Process and Fundamental Considerations of Selected Hydrometallurgical Systems

Martin C. Kuhn, Editor

Society of Mining Engineers
of
American Institute of Mining, Metallurgical, and Petroleum Engineers, Inc.

New York, New York • 1981

Printed in the United States of America by
Kingsport Press, Inc., Kingsport, Tennessee

Library of Congress Catalog Card Number 79-57685
ISBN 0-89520-282-4

V OHaldane
Nov 82

Editorial Committee

EDITOR

Martin C. Kuhn
Vice President
Mountain States Mineral Enterprises, Inc.
Tucson, Arizona

SECTION EDITORS

I. Hydrometallurgical Recovery From Ores

John L. Shafer
Staff Engineer
Minerals Division
EXXON Research and Engineering Company
Florham Park, New Jersey

II. Hydrometallurgical Recovery From Concentrates

William G. Sherwood
Section Supervisor
Extractive Research & Development, Inc.
AMAX
Golden, Colorado

III. Hydrometallurgical Processes for By-Product Recovery

Malcolm L. Jansen
Manager, Australian Operations
The Ralph M. Parsons Company
Sydney, Australia

IV. Fundamental Aspects of Solvent Extraction Phenomena

L.L. Tavlarides
Department of Chemical Engineering
Illinois Institute of Technology
Chicago, Illinois

Jan D. Miller
Professor of Metallurgy
Department of Metallurgy and
Metallurgical Engineering
University of Utah
Salt Lake City, Utah

V. Electrochemical Reactions and Solution Chemistry

Renato G. Bautista
Ames Laboratory USDOE and
Department of Chemical Engineering
Iowa State University
Ames, Iowa

T.W. Chapman
Chemical Engineering Department
University of Wisconsin
Madison, Wisconsin

FOREWORD

The collection of technical literature published herein represents the work of many individuals, companies, universities, and technical societies. The field of hydrometallurgy spans different industries, academic departments, and professional organizations and for this reason provides a difficult task for the industrial and scientific practitioner to keep abreast of the technical developments. A chronological listing of the sources of the literature presented herein is as follows:

"The Hydrometallurgical Recovery of Metals from Ores, Concentrates, and Secondary Sources," American Chemical Society Symposium, Anaheim, California, March 10, 1978, M.D. Kuhn, Program Chairman; J.L. Shafer, W.G. Sherwood, Session Chairmen.

"Fundamental Aspects of Hydrometallurgical Processing," Annual AIChE Meeting, San Francisco, California, November 25-29, 1979. L.L. Tavlarides, J.D. Miller, T.W. Chapman, R.G. Bautista, Session Chairmen.

"Hydrometallurgical Processes for By Product Recovery," Annual AIME Meeting, Las Vegas, Nevada, February 24-28, 1980, P.B. Queneau, M.C. Kuhn, Session Chairmen.

The ultimate objective of technology is the more efficient and economical use of our natural resources. We have in the various sections of this book attempted to highlight economically viable processes, emerging processes, and new technologies which may be the foundation of future processes.

This book has been made possible through the diligent efforts of Messrs. J.L. Shafer, W.G. Sherwood, M.L. Jansen, L.L. Tavlarides, J.D. Miller, R.G. Bautista, and T.W. Chapman. Ms. Marianne Snedeker, Manager of Publications of SME-AIME, deserves special thanks for making it a reality.

Martin C. Kuhn
Editor

TABLE OF CONTENTS

Section I

Hydrometallurgical Recovery From Ores

John L. Shafer, Section Editor

Chapter 1

RECOVERY OF COPPER FROM OXIDE ORES, TODAY'S PRACTICE

Gary A. Kordosky

Henkel Corporation
Mineral Industry Division

ABSTRACT

Winning copper from oxide ores usually involves hydrometallurgical techniques. The processes involved are the leaching of copper from the ore and the recovery of the leached copper from the aqueous solution.

RECOVERY OF COPPER FROM OXIDE ORES, TODAY'S PRACTICE

Gary A. Kordosky

Henkel Corporation
Minerals Industry Division

INTRODUCTION

Analysis of early copper tools, the region of their discovery and other evidence indicates that the recovery of copper from oxide ores, dating from about 4-5,000 B.C., was the first form of metallurgical practice known to man.(1) Specifically it was a form of pyrometallurgy, where the rich oxide ore was heated to about 800° C in a windswept wood fire or a pottery kiln, and then reduced by carbon monoxide to molten copper. This discovery was an important step in the development of civilization since it led to the use of copper for the tools of agriculture, art, construction, and war.

At that time oxide ores were the most important source of copper for several reasons; a) they were more amenable to early smelting practice, and, b) they were normally much easier to find and mine. Today the production of copper from oxide ores represents only about 10% of the total world copper production. This is a result of the decreased availability of oxide ores more than the recovery techniques employed, even though these can be important. Examples of orebodies composed completely of copper oxides are known, but normally copper oxide represents only a small fraction of the total ore present. However, the importance of copper recovery from oxide ores may be much greater than the 10% of the world production figures would indicate. Since it is increasingly more expensive to mine and process all ores, the economic recovery of copper from the oxide portion of the orebody may be a crucial factor in making an operation go or no go.

Arbiter (2) has defined mineral benefication in the following manner: "To make ore from waste". In a true sense this is what the first person who managed to smelt copper from an oxide ore did. Today the recovery of copper from oxide ores in many cases also represents "making ore from waste" as will be evident later in this discussion.

Most oxide ores as such are not amenable to present day flotation techniques nor are they rich enough for direct smelting, thus the winning of copper from these ores usually involves hydrometallurgical techniques. Normally two separate processes are involved; the leaching of copper from the ore, and the recovery of the leached copper from the leach solution.

The object of this paper will be to discuss these techniques in general and then to look closely at the operating data for several commercial operations.

LEACHING

The most important oxide ores and their general formulas are chrysocolla ($CuSiO_3 \cdot nH_2O$), azurite [$2CuCo_3 \cdot Cu(OH)_2$], and malachite [$CuCO_3 \cdot Cu(OH)2$].(3) In almost all instances aqueous sulfuric acid is used as the leachant because of its low cost and relative ease of handling. The dissolution rates of most copper oxide ores are very fast provided there is good contact between the acid solution and the oxide to be leached. Theoretically, the stoichiometric requirement of acid to copper leached is 2 moles of protons to one mole of copper ions released. The following equations are representative of the leaching process.

$$[CuCO_3 \cdot Cu(OH)_2] + 4H^+ \longrightarrow 2Cu^{2+} + CO_2(g) + 3H_2O \quad (1)$$

$$(CuSiO_3 \cdot nH_2O) + 2H^+ \longrightarrow Cu^{2+} + SiO_2(H_2O)x \quad (2)$$

However, in real practice acid consumption is much higher depending upon the gangue present and the method of leaching. The following generalizations can be made with respect to acid consumption with a given ore: (4)

1. A finer grind results in greater acid consumption;
2. The longer the ore-acid contact time the greater the acid consumption;
3. The lower the pH of the leachant the greater the acid consumption.

The chemical equation that describes the leaching process of a given oxide is very simple, and the dissolution of copper is very rapid and complete when good contact is made between the acid and the oxide. However, bringing about this required contact in an efficient yet economical fashion can present problems. Indeed, leaching has been described by E.E. Malouf (5) in the following manner, "Leaching as currently practied is empirical in nature and only recently has research effort been applied to convert the art of leaching into a science." Raymond Jones (6) has also termed "leaching an art as much as a science" and stated further that "each leaching operation has a personality all its own." The reasons for this are many, among them being a) the nature of the ore, b) the grade of the ore, c) the economic constraints on the leaching operation, and d) the type of leaching method used.

Leaching methods can be divided into 4 main categories:

1) in situ or in place leaching,

2) dump or heap leaching,
3) vat leaching,
4) agitation leaching.

1. With the in situ or in place leaching technique the ore is not mined (moved) in the conventional sense but rather leaching solution is caused to come into contact with the in place orebody and is then withdrawn. Old worked out mines, deep-lying orebodies, or small pocket-like orebodies might be leached in this manner, provided the geology of the area is such that there is no loss or deflection of the leach solution to the groundwater. This method holds real promise for the future as the grade and number of mineable orebodies falls. The % copper extracted should run 50 to 60% of the copper present.

2. Dump and heap leaching are similar techniques in that the ore is piled in a canyon or prepared area. Next the leachant is distributed on the surface of the ore pile, allowed to trickle through, and is then collected at the bottom of the ore pile, generally against a dam. Dump leaching differs from heap leaching in the grade of ore leached, the size of the ore pile and the recovery time cycle of the copper. In some cases heap leaching requires some crushing of the ore. The total copper recovery is around 60% and it may take months or even years to get this copper.

3. Vat leaching of oxidized copper ores is generally reserved for richer and cleaner ores. The crushed ore is submerged in 50 to 100 g/l H_2SO_4 and a solution carrying up to 50 g/l Cu results. Extraction efficiencies run very high, normally greater than 90% and sometimes close to 100% of the acid soluble copper oxide. Generally vat leaching is countercurrent and involves a series of leach-drain then wash-drain cycles, carried out over a period of 7-14 days. This insures complete copper leaching with minimum copper carried in the tails. As ore grades fall, vat leaching will become less important.

4. Agitation leaching is practiced when the nature of the ore is such that the other leaching methods do not work. The ore is crushed and ground into fine particles before contact with aqueous sulfuric acid in a series of agitated tanks. The leaching rate is very fast (2-5 hours) and complete (up to 100% of the acid soluble copper). Solid-liquid separation and washing of the leach residue follow. The pregnant leach solution may be high or low in copper depending upon the oxide leached. At the present time agitation leaching of old copper tailings is being practiced commercially and will probably become more important in the future.(7)

As previously stated, bringing about proper contact between the leachant and copper oxide to be leached is the most difficult part of leaching. These complications are mechanical as well as chemical in nature. For example, the leachant should move through the copper ore evenly, wetting all surfaces, and continue to do so over the life of the dump, heap or vat. However if too many fines are generated during the ore preparation these fines can cause plugging. The leach solution will then begin to channel around certain sections of the ore and leaching efficiency falls. Plugging problems can also be generated chemically. Gypsum may solubilize in one part of the ore mass being leached and precipitate in another. In some cases sulfate salts of other metal ions may also do the same thing. Again this leads to channeling and a resultant loss in leaching efficiency.

Studies (8) on the leaching of chrysocolla with acid have shown that the kinetics follow a shrinking core model. Initially the copper is released very fast because of chemical reaction at the exposed copper oxide surface but with time copper leaching slows. Two factors have been shown to be important: a) the diffusion of protons into, and copper ions out of, the silicate layer surrounding the unreacted chrysocolla and, b) diffusion of ions through the boundary layer at the silicate-solution interface. The rate of leaching was found to be inverseley related to particle size. But if the particle size is too fine the diffusion of ions into and out of the pile of particles tends to negate the rate increase due to greater oxide surface area. Leached chrysocolla will retain its original form as long as it is wet, but the loss of waters of hydration when it dries causes the particles to collapse under little or no pressure. Thus leaching - drying cycles may prove advantageous when dump or heap leaching chrysocolla. The kinetic study readily explains the time differences between the recovery of copper by the various leaching methods.

Data from three commercial leaching operations is given below. As the data is studied the reader will better understand: a) the leaching techniques involved and, b) the interplay of factors which dictated the type of leaching operations to be used.

Dump Leaching

The operation chosen is the dump leaching program of the Cyprus Bagdad Copper Company.(6) The ore leached is the oxide over-burden of a porphyry copper sulfide orebody with a copper sulfide cut off of 0.3% and containing $\sim$0.5% oxide copper, primarily as chrysocolla. The rock is piled into dumps as it is mined with no crushing or any other special preparations. Currently, Bagdad is leaching 3 dumps: Alum Creek (1960), Mineral Creek (1963), and Copper Creek (1968). Dump life is unknown, but the older dumps are showing some signs of their age, i.e., a significantly lower rate of copper leaching. The shallowest dump is 180-200 feet at the center while the deepest is $\sim$350 feet and growing. Only about 10% of the total dump area is under leach at any given time. Fresh ore is sprayed with about 7-9 g/l H_2SO_4 while older ore is leached with 4-5 g/l H_2SO_4. The first effluent from a new dump is nearly barren in Cu and Fe but after a short period of time the dump is conditioned and starts working, i.e., copper is leaching. Mr. Jones, of the Cyprus Bagdad Copper Company, states that only a small area of a new dump should be covered with solution at any given time so that the dump conditioning can be better controlled.

All dump areas are on a leach, drain, dry cycle. For example, when the copper production from a given area of a dump begins to fall, the application of leach solution is terminated, and this area of the dump is allowed to drain and dry. When leaching is again started at some later date, the copper production is higher than when leaching was terminated. This phenomenon is in keeping with the kinetic model previously discussed, i.e., when the chrysocolla dries the particles collapse exposing fresh ore for leaching.

If the acid concentration of the leaching solution is increased and the solution flow decreased, a good copper oxide dump will still provide the same amount of copper within reason. For example, a 10% decrease in solution flow coupled with a 10% increase in acid concentration should result in approximately the same total copper production.

At Bagdad the pregnant leach solutions from the various areas of the dumps under leach are blended to give a fairly constant feed solution containing 1.5-1.8 g/l Cu at a pH of 2-2.2.

Acid consumption at Bagdad is fairly low, requiring ~ 4.9 pounds of concentrated H_2SO_4 per pound of copper produced. Of this requirement, ~ 3.34 pounds are fresh acid addition and 1.54 pounds result from the solvent extraction process used for copper recovery. This aspect of the Bagdad operation will be discussed under the section on the recovery of copper from leach solutions.

Heap Leaching

Ranchers Exploration and Development Company has been heap leaching chrysocolla at its Bluebird Mine in Miami, Arizona since 1964.(9) Natural canyons are used as ore leaching areas with minimal land preparations. There is some scraping of vegetation and a 3 foot berm is built down the middle of the canyon. Solution collection is via perforated pipes which are laid along the berm and solution loss to the ground is very slight. If solution loss to the ground is expected to be a problem, soil cement can be used in the site preparation.

The ore is first ripped with a cat pulling a ripper and is then transported to the heaps using earth movers. No crushing or other special ore preparations are involved. The heaps are built in 20 foot lifts and have 90,000 square feet of surface area. A small berm is built around the leaching area, the top 3 feet of the surface is ripped and then plastic pipes are laid over the area for solution distribution. Leaching solution is applied according to the following schedule:

Leachant (g/l H_2SO_4)	Time
50	First 10 days
30	Next 20 days
20	Next 30 days
5-7 (Raffinate)	Rest of time

After the leaching of copper from a lift has slowed to a low level, solution application is stopped, the top of the heap is allowed to dry, and then the pipes are removed. After ripping the old heap to a 3 foot depth, a new lift is built on top and the leach cycle begins anew. Heads of up to 10 g/l copper are sometimes obtained from a new heap and these fall with time until the lift is retired from direct leaching. Leach solutions from the various heaps under leach are blended to provide a feed solution of ~2000 gal/min containing 2 g/l Cu and 2-3 g/l H_2SO_4. Acid usage runs about 5 pounds of concentrated H_2SO_4 per pound of copper produced. Eventual copper recovery is believed to be about 60% which is about the same recovery obtained using dump leaching. This is not surprising, because once there are 6 to 8 lifts stacked up, the ore pile is really a dump.

The main difference between heap and dump leaching of copper oxide ores is the time needed for the start of copper recovery. Whereas it may take a year or more to build a dump it only takes weeks to build a lift. For this reason, heap leaching is used as a primary recovery operation when the economics of the operation are dependent upon the production of copper from the oxide ore. Dump leaching is used when the oxide ore is considered as overburden and the building of a dump is part of a mining program aimed primarily at sulfide ore.

Agitation Leaching

The agitation leaching example is taken from Anamax Mining Company just south of Tucson, Arizona.(10) The leachable ore is principally chrysocolla with small amounts of carbonates, oxides and native copper. The non-leachable portion of the ore is a copper manganese silicate. The oxide ore at Anamax was not considered to be economically processable until only a few years ago, and in that sense it was not really ore but overburden even though it contained about 1% soluble copper. This was because of the high acid consuming nature of the ore. However once a cheap, long term supply of H_2SO_4 was secured this obstacle was removed. The oxide ore is first crushed, then milled to 15% + 65 mesh and finally slurried with sulfuric acid. Leaching takes place at elevated temperature in a series of 5 agitated leaching tanks where the rate of acid addition is controlled within certain pH ranges. The retention time in the leaching vessels is 5 hours. The leaching slurry then proceeds to the countercurrent decantation-clarification circuit.

Leaching results in 100% recovery of the acid soluble copper. The acid consumption ranges from 7 to 15 pounds of acid per pound of copper produced, depending upon the ore. The leach solution carries 2.5 to 3.0 g/l Cu at a pH of 1.8-1.9. Agitation leaching was chosen because of the type of ore (no other method would really work) and the high recovery of copper that is obtained. The time required for processing copper from ore to final product is about 11 days with the major portion of the holdup coming in the countercurrent decantation-clarification circuit.

Agitation leaching is the most capital intensive of the various types of leaching processes and usually has the highest operating costs per ton of ore treated. This is due to the crushing and milling required for the ore and the solids-liquid separation required for the leach solution. However, as the data indicate, high copper recoveries and short holdup times result.

RECOVERY OF COPPER FROM LEACH SOLUTIONS

There are three principal methods which are used to recover copper from acid leach solutions: a) cementation, b) direct electrowinning, and c) solvent extraction-electrowinning.

Cementation

Cementation can be described by the following equation:

$$Cu^{2+}_{Aq} + Fe^{\circ}(s) \longrightarrow Cu^{\circ}(s) + Fe^{2+}_{Aq} \qquad (3)$$

Any metal more active than copper could potentially be used as a precipitation agent. However, because of economics, shredded scrap iron is the only practical choice. This reaction has a driving force of 0.75 volts at standard conditions and proceeds to virtual completion. Whereas the stoichiometry of this reaction predicts an iron usage of 0.88 pound to precipitate one pound of copper, in actual practice, the usage rate is 2 to 3 times this amount. Two side reactions account for this large excess:

$$Fe^{\circ}(s) + 2Fe^{3+}_{Aq} \longrightarrow 3\ Fe^{2+}_{Aq} \qquad (4)$$

$$Fe^{\circ}(s) + 2H^{+}_{Aq} \longrightarrow Fe^{2+}_{Aq} + H_2(g) \qquad (5)$$

The cementation reaction has been well studied (11, 12) and the following generalizations observed:

1. The rate of copper precipitation is proportional to the copper concentration of the solution, the exposed iron area, and the specific rate constant.

2. The greater the ferric to cupris ion ratio, the greater the iron consumption in the cementation reaction.

3. Operating cementation at conditions that exclude oxygen from the system decreases iron consumption by decreasing the oxidation of Fe^{2+} to Fe^{3+}.

4. Cementation reactors which operate with turbulent flows and short retention times maximize the amount of copper recovered for the iron consumed.

The two most common systems for cementation are the open launders and the Kennecott cone. In the open launder system the copper bearing leach solution flows by gravity through a trough or series of troughs filled with scrap iron. The precipitated copper falls to the bottom of the launder, often times through a false bottom, and is eventually washed into a holding pond from whence it is recovered and dried.

The Kennecott cone system is shown in Figure 1. Pregnant leach solution is force swirled up through a bed of scrap iron. The copper which precipitates is carried upwards by the solution flow until the velocity of the solution has slowed. This allows the copper to settle from solution and pass down through the stainless steel screen into the copper settling zone. The reactor is designed for high flow rates, a high degree of solution turbulence, and short solution-scrap iron contact times relative to the typical launder system. As expected, these design features result in less iron consumption per pound of copper produced by up to 50% compared to a launder system. Copper recoveries of 90% or greater are common using cementation whether a launder or a Kennecott cone is used to produce the precipitate.

The cement copper produced via cementation is not of high purity and must be further refined.

Direct Electrowinning

Direct electrowinning has been used successfully for many years on a commercial scale to recover copper directly from high copper tenor leach solutions. An excellent study of this subject has been presented by Anderson and co-workers.(13)

Most feed solutions for direct electrowinning contain Cu^{2+}, Fe^{3+}, Fe^{2+} plus many other cations, principally Al^{3+}, and Mg^{2+}, in a sulfuric acid matrix. The electrowinning cell is composed of copper cathodes and inert anodes, normally of a lead alloy. The chemistry of this system can be described by the following electrochemical reactions:

$$\text{Anode:}\quad 2H_2O \longrightarrow 4H^{+} + O_2(g) + 4e^{-} \qquad (6)$$

$$Fe^{2+} \longrightarrow Fe^{3+} + e^{-} \qquad (7)$$

$$\text{Cathode:}\quad Cu^{2+} + 2e^{-} \longrightarrow Cu^{\circ} \qquad (8)$$

$$Fe^{3+} + 3^{-} \longrightarrow Fe^{2+} \qquad (9)$$

The cyclic oxidation - reduction of iron at the anode and cathode results in a steady state condition, whereby the concentrations of Fe^{2+} and Fe^{3+} remain constant. Thus, the net solution change in an electrowinning cell is the reduction of the Cu^{2+} concentration and an increase in acidity, i.e., equations 6 and 8. Hence the overall cell reaction is often written as:

$$Cu^{2+} + H_2O \longrightarrow Cu^{\circ} + 1/2\ O_2(g) + 2H^{+} \qquad (10)$$

$$\text{with } E^{\circ} = -.892V$$

The object of direct electrowinning is to produce a copper cathode of high quality at as low a cost as possible. Anderson (13) showed that cathode quality could be improved by: a) increasing the copper ion concentration, b) increasing the solution agitation, c) increasing the temperature, d) decreasing the viscosity of the solution (decreasing the inert salt concentration), and e) decreasing the applied current density.

The cost of producing copper by direct electrowinning is a function of the capital cost as well as the operating cost of the tankhouse. The capital cost per pound of copper produced per unit of time is directly related to the current density used in the tankhouse, i.e., the higher the current density the greater the amount of copper which can be produced in a given unit of time. The principal contributor to operating costs is the current efficiency defined as the current needed to deposit the copper produced in a unit of time divided by the total current used in electrowinning for the same period of time. The following factors were found to increase current efficiency: a) decreasing the temperature, b) increasing the inert salt concentration, c) decreasing the Fe^{3+} concentration, d) increasing the applied current density, and e) decreasing the agitation.(13)

A comparison of the factors which increase cathode quality with those which increase current efficiency, shows that there is no easy way to both, increase current efficiency and to increase cathode quality by changing tankhouse practices. If power costs were low, current efficiency would not be an important consideration. Also if refining costs were low, poor quality copper would not be important. However, both of these costs are accelerating at a rapid pace. The net result is that as the grades and quality of ores decline direct electrowinning will gradually diminish in importance as a method to recover copper.

A recent paper (14) reported on the development of direct electrowinning technology as applied to dilute copper leach solutions. Ferric ion decreases current efficiency and in addition the copper must be further refined. The authors claim some economic advantages over the cementation process and over solvent extraction for the recovery of copper from certain dilute leach solutions. However, there are currently no commercial operations employing this technology nor have plans to construct any been announced.

Solvent Extraction - Electrowinning

The goal of solvent extraction can be stated simply; to concentrate and purify the copper from a leach solution, so as to deliver to electrowinning a solution from which high quality copper can be won while operating at high current effieiency.

In the copper systems under discussion the equation below is a simplified representation of the chemistry involved:

$$[2\ R\text{-}H]_{Org} + [Cu^{2+}]_{Aq} \underset{\text{stripping}}{\overset{\text{extraction}}{\rightleftharpoons}} [R_2Cu]_{Org} + [2H^{+}]_{Aq}$$

In order to successfully apply solvent extraction technology to the recovery of copper from typical copper oxide leach solutions, a reagent must be available which will selectively extract copper from typical leach solutions yet will be strippable to give a solution amenable to electrowinning. Certain phenolic oximes of the general structure shown below have been used commercially in copper recovery operations for over ten years.

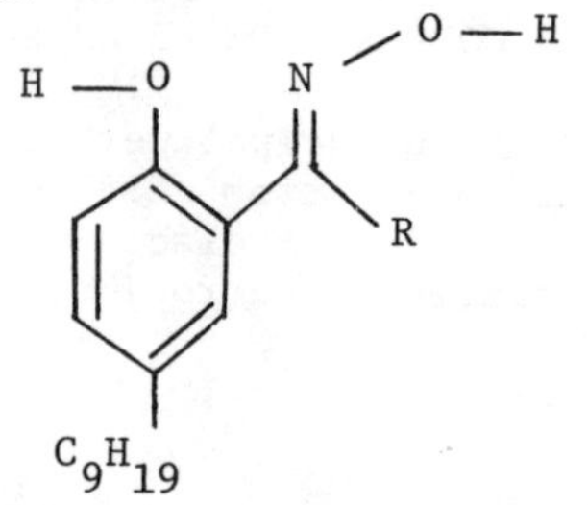

The actual stoichiometry of the extraction reaction is probably not constant at 2 mole of reagent to one mole of Cu over a wide pH range. However, the data developed is difficult to interpret.(15) In actual practice, though, the above equation does serve well to predict and explain behavior in commercial systems. A graphical depiction of this system is given in Figures 2 and 3. Figure 2 is an extraction equilibrium isotherm generated by equilibrating a stripped organic extractant dissolved in kerosene with a copper feed solution, at various organic to aqueous phase ratios. This isotherm is a function of, and unique to, the parameters of this system: a) the organic reagent and its concentration; b) the ratio of complexed to uncomplexed reagent in the stripped organic; c) the copper content of the feed; d) the pH of the feed; e) the other constituents of the aqueous phase; f) the temperature of the system, and g) the diluent (kerosene) used as the organic reagent carrier.

Figure 3 is a stripping equilibrium isotherm generated by equilibrating a loaded organic extractant with a typical tankhouse electrolyte, again at various organic to aqueous phase ratios. As in Figure 2, this isotherm is unique to the parameters of the system.

The extraction and stripping isotherms can be used to predict the staging requirements of a circuit. For example, consider the extraction isotherm (Figure 2) and suppose that the stripped organic coming into extraction contains 0.14 g/l Cu and that the organic and aqueous advance flow rates are equal. Knowing this, an operating line can be constructed by starting at the point at which the stripped organic value intersects the isotherm, and drawing a line to the right with a slope equal to the aqueous/organic flow rates, (one in this instance), until the operating line intersects the isotherm, and drawing a line to the right with a slope equal to the aqueous/organic flow rates, (one in this instance), until the operating line intersects the vertical line representing the copper content of the feed solution. Next, the horizontal lines to the isotherm curve and the vertical lines to the operating line are filled in completing the McCabe-Thiele diagram. Each triangle represents one counter-current stage of extraction. In this case under discussion a raffinate of 0.042 g/l

Cu and a loaded organic of 1.16 g/l Cu are predicted in two countercurrent stages of extraction. A similar operation on the stripping isotherm predicts a stripped organic of 0.08 g/l Cu in two stages when operating at an O/A advance flow ratio of 20/1, and building the pregnant electrolyte to ~50 g/l Cu.

The predicted values are the best values attainable under conditions of ideal mixing. In actual commercial operations, mixing efficiencies vary from 85 to 90% of theoretical. Thus, the actual results expected in a commercial operation can be predicted by working up the McCabe-Thiele diagram as previously described, except that the horizontal line is drawn to only 85% of the theoretical, or whatever mixing efficiency is expected. When this is done using Figure 2, a two stage raffinate of 0.10 g/l Cu and a three stage raffinate of 0.36 g/l Cu are predicted. These numbers are remarkably close to the numbers obtained in an actual circuit operation with the organic and feed under discussion.

A conceptual picture of SX-EW is shown in Figure 4. Copper ions, leached from the ore by sulfuric acid, are carried first by the leach solution, then by the organic, and finally by the electrolyte to the tankhouse where they are despoited as high quality copper. At the same time, an equivalent amount of hydrogen ions are carried from the tankhouse by the electrolyte, transferred to the leach solution via the organic, and carried to the ore pile to leach more copper. The leach cycle, organic cycle, and electrolyte cycle are normally closed loops such that there are no streams exiting to the environment.

Whereas the solvent extraction process looks relatively simple, and indeed operates in the same manner, the chemical requirements on the organic extractant are many:

1) It must load and strip the copper under conditions applicable to the industry;
2) It must be selective for copper under the conditions applicable to the industry;
3) It should have fast extraction and stripping kinetics;
4) It must exhibit good phase separation;
5) Its copper complex must be soluble in the the kerosene carrier;
6) It must be chemically stable to the circuit conditions;
7) It must have an acceptable cost.

Many times optimizing one desired property will cause a loss in another desired property. A good extraction reagent must have a balance of all the properties.

All commercial operations use mixer-settlers for effecting copper transfer and phase separation (Figure 5). The mixer is normally designed for a 3 minute residence time while the settler is designed such that the flow per unit area is 1.5 to 2.0 gal/min/ft^2. The staging is such that 90% or greater total copper recovery is realized.

As previously stated the goal of solvent extraction is to concentrate and purify the copper from a leach solution such that high quality copper can be electrowon while operating at high current efficiency. In actual practice, the final copper recovered by the SX-EW process ranges from good quality (suitable for alloying) to superior quality (suitable for wire bar). Current efficiencies vary from 85 to 93%.

An example of an efficient SX-EW plant is the Cyprus-Bagdad operation.(16) The plant was designed to produce 40,000 lbs of copper per day. Since its startup in August of 1970, this facility has produced slightly more than 40,000 lbs Cu/day with no major shutdowns. The copper produced is electrolytic quality and is registered as such on the Chicago Commodities Exchange. The plant has 4 modules, each with 4 extraction and 3 stripping stages. For the time period January to August, 1977, the feed to SX averaged 1.63 g/l Cu and a raffinate of 0.15 g/l Cu was obtained resulting in an average daily copper production of 41,000 pounds. Electricity usage for this time period was 0.90 kwh per pound of copper produced. The electrolyte bleed from the tankhouse to the leach solution was only 2% of the total electrolyte flow. These numbers are typical of the Cyprus-Bagdad operation.

The feeds to existing commercial solvent extraction plants vary between 1 and 5 g/l Cu depending upon the particular operation. However, plants now under consideration will expand this range from a low of 0.3 g/l Cu to a high of 20 g/l Cu.

The production capacity of SX-EW plants is also quite variable. Present day commercial plants vary in capacity from a low of 15 ton/day to a high of ~240 ton/day.

The future of solvent extraction as a technique for recovering copper from acid leach solutions is bright. Recent innovations in equipment (for example electrostatic separators) (17) and reagent synthesis (18) brighten the outlook even more.

BIBLIOGRAPHY

1. Poss, J.R., Stones of Destiny, Michigan Institute of Technology, Houghton, Mich., 1975 pp. 1-22.

2. Arbiter, N., "Mineral Benefication: A Perspective", AIME Centennial Volume, AIME, New York, 1971, pp. 85-118.

3. Biswas, A.K., and Davenport, W.G., "Hydrometallurgical Copper Extraction: Introduction and Leaching", Extractive Metallurgy of Copper, 1st Edition, Pergamon Press, Oxford, England, 1976, pp 254-293.

4. Dudas, L., Maass, H., and Bhappu, R., "Role of Minerology in Heap and In Situ Leaching of Copper Ores", Solution Mining Symposium 1974, Aplan, F.F., McKinney, W.A., and Pernichele, A.D., ed., AIME, New York, 1974, pp. 193-210.

5. Malouf, E.E., "Leaching as a Mining Tool", International Symposium on Hydrometallurgy, Evans, D.J.I., and Shoemaker, R.W., ed.,

AIME, New York, 1973, pp. 615-626.

6. Jones, R., Cyprus Bagdad Copper Company, Private Communication, August, 1977.

7. Holmes, J.A., Deuchar, A.D., Stewart, L.N., and Parker, J.D., "Design, Construction and Commissioning of the NChanga Tailings Leach Plant", Extractive Metallurgy of Copper, Vol. II, Hydrometallurgy and Electrowinning, Yannapoulas, J.C., and Agarwal, J.C., ed., AIME, New York, 1976, pp. 905-925.

8. Pohlmant, S.L., and Olson, F.A., "A Kinetic Study of Acid Leaching of Chrysocolla Using a Weight Loss Technique", Solution Mining Symposium 1974, AIME, New York, 1974, pp. 446-460.

9. Longwell, R., Bluebird Mine, Ranchers Exploration and Development Company, Private Communication, August, 1977.

10. Moyer, R., Oxide Plant, Anamax Mining Company, Private Communication, August, 1977.

11. Nadkarni, R.M., and Wadsworth, M.E., "Kinetic Study of the Cementation of Copper with Iron", Advances in Extractive Metallurgy, Institution of Mining and Metallurgy, London, 1968, pp. 918-941.

12. Schlitt, W.J., and Richards, K.J., "Chemical Aspects of Copper Cementation", Solution Mining Symposium 1974, AIME, New York, 1974, pp 401-421.

13. Anderson, T.N., Wright, C.N., and Richards, K.J., "Important Electrochemical Aspects of Electrowinning Copper From Acid Leach Solutions", International Symposium on Hydrometallurgy, AIME, New York, 1973, pp. 171-202.

14. Amman, P.R., Cook, G.M., Portal, C., and Sonstelie, W.E., "Direct Electrowinning Copper From Dilute Leach Solutions", Extractive Metallurgy of Copper, Vol. II, Hydrometallurgy and Electrowinning, Yannapoulos, J.C., and Agarwal, J.C., ed., AIME, New York, 1976, pp. 994-1008.

15. Virnig, M.J., Unpublished Observations, Henkel Corporation, Minneapolis, Minn, 1976-77.

16. Jones, R.L., "Cyprus Bagdad's Solvent Exchange Process", Mining Engineering, Vol. 29, Sept, 1977, pp. 38-42.

17. Warren, K.W., Prestridge, F., and Sinclair, B., "The Use of Electrostatic Separators to Replace Mixer-Settlers in Solvent Extraction Plants", Annual Meeting Arizona Section of AIME, Tucson, Arizona, Dec. 5, 1977.

18. Kordosky, G.A., Mackay, K.D., and Virnig, M.J., "A New Generation Copper Extractant" Trans. AIME, Vol. 262, No. 1, March, 1977, pp. 36-42.

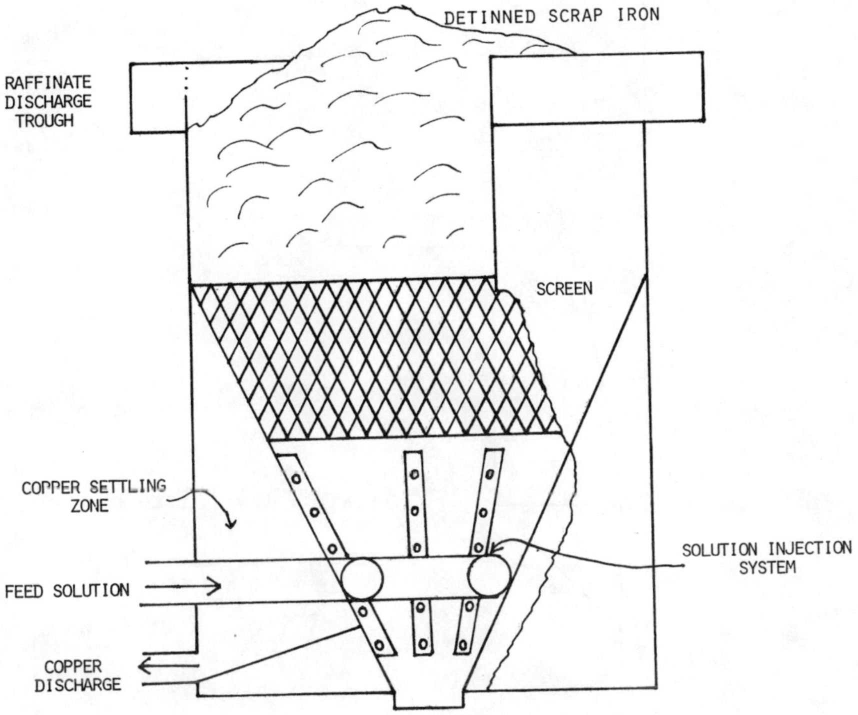

FIGURE 1 - KENNECOTT CONE PRECIPITATOR (12)

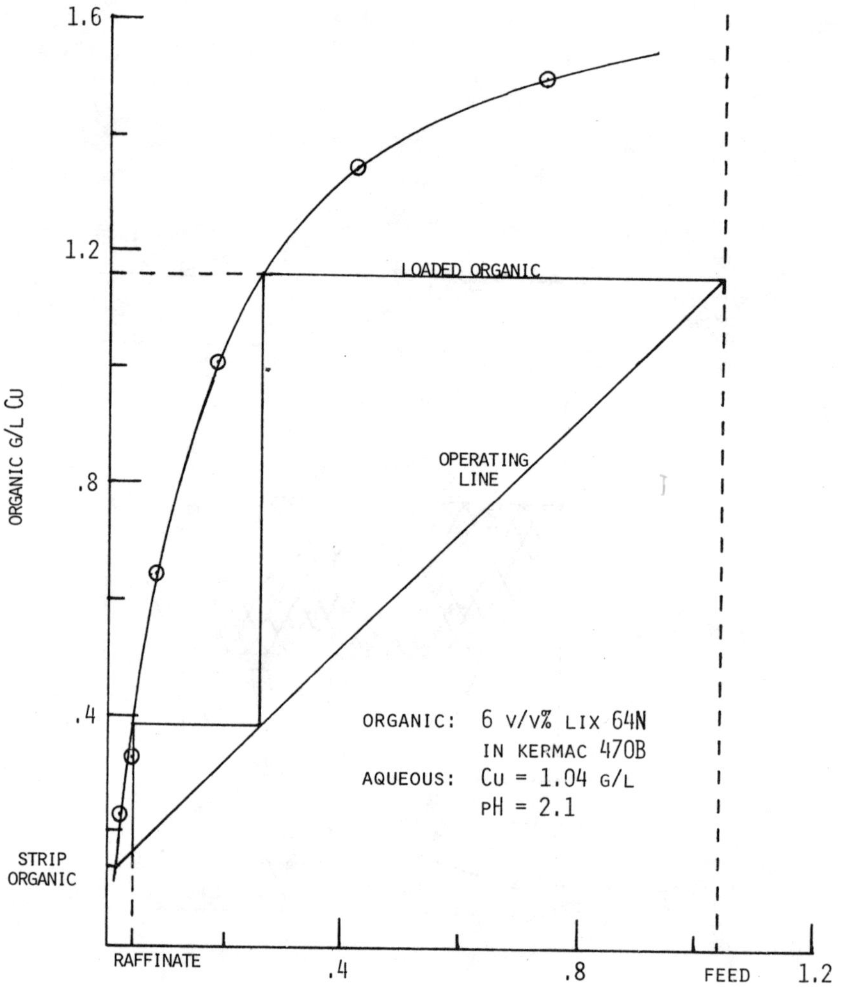

FIGURE 2 - EQUILIBRIUM EXTRACTION ISOTHERM

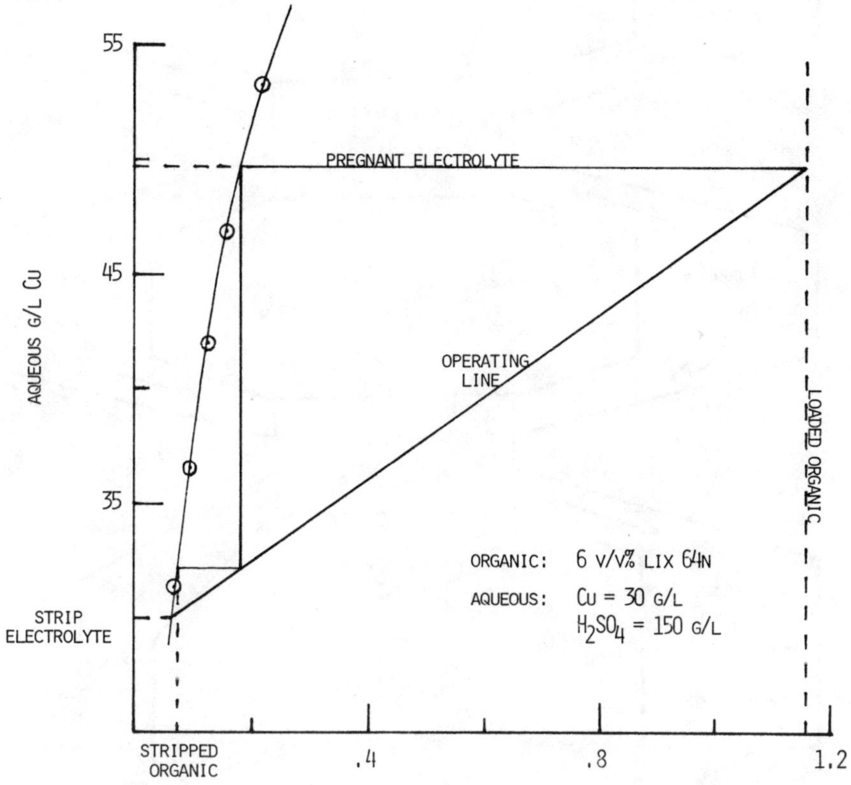

FIGURE 3 - EQUILIBRIUM STRIPPING ISOTHERM

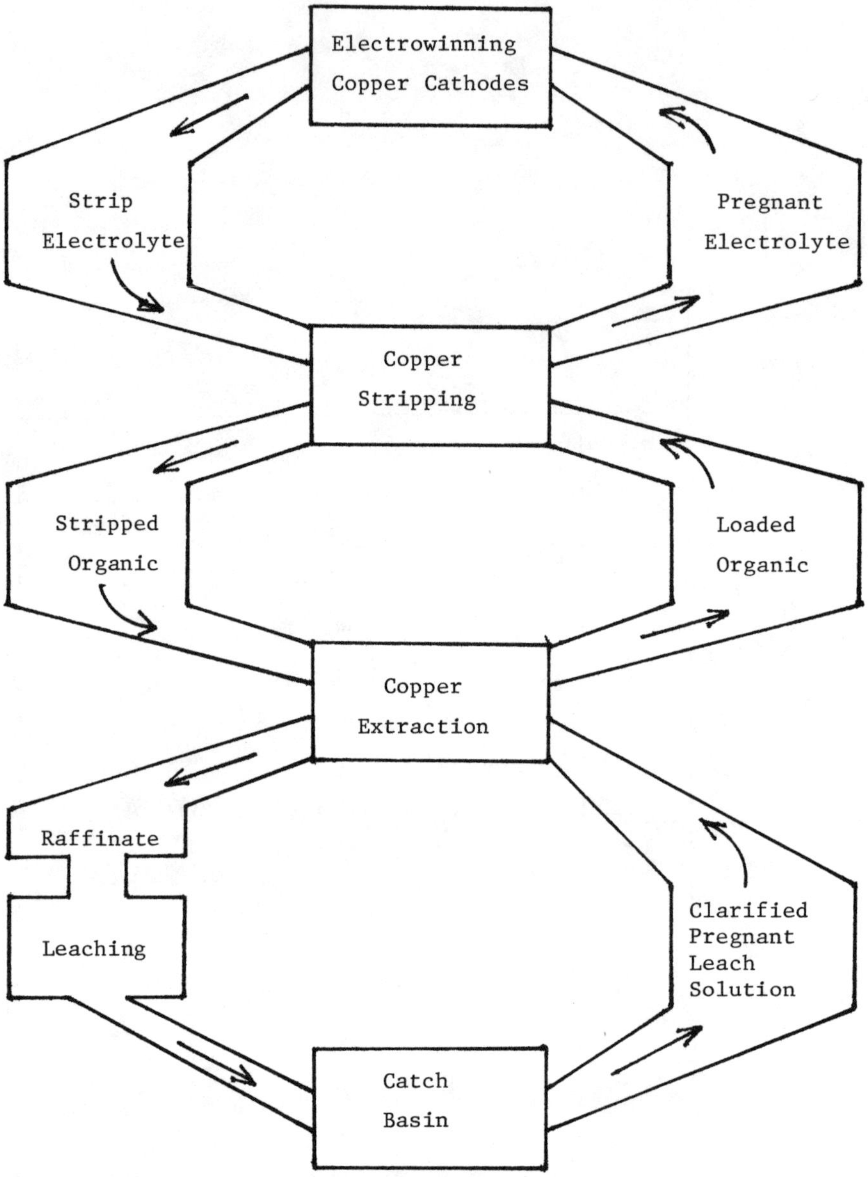

FIGURE 4 - SCHEMATIC OF THE SOLVENT EXTRACTION - ELECTROWINNING PROCESS

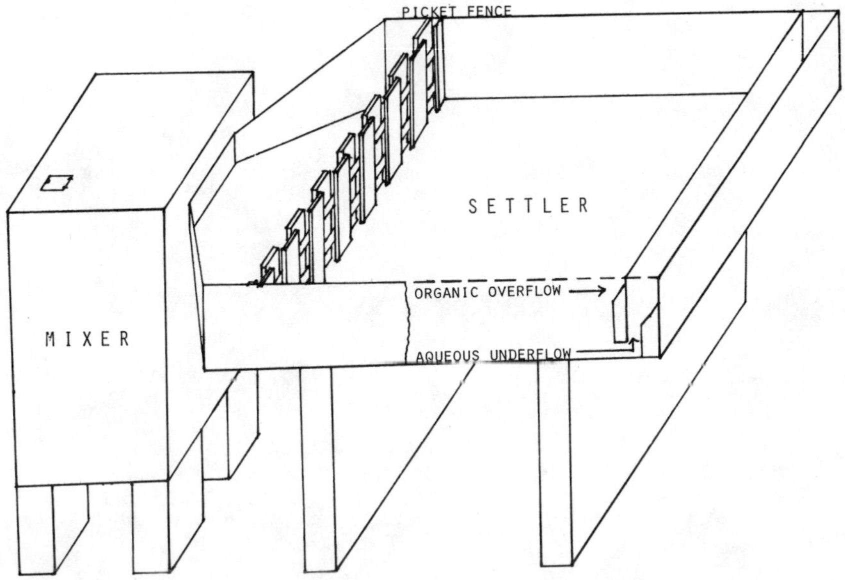

FIGURE 5 - TYPICAL MIXER SETTLER

Chapter 2

CHEMICAL ASPECTS OF THE OCCURRENCE AND EXTRACTION OF URANIUM

James K. Grunig

Anaconda Copper Company

ABSTRACT

As the real price of uranium increases, deposits of lower and lower grade will be treated. The various occurrences of uranium ore are described and the chemical aspects of extraction are detailed.

CHEMICAL ASPECTS OF THE OCCURRENCE AND EXTRACTION OF URANIUM

James K. Grunig

Anaconda Copper Company

Occurrence in Nature

Uranium, at 4 grams/ton, is not common in the earth's crust. It is about as abundant as beryllium and boron, and 40 times as abundant as silver. Table I shows, for some less abundant elements calculated concentrations in grams per ton (approximately parts per million) in the upper crustal rocks and concentrations in abnormally rich igneous rocks. Among the examples given, the ratio of the concentration in abnormally rich igneous rocks to the crustal concentration ranges from 10 to 100. Uranium is not concentrated to the extent of the noble metals or those which form stable sulfides.

Table I. Crustal Abundances and Relative Concentrations of Uranium and Some Other Elements.

Element	Concentration in Earth's Crust* g/ton	Concentration in Abnormally Rich Igneous Rocks g/ton	Concentration Ratio In Richest Rocks To Earth's Crust	Richest Rocks
Cr	200	3,400	17	Dunites
Ni	80	3,160	40	Peridotites
Cu	70	2,500	36	Gabbros
Co	23	240	10	Peridotites
Th	11.5	200	20	Peralkaline Granites
Be	6.0	700	12	Nephelene Syenites
U	4.0	100	25	Peralkaline Granites
B	3.0	30	10	Serpentines
Mo	2.5	80	32	Granite
Ag	0.1	10	100	Peridotites
Pd	.01	.2	20	Dunites
Pt	.005	.5	100	Dunites

*Crustal abundance data from Rankama, K. and Sahama, T. G. Geochemistry, U. of Chicago Press (1950).

Molybdenum and other elements such as vanadium, selenium, and tellurium tend to be concentrated along with uranium in reduced zones in sedimentary rocks. Pyrite and organic carbon also concentrate there.

Aqueous Uranium Chemistry. Uranium has a positive valence of either 4 or 6 under normal hydrometallurgical conditions or in natural solutions. The 3^+ state requires extremely reducing conditions and uranium 5^+ disproportionates in the presence of water to U^{4+} and $UO_2{}^{2+}$.

Uranium does not substitute in the common rock-forming silicate minerals. Thus, although it is found in complex accessory minerals in alkaline rocks along with the rare earths, its principal occurrence in such rocks is as the simple oxide uraninite. It does not concentrate with the base metal sulfides because it does not form a sulfide in the presence of water.

At least 90% of the uranium produced today comes from either sandstones less than 350 million years old, or from conglomerates which are about a billion years old.

The greatest production of uranium has come from uranium deposits in sandstone, which have been formed by the entrapment of uranium from circulating groundwaters by a rather complex biogeochemical localization mechanism, commonly exemplified by a configuration known as a roll front. (See Figure 1.) The front moves through the rock much as a chromatographic concentration band. The chemical and mineralogic features of such occurrences are well-known and have been duplicated in part by laboratory experiments. The front is driven by dissolved oxygen in circulating groundwaters and is fueled by sulfur and/or organic material. Chem-autotrophic bacteria such as Thiobaccillus Ferro-oxidans accelerate the oxidation of pyrite at the trailing edge of the front. They lower the pH and provide sulfate for sulfate reducers such as Desulfovibro Desulfuricans, which use organic material inside the ore zone to reduce aqueous sulfate ion to sulfide ion. This in turn reduces iron and precipitates it as pyrite. Uraniun borne in the groundwater is precipitated either as uraninite or cofinite, nominally the uranous oxide and silicate, respectively, and/or entrapped in the organic material which characterizes many roll front deposits. Sulfate is also provided by circulating groundwaters and the balancing calcium combines with carbonate ion released by increasing pH to precipitate calcium carbonate in the leading edge of the front. The roll front concept was crystallized by Harshman[4] and characteristics of the various zones have been described by Rubin[5].

Uranium is soluble in carbonate-bicarbonate waters as either U^{4+} or U^{6+} complexes above pH 7.5 or so.

Oxygen bearing bicarbonate groundwaters can carry uranium as a uranyl carbonate complex. The presence of Fe^{3+} hydroxides in the oxidized rock is consistent with this. The presence of pyrite in the ore zone requires a pH and Eh at which uranium is U^{4+} and carbonate complexes are unstable. The conditions upstream from the front allow transport and the conditions at the boundary between oxidized sandstone and ore allow some leaching of uranium. Thus, the roll front can concentrate uranium and other elements chromatographically from either the circulating groundwater or the host rock or from both. Table II lists the characteristics of the different zones of an idealized roll front.

Perhaps greater reserves of uranium occur in the conglomerate deposits than in sandstone. These deposits are worldwide in distribution, all occurring in rocks about a billion years old. Uranium in these deposits is associated with gold, iron sulfides, detrital heavy

Figure 1. Roll Front Showing Oxidized, Ore and Unoxidized Zones. After Harshman[4].

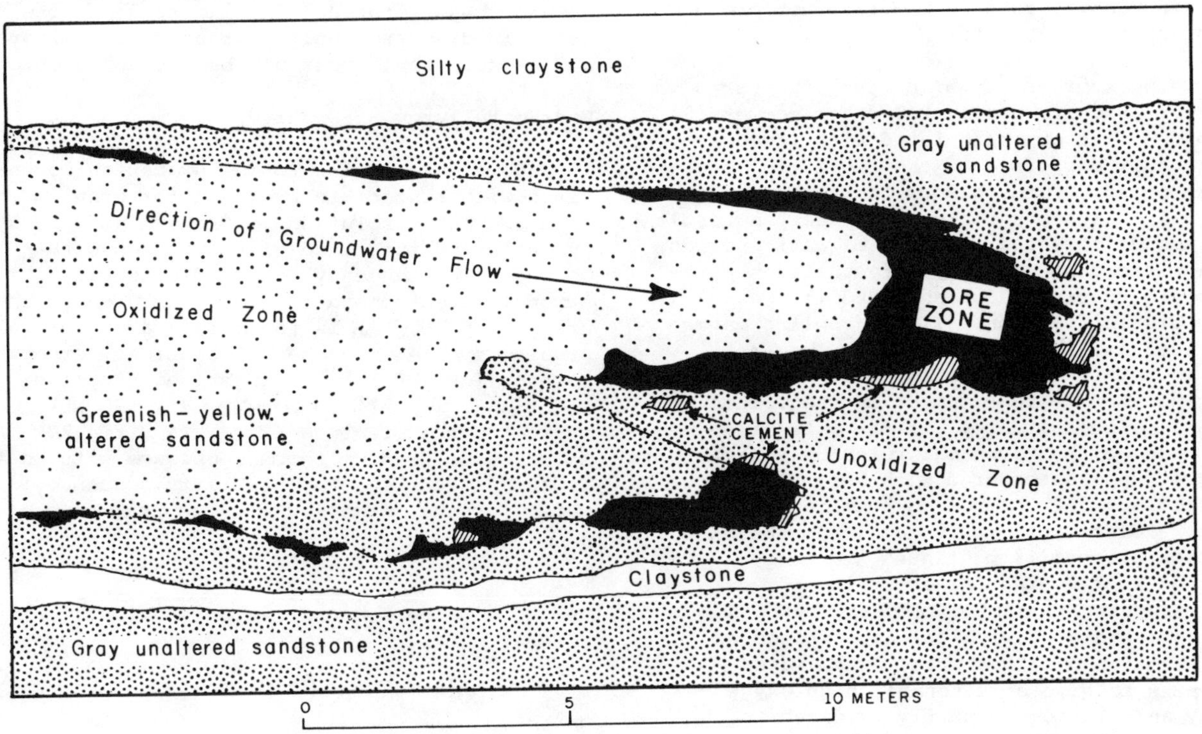

Table II. Characteristics of Roll Front Zones.

	Oxidized Zone	Ore Zone	Unoxidized Zone
Radiometric U/Chemical U	>1.0	≦1.0	>1.0
Uranium	.0002-.001%	.01-1.0%	.0004-.0008%
Molybdenum	.0001-.0002%	.001-.01%	.00005-.0002%
Vanadium	.0001-.0002%	.01-.5%	.0001-.0002%
Selenium	.002-.004%	.01-.3%	.00005-.0001%
Iron (Ferric)	.33-.66%	<.1%	.16-.33%
(Ferrous)	.16-.33%	1.0-3.0%	33-.66%
Sulfur (Sulfide)	<.002%	1.0-3.0%	<1.0
(Sulfate)	<.002%	.1-1.0%	<.002%
Carbon (Carbonate)	.01-.05%	.05-.6%	.2-.6%
(Organic)	.01%	.1-.6%	.03-.06
Typical Minerals	Clays, Hydrous Fe Oxides Feldspars, Hematite	Feldspars, Pyrite Calcite	Feldspars, Pyrite Calcite, Magnetite
pH	8.0±.3	6.0±.5	8.0±.3
Eh	>-.18	-.14 to -.28v	<-.27
Organic Compounds	Clay-organic Complexes Only	Coal-like and Higher Humates	High Molecular Weight Humates
Microbiota	Thiobacillus Ferrooxidans	Desulfovibrio Desulfuricans	
Groundwater Features	HCO_3^-, $SO_4^=$, Ca, Mg	HCO_3^-, $S^=$, SO_4 Ca, Mg	HCO_3^-, Ca, Mg
Color	Light Green to Buff	Gray to Black	Light Gray

minerals and with coal-like organic materials. In South Africa, uranium is a by-product of gold mining; in Canada, gold is minor to not present in major uranium deposits.

Excluding sources of which it is recovered as a minor by-product, nearly all other commercial uranium deposits are associated with igneous rocks, either dispersed in the rocks, in pegmatites associated with igneous rocks or in veins peripheral to igneous rocks which are the result of extreme differentiation of a magmatic series. As a magmatic mass cools and crystallizes, uranium is progressively concentrated in the remaining liquid fraction. If the cooling magma is confined and the late-stage liquors are forced to remain in the solid rock, the uranium is concentrated by containment in the volume of rock last to consolidate. If the late-stage liquors escape, hydrothermal veins or aplitic dikes are found associated with the magma after it has cooled. If the volume containing the late-stage differentiate is remelted, the concentration process is continued and large volumes of rock containing about 100 ppm of uranium may result. At the temperatures and pressures associated with the late stages of crystallization of a granitic magma, water dissociates to give a reasonably high oxygen activity and uranium is partly oxidized to the hexavalent state. Thus, instead of fully quadrivalent uraninite being present, uraninite which contains U^{6+} is always found. It is common for 33% of the uranium to be hexavalent giving rise to an empirical formula approximating U_3O_7. In such late-stage rock assemblages, uranium also occurs in accessory minerals such as rare earth oxides, phosphates and the oxides of tantalum, columbium and titanium. Indeed many of the ancient deposits of Australia and Western Canada are characterized by hydrated titanium-uranium-oxide minerals in vein or pegmatitic deposits.

Beneficiation of Uranium Ores

Most metallurgical treatment sequences contain a physical beneficiation or concentration step wherein a specific mineral assemblage containing the sought material is separated from the bulk of the ore as a prelude to hydro or pyro-chemical treatment. Concentration must pay for itself in reduced cost of subsequent processing.

Sandstone Ores. So far, the uranium in sandstone deposits has defied concentration because its minerals have neither a unique combination of physical properties nor a special set of surface parameters. The minerals containing uranium are generally very poorly crystallized and much of the uranium present occurs as extremely dilute dispersions in materials ranging from clays to coal-like organics. Concentrating such materials is usually not worthwhile because neither reagent usage nor plant costs are decreased appreciably. Moreover, when compared with base metal sulfides, flotation recoveries are unacceptably low.

Conglomerate Ores. These considerations have applied to conglomerate uranium ores, as well as sandstone ores, and even the uranium associated with late-stage magmatic phenomena is not usually beneficiated. The latter does oftentimes occur in discrete, well characterized, minerals and as such could possibly be concentrated.

Hydrometallurgy of Uranium

Figure 2 shows the formal oxidation potentials with respect to hydrogen for uranium in acid and basic solution[6]. Uranium is readily hydrated and this tendency increases with charge. The tendency to coordinate with strongly electronegative atoms is great. The uranyl ion ($UO_2{}^{2+}$) is the only stable hexavalent uranium species in aqueous solutions, except perhaps for $(UF_x)^{(6-x)+}$ in strong fluoride solutions. Reference[7], (P. 261, 60 Nb) indicates the presence of UF_6 in strong HF solutions. Uranium forms complexes with most ligands. For U^{4+}, the stability sequence in water for the formation of a neutral species is:

$$F^- > HPO_4{}^{2-} > SO_4{}^{2-} > NO_3{}^- > Cl^- > OH^-$$

For $UO_2{}^{2+}$, the sequence is only slightly different:

$$F^- > CO_3{}^{2-} > H_xPO_4{}^{(3-x)-} > SO_4{}^{2-} > Cl^- > NO_3{}^- > SO_3{}^{2-} > OH$$

These sequences are based on selected data shown in Table III. In spite of the relative

Figure 2 Formal oxidation potentials for Uranium, Iron, Manganese, and Nickel

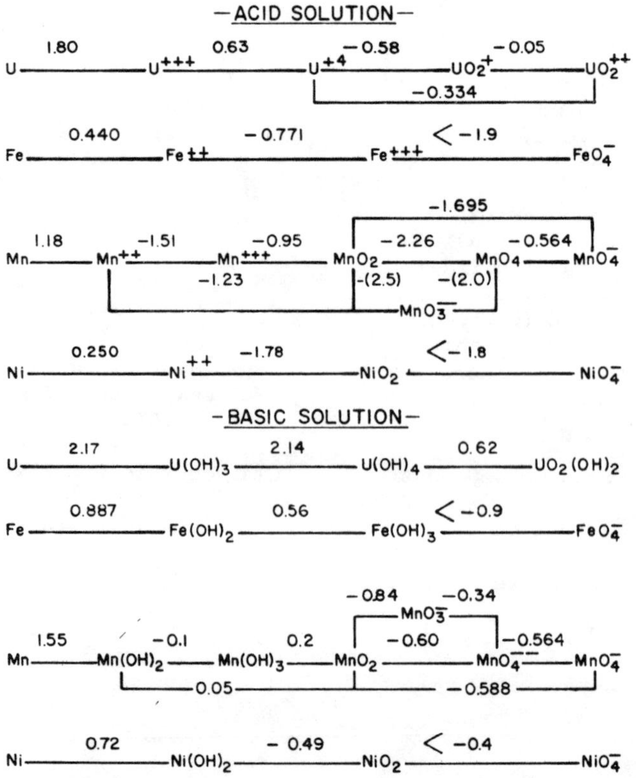

Uranium Data From Jones and Choppin(8), other Data from Latimer(4)

instability of hydroxyl complexes the insolubility of the hydrolysates allows precipitation of U^{4+} hydroxides and alkali diuranates near pH 7.0 even from carbonate solutions.

Dissolution of Uranium Minerals. Intrinsic in any dissolution technique for uranium is oxidation of quadrivalent uranium to hexavalent. This can be accomplished in a variety of solutions, the most common of which is dilute sulfuric acid at a pH ranging from 1.6 to 2.0. In such solutions, a chemical oxidant is generally used to oxidize iron to the ferric state, which in turn oxidizes uranium. At suitable pressures, gaseous oxygen can be used, a higher pressure being required to increase the solubility and therefore the activity of oxygen in the solution. In the case of the sulfuric acid dissolution, uranium goes into solution as the uranyl sulfate complex, the oxidation and solution of uranium being enhanced by the great stability of this complex ion species.

Similarly, the stability of the uranyl carbonate complexes is advantageous in the dissolution of uranium from ores by either sodium bicarbonate or ammonium bicarbonate solutions. Reasonably refractory ores containing a large amount of calcium carbonate (limestone) are treated using sodium bicarbonate solutions at elevated pressures and temperatures, where oxygen super-pressure is utilized for the oxidation/dissolution reaction. At the other end of the spectrum, uraninite (or possibly a naturally occuring chelate of uranium with a high molecular weight organic) is treated in-situ using bicarbonate solutions containing either dissolved oxygen or hydrogen peroxide as an oxidant.

Most uranium production comes from relatively low-grade ores, the usual sandstone mill feed being about 0.1% U_3O_8 , in which uranium occurs in mineralogic configurations ranging from the readily soluble uraninite to highly insoluble mixtures of either uranium and silicon or uranium and silicon or uranium and titanium oxides. For these ores, minimization of leaching costs requires that the lixiviant be tailored to react as little as possible with the host materials. Thus, the high temperature-high pressure sodium bicarbonate leach was devised for refractory uranium ores occurring in limestone; the sulfuric acid-ferric sulfate leach for uranium occuring in sandstone; and the bicarbonate leach for the in-situ leaching of uraninite in calcareous sandstone.

As the real price of uranium increases, deposits of lower and lower grade will be treated. The solubilization of uranium from these lower grade ores will require an increasing knowledge of the chemistry of the host rock, as well as of the uranium minerals, in order to minimize

Table III. Selected Stability Constants for Uranium Complexes in Aqueous Solutions

Cation	Ligand	Method*	Temp	Medium	K_1	K_2	K_3	K_4	β_1	β_2	β_3	β_4	Reference
U^{4+}	F^-	relxn	20°C	2.0 $HClO_4$					7.15	13.4	17.7	23.1	(12)
"	HPO_4^{2-}	sol	?	0.3 $HClO_4$					12.0	22.0	30.6	38.6	(14)
"	SO_4^{2-}	aix	25°C	1.0 H_2SO_4					2.27	3.87			(15)
"	NO_3^-	sp	26.S	2.0 $HClO_4$	0.20					0.17	-0.02	-0.46	(7)
"	NO_3^-	aix	25°C	1.0 HNO_3					0.79	-1.66			(15) (13)
"	Cl^-	sp	25°C	0.5 $NaClO_4$	-0.20								(7)
"	Cl^-	dis	25°C	2.0 $NaClO_4$	0.08	-0.02							(7)
"	$(OH)^-$	sp	25°C	0.5 $NaClO_4$	-1.47								(7)
"	$(OH)^-$	sp	25°C	2.0 $NaClO_4$	-1.68								(7)
UO_2^{2+}	F^-	pH	20°C	1.0 $NaClO_4$	4.59	5.34	2.56	1.35					(8)
"	CO_3^{2-}	sol-sp?	25°C ?	O Corr			3.78						(7)
"	CO_3^{2-}	E	25°C	0.1 $NaClO_4$	**$\beta_{1.1.1}$ 4.10, $\beta_{2.1.0}$ 16.16, $\beta_{3.1.0}$ 21.57								(9)
"	$H_2PO_4^-$	sol	25°C	1.0 $HClO_4$	1.58	1.18							(7)
"	PO_4^{3-}	sol	25°C	1.0 $HClO_4$	***$K_{1,2} = 2.3$								(7)
"	SO_4^{2-}	sp	20°C	1.0 $NaClO_4$	1.75	0.90							(7)
"	Cl^-	E	25°C	1.0 (NO_3^-)	-0.05								(11)
"	Cl^-	aix	25°C	O Corr	-0.1, -0.82, -1.70								(7)
"	NO_3^-	E	25°C	1.0 (ClO_4)	-0.72								(11)
"	SO_3^{2-}	E, sp, sol	25°C ?		-6.0	-3.3							(10)
"	$(OH)^-$	E	25°C	0.1 $NaClO_4$	****$\beta_{1.-1}$ -4.39, $\beta_{2.-1}$ -2.22 $\beta_{2.2}$ -6.09 $\beta_{3.-5}$ -15.64								(9)
"	$(OH)^-$	dis	25°C	0.1 $NaClO_4$	-4.2	-5.2							(7)

* Method abbreviations as follows: relxn = proton resonance, sol = solubility, aix = axion exchange, sp = spectophotometry, dis = distribution between two phases, pH = pH method not specified, E = E.M.F.

** $\beta_{1.1.1} = UO_2CO_3OH$, $\beta_{2.1.0} = UO_2(CO_3)_2$, $\beta_{3.1.0} = UO_2(CO_3)_3$

*** $K_{1,2}$ for $UO_2(H_2PO_4)_2H_3PO_4$

**** $\beta_{1.-1} = UO_2OH$, $\beta_{2.-1} = (UO_2)_2OH$, $\beta_{2.-2} = (UO_2)_2(OH)_2$, $\beta_{3.-5} = (UO_2)_3(OH)_5$

Stability constants reported as log (10) of equilibrium constant. K and β are defined below where M is a cation and L is a ligand.

$$k_1 = \beta_1 = \frac{[ML]}{[M][L]}, \quad k_2 = \frac{[ML_2]}{[ML][L]}, \quad k_3 = \frac{[ML_3]}{[ML_2][L]}, \quad \ldots K_n \frac{[ML_n]}{[ML_{n-1}][L]} \qquad \beta_2 = \frac{[ML_2]}{[M][L]^2}, \quad \beta_3 = \frac{[ML_3]}{[M][L]^3}, \ldots \beta_n = \frac{[ML_n]}{[M][L]^n}$$

and $K_1K_2 = \beta_2$, $K_1K_2K_3 = \beta_3$

reagent costs and the cost of disposing of solutions resulting from metallurgical activities.

Concentration of Uranium from Solution. In the early plants, which treated relatively high-grade ores, solutions resulting from dissolution of uranium from ore contained several grams of uranium per liter, and the most appropriate process used direct precipitation of uranium from such liquors by the addition of a base. Ammonium or magnesium hydroxide was added to acid sulfate liquors and sodium hydroxide to sodium bicarbonate liquors to hydrolyze uranium.

With the advent of commercial nuclear power, and the resulting requirements for additional uranium, the exploitation of lower-grade deposits was undertaken and a concentration step between leaching and precipitation was required. In 1949, the ability of strong base anion exchange resins to adsorb anionic uranyl complexes was discovered. This discovery has been developed sufficiently that now most uranium is recovered from solution by selective adsorption of an anionic uranyl complex either by a strong base anion exchange resin or by a long chain aliphatic tertiary amine dissolved in kerosene. The adsorbed uranium is eluted or stripped using a solution of high ionic strength containing a ligand which either forms a stable complex with the uranium or forms a stable complex with the ion exchanger. Table III shows examples of aqueous stability constants for several uranium complexes. These data indicate the preponderance of anionic complexes in carbonate and sulfate solutions compared with chloride and nitrate. Removal of uranium from solution by anionic exchangers is favored in sulfate and carbonate systems, and elution or stripping is favored in chloride or nitrate systems.

Table IV lists approximate selectivity constants for strong base resins for different anionic species in dilute (.001 to .01 N) solutions. The constant allows estimation of limits of distributions, for pairs of ions, between solution and resin. It has the form of the mass action constant but is used with expressions for fractional ionic strength such as are detailed in Appendix A. Because several simplifying assumptions are made and many factors in solution chemistry are not accounted for, the results must be used with caution. Nevertheless properly used, selectivity constants provide some predictive capability and illustrate some of the strengths and weaknesses of the ion exchange method.

The selectivity of a strong base resin for uranyl tricarbonate is high. On a weight basis, the concentration of uranium on loaded resin is on the order of 1000 times that of the aqueous concentration in 0.01 M ammonium bicarbonate solutions near neutrality. In 0.25 M sulfate-bisulfate plant solutions at pH 2.2, the concentration ratio is on the order of 75. These ratios are in line with the selectivity constants of Table IV.

Solute elution is effected by almost any anion if the concentration is high enough as witnessed by the data of Table IV. Usual elution solutions are 0.5 to 1.5 molar sodium or ammonium chloride solution which is 0.5 molar in sodium sodium bicarbonate. Other eluants include 1.0 to 1.5 molar ammonium carbonate and acidified 1.0 molar ammonium sulfate. Nitrate is a very effective eluant, but is not generally used because of cost, interference during loading, and pollution problems.

Table IV. Approximate Selectivities for Anions, Strong Base Resins.

Ion	$K^{A}_{Cl^-}$	Reference
$HPO_4{}^{2-}$	0.01	(17)
$CO_3{}^{2-}$	0.02	(17)
OH^-	.05*.07	(16)
F^-	0.1	(16)
SO_4^2	0.15	(16)
$HCO_3{}^-$	0.4	(17)
Cl^-	1.0	(16)
Bo^-	3.0	(16)
$NO_3{}^-$	4.0	(16)
I^-	8.0	(16)
$UO_2(SO_4)^{2-}$	8.0	**
$UO_2(CO_3)_3{}^{4-}$	100.0	**

* For type I resin.
** Calculations by author.

The long chain aliphatic amines, such as tri-n-dodecylamine or tri-caprylylamine, exhibit high selectivity for uranium from acid sulfate solutions. These reagents are dispersed in a special high flash point kerosene along with a long chain alcohol, such as isodecanol. They are contacted countercurrently with uranium bearing solutions in order to achieve maximum concentration of uranium in the organic phase and minimum concentration in the aqueous phase. Contact is made in agitated tanks with interstage separation in continuous settler-skimmers. The separated organic phase is then contacted with an aqueous stripping solution in countercurrent stages, to achieve a minimum uranium concentration in the organic and a maximum in the aqueous phase. In both the loading and stripping operations, concentration is limited by the allowable concentration distribution between the phases. Figure 3 shows the concentration distribution for extraction of uranium by a commercial amine from an operating mill solution[18]. The ratio between the concentration in the organic and aqueous phases ranges from 1000 at low aqueous concentrations to around 3.6 at near saturation of the organic. For molybdenum and vanadium the ratios are less than unity under mill conditions.

From acid sulfate leach solutions uranium is loaded as the uranyl disulfate $UO_2(SO_4)_2{}^{2-}$ or trisulfate $UO_2(SO_4)_3{}^{4-}$ complex[19]. The solvent is usually stripped with 1.5 molar ammonium sulfate at pH 4.2 or with 1.0 to 1.5 molar ammonium chloride at a pH near 2.0 as shown by the distribution diagram in Figure 4. Vanadium is stripped along with uranium but molybdenum is held by the solvent and a side stream of uranium-depleted organic is treated with sodium or ammonium carbonate solution to remove molybdenum.

Figure 3. Typical Extraction Distribution Diagram Showing Organic and Aqueous Concentrations of Uranium, Molybdenum and Vanadium. Courtesy Ashland Oil Co.[18]

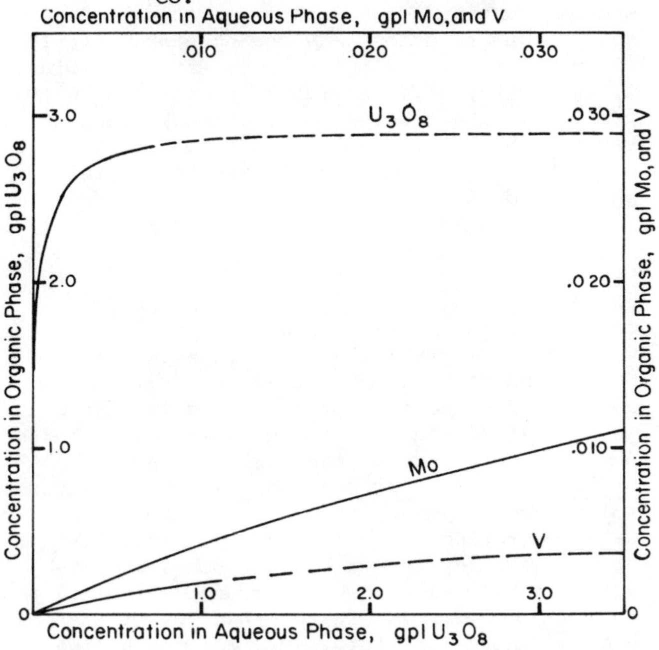

Aqueous Feed - .76 gpl U_3O_8, .035 gpl Mo, .011 gpl V, pH = 1.30, EMF = -.405 volts against calomel electrode.

Organic Phase - 2.5 W/V % triisooctylamine, 2.5 W/V % isodecanol balance is high flash point kerosene.

Uranium Precipitation

Uranium may be precipitated by reduction from U^{6+} to U^{4+} using H_2, SO_2 or Fe°. The U^{4+} phosphate is only slightly soluble and attempts have been made to utilize this fact for recovery of a uranium concentrate.

The alkali diuranates, for example $Na_2O \cdot 2UO_3$ are quite insoluble. The "yellowcake" of commerce consists of the yellow salts obtained by neutralization of acid solutions using either sodium, ammonium or magnesium hydroxide. Where bicarbonate is present a strong base must be added to hydrolyze the carbonate complexes. Typical reactions include: for sulfate complexes,

$$2(NH_4)_2UO_2(SO_4)_2 + 6NH_4OH \rightleftarrows (NH_4)_2O \cdot 2UO_3 + 4(NH_4)_2SO_4 + 3H_2O \quad (1)$$

for carbonate complexes,

$$2Na_4UO_2(CO_3)_3 + 6NaOH \rightleftarrows Na_2O \cdot 2UO_3 + 6Na_2CO_3 + 3H_2O \quad (2)$$

and for chloro complexes,

$$2Na_2UO_2Cl_4 + 6NaOH \rightleftarrows Na_2O \cdot 2UO_3 + 8NaCl = 3H_2O \quad (3)$$

Although these reactions are widely utilized they are not well-known. Because of the strong tendency of U^{6+} to be hexacoordinated, the precipitated compound almost certainly passes through a state of full hydration ($Na_2U_2O_3(OH)_8$ or $Na_2U_2O_7 \cdot 4H_2O$) before crystallizing to the anhydrous diuranate, $Na_2U_2O_7$.

Figure 4. Typical Strip Distribution Diagram Showing Uranium Concentrations in Organic and Aqueous Phases. Courtesy Ashland Oil Co.[18]

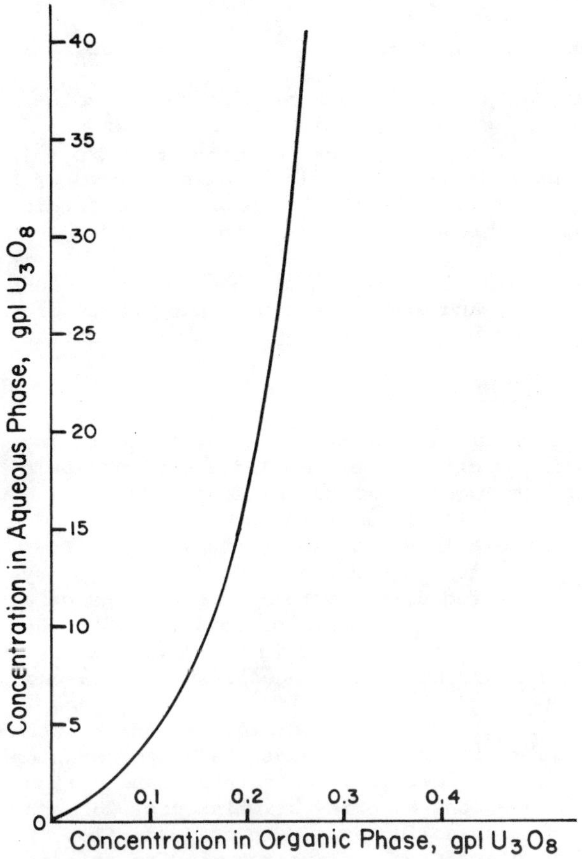

Aqueous Feed - 200 gpl $(NH_4)_2SO_4$ with H_2SO_4 to pH 4.3.

Organic Phase - 2.5 W/V % triisooctylamine, 2.5 W/V % isodecanol, balance is high flash point kerosene.

If sodium must be removed to meet marketing requirements, washing with an ammonium salt converts the sodium to the ammonium diuranate:

$$Na_2O \cdot 2UO_3 \cdot 4H_2O + 2NH_4OH \rightleftarrows (NH_4)_2O \cdot 2UO_3 \cdot 4H_2O + 2N_2OH \quad (4)$$

This conversion is not readily accomplished after drying.

Precipitates from sulfate solutions are usually contaminated by basic uranyl sulfates and may have some U^{6+} hydroxide in addition to the sulfate salts from the mother liquor.

Precipitations from chloride solutions are not contaminated to a great extent by basic chlorides but do contain chloride salts.

Precipitates from sodium carbonate solutions contain basic carbonates as well as sodium carbonate and hydroxide. These must be washed thoroughly, often with ammonium solutions, to obtain a marketable product. Because of the washing requirement, a coarse crystalline product is desired. Considerable art and close control of process conditions are required to meet this need.

In some plants, hydrogen peroxide is used to precipitate the insoluble peroxide. For example,

$$UO_2SO_4 + H_2O_2 + nH_2O \rightleftarrows UO_4 \cdot nH_2O + H_2SO_4 \quad (5)$$

Some plants treating ammonium carbonate solutions are stripping resin with 1.0 molar $(NH_4)_2CO_3$. This solution contains about 20 gpl U_3O_8 as a carbonate complex. When sparged with steam, NH_3 and CO_2 are driven off and a terminal product approaching $UO_3 \cdot 3H_2O$ is obtained.

Most plants dewater the washed product centrifugally and dry it in a multiple hearth or tray drier before packaging it for shipment.

Typical Uranium Mill

The chemical considerations outlined in the previous sections are exemplified by the Sohio Petroleum Company's L-Bar Mill near Seboyeta, N.M. The mill is illustrated in Figure 5. Mine-run ore is delivered by truck to Ore Receiving where it is either placed in temporary storage or fed directly to an ore bin whence a pan feeder and belt conveyor conduct it to the Grinding Circuit where it is mixed with warm water and ground semi-autogenously in a Cascade mill operated in closed circuit with cyclone classifiers. The mill operates as a high intensity attritioner. It cleans the softer uranium-bearing materials from the natural sand grains in the ore, the majority of which pass 28 mesh. The >28 mesh solids appear in the underflow of a cyclone classifier and are returned to grinding. The fine overflow (<28 mesh) passes through an automatic sampler to one of two slurry holding tanks. The fine slurry is metered through a mass recorder/totalizer to the Acid Leach Circuit where it is subjected to the attack of hot dilute sulfuric acid in the presence of sodium chlorate as an oxidizer. The slurry at about 50% solids passes serially through six tanks taking about one hour for each tank. A free acid level of about 20 gpl H_2SO_4 (pH 0.5 to 0.6), an EMF (platinum to calomel) $\geqq$500 MV, and a temperature of 50° to 60° is maintained in the leach. Liquid and solids are then separated in Counter Current Decantation where six continuous settler/decanters (thickeners) are used to wash the solids free of dissolved uranium. About 3.0 times as much raffinate from solvent extraction is added for washing as the water contained in the leach slurry. A pregnant overflow to the Clarification Circuit of 0.4 to 0.6 gpl U_3O_8 is produced while the solution discarded with the solids to tailings allows a bleed of impurities from the plant.

After removal of suspended solids to <10 ppm, the clarified pregnant solution goes to the Solvent Extraction Section where it is contacted with an organic solution of about 3.0% each of tricaprylylamine and tridecanol in high flash point kerosene. Contact is made in agitated tanks (mixers) from which the emulsion so formed goes to continuous decanters (settlers). Organic and aqueous solutions advance countercurrently. Pregnant organic is built up to between 2.0 and 2.5 gpl U_3O_8, while raffinate contains 5 to 10 ppm U_3O_8. The process is reversed in the strip section where a 1.0 molar ammonium sulfate solution at pH 4.0 is contacted with the organic phase to deplete it to <.10 gpl U_3O_8 while an aqueous phase containing from 15 to 30 gpl U_3O_8 is produced.

In the Precipitation Section the pH is increased to pH 7.0 with an air-anhydrous ammonium mixture. The first tank is held at about 60°C. The second tank is brought to pH 7.2. The ammonium diuranate so precipitated is washed in two settler/decanters (thickeners) before being pumped as a slurry to a centrifuge in the Drying and Packaging Section. The centrifuged solids are dropped directly into a rotating tray drier operating at 260° to 400°C. From the drier they drop into a storage bin for packaging in 55 gallon steel drums for shipment. Overall recovery is greater than 90% of the uranium in the ore into a product containing greater than 90% U_3O_8.

Literature Cited

1. Rankama, K. and Sahama, T.G., Geochemistry, University of Chicago Press (1950).
2. Wedepohl, K. G., Handbook of Geochemistry, SpringerVerlag (1969).
3. Merritt, R. C., The Extractive Metallurgy of Uranium, Colo. Sch. of Mines Resch. Inst. (1971).
4. Harshman, E. M., Uranium Ore Rolls in the United States, Uranium Exploration Geology, Intl., Atom. Enrg. Agncy., STI/PUB/ (1970), 277 (IAEAPL-391/4), 219231.
5. Rubin, B., Uranium Roll Front Zonation in the Southern Powder River Basin, Wyoming, Wyo. Geol. Assn. Earth Sci. Bull. (1970), 512.
6. Latimer, W. M., Oxidation Potentials, PrenticeHall (1952).
7. Sillen, L. E. and Martell, A. E., Stability Constants of Metal-Ion Complexes, Special Publication 17, The Chemical Soc. Burlington House, W. 1, London.
8. Jones, A. D. and Choppin, G. R., Complexes of Actinide Ions in Aqueous Soln., Actinides Rev., 1(4) (1969), 31136.
9. Tsymbal, C., Solution Chemistry of Uranium (VI), Lab. Chim. Nucl., C.E.N., Grenoble, Fr., Commis. Energ. At. (Fr), Rapp (1969) CEA-R-3476, pp 99.
10. Zakharova, F. A. and Orlova, M. M., Formation of Complexes Between the Uranyl and Sulfite Ions, Zh, Neorg. Khim. 12(11) (1967) 3016-22, Russ.
11. Ohashi, H. and Morozumi, T., Electrometric Determination of Stability Constants of Uranyl-Chloride and Uranyl-Nitrate Complexes with pCl-stat., Nippon Genshiryoku Gakkaishi, Feb. 9 (1967) pp 65-71.
12. Vdovenko, V. M., Romanov, G. A. and Shcherbakov, V. A., A Study of the Complexation

Figure 5. Sohio Petroleum Co.'s L-Bar Mill, Valencia Co. N. M., Courtesy Sohio Petroleum Co.[20]

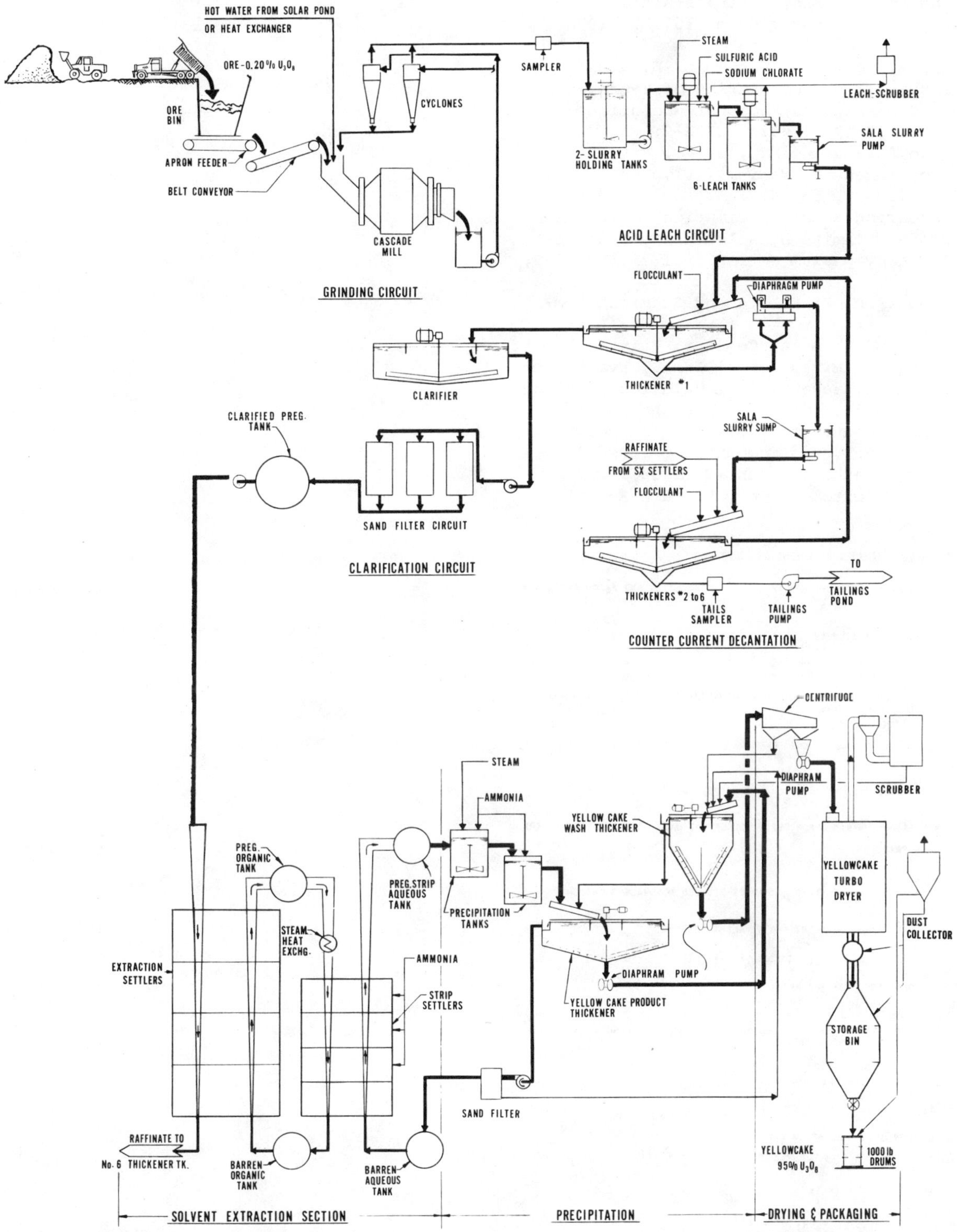

of Uranium (IV) by Fluoride Ions by the Proton Resonance Method, Radiokhimiya 5(5) (1963), 581-5.

13. Klygin, A. E., Kolyada, N. S. and Smirnova, I.D., Spectrophotometric Study of Complexing in the Uranyl Nitrate-Nitric Acid-Water System, Zh. Neorg. Khim. 15(12) (1970), 330 0-3.

14. Moskvin, A. E., Essen, L. N. and Bukhtiyarova, T. N., Complexing of Tetravalent Thorium and Uranium in Phosphate Solutions, Zh. Neorg. Khim. 12(12) (1967), 3390-2.

15. Rao, C. L. and Pai, S. A., Nitrate and Sulfate Complexes of Uranium (IV), Radiochem. Acta 12(3) (1969), 135-40.

16. Anderson R. E., Estimation of Ion Exchange Process Limits by Selectivity Calculations, AIChE Symposium Series, Vol 71, No. 152, 236-242.

17. Anderson, R. E. Diamond Shamrock Company.

18. Ashland Chemical Company, Private Report by Colo, Sch. of Mines Resch. Inst., Evaluation of Adogen Amine Reagents in Liquid Ion Exchange of Uranium (1973).

19. Ling Ch'iu, et. al., Yuan Tzu Meng, 3, 21532 (ch) (1965), via Communist Chinese Sci. Abstr., Phys. Math. 15, Jan. 2 (1966).

20. Sohio Petroleum Co., Private Report.

APPENDIX A - ION EXCHANGE DISTRIBUTION EQUATIONS

FOR THE EXCHANGE REACTION BETWEEN CATION A_S^+ IN SOLUTION AND B_R^+ ON THE RESIN:

$$A_S^+ + B_R^+ \rightleftharpoons A_R^+ + B_S^+$$

THE MASS ACTION EXPRESSION IS:

$$K_{B^+}^{A^+} = \frac{[A_R^+]\,[B_S^+]}{[A_S^+]\,[B_R^+]}$$

FOR SOLUTIONS WHERE C = TOTAL IONIC CONCENTRATION AND ONLY TWO EXCHANGABLE CATIONS ARE INVOLVED:

$$X_{A_S}^+ = \text{EQUIVALENT FRACTION OF TOTAL IONIC STRENGTH} = \frac{[A_S^+]}{C_S}$$

$$\text{AND, } X_{B_S}^+ = \frac{[B_S^+]}{C_S} \qquad \text{AND, } X_{B_S}^+ = X_{A_S}^+ = 1$$

SIMILAR EXPRESSIONS OBTAINED FOR RESIN CONCENTRATIONS. SUBSTITUTING IN THE MASS ACTION EXPRESSION:

$$\frac{X_{A_R}^+}{1-X_{A_R}^+} = K_{B^+}^{A^+} \frac{X_{A_S}^+}{1-X_{A_S}^+}$$

GOVERNS EXCHANGES BETWEEN MONOVALENT CATIONS.

FOR EXCHANGES BETWEEN MONOVALENT AND DIVALENT CATIONS:

$$2A_R^+ + B_S^{2+} \rightleftharpoons 2A_S^+ + B_R^{2+}$$

THE MASS ACTION EXPRESSION BECOMES:

$$K_{A^+}^{B^{2+}} = \frac{[A_S^+]^2 \quad [B_R^{2+}]/2}{[A_R^+]^2 \quad [B_S^{2+}]/2}$$

AND THE EQUIVALENT FRACTION FORM IS:

$$\frac{X_{B_R}^{2+}}{(1-X_{B_R}^{2+})^2} = K_{A^+}^{B^{2+}} \frac{C_R}{C_S} \frac{X_{B_S}^{2+}}{(1-X_{B_S}^{2+})^2}$$

THE GENERAL FORM OF THE EQUATION IS:

$$mA_S{}^n + nB_R{}^m \rightleftharpoons mA_R{}^n + nB_S{}^m$$

THE MASS ACTION EXPRESSION IS:

$$K_{B^m}^{A^n} = \frac{[A_R{}^n]^m \quad [B_S{}^m]^n}{[A_S{}^n]^m \quad [B_R{}^m]^n}$$

THE EXPRESSIONS FOR EQUIVALENT FRACTION OF TOTAL IONIC STRENGTH ARE:

$$X_{A_S}{}^n = \frac{[A_S{}^n]}{C_S} \text{ AND } X_{B_S}{}^m = \frac{[B_S{}^m]}{C_S} \qquad X_{A_R}{}^{n-} = \frac{[A_R{}^m]}{C_R} \text{ AND } X_{B_R}{}^{m-} = \frac{[B_R{}^m]}{C_R}$$

AND THE EQUIVALENT FRACTION FORM OF THE MASS ACTION EQUATION IS:

$$\frac{(X_{B_R}{}^m)^n}{(1-X_{B_R}{}^m)^m} = K_{A^n}^{B^m} \frac{C_R{}^m C_S{}^n}{C_R{}^n C_S{}^m} \frac{(X_{B_S}{}^m)^n}{(1-X_{B_S}{}^m)^m}$$

Chapter 3

THE CHEMISTRY OF GOLD AND SILVER RECOVERY FROM NATURAL RESOURCES

George M. Potter and Roshan B. Bhappu

Mountain States Research and Development
Tucson, Arizona 85731

ABSTRACT

The chemistry involved in the recovery of gold and silver from natural resources through hydrometallurgical and pyrometallurgical processes is reviewed. Process metallurgy for the cyanidation, activated carbon adsorption, carbon stripping, and electrowinning of gold and silver is discussed. The application of zinc sodium sulfide use for silver recovery from cyanide solutions is covered. Electrooxidation, other oxidation methods, and roasting processes for problem ores and concentrates containing organic carbon and impurities such as arsenic, antimony, and tellurides are summarized. The smelting of cathodes, precipitates, and other products, and the various wet and furnace steps in separating dore metal are also reviewed.

THE CHEMISTRY OF GOLD AND SILVER RECOVERY FROM NATURAL RESOURCES

George M. Potter and Roshan B. Bhappu

Mountain States Research and Development
Tucson, Arizona 85731

INTRODUCTION

Gold and silver have maintained their historical desirability for storing wealth and use in the arts but have also gained highly important industrial uses in the last century. Notably, gold, for example, has large applications in high-reliability electronics and silver is widely used in photography and electrical service. The world production of gold in 1975 was 38.6 million ounces and of silver was 294.3 million ounces. In the United States, 1975 newly-mined gold production was one million ounces, while the total demand for gold was four million ounces, about half for jewelry and arts and half for industrial and dental purposes. United States mine production of silver in 1975 was 34.9 million ounces and demand was 160 million ounces, of which about half was used in photography and the remainder in silverware, electrical equipment, and other industrial and arts applications (1).

Gold and silver production methods in the United States and other countries are similar. In 1975, 59 percent of the newly mined U.S. gold was produced by the cyanidation process, 38 percent as a by-product of base metal smelting, and only three percent by amalgamation and placer mining. As for silver, about 99 percent of the production in 1975 came from base metal smelting and only one percent from cyanidation, with negligible amalgamation production (1). However, new U.S. cyanidation production facilities on stream in 1977 should about double the U.S. primary silver production coming from cyanide-processed ore.

This paper focuses on the chemistry of the widely used processes: cyanidation, smelting, and subsequent refining. Renewed interest in hydrometallurgical processes for base metal concentrates also merits brief review of by-product gold and silver potential from such methods.

HYDROMETALLURGICAL EXTRACTION FROM ORES

Cyanidation - Leaching

The cyanidation leaching process for gold and silver ores began to replace the earlier chlorination and sodium thiosulfate leaching methods in about 1900, because cyanidation is generally cheaper, more effective, and more selective. The cyanidation process consists of percolation leaching of crushed ore or agitation of a ground slurry with dilute cyanide solution, generally less than 0.3 percent sodium cyanide ($NaCN$) in strength. The solution is alkaline to prevent hydrolysis of the cyanide with accompanying cyanide loss, and lime (CaO) is commonly used to attain the typical pH values of 10.0 to 10.5, although sodium hydroxide ($NaOH$) or sodium carbonate (Na_2CO_3) may be used. Any of a number of alkaline cyanide salts may be used effectively, and some organic nitriles also work. However, sodium cyanide, or sometimes calcium cyanide, is usually preferred. Dissolved oxygen is an essential ingredient in the leaching step. Dissolved gold and silver may be precipitated from clarified leaching solutions by metallic displacement, usually with zinc dust. Alternatively, unclarified solutions may be treated with granular activated carbon or resins to adsorb the precious metal values.

In recent years, comprehensive articles on cyanidation chemistry include reports by N. P. Finkelstein (2) and by Hedley and Tabachnick (3).

Gold in the form of clean, fine metallic particles is readily amenable to cyanidation. Gold occurring as selenide or telluride minerals must be roasted or chemically processed to make the gold satisfactorily leachable in dilute cyanide. According to Barsky (4) et al, pure gold is soluble at the rate of about three milligrams per centimeter squared per hour with proper aeration and agitation.

The main gold dissolution reaction, as reported by N. P. Finklestein (2) and Hedley and Tabachnik (3), is now indicated to be:

$$2Au + 4CN^- + O_2 + 2H_2O \longrightarrow 2Au(CN)_2^- + 2OH^- + H_2O_2 \quad (1)$$

Some of the hydrogen peroxide thus formed may be used in the reaction:

$$2Au + 4CN^- + H_2O_2 \longrightarrow 2Au(CN)_2^- + 2OH^- \quad (2)$$

A small part of the gold may be dissolved as shown by the classical Elsner equation:

$$4Au + 8CN^- + O_2 + 2H_2O \longrightarrow 4Au(CN)_2^- + 4OH^- \quad (3)$$

The maximum rate of gold dissolution takes place in cyanide solutions of not over 0.05 percent strength (as NaCN). Gold dissolution rate appears to peak at about 85 degrees celsius, as an optimum compromise between oxygen content of the solution and activity. However, heating solutions is not considered practical for ores.

Metallic silver also dissolves in dilute cyanide solutions in the presence of oxygen, at one-half the rate of gold, or about 1.5 milligrams per centimeter squared per hour. The equations are similar to those for gold. However, silver occurs in its ores most commonly as sulfide minerals. The reaction for silver sulfide is:

$$Ag_2S + 4CN^- \longrightarrow$$

$$2Ag(CN)_2^- + S^{-2} \quad (4)$$

The sulfide ion must be removed from action so that the dissolution can proceed, and such removal is by oxidation as follows:

$$S^{-2} + CN^- + 0.5\ O_2 + H_2O \longrightarrow$$

$$CNS^- + 2OH^- \quad (5)$$

$$2S^{-2} + 2O_2 + H_2O \longrightarrow$$

$$S_2O_3^{-2} + 2OH^- \quad (6)$$

and the thiosulfate so formed eventually further oxidizes:

$$S_2O_3^{-2} + 2OH^- + 2O_2 \longrightarrow$$

$$2SO_4^{-2} + H_2O \quad (7)$$

Silver halides do not require oxygen for cyanidation, for example:

$$AgCl + 2CN^- \longrightarrow$$

$$Ag(CN)_2^- + Cl^- \quad (8)$$

Current and historical price ratios for gold and silver require much higher silver ore weight content than gold for equivalent economy. The amount of cyanide consumed in actual silver dissolution from treatable ores is thus considerable, aside from cyanide consumed by impurities. Silver leach solutions may well contain as much as 0.3 percent NaCN.

Part of the alkaline cyanide used in leach solutions is lost by entrapment in residues, part by action of cyanicides (cyanide-consuming impurities), some as hydrogen cyanide (HCN) gas, and part as ammonia (NH_3) and nitrogen (N) compounds. Cyanide may react with carbon dioxide (CO_2) as follows:

$$CN^- + CO_2 + H_2O \longrightarrow$$

$$HCN + HCO_3^- \quad (9)$$

Also, cyanide may hydrolyze accordingly:

$$CN^- + H_2O \rightleftharpoons$$

$$OH^- + HCN \quad (10)$$

Losses by CO_2 and hydrolysis reactions are reduced by maintaining alkaline protection, usually through the use of lime (CaO) at pH 10.0 to 10.5. At a pH higher than 10.5 with lime, the gold and silver dissolution rates are appreciably slowed, whereas, at a pH of less than 10, loss of cyanide by hydrolysis becomes increasingly evident with lowering pH.

Cyanidation - Metal Recovery

Gold and silver are commonly recovered from clarified leaching solutions by precipitation with zinc dust. The effectiveness of the zinc is enhanced by de-aerating the solution prior to zinc addition as practiced in the Merrill-Crowe system by vacuum. The main reactions for gold are probably:

$$2Au(CN)_2^- + Zn \longrightarrow$$

$$2Au + Zn(CN)_4^{-2} \quad (11)$$

$$2Au(CN)_2^- + Zn + 3OH^- \longrightarrow$$

$$2Au + HZnO_2^- + 4CN^- + H_2O \quad (12)$$

The silver precipitation reactions are similar.

The precipitation of gold and silver by zinc is a cementation process, proceeding by local anode-cathode reaction and also resulting in a dissolution of zinc. By adding soluble lead salts, the reaction may in some cases be improved probably through assisting the electrochemical activity. The removal of oxygen under vacuum decreases the tendency of gold to redissolve in cyanide and cuts down the oxidative dissolution of the zinc in the cyanide. Reactions for the oxidative dissolution of zinc are given by N. P. Finkelstein (2), as follows:

$$Zn + 0.5\ O_2 + 4CN^- + H_2O \rightleftharpoons$$

$$Zn(CN)_4^{-2} + 2OH^- \quad (13)$$

$$Zn + 0.5\ O_2 + H_2O \rightleftharpoons$$

$$HZnO_2^- + H^+ \quad (14)$$

Aluminum is also an effective precipitant for gold and silver from de-aerated cyanide solutions, although it has been little used commercially. The reaction is an anodic-cathodic cementation, and the solution must be alkaline with NaOH to prevent the formation of insoluble calcium aluminates. Hamilton (5) gives the reaction:

$$2NaAg(CN)_2 + 4NaOH + 2Al \longrightarrow$$

$$4NaCN + 2Ag + Na_2Al_2O_4 + 4H^+ \quad (15)$$

The gold reaction is similar.

Renewed interest in a neglected process for silver precipitation has recently been shown by the precious metal industry, namely the use of sodium sulfide to precipitate silver sulfide and regenerate cyanide, as in the reaction: (6)

$$2NaAg(CN)_2 + Na_2S \longrightarrow$$

$$Ag_2S + 4NaCN \quad (16)$$

The precipitated sulfide would be recovered by filtration. Gold is not precipitated, but could be recovered in a subsequent activated carbon adsorption step, as described later in this paper.

As an alternative to the liquid-solid separation and zinc dust precipitation of gold and silver from cyanide solutions, the application of ion exchange resins or activated carbon to either clear solutions or slurries may be considered. The use of ion exchange resins was proposed by A. J. Gilmore (7), who presented the following reaction:

$$RCl + NaAu(CN)_2 \longrightarrow$$

$$RAu(CN)_2 + NaCl \qquad (17)$$

where R is an anion exchanger.

The stripping reaction with Amberlite IRA-400 resin (Rohm and Haas) was given as:

$$RAu(CN)_2 + NH_4SCN \longrightarrow$$

$$RSCN + NH_4Au(CN)_2 \qquad (18)$$

The stripped gold in solution was then electrowon as follows:

$$NH_4Au(CN)_2 \longrightarrow$$

$$Au\downarrow + NH_4^+ + 2(CN)^- \qquad (19)$$

Both gold and silver in cyanide leach solutions are successfully recovered commercially by adsorption on granular activated carbon. The mechanism of gold adsorption on activated carbon was discussed by R. J. Davidson (8), and other definitive work on gold and silver adsorption by Cho, Dixon, and Pitt has been reported (20), (21), (22). The Freundlich isotherm relationship best describes the adsorption:

$$\frac{X}{M} = kc^{1/n} \qquad (20)$$

where: X = mass of solute adsorbed
M = mass of adsorbent
c = concentration of solute remaining in solution
k and n are constants

Calcium ions present in gold and silver cyanide solutions favor the adsorption of the auro- or argento-cyanide complex. whereas sodium ions inhibit the adsorption, particularly above pH 10. A practical process for the stripping of gold and silver values from activated carbon was described by the Bureau of Mines (9). A one percent NaOH, 0.1 percent NaCN strip solution at near-boiling temperature is used to strip precious values from carbon typically loaded from 200 ounces to 600 ounces of combined gold and silver per short ton of carbon. Stripping takes from 40 to 60 hours. Strip solutions are electrolyzed to recover gold and silver on steel wool cathodes. Carbon is generally reused, after periodical reactivation by heat treatment at 600 degrees celsius. An elevated temperature of 160 degrees celsius cuts stripping time to four to six hours (10), as does the use of an ethanol or methanol-caustic-cyanide aqueous mixture at 60 to 80 degrees celsius (6).

Cyanidation - Impure Ores and Treatment

Some minerals and their soluble oxidation products occurring in gold and silver ores are so-called cyanicides, substances that consume cyanide without benefit to the process and that may also inhibit the dissolution of gold and silver. Notable examples are iron, copper, arsenic, antimony, and zinc minerals.

Iron minerals oxidize to form ferrous iron that forms ferrocyanides in alkaline solution:

$$Fe^{++} + 6CN^- \longrightarrow$$

$$Fe(CN)_6^{-4} \qquad (21)$$

Numerous other complex ferric and ferrous salts are also formed. All are either very weak or completely ineffective precious metal solvents. Also, even slight amounts of sulfide ion derived from iron sulfide minerals may passivate gold surfaces and inhibit cyanidation. Aeration in alkaline circuit generally controls iron problems prior to cyanidation.

Copper sulfide and oxide minerals are soluble in dilute alkaline cyanide solutions, and from about five to more than 95 percent of the minerals may dissolve, thus

$$2CU^{++} + 7CN^- + 2OH^- \longrightarrow$$

$$2Cu(CN)_3^{-2} + CNO^- + H_2O \qquad (22)$$

The copper cyanogen complexes represent a loss of cyanide and are not satisfactory gold and silver solvents.

In cyanidation, arsenic and antimony minerals, particularly As_2S_3 and Sb_2S_3, are increasingly soluble as alkalinity increases and seriously inhibit cyanidation, as shown by Hedley and Tabachnik (3) as follows:

$$2As_2S_3 + 6Ca(OH)_2 \longrightarrow$$

$$Ca_3(AsO_3)_2 + Ca_3(AsS_3)_2 + 6H_2O \qquad (23)$$

$$Ca_3(AsS_3)_2 + 6Ca(OH)_2 \longrightarrow$$

$$Ca_3(AsO_3)_2 + 6CaS + 6H_2O \qquad (24)$$

$$Ca_3(AsS_3)_2 + 6NaCN + 3O_2 \longrightarrow$$

$$6NaCNS + Ca_3(AsO_3)_2 \qquad (25)$$

The detrimental effect is believed to be caused by relatively stable $S^=$. The addition of lead nitrate seems to accelerate the oxidation of the sulfide ion.

Preliminary roasting or caustic leaching may be used to prepare ores for cyanidation, if economics permit.

Zinc minerals dissolve in alkaline cyanide solutions in varying degrees. The resulting zinc complex cyanides are not active solvents for gold and silver, but do not inhibit cyanidation if free cyanide is present.

Carbon occurring in various forms in carbonaceous ores may seriously inhibit gold and silver dissolution by preventing physical access of the solution to included mineral grains, but may also adsorb dissolved precious metals and cause losses to solid tailings. Carbonaceous ores may be roasted and cyanided if costs are favorable. The U.S. Bureau of Mines studied carbonaceous ores and found that alkaline hypochlorite pretreatment generally made the slurries amenable to subsequent cyanidation (11). The hypochlorite might be added as such or generated by adding gaseous chlorine to alkaline slurries. Alternatively, the hypochlorite may be generated in place by electrolysis of salt brine slurries.

Chlorination

The chlorine process for gold recovery was used widely before cyanidation displaced it around 1900. Chlorine in the presence of water attacks fine accessible gold rapidly, as follows:

$$Au + 3/2Cl_2 \longrightarrow AuCl_3 \text{ (water soluble)} \quad (26)$$

The soluble gold, after solids separation, was generally precipitated with ferrous sulfate or sulfur dioxide. Ore containing appreciable silver was then treated by thiosulfate (hyposulfite) leaching to dissolve silver chloride, as later described. More recently, Finkelstein, et al (12) and van Zyl and others (13) reviewed the chlorination process, reported on pilot work, and suggested the method for treating Merrill slimes (zinc dust precipitation) and gravity concentrates.

Chlorination - Leaching of Base Metal Ores

In recent years, great interest has been focused upon the processing of base metal ores by hydrometallurgical means as a substitute for pyrometallurgical reduction. To the extent that the new hydrometallurgy is commercially adopted, precious metals will be produced through these systems. For example, the Duval "CLEAR" process is described in U.S. Patent 3,879,272 (April 22, 1975, Atwood et al). The method comprises ferric chloride attack of chalcopyrite to form cupric chloride, reduction of the cupric chloride to cuprous chloride, recovery of metallic copper (electrolysis preferred), and regeneration of ferric chloride by oxidation, concurrently purging iron, sulfate, and other impurities from the system.

Equations as given in the patent are:

Reduction Stage:

$$2CuFeS_2 + 3CuCl_2 \longrightarrow 4CuCl + FeCl_2 + 2S + CuFeS_2 \quad (27)$$

($4CuCl + FeCl_2$ → to electrolysis; $2S + CuFeS_2$ → to combined oxidation, regeneration-purge)

Electrolysis Stage:

$$4CuCl + FeCl_2 \longrightarrow 2Cu + 2CuCl_2 + FeCl_2 \quad (28)$$

($2Cu$ → Product; $2CuCl_2 + FeCl_2$ → to combined oxidation, regeneration-purge)

Combined Oxidation, Regeneration - Purge Stage:

$$CuFeS_2 + 2S + 2CuCl_2 + FeCl_2 + 3/2O_2 + 3H_2O \longrightarrow 3CuCl_2 + 2Fe(OH)_3 + 4S \quad (29)$$

($3CuCl_2$ → to reduction; $2Fe(OH)_3 + 4S$ → residue)

Gold and silver in the Duval "CLEAR" process probably dissolve to differing degrees, and the soluble precious values are recovered together with the copper by electrolysis. The rest of the gold and silver remains in leach residues for processing by cyanidation or other known methods.

Other hydrometallurgical processes have also been pioneered and operated for treating base metal concentrates, mostly copper. The Arbiter Anaconda process (14) employs an ammoniacal sulfate leaching step, followed by solvent extraction and electrowinning. Precious metal values remain in the residue for further treatment.

Other ferric chloride leaching methods have been developed by the Cymet group (15) for copper concentrates and the U.S. Bureau of Mines (16) for lead.

THIOSULFATE LEACHING

Silver chloride, whether a natural mineral or created by chloridizing roasting, is rapidly dissolved in sodium thiosulfate ($Na_2S_2O_3 \cdot 5H_2O$, the hyposulfite of photography) as follows:

$$Ag^+ + 2(S_2O_3^{-2}) \longrightarrow Ag(S_2O_3)_2^{3-} \quad (30)$$

The silver was generally precipitated as sulfide by adding a soluble sulfide. Although the process is not applied much at present, it might find a revived role in a post-chlorination step in a combination process.

PYROMETALLURGICAL PROCESSING OF GOLD

Aside from hydrometallurgy, chiefly cyanidation at present, smelting processes account for much of the gold and silver production into refinable materials. Copper (or copper-nickel) and lead ores account for a large part of the by-product gold and silver production, usually via froth flotation concentrates.

LEAD ORES

Lead ores and concentrates are commonly smelted in the blast furnace (17). Lead sulfides, exemplified by galena (PbS), are roasted to oxide as follows:

$$2PbS + 3O_2 \longrightarrow$$

$$2PbO + 2SO_2 \qquad (31)$$

The oxide sinter formed in roasting is then charged to the blast furnace with suitable fluxes and reduced to metallic lead, which "collects" or dissolves most of the precious metals, thus

$$2PbO + C \longrightarrow$$

$$2Pb + CO_2 \qquad (32)$$

Impurities form iron and calcium silicate slags, which are drawn off separately. Gold and silver may be removed from molten lead in a separate refining step by stirring in metallic zinc and skimming off a resulting precious metal crust for further refining. Also, lead may be electrorefined, leaving gold and silver in an insoluble slime.

COPPER-NICKEL ORES

Copper ores and sulfide concentrates are generally smelted to an iron and copper sulfide called matte, a good collector for gold and silver, and a discardable siliceous slag. Smelting usually takes place in reverberatory or autogenous furnaces, rarely in a blast furnace. The molten matte from smelting is blown with air in a converter to burn off sulfur and slag off iron, leaving gold-silver bearing copper metal for casting into electrolytic refining anodes. Copper matte is a mixture of Cu_2S and FeS in composition and the reactions are:

$$Cu_2S + O_2 \longrightarrow$$

$$2Cu + SO_2 \qquad (33)$$

$$2FeS + 3O_2 \longrightarrow$$

$$2FeO + 2SO_2 \qquad (34)$$

The process is exothermic.

Copper-nickel and nickel ores and concentrates are smelted in a process similar to that of copper. However, the converting step is halted when the iron has been slagged off and the molten material is copper-nickel sulfide containing any precious metals. Subsequent processing is by special hydrometallurgical, smelting, and electrolytic treatment. Finally, the gold and silver are contained in an anode slime or mud product from electrolysis, as in the case of copper.

SOME REFINING PROCESSES

The electrorefining of base metals, notably copper, plays an important, if indirect, part in the production of precious metals from base metal ores. As much as 90 percent of the world's new copper production is made by electrolytic refining, the only practical process for recovering gold, silver, and platinum group metals from copper.

Copper anodes for electrolytic refining usually contain from 99.0 to 99.5 percent copper. They are electrolyzed in a solution containing about 40 grams of copper and 200 grams of free sulfuric acid per liter. Current density may be 300 amperes per square foot or more provided that special agitation and current reversal methods are used. The reaction at the anode is:

$$Cu^o - 2e \longrightarrow Cu^{++} \qquad (35)$$

and at the cathode

$$Cu^{++} + 2e \longrightarrow Cu^o \qquad (36)$$

The power consumption is generally 0.08 to 0.20 KW hour per pound copper. The precious metals remain insoluble and are collected as a sludge or slime from the bottom of the refining cells. The refining of the slimes is discussed separately.

Although lead bullion may be refined by an alternative pyrometallurgical process (Parkes), crude lead may also be refined electrolytically by the Betts process. A lead fluosilicate ($PbSiF_6$) and free fluosilicic acid (H_2SiF_6) electrolyte is used with a current density of 14 to 16 amperes per square foot. Precious metals remain as insoluble slimes.

Typical anode slime from copper electrorefining contains precious metals and copper and impurities insoluble in dilute sulfuric acid. The anode slime generally ranges in size from minus 300-mesh to less than one micron. A typical analysis of slime, in percent, is: copper, 30; selenium, 12; tellurium, 3; lead, 2; arsenic, 2; antimony, 0.5; silver, 8.5; gold, 0.85 (18). Copper occurs as metallic particles and combined as copper selenide. Silver is also combined with selenium. Inasmuch as the selenides and tellurides are difficult to dissolve in dilute sulfuric acid, processing methods include a step to decompose them. Oxidizing roasting, sodium bisulfate fusion, or acid ferric solution leaching may be used to break down selenides and tellurides for further processing.

Sulfuric acid electrolyte (140 to 200 grams H_2SO_4 per liter) leaching is effective in dissolving that part of the copper in the slimes that is not tied up as selenides. The filtrate from the leaching is recycled to the electrorefining tank house. In one process the leached slimes are then given an oxidizing roast (700 to 800 degrees celsius) to recover selenium as the volatile selenium dioxide. The oxidized residue is further treated in an electric furnace under oxidizing conditions and then cupelled to obtain precious metal (doré) bullion. The roasting equation is:

$CuSe + 3/2O_2 \longrightarrow$

$CuO + SeO_2$ (37)

An alternative process, sodium bisulfate fusion, is used by the Kennecott Copper Corporation at the Utah Refinery (18). The fusion of dried slime with bisulfate takes place in a rotary kiln at 540 to 650 degrees celsius. Selenium and tellurium compounds decompose. Selenium volatilizes, according to the equations:

$Cu + 2H_2SO_4 \longrightarrow$

$CuSO_4 + SO_2 + 2H_2O$ (38)

$Ag_2Se + 2H_2SO_4 + O_2 \longrightarrow$

$Ag_2SO_4 + SeO_2 + SO_2 + 2H_2O$ (39)

$CuSe + 2H_2SO_4 + O_2 \longrightarrow$

$CuSO_4 + SeO_2 + SO_2 + 2H_2O$ (40)

$AuTe + 2H_2SO_4 \longrightarrow$

$Au + TeO \cdot SO_3 + SO_2 + 2H_2O$ (41)

The volatilized selenium is recovered in water spray scrubbers and electrostatic precipitators. The molten bisulfate fusion is granulated and leached with hot water to dissolve most of the copper sulfate and sodium sulfate. Some silver sulfate dissolved in the process is reprecipitated by stoichiometric addition of copper powder. The slurry is then dewatered, filtered, and dried. Dry filter cake is melted in a reverberatory furnace and the first slag is removed. A soda slag is then made with sodium carbonate additions to take off such impurities as arsenates, selenates, tellurates, and antimonates. The remaining copper and impurities are removed with a niter slag. The resulting doré bullion is cast into anodes for the Moebius electrolytic process to recover pure silver. Insoluble gold from the Moebius cells is then treated by the Wohlwill electrolytic process to purify the gold.

In the oxidation roasting and the bisulfate fusion processes, silver, selenium, and tellurium tend to dissolve and are difficult to recover. To overcome this disadvantage, the Hitachi Smelter and Refinery, Japan (19), developed a new selective leaching process using ferric sulfate as the oxidizer, based upon the equation:

$2Fe^{3+} + Cu \longrightarrow$

$2Fe^{2+} + Cu^{2+}$ (42)

A preliminary leach of the slimes with sulfuric acid solution without ferric iron removes some of the copper. Then a leaching step with three to eight grams per liter Fe^{3+} and 150 grams per liter H_2SO_4 selectively removes the remainder of the copper. Subsequent electrolysis of the leaching solution in a nondiaphragm cell regenerates sufficient ferric iron for recycling to the leaching step, even though the ferric regeneration efficiency is only 40 to 60 percent. Copper is electrowon with 60 to 80 percent efficiency. The reactions are:

Anode Reaction: $Fe^{2+} \longrightarrow Fe^{3+} + e^-$ (43)

Cathode Reaction: $Cu^{2+} + 2e \longrightarrow Cu^o$ (44)

Electrolytic processes are extensively applied for refining gold and silver, as described in the "Gold Metallurgy in South Africa", volume (2). Relatively pure silver anodes cast from leached slimes and other sources are refined in a vertical electrolytic unit, the Moebius cell, with anode bags and cathode scrapers. Silver anodes generally containing more than about five percent gold are preferably treated in the horizontal Thum cell, which has more sludge fall-out capacity than the Moebius cell. Also, the Thum cell is better adapted to processing anode fragments. The composition of the electrolyte for each type of cell is given in the following table (2).

	Vertical	
	Fresh	Discard
Silver g/l	30	30 to 40
Copper g/l	2 to 4	70
Lead g/l		2
Potassium nitrate g/l	15	15
Nitric acid g/l	10	10

	Horizontal	
	Fresh	Discard
Silver g/l	30	100
Copper g/l	5 to 10	80
Lead g/l		10
Nitric acid g/l	20	20

The role of copper in the fresh silver refining electrolyte is to increase conductivity. Copper may be built up to 40 to 100 grams per liter before stripping from the electrolyte. The reactions are:

Anode:

$Cu^o - 2e \longrightarrow Cu^{++}$ (45)

$Ag^o - e \longrightarrow Ag^+$ (46)

Cathode:

$Cu^{++} + 2e \longrightarrow Cu^o$ $E_o = + * 0.34V$ (47)

$Ag^+ + e \longrightarrow Ag^o$ $E_o = + * 0.80V$ (48)

Thus copper does not electrically plate with the silver. Gold remains insoluble at the anode and is recovered as a sludge.

Gold electrorefining is done in the Wohlwill cell in a gold chloride solution. The electrolyte may be prepared by dissolving gold with chlorine gas in hydrochloric acid:

* American convention as to sign.

$$2Au + 2HCl + 3Cl_2 \rightarrow 2HAuCl_4 \quad (49)$$

Main electrolytic reactions are:

Anode: $$Au - 3e \rightarrow Au^{3+} \quad (50)$$

$$Au^{3+} + 4Cl^- \rightarrow AuCl_4^- \quad (51)$$

Cathode: $$AuCl_4^- \rightarrow Au^{3+} + 4Cl^- \quad (52)$$

$$Au^{3+} + 3e \rightarrow Au \quad (53)$$

The Miller (chlorine) process is probably the most important currently applied gold refining process. The method is capable of producing 99.99 percent or more pure gold, satisfactory for many purposes. The process consists of passing chlorine gas into molten impure bullion. Various impurities form chlorides, rise to the surface of the melt, and are either skimmed off as molten chlorides or volatilize. Some of the reactions are:

$$2Ag + Cl_2 \rightarrow 2AgCl \quad (54)$$

$$2Cu + Cl_2 \rightarrow 2CuCl \quad (55)$$

$$Fe + Cl_2 \rightarrow FeCl_2 \quad (56)$$

$$Pb + Cl_2 \rightarrow PbCl_2 \quad (57)$$

$$Zn + Cl_2 \rightarrow ZnCl_2 \quad (58)$$

$$2Fe + 3Cl_2 \rightarrow 2FeCl_3 \quad (59)$$

Silver chloride skimmed from Miller process crucibles after cooling may be purified by leaching with hydrochloric acid and sodium chlorate and then treated with zinc dust to produce metallic silver.

REFERENCES

(1) U.S. Department of the Interior, Bureau of Mines, Minerals Yearbook (1975), Gold, Silver preprint chapters

(2) Finkelstein, N.P. Gold Metallurgy in South Africa, edited by R.J. Adamson, Chamber of Mines of South Africa (1972), Chapter 10 - The Chemistry of the Extraction of Gold from Its Ores, pp. 284-351

(3) Hedley, Norman and Howard Tabachnick, Chemistry of Cyanidation, American Cyanamid Company, Mineral Dressing Notes, No. 23, (December 1968)

(4) Barsky, G., S.J. Swainson and N. Hedley, "Dissolution of Gold and Silver in Cyanide Solutions", Trans. AIME, Vol. 112, (1934), pp. 660-667

(5) Hamilton, E.M., Manual of Cyanidation, McGraw-Hill Book Co., (1920)

(6) Heinen, H.J., D.G. Peterson; and R.S. Lindstrom, (Reno Metallurgy Research Center, U.S. Bureau of Mines) "Silver Extraction from Marginal Resources"; Oral Presentation -- The 104th TMW-AIME Annual Meeting, New York, New York, February 16-20, 1975

(7) Gilmore, A.J., "A Proposed Use for Ion Exchange in Gold Cyanidation", Canadian Mining Journal, (May 1967), pp. 63-65

(8) Davidson, R.J., "The Mechanism of Gold Adsorption on Activated Charcoal", Journal of the South African Institute of Mining and Metallurgy, (November 1974), pp. 67-76

(9) Zadra, J.B., A.L. Engel, and H.J. Heinen, "Process for Recovering Gold and Silver from Activated Carbon by Leaching and Electolysis", U.S. Bureau of Mines, R.I. 4843, (February 1952)

(10) Ross, J.R., H.B. Salisbury, and G.M. Potter, "Pressure Stripping Gold from Activated Carbon", AIME Annual Meeting, Chicago, Illinois (February - March 1973)

(11) Scheiner, B.J., R.E. Lindstrom, and T.A. Henrie, "Oxidation Process for Improving Gold Recovery from Carbon-Bearing Gold Ores", Report of Investigations 7573, U.S. Bureau of Mines (1971)

(12) Finkelstein, N.P., R.M. Hoare, G.S. James, and D.D. Howat, "An Aqueous Chlorination Process for the Treatment of Merrill Slimes and Gravity Concentrates from Gold Ores", Part 1 - A study of the chemistry of the process and a report of laboratory test work, Journal of the South African Institute of Mining and Metallurgy, (December 1966), pp. 196-215

(13) van Zyl, J.J.E., N.P. Finkelstein, H.J. Borey, and D.D. Howat, Reference Title

REFERENCES (continued)

(9) Part II - The design and operation of the pilot plant at Western Holdings, Ltd., Welhom, Journal of the South African Institute of Mining and Metallurgy, (December 1966), pp. 216-245

(14) Kuhn, M.D., N. Arbiter, and H. Kling, "Anaconda's Arbiter Process for Copper", Canadian Mining and Metallurgical Bulletin, (February 1974)

(15) "The Cyprus Copper Process", Cyprus Mines Corporation Brochure, October, 1977

(16) Haver, F.P., and M.M. Wong, "Ferric Chloride Brine Leaching of Galena Concentrate", U.S. Bureau of Mines, R.I. 8105, (1976)

(17) Newton, Joseph, Extractive Metallurgy, John Wiley & Sons, (1959)

(18) Leigh, Arthur N., "Precious Metals Refining Practice", AIME International Symposium on Hydrometallurgy, (1973)

(19) Soto, Junichi; Imamura, Tatsuo; Hojo, Mamoru; and Suzuki, Takashi, "New Process for Copper Anode Slime Treatment at the Hitachi Smelter and Refinery", World Mining and Metals Technology, AIME, (1976)

(20) Cho, E.H., and C. H. Pitt, Metallurgical Transactions B, Vol. 10B, No. 2, "The Adsorption of Silver Cyanide on Activated Charcoal", p. 159

(21) Cho, E.H., and C.H. Pitt, Metallurgical Transactions B, Vol. 10B, No. 2, "Kinetics and Thermodynamics of Silver Cyanide Adsorption on Activated Charcoal", p. 165

(22) Cho, E.H., S.N. Dixon, and C.H. Pitt, Metallurgical Transactions B, Vol. 10B, No. 2, "The Kinetics of Gold Cyanide Adsorption on Activated Charcoal", p. 185

Chapter 4

THE RECOVERY OF METAL VALUES FROM NICKEL-BEARING LATERITE ORES BY REDUCTIVE ROAST/AMMONIA LEACH TECHNOLOGY

J. Mark Richardson, Lawrence G. Stevens, Martin C. Kuhn

Mineral Sciences Division
UOP Inc.
Tucson, Arizona 85706

ABSTRACT

The majority of the world's known and anticipated reserves of nickel ores are lateritic deposits. At present, there are several different process alternatives for recovering nickel and other desired metal values from laterite ores. An overview is given of the selective reduction roast/ammoniacal leach process, one of the processes in commercial operation.

THE RECOVERY OF METAL VALUES FROM NICKEL-BEARING LATERITE ORES BY REDUCTIVE ROAST/AMMONIA LEACH TECHNOLOGY

J. Mark Richardson*, Lawrence G. Stevens, Martin C. Kuhn**

Mineral Sciences Division
UOP Inc.
Tucson, Arizona 85706

INTRODUCTION

The majority of the world's known and anticipated reserves of nickel ores are lateritic deposits. Laterites are near-surface deposits of oxidized material formed by extreme weathering (or "laterization") of parent ultramafic rock. In contrast, the majority of the world's present production of refined nickel is from sulfide deposits, which tend to extend deep into the earth. Although sulfide ores are relatively easy to process (usually by a combination of ore-dressing and pyrometallurgical techniques), mining costs are high and many established mines are getting deeper and obtaining lower grade ore. Laterites, however, are quite abundant and are much less expensive to mine. It is true that processing costs for laterites are higher than sulfide ores because of the necessity of chemical and/or thermal treatment of the entire ore. However, as processing technology improves, sulfide reserves shrink, and world demand for nickel increases, the importance of laterite ores will continue to grow.

At present, there are several different chemical process alternatives for recovering nickel and other desired metal values from laterite ores. Those which are in commercial operation around the world include pressure acid leaching, matte smelting, production of ferronickel in an electric or blast furnace, and selective reduction roast/ammoniacal leach. The purpose of this paper is to give an overview of the selective reduction roast/ammoniacal leach process route.

MINERALOGY OF LATERITE ORE DEPOSITS

Nickel laterite deposits can be typified because of overall similarities. However, because of the prevalence of slightly different conditions during the weathering process, or because of a difference in parent rock characteristics, each deposit has unique features. The parent ultramafic rocks from which laterites can be formed include ferromagnesian rocks such as peridotites, dunites and serpentinites. The parent rock usually contains about 0.2 to 0.3% nickel. All laterites are formed by weathering of the parent rock in areas of abundant rainfall. This means that most deposits are found in tropical or subtropical climates, although some are known to exist in more temperate zones. The nickel in the ultramafic rock usually occurs as an ion replacement in the silicate lattice or as an associated sulfide. Mildly acidic solutions are formed by the action of rainwater on decomposing organic surface material. These solutions attack and leach nickel out of the ultramafic rock, carrying it downward. Reprecipitation of the nickel deeper in the ore deposit causes an enrichment of nickel values that range from 5 to 20 times greater than the content in the parent rock. The same process also results in concentration of magnesium and silica in the lower zones. Iron oxides and aluminum oxide are not leached and remain behind in the upper zones as residual accumulations. The end result of the "laterization" process is the formation of three distinct zones overlying each other and varying in chemical and mineralogical composition.

The three zones of a typical laterite deposit are: the leached or canga zone, the iron oxide zone and the saprolitic or magnesium-silicate-rich zone. Below the saprolitic zone lies the remaining unaltered parent rock. Any of the three zones may be extremely thin or absent, while the thickness of individual zones may vary as much as 30-plus meters. Some deposits may even be composed entirely of almost one zone.

The leached canga zone contains little or no enriched nickel and is thus not considered as ore to be processed. The canga is mainly composed of iron oxides and alumina with lesser quantities of silica. In the iron oxide zone, nickel oxide is finely disseminated among a mixture of several different iron oxides (among them goethite, limonite, magnetite and hematite). The lower, or saprolitic zone is composed largely of hydrated magnesium silicates in which some of the magnesium in the lattice structure is replaced by nickel. In an economically feasible deposit, nickel content in the iron oxide zone might vary from 0.8% to 1.5% with iron content ranging from 20% to 50%. In this zone, only small amounts of magnesium and silica, usually a few percent or less, are present. In the saprolite zone, nickel usually tends to average about 1% to 2%, but may reach 5 or more percent, while iron content drops dramatically to 20% or less. Conversely, magnesium and silica contents in this zone can increase to a level of 30% and 40%, respectively. Laterite deposits also usually contain cobalt, mainly in the iron oxide zone, in concentrations up to a few tenths of a percent. Like the

* Currently with Fluor Mining and Metals, San Mateo, CA.
** Currently with Mountain States Mineral Enterprises, Tucson, AZ.

nickel, the cobalt is disseminated among the iron oxides. Copper has also been known to occur in laterite deposits. A transition zone, containing intermediate values of the major constituents, and in which the iron oxide ore is usually intermingled with the saprolitic ore, often exists directly below the iron oxide zone. Figure 1 depicts the variation in chemical composition and ore type with depth for a "typical" deposit. (1, 2)

LATERITE PROCESSING METHODS - GENERAL

Several commercial alternatives exist for the treatment of nickel laterite ores. The choice of which process to employ depends on several factors, including physical and chemical nature of the ore body, and geographical location, which is related to supply and cost of necessary raw materials and energy. Different processes also result in different end products and the marketability of such products may depend on geographical location. Processes currently in operation include pressure acid leaching, matte smelting, production of ferronickel by electric or blast furnace, and selective reductive roast/ ammoniacal leach. A brief description (3) of each follows:

Pressure Acid Leaching

A typical flowsheet for acid leaching is shown in Figure 2. (3) After crushing and grinding, the ore is slurried and leached with acid (usually sulfuric acid) at high temperature and pressure, usually in excess of 522.04°K (480°F) and 4.137×10^6 Pa (600 psig). The resulting metal-containing solution is then separated from the solid residue by means of dewatering thickeners and the metal values precipitated by sulfide addition (usually H_2S). An alternative route to product recovery is solvent extraction of the pregnant leach liquor followed by electrowinning. Recoveries of desired metal values by acid leach processing are usually very high. On the other hand, acid leaching is not very selective. Magnesia and alumina consume excessive amounts of H_2SO_4, meaning that the feed ore must be low in these compounds or operating costs may become prohibitive. Therefore, the application of acid leaching is limited to a narrow range of ore types.

Matte Smelting

A typical flowsheet for matte smelting is shown in Figure 3. (4) In this process, ground ore is charged to a blast furnace in the presence of coke, gypsum, and limestone. The coke is present as a reducing agent to convert nickel and iron oxides to the metals. The gypsum is reduced to calcium sulfide, which reacts with the nickel and iron to produce a metal sulfide matte. In addition to the molten matte, the high temperatures of the process result in the formation of a molten slag by the remainder of the ore and limestone. The immiscibility of the slag and matte facilitates their physical separation. The molten nickel-iron matte is then charged into converters and blown with air to preferentially oxidize the iron. Addition of silica flux produces a secondary, iron-rich slag which can be easily separated from what is now the nickel matte. The nickel sulfide matte can be marketed as such or can be submitted to further refinement by oxidation roasting followed by reduction to nickel metal in a retort furnace. Due to the inherent selectivity, the technique of matte smelting can be used on a wide variety of ores with moderate to high recovery of nickel. However, the process is very energy intensive and may be ruled out in certain geographical locations because of both economical and political concerns. (Sulfur dioxide gas emission is an unwanted, but unavoidable, by-product of this process.)

Production of Ferronickel

Ferronickel is a desired raw material used in the production of nickel alloy steels. The growing demand for ferronickel in this century has led to the presently employed techniques of production, which are a variation of the matte smelting process. A typical flowsheet for the production of ferronickel is shown in Figure 4. (4) The major difference between ferronickel production and nickel matte production is that in ferronickel production, no sulfiding agent is added, with the initial reduction step resulting in molten crude ferronickel and silicate slag. The crude ferronickel is then usually oxidized in a converter to remove impurities such as carbon, silicon, chromium, and phosphorus in a secondary slag. Again, a wide range of ores may be treated, but those high in silicates are desirable because of the ability to form a good primary slag. This process is also very energy intensive, particularly if electric smelting is used to melt a high-silica ore (high-silica ores have higher temperatures of fusion and require the accurate temperature control that can be provided by electric furnaces).

Selective Reduction Roast/Ammoniacal Leach

A typical flowsheet for the selective roast/ leach process is given in Figure 5. The main objective in using the selective roast/leach route is to reduce the desired metal oxides in laterite ores to the respective metals, using a gaseous reductant and a moderately high temperature. The metals are oxidized in an ammoniacal ammonium carbonate solution. An important precaution is to ensure that the iron oxides in laterites are not reduced to a form that is soluble in the leach, or that might affect the oxidation and subsequent solubilization of the nickel and cobalt. Furthermore, reduction to metallic iron results in excessive consumption of costly reductants. Hence the reductive roast must be selective. After the desired metals are solubilized, several established methods of recovering them from solution are available.

All selective reduction roast/ammoniacal leach processes in current practice are similar in the reduction of certain constituents of a laterite ore so that only the desired metals are leached. Where these processes vary slightly is in how they accomplish the reduction step. The amount of metal solubilized by leaching is greatly dependent on the conditions of the roast that allow the metal to be liberated. In general, ore from the oxide zone tends to be the easiest to reduce obtaining high recoveries and requiring little in the way of specialized treatment. On the other hand, the transition and saprolitic ores are often somewhat resistant to straightforward reductive roasting. They may require special chemical additives or increased reductant strength to aid in liberating the desired metals from the magnesium-silicate lattice and to aid in preventing unwanted side reactions which produce insoluble metal compounds. Although the composition and mineralogy of laterites can generally be typified as in Figure 1, individual ore bodies may vary widely in the amount of oxide, transition, and saprolitic ores contained. Thus, the roast treatment of the ore must be tailored for the deposit in question.

The "standard" reductive roast/ammoniacal leach process which has proven satisfactory in treating most oxide ores, was defined by the work of M. H. Caron (5) during the period from the 1920's to the 1950's. A patent issued to Caron in 1924 was the basis for a plant built at Nicaro, Cuba, during World War II. In the Caron process, ground and dried ore is subjected to a reducing roast, followed by cooling under non-oxidizing conditions, oxidative leaching in an ammoniacal ammonium carbonate solution, and steam stripping of the pregnant solution to recover nickel and cobalt as a precipitate and to recycle ammonia. The precipitate is a basic nickel carbonate that is then thermally decomposed to a high-purity metal oxide. All existing commercial reductive roast/ammoniacal leach technology is a variation in some manner of Caron's original process. Accordingly, it is this process (as embodied by the operation at Nicaro) that will be drawn upon most heavily in the discussions of unit operations and chemistry that follow.

In the past decade, other selective reductive roast plants have gone onstream, most notably Greenvale, in Australia, and Marinduque, in the Philippines. The Greenvale operation (6) differs from the Caron-Nicaro technique mainly in the patented direct addition of sulfur-containing fuel oil to the ore prior to roasting. Sulfur is an additive widely held to be beneficial in increasing metal extraction from the more refractory laterite ores. One hypothesis is that mixing fuel oil (which will volatilize during the roast) directly into the ore helps to provide greater interphase contacting between solid oxide and gaseous reductants. This, in turn, leads to improved reaction kinetics and thus, greater reduction to soluble metal in unit time. Product recovery in the Greenvale operation is much the same as Nicaro, except that cobalt is precipitated as sulfide, prior to nickel removal by steam-stripping. Not as much is known about the Marinduque operation other than that it uses more highly concentrated gaseous reductants than specified by Caron-Nicaro. (7)

Other reductive roast/ammonia leach processes currently in advanced stages of development, but not in commercial operation, include the U.S. Bureau of Mines (USBM) process (8) and the UOP additive process. (1) These two technologies have in common the option of using solvent extraction followed by electrowinning as a method of product recovery from solution. The Bureau of Mines process, which appears to be suited mainly for oxide and transition ores, employs pure carbon monoxide (rather than a mixture of CO and hydrogen as at Nicaro) and lower-than-normal roast temperatures. Sulfur, usually in the form of pyrite, is also employed as an additive. The USBM uses ammoniacal-ammonium sulfate solution for leaching.

The UOP process relies on the patented use of various additives in the roast step to handle a wide range of ore types, including the more refractory saprolitic ores. The additives are sulfur forms (gaseous or solid) and chlorides. As specified in three patents, (9, 10, 11), the additives can be used in various combinations, depending on the degree of difficulty of the ore being treated. The possible effects of such additives are also discussed further in this paper.

SELECTIVE REDUCTIVE ROAST/AMMONIACAL LEACH PROCESS DESCRIPTION

In order to obtain the desired metal values from a mined ore, an integrated selective reduction roast/ammoniacal leach process must comprise the following unit operations:

- Ore preparation (mining, size reduction and moisture removal)
- Reduction roasting
- Ammoniacal leaching
- Product liquor recovery
- Product recovery (via the options of steam stripping, sulfide precipitation, or solvent extraction)

Such an integrated process is shown in Figure 5. To better understand what is accomplished by the overall process, a more detailed look at the individual unit operations, including the chemistry is helpful.

Ore Preparation

Mining and Blending. Before any processing takes place, the ore must be mined from the deposit. Because of its proximity to the surface (most deposits do not exceed 90 meters in depth), mining can be accomplished by conventional stripping techniques. Before mining, the deposit is well-characterized with respect to ore type and composition by an extensive drilling and test pitting program. A mining plan is then established which allows either for the selective mining of a particular ore zone or a specified blend of several zones. The mining plan followed depends on the process to be employed. At Nicaro, (3) effort was directed toward obtaining a blend of ore that was largely oxide in nature, as the process was not capable of satisfactorily handling more saprolitic ore types. In the case of some operations, the various zones are mined individually and the ore types are blended at the plant site to the desired composition.

Size Reduction and Moisture Removal. Size reduction of the ore is required for several reasons:

- Bulk handling of the ore through the various unit operations is facilitated if the particles are uniformly small.
- Increased specific surface area is important to successful roasting and leaching. (Reaction rates often become less mass transfer-and-diffusion-limiting if more surface area is available.)
- Grinding the ore facilitates the decomposition of the normal structure of the ore. The surfaces of nickel-containing minerals are exposed to chemical attack, whereas, without size reduction, they might remain locked within other inert minerals not participating in any reactions.

Laterite ores are very friable and do not require the extensive crushing and grinding circuits needed for harder sulfide ores. Usually after mining, the wet ore (containing 20-50% moisture) is crushed to less than 1 inch in size, then dried in a rotary kiln to less than 1-3% moisture. The moisture is removed to facilitate dry grinding in a ball mill and allow storage and transport. In a typical operation, the ore is ground to 100% passing 65 mesh (210 microns), but coarser or finer grinds may be required, depending on ore type.

Reduction Roasting

In reduction roasting, the dried, ground ore is gradually heated to a specified temperature under reducing atmosphere. The ore is usually fed to a multiple hearth roaster, falling by gravity from one hearth to the next, with temperature increasing from the top to the bottom of the roaster. In the upper, cooler hearths the ore is preheated and bound waters of hydration are driven off, thus beginning the process of decomposition of the nickel-containing lattice structures. Reduction of metal oxides mainly takes place on the lower, hotter hearths, usually at temperatures in the range of 1300° - 1650°F. After reduction, the ore is cooled under inert atmosphere to prevent re-oxidation of the nickel and other desired metals, if any. Additives are blended in with the ore before roasting. A flowsheet example of a typical roasting operation is given in Figure 6.

A typical oxide zone ore might have the following composition:

Ni	1.26%
Co	0.1 %
Fe	46.5 %
Mg	0.1 %
SiO_2	1.9 %

while a typical saprolite ore might look as follows:

Ni	2.2 %
Co	0.01 - 0.05%
Fe	15 - 20%
Mg	15 - 20%
SiO_2	40%

In either case, or in a blend of both ore types, the primary reactions taking place during roasting are those affecting nickel and iron: (1, 3)

$$NiO + H_2 = Ni + H_2O \quad (1)$$

$$3Fe_2O_3 + H_2 = 2Fe_3O_4 + H_2O \quad (2)$$

$$Fe_3O_4 + H_2 = 3FeO + H_2O \quad (3)$$

$$FeO + H_2 = Fe + H_2O \quad (4)$$

$$NiO + CO = Ni + CO_2 \quad (5)$$

$$3Fe_2O_3 + CO = 2Fe_3O_4 + CO_2 \quad (6)$$

$$Fe_3O_4 + CO = 3FeO + CO_2 \quad (7)$$

$$FeO + CO = Fe + CO_2 \quad (8)$$

$$Ni + Fe = FeNi \quad (9)$$

Similar reactions occur for cobalt and copper.

To reduce a maximum of nickel oxide to the metal, most of the iron present must also be reduced. A substantially large portion of the nickel in laterite ores is tied up in solid solution with iron oxides. The iron oxides must first be attacked and reduced to liberate nickel oxide for its own reduction. However, the reduction of iron all the way to the metallic state is neither necessary nor desirable. The avoidance of the production of metallic iron is what makes reduction roasting of nickel laterite ores a selective process. There are several reasons for obtaining maximum selectivity:

(1) The reduction to metallic iron consumes costly reductant unnecessarily. A combination of equations 2 and 3, or equations 6 and 7, shows that one mole of reductant (either H_2 or CO) is sufficient to reduce two moles of iron as hematite (Fe_2O_3) to wustite (FeO). However, to reduce wustite to metallic iron, an additional mole of reductant is required to reduce only one mole of iron.

(2) Metallic iron forms various nickel alloys, as suggested by Equation 9, some of which are insoluble in ammoniacal ammonium carbonate solution, depending on the ratio of iron to nickel.

(3) In the ammoniacal leach, free metallic iron first oxidizes to the ferrous state to dissolve and form an instable ammine complex. The ferrous then oxidizes further to form $Fe(OH)_3$, which

is insoluble, gelatinous, and has a tendency to coat reduced ore particles with a film that interferes with dissolution of nickel.

Having established the undesirability of producing metallic iron in the reduction, the problem remains to find those conditions that will limit the production of metallic iron while permitting the production of metallic nickel. This problem can be resolved by considering the thermodynamics of the system. Experimental work by many investigators has shown that the active mass of the solid phase in laterite reductive roast systems can be considered as constant, and as such, may be neglected in formulating equilibrium constants for the pertinent reactions. This means that K, the equilibrium constant for the Reactions 1 through 4 may be expressed as:

$$K = \frac{H_2O}{H_2} \qquad (10)$$

For Reactions 5 through 8, K is given by:

$$K = \frac{CO_2}{CO} \qquad (11)$$

Thermodynamic equilibrium data for Equations 1 - 8 are available in the form of free energies of formation, expressed as functions of temperature. These data can be used to calculate equilibrium values of K, as a function of temperature for each reaction. The resulting equilibrium curves graphically illustrate the conditions under which desired reactions will proceed (and under which undesired reactions will not).

The data in Figure 7 (12, 13) give the equilibrium curves for the systems Fe:H:O and Ni:H:O, while Figure 8 (13, 14) illustrates equilibria for the systems Fe:C:O and Ni:C:O. From the figures, it can be observed that relatively high K values will permit the reduction of NiO to proceed, particularly at higher temperatures in the case of the Ni:H:O system. A calculation of the heat of reaction (15) for the reduction of NiO reveals about -10,467 joules (-2500 calories) per mole of NiO at 800°K (980°F) and approximately -12,560 joules (-3000 calories) per mole at 1000°K (1340°F), meaning that the reaction is slightly, but increasingly, exothermic at higher temperatures. Unfortunately, the data in Figure 7 also reveal that much lower K values are required for the reduction of iron oxides. Since nearly all laterite ores contain a significant amount of nickel oxide in solid solution with iron oxides, some reduction of iron must be achieved to liberate nickel oxide for its own reduction. If similar heat of reaction calculations are made for the reduction of Fe_3O_4 to FeO, it is found that about +58,615 joules (+14,000 calories) per mole Fe_3O_4 at 800°K (980°F) and about +46,055 joules (+11,000 calories) per mole Fe_3O_4 at 1000°K (1340°F) are required. This signifies a fairly endothermic reaction. Thus, where nickel oxide alone would require little heat and reductants once a high initial temperature was attained, the need to reduce some iron requires large amounts of heat and reductants. Fortunately, operation in the region where Fe_2O_3 and Fe_3O_4 reduce, provides very fast reaction rates for the nickel reduction reaction. From the figure, it is seen, for example, that operation in the vicinity of 1000°K (1340°F) will require a maximum K value of no greater than about 1.5 (so that Fe_3O_4 is reduced), and a minimum of about 0.45 to avoid metallic iron production. Very simply, this means that water vapor in the gas phase at equilibrium should be no greater than 1.5 times the volume concentration of hydrogen and no less than about half. Other temperatures, of course, will result in different limitations on K, with temperature used depending heavily on ore type.

Temperature also has an effect on reaction kinetics, with higher temperatures giving faster reactions. For this reason, a practical lower limit of about 866°K (1100°F) is used in most operations. An upper limit of about 1144° to 1172°K (1600° to 1650°F) is observed for most ores as well, primarily to avoid the sintering that occurs at higher temperatures. Furthermore, an ore that contains magnesium and silica in any quantity, will tend, at higher temperatures to form the forsterite structure ($MgSiO_3$). Forsterite, when it contains nickel in the lattice, is extremely inert to any reduction of the nickel. Finally, the higher the temperature, the greater will be the fuel consumption required for heating, as the overall reduction reaction is usually endothermic (as was seen by the example of Figure 7 and the related thermodynamic calculations).

A similar review of Figure 8 for the Ni:C:O and Fe:C:O systems can be made, with similar conclusions being drawn. Thermodynamic calculations for the reduction of Fe_3O_4 by CO reveal an endothermic reaction ∿ +20,934 joules (+5000 calories) per mole Fe_3O_4 at 800°K (980°F) and ∿ +12,560 joules (+3000 calories) per mole Fe_3O_4 at 1000°K (1340°F).

The use of Figures 7 and 8 is not the most exacting method, since a single reductant is assumed, when in fact most commercial systems use mixture of H_2, CO, H_2O and CO_2 diluted in nitrogen. However, the equilibrium diagrams do allow an estimation of conditions. Actual exact operating conditions are usually then pinned down by trial and error laboratory testing in the regions suggested by the diagrams. It is possible, though, to approach reality a little closer through the use of a Chaudron diagram, which is an equilibrium diagram based on reducing power. Reducing power is defined as total volumetric concentration of reductants divided by total active gas concentration:

$$P = \frac{CO + H_2}{CO + H_2 + CO_2 + H_2O} \qquad (12)$$

Figure 9 (16) is a Chaudron diagram for the system Fe:C:H:O. From Figure 9, it can be seen that in order to prevent metallic iron from forming in the temperature range 1100° - 1650°F, the reducing power must be less than 65%. To accomplish the minimum reduction of iron, however, at least 10 to 50% reducing power is necessary.

Additives are used in roasting to modify normal system behavior. Although no postulated mechanisms which show how additives affect reduction chemistry have been proven conclusively, evidence of the positive effects of additives is abundant. Consequently, speculation is easy on how these additives might interact with the system. For instance, pugging fuel oil into the ore provides a source of reductant that is intimately associated with a solid particle. This situation improves the interfacial transfer of the reactants and products, and thus, may dramatically alter the overall rate of reaction and possibly lower the temperature required to achieve reduction. One theory (1) holds that sulfur, in many forms such as pure powder, pyrite, or H_2S gas, is beneficial because it inhibits the formation of some insoluble nickel-iron alloys and promotes the formation of soluble alloys. Further speculation indicates that in more saprolitic ores, sulfur prevents the re-combination of thermally decomposed magnesium silicates with nickel oxide. If the recombination occurs before the oxide can be reduced to the metal, no solubilization of nickel will be possible. Halide additives (1, 9, 10, 11) such as chloride forms, are thought to be beneficial because they aid in breaking down magnesium silicate structures containing NiO, allowing the reduction to proceed. In fact, improved reaction kinetics have been noted for many saprolitic ores when roasted in the presence of halides, lending credence to the theory that the halide additives speed up the thermal decomposition process caused by roasting. Finally, halide additives are thought, because of supporting DTA* evidence (1), to inhibit or even eliminate the formation of the forsterite structure in the higher temperature range. This allows use of higher temperatures on saprolitic ores, which in turn promotes faster reaction.

Ammoniacal Leaching

After roasting, the now-calcined ore must be cooled and sent to leach. To achieve this in a typical operation such as Nicaro, the ore is first dropped onto a final hearth or hearths for which the temperature is much lower than the maximum reduction temperature. The product is then passed through an externally cooled rotary tube that contains a non-oxidizing (and usually inert) atmosphere. After cooling to about 300°F, the ore is discharged to a stirred quench tank containing ammoniacal leach liquor. A typical roasted oxide ore after losing 20% of weight, might have the following composition:

Ni	1.58%
Co	0.13%
Fe	58.1 %
Mg	0.13%
SiO_2	2.38%

Until the calcined ore reaches the leach liquor, it must not come into contact with any oxidants (the slightest amount of air is detrimental) or re-oxidation of the nickel metal will occur. Re-oxidation, or passivation as it is commonly called, occurs easily at elevated temperatures and renders the nickel insoluble.

* Differential Thermal Analysis

Once in the leach liquor, however, the object is to oxidize. The following equations describe the idealized overall leach reactions that take place:

$$Ni + \tfrac{1}{2}O_2 + 8NH_3 + 2CO_2 + H_2O = Ni(NH_3)_6^{++} + 2NH_4^{+} + 2CO_3^{--} \quad (13)$$

$$FeNi + O_2 + 12NH_3 + 3CO_2 + H_2O = Ni(NH_3)_6^{++} + Fe(NH_3)_4^{++} + 2NH_4^{+} + 3CO_3^{--} \quad (14)$$

$$Fe + \tfrac{1}{2}O_2 + 6NH_3 + 2CO_2 + H_2O = Fe(NH_3)_4^{++} + 2NH_4^{+} + 2CO_3^{--} \quad (15)$$

As will be seen from thermodynamic data presented further on in this paper, the iron-ammine complex is stable only under very limited conditions. In oxidizing conditions, the complex decomposes rapidly, with ferrous iron oxidizing rapidly to ferric, forming the insoluble hydroxide:

$$4Fe^{++} + O_2 + 2H_2O + 8OH^{-} = 4Fe(OH)_3 \quad (16)$$

Leaching takes place in a countercurrent mixing and decantation system which builds the concentration of dissolved metal values to a level necessary for efficient product recovery from the liquor (usually 10 to 12 gpl Ni is considered sufficient). Air is pumped into the leach tanks to provide the necessary oxygen; high-speed impellers ensure that efficient mixing and oxygen mass transfer take place. Oxygen mass transfer from the gas phase to the liquid phase is essential for the leach reaction to proceed, since the crucial reaction takes place at the liquid/solid particle interface.

Reduced and cooled calcine is commonly leached at ambient temperatures and pressures. An ammonia-ammonium carbonate leach solution is intimately mixed with the reduced ore in the presence of oxygen to oxidize and solubilize the metal values of commercial interest, i.e., nickel, cobalt and copper. Once leached, the dissolved metal values are separated and recovered from the pulp in a liquid solid separation. Countercurrent decantation thickening is used to accomplish both the recovery of metals and the reagents as well as to increase the concentration of the metal species. A typical product pregnant liquor from the leach circuit might have the following composition:

Ni	10 - 12 gpl
Co	0.1 - 0.2 gpl
Fe	0.1 gpl
Mg	0.01 gpl
SiO_2	Not Detected

The oxidizing ammonia-ammonium carbonate leach system is utilized principally because of its iron selectivity. The most obvious advantages are the rejection of reduced iron to the leach residue and the minimized corrosion problems in ammonium carbonate systems. The high temperature reduction of laterite ores tends to reduce to the metallic state the nickel, cobalt, copper and some iron. The leach system must therefore be capable of oxidizing the metals in question and also be capable of complexing the metals of interest to allow for their recovery. An ideal ligand for this system is ammonia. The chemistry of leaching can best be portrayed in terms of aqueous oxidation, reduction and pH. Solution potential - pH diagrams have been constructed which represent the Ni - NH_3 - H_2O, Co - NH_3 - H_2O, Cu - NH_3 - H_2O and Fe - NH_3 - H_2O systems. These diagrams are shown in Figures 10 - 13. The thermodynamic data used in the construction of the diagrams were taken from Latimer (17), Sillen and Martell (18), Butler (19) and Lange's Handbook of Chemistry (20).

Shown in Figure 11, is the solution potential - pH diagram describing the Co - NH_3 - H_2O system. Unlike the nickel system, the cobalt undergoes oxidation from cobaltous to cobaltic at moderately oxidizing conditions when the cobalt exists as the relatively stable hexammine complex over a very limited pH range of about 1 pH unit. If the pentammine complex of cobaltous were shown, this stability of field would be slightly expanded since Butler (19) reports a K_6 of -0.62. As the solution potential becomes more oxidizing, the hexammine complex of cobaltic exhibits a much expanded stability field with respect to pH. At Eh values in excess of 0.7 volts, the field is about 5 pH units wide. Insofar as leaching is concerned, the solution pH is governed by the pH limits established for the nickel system regardless of potential. The leaching behavior of cobalt has been studied and reported by Vu and Han. (21)

The solution potential - pH relationships in the copper system are shown in Figure 12. Both cuprous and cupric copper complex with ammonia. The cuprous diammine and the cupric tetraamine complexes are shown in the Figure. Of major interest in the copper system, is the necessity to operate at higher solution potential values in order to effect the oxidation and dissolution of copper. Of all the metals of economic interest, copper provides the largest stability fields for the respective ammine complexes and therefore places the least stringent conditions on the leach.

Often the fact that a ferrous tetraamine complex is stable at certain Eh - pH values is overlooked. Therefore, Figure 13 is offered to show the limited stability field of the ferrous tetraamine complex. Even though the field is small and at low potentials, the introduction of freshly reduced ore to leach solution can result in a portion of the leach taking place in this stability field. If considerable iron has been reduced to an easily leachable form, an oxygen mass transfer-limited leach system might be observed which can seriously affect the desired leach kinetics. Since one of the major objectives of the leach is to reject iron to the residue sufficient oxygen mass transfer must be available to operate at Eh values greater than -0.2 volts. (All Eh values are given in the European System.) (Note that the given diagrams represent equilibrium only and do not reflect nor even suggest the leach kinetics.)

Product Liquor Recovery

The advantage of the ammoniacal ammonium carbonate leach system is that a closed-loop process can be built which enables the continuous recycle of most of the ammonia and carbon dioxide originally present. Figure 14 illustrates a typical liquor and reagent recovery system. After nearly all of the desired metal values that are extractable have been leached from the calcined ore, the residue is washed in a countercurrent decantation train of thickeners to remove any solubilized metal that may be entrained in the waste slurry. The final washed slurry goes to a steam stripping tower (or train of such towers) where countercurrent contacting of the slurry with steam causes volatilization of the ammonia and carbon dioxide. The off-gasses are vented to the reagent recovery system, while the tailings slurry, containing only calcined ore and water, is pumped to a disposal pond. The vented ammonia and carbon dioxide from this operation are passed to the reagent recovery section, which is a series of condenser-absorber and scrubber towers designed to re-solubilize the reagent with water. Also recovered in this section are the ammonia and carbon dioxide vapors given off from the leaching and washing circuits. All tanks and thickeners in those circuits are covered and vented for this purpose. If steam stripping of the product liquor to recover metal carbonates is used as the product recovery alternative, then the volatilized ammonia and carbon dioxide from that operation are also recovered. The resulting underflow from the product recovery section is fresh leach solution, the strength of which is controlled by adding necessary amounts of fresh make-up ammonia and carbon dioxide during the recovery operations. This fresh leach solution is re-introduced to the leaching operation by feeding it to the final thickeners of the wash circuit.

Product Recovery

Once the metal values have been solubilized, several options are available for final product recovery from the pregnant liquor. The choice of which option to use depends purely on economical considerations such as price and availability of raw materials and/or desirability and marketability of the final product. The options include steam stripping, sulfide precipitation, and solvent extraction. Each of these options is independent of ore type fed to the roast, due to the homogenous and fairly constant nature of the product liquor. The basic principles and chemistry of each are as follows:

Steam Stripping. The clarified pregnant liquor from the leaching operation is advanced to a countercurrent steam distillation or stripping tower in which the complexed ammonia and carbonate are thermally driven off. The pregnant liquor is charged to the top of the stripper and steam is injected at the bottom. The descending solution is gradually depleted of ammonia until the nickel and cobalt begin to precipitate. This usually occurs when total ammonia concentration drops below about 2%. As heating continues, the following reactions take place:

$$Ni(NH_3)_2^{++} = Ni^{++} + 2NH_3 \qquad (17)$$

The nickel ion combines with hydroxyl and carbonate ions in solution to form the insoluble basic carbonate:

$$5Ni^{++} + 6OH^- + 2CO_3^{--} = 3Ni(OH)_2\ 2NiCO_3 \qquad (18)$$

Ammonia vapor from the top of the still is sent to reagent recovery, while the basic nickel carbonate is discharged as a slurry from the bottom of the still. The slurry is then thickened and filtered to about 30% moisture content. The nickel carbonate cake is fed to a furnace (usually a rotary kiln) for thermal decomposition at about 1600°F. This reaction, which takes place in an oxidizing atmosphere is defined by the equation:

$$3Ni(OH)_2\ 2NiCO_3 = 5NiO + 3H_2O + 2CO_2 \qquad (19)$$

The nickel oxide thus produced is known as "Sinter 75" for the approximate content of nickel and can be marketed "as-is", or can be reduced to higher nickel content. Typical sinter product compositions are:

	Sinter 75	Sinter 90
Ni	75.0%	89.0%
Co	1.0%	1.2%
Cu	0.9%	0.3%
Fe	0.5%	0.6%
S	0.02%	0.02%

Sulfide Precipitation. In a typical operation, hydrogen sulfide is bubbled through the clarified pregnant liquor, causing the following reaction:

$$H_2S + Ni(NH_3)_6^{++} = NiS + 6NH_3 + 2H^+ \qquad (20)$$

Unfortunately, similar reactions take place for any other dissolved metals, with purity of the nickel sulfide depending on solution content of these other metals. Further refining of the nickel sulfide is required.

Solvent Extraction - Electrowinning. The solvent extraction and electrowinning of nickel from ammoniacal sulfate or ammoniacal carbonate is well known technology. Nickel extraction from ammonium sulfate solutions has been demonstrated on a commercial scale in Texas, since 1972. (22) Additional laboratory and pilot plant data on nickel extraction from ammonium carbonate solutions have been published by Merigold and Sudderth (23) and Agarwal, et al. (24)

Nickel is typically extracted from ammoniacal solutions with an organic extractant, such as Henkel Corporation's LIX 64N. LIX 64N is usually diluted in an inert organic carrier, such as kerosene, to a concentration appropriate to extract the amount of nickel present in the feed liquor. Generally, 35-40 volume percent LIX 64N is the maximum amount of active extractant which can be used.

The overall chemical reaction for extraction can be expressed as:

$$[Ni(NH_3)_4^{+2} + CO_3^{-2}]_{AQ} + 2RH_{ORG} + H_2O = \qquad (21)$$

$$R_2Ni_{ORG} + [(NH_4)_2CO_3 + 2NH_4OH]_{AQ}$$

where R = organic molecule (LIX 64N)

Note from this reaction that only the metal ion is extracted from the aqueous phase. In actuality, a small amount of ammonia is extracted by LIX 64N in addition to the metal ion.

A typical nickel extraction isotherm is shown in Figure 15 with a McCabe-Thiele stage construction diagram for 90 percent stage efficiency. This diagram shows that three countercurrent stages are needed to extract substantially all of the nickel.

The ammonia that is co-extracted with the nickel on LIX 64N must be scrubbed off the solvent prior to stripping the nickel. A dilute sulfuric acid scrub is usually employed to remove the extracted ammonia.

The nickel is stripped from LIX 64N with a weak sulfuric acid solution. The concentration of sulfuric acid in the stripping solution must be equivalent to the amount of nickel on the loaded organic and this acid must be consumed during stripping. For example, if a loaded organic contained 5 g/l nickel, then the equivalent acid concentration of the electrolyte from the SX stripping system must have a pH of approximately 3, in order to effectively electrowin the nickel from the electrolyte. Nickel is electrowon from the electrolyte according to commercial practice using insoluble Ca - Pb anodes and diaphragm type cells. (25) The principal cell reactions are as follows:

$$\text{Anode: } 2H_2\downarrow\ ®\ \downarrow_2 + 4H^+ + 4e^- \qquad (22)$$

$$\text{Cathode: } 2(Ni^{++} + 2e^- = Ni^o) \qquad (23)$$

$$\text{Net: } 2Ni^{++} + 2H_2O = 2Ni^o + 4H^+ + O_2 \qquad (24)$$

$$2NiSO_4 + 2H_2O = 2Ni^o + 2H_2SO_4 + O_2 \qquad (25)$$

The sulfuric acid generated at the anode is equivalent to the amount of nickel plated out and is utilized by recirculation back to the nickel stripping section in SX.

Figure 16 is a block flowsheet illustrating the combined solvent extraction electrowinning unit process as used for the UOP nickel process.

The primary advantage of this product recovery route is the selectivity afforded. By controlling pH, the same organic solvent can be used to load both nickel and copper simultaneously, and strip them separately. Cobalt is rejected by LIX 64N and remains with the raffinate. The cobalt may be removed as a separate product either by sulfide precipitation or resin ion exchange followed by electrowinning. Purity of nickel cathode product is well in excess of 99%.

***

REFERENCES

(1) Stevens, L.G., Goeller, L.A., Miller, M., "The UOP Nickel Extraction Process - An Improvement in the Extraction of Nickel from Laterites", paper presented at the Canadian Institute of Mining and Metallurgy, 14th Annual Conference of Metallurgists, Edmonton, Alberta, Canada, August 26, 1975.

(2) Weege, R.J., Beres, W.J., "The Exploration and Development of Nickel Laterites", paper presented at AIME Annual Meeting, Denver, Colorado, February 27, 1978.

(3) Boldt, J.R., Queneau, P., The Winning of Nickel, D. Van Nostrand Co., Princeton, N.J., 1967.

(4) Simons, C.S., "Nickel Pyrometallurgy in Perspective", paper presented at AIME Annual Meeting, New York, March, 1971.

(5) Caron, M.H., U.S. Patent 1,487,145 issued 1924.

(6) Private Communication

(7) Private Communication

(8) Seimens, R.E., Corrick, J.D., "Process for Recovery of Nickel, Cobalt, and Copper from Domestic Laterites", Mining Congress Journal, January, 1977, pp. 29-34.

(9) Stevens, L.G., Miller, M., Goeller, L.A., U.S. Patent 3,772,423, issued Nov. 13, 1973.

(10) Stevens, L.G., Miller, M., Goeller, L.A., U.S. Patent 3,772,424, issued Nov. 13, 1973.

(11) Miller, M., Goeller, L.A., Stevens, L.G., U.S. Patent 3,845,189, issued Oct. 29, 1974.

(12) Eastman, E.E., "Equilibria in the Systems Iron: Carbon: Oxygen and Iron: Hydrogen: Oxygen, and the Free Energies of the Oxides of Iron", Journal of American Chemical Society, 1922, pp. 975-998.

(13) Emmett and Schulz, Journal of American Chemical Society, 1938.

(14) Bogatski, D.P., "Reduction of Nickel Oxides with Carbon Monoxide", Metallurgy, Vol. 13, No. 1, 1938, pp. 84-90.

(15) Wicks, C.E., and Block, F.E., "Thermodynamic Properties of 65 Elements - Their Oxides, Halides, Carbides, and Nitrides", USBM Bulletin 605, 1963.

(16) Bonnivard, G., "Selective Reduction of Nickel Ores", paper presented at the IXth International Congress of Metallurgy, Prague, Czechoslovakia, 1970.

(17) W.M. Latimer, Oxidation Potentials, Prentice - Hall, Inc., Englewood Cliffs, N.S., 1952, p. 384.

(18) Sillen, L.G., Martell, A.E.: Stability Constants of Metal - ion Complexes, The Chemical Society, Burlington House, W.I., London, 1964, pp. 149-158.

(19) Butler, J. H., Ionic Equilibrium - A Mathematical Approach, Addison - Wesley Publishing Company, Inc., Reading, Massachusetts, 1964, p. 304.

(20) Lange's Handbook of Chemistry - Dean, D.A. - Editor: Eleventh Ed., McGraw - Hill Book Company, N.Y., N.Y., 1973, pp. 9-17, 9-18.

(21) Vu, C., and Han, K.H., "Leaching Behavior of Cobalt in Ammonia Solutions", Transactions/ Section C, Inst. of Mining and Metallurgy, Vol. 86, 1978 pp. C119-125.

(22) Eliasen, R.D. and E. Edmunds Jr., "The S.E.C. Nickel Process", CIM Bulletin, February, 1974, pp. 82-86.

(23) Merigold, C.R. and R.B. Sudderth, "Recovery of Nickel by Liquid Ion Exchange Technology", International Symposium on Hydrometallurgy, edited by D.J.I. Erans and R.S. Shoemaker, AIME, New York, 1972, pp. 552-558.

(24) Agarwal, J.C. et al., "A New FIX on Metal Recovery From Sea Nodules", Engineering and Mining Journal, December 1976, pp. 74-78.

(25) Toivahen and Gronqvist, P., "Nickel Refining in Finland", paper presented at Annual General Meeting of CIM, Montreal, Canada, April, 1964.

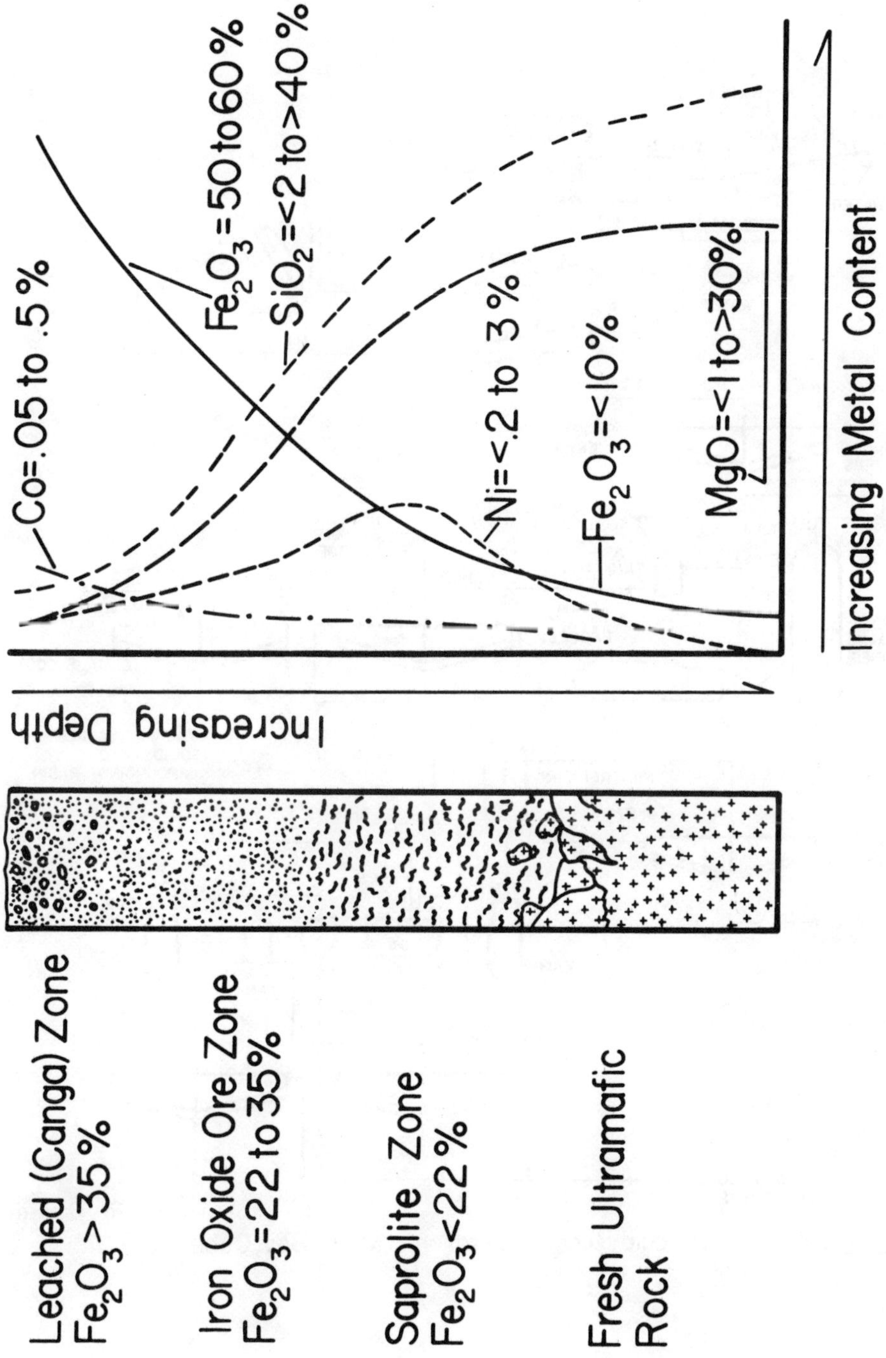

Figure 1: Chemical composition of nickel laterite ore zones.(2)

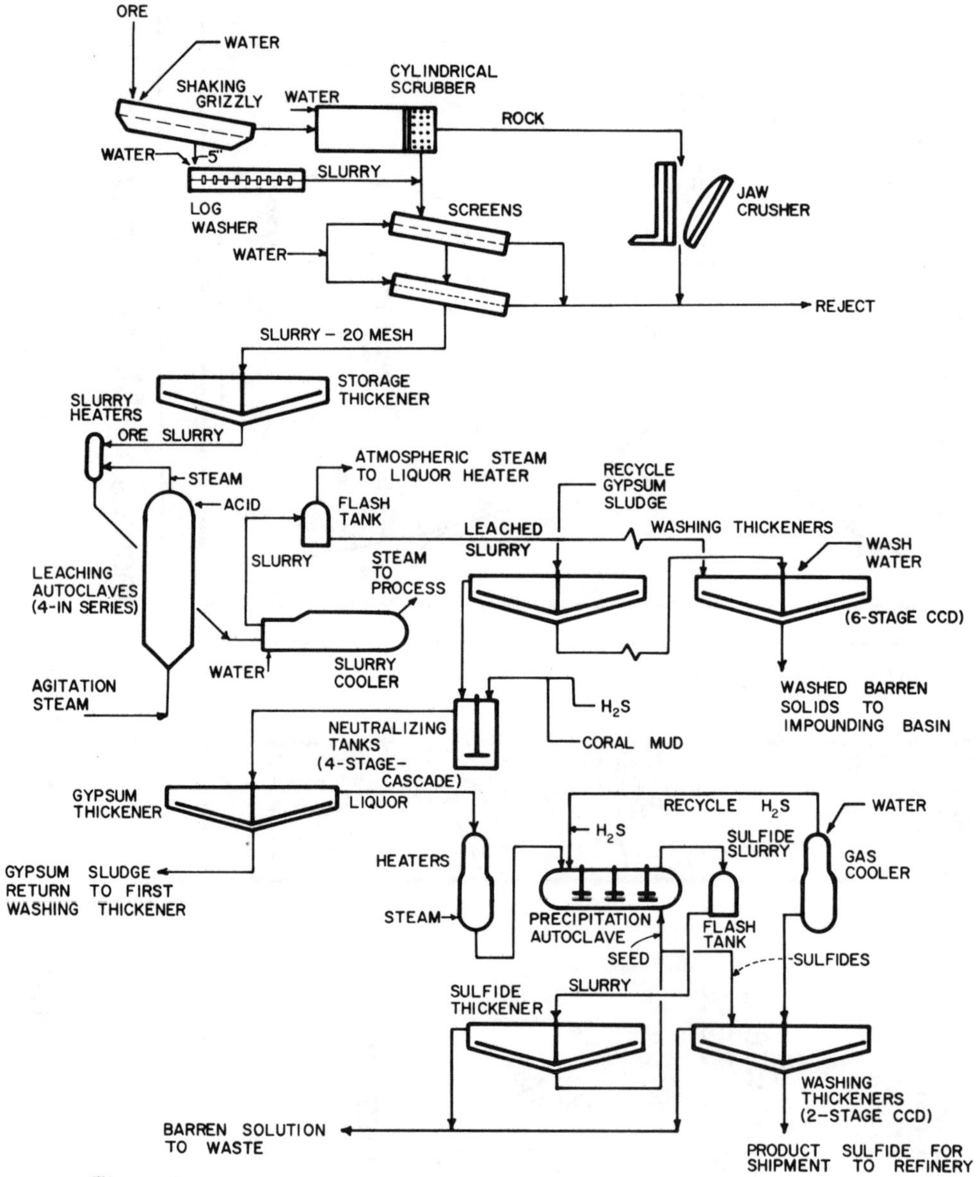

Figure 2: Typical acid leach operation – Moa Bay, Cuba.

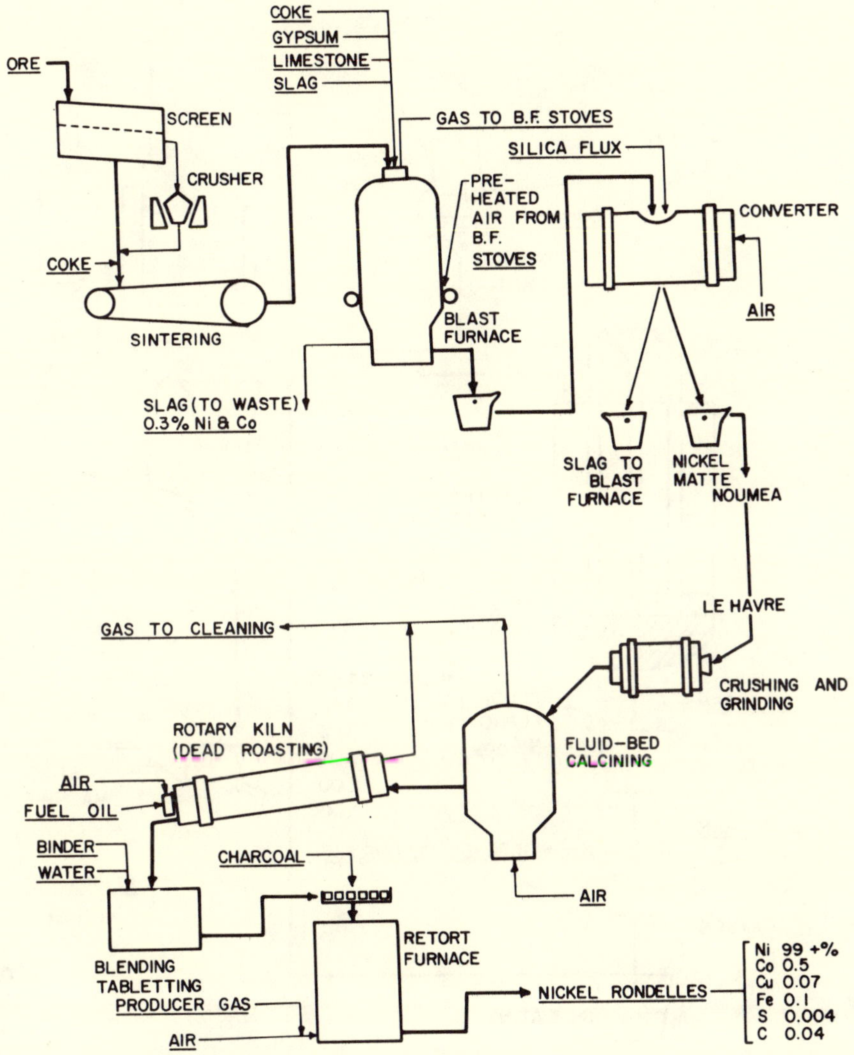

Figure 3: Matte smelting of garnierite (SLN) (4)

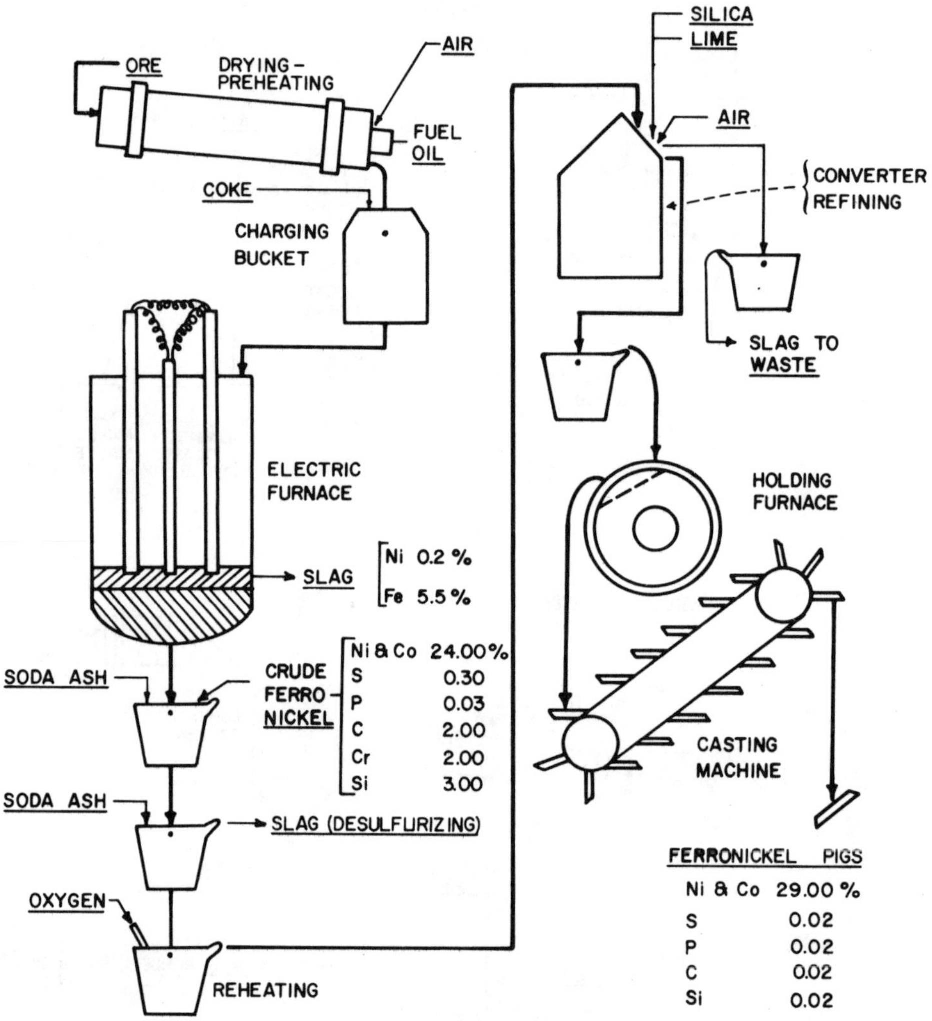

Figure 4: Ferronickel from garnierite (SLN) (4)

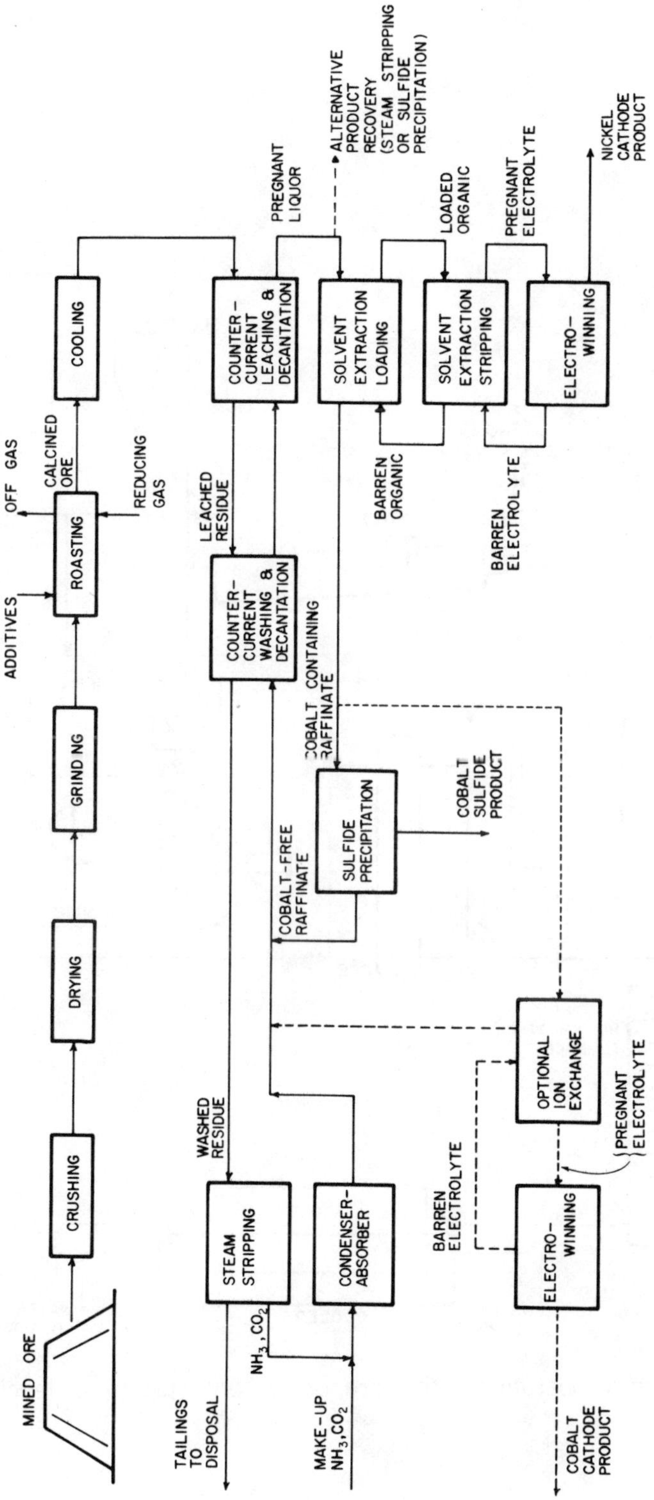

Figure 5: Simplified block flow diagram of selective reductive roast/ammoniacal leach. (Showing solvent extraction product recovery route.)

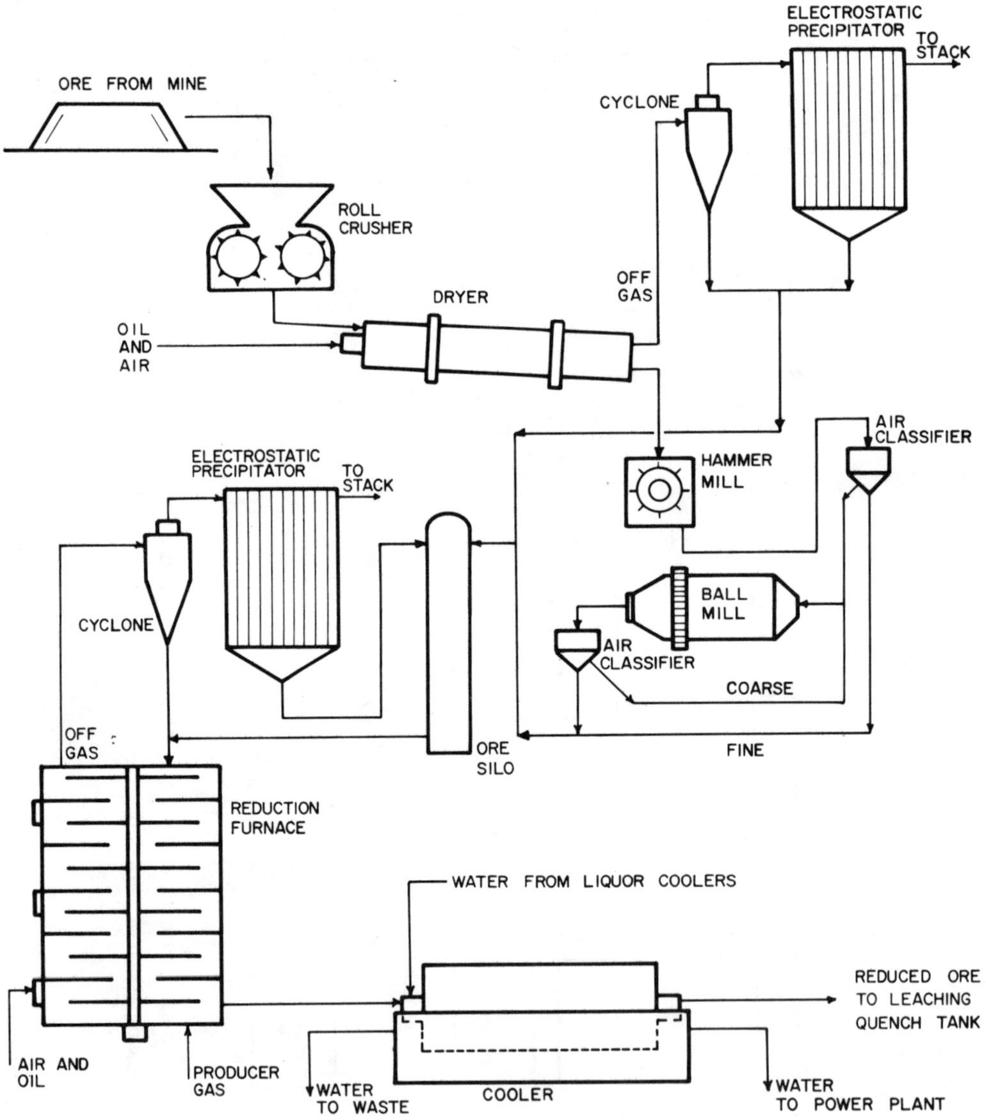

Figure 6: Nicaro operations for preparing ore prior to leaching.

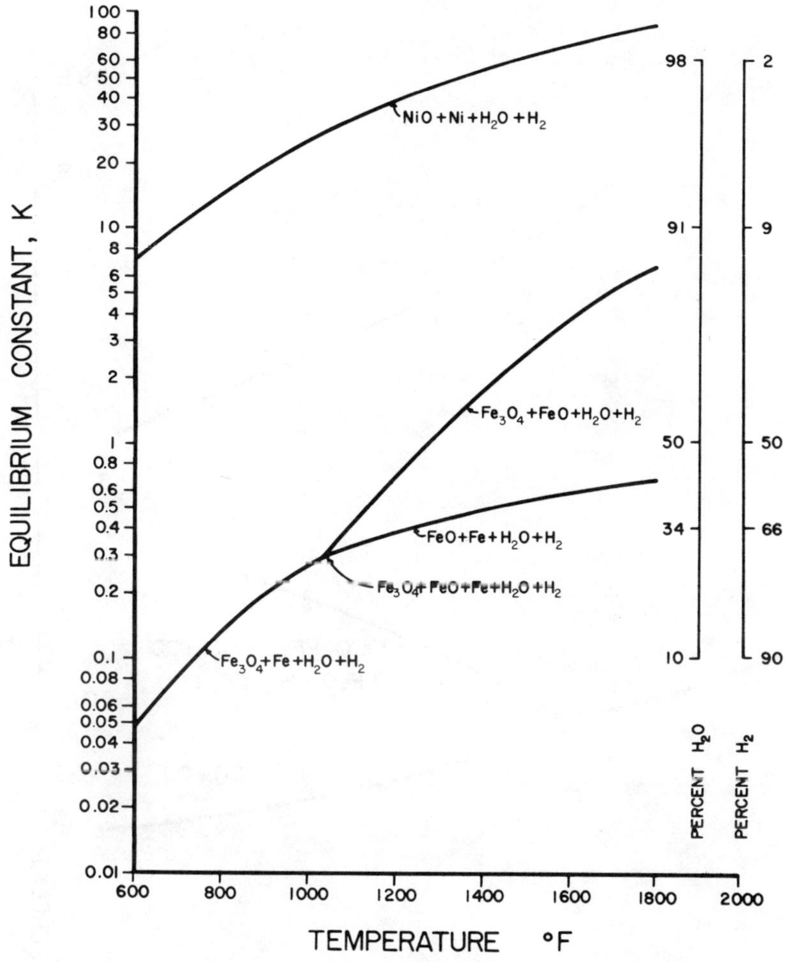

Figure 7: EQUILIBRIA IN SYSTEMS
Fe:H:O and Ni:H:O
Equilibrium Constant, $K = \frac{H_2O}{H_2}$ Vs. Temperature

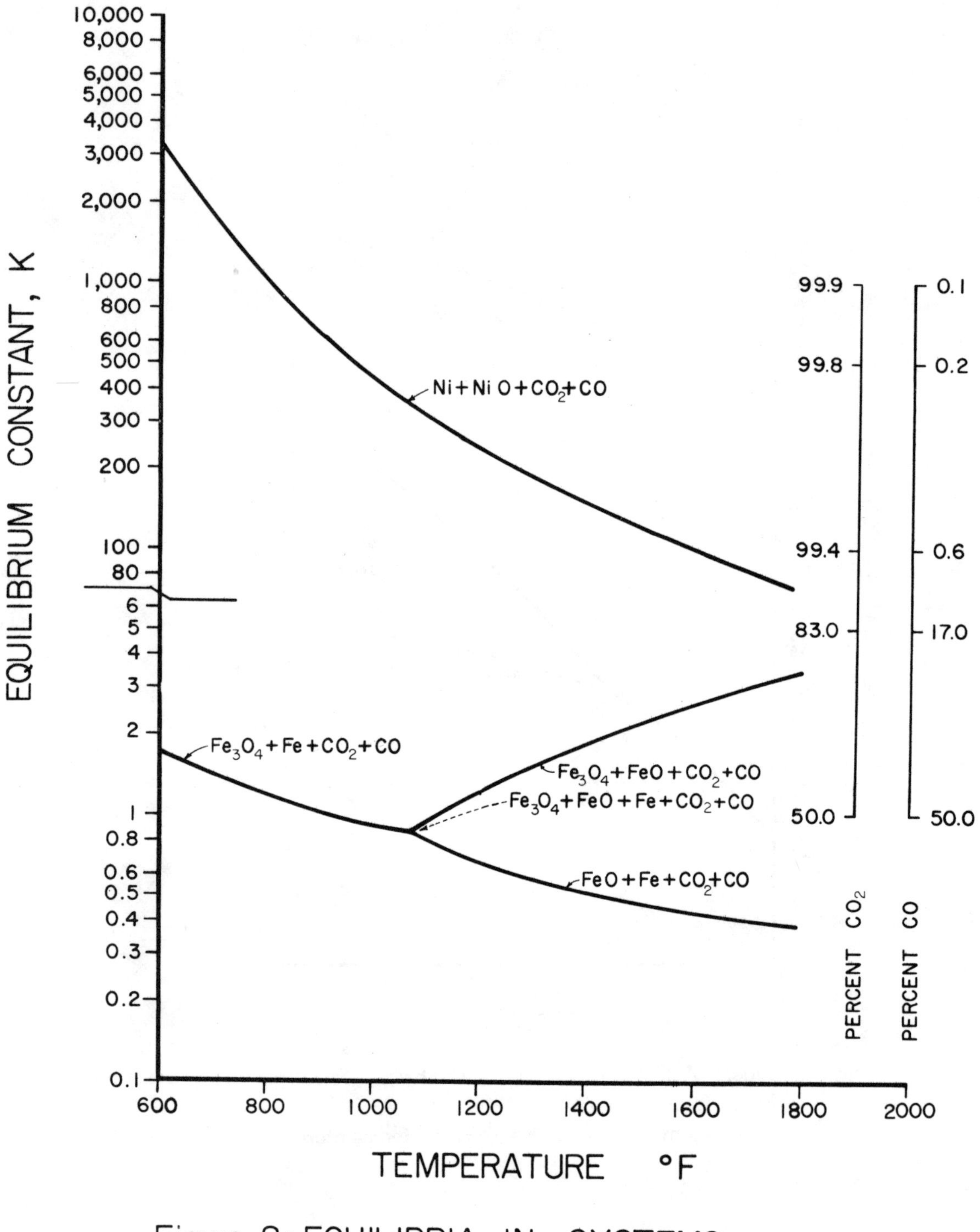

Figure 8: EQUILIBRIA IN SYSTEMS Fe:C:O and Ni:C:O Equilibrium Constant, $K = \frac{CO_2}{CO}$ Vs. Temperature

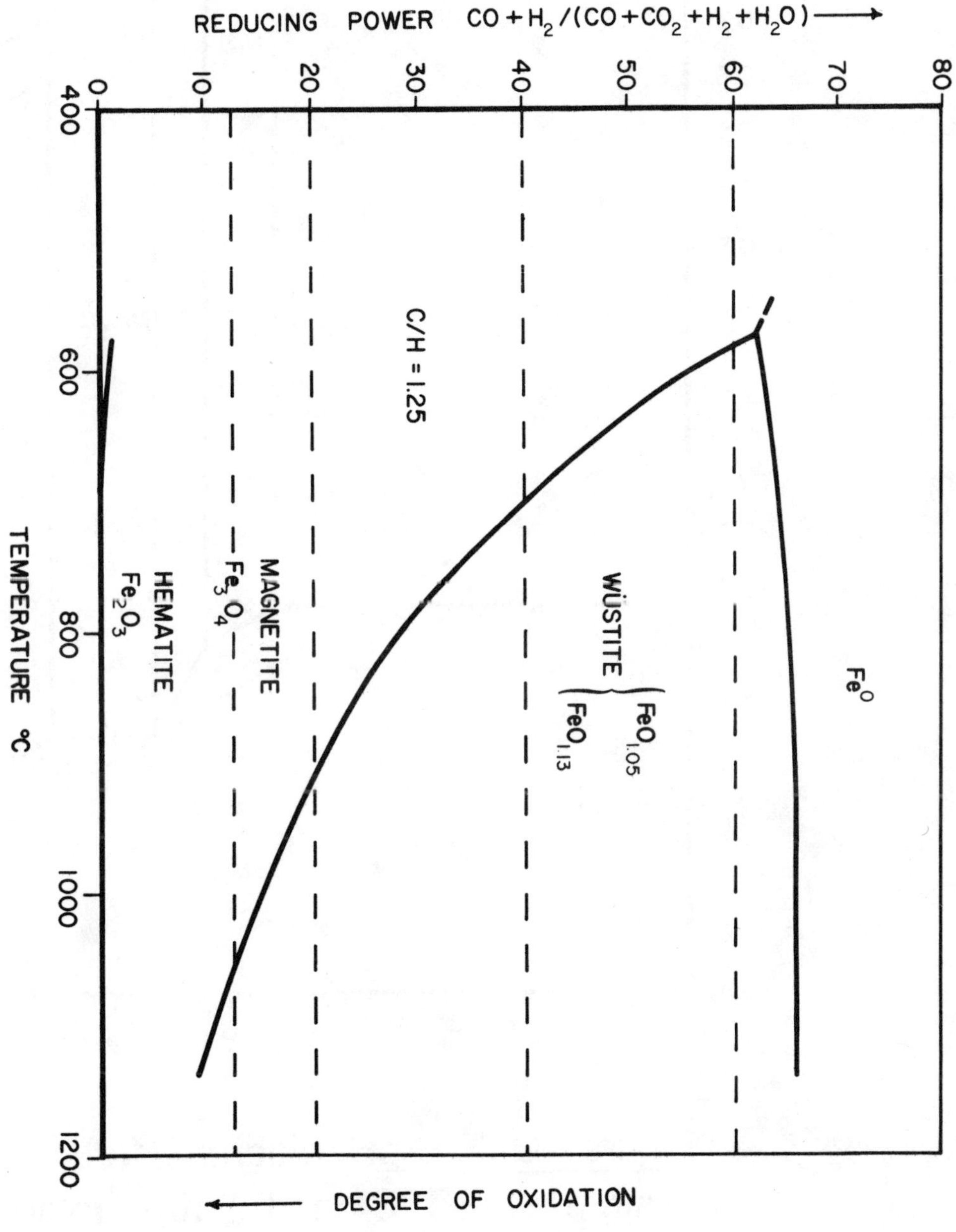

Figure 9: Equilibria of the oxides of iron with a mixed gas.

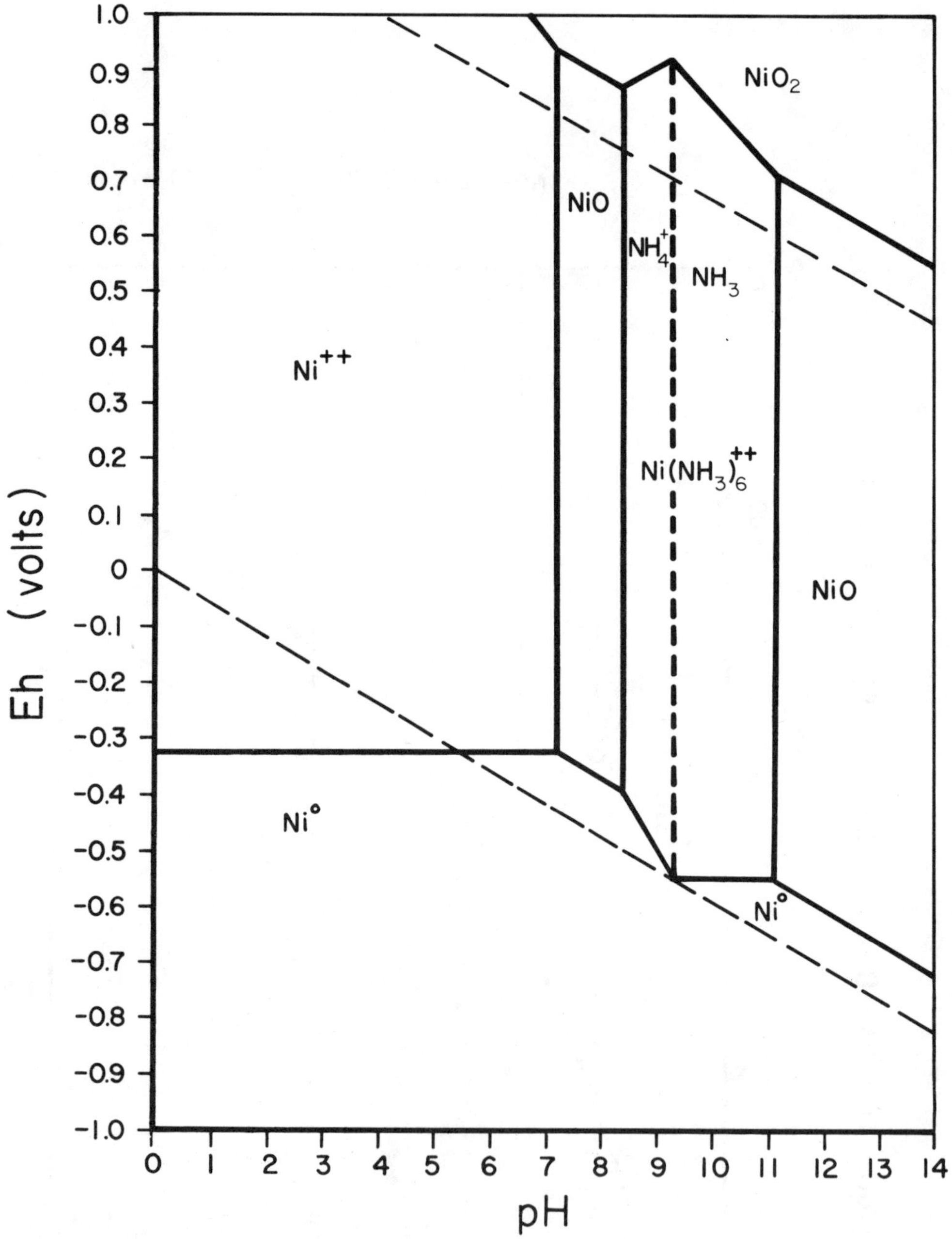

Figure 10: Eh–pH diagram for $Ni\text{-}NH_3\text{-}H_2O$ system at 25°C and 1 atm. total pressure. The activities of NH_3 above pH 9.25 and NH_4^+ below 9.25 are unity. The ionic nickel species is 10^{-2} M.

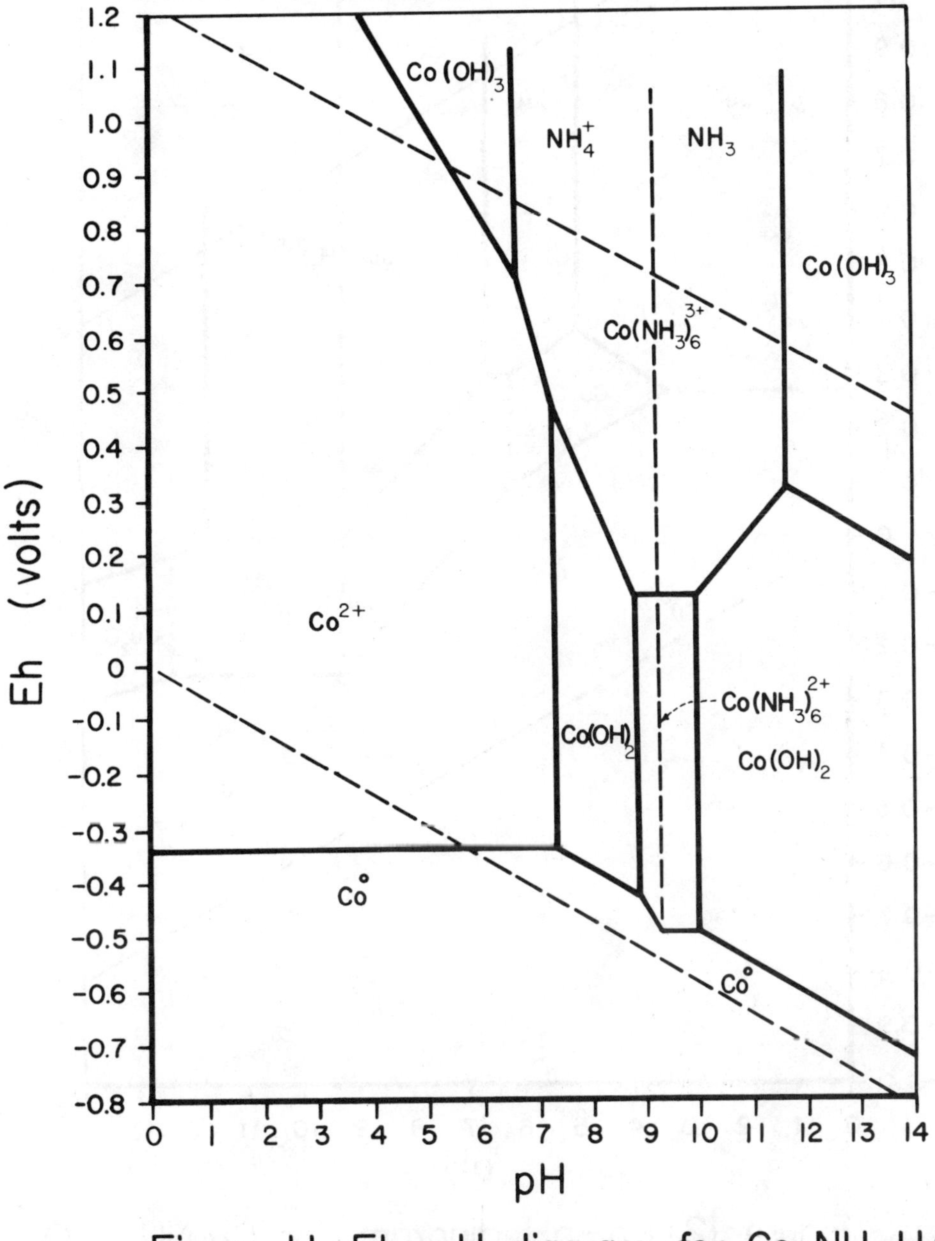

Figure 11 : Eh−pH diagram for Co-NH_3-H_2O system at 25°C and 1 atm. total pressure. The activities of NH_3 above pH 9.5 and NH_4^+ below 9.25 are unity. The ionic cobalt species is 10^{-2} M.

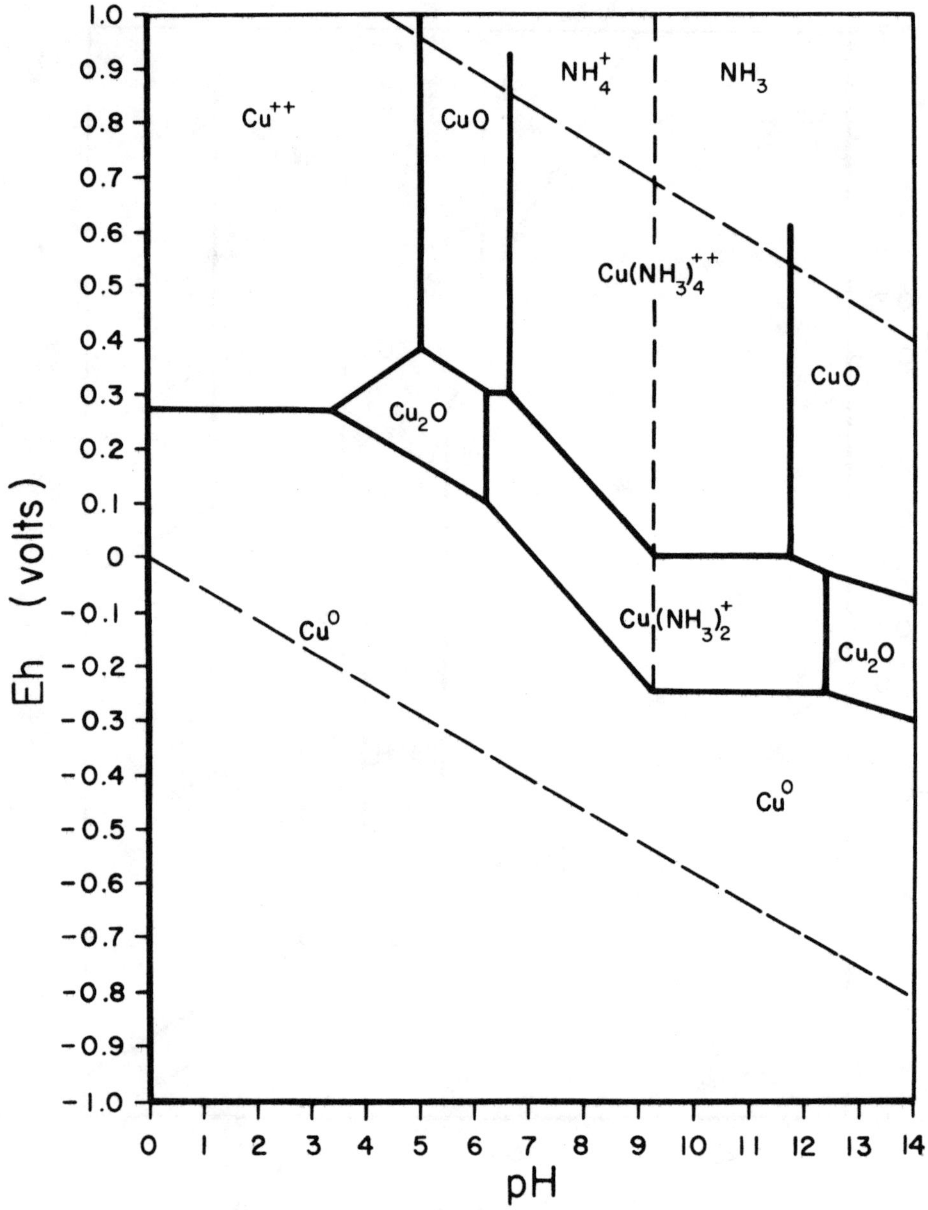

Figure 12: Eh−pH diagram for $Cu\text{-}NH_3\text{-}H_2O$ system at 25°C and 1 atm. total pressure. The activities of NH_3 above pH 9.25 and NH_4^+ below 9.25 are unity. The ionic copper species is 10^{-2} M.

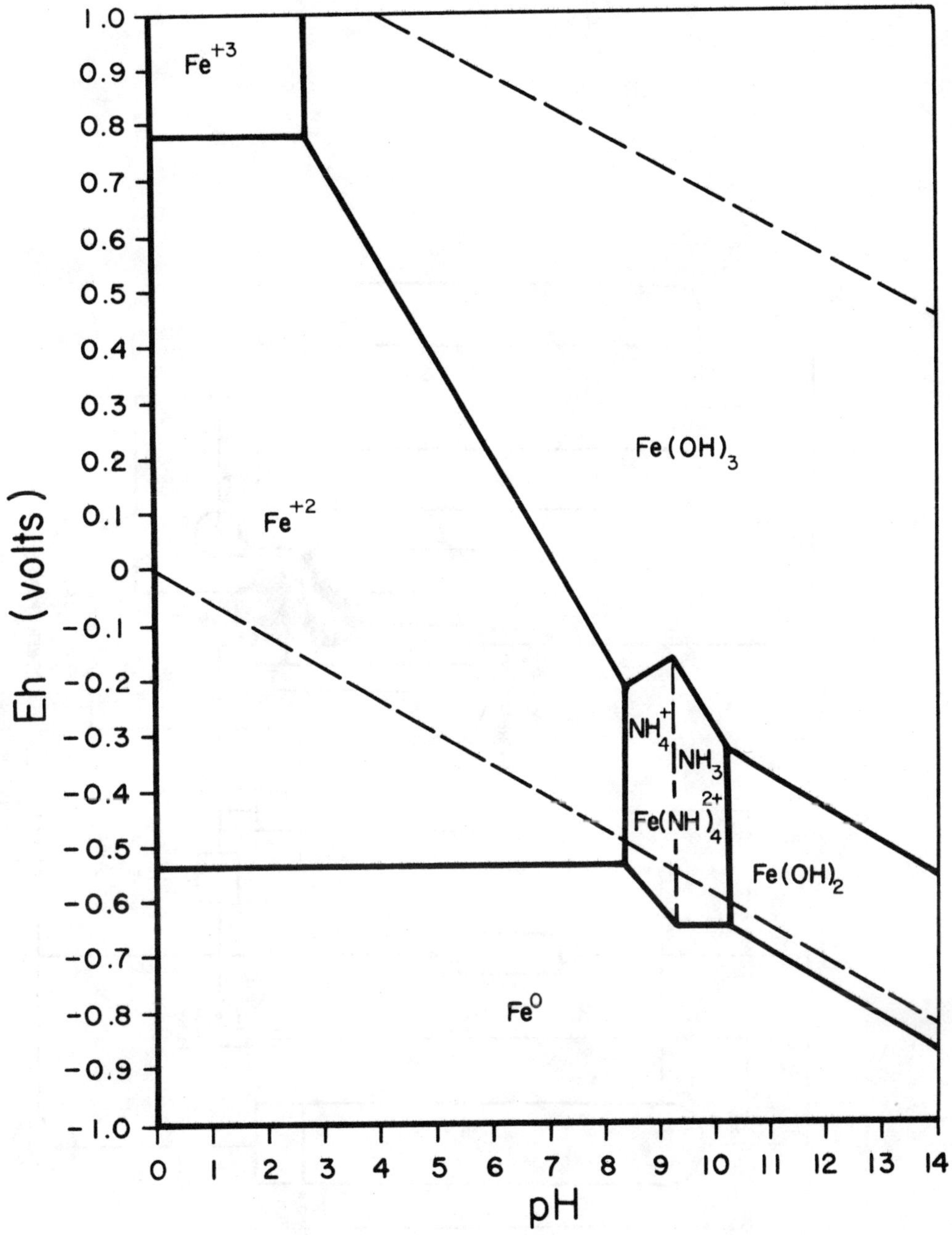

Figure 13: Eh–pH diagram for $Fe-NH_3-H_2O$ system at 25°C and 1 atm. total pressure. The activities of NH_3 above pH 9.25 and NH_4^+ below 9.25 are unity. The ionic iron species is 10^{-3} M.

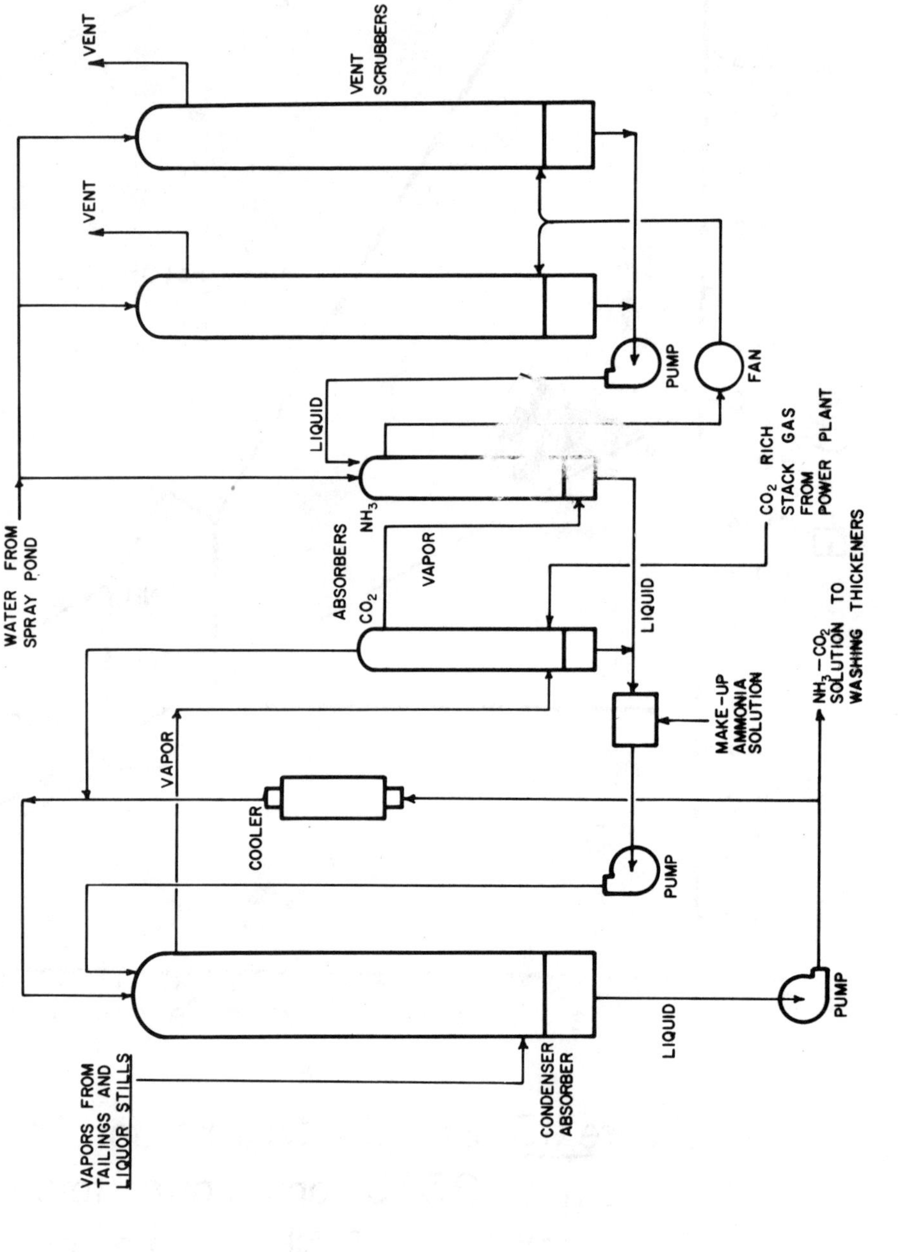

Figure 14: Ammonia and carbon dioxide recovery system at Nicaro.

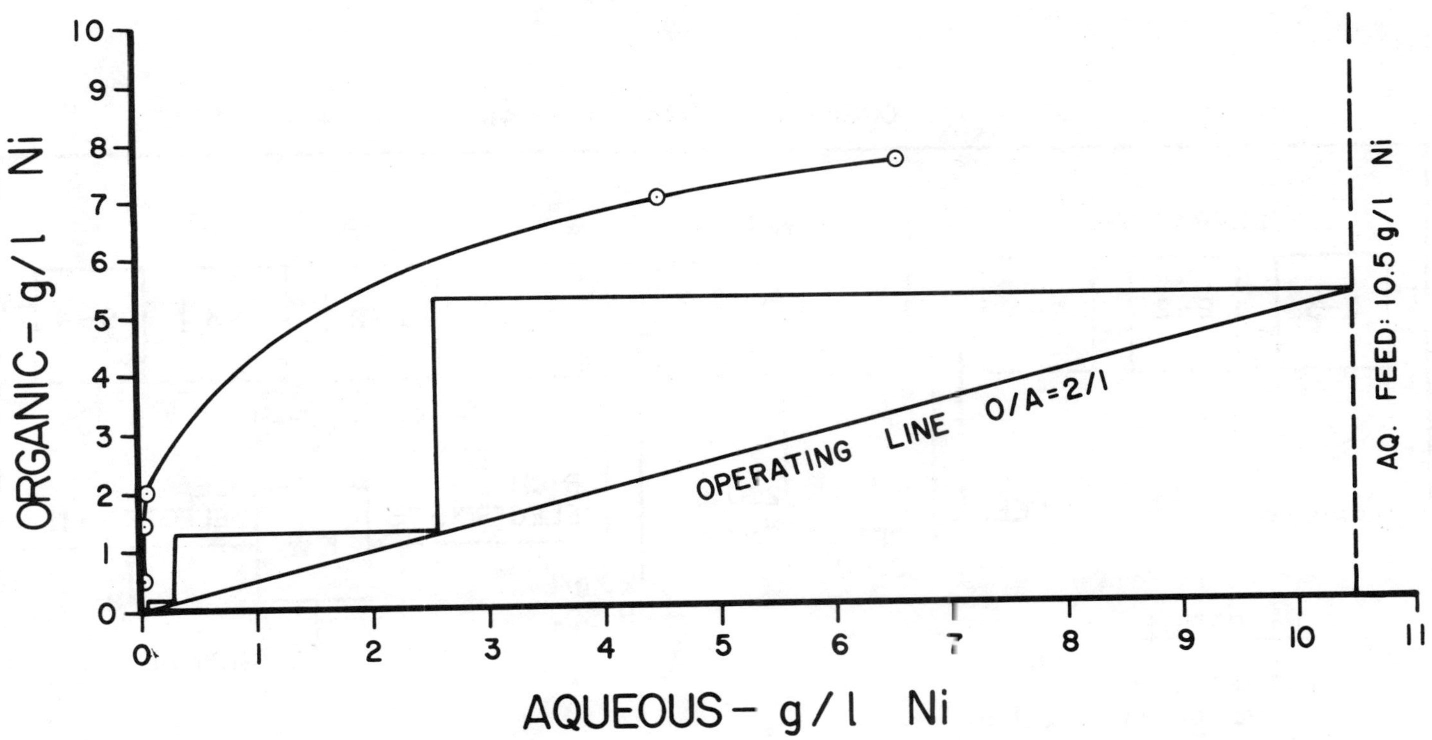

Figure 15: Extraction isotherm for LIX 64N Nickel in ammonium carbonate solution – pH = 9.5 T = 75 °F McCabe – Thiele construction for 90% stage efficiency.

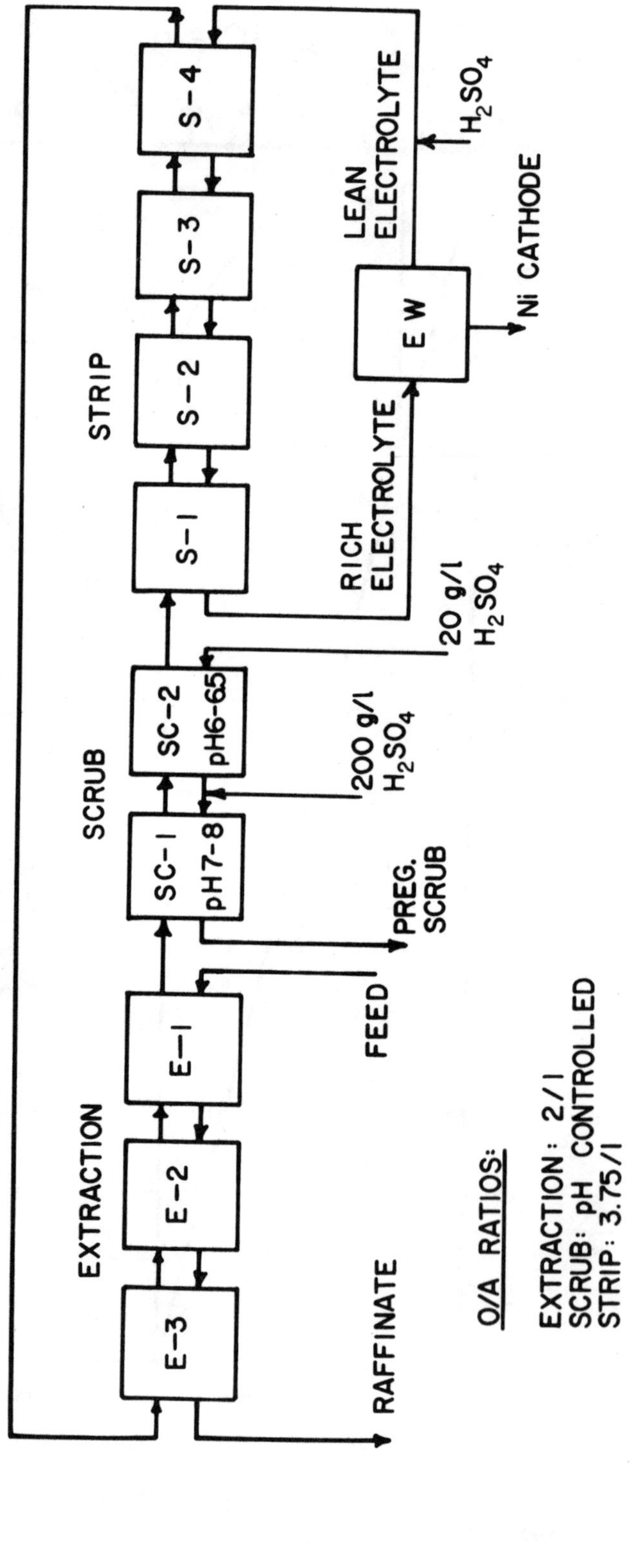

Figure 16: Block flowsheet for UOP nickel process solvent extraction – electrowinning system.

Chapter 5

MICROBIOLOGICAL PROCESSES IN RECOVERY OF METAL FROM ORES

Corale L. Brierley and James A. Brierley

New Mexico Bureau of Mines and Mineral Resources and Department of Biology
New Mexico Institute of Mining and Technology
Socorro, New Mexico 87801

ABSTRACT

Microorganisms have been considered as agents which may be of importance in catalyzing hydrometallurgical processes for the extraction of metals from low-grade ore minerals. Various microorganisms and their role in bacterial leaching are described. Bacterial leaching as applied to copper, uranium, zinc, lead, and coal is discussed.

MICROBIOLOGICAL PROCESSES IN RECOVERY OF METALS FROM ORES

Corale L. Brierley and James A. Brierley

New Mexico Bureau of Mines and Mineral Resources, and Department of Biology
New Mexico Institute of Mining and Technology
Socorro, NM 87801

INTRODUCTION

Microorganisms have been considered as agents which may be of importance in catalyzing hydrometallurgical processes for extraction of metals from low-grade ore minerals. Most research has dealt with the bacteria of the genus Thiobacillus and their role in leaching of copper and uranium. There is a rather voluminous literature covering the topic including recent reviews (1,2,3,4), symposia proceedings (5,6) and one book (7).

Microbial activities may also aid in restoration of in-site leaching environments by transforming nitrogen ions for removal from leached formations.

MICROORGANISMS

The microbes, Thiobacillus ferrooxidans and Thiobacillus thiooxidans are well recognized components of bacterial leaching systems. Their role in dump leaching was first recognized in 1957 (8). Most hydrometallurgical processes and studies dealing with biological processes have utilized these two microbes. However, the leach dump environment is a complex ecosystem containing yeasts (9), protozoa (10), and algae (Chlamydomonas is present in the leach dump effluent from Chino Mines Leach Dump, Kennecott Copper Corporation, Hurley, N.M.) in addition to the thiobacilli. Recently, facultative-thermophilic microorganisms, capable of iron oxidation at 50°C, have also been found in samples from copper-leaching dumps (11). It is quite likely that other interesting microorganisms not presently identified may also exist in leaching environments. Much more research is needed to determine what other types of microorganisms inhabit leaching systems and how these microbes relate to the process of metal's extraction.

The thermophilic (grow at 80°C) and acidophilic (require pH 3.5 or less) bacteria of the genus Sulfolobus (12,13) can oxidize reduced sulfur and iron compounds and catalyze the leaching of chalcopyrite and molybdenite (12,14,15, 16,17). Although these microbes show promise for catalysis of metal's leaching in the laboratory, they have not yet been found in large-scale leaching operations.

The facultative-thermophilic, iron-oxidizing bacteria found in leaching systems (11,17,18) can apparently grow over a range of temperatures from 30 to 50°C. However, for these microbes to grow, they require the presence of organic compounds (e.g., cysteine, glutathione) since they do not fix carbon dioxide as do the Thiobacillus species (19,20,21). The T. ferrooxidans found in leach dumps may be supplying the organic compounds required by the facultative-thermophilic microbes (20). Initial research indicates that this group of thermophilic microbes may have an active role in metal's solubilization (19,22).

An iron-oxidizing microorganism, Leptospirillum ferrooxidans, has been shown to form a consortium with sulfur-oxidizing microorganisms such as Thiobacillus organoparus to bring about metal's dissolution (22). This is the first indication that in a leaching system a complex association of microorganisms, and not necessarily a single microbe, may facilitate metal's-extraction. In a recent study (23) iron-oxidizing thiobacilli were found to oxidize chalcopyrite with liberation of soluble copper and concurrent formation of jarosite; the doubling time of these organisms on this substrate was 38.4 ± 2.9 h. These same organisms oxidized a galena concentrate, exhibiting a multiplication rate of 39.6 ± 2.7 h; x-ray diffraction analysis indicated that galena was converted to anglesite. The microbial oxidation of these two solid substances was believed to be the result of a mixed culture of iron-oxidizing thiobacilli. One strain was thought to be responsible for the oxidation of lead sulfide, and the other strain oxidized chalcopyrite and pyrite. There is also a strong possibility that many biological leaching studies using T. ferrooxidans may have in fact involved a microbial consortium rather than a pure culture. A mixotrophic contaminant, which can use sulfur or organic compounds for energy and identified as Thiobacillus acidophilus, has been found in several cultures of T. ferrooxidans (24,25,26). It is possible that commercial metal's leaching is facilitated by mixed cultures of more than one microorganism.

BACTERIAL LEACHING

Copper

Dump leaching of copper is extensively practiced in the western United States and accounts for about 11.5% of the total copper production of this country (27). The microorganisms assumed to be the principals involved in the metal's extraction process are Thiobacillus ferrooxidans and Thiobacillus thiooxidans. These organisms require oxygen, an acid environment of about pH 1.5 to 3.0, and they utilize carbon dioxide as their sole source of carbon. The microbes function in the temperature range of 18° to 40°C. T. ferrooxidans derive energy for growth from the oxidation of ferrous iron.

$$2FeSO_4 + \tfrac{1}{2} O_2 + H_2SO_4 \rightarrow Fe_2(SO_4)_3 + H_2O \quad (1)$$

T. ferrooxidans increase the rate of reaction (1) by 500,000 times the spontaneous oxidation rate in air (28). The T. ferrooxidans are present at leach-dump surfaces at concentrations of 10^{7} per gram, and the numbers decline with depth into the dump (29). T. thiooxidans oxidize sulfur, accordingly:

$$S^o + 3/2\ O_2 + H_2O \rightarrow H_2SO_4 \quad (2)$$

These organisms are also found in large numbers, up to 10^7 per gram of ore, in copper dump leaching operations. The ferric iron produced by T. ferrooxidans is directly responsible for the dissolution of metals (M) from metallic sulfide (MS) minerals,

$$MS + 2Fe^{3+} \rightarrow M^{2+} + 2Fe^{2+} + S^o \quad (3)$$

and the elemental sulfur produced by this reaction is oxidized by T. thiooxidans.

Several specific metal dissolution reactions are:

$$CuFeS_2 + 2Fe_2(SO_4)_3 \rightarrow CuSO_4 + 5FeSO_4 + 2S^o \quad (4)$$

(Chalcopyrite)

$$Cu_2S + 2Fe_2(SO_4)_3 \rightarrow 2CuSO_4 + 4FeSO_4 + S^o \quad (5)$$

(Chalcocite)

The sulfur generated by these chemical reactions can passivate the mineral surface and inhibit further leaching. One function of T. thiooxidans is believed to be the oxidation of the sulfur to soluble sulfuric acid thus allowing leaching to continue. The involvement of bacteria in such metal dissolution reactions has been called "indirect" since the bacteria generate the oxidizing agent, ferric iron, and provide a suitably acid environment to maintain the oxidized metal species in solution.

The thiobacilli also directly effect metal dissolution. The mechanism of attack has not been elucidated, but it is widely believed that the metal moiety itself is oxidized by the microbes. Some specific reactions in which the bacterial leaching is "direct" include:

$$2Cu_2S + O_2 + 2H_2SO_4 \rightarrow 2CuS + 2CuSO_4 + 2H_2O \quad (6)$$

$$4FeS_2 + 15O_2 + 2H_2O \rightarrow 2Fe_2(SO_4)_3 + 2H_2SO_4 \quad (7)$$

$$4CuFeS_2 + 17O_2 + 2H_2SO_4 \rightarrow 4CuSO_4 + 2Fe_2(SO_4)_3 \quad (8)$$

$$2H_2O$$

A recent paper (26) has indicated that T. ferrooxidans can enzymaticaly oxidize the sulfide moiety of pyrite. However, the extent and importance of the direct oxidation mechanism in the leaching system are not known.

The development of in-situ copper leaching processes will require study of the response of the microbes to the physico-chemical environment will develop. In-situ leaching will occur at increased hydrostatic pressure and elevated temperatures. A preliminary report (30) has indicated that hydrostatic pressure (690 kPa) does affect oxidation of chalcopyrite by T. ferrooxidans; increased atmospheric pressure, which increases oxygen partial pressure, is deleterious to the organisms (S. Zeiger, unpublished data). Studies are presently in progress at New Mexico Institute of Mining and Technology to determine if microbes can tolerate in-situ conditions for leaching of copper and uranium.

Uranium

The bacterial leaching of uranium is an indirect process. T. ferrooxidans oxidize ferrous iron and pyrite according to reactions (1) and (7) to produce the oxidant, ferric iron, which reacts with reduced tetravalent uranium

$$UO_2 + Fe_2(SO_4)_3 \rightarrow UO_2SO_4 + 2FeSO_4 \quad (9)$$

Ferric iron is regenerated by the bacteria, and the sulfuric acid necessary to maintain the oxidized uranium in solution is added and/or generated by the bacterial oxidation of pyrite (reaction 7). The microbial process has been used as a scavenger of uranium values from mined-out areas and low-grade stopes in the Elliot Lake region of Canada. Bioextraction is now applied as the principal means of recovering uranium from heaps and fractured stopes at one Elliot Lake Mine (31). Acid leach solutions are percolated through the underground ore, and the oxidizing agent, ferric iron, is regenerated biologically. The uranium ores of northern Ontario are amenable to bacterial leaching because of the substantial pyrite concentrations. However, the uranium ores of the Rocky Mountain region and south Texas are secondary uranium formations, deposited in the roll front configuration, and they lack the large concentrations of pyrite essential for effective bacterial leaching. These deposits are amenable to in-situ, or bore-hole mining, by either acid or carbonate leaches (32). Little research has been conducted on the bacterial leaching of these rollfront ores, but preliminary investigations on ore from the Anaconda Company Jackpile-Paguate mines of New Mexico indicated that suitable energy sources are unavailable for the bacteria, and that the organisms lose viability when in protracted contact with the ore (33).

In the uranium leaching environment bacteria may catalyze other reactions which are not directly involved in the extraction but may influence the mining operation process. The possible use of micro-organisms, specifically those which either oxidize or reduce nitrogenous compounds, for restoration of in-situ uranium mines has recently been reviewed (2). The use of leach solutions containing ammonium carbonate or ammonium bicarbarate can result in increased levels of ammonium ion adsorbed to base exchange sites of clays. Later, oxidation to nitrite and nitrate and resulting mobilization may be responsible for ground water quality degradation. Bacteria capable of ammonium ion oxidation could be used under controlled conditions during formation restoration for mobilization of the bound ammonia. Species of the genus Nitrosomonas oxidize ammonium to nitrite,

$$NH_4^+ + 3/2\ O_2 \rightarrow 2H^+ + H_2O + NO_2^- \quad (10)$$

which is further oxidized by Nitrobacter.

$$NO_2^- + 1/2\ O_2 \rightarrow NO_3^- \quad (11)$$

These two organisms are always found together and accomplish the oxidation of ammonia to nitrate (the processes of nitrification). The nitrate has little affinity for minerals and will remain in solution. The nitrate could be flushed from a leach site. By the bacterial process of denitrification nitrate can be converted to nitrogen gas. This occurs in the absence of oxygen. Many diverse types of microbes can reduce nitrate,

$$2NO_3^- + 10(H^+ + e^-) \rightarrow N_2 + 4H_2O + 2OH^- \quad (12)$$

if they are provided with an oxidizable energy source, usually an organic substrate, for growth. Although the nitrogen gas may appear innocuous, it could lead to an airlock in the formation and blockage of solution flow. However, diatomic nitrogen gas would not result in toxicity problems which are assoicated with nitrate and nitrite.

Zinc and Lead

Microbial solubilization of zinc has been extensively reported (34,35,36,37,38). The T. ferrooxidans are effective at oxidation of zinc sulfide and tolerating high concentrations of soluble zinc (more than 10 g l^{-1} (39)). The microbial facilitated oxidation of zinc sulfide may be the result of direct microbial oxidation of the mineral and indirect oxidation by microbially-generated ferric iron. Conditions which favor growth of T. ferrooxidans (e.g., increased CO_2 concentration) facilitate zinc leaching (35). Increase of surface area of the metal sulfide also enhances extraction of a sphalerite concentrate (36).

T. ferrooxidans have potential for use in processes for selective extraction of some metals with concurrent enrichment of others in precipitate form. Copper, zinc and cadmium can be leached from a lead-sulfide concentrate (40). Lead-sulfide was found to be partly oxidized (40 to 60%) with the resulting formation of lead sulfate. The T. ferrooxidans microbes have also been used for leaching of lead-zinc (galena and sphalerite) ore with extraction of 40% zinc and 60% lead in 271 to 286 days (37). Percolation columns were used in the experimentation. The zinc was solubilized and the lead remained in the column as a lead-sulfate precipitate which could be dissolved with a sodium chloride solution. The maximum rate of galenite oxidation was reported to occur at a temperature between 50 to 65°C with a soluble ferric iron concentration greater than 20 g l^{-1}. This temperature range is beyond the range of growth of T. ferrooxidans. The microbial leaching was either a consequence of undefined thermophilic microbial activity or due solely to oxidations by ferric iron.

Coal

Microbial processes may also be involved in benefication of coal. Pyrite can be oxidized by a T. ferrooxidans plus T. thiooxidans mixture or Sulfolobus in order to decrease coal sulfur content (41,42,43,44,45). The thermophilic Sulfolobus has shown potential for economic removal of pyritic-sulfur from coal (41). The problem of removal or decrease of organic-sulfur content by microbiological means has not yet been solved. A mixed population of unidentified microbes capable of degradation of dibenzothiophene was able to reduce, by almost 20%, the organic-sulfur content of a coal sample (41). Development of microbiological processes for removal of sulfur (including pyrite and organic sulfur) will provide better understanding of the microbial aspects for recovery of metal values from diverse substrates.

REFERENCES

1. Brierley, C.L., "Bacterial Leaching," CRC Critical Reviews in Microbiology, Vol.6, Nov. 1978, pp. 207-262.
2. Brierley, C.L., "Microbiology and Mining--Practices and Problems," New Mexico Geology, Vol.1, No.3, Aug. 1979, pp. 33-36.
3. Kelly, D.P., Norris, P.R., and Brierley, C.L., "Microbiological Methods for the Extraction and Recovery of Metals," Microbial Technology: Current State, Future Prospects, Bull, A.T., Ellwood, D.C., and Ratledge, eds. Cambridge University Press, Cambridge, 1979, pp. 263-308.
4. Torma, A.E., "The Role of Thiobacillus ferrooxidans in Hydrometallurgical Processes," Advances in Biochemical Engineering, Vol.6, 1977, pp. 1-37.
5. Schwartz, W., ed., Conference Bacterial Leaching 1977, Verlag Chemie, New York, 1977, 270 pp.
6. Murr, L.E., Torma, A.E., and Brierley, J.A., eds., Metallurgical Applications of Bacterial Leaching and Related Microbiological Phenomena, Academic Press, New York, 1978, 526 pp.
7. Karavaiko, G.I., Kuznetsov, S.I., and Golonizik, A.I., The Bacterial Leaching of Metals from Ores, Technicopy Limited (english translation), England, 1977, 205 pp.
8. Bryner, L.C., and Anderson, R., "Microorganisms in Leaching Sulfide Minerals," Industrial and Engineering Chemistry, Vol.49, 1957, pp. 1721-1724.
9. Ehrlich, H.L., and Fox, S.I., "Copper Sulfide Precipitation by Yeasts from Acid Mine-Waters," Applied Microbiology, Vol.15, No.1, Jan. 1967, pp. 135-139.
10. Ehrlich, H.L., "Microorganisms in Acid Drainage from a Copper Mine," Journal of Bacteriology, Vol.86, No.2, Aug. 1963, pp. 350-352.
11. Brierley, J.A., "Thermophilic Iron-Oxidizing Bacteria Found in Copper Leaching Dumps," Applied and Environmental Microbiology, Vol.36, No.3, Sept. 1978, pp. 523-525.
12. Brock, T.D., Brock, K.M., Belly, R.T., and Weiss, R.L., "Sulfolobus: A New Genus of Sulfur-Oxidizing Bacteria, Living at Low pH and High Temperatures," Archives of Mikrobiologia, Vol.-84, 1972, pp. 54-68.
13. Brierley, C.L., and Brierley, J.A., "A Chemoautotrophic and Thermophilic Microorganism Isolated from an Acid Hot Spring," Canadian Journal of Microbiology, Vol.19, No.2, 1973, pp. 183-188.
14. Brierley, C.L., "Thermophilic Microorganisms in Extraction of Metals from Ores," Developments in Industrial Microbiology, Vol.18, 1977, pp. 273-284.
15. Brierley, J.A., and Brierley, C.L., "Microbial Leaching of Copper at Ambient and Elevated Temperatures," Metallurgical Appli-

cations of Bacterial Leaching and Related Microbiological Phenomena, Murr, L.E., Torma, A.E., and Brierley, J.A., eds., Academic Press, Inc., New York, 1978, pp. 477-489.

16. Brierley, C.L., "Molybdenite-Leaching: Use of a High-Temperature Microbe," Journal of the Less-Common Metals, Vol.36, 1974, pp. 237-247.

17. Brierley, C.L., Brierley, J.A., Norris, P.R., and Kelly, D.P., "Metal-tolerant Microorganisms of Hot, Acid Environments," Society for Applied Bacteriology, Technical Series Publication, No.14, in press.

18. Brierley, J.A., and Lockwood, S.J., "The Occurrence of Thermophilic Iron-Oxidizing Bacteria in a Copper Leaching System," FEMS Microbiology Letters, Vol.2, 1977, pp. 163-165.

19. Brierley, J.A., Norris, P.R., Kelly, D.P., and LeRoux, N.W., "Characteristics of a Moderately Thermophilic and Acidophilic Iron-Oxidizing Thiobacillus," European Journal of Applied Microbiology and Biotechnology, Vol.5, 1978, pp. 291-299.

20. Brierley, J.A., Lockwood, S.J., Kelly, D.P., and Norris, P.R., "Requirement of Organic Compounds by Iron-Oxidizing Facultative-Thermophilic Thiobacilli," Abstracts of the Annual Meeting of the American Society for Microbiology, 1979, p. 96.

21. Norris, P.R., Brierley, J.A., and Kelly, D.P., "Physiological Characteristics of Two Facultatively Thermophilic Mineral-Oxidizing Bacteria," FEMS Microbiology Letters, Vol.7, 1980, pp. 119-122.

22. Norris, P.R., and Kelly, D.P., "Dissolution of Pyrite (FeS_2) by Pure and Mixed Cultures of Some Acidophilic Bacteria," FEMS Microbiology Letters, Vol.4, 1978, pp. 143-146.

23. Kingma, J.G., and Silver, M., "Growth of Iron-Oxidizing Thiobacilli in the Presence of Chalcopyrite and Galena," Applied and Environmental Microbiology
Environmental Microbiology, Vol.39, No.2, Mar. 1980, pp. 635-641.

24. Guay, R., and Silver, M., "Thiobacillus acidophilus sp.nov.; Isolation and Some Physiological Characteristics," Canadian Journal of Microbiology, Vol.21, 1975, pp. 281-288.

25. Tuovinen, O.H., Kelly, D.P., Crawford, S.D., and Eccleston, M., "Metabolic Transitions in Cultures of Acidophilic Thiobacilli," Metallurgical Applications of Bacterial Leaching and Related Microbiological Phenomena, Murr, L.E., Torma, A.E., and Brierley, J.A., eds., Academic Press Inc., New York, 1978, pp. 61-81.

26. Arkesteyn, G.J.M.W., "Pyrite Oxidation by Thiobacillus ferrooxidans with Special Reference to the Sulfur Moiety of the Mineral," Antoni van Leeuwenhock Journal of Microbiology, Vol.45, No.3, 1979, pp. 423-435.

27. Wadsworth, M.E., "Physico-Chemical Aspects of Solution Mining," Symposium on Inplace Leaching and Solution Mining, Kim, Y.S., ed., 1975, Chap.8.

28. Lacey, D.T., and Lawson, F., "Kinetics of the Liquid-Phase Oxidation of Acid Ferrous Sulfate by the Bacterium Thiobacillus ferrooxidans," Biotechnology and Bioengineering, Vol. 12, 1970, pp. 29-50.

29. Bhappu, R.B., Johnson, P.H., Brierley, J.A., and Reynolds, D.H., "Theoretical and Practical Studies on Dump Leaching," Trans. SME/AIME, Vol.244, No.3, Sept. 1969.

30. Bosecker, K., Torma, A.E., and Brierley, J.A., "Microbiological Leaching of a Chalcopyrite Concentrate and the Influence of Hydrostatic Pressure on the Activity of Thiobacillus ferrooxidans," European Journal of Applied Microbiology and Biotechnology, Vol.7, 1979, pp. 85-90.

31. McCreedy, H.H., "Bacterial Leaching to be Applied to a New Canadian Uranium Mine," Coordinated Research Committee Meeting, International Atomic Energy Agency, Coventry England, Dec. 13-15, 1976.

32. Shock, D.A., and Conley, F.R., "Solution Mining--Its Promise and Its Problems," Solution Mining Symposium, Aplan, F.F., McKinney, W.A. and Pernichele, A.D., eds., The American Institute of Mining, Metallurgical, and Petroleum Engineers, New York, 1974, pp. 1-37.

33. Brierley, C.L., "Biogenic Extraction of Uranium From Ores of the Grants Region," Metallurgical Applications of Bacterial Leaching and Related Microbiology Phenomena, Murr, L.E., Torma, A.E., and Brierley, J.A., eds., Academic Press, Inc., New York, 1978, pp. 345-362.

34. Torma,A.E., "Microbiological Oxidation of Synthetic Cobalt Nickel and Zinc Sulfides by Thiobacillus Ferrooxidans," Revue Canadianne de Biologie, Vol.30, No.3, Sept., 1971, pp. 209-216

35. Torma, A.E., Walden, C.C., Duncan, D.W., and Branion, R.M.R., "The Effect of Carbon Dioxide and Particle Surface Area on the Microbiological Leaching of a Zinc Sulfide Concentrate," Biotechnology and Bioengineering, Vol. 14, 1972, pp.777-786.

36. Torma, A.E., and Guay, R., "Effect of Particle Size on the Biodegradation of a Sphalerite Concentrate," Naturaliste Canadianne, Vol.103, 1976, pp.133-138.

37. Ilyaletdinov, A.N., Kamalov, M.R., Stukanov, V.A., "Microbiological Leaching of Zinc and Lead from Ores of the Tekeli Deposit," Microbiology, Vol.46, No.5, Sept.-Oct., 1977, pp. 696-703.

38. Torma, A.E., and Sakaguchi, H., "Relation Between the Solubility Product and the Rate of Metal Sulfide Oxidation by Thiobacillus ferrooxidans, "Journal of Fermentation Technology, Vol.56, No.3, 1978, pp. 173-178.

39. Tuovinen, O.H., Niemela, S.I., and Gyllenberg, H.G., "Tolerance of Thiobacillus ferrooxidans to some Metals," Antonie van Leeuwenhock Journal of Microbiology, Vol.37, No.4, 1971, pp. 489-496.

40. Torma, A.E., and Subramanian, K.N., "Selective Bacterial Leaching of a Lead Sulphide Concentrate," International Journal of Mineral Processing, Vol.1, 1974, pp. 125-134.

41. Detz, C.M., and Barvinchak, G., "Microbial Desulfurization of Coal," Mining Congress Journal, July 1979, pp. 75-86.

42. Chandra, D., Roy, P., Mishra, A.K., Chakrabarti, J.N., and Sengupta, B., "Microbial Removal of Organic Sulphur From Coal," Fuel, Vol.58, July 1979, pp. 549-550.

43. Ford, W.H., Roffman, H.K., and Beimborn, W.A., "Bacterial Removal of Sulfur From Coal," Combustion, August 1977, pp. 36-38.

44. Capes, C.E., McIlhinney, A.E., Sirianni, A.F., and Puddington, I.E., "Bacterial Oxidation in Upgrading Pyritic Coals," The Canadian Mining and Metallurgical Bulletin, Nov. 1973, pp. 88-91.

45. Silverman, M.P., Rogoff, M.H., and Wender, I., "Removal of Pyritic Sulphur From Coal by Bacterial Action," Fuel (London), Vol.42, 1963, pp. 113-124.

Section II

Hydrometallurgical Recovery from Concentrates

William G. Sherwood, Section Editor

COPPER RECOVERY FROM SULFIDE CONCENTRATES BY THE U.B.C. - COMINCO FERRIC CHLORIDE LEACH ROUTE

E. Peters*, G. M. Swinkels+ and A. Vizsolyi**

*Professor, Metallurgical Eng. Department,
University of British Columbia,
Vancouver, B. C.

+Asst. Manager, Tecnical Research,
Cominco, Ltd., Train, B.C.

**Sr. Research Officer, B. C. Research,
3650 Wesbrook Mall, Vancouver, B.C.

ABSTRACT

A process is described utilizing ferric chloride leaching followed by reduction of the cupric content to cuprous by cement copper, with CuCl obtained thereafter by crystallization. The residual liquor is cemented with sponge iron to produce cement copper and a copper free ferrous chloride solution. The latter is oxidized in an autoclave with oxygen to regenerate ferric chloride and to precipitate excess iron as a granular, filterable hematite. Cuprous chloride crystals are reduced to high purity metal with hydrogen, while sulphur reports almost quantitatively in residues in the elemental form. Pyrite and molybdenite are not leached.

This paper describes the important variables in the process, as well as the deportment of minor elements as silver, gold, selenium, tellurium, bismuth, nickel, lead, zinc, magnesium, etc. An elegant method of isolating silver from CuCl crystals is described, as well as recovery of gold from residues by a chloride reagent. The process is described as non-polluting and capable of producing refined copper, ore grade iron oxide, elemental sulphur, and yields of minor metals as good or better than obtainable by conventional smelting, without the use of solvent extraction or electrochemistry.

COPPER RECOVERY FROM SULFIDE CONCENTRATES BY THE U.B.C. - COMINCO FERRIC CHLORIDE LEACH ROUTE

E. Peters*, G.M. Swinkels+ and A. Vizsolyi**

*Professor, Metallurgical Eng. Dept.,
University of British Columbia,
Vancouver, B.C.,
+Asst. Manager, Technical Research,
Cominco, Ltd., Trail, B.C.
**Sr. Research Officer, B.C. Research,
3650 Wesbrook Mall, Vancouver, B.C.

Ferric chloride leaching of copper concentrates must now be regarded a state-of-art, with the operation of a small production plant by DUVAL Corporation (1) and a successful pilot plant in regular operation by Cyprus Metallurgical Corporation (2). It seems almost too late to publish a report on the earlier work done in our U.B.C. and Cominco laboratories, which eventually led to a patent (3). Nevertheless, this report is of value because it deals with most metallurgical questions associated with the hydrometallurgy of copper concentrates by the ferric chloride leaching route, in order to show that the regular criticisms of the conventional smelting protagonists can be met, when the problems are studied seriously.

In order to compete successfully with smelting and meet environmental and conservation demands, the following objectives should be met by an acceptable hydrometallurgical copper process:

(1) High copper recoveries; at least as high as conventional smelting (>98%).
(2) High conversion of sulfur to a marketable elemental form.
(3) Recovery of precious metals (silver and gold).
(4) Production of refined copper (without separate refining).
(5) Recovery or safe disposal of minor metals (such as tellurium, selenium, arsenic, antimony, bismuth, lead, zinc, molybdenum, nickel and cobalt).
(6) Complete closure of aqueous circuits, with control of purged material and residues as innocuous solids.
(7) Separation of iron as a marketable product.
(8) Low energy consumption, by avoidance of any electrochemical steps.
(9) Selection of satisfactory affordable containment materials.

It is the object of this report to describe how these objectives were met, in terms of the chemistry and the test results.

The Process

The U.B.C. - Cominco process as patented is shown in Fig. 1. The essential features of this process can be summarized as follows:

1. Leaching takes place by control of lixiviant and concentrate ratio in such a way that the copper in solution is divided about evenly between cupric and cuprous. The leaching stoichiometry is approximated by:

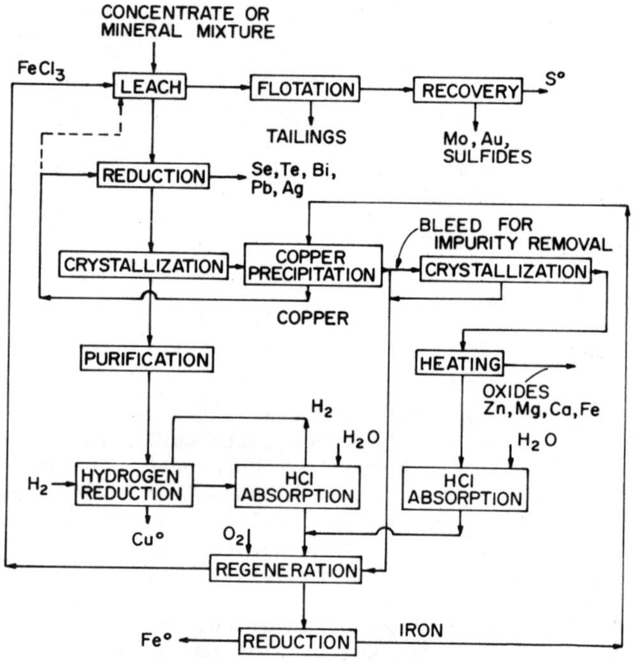

Fig.1: The Ferric Chloride Copper Process as Patented by Milner et al (3).

$$3.5FeCl_3 + CuFeS_2 \rightarrow 0.5CuCl_2 + 0.5CuCl + 4.5FeCl_2 + 2S° \quad (1)$$

In general, all the iron (from chalcopyrite and other copper minerals) as well as copper and silver is leached, along with certain minor metals. Minerals remaining unleached are: silicious material; pyrite, FeS_2; molybdenite, MoS_2, and contained gold. Silicious material is rejected by flotation, and elemental sulfur is separated from unleached sulfides by filtering liquid sulfur.

2. The cupric component in the hot leach liquor is reduced to cuprous by means of cement copper (produced in a subsequent step). Typically, the pregnant solution containing more than 50 g/l total copper is treated with more than 25 g/l cement copper to yield a solution containing 75 g/l total copper, essentially as cuprous. A residue contains excess copper and noble impurities (tellurium, bismuth). Silver is separated if present in excessive quantities, but, in general, when silver/copper ratios are less than about 1/100, silver remains quantitatively in solution.

3. The CuCl containing solution is cooled to ambient temperature and CuCl crystals are separated.

The supernatant liquor is typically 120-200 g/l in Fe^{II} and > 25 g/l Cu^{I}. The crystals contain $FeCl_2$ and AgCl in small quantities. If the Ag/Cu ratio in the ore is less than 1/100, all the silver will report with the CuCl crystals in a steady-state operating plant.
4. The CuCl crystals are purified by a process that is not described in the U.B.C.-Cominco patent (3). In fact, this process involves recrystallization from NH_4Cl saturated brines, and a quantitative silver separation is achieved, as described later.
5. The supernatant liquor from crystallization is treated by cementation with sponge iron, to obtain a solution low in copper and containing mainly $FeCl_2$. This cement copper is used in the second step. The sponge iron is a product of the process, as described below.
6. The low-copper $FeCl_2$ solution is oxidized with oxygen at 160°C to produce $FeCl_3$ and a filterable hematite. The reaction stoichiometry is:

$$6FeCl_2 + 3/2O_2 \rightarrow 4FeCl_3 + Fe_2O_3 \qquad (2)$$

This hematite is reduced to sponge iron by any suitable sponge iron process. The hematite is rather pure and low in bound water. Some of the sponge iron is recycled to cementation; the remainder is for sale.
7. CuCl crystals are reduced with hydrogen to produce copper and HCl, which is recycled. The copper is pure enough to meet wire-bar specifications, if made from purified CuCl crystals.
8. Gold can be recovered from residues by means of a $FeCl_3$ leach, not mentioned in the patent. This will be described.
9. A conventional technology is used to treat any bleed streams, converting metal content to oxides, and recycling contained chloride as HCl.

The Leach: Theory and Performance

While equation (1) represents the approximate stoichiometry of the leach, it is useful to focus on the essential thermodynamics, which define (in principle) the character of a leach solution that has equilibrated with chalcopyrite (the most important and most refractory of the copper minerals in ordinary concentrates). At equilibrium the most useful of the various equilibria in the system can be written:

$$CuFeS_2 + 3Cu^{++} + 8Cl^- \rightleftarrows 4CuCl_2^- + Fe^{++} + 2S° \qquad (3)$$

$$\log[CuCl_2^-] = 1/4 \log K_3 + 3/4 \log A_{Cu^{++}} + 2 \log A_{Cl^-} - 1/4 \log A_{Fe^{++}} - \log \gamma_{CuCl^-} \qquad (3a)$$

Since the calculated value of log K_3 is about +1.8, under standard conditions the $CuCl_2^-$ concentration should not be much higher than that of Cu^{++} (and for that matter, CuCl is not very soluble). But in strong brines the activity of chloride ions rises very rapidly; for example, it is estimated that a 3M.$FeCl_2$ solution of total ionic strength ≃ 10 will have a value of A_{Cl^-} of about 50 (4). Thus the term [2 log A_{Cl^-}] has a value of about 3.4, and this permits the $CuCl_2^-$ value to be 1M. (assuming $\gamma_{CuCl_2^-} \simeq 1$) at a Cu^{++} concentration of only about 1.3 x 10^{-5}M. In other words, the Cu^{II}/Cu^{I} ratio at equilibrium in 3M.$FeCl_2$ solution is almost 10^{-5}, while under standard conditions (say 1/2M.$FeCl_2$) it is only about 0.2. Strong brines therefore yield significant improvements in leaching thermodynamics over brines of approximately unit ionic strength.*

Leaching kinetics become rather slow, however, when Cu^{II}/Cu^{I} values fall below about unit value, even in strong brines. DUVAL patent assignors Atwood and Curtis (5) made claims that concentrates (i.e. chalcopyrite) could be used to reduce essentially all the cupric to cuprous (as, indeed, the above thermodynamic reasoning supports) but actual test work reported in their patent indicates that the process was too slow in unpressurized equipment (up to 105°C) and yielded a copper precipitating side reaction at 140°C, viz.

$$CuFeS_2 + CuCl_2 \rightarrow 2CuS + FeCl_2 \qquad (4)$$

$$\text{or } CuFeS_2 + 2CuCl \rightarrow Cu_2S + CuS + FeCl_2 \qquad (5)$$

Leaching tests done in U.B.C. and Cominco laboratories may be summarized with the help of Figs. 2 and 3 and Table I as follows:
1. Leaching rate curves for concentrates of two general types** are shown in Fig. 2.

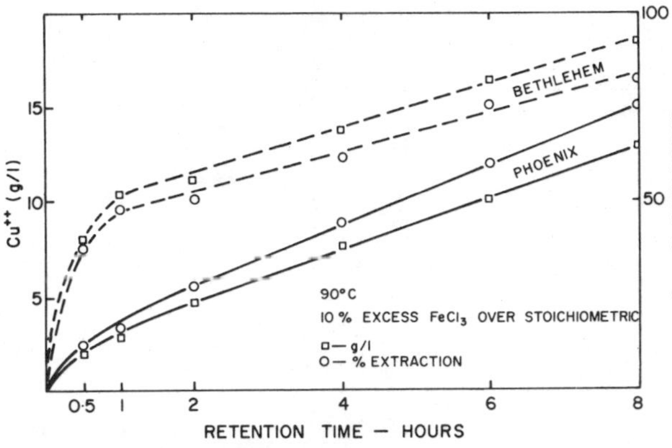

Fig. 2: Leaching Rate Curves for Bornite-containing (Bethlehem and Pyrite-containing (Phoenix)Chalcopyrite Contentrates.

Phoenix concentrates leach more slowly initially than Bethlehem concentrates, probably because the latter contains the more reactive mineral, bornite (Cu_5FeS_4) to the extent of 10-15%, and accounting for 20-30% of the contained copper. The extrapolated residence time required for complete extraction of copper remains similar for the two concentrates, indicating that chalcopyrites from the two different mineral assemblages have similar reactivities to the lixiviant. Maximum copper

* In fact, this shift in thermodynamics with ionic strength in chloride media leads to Cu^{II} becoming an effective oxidizing agent to chalcopyrite in strong brines.

**In general, the Highland Valley in British Columbia yields concentrates containing chalcopyrite and bornite mineralization, and pyrite is almost absent (Valley and Bethlehem mines). The Phoenix mine at Grand Forks, B.C. contains little or no bornite, but pyrite is present in the ore and comprises about 10% of the concentrates.

extractions demonstrated by experiment were in the range of 99.6 to 99.9% (under slightly different conditions: grinding to >88% -325 mesh and leaching for 4 to 5 hours at 100°C).

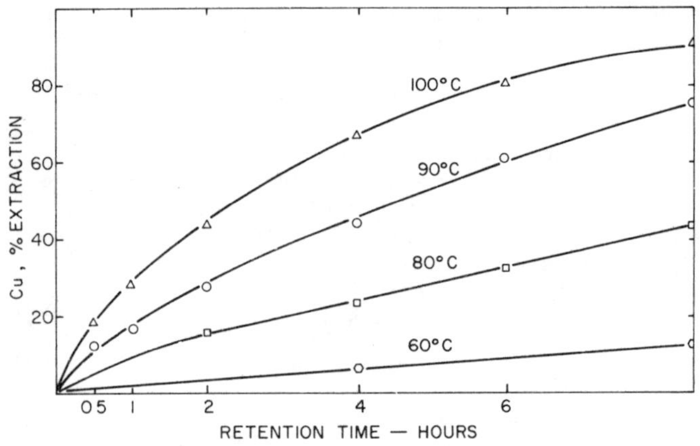

Fig. 3: Leaching Rate Curves for Phoenix Concentrates at Several Temperatures.

TABLE I

FERRIC CHLORIDE LEACHING OF PHOENIX COPPER CONCENTRATE* in TWO STAGES

Material	Wt. g.	Vol. cm^3.	g/l or % Cu(e)	Fe	S
(a) Small Scale Test Concentrate (a)	60	-	28.4	28.6	31.2
Stage I Lixiviant	-	1000	-	$76Fe^{III}$	-
Stage I Residue	24.5	-	6.1	11.8	75.7
Stage I Pregnant	-	1000	16.2	95.2	-
Stage II feed (c)	7.9	-	(17)	(32.5)	(31.8)
Stage II Lixiviant	-	100	-	$76Fe^{III}$	-
Stage II Residue	5.1	-	0.3	23.5	49.0
Stage II Pregnant	-	100	15.9	n.a.	_
(b) Larger Scale Test Concentrate (b)	5,100	-	28.1	28.3	30.6
Stage I Lixiviant	30,000	-	-	$135Fe^{III}$	-
Stage Residue	-	(2700)	9.0	15.8	57.8
Stage I Pregnant	30,000	-	42	202	-
Stage II Feed (d)	-	(2700)	9.0	15.8	57.8
Stage II Lixiviant	30,000	-	-	$135Fe^{III}$	-
Stage II Residue	-	(2360)	0.10	12.2	66.0
Stage II Pregnant	30,000	-	17(?)	194	-

* (a) as supplied Granby Mining Company
(b) ground to 98% - 325 mesh.
(c) Stage I residue with sulfur removed by CS_2
(d) Stage I residue (wet). Dry weight estimated
(e) Total Cu only; for Cu^I and Cu^{II} distribution in pregnant solution, see on "Reduction and Crystallization."

2. The temperature dependence of the leach, shown in Fig. 3, is consistant with an Arrhenius activation energy of about 18 kcals per mole (75 k joules per mole) indicating that the reaction rate is chemically controlled and that high temperatures (to 100°C) are desirable for minimum residence time.

3. Two stage leaches at two different scales of testing, shown in Table I for Phoenix concentrates indicate that copper extractions can be very high (>99.8% and that the residues contain an iron sulfide mineral (pyrite, FeS_2*). Elemental sulfur yield is very high, although sulfur from pyrite remains unchanged. Sulfate formation is detectable, but accounts for less than 2% of the chalcopyrite sulfur.

*The iron-sulfur ratio remaining when elemental sulfur is deducted seems too high for pyrite; however, elemental sulfur analyses are not precise enough to justify any alternate conclusions.

Similar experiments with Valley copper concentrates (high in bornite) yielded residues containing .03% Cu, 1.8 to 3.6% Fe, and 47.4 to 47.8% total sulfur (mainly elemental) in single stage leaches. The copper extraction was even higher (99.96%) than for similar leaches with Phoenix concentrates. A second stage leach was evidently not necessary because of a fine grind (98% -325 mesh) and the lower ferric iron consumption by bornite, leading to a higher excess of Fe^{+++} + Cu^{++} in these tests than in equivalent tests with Phoenix concentrates.

Although these leaching tests were not done with strict material balances for silver, gold, and other minor metals, there was evidence that contained silver was leached (in excess of 90% with Phoenix concentrates) along with tellurium, bismuth, arsenic, and antimony, as indicated by their appearance in leach solution and low levels in leach residues.

These leaching results therefore verify that the ferric chloride system is capable of achieving very high extractions of copper on two kinds of concentrates that are representative(in their basic mineralization)of concentrates shipped from most mines. Unfortunately, no data were collected on the all important arsenic-and antimony-containing minerals, enargite (Cu_3AsS_4), tetrahedrite (Cu_3SbS_3), and tennantite (Cu_3AsS_3). Although these minerals should decompose according to thermodynamic forces, they may be more refractory than chalcopyrite. An indication of this is found in the very slow leaching rates encountered for enargite in ferric sulfate solutions (6).

Reduction and Crystallization

Leach solutions were prepared in some cases that were high in Cu^I/Cu^{II} ratio, although the values were obscured in these very strong liquors by the fact that Fe^{+++} and Cu^+ were both complexed by chloride ions, while Fe^{++} and Cu^{++} ions were complexed only weakly, if at all. Thus the common assumption that Fe^{III} and Cu^I react to completion is not completely valid when chloride ions are present at a concentration of 6 to 8N and at an activity of close to 50. For example, analysts reporting on the composition of Leach I pregnant solution in Table I stated 40 g/l Cu^I, 2 g/l Cu^{II}, 181 g/l Fe^{II}, and 21 g/l Fe^{III}.* An equally plausible analysis may have been 14 g/l Cu^I, 28 g/l Cu^{II}, 202 g/l Fe^{II}, and nil Fe^{III}. The reaction between Fe^{III} and Cu^I is so reversible that any dilution by the analyst would change the results. In any case, it is unimportant, because all the Fe^{III} and Cu^{II} must be reduced, to create conditions under which a maximum yield of CuCl can by crystallized.

CuCl solubility curves in $FeCl_2$ - based brines are shown in Fig. 4, and these data are converted

*The determination is made by titrating the diluted sample to two successive REDOX end points with an oxidizing reagent (i.e. $K_2Cr_2O_7$). The effect of Cl^- concentration on these end points is significant, but normally neglected.

into CuCl/Cl$^-$ ratios in Fig. 5. Superficial examination of equation (1), i.e.

$$CuFeS_2 + 3.5FeCl_3 \rightarrow 0.5\ CuCl + 0.5CuCl_2 + 4.5FeCl_2 + 2S° \quad (1)$$

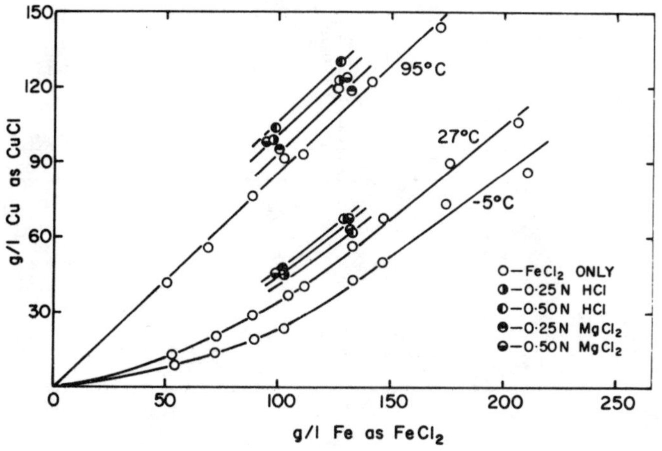

Fig. 4: Cuprous Chloride Solubilities in Ferrous-Chloride-based Brines at Several Temps.

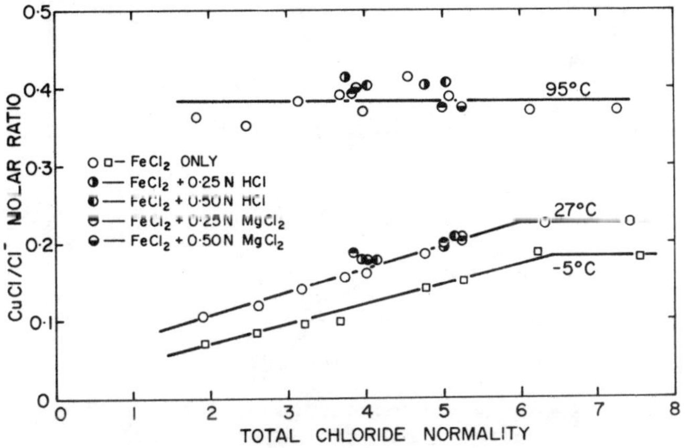

Fig. 5: CuCl/Cl$^-$ Ratios in CuCl-saturated $FeCl_2$-based Brines.

indicates that the molar ratio of CuCl to Cl$^-$ (from $FeCl_2$ and $CuCl_2$) is only 0.05, well below CuCl saturation levels indicated in Fig. 5. Therefore, it is not possible to crystallize CuCl from pure ferric chloride leach solutions resulting from this particular stoichiometry. If bornite is present in the concentrate, the leaching stoichiometry (yielding a 1:1 Cu^I/Cu^{II} ratio) is

$$Cu_5FeS_4 + 9.5FeCl_3 \rightarrow 2.5CuCl + 2.5CuCl_2 + 10.5FeCl_2 + 4S° \quad (6)$$

and leads to a CuCl/Cl$^-$ ratio in the pregnant solution of about 0.10. In fact, even an iron free mineral such as chalcocite (Cu_2S) or covellite (CuS) will not yield higher CuCl/Cl$^-$ ratios than 0.14, if the Cu^I/Cu^{II} ratio in the pregnant solution is 1:1. Clearly, there is a need for another way of obtaining higher CuCl/Cl$^-$ ratios in preparation for CuCl crystallization, especially if the concentrates being leached are mainly chalcopyrite. There are, in fact, three methods of enhancing the CuCl/Cl$^-$ ratio available and these can be expressed as follows:

(a) Reduction of all Cu^{II} with cement copper in a separate reduction step:

$$CuCl_2 + Cu° \rightarrow 2CuCl \quad (7)$$

In this case, the overall leach-reduction two-step stoichiometry is:

$$CuFeS_2 + 0.5Cu° + 3.5FeCl_3 \rightarrow 1.5CuCl + 4.5FeCl_2 + 2S° \quad (8)$$

The ratio of CuCl: Cl$^-$ is now 0.167, and crystallization is possible from slightly diluted chloride brines, less than about 4N. according to Fig. 5. It is also possible to extend this further by using more cement copper (and, of necessity, more $FeCl_3$) to enhance the CuCl: Cl$^-$ ratio, i.e.

$$CuFeS_2 + xCu° + (3 + x)\ FeCl_3 \rightarrow (1 + x)\ CuCl + (4 + x)FeCl_2 + 2S° \quad (9)$$

The CuCl/Cl$^-$ ratio is then (1 + x): 2(4 + x) and clearly increases with the value of x until rystallization conditions are optimized. The value of x necessary is smaller if there is some bornite in the copper concentrates.

(b) Substitution of $FeCl_3$ by $CuCl_2$ recycled to the leach:

$$CuFeS_2 + 7CuCl_2 \rightarrow 4CuCl + 4CuCl_2 + FeCl_2 + 2S° \quad (10)$$

If all the $FeCl_3$ is substituted by $CuCl_2$, the CuCl/Cl$^-$ ratio is 0.4, even when the Cu^I/Cu^{II} ratio is 1:1. Quite clearly, this route is feasible if crystallization from leach solutions is contemplated. It is the route chosen by Cyprus Metallurgical Corporation (2).

(c) Substitution of $FeCl_3$ by oxygen and HCl:

$$CuFeS_2 + 3.5HCl + 0.875\ O_2 \rightarrow 0.5CuCl + 0.5CuCl_2 + FeCl_2 + 2S° + 1.75H_2O \quad (11)$$

This particular route would produce a value of 0.167 for the CuCl/Cl$^-$ ratio on the basis of the stoichiometry of equation (11). It has not been tested explicitly, and suffers from a need to generate HCl in substantial excess of that produced by hydrogen reduction of CuCl. In fact, this route would work best if oxygen and HCl is used to generate $CuCl_2$ from CuCl, and the resulting $CuCl_2$ is used according to equation (10). This is because there is indirect evidence that only Cu^I (of all the reduced species present in the system) is likely to have a rapid reaction with oxygen.

We have chosen option (a) for the UBC-Cominco process because cement copper would be available from an iron cementation on the crystallization mother liquor. In general, it is desired to have a CuCl/Cl$^-$ ratio of about 0.3 after the reduction step, but this proved difficult to achieve for chalcopyrite concentrates. Even a value of 0.25 would require a circulating load of cement copper that was unacceptably high, except under conditions of high bornite content in the concentrates. Since the first application of the process was anticipated to be for Valley Copper concentrates (50% of the copper as bornite) the crystallization route with cement copper was feasible, although the amount of cement copper returned from the reduction step was determined to be rather high at high chloride concentrations (2.3 lbs. cement copper per lb. copper leached). It was concluded that this ratio could be improved by either (a) operating at lower brine concentrations to improve CuCl crystallization at lower $CuCl/Cl^-$ ratios, or (b) recirculating some $CuCl_2$ to the leach, in order to improve the $CuCl/Cl^-$ ratio partially by the method of equation (10). However, the cement copper route was used as far as possible, in order to generate copper-free $FeCl_2$ solutions for oxydrolysis.

Cuprous chloride crystals from crystallization, obtained from solutions containing 120 to 200 g/l iron as $FeCl_2$, contained about 0.1% Fe after washing, and retained silver, presumably as AgCl in solid solution. Although some fractional separation of AgCl from CuCl was evident in crystallization, it was not a useful separation route in a steady-state operation where silver recovery is desired. Therefore, a purification method was desired that would yield truly pure CuCl crystals and provide for silver recovery.

Cement copper recirculated for the reduction step contained some impurities, and the more noble of these (especially those that form intermetallic phases with copper, i.e. arsenic, antimony, bismuth, tellurium, and selenium) as well as noble metals as silver (if present in large amounts in the leach liquor) could be obtained as a residue in the reduction steps. Silver was not obtained in useful amounts at this point if its concentration in the concentrates was rather low (say, less than 1 part silver per 100 parts copper).

Purification of CuCl Crystals

The objective of CuCl purification was to recover silver, as well as to obtain sufficiently pure CuCl so that, after hydrogen reduction, it would not require refining. A unique process was developed, based on the following reasoning and observations:

Fig. 6 shows an E_h-pH diagram at 1M. NH_4^+ or NH_3 for the $Cu-Cl-NH_3-H_2O$ system. Also shown is the line representing the AgCl/Ag potential for an AgCl saturated and a 0.3M. AgCl solution. It is of interest that the cuprous ammine, $Cu(NH_3)_2^+$, becomes a reducing agent as it forms and that the standard $Cu(NH_3)_4^{++}$ - $Cu(NH_3)_2^+$ potential becomes low enough to reduce AgCl to metal. It turns out that silver ammines are not expected to form in strong chloride solution on the basis of thermodynamic data. The method for separating silver from copper by addition of ammonia can therefore be predicted: silver would be precipitated by the reaction:

$$Cu(NH_3)_2^+ + 2NH_3 + AgCl_2^- \rightarrow Ag^\circ + Cu(NH_3)_4^{++} + 2Cl^- \quad (12)$$

The experimental results are demonstrated in Fig. 7. CuCl was dissolved in $4M.NH_4Cl$ and ammonia was added by increments. Silver levels, beginning at 0.56 g/l, fell below 0.1 g/l at about pH - 7.5 and below 0.01 g/l at about pH=10. The silver produced was a pure, white, filterable crystal.

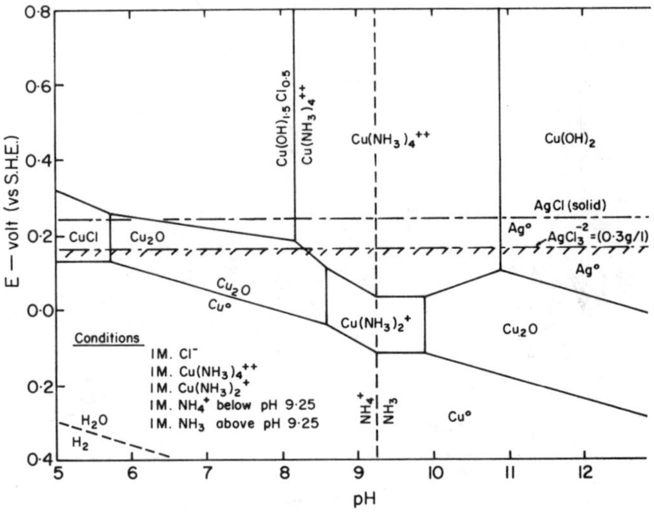

Fig. 6: E_h-pH Diagram for the $Cu-NH_3-Cl$ System.

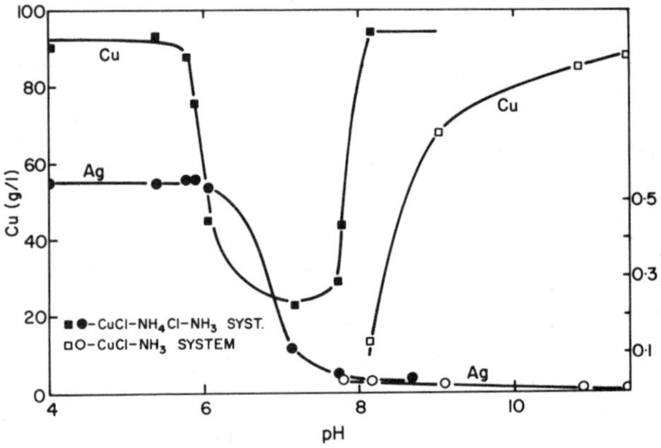

Fig. 7: Ag- and Cu-contents of Ammoniacal Cuprous Chloride Solutions as a Function of pH.

The supernatant solution was steam stripped back to a pH of about 7, and yielded crystals that proved to have the composition of $CuNH_3Cl.H_2O$. At the time we did this work, we did not find any thermodynamic data on this compound and so failed to incorporate it into the E_h - pH diagram. We rather expected to find Cu_2O. However, our strong brines (more than 5M. in total chloride) presumably closed the Cu_2O zone calculated at $1M.Cl^-$ and shown in Fig. 6, and permitted the crystallization of this salt. It gave up its ammonia and reverted to CuCl at modest temperature (less than 250°C) or under vacuum at ambient temperature and was free of silver. A cyclic silver recovery process is suggested in the schematic flow sheet of Fig. 8.

Other silver-copper separations applicable to CuCl crystals or CuCl - containing brines are also potentially applicable in this process (7). However, these will not be discussed here.

Cementation of Copper

Cementation of copper from CuCl - containing

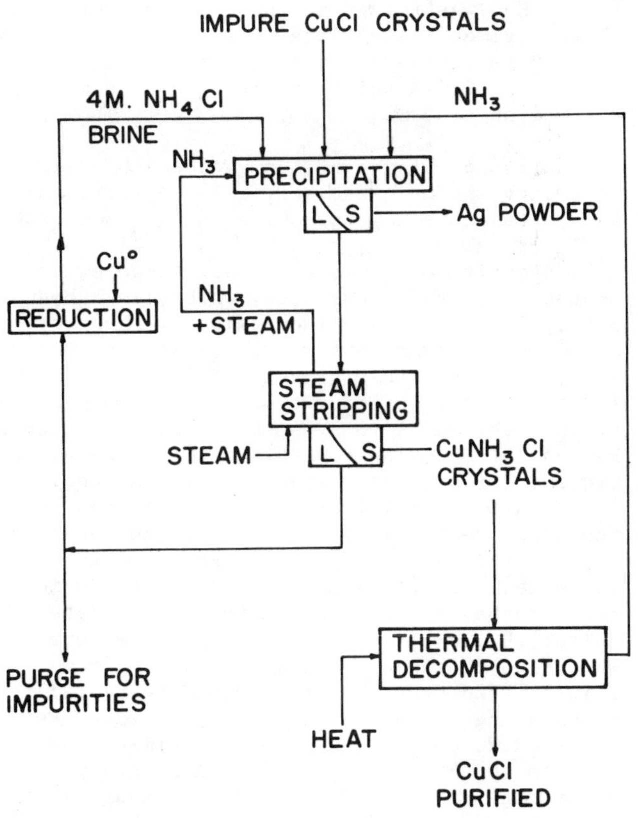

Fig. 8: Schematic Flow Sheet of a CuCl Purification Scheme that Includes Silver Recovery.

brines is a relatively straight forward procedure and in the laboratory it is easily accomplished with either small finishing nails as a standard iron source, or commercial sponge iron. Results of cementation tests done on a solution that had not been subject to prior CuCl crystallization are presented in Table II for purposes of indicating iron consumption rates and solution composition changes. Analyses of the six successive batches of cement copper are shown in Table III.

TABLE II

DISPLACEMENT OF COPPER FROM CUCL* - CONTAINING SOLUTIONS WITH SPONGE IRON+ BY STAGES

SOLUTION VOLUME FOR EACH STAGE = 15.9L

	Iron gms.	Solution g/l		Precipitate			
		Cu	Fe	gms.	Cu %	Fe %	Other %
Initial	-	57.7	151	-	-	-	-
Stage I	207	48.5	164	153	95.7	0.1	1.7
Stage II	78	38.1	169	164	97.6	<0.5	0.31
Stage III	75	28.4	173	156	98.8	<.05	0.23
Stage IV	72	18.9	178	152	98.3	<.05	0.18
Stage V	69	10.1	186	138	98.9	<.05	0.16
Stage VI	60	1.5	188	137	91.1	5.6	0.14

* 37.1 Cu^{+}, 57.7 Cu_t, 151 Fe^{++}, 262 Cl^{-}, 2.4 S/SO_4 (g/l)
\+ Fe 96.0%, C <.02%, S .005%, P .012%, Mn <0.6%, Insol <.05, Hydrogen loss <1.2%, -150 mesh.

These results indicate that (a) once any residual $CuCl_2$ is reduced to CuCl, the iron consumption is about 0.48 lbs. (contained iron) per lb. (contained Cu in ppte.), about 10% in excess of theorectical for cuprous. Typically, solutions could be brought down to 1 g/l copper or less, if excess iron is used. Spectrographic assays showed that in six successive increments of cementation going from 56 g/l Cu to 1 g/l Cu, minor impurities followed the patterns indicated:

Table III

Spectrographic Analyses of Cement Copper

ppm except where % is shown

Batch	I	II	III	IV	V	IV
Al	200	100	100	40	70	100
Sb	.2%	10	-	-	-	50
As	.1%	.1%	500	200	50	50
Bi	20	100	100	200	100	7
B	-	5 erratic	-	-	-	-
Ca	10	5	7	3	5	5
Cr	200	200	200	200	200	200
Co	-	-	-	-	-	3
Ga	.2	.2	.1	.1	.1	.1
Ge	-	-	-	1	2	20
Fe	.1%	700	300	500	500	2%
Pb	5	3	2	10	7	20
Mg	30	20	20	20	20	20
Mn	300	300	200	200	200	200
Hg	10	-	-	-	-	-
Mo	20	20	20	20	20	20
Ni	5	5	7	7	7	70
Si	.1%	700	300	500	500	300
Ag	500	500	500	300	300	200
Na	10	5	-	5	-	5
Te	500	-	-	-	-	-
Th	.3	-	-	-	-	-
Sn	50	40	40	30	30	40
Ti	10	4	2	5	5	2
V	30	20	20	7	3	2
Zn	30	30	5	10	5	10

(a) Decreasing levels: Al, Sb, As, Hg, Ag, Te, Tl, V, Zn
(b) Even distribution: Bi, Ca, Cr, Ga, Mg, Mn, Mo, Si, Sn, Ti
(c) Increasing levels: Ge, Pb, Ni

Some of these impurities, particularly most of those with an even distribution, came from the sponge iron. Impurities occurring in decreasing levels are those more noble than copper and those forming intermetallic compounds more noble than copper. Elements having increasing levels are less noble than copper, but are noble enough to be cemented by an excess of iron, and so are capable of being purged from solution by this means.

Some impurities in the above list, while present in the cement copper in spectrographic amounts, would nevertheless build up in solution and require a purge treatment. These are: zinc, calcium, magnesium, aluminium, and perhaps manganese.

Solution Regeneration

After cementation, the barren solution contains $FeCl_2$ and perhaps some tramp chlorides that have not been purged by the cementation process. The copper content of these solutions as contemplated in the process was to be less than 1 g/l, although levels as high as 10 g/l were tested.

Some laboratory results of solution regeneration conducted in the temperature range of 130°C to 160°C are shown in Table IV and in Fig. 9, using copper-free synthetic stock solution containing $CaCl_2$, $MgCl_2$, $ZnCl_2$ and KCl in quantities shown. The results show that virtually complete oxidation to ferric chloride is possible, with hydrolysis of excess iron, as follows:

TABLE IV

OXIDATION PRODUCTS OF $FeCl_2$ SOLUTIONS

STARTING SOLUTION (g/1) 204.5 Fe^{++}; 6Ca^{++}; 3.6Mg^{++}; 10Zn^{++}; 12K^{+}; 300Cl^{-}; nil Fe^{+++}; nil Cu^{++}, Cu^{+}

Experimental Conditions*	Final Solution g/l Fe^{II}	Fe^{III}	HCl(free)	Washed Final Solids % Fe_2O_3	Cl	Wt loss**	Color
130°C 40 min.	56.2	92.3	15.7	75.2	1.26	20.9	Ochre
130°C 120 min.	1.6	126	19.3	87.0	1.1	11.3	brn/red
140°C 40 min.	33.7	110	17.5	90.0	1.55	9.5	lt.grey
140°C 120 min.	nil	138	22.6	90.8	1.05	5.8	dk grey
150°C 20 min.	74.5	112	26.6	90.8	1.18	10.2	lt.grey
150°C 80 min.	0.05	141	23.4	93.0	0.80	7.6	dk grey
160°C 20 min.	39.1	112	19.7	92.7	1.02	4.2	dk.grey
160°C 60 min.	nil	116	19.7	93.0	0.60	2.4	dk.grey

* 110 psig oxygen; shaking autoclave (120 cm^3)

** 925°C

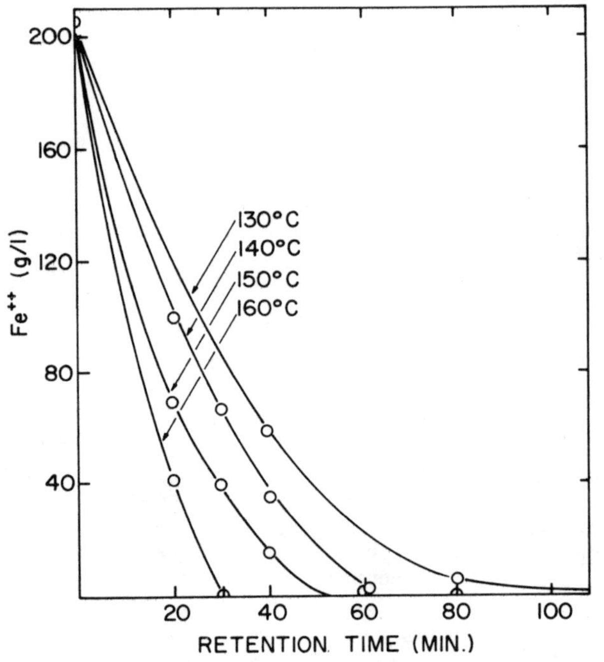

Fig. 9: Rate-Curves for Oxidation of $FeCl_2$-containing Brines with Oxygen; PO_2 = 110 psia (760 kPa).

$$6FeCl_2 + \frac{3}{2} O_2 + xH_2O \rightarrow (4 - \frac{2x}{3})FeCl_3 + (1 - \frac{x}{3}) Fe_2O_3 + 2xHCl \qquad (13)$$

Titrations indicate that free HCl ranges from 0.5 to 0.7M. beginning with 204.5 g/l Fe^{++} while final solutions range between 115 and 140 g/l Fe^{+++}. The theoretical value of Fe^{+++} is 136 g/l when x in equation (13) is zero, indicating that analytical results are in approximate agreement.

Of special interest is the nature of the precipitate. It proved to be hematite (Fe_2O_3) only at the highest temperatures (150 - 160°C) or after prolonged digestion at slightly lower temperature (140°C). In general, goethite ($Fe_2O_3.H_2O$)was formed at lower temperature and shorter times. Residence times of 20 minutes at 160°C were adequate for both complete oxidation of iron to $FeCl_3$ and precipitation of hematite in a filterable form.

Calcination of the solids yielded the weight loss data shown in Table IV. The weight loss can be taken mainly as H_2O, although above 200°C $FeCl_3$ can also be expected to volatilize. The grey hematite obtained at higher temperatures proved to contain less total water, less bound water, and lower chloride residuals than material obtained at lower temperatures. Washing tests showed that H_2O repulping could lower Cl^- from 1.28 to 0.42%, while a wash with 2% ammonia dropped the value to 0.05% Cl^-, indicating that removal of chloride presents no difficulty. Subsequent tests with the same solutions supplemented with ∿ 2 g/l sulfate sulfur indicated that the hematite precipitate would collect at least half of this sulfur, probably as potassium jarosite. Solutions not containing potassium may also be expected to precipitate sulfate sulfur, but probably not to such a low level. It should be possible to isolate sulfate from solution and prepare a pure hematite by (a) recirculating a small amount of regenerated $FeCl_3$ solution to the $FeCl_2$ liquor containing the sulfate (b) heating this to boiling and precipitating a jarosite type of salt and (c) separating this from the mother liquor before feeding the liquor to the reoxidation autoclaves.

Oxidation of $FeCl_2$ solutions was conducted in titanium autoclaves, and there was no evidence of corrosion after upwards of 20 tests. Corrosion literature indicate (8) that Fe^{III} is a corrosion inhibitor for the HCl corrosion of titanium. It would probably be necessary to recycle ∿ 5% of the product $FeCl_3$ liquor to the front end of the reoxidation autoclaves in a plant, to protect titanium pumping equipment.

Hydrogen Reduction of Purified CuCl Crystals

It is thermodynamically possible to reduce CuCl with hydrogen according to the equation:

$$CuCl + \frac{1}{2} H_2 \rightarrow Cu + HCl_{(g)} \qquad (14)$$

Experiments conducted in a tube furnace operating at 400°C proved this feasibility. However, calculations show that (a) the reaction is significantly endothermic (+ 10.43 kcals/mole at room temperature) and the physical properties of CuCl are not promising: a low melting temperature (422°C) and relatively high vapour pressures (1mm at 546°C). Furthermore, the vapour has a high molecular weight (Cu_2Cl_2 or Cu_3Cl_3) to make volatilization at higher temperatures a severe fuming problem. To provide 10.43 kcals of endothermic energy by superheating the hydrogen would require excess temperatures of about 3000°C (for a stoichiometric amount of hydrogen) or 300°C (for a ten-fold excess over stoichiometric hydrogen.) Even the latter would lead to melting and fuming problems.

Options that we considered were to oxidize the CuCl to a mixture of basic chloride and cupric chloride and (a) reduce the mixture or (b) reduce only the basic cupric chloride, recycling the water-soluble cupric chloride. The oxidation stoichiometry is:

$$3CuCl + \frac{3}{4} O_2 + \frac{3}{2} H_2O \rightarrow 2Cu(OH)_{1.5}Cl_{0.5} + CuCl_2 \quad (15)$$

Hydrogen reduction of this mixture yields an exothermic reaction heat of about 0.66 kcals per mole of copper reduced. If it is first dried to a CuO - $CuCl_2$ mixture, the exothermic reaction heat is increased to 6.26 kcals per mole. If only the $Cu(OH)_{1.5}$ $Cl_{0.5}$ basic salt is hydrogen reduced, the exothermic reaction heat is about 5.1 kcals per mole copper reduced. Thus, it is clear that the endothermic reaction of CuCl with hydrogen can be offset by oxidizing it to a divalent form, especially if the oxidation product is dehydrated or if $CuCl_2$ is first leached out.

We did not make a final choice between these options, but assumed that final hydrogen reduction of CuCl would be by a route involving oxidation to cupric mixtures first. A choice was made by Cyprus Metallurgical Corporation in favour of a patented (9) fluid bed reactor in which CuCl crystals were reduced by superheated hydrogen at a temperature of about 500°C. Clean silica sand is added as a fluidizing facilitator by preventing the sticking of copper particles, and is consumed at a rate of .08 lbs sand per lb. copper. Data on the heat balance are not sufficiently detailed in published works (10,11) to permit evaluation, but it appears that hydrogen is used as a fluidizing gas at about a five-fold excess over stoichiometric, and that the CuCl crystal and sand mixture is injected into the middle of the bed. The endothermic heat requirements may also be reduced modestly by partial oxidation of the CuCl crystals during storage in the presence of air. The exhaust gases of the fluid bed are scrubbed free of HCl (and, possibly, CuCl fumes) in two stages by water (yielding aqueous HCl) and caustic soda (yielding NaCl), and the remaining hydrogen is recycled after preheating.

Residue Treatment and Recovery of Gold

The ferric chloride leach yielded copper recoveries so complete that there was no evidence of copper minerals in residues. However, the residues contained pyrite (FeS_2), molybdenite (MoS_2) and silica (SiO_2) as well as elemental sulfur, and gold in an undetermined form. Direct recovery of gold was possible, in principle, from these residues, using ferric chloride solutions sufficiently low in ferrous chloride, according to the following leaching stoichiometry:

$$Au + 3Fe^{+++} + 4Cl^- \rightarrow AuCl_4^- + 3Fe^{++} \quad (16)$$

In fact, the method chosen for this was to use a small amount of chlorine gas to treat regenerated ferric chloride solution to the extent of eliminating any significant ferrous iron and allowing a few p.p.m. excess aqueous chlorine. The solutions are sufficiently acid to avoid formation of other oxidizing species such as hypochloric acid (HClO). Excessive amounts of chlorine could also not be used because of the danger of forming sulfur chloride (S_2Cl_2) with the elemental sulfur present in the residues.

Control of the leach would be achieved by maintaining the potential between two critical values represented by the equations:

$$AuCl_4^- + 3e \rightleftarrows Au + 4Cl^- \quad (17)$$

$$E_{17} = E_{17^\circ} - \frac{2.3RT}{F} (\frac{4}{3}) \log A_{Cl^-} + \frac{2.3RT}{F} (\frac{1}{3}) \log A_{AuCl_4^-} \quad (17a)$$

$$S_2Cl_2 + 2e \rightleftarrows 2S^\circ + 2Cl^- \quad (18)$$

$$E_{18} = E^\circ_{18} - \frac{2.3RT}{F} \log A_{Cl^-} + \frac{2.3RT}{F} (\frac{1}{2}) \log A_{S_2Cl_2} \quad (18a)$$

The values of $\overset{\circ}{E}_{17}$ and $\overset{\circ}{E}_{18}$ are 1.00 and 1.23 volts respectively. If the chloride ion activity is taken as 50 and an acceptable pregnant solution for gold is 10^{-4}M. (∿ 20 mg/liter), while the S_2Cl_2 maximum permitted activity is, say 0.01, then the values of E_{17} and E_{18} are, respectively, 0.79 and 1.07 volts. The Fe^{++}/Fe^{+++} ratio at 0.79 volts is about 0.03 at 1M. Cl^- and will be much lower in the strong ferric chloride leaching brines because Cl^- complexes Fe^{+++} much more than Fe^{++}. Therefore, maintaining a potential in the range of the safe window between 0.79 and 1.07 volts will require only a trace of chlorine, about 0.2 p.p.m. maximum, in excess of that required to oxidize any Fe^{++} and other reducing agents present.

To confirm the ability of ferric chloride solutions containing a trace of chlorine to dissolve gold, some experiments were done with gold foil, and these are reported in Fig. 10. Although initial potentials were too high (1.35 volts) chlorine was consumed almost quantitatively by the gold, with final solution potentials as shown (1.00 to 1.02 volts). Final solutions in these experiments contained 20 to 200 mg/l gold. The chlorine consumption was less than predicted by equation (17) and suggested that, in fact gold dissolves in these dilute solutions mainly in the monovalent form, probably as the complex $AuCl_2^-$.

Experiments were also conducted on ferric chloride leach residues containing around 1 oz/ton gold. At a solids loading of 300 g/l and 60°C, with chlorine addition of 15 ml. per minute per liter of solution, gold extractions of 91.5% were obtained, in two stages, with a solution loading of 5.3 mg/l in the first stage and 0.9 mg/l Au in the second stage. Up to 6.4 g/l sulfate was formed at the same time, presumably from the reaction sequence*

$$2S^\circ + Cl_2 \rightarrow S_2Cl_2 \quad (19)$$

$$2S_2Cl_2 + 2H_2O \rightarrow 3S^\circ + SO_2(aq) + HCl \quad (20)$$

$$SO_2(aq) + 2H_2O + Cl_2 \rightarrow H_2SO_4 + 2HCl \quad (21)$$

Thus sulfur would have become a substantial consumer of chlorine.

*FeS_2 may also have been attached by chlorine to yield sulfate.

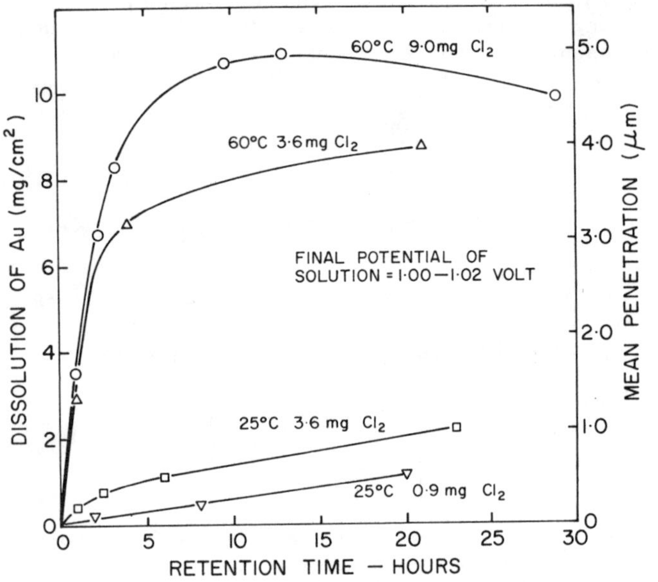

Fig. 10: Gold Dissolution Rate Curves in Chlorinated $FeCl_3$-containing Brines

TABLE V

FLOTATION AND SULFUR FILTRATION RESULTS ON LEACH RESIDUES

		%Cu	%Fe	$\%S_T$	%S°
Leach Residue:	Phoenix	0.30	7.9	66.2	(60)
	Valley	0.32	2.85	48.9	(46)
Leach Residue: Flotation Concentrate*	Phoenix	0.40	6.8	98.2	(83.1)
	Valley	0.53	0.98	89.3	(88.1)
Flotation Tailing:	Phoenix	0.07	10.4	8.9	n.a.
	Valley	0.09	4.9	4.9	n.a.
Liquid Sulfur Filtrate**:	Phoenix	-	-	~100	~100
	Valley	-	-	~100	~100
Filter Cake:	Phoenix	1.14	20.2	62.0	40.1
	Valley	2.4	4.5	47.8	42.0

* Flotation of 400 gms leach residues in 2 liters water at pH =3.5 with pine oil frother and diesel oil collector.

** Filtered in the presence of water under 50 psi N_2 pressure at 125°C.

These results were sufficiently conclusive to permit the assumption that a ferric chloride-chlorine leach for gold could be worked out. Recovery of gold from these solutions would probably be by means of cementation on a small quantity of copper concentrates expressly selected as a gold concentrate for shipment to a smelter. The disappearance of free chlorine and appearance of modest amounts of Fe^{++} would be enough to guarantee precipitation of gold, and copper concentrate is suitable both as a consumer of oxidant and as a collector of the precipitate.

Leach residues consist of elemental sulfur, unreacted sulfide minerals, and silicious material as well as unleached gold. The unleached sulfides consist mainly of pyrite and molybdenite. Residues from both Phoenix and Valley concentrate leaches were tested by flotation, followed by molten sulfur filtration (under pressure) of the floated material. The results of these tests are shown in Table V.

Flotation led to a sulfur concentrate and a tailing containing virtually no residual copper. Liquid sulfur filtration led to a relatively pure sulfur and a filter cake still containing 40% elemental sulfur. This cake should contain most of the molybdenite and pyrite, and represents less than 10% of the original weight of Phoenix concentrates and about 5% of the original weight of Valley Copper concentrates.

Purge Liquor Treatment

Certain impurities, primarily magnesium and zinc, will be dissolved from concentrates but will not separate from solution in any steps mentioned so far. These are not particularly detrimental impurities, but they cannot build up in solution indefinitely. It was contemplated that recirculating leach liquor should be purged at a rate suitable for limiting the level to which these impurities should rise in an operating plant. The contemplated treatment is a state of art; the purge solution is boiled down until most of the salts can be crystallized on cooling. The hydrated salts, containing mainly $FeCl_2$ with impurities as zinc, calcium, and magnesium, are then decomposed thermally in a fluid bed or rotary kiln, in the presence of steam, according to the stoichiometry

$$MCl_2 + H_2O \rightarrow MO + 2HCl(g) \qquad (22)$$

The oxide residue is disposed of in a land fill and the HCl is absorbed and recycled to the circuit. The small amount of supernatant liquor from the crystallization is passed back to the circuit to eliminate copper losses in the purge stream.

Conclusions

The ferric chloride leach circuit described above was able to obtain extraordinarily high copper recoveries, and dispose of iron and sulfur in salable forms, as well as to permit recovery of precious and minor metals commonly present in copper concentrates. The flowsheet was not operated on a continuous or integrated basis; however, every unit operation was tested on real solutions obtained and stored from previous units, as well as from synthetic solutions. Some leach tests were carried out on a scale of up to about 30 liters, while other tests were generally on a small bench scale averaging 1 to 4 liters. If fully integrated in a plant, the overall flowsheet as studied should permit complete closure of solution circuits, so that there should be no atmospheric or groundwater pollution; only innocuous solids would be sent to waste, and these in remarkably small tonnages by comparison to concentrate tonnages treated.

Titanium was found to be suitable as a container material in the most severe environments studied, as long as several grams per liter of ferric iron was present. Fiber-reinforced plastics were contemplated for most unpressurized unit operations, with acid-proof brick lining to resist abrasion. However, ferric chloride solutions were found to corrode all steels rapidly, including the best stainless steels, and so it was recognized that plant construction would require an on-site corrosion engineer to assist in avoiding mistakes in materials selection for critical components.

We cannot claim to know all there is to know about the ferric chloride processing of copper concentrates; nor have we necessarily solved all the problems in the most economical way or optimized the circuits. However, it has become clear that ferric chloride offers some unique capabilities, and it is not surprising to us that others have pursued similar chloride routes to pilot plant and production scales.

Acknowledgement

This paper is based on research and development work at both U.B.C. and Cominco laboratories. All the U.B.C. work was supported by Cominco Ltd., through a contract with the University, and is gratefully acknowledged, as is permission to use selected data from Cominco's laboratories.

References

1. Anon, Mining J., June, 1973, p. 457, E. & M. J., July 1973, p. 123, Chemical Engineering, Oct. 1, 1973, p. 73.
2. McNamara, J.H., Ahrens, W.A., and Franek, J.G., presented at the AIME Annual Meeting, Denver, Colorado, Feb. 1976.
3. Milner, E.F.G., Peters, E., Vizsolyi, A.I., and Swinkels, G.M., Canadian Patent No. 938793, Dec. 25, 1973, and U.S. Patent No. 3, 798, 026, March 9, 1974.
4. Peters, E., Chloride Hydrometallurgy Proceedings, Benelux Metallurgie, Sept. 26 to 28, 1977.
5. Atwood, G.E., and Curtis, C.H., U.S. Patent No. 3, 785, 944, Jan. 15, 1974.
6. Dutrizac, J.E., and McDonald, R.J.C., Can. Met. Quarterly 11, 3 (1972) pp. 469-476.
7. Peters, U.S. Patent 4, 101, 315, Peters, U.S. Patent 4, 124, 379, Swinkels, G.M., U.S. Patent 4, 097, 271.
8. Cobb, J. and Uhlig, H., J. Electrochem. Soc., 99, 13 (1952).
9. Stethans, F.M., Blair, J.C., U.S. Patent 4, 039, 324.
10. McNamara, J.H., Ahrens, W.A., and Franek, J.G., C.I.M.M. Bulletin 73, 201-204, (1980).
11. Goens, D.N. and Kruesi, P.R., U.S. Patent 4, 013, 457.

Chapter 7

A NOVEL OXIDANT FOR NICKEL HYDROMETALLURGY

V. A. Ettel, M. A. Mosoiu, E. A. Devuyst

INCO Metals Company
J. Roy Gordon Research Laboratory
Mississauga, Ontario

ABSTRACT

Nickelic hydroxide is an important metallurgical reagent used for precipitating cobalt from nickel sulfate solutions.

Existing methods of preparation of nickelic hydroxide involve electrolytic oxidation of $Ni(OH)_2$ or the use of strong oxidizing agents such as chlorine, ozone and persulfate. Although the oxidizing properties of the $SO_2 + O_2$ system have been known for some time, e.g., detection of SO_2 and oxidation on iron in acidic solution, the application of this system for the preparation of nickelic hydroxide as a process was not considered in the past.

The present paper describes a new method for chemical preparation of nickelic hydroxide which has been developed in INCO laboratories. The technique consists of reacting nickelous hydroxide suspensions with a slurry of calcium sulfite in the presence of air or oxygen. The main advantage of the process is the cost of reagents and an absence of soluble by-products (e.g., Cl^-).

The chemistry of this process is unusual since the redox potential obtained during the reaction exceeds that of the oxygen electrode. Therefore, the effect of $CaSO_3$ is not just a catalytic acceleration of the reaction $Ni(OH)_2 + O_2$. The $SO_3^=$ must be involved in the stoichiometry of the overall reaction producing in the initial phase some strongly oxidizing intermediate species.

A NOVEL OXIDANT FOR NICKEL HYDROMETALLURGY

V.A. Ettel, M.A. Mosoiu, E.A. Devuyst

INCO Metals Company
J. Roy Gordon Research Laboratory
Mississauga, Ontario

INTRODUCTION

Nickelic hydroxide, referred to as nickel black in view of its appearance, is the high valency hydroxide of nickel used for removing cobalt from nickel sulfate electrolytes or leach liquors. Nickel black can oxidize cobaltous ions in such solutions to precipitate insoluble cobaltic hydroxide (1-8):

$$Ni(OH)_3 + Co^{2+} + 2OH^- \rightarrow Ni(OH)_2 + Co(OH)_3$$

which can subsequently be removed by filtration. Nickelic hydroxide can be made by reaction of strong oxidizing agents with solutions of nickel salts or suspensions of nickelous hydroxide. Ample information is available on methods of preparation of nickel black using sodium hypochlorite or chlorine (1, 4, 5, 8, 9, 10), sodium persulfate (6, 11) and ozone (12, 13, 14, 15, 16) as the oxidizing agents. Chlorine or sodium hypochlorite have been used commercially to make nickel black, but they have the disadvantage of introducing chloride ions into the nickelic hydroxide product which therefore cannot be used for the removal of cobalt from nickel sulfate electrowinning electrolytes. In contrast, ozone can be considered to be a very clean reagent but it is much more expensive than chlorine. There is no known metallurgical plant using ozone for the production of nickel black. Alternatively, nickel black can be obtained electrolytically by anodic oxidation of nickelous hydroxide slurries. The electrochemical method for preparing nickel black was developed by Outokumpu and is used in several nickel electrowinning plants. This process too is rather expensive to run, because of low current efficiency. However, it is currently the only applied method for making chloride free nickel black which is suitable for cobalt removal from nickel sulfate electrolytes.

PRINCIPLE OF THE NEW METHOD

The method of preparation of nickel black which was developed in INCO laboratories (17) consists of oxidizing nickelous hydroxide with air in the presence of calcium sulfite.

The overall chemical process can be described by two parallel reactions:

$$Ni(OH)_2 + \tfrac{1}{4}O_2 + \tfrac{1}{2}H_2O \rightarrow Ni(OH)_3$$

$$\Delta G_{25°C} = +25.5 \text{ kcal/mole} \qquad (1)$$

$$CaSO_3 + \tfrac{1}{2}O_2 + 2H_2O \rightarrow CaSO_4 \cdot 2H_2O$$

$$\Delta G_{25°C} = -98.9 \text{ kcal/mole} \qquad (2)$$

However, reaction (1) is not thermodynamically favourable and the feasibility of the process cannot be, therefore, explained by a catalytic or inductive effect of reaction (2) upon reaction (1). The explanation of this paradox is not simple and we will first discuss the practical, metallurgical results with the new process before trying to speculate why it works.

RESULTS OF BATCH PREPARATION OF NICKELIC HYDROXIDE

Nickelic hydroxide is obtained by treating a nickelous sulfate solution containing 10 to 40 g/l nickel with lime to raise the pH of the solution to at least 8 and precipitate substantially all nickel as a basic nickelous precipitate and introducing into the resulting slurry a mixture of air and freshly prepared calcium sulfite suspension. Calcium sulfite suspension is prepared by sparging sulfur dioxide through a 1.5 molar calcium hydroxide suspension until a pH of about 5 is obtained. The calcium sulfite suspension at pH 5 is reneutralized with lime to pH 11 to stabilize the sulfite. The amount of O_2 supplied by sparging air into the reacting nickel hydroxide-calcium sulfite mixture, should be in excess of the stoichiometric amount needed to oxidize the calcium sulfite and the rate of addition of calcium sulfite should not exceed its rate of consumption (about 0.1 mole $CaSO_3$ per litre and per hour). The reaction can be carried out conveniently at room temperature.

The results of a typical example of a batch oxidation test carried out in a 1 ℓ capacity stirred vessel, (Fig. 1), with air introduced through a fritted glass filter at a rate of 10 ℓ of air per litre of slurry and per minute are plotted on Fig. 2. The nickelous hydroxide slurry to be oxidized contained 35 g/l nickel and was at pH 8. A freshly prepared 1.2M $CaSO_3$ suspension at pH 11.5 was added to the above slurry at room temperature and at a rate of 0.15 mole $CaSO_3$ per litre of slurry and per hour. The pH, oxidation-reduction potential versus the standard hydrogen electrode potential and percent of the nickel as Ni(III) were monitored throughout the experiment. The pH of the suspension fell from 8 to 7.1 in the course of the oxidation process. As is shown on Fig. 2, the rate of formation of nickelic hydroxide was approximately constant at a value of 4g Ni(III)/ℓ/h up to oxidation of 65% of the Ni to Ni(III). The efficiency of the process, ex-

FIGURE 1. Batch Preparation of Nickelic Hydroxide

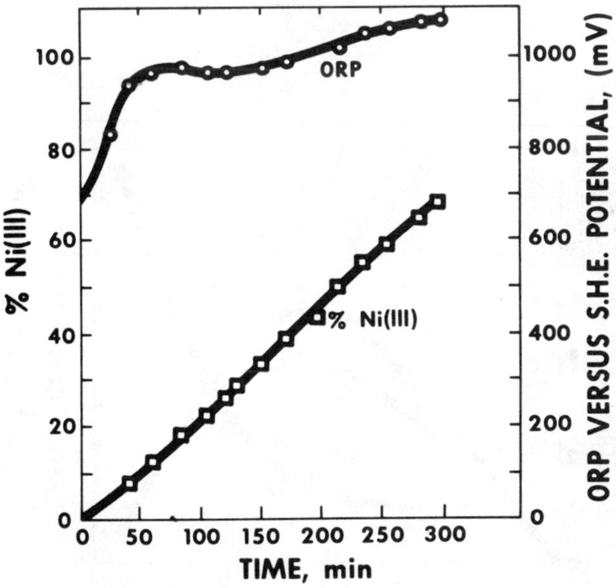

FIGURE 2. Percent of Total Nickel Oxidized and Oxidation-Reduction Potential Versus Time

pressed in terms of moles of $CaSO_3$ consumed per mole of Ni(III) produced, was about 2. The oxidation-reduction potential versus the standard hydrogen electrode potential was seen to increase rapidly from +640 mV to +940 mV in the first hour and more slowly thereafter, from +940 mV to +1,100 mV.

A series of comparative batch tests were carried out in which the pH of the starting nickelous hydroxide slurries to be oxidized was varied between 6.6 and 10. Three different sulfur dioxide incorporating reagents were tested, respectively $CaSO_3$, SO_2 and Na_2SO_3 (see Table 1). The results clearly show that below the precipitation pH of nickelous hydroxide, no nickelic hydroxide was formed (tests B2 and B4). At pH 7, where some nickelous hydroxide is precipitated from solution, a small percentage of the nickel was oxidized to nickelic hydroxide, but at the expense of a large consumption of SO_2 (test B5). At pH 8.3 and above, where virtually complete precipitation of nickel as a basic nickel salt is obtained, large amounts of nickelic hydroxide were formed (tests B3, B6 and B7). The above results demonstrate the superior efficiency of the process of nickel oxidation to Ni(III) when treating nickelous hydroxide suspensions rather than clear nickel solutions as has been proposed in the patent literature (18).

The results shown in Table 1 indicate that $CaSO_3$ is a superior SO_2 incorporating reagent to both SO_2 or Na_2SO_3. Indeed, at a given pH, the consumption of SO_2 in the form of $CaSO_3$ per mole of Ni(III) formed was at least 10 times smaller than that obtained with SO_2 or Na_2SO_3. Other sulfite incorporating reagents such as $MgSO_3$ and $BaCO_3$ were briefly tested, but also with limited success when compared with $CaSO_3$. Thus, in the work which is further discussed here, only $CaSO_3$ will be considered as the SO_2 incorporating reagent for the oxidation process of nickelous hydroxide.

The effect of the total nickel concentration in the nickelous hydroxide suspension to be oxidized, in effect the % of solids in the reacting slurry, is shown in Fig. 3. The dilute suspension was oxidized more efficiently than the more concentrated suspension. The pH in those tests was maintained at about 8 and $CaSO_3$ was added in both cases at a rate of 0.29 moles of $CaSO_3$ per mole of nickel and per hour. The results also show that in both tests, over 80% of the Ni can be oxidized.

The temperature of the nickelous hydroxide slurry has a relatively small effect on the efficiency of oxidation as shown in Fig. 4. The overall efficiency in terms of moles of $CaSO_3$ consumed per mole of Ni(III) formed was about 30% better at 55°C than at 25°C.

A series of tests were carried out at 25°C and 55°C in which the rate of addition of $CaSO_3$ was varied between 0.04 moles and 0.19 moles $CaSO_3$/ℓ/h. The results on Fig. 5, show that the rate of formation of Ni(III) passes through a maximum with increasing rate of addition of $CaSO_3$. This maximum occurs at a $CaSO_3$ addition rate of approximately 0.1 mole/ℓ/h both at 25°C and 55°C. Thus, if $CaSO_3$ is introduced at a rate higher than its rate of utilization, the result is not only a much poorer ratio of reducing agent

TABLE 1. Effect of pH upon % of Total Nickel Oxidized and Reducing Agent Consumption

TEST	pH OF SOLUTION OR SUSPENSION	REDUCING AGENT	AMOUNT OF REDUCING AGENT ADDED $(\frac{\text{mole}}{\text{mole Ni}})$	% OF TOTAL NICKEL OXIDIZED	REDUCING AGENT CONSUMPTION $(\frac{\text{moles } SO_2}{\text{moles Ni(III)}})$
B2	6.6	$CaSO_3$	0.79	0	∞
B3	8.3	$CaSO_3$	0.79	36.7	2.15
B4	6.6	SO_2	0.72	0	∞
B5	7.0	SO_2	1.05	1.05	100
B6	8.3	SO_2	0.79	3.20	24
B7	10.0	Na_2SO_3	5.00	24.0	21

CONDITIONS: ROOM TEMPERATURE, NICKEL TOTAL: 15 g/l, pH CONTROLLED WITH LIME, RATE OF ADDITION OF REDUCING AGENT: 0.2 mole/l.h

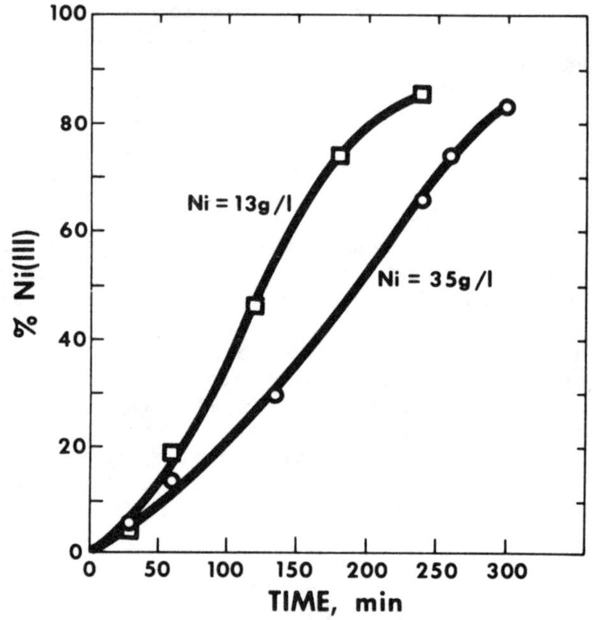

FIGURE 3. Effect of Total Nickel on Rate of Oxidation of $Ni(OH)_2$. (25°C, pH 8, $CaSO_3$ Addition Rate: 0.29 mole/mole Ni·h)

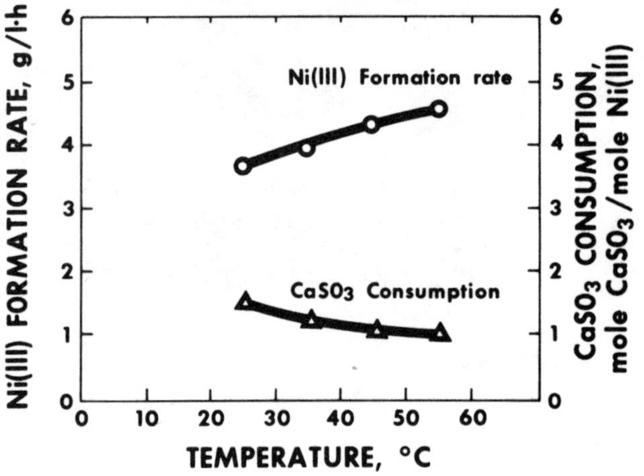

FIGURE 4. Effect of Temperature on Ni(III) Formation Rate and $CaSO_3$ Consumption. (pH 8, Ni total: 13.5 g/l; $CaSO_3$ Addition Rate: 0.09 mole/ℓ·h·)

used to nickelic hydroxide formed, but also a much lower rate of formation of the nickelic hydroxide. The relative position of the two curves shows that reaction yield per mole of $CaSO_3$ was better at 55°C than at 25°C.

The rate of air sparging had no apparent effect on the rate of Ni(III) formation, provided excess air was available for the reaction stoichiometry.

RESULTS OF CONTINUOUS PREPARATION OF NICKELIC HYDROXIDE

Continuous nickelous hydroxide oxidation tests were carried out in a 4 liter capacity Wemco flotation unit shown on Fig. 6. The nickelous hydroxide and calcium sulfite suspensions were added independently at the desired flowrates by means of a multichannel Watson Marlow pump. Air was sparged through the slurry at a rate of 6 to 8 liters per liter of reactants and per minute. The slurry was stirred at 2,400 rpm. The pH of

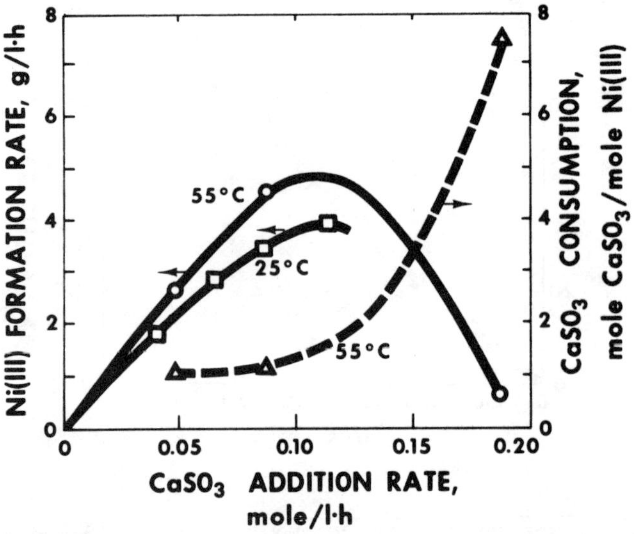

FIGURE 5. Effect of $CaSO_3$ Addition Rate on Ni(III) Formation Rate and $CaSO_3$ Consumption (pH 8, Ni total: 13.5 g/l)

FIGURE 6. Continuous Preparation of Nickelic Hydroxide

the reacting slurry was controlled by means of a Radiometer titrator. Only small additions of lime were necessary to keep the pH at the desired value. The nickelic hydroxide product was collected by overflow into a bulk reservoir.

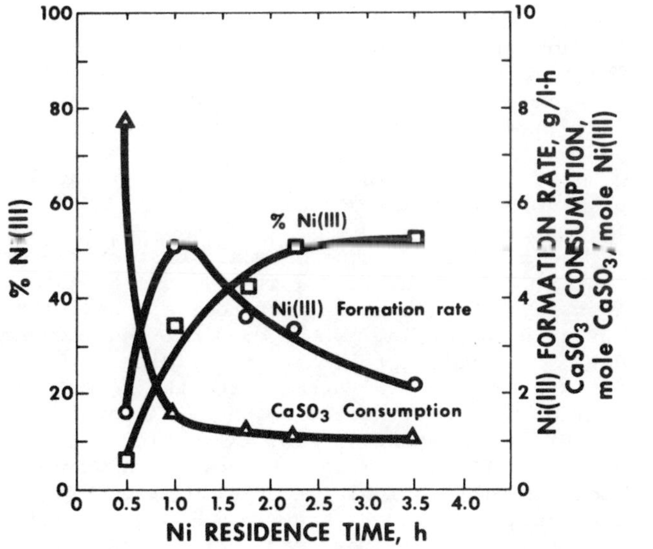

FIGURE 7. Effect of $Ni(OH)_2$ Slurry Residence Time on % of Total Nickel Oxidized, Ni(III) Formation Rate and $CaSO_3$ Consumption. (25°C, pH 8, Ni total: 15 g/l, $CaSO_3/Ni(OH)_2$: 0.55 mole/mole)

The effect of varying the residence time of nickelous hydroxide between 0.5 and 3.5h at a fixed $CaSO_3/Ni(OH)_2$ feed ratio of ∿0.55 mole/mole upon the rate of formation of Ni(III), the % of total Ni oxidized to Ni(III) and the consumption of $CaSO_3$ is shown in Fig. 7. With long residence times, corresponding to slow $Ni(OH)_2$ feed rates, the conversion of Ni(II) to Ni(III) reached a maximum of 52%. Decreasing the residence time by increasing the $Ni(OH)_2$ feed rate logically increased the rate of production of Ni(III), but the % of total Ni oxidized to Ni(III) started to decrease. At a $Ni(OH)_2$ feed rate corresponding to a residence time of 1h, a maximum production rate of 5g Ni(III)/ℓ/h was obtained, and further increase of the $Ni(OH)_2$ feed rate resulted in a drastic drop in the Ni(III) formation rate. At those high $Ni(OH)_2$ feed rates, corresponding to about 30 minutes residence time or less, the redox potential versus the standard hydrogen potential dropped from +1,100 mV to +250 mV (SHE). The consumption of $CaSO_3$ was approximately of 1 mole $CaSO_3$ per mole of Ni(III) formed at low $Ni(OH)_2$ feed rates but increased sharply when reaching $Ni(OH)_2$ feed rates corresponding to 30 minutes residence time.

The effect of varying the $CaSO_3/Ni(OH)_2$ feed ratio between 0.5 and 2 mole/mole at a constant residence time of nickelous hydroxide of 2h upon the rate of formation of Ni(III), the % of the nickel oxidized to Ni(III) and the consumption of $CaSO_3$ per mole of Ni(III) formed is plotted on Fig. 8. The results show that both the % of

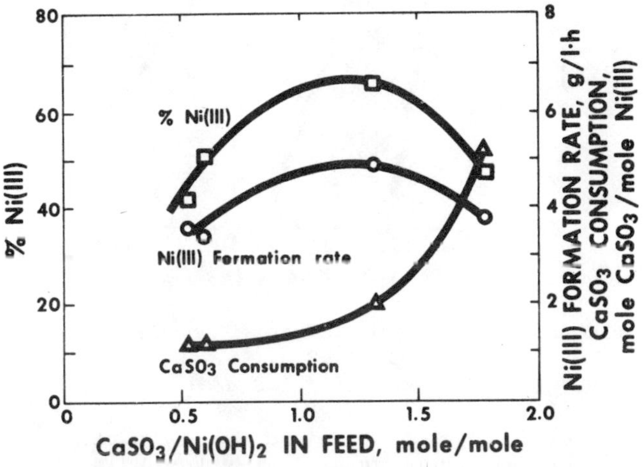

FIGURE 8. Effect of $CaSO_3/Ni(OH)_2$ in Feed on % of Total Nickel Oxidized, Ni(III) Formation Rate and $CaSO_3$ Consumption. (25°C, pH 8, Ni Total: 15 g/l, $Ni(OH)_2$ Residence Time : 2 h)

total Ni oxidized to Ni(III) and the formation rate of Ni(III) passed simultaneously through a maximum value at a $CaSO_3/Ni(OH)_2$ molar feed ratio of about 1.3. At lower $CaSO_3/Ni(OH)_2$ molar feed ratios, both the Ni(III) formation rate and conversion dropped in proportion to the feed rate of $CaSO_3$. At higher $CaSO_3/Ni(OH)_2$ molar feed ratios, the Ni(III) formation rate and conversion decreased due to the sharp increase in $CaSO_3$ consumption per mole of Ni(III) formed. In the example given in Fig. 8, the maximum conversion of Ni(II) to Ni(III) in the product was 70% with a corresponding maximum Ni(III) formation rate of about 5 g/ℓ/h and a consumption of about 1.3 moles of $CaSO_3$ per mole of Ni(III) formed.

RESULTS ON COBALT REMOVAL FROM NICKEL-COBALT SULFATE CONTAINING SOLUTIONS

Nickel black prepared with $CaSO_3$ and air shows the same activity for cobalt removal from nickel containing liquors as nickel black prepared by other methods. Examples of cobalt removal from a typical dilute nickel sulfate leach liquor containing 6.3 g/l Ni and 0.28 g/l Co (Ni/Co of ∿20) are shown on Fig. 9. Both the rate of cobaltic

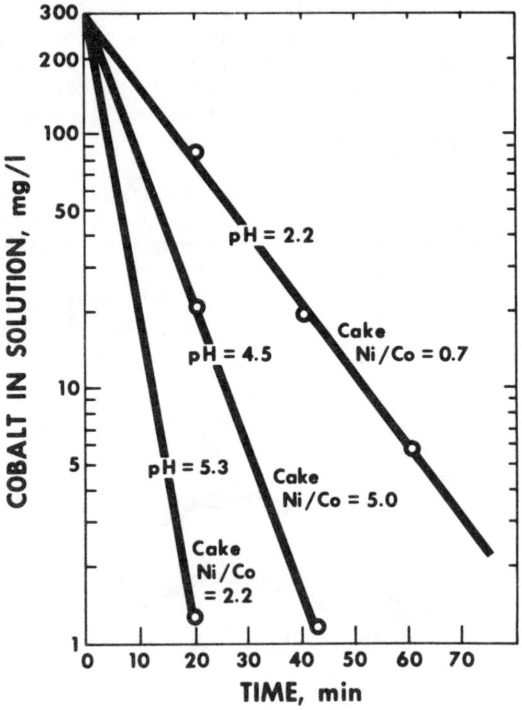

FIGURE 9. Effect of Initial pH on Rate of Cobalt Precipitation. (6.8 g/l Ni, 0.28 g/l Co; 80°C; Ni(III)/Co(II) : 1.3 g/g; 31% Ni(III))

hydroxide precipitation and the Ni/Co ratio in the washed cake decreased with decreasing initial solution pH. However, cobalt was removed to well below 10 mg/l using a 30% excess of Ni(III) in less than one hour for initial solution pH's in the range 2.2 to 5.3.

Cobalt removal from nickel electrowinning type electrolyte, containing 93 g/l Ni, 100 g/l Na_2SO_4, 5 g/l H_3BO_3 and 0.58 g/l Co was tested in a batch experiment using $CaSO_3$-air prepared nickel black. The result is plotted on Fig. 10. Virtually all of the cobalt was precipitated from the electrolyte after 90 minutes by adding to the solution 70% excess Ni(III) with respect to the cobalt, at 75°C and keeping the pH of the solution under treatment at about 5.3 with the aid of lime. The cobaltic hydroxide precipitate had a Ni/Co ratio of about 8 and was further upgraded at pH 3 with a solution of water and sulfuric acid. The upgraded cake had a Ni/Co ratio of about 1.

An example of cobalt removal from leach liquor is given in Table 2. A final acid washed cobaltic hydroxide cake with Co/Ni of 3 was obtained. The precipitates showed excellent settling characteristics.

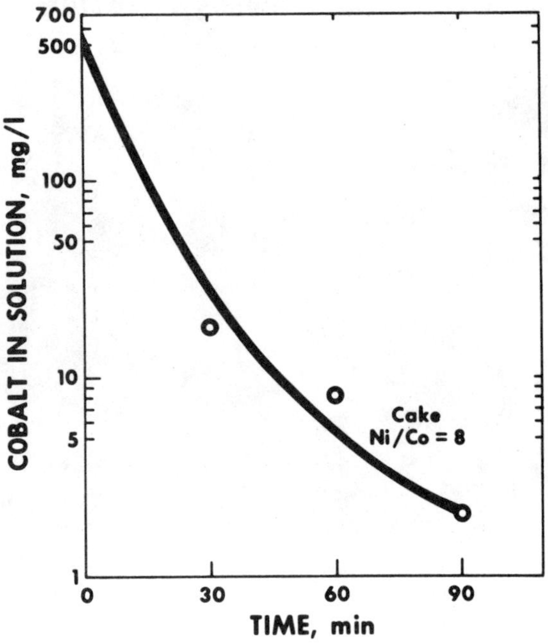

FIGURE 10. Cobalt Removal from Nickel Sulfate Electrowinning Electrolyte. (Ni : 93 g/l; Co : 0.53 g/l; Na_2SO_4 : 100 g/l; H_3BO_3 : 5 g/l; 75°C; pH 5.3; Ni(III)/Co(II) : 1.7 g/g;)

WHAT IS THE OXIDANT?

Having established that the new process for producing nickelic hydroxide is feasible and attractive, it would be interesting to find out why it works and in particular, what is the oxidant and what is the role of $CaSO_3$. It is known that only strong oxidants such as Cl_2, Br_2, $S_2O_8^{2-}$ and O_3 can oxidize $Ni(OH)_2$. H_2O_2 is for example, without action and so is O_2, even at elevated pressures and temperatures.

The simplest assumption would be that the oxidant is O_2 and that $CaSO_3$ simply acts as a catalyst. This explanation must however, be ruled out since according to the credible thermodynamic data, O_2 should not be able to oxidize $Ni(OH)_2$ to $Ni(OH)_3$ (Fig. 11). Furthermore, the redox potential observed during the batch oxidation of $Ni(OH)_2$ (also shown in Fig. 11) varied between 980 and 1,180 mV (SHE) at pH 7-8, which is well above the theoretical potential of the O_2/H_2O electrode in this pH range (850 - 930 mV). It must be, therefore, assumed that the oxidation of $CaSO_3$ is part of the reaction stoichiometry rather than being a parallel reaction. This conclusion can be supported by the observation that the stoichiometric ratio of $CaSO_3$ used to $Ni(OH)_3$ produced converged to 1 as the $CaSO_3$ addition rate was reduced or as the temperature was increased. In the continuous tests, the best $Ni(OH)_3$ formation rate was achieved when the molar ratio of $CaSO_3$ and $Ni(OH)_2$ in the feed was slightly above 1.

The simplest reaction sequence involving $CaSO_3$ in the stoichiometry is based on the assumption that a salt of Caro's acid is produced as the active intermediate:

TABLE 2. Cobalt Removal from Leach Liquor
[Cobalt Removal: 90 Min at 75 C; Cobalt Upgrading 90 Min at 75°C]

STREAM	VOLUME OR WEIGHT	ASSAYS (wt %) or (g/l) Ni	Co	DISTRIBUTIONS (%) Ni	Co	Ni/Co (g/g)	SETTLING % SOLIDS IN 1 h
FEED LIQUOR	4ℓ	31.8	0.57	100	100	56	-
PURIFIED LIQUOR	3.7ℓ	32.4	0.05	94	9	650	-
PRIMARY COBALT CAKE (WET)	97.5g	9.68	2.10	7	91	4.6	~15
UPGRADED COBALT CAKE (WET)	27.3g	2.25	7.00	0.5	84	0.32	~15

REAGENT	CONSUMPTION g/g Co REMOVED
Ni(III)	1.5
$Ca(OH)_2$	5.0
H_2SO_4	5.0

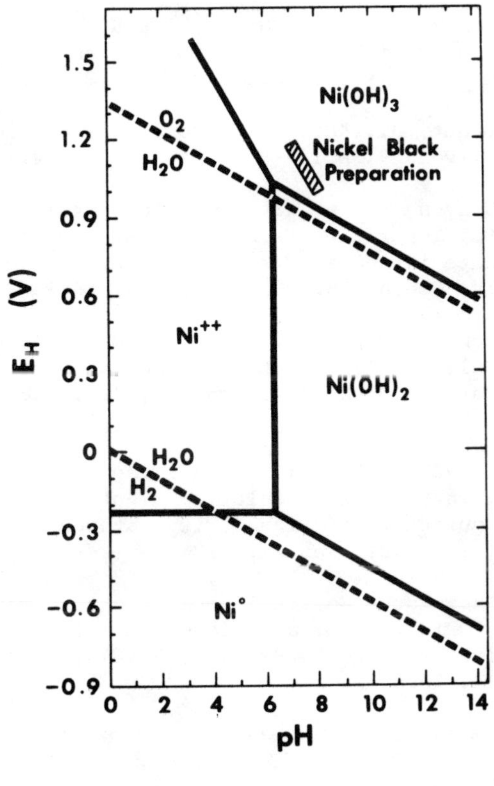

FIGURE 11. Eh-pH Diagram for Ni-H_2O (23, 24)

$$CaSO_3 + O_2 \rightarrow CaSO_5 \quad (3)$$

$$CaSO_5 + 2Ni(OH)_2 + H_2O \rightarrow CaSO_4 + 2Ni(OH)_3 \quad (4)$$

The theoretical stoichiometric yield of this reaction is, however, 2 mole $Ni(OH)_3$/mole and one would have to assume that even under the most favourable conditions tested, ~50% of $CaSO_3$ was wasted by a parallel oxidation of $CaSO_3$ to gypsum:

$$2CaSO_3 + O_2 \rightarrow 2\ CaSO_4 \quad (5)$$

This scheme is favourable thermodynamically and Caro's acid was found to oxidize $Ni(OH)_2$ efficiently. Attempts to produce analytical quantities of $CaSO_5$ by reaction (3) in the absence of $Ni(OH)_2$ have, however, failed. Attempts to produce $CaSO_5$ by adding H_2SO_5 to a slurry of $Ca(OH)_2$ at pH 12 have also failed due to a rapid decomposition to $CaSO_4$ and O_2. A similar experiment using NaOH led to the sodium salt of Caro's acid. If $CaSO_5$ is the oxidant, it seems to be too unstable to be detected by simple means. The observation (19) that $CaSO_3$ can catalyze the otherwise slow reaction of SO_2 with O_2 could possibly also be explained by assuming the $CaSO_5$ intermediate:

$$CaSO_3 + O_2 \rightarrow CaSO_5 \quad (6)$$

$$CaSO_5 + 2SO_3^{2} \rightarrow CaSO_3 + 2SO_4^{2-} \quad (7)$$

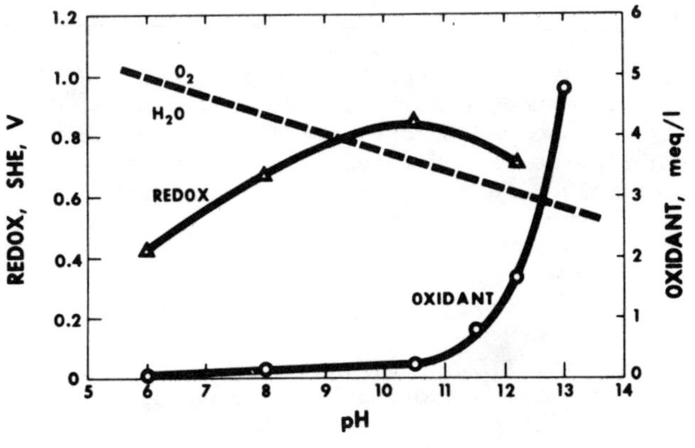

FIGURE 12. Oxidation of Na_2SO_3 with Air. (1 g/l Na_2SO_3, 25°C, 30 minutes)

The stability exhibited by the sodium salt of Caro's acid led us to look at the oxidation of SO_2 (or Na_2SO_3) with air, in the absence of nickel. As shown in Fig. 12, analytical quantities of oxidant were indeed obtained, but only at pH's >8. The oxidation-reduction potential of the oxidized solutions was above the theoretical O_2-H_2O potential in the pH region where oxidant was made. The oxidant was also found to be stable over a period of several hours and readily oxidized $Ni(OH)_2$ to $Ni(OH)_3$ in the absence of air. The above observations may suggest that the calcium ion in the $CaSO_3$-air system merely keeps the amount of SO_2 species in solution low enough to prevent back reduction of $Ni(OH)_3$.

The well known observation that $Ni(OH)_2$ alone is an effective catalyst for oxidation of SO_2 with O_2 (20,21,22) suggests that some kind of $Ni(OH)_2$-SO_3^{2-} complex could be formed initially which can react rapidly with molecular oxygen. This reaction could lead to a number of oxidizing species e.g. the previously assumed SO_5^{2-} but also $S_2O_8^{2-}$ or even free radicals such as HO_2 or OH. Our observations however also indicate that such species can be formed in the presence of only traces of Ni.

However interesting, the problem of what is the oxidant or how to write the chemical reactions of the process, is obviously too complex for us to study. We hope, that somebody else will be able to resolve the mysterious chemistry while we will try to develop the process for commercial use.

CONCLUSIONS

We have described in this paper a new, inexpensive method of production of nickelic hydroxide. The only reagents required are lime (or even limestone), sulfur dioxide and air.

The nickel black prepared by this method is a suitable reagent for the removal of cobalt from metallurgical nickel sulfate liquors.

The redox potential obtained during the reaction exceeds that of the oxygen electrode and this suggests production of some strongly oxidizing intermediate species. It is not clear at this time whether such species are formed at the surface of the $Ni(OH)_2$ or in solution phase nor is it known which species are involved. We speculate that a salt of Caro's acid, e.g. $CaSO_5$, may be involved in the overall reaction, but other evidence indicates that $Ni(OH)_2$ may itself be involved in the primary reaction with molecular oxygen.

REFERENCES

1. Wallis, A.E., and Pincott, J., "Production of Pure Cobaltic Hydroxide", U.S. Pat. 2,726,144, Dec. 6, 1955.

2. Toivanen, T., and Grongvist, P.O., "Nickel Refining in Finland", The Canadian Mining and Metallurgical Bulletin, Vol. 626, No. 6, April 1964, pp. 653-658.

3. Gran, T.V., "Oxidation of Cobalt by Black Nickel Hydrates during Repulping of Cobalt-Nickel Cakes", Zhurnal Prikladnoi Khimii, Vol. 37, No. 3, March, 1964, pp. 487-492.

4. International Nickel Ltd. "Oxidative Removal of Cobalt from Nickel Salt Solutions" Fr. Demande 2,055,769, June 4, 1971.

5. Gandon, L., Bozec, C. and Lenoble, P., "Basic Ni(III) Carbonate", Ger. Offen. 2,038,754, Feb. 18, 1971.

6. Lever, F.M., MacGregor, J.J., and Westwood, W., "Preparation of Nickel Hydroxide" Can. Pat. 965,931, Jan. 11, 1975.

7. Outokumpu, Oy., "Separating Nickel, Cobalt and Chromium from Iron in Metallurgical Products such as Ferronickel", Fr. Demande 2,183,166, Jan. 18, 1974.

8. Gandon, L. et al, "Removal of Cobalt from Nickel Salt Solutions", U.S. Pat. 2,890,243, June 17, 1975.

9. Meidert, F., "Nickelic Hydroxide", Can. Pat. 284,237, Oct. 23, 1928.

10. Feitknecht, W., Christen, H.R., and Studer, H., "Higher Nickel Hydroxides", Z. anorg. u. allgem. Chem., Vol. 283, 1956, pp. 88-95.

11. Labat, J., "Nickel Hydroxides", Annales de Chimie, Vol. 9, No. 7-8, 1964, pp. 399-427.

12. Henych, R. et al, "Removing Metal and Nonmetal Impurities from Technical Nickel Carbonate in the Manufacture of Pure Cathodic Nickel", Czech. 102,895, March 15, 1962.

13. Henych, R.,et al, "Removing Metal Impurities from Sulfate Electrolytes in the Manufacture of Pure Cathode Nickel", Czech 106, 496, Feb. 15, 1963.

14. Henych, R. et al, "Separating Cobalt from Nickel in Sulfate or Chloride Electrolytes in the Manufacture of High Purity Cathode Nickel", Czech. 106, 524, Feb. 15, 1963.

15. Rheinische Westfaelisches Elektrizitaetswerk A.G., "Oxidation of Nickelous Hydroxide for Alkaline Batteries", Brit. 1,050,876, Dec. 14, 1966.

16. Faber, P., "Nickel(III) Hydroxide for Alkaline Batteries", Ger. Offen. 2,122,165, Nov. 30, 1972.

17. Ettel, V.A. and Mosoiu, M.A., "Preparation of Nickel Black", U.S. Pat. 4,006,216, Feb. 1, 1977.

18. Kniprath, E., "Redox Precipitation of Non-Ferrous Heavy Metals from Aqueous Solutions", Can. Pat. 935,660, Oct. 23, 1973.

19. Haber, F., and Bran, F., "Aber die Autoxyolation", Z. Phys. Chem., Vol. 84, 1900, p. 35.

20. Reinders, W., and Vless, S.I., "Reaction Velocity of Oxygen with Solutions of Some Inorganic Salts", Receuil de Traveaux Chimiques des Pays Bas Belges, Vol. 44, 1925, pp. 249-268.

21. Titoff, A., "Beiträge zur Kenntnis der negativen Katalyse im homogenen System", Z. Phys. Chem. (Leipzig), Vol. 45, 1903, pp. 645-683.

22. Böttger, W., and Thoma, E., "Clarification of the Wicke Reaction", J. Prakt. Chem., Vol. 147, 1936, pp. 11-21.

23. Karapet'Yants M.Kh., "Handbook of Thermodynamic Constants of Inorganic and Organic Compounds", Ann. Arbor-Humphrey, London, 1970, p. 189.

24. Latimer, W.M., "Oxidation Potentials", Prentice-Hall, New York, 1952, p. 199.

Chapter 8

HYDROMETALLURGICAL NICKEL AND COBALT RECOVERY FROM SULPHIDE CONCENTRATES

B. Meddings

Analytical Services and Quality Control
Sherritt Research Centre
Sherritt Gordon Mines Limited
Fort Saskatchewan, Alberta

ABSTRACT

The technology involved in the Sherritt process for the recovery of nickel and cobalt from sulphide concentrates and mattes is presented in a manner which emphasises the chemistry of the process. The hydrometallurgical process takes advantage of the differing stabilities of the various oxy-acids of sulphur from thionates through thiosulphates to sulphate. Some separations are based upon stability differences, some upon kinetic differences and some upon valence state differences and similarities. The engineering and economic factors involved in hydrometallurgy are not discussed in either detail or depth.

HYDROMETALLURGICAL NICKEL AND COBALT RECOVERY FROM SULPHIDE CONCENTRATES

B. Meddings

Analytical Services and Quality Control
Sherritt Research Centre
Sherritt Gordon Mines Limited
Fort Saskatchewan, Alberta

Most of the nickel produced today is derived from sulphide ore bodies. Most of that production comes from a combination of pyrometallurgy and electrolytic refining. In contrast, this paper deals with the production of refined nickel and cobalt from sulphide concentrates by a purely hydrometallurgical process developed by Sherritt Gordon Mines Limited. This process was developed, piloted and brought into full scale commercial production between 1948 and 1954. The information given is therefore not new but it is deliberately presented in such a manner as to emphasise the chemistry of the process.

A very much simplified flowsheet of the Sherritt process is shown in Figure 1 and this emphasises the three unit processes which must be carried out in any hydrometallurgical metal refining process:

1. Leaching
2. Solution Purification
3. Metal Recovery

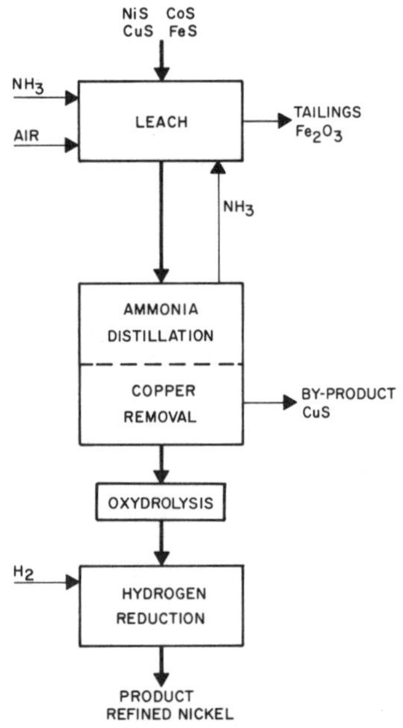

FIGURE 1. SIMPLIFIED FLOWSHEET

1. Leaching

The Sherritt process was originally developed to process the sulphide concentrate from the Lynn Lake mine in northern Manitoba. Over the years, the process has been modified to deal with a variety of sulphide concentrates and mattes from all over the world. A typical nickel sulphide concentrate has the following composition:

Iron	40.5% wt.	
Cobalt	0.30%	
Nickel	12.0%	
Copper	1.55%	
Sulphur	32.0%	
Pyrrhotite	FeS	0.62 moles
Pentlandite	$FeNiS_2$	0.28 moles
Pyrite	FeS_2	0.07 moles

The four metals are the last four elements of the first transition series, having atomic numbers 26 through 29, and chemists would expect these four sulphides to have very similar chemical behavior. Because nickel is refined on a considerably smaller scale than steel, it is not economically practical to recover iron during the refining of nickel. The first major objective of all methods of nickel refining is, therefore, the rejection and removal of iron from the other three metals.

In the Sherritt process, advantage is taken of the fact that, while all four metals form soluble ammine complexes, there is a marked difference in their resistance to hydrolysis and oxidation. The ammine complexes of iron are easily hydrolysed, and then oxidi ed, in ammoniacal solutions where the ammines of cobalt, nickel and copper are quite stable. The Sherritt process, therefore, starts by leaching the sulphide concentrate in an ammoniacal solution of ammonium sulphate into which air is sparged. The sulphides are oxidised to sulphates, the metals pass into solution and the iron is hydrolyzed out and oxidized to hydrated ferric oxide which is filtered off and discarded to the tailings pond. One can represent the overall reaction by the following series of equations:

$$NiS + 2NH_3 + 4O \longrightarrow [Ni(NH_3)_2]^{++} + SO_4^{=}$$

$$CoS + 2NH_3 + 4O \longrightarrow [Co(NH_3)_2]^{++} + SO_4^{=}$$

$$CuS + 2NH_3 + 4O \longrightarrow [Cu(NH_3)_2]^{++} + SO_4^{=}$$

$$FeS + 2NH_3 + 4O = 1/2O + H_2O \quad 1/2Fe_2O_3 + 2NH_4^{+} + SO_4^{=}$$

Because of requirements in the metal recovery operation, the desired nickel ammine in the Sherritt process is the diammine. However, freshly precipitated hydrated iron oxide is a very efficient absorber for metal ions, and if this leach step is carried out with only sufficient ammonia to give an ammonia to total metals (i.e., Ni+Co+Cu) molar ratio of 2:1, the discarded iron oxide will contain a large proportion of the valuable metals. This absorption can be decreased to acceptable levels by increasing the ammonia level to an ammonia/metals ratio of 6 or 7 to 1.

When the Sherritt ammonia leach process is used for matte instead of concentrate, this ammonia to metals ratio can be maintained at about 2/1. The reason is that typical mattes contain about 70-75% nickel and less than 1% iron so that the absorption losses on the small amount of iron residue is negligible.

The path of the sulphur as it passes through the leaching step must now be considered in more detail. To a first approximation, the sulphides are oxidized to sulphates

$$S^{=} + 2\,O_2 \longrightarrow SO_4^{=}$$

However, this is an oversimplification of what turns out to be a rather complicated series of reactions.

If the leach solution is analysed for both total dissolved sulphur and sulphur present as sulphate, it is found that the total sulphur always exceeds the sulphate sulphur. In the early stages of the leach, the sulphate sulphur will be as low as 60% of the total sulphur. Complete analysis of the leach solutions shows that the sulphur is present in solution in four major species:

Sulphate $SO_4^{=}$ or 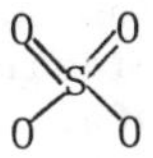

Thiosulphate $S_2O_3^{=}$ or 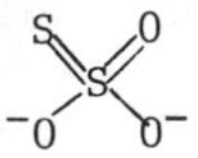

Sulphamate $SO_3NH_2^{-}$ or 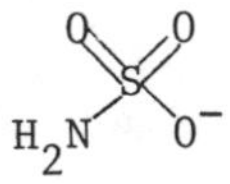

and Trithionate $S_3O_6^{=}$ or 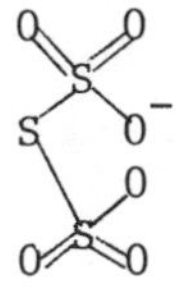

The full details of all the interactions possible between these species under the range of permissible leach conditions have not been completely elucidated. However, there is some evidence that thiosulphate is the initial product of oxidation of the sulphide matrix of the minerals. Under oxidising conditions in the ammoniacal solutions employed, there is then further oxidation to trithionate and sulphate. The sulphamate appears to be produced by a side-reaction between sulphate and ammonia.

Figure 2 shows a typical laboratory batch leach curve and the changes in the sulphur species as the leach progresses and nickel and copper pass into solution. The composition of this leach solution must be controlled in the full scale commercial plants because the amount of thiosulphate and trithionate passing from the leach section to the solution purification section turns out to be critical.

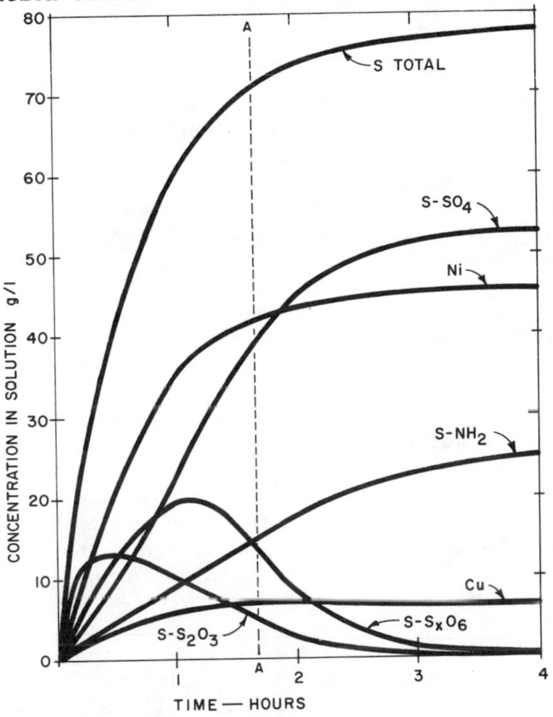

FIGURE 2. BATCH LEACHING CURVE

2. Solution Purification

The major function of the solution purification section of the Sherritt process is the removal of copper and the adjustment of the ammonia to metals ratio.

The adjustment of the ammonia to metal ratio is necessary because the nickel reduction section requires this ammonia to metal ratio to be rather exactly 2 to 1. The most direct method of removing ammonia from the leach solution is by distillation of the ammonia and recycling this ammonia back to the leach autoclaves. During the early development work on the Sherritt process, it was discovered that this distillation step produced a black precipitate. Analysis of this material showed it to be a mixture of cupric and cuprous sulphides. Moreover, it was remarkably free of nickel and cobalt contamination.

Investigation of this reaction revealed that the stability of the thionate and thiosulphate species decreased as the ammonia content of the solution fell and these species, in effect, disproportionated to sulphate and sulphide. This homogeneous generation of sulphide led to the precipitation of the most insoluble metal sulphide, namely copper sulphide. The combination of ammonia distillation and copper precipitation is now known as the "Copper Boil."

A series of equations can be written to show this disproportionation to sulphate and sulphide and the concurrent reduction of some cupric ions to cuprous:

$$S_2O_3^{=} + H_2O \longrightarrow H_2SO_4 + S^{=}$$

$$S_3O_6^{=} + 2H_2O \longrightarrow 2H_2SO_4 + S^{=}$$

$$=2S_2O_3^{=} + 4H_2O + 8Cu^{++} \longrightarrow S_3O_6^{=} + H_2SO_4 + Cu^{+} + 6H^{+}$$

This serendipitous precipitation of copper during the removal of ammonia from the leach solutions was found to go furthest to completion when the leach solutions contained roughly equal weights of copper in solution and of sulphur in the thionate and thiosulphate forms. However, the removal of copper was not entirely complete, copper usually decreasing from 5-6 gpl in the leach solution to about 0.1-0.5 gpl in the copper boil filtrate. This residual copper is removed by injecting hydrogen sulphide gas which effectively lowers the copper strip residue, is always contaminated with nickel and cobalt, and is recycled to each of the autoclaves.

The operating requirement that the amount of sulphur present as thionate and thiosulphate in the leach solution should equal the copper in solution can now be related to the leaching curve shown in Figure 2.

The line marked A-A represents the moment in time when the optimum conditions for ammonia distillation and copper removal are obtained. In the commercial plants, the leaching autoclaves are discharged at this moment and the solids and liquid separated. The clear solution goes to the copper boil section (i.e., ammonia distillation and copper removal). The incompletely leached solids are charged into another set of leaching autoclaves with fresh leach solution and leached until all the recoverable nickel is brought into solution. This section is called the final leach and the filtered and washed solids from this section are discarded to the tailings pond, while the leach solution passes to the first leaching autoclaves. The first leaching train is known as the Adjustment Leach because it is here that the Cu/S ratio is adjusted to give the optimum ratio for copper removal. It may be emphasi ed that the reagent required for the copper removal is generated from the ore itself during leaching.

The original very simplified flow diagram of Figure 1 can now be expanded to that shown in Figure 3.

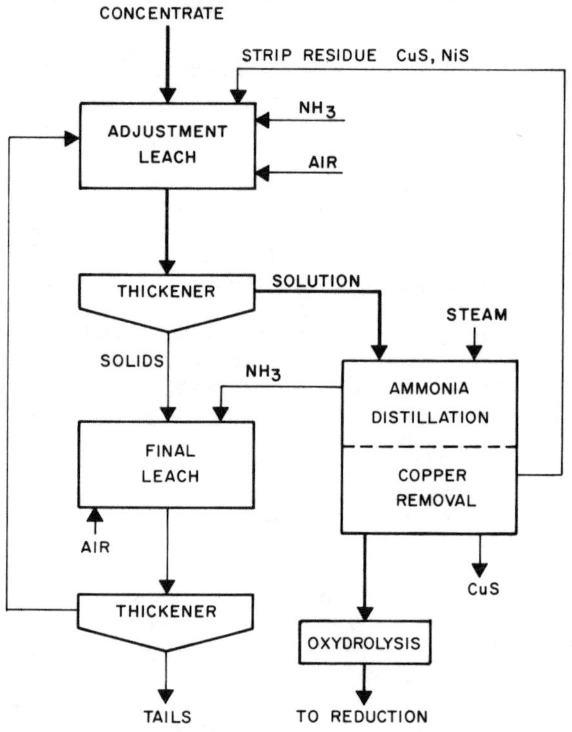

FIGURE 3. FLOWSHEET BEFORE REDUCTION

After the ammonia distillation and copper removal steps, the solution contains only nickel and cobalt metals together with ammonium sulphate, free ammonia and the sulphamate produced during the oxidation of the sulphide minerals. There is also usually some 1-2 gpl of sulphur in the form of residual thiosulphate and thionate. If this solution passes directly to the hydrogen reduction section for nickel production, some of the thiosulphate and thionate is reduced to sulphide which contaminates the nickel product.

The solution after reduction would also contain sulphamate, thereby contaminating the ammonium sulphate which is produced from the reduction end solution and sold as fertiliser. Since the ammonium sulphamate is a herbicide, this is not desirable.

For these reasons, the last operation in the solution purification section is to convert the thiosulphate, thionate and sulphamate to sulphate. The thiosulphate and thionate can easily be oxidised with air to sulphate. However, sulphamate is comparatively stable in solution and the only practical method of dealing with it is to hydrolyse it to sulphate at elevated temperatures. The copper-free solution is therefore oxidised with air in autoclaves at 450°F. This operation combines the oxidation and hydrolysis steps into one reaction and is therefore termed "Oxydrolysis."

$$S_2O_3^{=} + H_2O + 4O \longrightarrow 2HSO_4^{-}$$

$$S_3O_6^{=} + 2H_2O + 4O \longrightarrow 2H_2SO_4^{=} + SO_4^{=}$$

$$NH_4 \cdot SO_3 \cdot NH_2 + H_2O \longrightarrow (NH_4)_2SO_4$$

We can, therefore, summarise the chemistry of the solution purification section of the Sherritt process as:

1. Distillation of ammonia to give an ammonia/nickel and cobalt molar ratio of 2:1.
2. Simultaneous disproportionation of trithionate and thiosulphate to sulphate and sulphide with the resultant homogeneous precipitation of copper sulphide.
3. Final stripping of the last traces of copper by the injection of H_2S.
4. Simultaneous oxidation of residual trithionate and thiosulphate and hydrolysis of sulphamate to sulphate in the "oxydrolysis" section.

3. Metal Recovery

In the Sherritt process, the metal is recovered by hydrogen reduction of the aqueous nickel solution.

When considering the possibility of using gaseous hydrogen to reduce an aqueous solution of a metal salt to the elemental metal, the thermodynamics of the situation should be examined. The general case is:

$$Me^{n+} + \frac{n}{2} \cdot H_2 \longrightarrow Me + n.H^+$$

This can be viewed as two opposing reactions:

$$Me \quad Me^{n+} + n.e \quad \text{Electrode potential} = E_{Me} \qquad (2)$$

$$H_2 \quad 2H^+ + 2.e \quad \text{Electrode potential} = E_{H_2} \qquad (3)$$

Clearly, if the metal potential E_{Me} exceeds the hydrogen potential, then the metal will pass into solution and liberate hydrogen from water.

Conversely, if E_{H_2} exceeds E_{Me}, then hydrogen will pass into solution and reverse reaction (2) and precipitate metal Me from solution.

The potentials E_{H_2} and E_{Me} are single electrode potentials and are defined in the usual electrochemical nomenclature by:

$$E_{H_2} = E^o_{H_2} - \frac{2.303\ RT}{F} \cdot \log_{10} a_{H^+} \qquad (4)$$

$$E_{Me} = E^o_{Me} - \frac{2.303\ RT}{n.F} \cdot \log_{10} a_{Me^{n+}} \qquad (5)$$

The potential of the hydrogen electrode is governed by the concentration of hydrogen ions and hydrogen molecules in solution in contact with the electrode and so equation (4) should be rewritten as

$$E_{H_2} = E^o_{H_2} - \frac{2.303\ RT}{F} \cdot \log_{10} a_{H^+} + \frac{2.303\ RT}{2F} \cdot \log_{10} P^*_{H_2} \qquad (6)$$

where $P^*_{H_2}$ is the fugacity of the hydrogen. However, between 1 and 100 atmospheres, the use of fugacity in place of pressure only alters the last term in equation (6) by no more than 1%, and so we may use pressure (in atmospheres) in equation (6).

Since $-\log_{10} a_{H^+}$ is defined as pH and the standard electrode potential of hydrogen, $E^o_{H_2}$, is defined as zero, then equation (6) reduces to

$$E_{H_2} = 0.0592\ pH + 0.0296\ \log_{10} P_{H_2}$$

at 25°C, where P_{H_2} is the partial pressure of hydrogen in atmospheres.

The expression for the metal potential can be simplified in a similar manner.

The activity of the metal ion is defined as

$$a_{Me^{n+}} = f_{Me^{n+}} \cdot [Me^{n+}]$$

where Me^{n+} is the analytical concentration of the metal ion, Me^{n+}, and $f_{Me^{n+}}$ is the activity coefficient of that ion. Equation (5) can, therefore, be written as:

$$E_{Me^{n+}} = E^o_{Me^{n+}} = \frac{0.0592}{n} \cdot \log_{10} [Me^{n+}] - \frac{0.0592M}{n} \cdot \log_{10} f_{Me^{n+}} \qquad (7)$$

In hydrometallurgical operations, the range of concentrations usually encountered runs from 1 molal down to about 1×10^{-3} molal.

In the former case, the middle term of equation (7) becomes zero and the activity coefficient of most divalent metal sulphates is about 0.05 so that equation (7) reduces to

$$E_{Me^{n+}} = E^o_{Me^{n+}} + 0.04$$

In the latter case, at 1×10^{-3} molal, the last term of equation (7) becomes zero and we have

$$E_{Me^{n+}} = E^o_{Me^{n+}} + 0.09$$

So, between 1.0 and 1.0 x 10^{-3} molal, the electrode potential of most divalent metal sulphates lies between (E^o+0.04) and (E^o+0.09) volts where E^o is the standard electrode potential of the metal.

We can now plot the variation of the hydrogen electrode potential with pH over a practical range and use the ordinates to plot the metal potentials for a number of metals at 1.0 and 1.0 x 10^{-3} molal. This is shown in Figure 4.

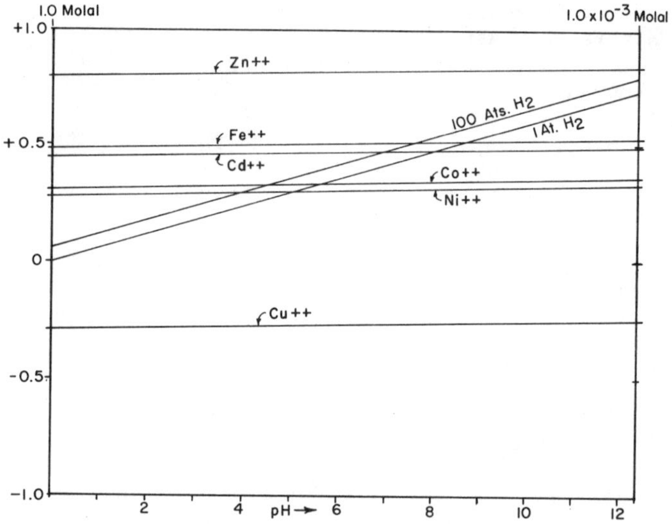

FIGURE 4. HYDROGEN AND METAL ELECTRODE POTENTIALS

This shows clearly that it is thermodynamically possible to reduce a metal from solution if the E_{H_2} line in Figure 4 lies above the $E_{Me^{n+}}$ line.

Clearly it is possible to reduce cupric ions to copper at all pH values but to reduce nickel and cobalt it is necessary to operate above a pH of about 5.5.

In practice the kinetic and chemical factors are of more concern than the thermodynamic considerations. The actual reduction can be viewed as

$$Ni^{++} + H_2 \qquad Ni^o + 2H^+$$

Clearly the acid produced will eventually reverse the reaction and so the acid produced must be neutralised. The easiest manner in which to do this is to incorporate a base into the nickel solution before reduction and this is the function of the ammonia in the solution produced from the ammonia distillation process.

Because there are two protons produced for each nickel atom produced the ideal ratio of ammonia to nickel is 2 to 1. If this ratio is achieved then the pH of the nickel solution will not change during reduction and the reduction can be carried to completion. However, if the ammonia to nickel ratio is less than 2/1 then the raction will cease before completion.

Thus if the ratio is 1.5/1 then all the base will be consumed when only 75% of the metal is reduced and the reaction will cease very shortly after this because the extra acid produced will lower the pH and so decrease the hydrogen potential. Conversely, if the ammonia to nickel ratio is higher, say 2.5/1 then when 80% of the nickel is reduced the NH_3/Ni ratio remaining in solution will be 4.5/1. At this ratio the nickel potential is so raised by the formation of nickel ammines that the potential is approximately 0.4 volts compared to 0.24 v for the standard electrode potential for nickel. At this high potential the reduction rate becomes uneconomically slow and the reduction is stopped when some 15-20% of the nickel remains in solution.

So in the Sherritt process the ammonia to metal ratio must be controlled rather closely at 2/1. There is another reason for this close control. The solution fed to the reduction section contains about 50 gpl nickel and 1-1.5 gpl cobalt. If both metals were reduced at the same rate the metal product would contain 2-3% cobalt. However, the usual specification for refined nickel metal sets a maximum limit of 0.1% cobalt. Sherritt nickel typically contains 0.06-0.08% cobalt. The control required to achieve this low cobalt content is partly achieved by control of the ammonia/metal ratio. During the reduction step cobalt is not reduced until the nickel concentration has been reduced to about that of cobalt. If the reaction can be stopped at this point then very little cobalt will appear in the nickel product. The reaction is stopped at this point by using an ammonia to metal ratio of less than 2/1. Depending upon the cobalt content of the reduction feed solution the NH_3/Ni+Co ratio is held at 1.9-1.95/1 and this is sufficient to stop the reaction before significant cobalt reduction has occurred. Another perspective of this control can be given by stating that use of a 1.9 - 1.95/1 ammonia to metal ratio results in reduction of 97% of the nickel in solution while only 2% of the cobalt is reduced.

The reduction of nickel ions by hydrogen is a heterogeneous reaction which only proceeds at useful rates upon suitable solid nuclei. If a nickel solution is subjected to hydrogen reduction in an autoclave the nickel will be deposited as a plating upon the autocalve walls - which is not economically recoverable. In commercial practise a nickel solution is nucleated for hydrogen reduction by the addition of small amounts of ferrous sulphate before hydrogen injection. This results in a very rapid reduction and the formation of very fine nickel powder. This powder is then allowed to settle in the autoclave, the reduced solution is discharged and fresh reduction feed solution is admitted to the autoclave.

The agitators are re-started to re-suspend the fine nickel powder, and hydrogen is again fed into the autocalve thereby reducing the nickel out of solution on to the fine nickel seed powder. The sequence of settling, discharging, re-filling and reduction is then repeated many times over. Because this series of operations causes an increase in the apparent density of the nickel powder each single reduction is called a "densification." In commercial practise about 50 to 75 such reductions are carried out in each densification series. As the nickel powder becomes denser and coarser it becomes less active as a catalyst and the time of reduction gradually lengthens from about 5 minutes in the early densifications to about 30-40 minutes in the last densifications. At this stage all the nickel powder and reduced solution is discharged from the autoclave. The powder is separated from the solution, washed with water and dried in a rotary kiln dryer.

The powder so produced is either sold as such or briquetted and sintered before sale. The typical analyses of these two products are as follows:

	Powder	Briquettes
Cobalt	0.06%	0.06%
Copper	0.004%	0.004%
Iron	0.006%	0.009%
Sulphur	0.020%	0.005%
Carbon	0.003%	0.003%

The slight increase in iron content observed in briquettes is due to physical pick-up from the steel conveyor belt in the sintering furnace. The decrease in sulphur is due to desulphurisation in the reducing atmosphere during sintering.

The solution discharged from the autoclaves after reduction contains about 1.0-1.5 gpl each of nickel and cobalt and 500 gpl of ammonium sulphate. The nickel and cobalt are removed by injecting hydrogen sulphide gas and filtering off the precipitated nickel and cobalt sulphides. The metal-free solution contains only 0.002 gpl of nickel and cobalt and is fed to an evaporator and crystallisation circuit which produces crystalline ammonium sulphate for sale as fertiliser.

The mixed nickel and cobalt sulphides are further processed to produce nickel and pure cobalt metal.

Recovery of Cobalt

The mixed nickel and cobalt sulphides are leached in aqueous solution in autoclaves at about 250°F and 100 psig of air with sufficient sulphuric acid added to give a pH of 1.5-2.5. Under these conditions the sulphides are rapidly oxidised to sulphates without any of the intermediate species formed in the ammoniacal leaching concentrates.

The solution of nickelous and cobaltous sulphate so produced contains some iron and this is removed by raising the pH to about 5.5 with ammonia and sparging with air to precipitate ferric hydroxide. This is removed by filtration when the clarified nickel-cobalt solution is passed to the oxidation circuit.

In the oxidation circuit the solution is made strongly ammoniacal and agitated under air pressure in an autoclave in order to oxidise the cobalt to the cobaltic state. Advantage can then be taken of the different chemistries of the cobaltic ammines and the nickelous ammines to effect a separation between the two metals. The amount of ammonia added is sufficient to give a molar ratio, NH_3/Ni+Co, of about 5.5/1. When this solution is reacted at 160°F under 100 psig of air almost all the cobalt is oxidised to the trivalent state. The solution discharged from the oxidation circuit usually contains about 35 gpl of nickel, 1-1.5 gpl of cobaltous and 34 gpl of cobaltic, with a free ammonia concentration of about 60-70 gpl. This solution is then passed to the nickel removal circuit.

In the nickel removal circuit the oxidised ammoniacal solution is carefully acidified with sulphuric acid to about pH 2 - 2.5 and cooled. Under these conditions nickel ammonium sulphate crystallises out and is filtered off. Any cobalt that is left in the cobaltous state at the end of the oxidation stage also precipitates as cobaltous ammonium sulphate and is removed with the nickel double salt. However, all the trivalent cobalt remains in solution as stable cobaltic ammines. The average ratio of ammonia to cobalt in the remaining clear solution is about 5.1/1 and and major cobalt species in solution is about 5.1/1 and the major cobalt species in solution is the cobaltic pentammine, $[Co(NH_3)_5]^{+++}$.

The removal of nickel is actually carried out in two stages. Between the two stages the solution is evaporated to decrease its volume by about 25%, then cooled, acidified to pH 2.0 and about 1-2 gpl of cobalt powder is added. The effect of the cobalt powder is to reduce some of the cobaltic cobalt to the cobaltous state,

$$Co + 2Co^{+++} \longrightarrow 3Co^{++}$$

and this is precipitated as cobaltous ammonium sulphate. This second precipitation of cobaltous ammonium sulphate carries down the last traces of nickel so that the resultant clarified solution contains about 40 gpl cobalt and 0.02-0.03 gpl nickel.

The purified cobalt solution must be reduced to the cobaltous state before being precipitated as pure cobalt powder by hydrogen reduction. This pre-reduction step is necessary for two reasons. Firstly, heating of the cobaltic solution to the temperature required for hydrogen reduction causes some hydrolysis and the precipitation of cobaltic oxide which contaminates the cobalt powder. Secondly, the ammonia to cobalt ratio in the cobaltic solution is 5/1 and so there is excess ammonia which is liberated during reduction and this very soon slows the reaction rate down to uneconomical rates.

The pre-reduction step is very conveniently carried out by adding fine cobalt powder to the cobaltic solution. Some sulphuric acid is also added to adjust the free ammonia to cobalt ratio to close to 2/1.

$$Co + 2Co(NH_3)_5^{+++} \longrightarrow 3Co^{++} + 10NH_3$$

The optimum NH_3/Co ratio for hydrogen reduction is about 2.2-2.4/1 which is somewhat higher than that used for nickel reduction. However, in the reduction of the pure cobalt solutions there is no need to stop the reduction before completion. The actual hydrogen reduction operation is carried out in the same manner as the nickel reduction using 300 psig of hydrogen pressure at 350°F. The average reduction time is about 10-15 minutes and a series of about 60 densifications is carried out after each nucleation.

Conclusion

We can now summarise the chemistry of the Sherritt hydrometallurgical refining of nickel/cobalt sulphide concentrates and mattes.

Leaching

1. Oxidation of the sulphide matrix to liberate the metals into solution.
2. Oxidation and hydrolysis of the iron ammines to precipitate and remove iron leaving only the valuable metals, nickel, copper and cobalt in solution as ammines.

Solution Purification

3. Simultaneous removal of ammonia by distillation and de-stabilisation of the thionate and thiosulphate species to cause disproportionation into sulphide and sulphate with the homogeneous precipitation of copper sulphide.
4. Final stripping of copper from solution by injection of hydrogen sulphide.
5. Removal of sulphamate and traces of thionate and thiosulphate by "oxydrolysis," a combination of oxidation and hydrolysis.

Metal Recovery

6. Recovery of pure nickel powder by hydrogen reduction which reduces 97% of the nickel from solution and only 2% of the cobalt, thereby leaving a 1/1 ratio of Ni/Co in the solution after reduction.
7. Recovery of the residual nickel and cobalt from solution by precipitation with hydrogen sulphide.
8. Recovery of ammonium sulphate from the barren solution and sale as fertiliser. In this way, all the sulphur from the original ore is converted to useful fertiliser and none is emitted as a pollutant in any form.

Cobalt Recovery

9. Dissolution of the mixed nickel/cobalt sulphides under acid conditions which converts sulphide directly to sulphate without intermediate species.
10. Oxidation and pH adjustment to hydrolyse iron and remove iron from the nickel/cobalt solution.
11. Oxidation under highly amoniacal conditions to bring the cobalt to the trivalent state so that the differences in chemistry between Co^{III} and Ni^{II} can be exploited.
12. Stabilisation of the cobalt as cobaltic pentammine while nickel is precipitated as nickelous ammonium sulphate at pH 2.0.

13. Conversion of cobaltic to cobaltous by adding elemental cobalt powder to the cobaltic solution.

14. Hydrogen reduction of the cobaltous solution to recover pure cobalt powder.

15. Recovery of residual cobalt from the reduction and solution by precipitation with hydrogen sulphide and recovery of ammonium sulphate from the barren solution by evaporation.

Chapter 9

COPPER RECOVERY FROM SULFIDE CONCENTRATES VIA ROAST/LEACH/ELECTROWINNING

Matthews, Dale C., Jr.
Principal Staff Engineer, Morrison-Knudsen Company, Inc.
Boise, Idaho

Simpson, Donald L.
General Manager, Smoky Valley Mining Division Copper Range Company
Box 480 Round Mountain, NV

ABSTRACT

Hecla Mining Company's Lakeshore Mine and Metallurgical complex near Casa Grande, Arizona, is the first plant in North America to utilize the roast/leach/electrowinning process for production of cathode copper from sulfide concentrates. Oxide copper ores are also mined at Lakeshore. Using sulfuric acid and sponge iron by-products of the roast/leach operation, the oxide ore is leached in vats, the copper is precipitated with iron and recovered as cement copper.

This paper will describe the roast/leach facilities and discuss the operating conditions at which more than 95 percent of the copper in concentrate is made soluble in acid solutions while less than 5 percent of the iron is rendered soluble. More than 97 percent of the sulfur contained in concentrate is captured either as strong sulfuric acid or as soluble sulfates in leach liquors. The iron rich residue after leaching is pelletized and reduced to sponge iron in an SLRN reduction kiln.

"COPPER RECOVERY FROM SULFIDE CONCENTRATES VIA ROAST/LEACH/ELECTROWINNING"

Matthews, Dale C. Jr.
Principal Staff Engineer, Morrison-Knudsen Company, Inc.
Boise, Idaho

Simpson, Donald L.
General Manager, Smoky Valley Mining Division Copper Range Company
Box 480 Round Mountain, Nv.

I. INTRODUCTION

The Lakeshore Mine complex, a joint venture of Hecla Mining Company and El Paso Natural Gas Company, is located approximately 35 miles south of Casa Grande, Arizona on the Papago Indian Reservation. Hecla Mining Company is the operating partner.

The Lakeshore ore body contains three distinct types of ore - two sulfide and one oxide. Tactite is a relatively high grade sulfide orebody with limited tonnage. Porphyry or massive disseminated ore is lower grade but represents a very large tonnage. Oxide ore is relatively high grade and generally overlies the porphyry ore.

The selection of a processing scheme for the Lakeshore property recognized the general shortage of smelting capacity in Arizona - this shortage was and still is being aggravated by environmental pressures. It was considered necessary to produce as much marketable product as possible at the mine site. Roast-Leach-Electrowinning of sulfide copper plus vat leaching and cementation of oxide copper appeared to meet the essential criteria for process selection:

(1) All pollution standards proposed by federal and state agencies could be met.
(2) The processes were not new but had been demonstrated in commercial practice.
(3) The processes produced standard marketable shapes.
(4) The processes take economic advantage of byproduct iron and sulfuric acid.
(5) The processes are efficient and economic at relatively small scale operation.

Even though the processes were not new, an extensive program of laboratory and pilot plant testing was required to confirm their suitability for treatment of Lakeshore ores and to provide specific information for design of the commercial plant.

Two years of commercial operation have proven the validity of the pilot plant work. The roast/leach/electrowinning process is a viable and economical alternative to smelting and refining treatment of sulfide copper concentrates.

While the vat leaching and cementation plants have performed as expected, toll smelting charges for cement copper have increased to the point that it is now economically justifiable to install solvent extraction and additional electrowinning facilities to produce only cathode product from both sulfide and oxide ores. This project is now in the engineering phase.

II. GENERAL PLANT DESCRIPTION (FIGURE 1)

Fig. I- LAKESHORE PLANT FLOWSHEET

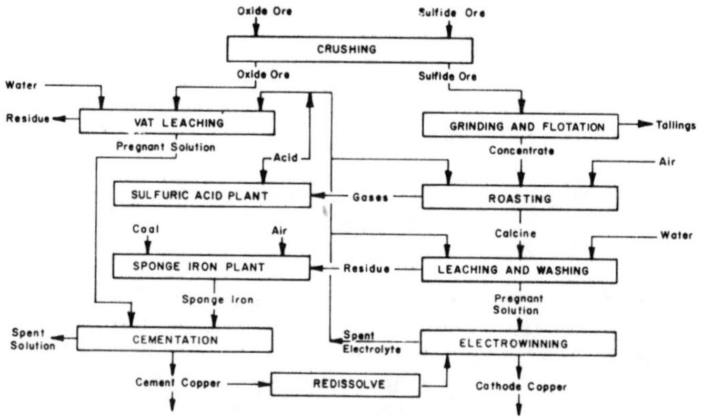

Approximately 11,000 tons of sulfide ore and 6,500 tons of oxide ore are mined daily. The principal mineral in the sulfide ore is chalcopyrite, a copper-iron sulfide. The principal oxide mineral is chrysocolla, a hydrated copper silicate.

Sulfide concentrates are produced in a conventional crushing plant and concentrator. These concentrates are roasted, producting calcine and SO_2 gasses suitable for conversion to sulfuric acid in a single pass contact acid plant. The calcine is leached in acid solution to produce pregnant solution for the electrowinning plant and an iron-rich residue suitable for production of sponge iron.

Cathode copper is produced from the pregnant solution in an electrowinning tankhouse. Spent electrolyte is recycled to the roast/leach plant to provide the acid requirement for calcine leaching and a bleed stream goes to vat leaching.

Oxide ore is crushed in the same crushing plant as the sulfide ore and is leached in vats using the acid in the tankhouse spent electrolyte and strong acid to solubilize the copper.

The residue produced from leaching the roaster calcine is approximately 50 percent iron. This material is pelletized, indurated and then reduced to sponge iron in an SLRN reduction kiln. The sponge iron pellets are ground and used to precipitate cement copper from the vat leach pregnant solutions. This cement copper is pelletized and shipped to a smelter.

A portion of the cement copper may be redissolved in sulfuric acid solutions to provide additional feed to the electrowinning plant.

More than 97 percent of the sulfur in the copper concentrates is captured either as strong acid or as soluble sulfates in leach liquors. Lakeshore has an unconditional operating permit issued for primary copper production by the Bureau of Air Quality, Arizona State Health Department.

III. Detailed Discussion of Roast-Leach-Electrowinning Plant (Figure 2)

a) Feed Preparation

Concentrate slurry is received from the concentrator at 20% solids and at the rate of about 400 tons per day. The concentrate is thickened to 65% solids and the clear thickener overflow is returned to the concentrator. The thickened slurry is stored in four agitated tanks with a total capacity of about 4,000 tons of concentrate. As each storage tank is filled, the density is adjusted to 65% solids and the sulfur is determined by X-ray analysis of a sample. Slurry of known composition is then pumped to a 150-ton capacity roaster feed tank. Tramp oversize material larger than 10 mesh is removed by screening ahead of the feed tank to avoid potential blockage of the roaster feed system.

The roaster feed typically contains about 26 percent copper, 26 percent iron, 30 percent sulfur and 15 percent insoluble material, mostly SiO_2. Other metals commonly present in minor amounts are zinc, molyhdenum, calcium, magnesium and aluminum.

Fig. 2 - FLOWSHEET ROAST-LEACH-ELECTROWINNING PLANT

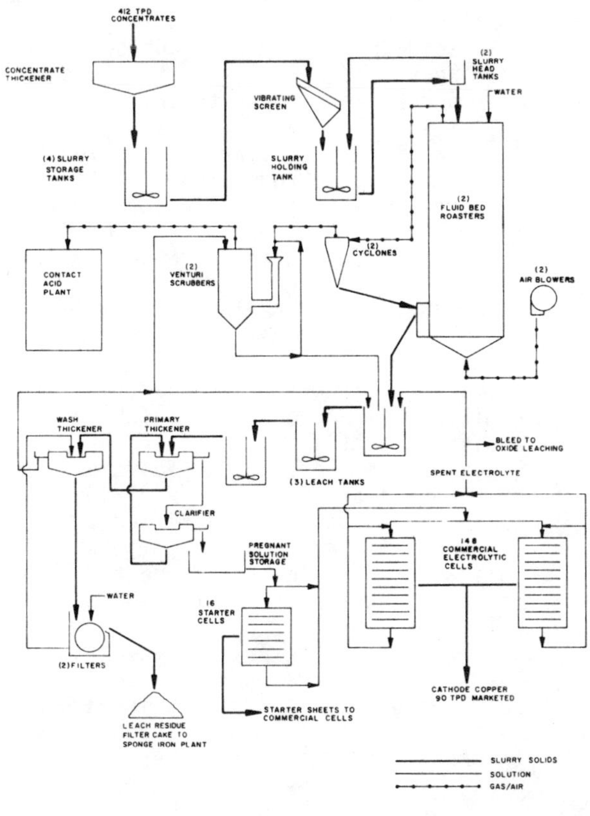

b) Roasting

Concentrates are roasted in two fluidbed roasters. Each roaster is 25 feet high with a 20 foot 9 inch hearth diameter. The freeboard area is expanded to 26 foot diameter. Fluidizing air enters the roaster through 1,000 tuyeres on the flat bottom.

Concentrate slurry is pumped from the slurry holding tank to a steady head system with an overflow returning to the holding tank. Slurry density is monitored with a nuclear density meter mounted on the pump line. Feed to each roaster is automatically controlled to a preset volume by use of magnetic flowmeters and air operated control valves. Cooling water and atomizing air are mixed with the concentrate slurry in the roaster from a spray head on the feed pipe. The spray is located in the center of the roaster and about three feet above the top of the fluidized bed.

Gasses exit the top of the roaster to a dry cyclone that removes about 95 percent of the dust load. This dust is returned to the roaster in the roaster discharge area. Gasses are further cleaned in a high-energy venturi scrubber to remove all of the roaster calcine dust. Scrubber solutions containing the dust are pumped to the calcine leach tanks. The clean gas, now containing about 5 percent SO_2, continues to a conventional contract sulfuric acid plant. The plant is designed to be autothermal at 4.5 percent SO_2 in the gas.

The important points in roaster operational control are:

(1) Space rate - that is, the theoretical upward velocity of the fluidizing air in the roaster bed. Normally the roasters operate with space rates between 1.75 and 2.15 feet per second.
(2) Sulfur/air ratio - that is the pounds of sulfur in the slurry feed per pound of fluidizing air. Normally the roasters operate with a sulfur/air ratio between 0.11 and 0.13.
(3) Temperature - the temperature in the roaster bed is most critical to the metallurgical performance of the system. The chemical reactions in the roaster are exothermic and it is possible to maintain the temperature within 2 degrees celcius by the addition of cooling water with the slurry feed. Normal operating temperature is between 680° C and 700° C.
(4) Roaster bed agglomeration - the concentrate in the slurry feed is very fine - more than 80 percent minus 400 mesh. It is necessary that some agglomeration take place in the bed; otherwise, it would all go with the gasses to the cyclone. If too much agglomeration occurs, the bed loses fluidization and shuts the plant down. Samples of the bed material are taken hourly and the feed gun spray is adjusted to maintain 3% to 5% plus 8 mesh material in the bed.

The chemical reactions in roasting are not completely defined, however, pilot and commercial plant experience have indicated the following:

(1) Under optimum conditions, more than 97 percent of the copper and less than 5 percent of the iron is rendered souble.

(2) Within the normal operating temperature range of 680-700° C, most of the copper is sulfated and is water soluble. At the higher temperature, there is less sulfation and acid is required for leaching.

The principal copper mineral in the concentrate is chalcopyrite, $CuFeS_2$. Some pyrite, FeS_2, is nearly always present. Some of the more important reactions are:

$$4\ CuFeS_2 + 13\ O_2 \text{ ---- } 4\ CuO + 2\ Fe_2O_3 + 8\ SO_2$$

$$4\ FeS_2 + 11\ O_2 \text{ ---- } 2\ Fe_2O_3 + 8\ SO_2$$

$$SO_2 + O_2 \text{ ---- } 2\ SO_3$$

$$CuO + SO_3 \text{ ---- } CuSO_4$$

$$Fe_2O_3 + SO_3 \text{ ---- } Fe_2(SO_4)_3$$

Temperature control is most critical in roaster operation. At higher temperatures, both copper and iron solubility decrease. The color of the leach residue is normally hematite red but at higher temperatures it is black, indicating formation of magnetite, Fe_3O_4. Copper is probably being tied up as insoluble copper ferrites - $CuFe_2O_3$ or $CuFeO_2$. Other elements present in the concentrate that form soluble sulfates include zinc, molyhdenum, magnesium and aluminum.

Fig. 3 – The effects of variations in roaster bed temperature and in sulfur to air ratio on % Cu dissolved.

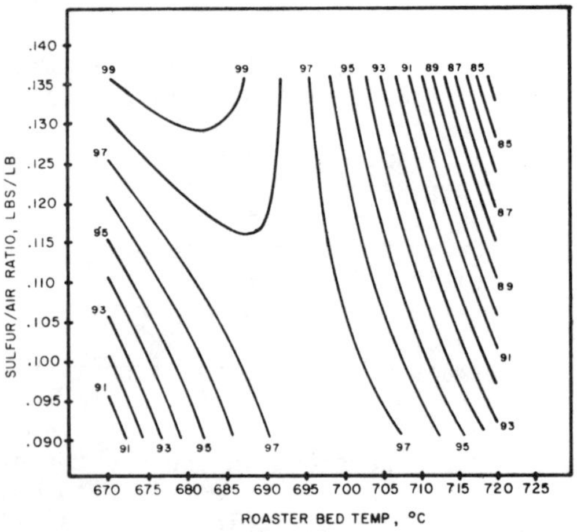

Fig 4. — The effects of variations in roaster bed temperature and in S to air ratio on % Fe dissolved

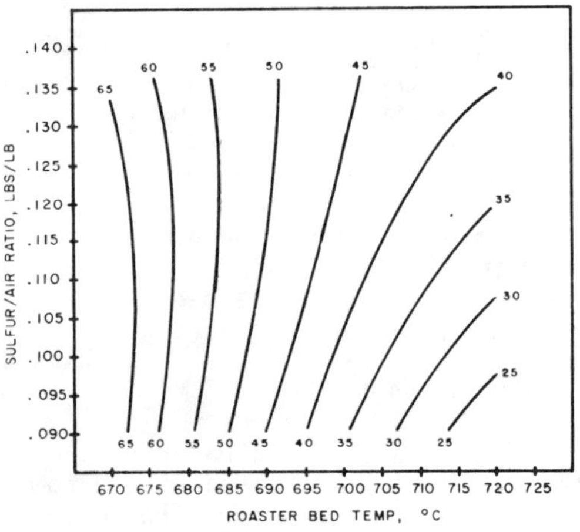

Variations in the sulfur to air ratio, while not as critical as temperature variation, also has an effect on roaster performance. The interreaction and effect of varying temperature and sulfur to air ratio on copper solubility, iron solubility and SO_2 content of roaster gasses are illustrated on response surfaces, Figures 3, 4 and 5. As can be seen, copper and iron solubility and SO_2 content of roaster gasses are all affected by variation in temperature.

Fig. 5 – The effects of variations in roaster bed temperature and in S to air ratio on % SO in exit gas L

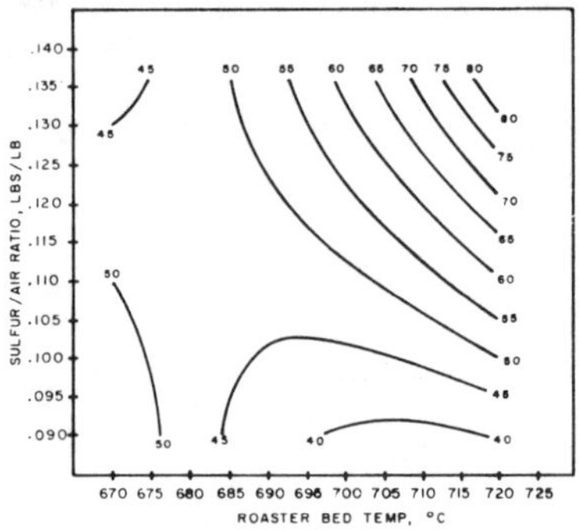

Copper solubility is slightly increased at higher sulfur to air ratios as is SO_2 content of roaster gas. However, at higher sulfur to air ratios, bed agglomeration and potential loss of fluidization become difficult to control.

(c) Leaching

Calcine from the roaster bed discharges into the first of three agitated leach tanks which perform the dual function of quenching the calcine and dissolving the soluble copper. Leach solutions come from three sources:

(1) Sufficient spent electrolyte from the electrowinning plant is recycled to provide the acid necessary for leaching and maintain about 15 gpl free acid in the new pregnant solution.

(2) Weak acid slurry from the venturi scrubbers reports to the leach tanks.

(3) Wash solutions in the leach residue washing circuit are controlled so that the introduction of this wash volume into the leach tanks will maintain the copper assay of the new pregnant solution at 55 gpl.

(d) Liquid/Solids Separation

Slurry from the leach tanks is pumped to a primary thickener. Flocculant is added to a mixing box ahead of the thickener to assist in settling. The overflow from the thickener is the pregnant solution and goes to a clarifier where additional flocculant is added. The clarifier overflow will contain less than 5- ppm suspended solids and goes to the pregnant solution storage tank at the electrowinning plant.

Underflow from the primary thickener contains the leach residue in a 50 percent solids slurry. This slurry is washed to remove all soluble copper through a wash thickener and two wash filters. Filter cake, containing about 50 percent iron and less than 2 percent copper is feed for the sponge iron plant. Filter wash water goes to the wash thickener. The overflow from the wash thickener is used in the roaster gas venturi scrubbers with the excess going to the calcine leach tanks.

Electrowinning

The electrowinning plant is designed to produce about 90 tons of cathode copper per day.

There are a total of 164 electrowinning cells in the tankhouse. The cells are cast concrete with plastisol PVC liners. Each cell has 73 insoluble lead anodes and 72 cathodes. Cathode size is 36" X 40".

Starting sheets are made on stainless steel blanks in 16 cells. About 2300 sheets weighing 10 pounds each are stripped daily. A portion of the sheets are cut up to make loops for the starting sheet assemblies.

The commercial section of the tankhouse has 148 cells. Cathodes weighing about 120 lbs. each are pulled on a 7-day cycle.

DC power is supplied from two silicon diode rectifiers. The unit for starting sheets supplies 29,000 amperes at 40 volts. The commercial cell rectifier supplies 29,000 amperes at 390 volts.

The pregnant solution feed from the Roast-Leach Plant contains about 55 gpl copper, 3 gpl total iron and 15 gpl free acid. A portion of the solution feeds the starting sheet cells and the spent electrolyte from these cells combines with the balance of new pregnant solution to make up the feed to the commercial cells.

All commercial cells are connected in series electrically. Solution flow is in three cascades. Cell conditions for starting sheet and commercial sections are as follows:

	Starting Sheet Cells	Commercial Cascades I	II	III
No. Cells	16	74	37	37
Cu - gpl	35	35	25	15
Free Acid - gpl	45	45	60	75
Temperature - °C	35	40	40	45
Current Density, Amps/ft^2	20	15	15	15

Animal protein glue is added to the cells at the rate of about 0.6 pounds per ton of cathode produced to inhibit dendritic copper crystal growth. A demisting agent is used to produce a stable foam cover on the cells that contain the mist caused by oxygen generation at the anodes.

Approximately 80 percent of the copper contained in the sulfide concentrates is recovered as cathode copper. The remaining copper is recovered as cement copper.

The cathodes meet all specifications for electrolytic copper and are shipped to a continuous copper rod casting plant.

Process Economics

The roast-leach-electrowinning process plant is environmentally acceptable with no capital intensive extra equipment requirements. This, coupled with the negligible fuel requirement for plant operation, makes the process competitive with other treatment alternatives.

Literature Cited

1. Griffith, W.A., Day, H.E., Jordan, T.S., Nyman, V.C., "Development of the Roast-Leach-Electrowin Process for Lakeshore", Journal of Metals, Vol, 27, No. 2, pp 17-25, February, 1975.

Chapter 10

The Recovery of Lead from Sulfide Concentrates Using a Chlorination/Brine Leach/Electrolysis Process

R. T. Um
W. K. Tolley
J. C. Stauter

UOP, Inc.
Del Plaines, Illinois

ABSTRACT

A hydrometallurgical process for the recovery of lead from sulfide concentrates is presented. The process combines the unit operations of selective dry chlorination, brine leaching and fused salt electrolysis as key steps. An overview of the chemistry involved in the process is discussed as it pertains to overall recovery.

INTRODUCTION

There is presently a great effort expended to find alternative methods of treating lead sulfide concentrates, chiefly because of increased environmental and health concerns. Much of this increased effort is directed toward hydrometallurgical processes because air pollution is essentially eliminated.

One of the more widely known examples of hydrometallurgical treatment for lead concentrates is the acidic ferric chloride leach process proposed by the United States Bureau of Mines (1,2,3) in which lead sulfide is converted to lead chloride and elemental sulfur in an aqueous medium. Lead chloride is then solubilized in hot brine solution followed by the precipitation of high-purity lead chloride crystals upon cooling. The solution contains ferrous species and impurities and is treated for recycle. The precipitated lead chloride undergoes fused salt electrolysis to produce pure lead and reclaim process chlorine.

The great reactivity of chlorine with the base metal sulfides was realized early in the 1900's and was proposed as an oxidizing agent which could be used in a gaseous phase treatment. Early descriptions of dry gaseous chlorination processes for complex sulfides were given by Malm (4) in 1914, and Ionides (5) in 1916. In these processes, chlorine, acting either by itself or by the formation of volatile sulfur chlorides and some metal chlorides, assured the conversion of lead and zinc compounds to the chlorides, which could then be treated for recovery through simple aqueous or brine dissolution. Recently, a patent issued to Cyprus Metallurgical Process Corporation (6) and a paper by Reynolds, et al, described a low temperature gaseous chlorination process for the treatment of lead sulfide concentrates in which the lead is converted to $PbCl_2$ with sulfur remaining in elemental form.

Low temperature, non-aqueous chlorination of lead sulfide concentrates, as done in the UOP laboratory, was found to be much more selective toward lead and to produce less conversion of impurities than ferric chloride processes. The generation of sulfates in the leach step also was diminished when using the dry chlorination method.

Since chlorine gas was found to react rather selectively toward lead sulfides at low temperature, a program to investigate its use in a hydrometallurgical lead process was begun by UOP, Inc. This paper reports some of those results along with an overview of the treatment of lead sulfide concentrates by chlorination.

THE PROCESS

A Simplified flowsheet showing the chlorination process is shown in Figure 1. Lead sulfide concentrates are chlorinated in a reactor controlled at 80-120°C into which a mixture of air and chlorine gas is passed. The resultant lead chloride is then brine leached to solubilize the lead chloride and separate the sulfur and gangue portions. The hot, pregnant brine solution is cooled to crystallize pure lead chloride. The lead chloride is dried and dissolved in a mixture of fused salts, preferably with sodium chloride, which is then electrolyzed to produce a pure lead product and chlorine for recycle.

FIGURE 1

HYDROMETALLURGICAL PROCESS TO TREAT LEAD SULFIDE CONCENTRATES

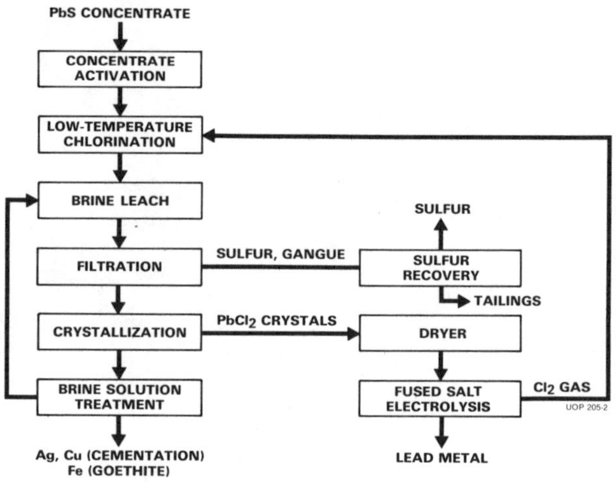

PROCESS CHEMISTRY

Chlorination

The low temperature gasious chlorination step can be expressed simply as:

$$PbS + Cl_2 \qquad PbCl_2 + S^o \qquad (1)$$

The reaction is exothermic, and at 100°C (373°K) has an enthalpy of -62.8 kcal/mole PbS converted to $PbCl_2$.

Chlorination from side reactions such as:

$$2S + Cl_2 \qquad S_2Cl_2 \qquad (2)$$

and

$$PbS + S_2Cl_2 \qquad PbCl_2 + 3S \qquad (3)$$

is minimized by maintaining the temperature in the range 80-100°C.

Lead sulfide selectivity for different chlorination processes at a temperature of 100°C is shown in Table 1. A lead concentrate of the composition shown in Table 2 was used for the comparison. The major differences in the process methods were in copper and zinc conversion with dry chlorination showing

TABLE 1

METAL SULFIDE CHLORINATION BY VARIOUS METHODS AT 100°C

METAL SULFIDE	$FeCl_3$ LEACH (CONVERSION %)	S_2Cl_2 LEACH (CONVERSION %)	DRY, GASEOUS CHLORINATION (CONVERSION %)
Pb	99.7	75.0	95.0
Zn	42.4	29.0	2.6
Fe	13.3	25.0	8.8
Cu	26.1	51.0	<1.0

UOP 205-9

TABLE 2

ASSAY OF LEAD SULFIDE CONCENTRATE

Element	Concentration (wt-%)
Pb	64.8
S	17.5
Fe	4.4
Cu	0.85
Zn	5.5

less reaction to impurities. The improvement in lead selectivity was also noted by Reynolds and co-workers (7) as reaction temperature was decreased from 150°C to 40°C.

Lead sulfide selectivity over conversion of impurity sulfides was most evident for tests maintained below 120°C. This increase in selectivity appears to correspond to a decrease in the amount of sulfur chlorides produced. To test the reactivity of sulfur to chlorination, samples of granulated sulfur were reacted under identical chlorination conditions to those used for lead concentrates. The results showed that elemental sulfur reacted very slowly with chlorine at 100°C, whereas at temperatures above 120°C the amount of sulfur chlorides produces increased rapidly.

The use of an activation treatment on the concentrates was an added factor in promoting chlorination selectivity. The activation involved roasting the concentrates in a neutral or reducing atmosphere at 400-600°C for 5-15 minutes. Chlorination of unactivated concentrates produced more sulfur chlorides and higher impurities conversion than concentrates tested following an activation treatment. Copper conversion, for example, dropped from 4-5% to less than 1%.

The heat treatment removes hydroxyls, oxides, sulfates and organics such as flotation reagents from the sulfide surfaces and, thus, exposes a higher lead surface area for active chlorination. Surface analysis by Electron Scan Chemical Analysis (ESCA) of concentrates before and after such treatment showed the marked surface concentration changes. X-ray diffraction, however, showed no change in crystal structure, an example of such ESCA findings is shown in Figure 2 for the surface removal of sulfate.

FIGURE 2

RESULTS OF ESCA SCAN SHOWS SULFATE REMOVAL FROM PbS

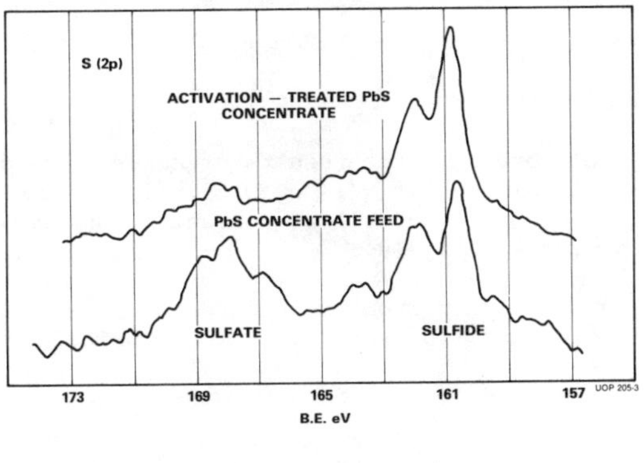

The activation pretreatment produced a change in the apparent activation energy in the chlorination step from +13.8 kcal/mole to +2.5 kcal/mole PbS. A plot showing the two activation energies is given in Figure 3.

FIGURE 3

ARRHENIUS PLOT FOR CHLORINATION OF PbS WITH GASEOUS CHLORINE

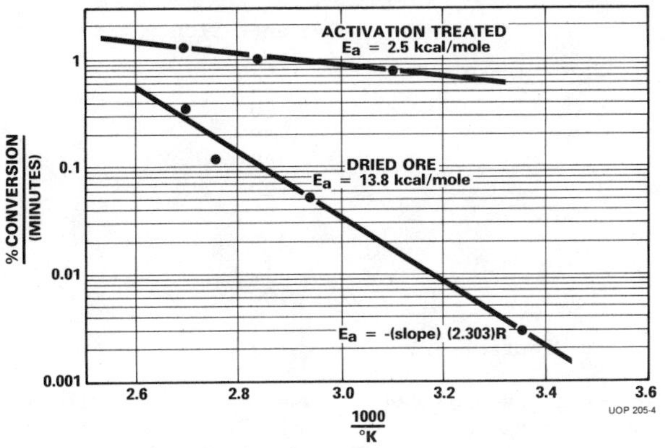

Chlorination of the activated concentrate at the same chlorine partial pressure which had been used with the unactivated concentrate resulted in an uncontrollable temperature rise and fusion of the concentrate. Chlorination under these experimental conditions was not selective for lead. However, by lowering the chlorine partial pressure in the activated case, good selectivity was achieved.

Brine Leaching

The lead chloride produced in the chlorination step was separated from the gangue, sulfur and unreacted metal sulfides by leaching in a brine solution made up of approximately 300 g/l NaCl (25 wt. % NaCl). The solubility of lead as a chloride complex in such a brine is temperature dependent and typical solubility relationships for the $PbCl_2$-NaCl-H_2O system are shown in Figure 4. It is interesting to note that there is a decrease in $PbCl_2$ solubility in the brine above approximately 26% NaCl because of the formation of some form of $PbCl_2$-NaCl double salt.

FIGURE 4

SOLUBILTY CURVES FOR THE $PbCl_2$ -NaCl-H_2O SYSTEM

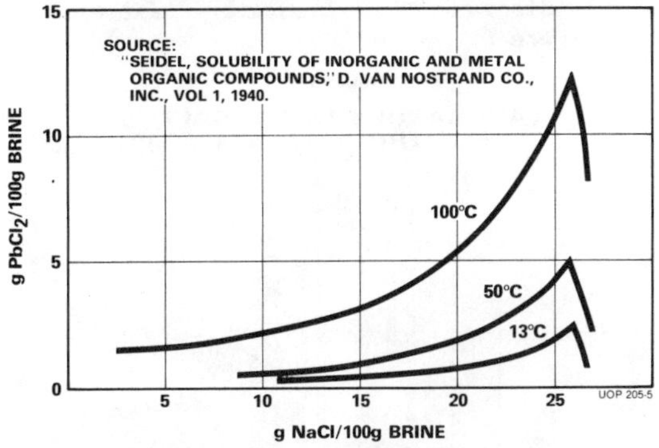

Lead chloride is dissolved in the brine at 100°C; the solution is filtered to remove solids and then cooled to crystallize pure $PbCl_2$, which goes to drying and fused salt electrolysis. The resultant lead chloride crystallized from brine solution had an assay as shown in Table 3 for the laboratory test series.

TABLE 3

ASSAY OF LEAD CHLORIDE CRYSTAL USED AS ELECTROLYSIS FEED

Element	Concentration (ppm)
Cu	20
Zn	30
Fe	70
Ag	20
Cd	20
As	100
Bi	80
Sb	100

Fused Salt Electrolysis

Pure lead was produced by fused salt electrolysis in a melt containing a mixture of 70 mole % $PbCl_2$ (bal. NaCl; or 92 wt. % $PbCl_2$). Bath temperature was 500°C to ensure adequate fluidity. A phase diagram depicting the salt system is shown in Figure 5.

FIGURE 5

$NaCl$-$PbCl_2$ PHASE DIAGRAM

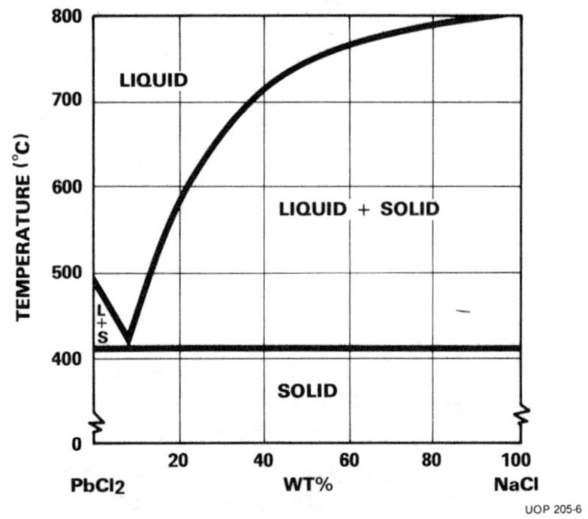

A schematic diagram of the laboratory test cell is shown in Figure 6.

FIGURE 6

LABORATORY FUSED SALT ELECTROLYSIS CELL

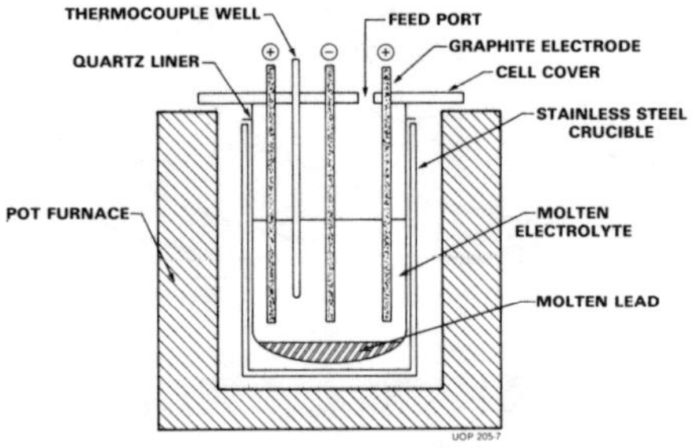

A binary salt system of $PbCl_2$ and NaCl was chosen so that impurity buildup in the fused salt bath could be controlled by recycling a portion of the electrolyte to brine leaching, thereby recovering both lead and NaCl.

Typical cell operation parameters and product purities are shown in Tables 4 and 5.

TABLE 4

TYPICAL OPERATING PARAMETERS FOR LABORATORY FUSED SALT ELECTROLYSIS

Voltage	2.2 V.
Current Density	720 A/ft^2
Current Efficiency	97.6%
Power Consumption	0.26 Kwh/lb Pb
Cathode Area	2 in^2
Electrode Spacing	1 in
Temperature	550°C

TABLE 5

PURITY OF ELECTROLYZED LEAD METAL PRODUCT

Element	Concentration (ppm)
Cu	5
Zn	1
Fe	5
Sb	1
Ag	10
Bi	0
As	0
S	3
Cl	10-20
Cd	4

Impurity Removal

Regardless of how selective the chlorination may be, some chlorination of impurities will occur, resulting in impurities being solubilized in the leach brine. Test studies were performed to remove silver, copper and iron impurities from the brine leach solution.

Silver and Copper

Silver and copper were removed from the recycle leach brine by cementation onto the lead powder. The reactions, ignoring complex chloride ions, can be written:

$$2Cu^{+} + Pb \qquad 2Cu + Pb^{+2} \qquad (4)$$

$$2\ Ag^{+} + Pb \qquad 2Ag + Pb^{+2} \qquad (5)$$

Figure 7 presents the results of cementation for a brine containing 1.8 g/l Cu, 0.25 g/l Ag. In 25 minutes the copper and silver concentrations were 11 and 3 ppm, respectively.

Cementation with iron was not tested but could be considered since iron removal from the brine was not difficult.

FIGURE 7

CEMENTATION ONTO LEAD POWDER

COPPER AND SILVER AT 75°C

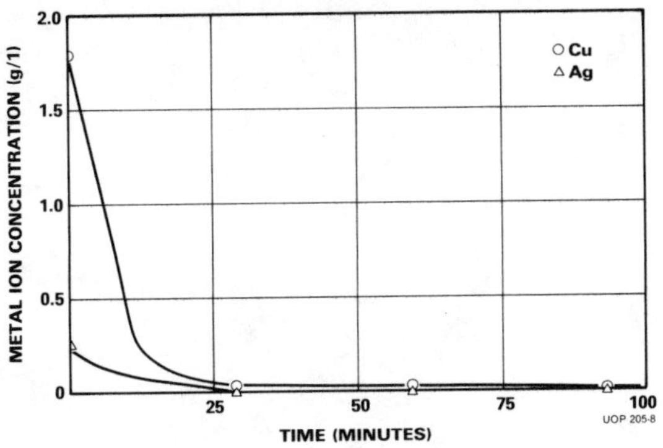

Iron

Iron was removed from the process brine using the "goethite" process for chloride solutions as described by Davey and Scott (8).

Iron chlorinated in the process was present as ferrous chloride in the solution during leaching. The unreacted sulfides in the brine pulp maintained the iron in the ferrous state. Because the iron was present in a reduced oxidation state, the "Vielle Montagne" process for producing goethite was used. In this process iron is oxidized to Fe(III) by air sparging at 55°C and precipitated at a controlled rate according to the equation

$$3H_2O + Fe^{+2} + \tfrac{1}{2}O_2 \quad 2FeOOH + 4H^+ \quad \text{(Goethite)} \qquad (6)$$

The goethite produced in our test solutions was a fine, easily filtered powder.

Optimum conditions for iron removal from the leach brine using this process were obtained at a temperature of 55°C and pH between 2 and 3.5, the lower being dictated by FeOOH solubility in acid, and the upper limit controlled by the solubility of $Fe(OH)_3$ which must be kept from precipitating to maintain good filterability.

Iron levels in the treated brine were maintained at concentrations of 0.8 g/l or less. This level was found to be satisfactory in further recycle tests.

Sulfur Recovery

Sulfur was extracted from the brine leach residue by solubilization with ammonium sulfide to form ammonium polysulfide. The ammonium polysulfide was then heated to vaporize ammonium sulfide, leaving solid, elemental sulfur. Sulfur extractions of 99.2% were achieved in this manner.

SUMMARY

A hydrometallurgical process for the production of pure lead from lead concentrates was presented.

Gaseous chlorination at controlled, low temperatures was found to be highly selective for lead sulfide, and was improved even more when an activation treatment was done prior to chlorination.

When fused salt electrolysis is included in the process, pure lead metal may be produced.

REFERENCES

1. Murphy, J. E., Haver, F. P., and Wong, M. M., USBM Report of Investigation 7913, "Recovery of Lead from Galena by a Leach Electrolysis Procedure" 1974.

2. Haver, F. P., Wong, M. M., USBM Report of Investigation 8105, "Ferric Chloride - Brine Leaching of Galena Concentrate" 1976.

3. Haver, F. P., et al, USBM Report of Investigation 8166, "Recovery of Lead from Lead Chloride by Fused Salt Electrolysis" 1976.

4. Malm, J. L., "Dry Chlorination of Ores" Metallurgical and Chemical Engineering," V. 12, No. 2, p.

5. Ionides, S. A., "The Dry Chlorination of Complex Ores", Mining and Scientific Press, V. 112, p/ 781-787, May 27, 1916.

6. Coltrinari, E. L., Reynolds, J. E., "Process for Separation and Recovery of Metal Values from Sulfide Ore Concentrates", US Patent 4,011,146, Mar. 8, 1977.

7. Reynolds, J. E., Goens, D. N., Kenney, C. W., "Pilot Plant Development of Chloride Processes for Lead-Zinc Concentrates", Lead-Zinc Update, D. O. Rausch, ed., SME-AIME, 1977.

8. Davey, P. T., Scott, T. R., "Removal of Iron from Leach Liquors by the "Goethite" Process", Hydrometallurgy, V. 2, p. 25-33, 1976.

Chapter 11

THE AMAX PORT NICKEL REFINERY PROCESS
FOR COBALT PURIFICATION AND RECOVERY

C. Nikolic*, J.L. Blanco**,
R. Crnojevich***, and W.G. Sherwood****

ABSTRACT

This paper describes the process used for cobalt purification and recovery at the AMAX Nickel Refining Co. plant in Port Nickel, Louisiana. In the first stage, cobaltic precipitate produced in the Outokumpu electrolytic nickel cobalt separation process is dissolved using methanol and sulfuric acid. Liquor produced in this step is converted to an ammoniacal complex, and sufficient ammonium sulfate is added to facilitate cobalt oxidation to cobaltic pentammine. Nickel is then selectively precipitated as nickel-ammonium sulfate double salt at pH 2.5. Final purification of the pentammine liquor is effected by pH adjustment with ammonia and removal of divalent nickel by ion exchange. Purified liquor is then advanced to hydrogen reduction to produce cobalt metal. Although similar to other cobalt refining processes used in the nickel industry, the Port Nickel process is unique because of a single point chemistry adjustment for a number of unit operations that occur in succession.

*Manager, Technical Services, AMAX Nickel Refining Co., Braithwaite, LA.

**V.P. Operations, AMAX Nickel Refining Co., Braithwaite, LA, USA.

***Superintendent, Process and Quality Control, AMAX Nickel Refining Co., Braithwaite, LA, USA.

****Section Supervisor, Extractive Metallurgy Research Department, AMAX Extractive Research and Development, Inc., Golden, Colorado, USA.

INTRODUCTION

The AMAX Port Nickel Refinery evolved from the installation built by Freeport Nickel Company in 1958-59 to refine nickel-cobalt sulfide "concentrates" from their laterite extraction plant at Moa Bay, Cuba. The present matte feed is entirely different from the nickel-cobalt precipitates for which the plant was designed, and the flowsheet has undergone substantial modification (see Figure 1). The electrolytic Outokumpu process for nickel-cobalt separation was installed to provide liquor of sufficient purity to produce Class I nickel. Cobaltic precipitate generated during the Outokumpu process is further refined (Figure 2) to provide low-nickel cobalt liquor, which is advanced to hydrogen reduction to produce cobalt metal.

PORT NICKEL
GENERAL FLOW DIAGRAM

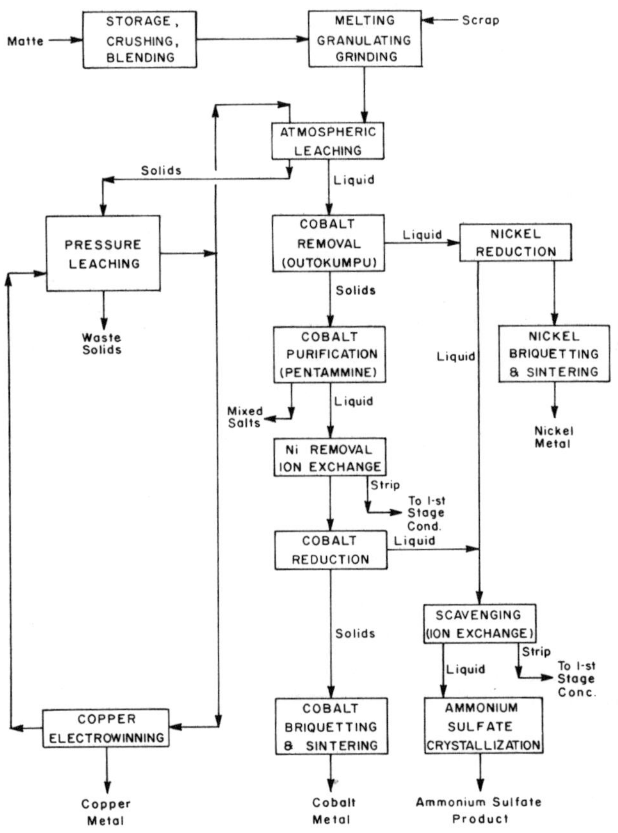

Figure 1

Although a detailed description of the nickel-cobalt separation process has been provided elsewhere[1], for better understanding of process requirements for recovering cobalt, several aspects of the separation process will be reviewed. This introductory discussion will be followed by a description of past and present Port Nickel cobalt refining practices and the laboratory research and development on which the processes were based.

Each stage of the cobalt recovery process performed on the cobaltic precipitate will be discussed, with particular emphasis on conversion of the precipitate to soluble cobaltic pentammine/nickel diammine complexes. Both laboratory and plant scale experiments have been performed using a direct ammoniation - autoclave dissolution approach and an acid-reductant pretreatment.

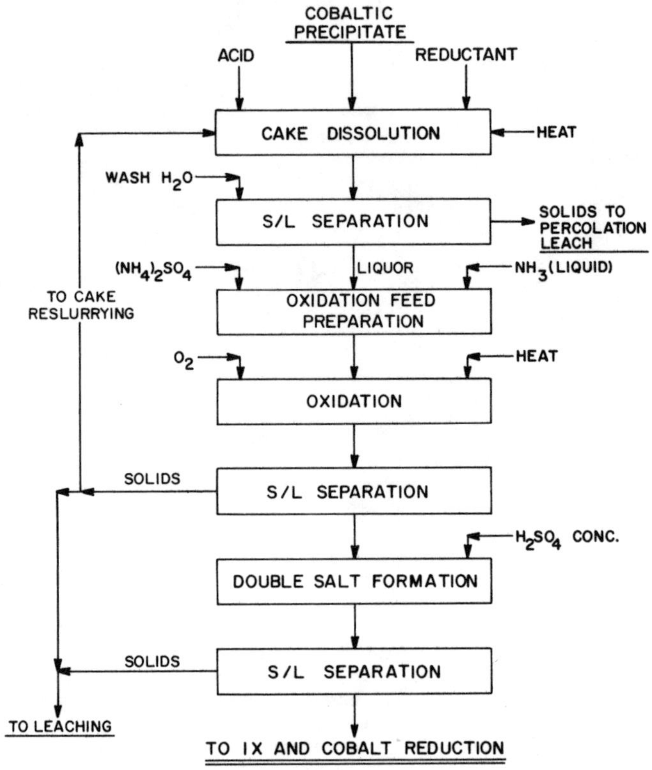

Figure 2. Cobalt Refining Flowsheet

A brief description of the nickel removal process, which utilizes selective nickel ammonium sulfate double salt precipitation and divalent cation exchange, is also given.

Finally, the chemistry and operating conditions for reduction of cobaltic pentammine to cobalt metal are described.

RESULTS AND DISCUSSION

The composition of the Outokumpu cobaltic precipitate is critical to the subsequent cobalt refining process. The chemistry of the precipitate formation and typical compositional analyses are described in the next section.

1. Cobaltic Precipitate Formation

Nickel sulfate liquor from the Port Nickel leach circuit, containing 100 gpl nickel and 1.5 gpl cobalt, is contacted with electrolytically generated nickel hydroxide at 65°C and pH 5.8, to yield the following general chemical reaction:

$$Co^{++} + Ni(OH)_3 \longrightarrow Co(OH)_3 + Ni^{++} \qquad (1)$$

In order to optimize the rate of cobalt precipitation, minimize consumption of nickelic hydroxide, and enhance product filterability, reactor pH and temperature must be carefully controlled. Several reactions take place simultaneously in the precipitation tanks:

$$Co^{++} + Ni^{+++}(solid) \longrightarrow Co^{+++}(liq) + Ni^{++}(solid) \quad (2)$$

$$Co^{+++} + 3\ H_2O \longrightarrow Co(OH)_3 + 3\ H^+ \quad (3)$$

$$NiSO_4 + 2NH_3 + 2H_2O \longrightarrow Ni(OH)_2 + (NH_4)_2SO_4 \quad (4)$$

$$m\ NiSO_4 + nH_2O \longrightarrow 2n\ Ni(OH)_2(m-2n)NiSO_4 + 2nH^+ \quad (5)$$

Reactions 2,4 and 5 can occur to the extent that the cobaltic precipitate contains up to 8 moles of nickel per mole of cobalt. Filter aid used in filtration of nickelic hydroxide also reports to this cake[1], as do iron, copper and lead.

With the exception of nickel, all impurities present in this cake are in the form of the metal's highest valent compound. Nickel is present in both the trivalent state, as a consequence of incomplete reduction of $Ni(OH)_3$ (Reaction 2), and in the divalent state due to hydrolysis reactions 4 and 5. The cobaltic precipitate has a tendency to incorporate large quantities of liquor, therefore, subsequent acid digestion and filtration are necessary to minimize the Ni/Co ratio in the precipitate that is advanced to the cobalt recovery circuit. After repulping in a bleedstream of return liquor from the precipitate dissolution process, the slurry has the following total composition:

Constitutent:	Co	Ni	Fe	Cu	Insoluble
Concentration (gpl):	60	20-60	0.50	0.1	50-80

After acid digesting the primary cobaltic precipitate, the majority of the unreacted nickelic oxidant has been removed and all nickel in the cobaltic slurry is in the entrained liquor as $NiSO_4$ and as basic nickel sulfate.

2. Cobaltic Precipitate Dissolution

General Considerations

Several major steps are involved in conversion of cobaltic cake to cobalt powder (Figure 2), the first of which is complete dissolution of cobalt and nickel. The solubility of cobaltic hydroxide has been studied in both acid[2] and ammoniacal[3,4,5,6] media; cobaltic ion is stable in both systems, as shown by the E_h-pH diagram in Figure 3. Based on thermodynamic considerations, conditions should exist under which cobaltic precipitate could be redissolved while avoiding cobalt reduction to the divalent state.

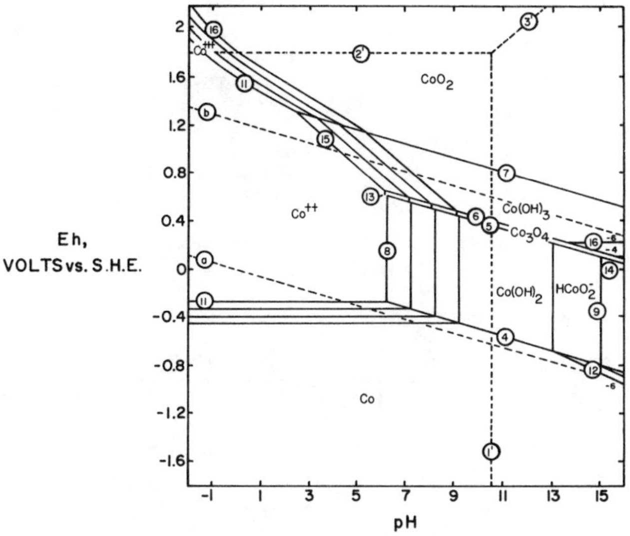

Figure 3. Cobalt E_h-pH Diagram

Dissolution in sulfuric acid yields cobaltic sulfate, as shown by Reaction (6). Dissolution in ammoniacal solutions, on the other hand, should yield a cobaltic ammine as shown by Reaction (7). When the dissolution is accomplished by reduction with, for example, an appropriate metal, acid must be added to the system, as shown by Reaction (8).

$$2\ Co(OH)_3 + 3H_2SO_4 \longrightarrow Co_2(SO_4)_3 + 6H_2O \quad (6)$$

$$2\ Co(OH)_3 + 4NH_3 + 3(NH_4)_2SO_4 \longrightarrow [Co(NH_3)_5H_2O]_2\ (SO_4)_3 + 4H_2O \quad (7)$$

$$2\ Co(OH)_3 + Co^o + 3\ H_2SO_4 \longrightarrow 3\ CoSO_4 + 6\ H_2O \quad (8)$$

Several reducing agents have been tested on a laboratory scale, including SO_2, H_2S, Ni^o, Co^o and CH_3OH. Cathodic reduction of cobaltic cake is also a viable alternative. Ammonia, ammonium sulfate and H_2SO_4 solutions have been tested as supporting electrolytes to perform the dissolution. Several approaches to dissolution of cobaltic precipitate have also been evaluated by AMAX on a plant scale:

1. Direct dissolution in ammonia-ammonium sulfate.
2. Dissolution using SO_2 as reducing agent.
3. Dissolution using methyl alcohol (methanol) as a reducing agent.

Direct Ammonia/Ammonium Sulfate Dissolution

Direct dissolution in an ammonia-ammonium sulfate solution is a simplistic approach to processing cobaltic precipitate.[7] Preliminary study of this system revealed that temperature, ammonia and ammonium sulfate concentrations are the major variables governing the efficiency of dissolution. Graphical presentations of the effects of these variables on cobalt extraction are given in Figures 4 and 5. Constant conditions in all tests incorporated in these graphs were:

Cobaltic cake	- Obtained from the secondary sulfuric acid step. 8.23% Ni, 8.24% Co, 8.6% filter aid and 49.2% moisture.
Test Charge	- Initial total cobalt concentration in the slurry was 20 gpl; 121 gms of cake in 500 mls.
Reaction Time	- 30 minutes to heat-up; 2 hours at temperature.
Oxygen Overpressure	- 7.0 Kg/cm^2
Ammonia Addition	- 2 M NH_3/M Ni

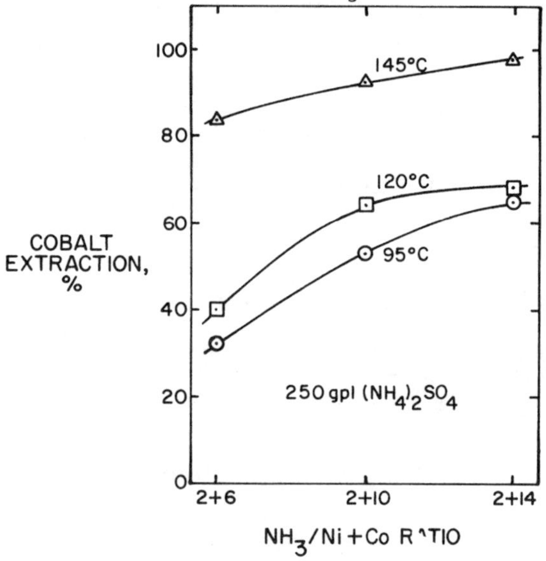

Figure 4. The Effect of Ammonia to Metals Ratio on Cobalt Extraction at 95, 120 and 145°C and 250 gpl Ammonium Sulfate.

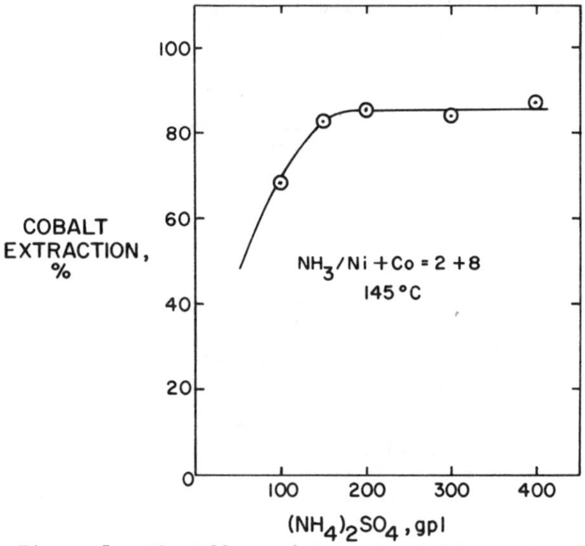

Figure 5. The Effect of Ammonium Sulfate Concentration on Cobalt Extraction.

Temperature is the variable having the most decisive effect on the direct dissolution process. Extractions of 90% or greater were observed only at temperatures of 145°C.

The ammonia to metals ratio is the second most important variable in determining Co recovery. An addition of ammonia corresponding to the molar ratio NH_3/Ni + Co of 2 + 10 (i.e. NH_3/total metals=6) should be adequate and is about the middle point of the optimum range. Increasing the molar ratio of ammonia to metals yields higher extractions at all temperatures.

The extraction of cobalt is not affected by increasing the ammonium sulfate concentration above 200 gpl. Below this level, cobalt extractions drop with decreasing concentrations of ammonium sulfate.

A full factorial experimental design program was performed with the three variables, and the results are shown in Figure 6.

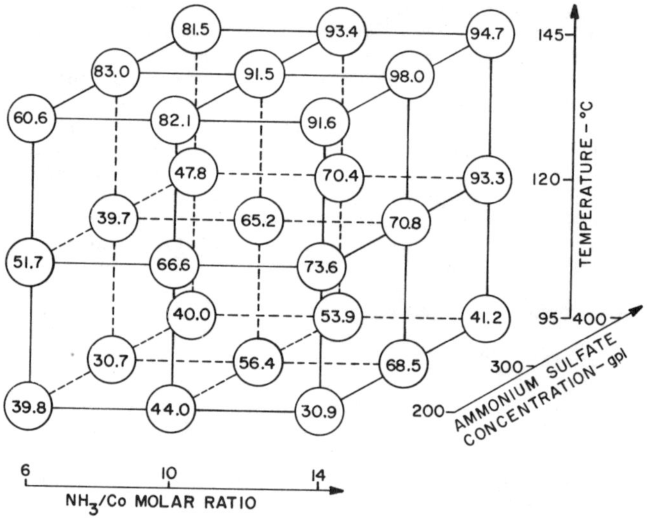

Figure 6. Tri-Dimensional Representation of Cobalt Extraction as a Function of Temperature, Ammonia and Ammonium Sulfate Concentration. Circled Numbers Represent % Cobalt Extraction.

Experimental results showed nickel extraction to be significantly greater than cobalt extraction. Extractions of 95% or higher were achieved for both metals in practically all tests. Filterability of the residues, obtained from tests in which 90% cobalt was leached, was adequate, but deteriorated rapidly with decreasing extraction levels. Dissolution of cobaltic precipitate occurs rapidly; 30% of the cobalt is dissolved by the time the system reaches temperature (Table I) and the reaction is essentially complete in about 30 minutes.

TABLE I

COBALT EXTRACTION AS A FUNCTION OF TIME

Temperature - 135°C, NH_3/Ni + Co = 2 + 10; $AmSO_4$ - 300 gpl

Initial cobalt concentration in the slurry - 20 gpl Co

Reaction Time (Minutes)	Test Results				
	Residue			Extraction	
	wt.(gm)	Ni(%)	Co(%)	Ni(%)	Co(%)
0	33.0	3.2	20.3	89.5	33.0
30	18.1	1.0	6.7	98.2	87.9
60	16.5	1.1	5.8	98.2	90.4
120	15.9	1.2	6.0	98.2	90.5

The effect of initial slurry cobalt concentration on the extent of cobalt extraction was studied in the range of 20-40 gpl Co. Under otherwise identical conditions (135°C; NH_3/Ni + Co = 2 + 10; 300 gpl, $AmSO_4$; 2 hours), feed cobalt concentration has no effect on the extent or rate of extraction.

Reduction/Dissolution with SO_2

The chemistry of dissolution of cobaltic precipitate with SO_2 as a reducing agent is identical to that in the presence of metallics or any other reducing agent:

$$2\ Co(OH)_3 + SO_{2(g)} + H_2SO_4 \longrightarrow 2\ CoSO_4 + 4H_2O \quad (9)$$

Controlled addition of both SO_2 and sulfuric acid is essential for proper oxidation of the resulting liquor in the subsequent step, which is identical to that described in the previous section. Filtration of the resulting slurry of $CoSO_4$ with filter aid, and corrosion hazards normally accompanying systems using SO_2 require accurate control of excess SO_2. The dissolution reaction can be brought to completion without H_2SO_4 addition because of the sulfurous acid that can be formed by reaction 10, which requires excess SO_2:

$$SO_2 + H_2O \longrightarrow H_2SO_3 \quad (10)$$

At least partial conversion of Co and Ni to sulfites is unavoidable under such conditions. High residual SO_2 in solution, poorer reaction rate and difficulties with handling subsequent unit operations are the major disadvantages of the SO_2 approach.

When sulfuric acid is added to the system simultaneously with SO_2 as shown by Reaction 9, the above problems are essentially eliminated. Dissolution rate, in the presence of sufficient sulfuric acid, is practically instantaneous and is only dependent on the rate of SO_2 addition. The theoretical requirement of SO_2 is 0.54 pounds SO_2 per pound Co^{3+}. The changes in pH during the course of dissolution as a function of SO_2 addition at various concentrations of sulfuric acid are shown in Figure 7.

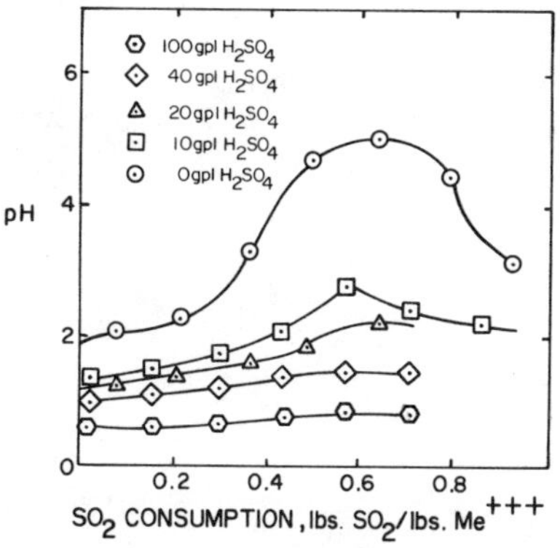

Figure 7. pH Change as a Function of Sulfuric Acid Concentration and Sulfur Dioxide Addition.

As shown in Figure 8, the redox potential can be used to determine the SO_2 reduction end point. A potential drop of about 600 mV occurs when the reaction is complete. If the proper quantity of H_2SO_4 is used, SO_2 consumption is stoichiometric, as shown in Reaction 9.

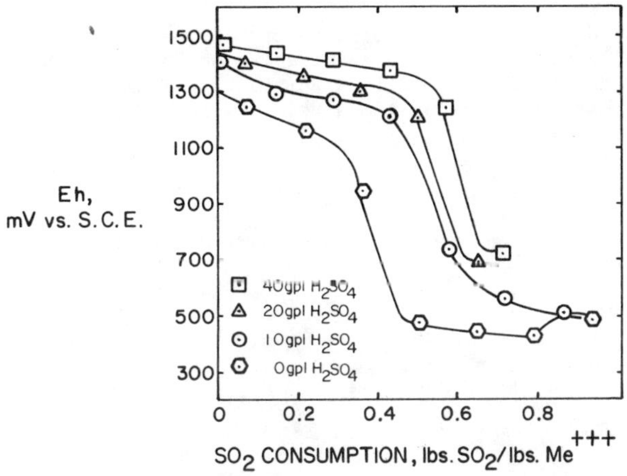

Figure 8. Redox Potential as a Function of the Consumption of Sulfur Dioxide.

Consumption of SO_2 per pound of trivalent feed metals at various levels of H_2SO_4 is shown in Table II. Cobalt losses to the filter aid residue occur as a result of incomplete dissolution caused by the use of SO_2 only or insufficient sulfuric acid to satisfy the stoichiometry of Reaction 9.

However, the problem of corrosion of downstream equipment, which was originally designed to handle ammoniacal liquors, could not be eliminated. In addition, residual SO_2 yields numerous sulfur species in various oxidation states. These compounds inhibit subsequent cobalt reoxidation (see Section 3) and also produced degradation of the divalent cation exchange resin used for nickel scavenging (see Section 4). Because of these difficulties, alternate reductants were evaluated and the SO_2 process was abandoned in April of 1977.

TABLE II

THE EFFECT OF ACID ADDITION ON SO_2 CONSUMPTION

Sulfuric Acid Addition	Initial pH	Final SO_2 pH	Consumption $g\ SO_2/g(Ni^{+++} + Co^{+++})$
gpl H_2SO_4			
0	1.9	5.0	0.78
10	1.2	2.8	0.63
20	1.1	1.9	0.55
40	1.0	1.5	0.55
100	0.3	0.8	0.55
Stoichiometric			0.544

Dissolution of Cobaltic Precipitate Using Methanol or Formaldehyde

The chemistry of cobaltic precipitate dissolution with methyl alcohol has been studied extensively in the Port Nickel Research and Development Laboratory.[8] We believe that the dissolution proceeds according to Reactions 11, 12 and 13.

$$2Co(OH)_3 + CH_3OH + 2H_2SO_4 \rightleftharpoons H_2CO + 2CoSO_4 + 6H_2O \quad (11)$$

$$2Co(OH)_3 + H_2CO + 2H_2SO_4 \rightleftharpoons HCO_2H + 2CoSO_4 + 5H_2O \quad (12)$$

$$2CO(OH)_3 + HCO_2H + 2H_2SO_4 \rightleftharpoons CO_2 + 2CoSO_4 + 6H_2O \quad (13)$$

Based on these reactions, about 0.09 pounds of methanol would be required to dissolve one pound of cobalt in the form of cobaltic precipitate. Table III is a summary of reaction times required as a function of methanol dose.

TABLE III

COBALTIC PRECIPITATE DISSOLUTION WITH METHYL ALCOHOL

Total Retention Time - 60 minutes

Test Conditions		Test Results				
Methanol Addition (lbs Methanol per lb Co)	Time Needed* to Complete Dissolution (min)	Residue Wt. g	Residue Ni %	Co %	Extraction Nickel %	Extraction Cobalt %
0.09	50	22.5	.38	.30	>99	99.3
0.18	40	19.6	.47	.37	>99	99.3
0.36	30	15.5	1.23	.97	>99	98.5
0.72	20	11.6	.60	.48	>99	99.4

*Residual filter aid turns white.

Similar dissolution tests performed with formaldehyde (Table IV) indicate that formaldehyde is as effective as methanol as a reductant; stoichiometry was also maintained in the case of formaldehyde, (0.14 lbs H_2CO/lb Co). When used directly, formaldehyde may be lost to the vapor space because of its very low boiling point (-21°C vs. 65°C) for methanol.

TABLE IV

COBALTIC PRECIPITATE DISSOLUTION WITH FORMALDEHYDE

Total Retention Time - 30 minutes

Test Conditions		Test Results						
Formaldehyde Addition g Formaldehyde per g Co^{+++}	Time Needed* to Complete Dissolution (min)	Residue Wt. g	Residue Ni %	Residue Co %	Residue Cu %	Residue Fe %	Extraction Ni %	Extraction Co %
0.06	30	18.2	1.48	9.27	.02	0.08	97.8	83.1
0.13	20	13.0	.46	.89	< .01	0.04	99.5	98.8
0.26	4	13.1	.09	.08	< .01	0.04	99.9	99.0
0.51	3	12.6	.08	.06	< .01	0.04	99.9	99.3

* Residual filter aid turns white.

When precipitate dissolution is complete, the resulting slurry, containing the filter aid as a solid phase, is filtered on a string filter. The cake from this step is reslurried and filtered on a drum filter to increase the recovery of cobalt. Final residue-filter aid is sent to a tailings pond where entrained cobalt is dissolved and recycled to further improve the recovery. Filtrate is returned (as required) to the Ni/Co separation area for repulping of cobaltic cake; the major portion of the filtrate is advanced to cobalt oxidation feed preparation (see next section).

3. Cobaltic Pentammine Formation

In the process of direct cobaltic cake dissolution, cobaltic pentammine is a reaction product. In any of the other process variations involving reduction of cobaltic cake for dissolution, reoxidation of cobaltous ion to cobaltic pentammine is required.

Following filter aid removal, as described above, the cobalt/nickel sulfate solution (typical composition shown in Table V) is treated by ammonia and ammonium sulfate addition in preparation for pentammine formation. Ammonia is added in quantities to satisfy the following molar ratios $NH_3/Co = 5$ and $NH_3/Ni = 2$. Ammonium sulfate is added according to the stoichiometry of the pentammine formation reaction:

$$2CoSO_4 + 8NH_3 + (NH_4)_2SO_4 + \tfrac{1}{2}O_2 \longrightarrow [Co(NH_3)_5H_2O]_2(SO_4)_3 \quad (14)$$

There is no ammonium sulfate addition for nickel; it is allowed to form the ammonium sulfate from its own ammine complex in the double salt formation step.

$$Ni(NH_3)_2SO_4 + H_2SO_4 + nH_2O \longrightarrow NiSO_4(NH_4)_2SO_4nH_2O \quad (15)$$

Excess ammonium sulfate required for stabilization of the cobaltic pentammine as well as suppression of nickel to <1 gpl in the double salt crystallization step is also added at this point (Figure 2). The target is always set to provide 200-250 gpl $(NH_4)_2SO_4$ in the cobaltic pentammine solution after the doublt salt crystallization step, as determined from the previously described factorial design experiments.

The Port Nickel process uses an NH_3/Co ratio in the oxidation step of 3.6-4.2 depending on the Ni/Co ration in the cobaltic precipitate. Oxidation of cobaltous ion to cobaltic pentammine is accomplished by oxygen addition in an autoclave. Because of the specific ion exchange step used in the final nickel removal from the cobaltic pentammine solution, oxidation has to be complete. The requirement is to oxidize 99.5% of the cobalt. The unoxidized cobalt will either crystallize as a cobalt-ammonium double salt with the nickel double salt, or it will be loaded on the IX resin. There is an increase in cobalt recycle to the matte leach circuit in either case.

In the process of oxidation, some cobalt precipitates out with iron and other minor impurities. After these impurities are removed on pressure-leaf filters, the solution is ready for double salt removal.

TABLE V

COBALT PROCESS STREAM COMPOSITIONS
(gpl or %)

Element	Feed Slurry	Filtrate	Oxidation Autoclave Feed	Double Salt Filtrate	Double Salt
Co	60	50	40	35	1.1%
Ni	40	35	28	.600	12.5%
N	2.5	-	120	-	-
Fe	.5	-	-	-	-
pH	2.5	1.0	-	2.6	-

4. Nickel Removal

Nickel is separated from the pentammine liquor in two steps. In the first step, the solution is acidified with H_2SO_4 to a pH of 2.4-2.6 and then cooled to about 32°F to crystallize out the nickel as nickel-ammonia sulfate double salt (Reaction 15).[9,10,11] When the nickel concentration drops to about 600 to 800 ppm, the slurry is centrifuged, and the centrate is advanced to the second nickel removal step. The crystalline double salt is recycled to the matte leaching circuit. The resultant cobaltic pentammine solution, containing less than 800 ppm Ni, (as shown in Table V) is ammoniated to pH 8, which is suitable for IX removal of divalent metal ions. The resin used is a Rhom and Haas product (XE-318) that is highly selective for divalent ion. Metal values are stripped with a sulfuric acid solution, which is returned to the leaching section. The purified cobaltic pentammine solution is advanced to the cobalt reduction area.

5. Cobalt Reduction

Cold cobaltic pentammine solution is advanced to the hydrogen reduction autoclaves to be heated with steam prior to introducing hydrogen gas. In the process of heating the solution, pre-reduction of the cobaltic complex with cobalt powder occurs according to the generalized reaction:

$$2Co(NH_3)_5H_2O^{+++} + Co^o \longrightarrow 3Co(NH_3)n^{++} + (10-3n)NH_3 + H_2O \quad (16)$$

The pre-reduction is practically complete before preheating is finished (i.e. within 45 minutes).

At this point, hydrogen gas is introduced into the autoclave and the reduction takes place. Although the cobalt reduction rate is slowed by the excess free ammonia present, the approach using rapid pre-reduction in the autoclave still offers the advantage of extending the operating times between cobaltic precipitate leaches.

Nucleation runs are performed with a normal autoclave charge of $9.5m^3$. The reduction conditions used are:

Temperature	-	190 - 205°
Pressure	-	32 Kg/cm^2
NH_3	-	not controlled in this step

Cobalt powder produced at Port Nickel is somewhat coarser (+40 mesh) than that produced in other processes, due to simultaneous preheating and pre-reducing in the autoclave. After washing and drying under a H_2 atmosphere, the cobalt powder is packaged in 250 kg drums for marketing, primarily to the chemical industry.

LITERATURE

1. Simons, C. S., Nikolic, C., Saarinen, H., Sherwood, W. G., Cobalt-Nickel Separation by the Outokumpu Process; Operations at the Port Nickel Refinery of AMAX Nickel Refining Company, 7th Annual Hydrometallurgical Meeting Conference of Metallurgists of CIM Vancouver. (Unpublished)
2. Pourbaix, M., Atlas of Electrochemical Equilibria in Aqueous Solutions, 1974, NACE, page 325.
3. Leshch, I. Y., Frumina, L. M., Bernovskaya, F. M., Scneerson, Y. M., J. of Applied Chem. of USSR, 43, 1970, p. 1683-7.
4. Shauffelbergh, F. A., U.S. Patent 2767054 (Oct. 16, 1956).
5. Kunda, W., et al, Canadian Institute Min. Met, 55 25 (1962).
6. Van Hare, G. F., U.S. Patent 2728636 (Dec. 27, 1955).
7. Nikiloc, C. B., Zundel, W. P., Rickard, R. S., U.S. Patent 3933975, Jan. 20, 1976.
8. Crnojevich, R., Wiewiorowski, E. D., Wilkinson, D. H., U.S. Patent 4151258 (March 6, 1978).
9. McGauley, P. J., U.S. Patent 2647828 (Aug. 4, 1953)
10. Forward, F. A., U.S. Patent 2647820 (Aug. 4, 1953)
11. Benoit, R. L. U.S. Patent 2822264 (April 11, 1956)

Section III

Hydrometallurgical Process for By-Product Recovery

Malcolm L. Jansen, Section Editor

Chapter 12

THE ELECTROSLURRY PROCESS FOR COPPER RECOVERY FROM SMELTER AND REFINERY WASTES

F.A. Baczek
Program Manager

B.C. Wojcik
Development Engineer

D.M. Lewis
Development Engineer

R.C. Emmett
Sr. Process Consultant

Eimco PMD Division
Envirotech Corporation
Salt Lake City, Utah

ABSTRACT

The ElectroSlurry™ electrowinning concept has been incorporated into a process flowsheet for recovering copper from smelter flue dusts. The process has averaged 96% recovery of the copper as high-grade electrowon cathode, while at the same time removing a majority of the impurities that degrade drawn copper as a result of recycling flue dust and liberator cathodes. The dust is processed through several stages consisting of leaching, copper precipitation, ElectroSlurry™ electrowinning, and waste liquor treatment. The process may be integrated into a smelter operation and become an effective and feasible means of maintaining good product purity.

THE ELECTROSLURRY* PROCESS FOR COPPER RECOVERY FROM SMELTER AND REFINERY WASTES

F. A. Baczek
Program Manager

B. C. Wojcik
Development Engineer

D. M. Lewis
Development Engineer

R. C. Emmett
Sr. Process Consultant

Eimco PMD Division
Envirotech Corporation
Salt Lake City, Utah

INTRODUCTION

During the past four years, the Eimco PMD Division of Envirotech Corporation has been developing a new approach for the hydrometallurgical recovery of copper from concentrates (1). The product is an electrowon cathode produced at 1076 A/m^2 (100 A/ft^2).

A unique feature of the process is the electrowinning cell which handles a slurry of copper bearing solids and electrolyte. Within the cell, leaching of copper solids (by means of an oxidant continuously regenerated at the anode) and electrowinning take place simultaneously. Thus, there is use of the energy at the anode which is normally lost to oxygen generation. There are no diaphragms separating an anolyte and catholyte, and solids are maintained in suspension by recirculation of the slurry. Under proper operating conditions, the cathode deposit produced is smooth and virtually free of occluded or plated impurities. Current efficiencies are in the range of 70-75% and power consumption averages 2.86 kwh/kg Cu (1.35 kwh/lb Cu).

The electrolyte is composed of copper sulfate, ferrous and ferric sulfates, sulfuric acid, and minor amounts of chloride. The concentration of these components can be varied over preferable operating ranges which are shown in Table 1. The solids feed into the cell and current density are controlled so that the copper leach and deposition rates are equal; thus, the component concentrations in the electrolyte do not change significantly. After a suitable retention time, the slurry exits the cell circuit and goes to a liquid-solid separation step. The electrolyte is returned to the feed end of the cell circuit and the residual solids go to by-product recovery. The cell operation is shown schematically for covellite in Figure 1.

Additional work has been directed toward developing other processes for recovering copper which incorporate the slurry electrowinning concept. Flowsheets have been defined and tested for chalcopyrite, simple sulfide concentrates (such as chalcocite), cement copper precipitates, and copper smelting flue dusts.

Investigations into processing flue dusts were begun when several copper companies expressed a need for alternative methods of handling the dusts which are an environmental and operating nuisance. The dusts easily become airborne and, in addition to copper, they contain various elements which are hazardous or degrade the final copper product. An alternative would be preferred that would recover the copper, remove the impurities, and condition the residues to be suitable for disposal.

Some present operations recycle the dusts to the smelter to recover the copper, however this also returns impurities such as arsenic, lead, antimony, and bismuth. This increases the probability that they will eventually show up in the end product. Also, large concentrations of dust in the flue gases will tend to foul waste heat boilers. Storage of rejected dust requires that provisions be made for containment to prevent the dusts from becoming airborne or leached into ground water. Selling of dusts to an independent processor has rarely been a satisfactory solution.

*"ElectroSlurry" is an Envirotech Corporation service mark for the design, engineering, production, technical knowhow and licensing of the hydrometallurgical treatment facilities.

Hydrometallurgical treatment of the dusts is an approach being investigated by many copper smelters. This can be effective, not only for removing the dusts from the smelting process and recovering copper, but also for purging impurities. Eimco PMD has developed and investigated techniques for precipitating copper from leach liquors that achieve good separation of copper from other elements such as iron, zinc, arsenic, antimony, bismuth, and molybdenum.

These precipitation methods are not only applicable to recovery of copper from flue dusts, but also the recovery of copper from tankhouse bleed electrolytes. For example, copper can be removed from partially stripped electrolyte by precipitation onto chalcopyrite green feed or elemental sulfur rather than deposited to form a highly impure liberator cathode. This approach, by eliminating electrolytic stripping avoids formation of arsine.

PROCESS DESCRIPTION AND CHEMISTRY

The basic process flowsheet, shown in Figure 2, consists of four main sections:

1) A leach section for the dissolution of copper,
2) A copper precipitation section for the separation of copper from the other components in the leach liquor,
3) A slurry electrowinning section for recovery of copper as an electrowon cathode,
4) A waste treatment section for treating waste streams for disposal.

The Leach Section

The purpose of the leach section, shown in Figure 3, is to dissolve copper from the dust. Three-stage, countercurrent sulfuric acid leaching is used because the majority of the dusts tested require a strong acid concentration (200 g/l) to achieve a copper recovery greater than 70%. Copper recoveries above 95% were obtained when ferric sulfate was added to the third stage.

The first stage is a weak acid leach of the readily soluble compounds, including copper sulfates and oxides. The second stage is a strong acid leach which consumes excess ferric sulfate from the third stage. The third stage is an oxidizing leach for that portion of the dust which consists of copper sulfides and the more refractory copper compounds. The average operating conditions are shown in Table 2 along with the results obtained for each stage. Slurry concentrations throughout the section average 15-20% solids. Sulfuric acid can be supplied from tankhouse bleed electrolytes or other sources.

When liberator electrolyte is used to supply acid to the process, the copper it contains is recovered in the subsequent copper precipitation section. Consequently, impurities which would co-deposit with copper in the liberator cathode are purged from the smelter circuit. Ferric sulfate can be provided by in-situ oxidation of the ferrous sulfate, normally in the range of 5-10 g/l of Fe^{+2} in the third stage, by oxygen sparging. The reaction proceeds according to Equation 1.

$$4FeSO_4 + O_2 + 2H_2SO_4 \rightarrow 2Fe_2(SO_4)_3 + 2H_2O \qquad (1)$$

Oxygen use efficiency is close to 95% as long as the Fe^{+3}/Fe^{+2} mole ratio is no greater than 0.8. If this ratio is not exceeded throughout the third stage, oxygen and acid requirements for ferric sulfate generation are about 0.15 kg O_2 and 0.88 kg H_2SO_4 per kg Fe^{+3} required.

Other metals and elements in the flue dust are leached along with the copper. Table 3 shows the leach of these other components for three typical flue dusts. Precious metals such as gold and silver contained in the feed dust are not extensively leached and concentrate in the solid leach tailings. Lead is only slightly soluble in the sulfate system. Typical analyses of the leach residues are shown in Table 4.

Underflow from the third stage thickener is easily filtered and washed on a horizontal belt filter. The filtrate and wash are combined with the thickener overflow and go to the second stage acid leach. The filtered leach solids can be disposed or processed further for recovery of additional values.

The Copper Precipitation Section

The purpose of the precipitation section, shown in Figure 4, is to separate copper from the other components of the leach liquor and to produce a feed material that is suitable for cell feed. Precipitation of CuS and Cu_2S with sulfur and sulfur dioxide accomplishes this requirement (2, 3, 4, 5). Equations 2 and 3 show the reactions.

$$CuSO_4 + S^o + SO_2 + 2H_2O \rightarrow CuS + 2H_2SO_4 \qquad (2)$$

$$CuSO_4 + CuS + SO_2 + 2H_2O \rightarrow Cu_2S + 2H_2SO_4 \qquad (3)$$

X-ray diffraction analysis of the product shows reaction 2 to be predominant.

The leach liquor feed to the precipitation section contains 15-45 g/l Cu^{+2} and 10-15 g/l H_2SO_4. Elemental sulfur, which is recycled from the slurry electrowinning circuit, is added in an amount equal to 95 to 100% of the stoichiometric requirement shown in Equation 2. Sulfur dioxide gas is added to provide good gas dispersion. Any excess SO_2 contained in the final waste liquor can be recovered by sparging with live steam and directing the resulting off gas back to the reactors. Air stripping cannot be used because the presence of oxygen with SO_2 slows the reaction rate and increases the SO_2 requirements. SO_2 removal is also facilitated by the high acid concentration of the waste liquor, which lowers the SO_2 solubility.

The reaction is carried out at an initial maximum pH of 1.0 to keep the precipitation selective for copper and leave impurities such as arsenic, bismuth, antimony, iron, and zinc in solution. The behavior of various impurities is shown in Table 5. Sulfuric acid, a product of the reaction, causes a slow rate of reaction as its concentration increases, and it is difficult to reach completion. This is especially true when initial copper concentrations in the area of 45 g/l results in a final sulfuric acid concentration of 150 g/l. To maintain a good rate of reaction and obtain complete copper recovery, a portion of the CuS precipitate product is recycled back to the reactors. The precipitate is recycled to provide a ratio of 2 grams of precipitate per gram of copper in solution. The reaction is carried out at 90°C and requires a retention time of 30 to 45 minutes. The effects of recycled CuS on the rate of reaction is shown in Figure 5. X-ray diffraction analysis of the product produced with recycle shows CuS is the major product with minor amounts of S^o present. The product averages 58% copper.

The product slurry leaving the reactors is thickened and filtered. Filtration is very rapid and therefore a good wash is possible. A final cake liquor content of 10 to 12 wt% should be obtained after a short air dry period. The filtrate and wash are combined and go to the waste neutralization section and the cake is repulped with electrolyte and sent to the slurry electrowinning section.

The Slurry Electrowinning Section

The flowsheet for the slurry electrowinning section is shown in Figure 6.

The repulped CuS precipitate cake is further diluted to 6-7% solids with recycle electrolyte and fed to the primary cells. The current density in this section is maintained at 1076 A/m^2. Retention time is approximately 2.5-3.0 hours. Overflow from primary cells goes to secondary cells operated at 860-915 A/m^2 where retention time is approximately 1.0-1.5 hours. Residual solids exiting the secondary cells contain 10% copper and 90% sulfur. Electrowon copper cathodes are produced at 75% current efficiency and require 2.97 kwh/kg Cu.

In a commercial operation, each cell will produce an average of 1.8 metric tons/day of copper with an operating cycle of 36 to 48 hours. Table 6 shows the average analysis of cathodes produced during bench and scale-up testing. The lead concentration is high due to the non-continuous operation of test cells. The shutdown of the cells between tests causes the formation of lead sulfate on the anode surface which contaminates the copper deposit in subsequent tests. The lead concentration is expected to decrease in continuous operation. The high levels of sulfur and oxygen represent tests which studied the effect of high current densities during the initial 1/2 hour of deposition. In actual operation, the current density is gradually increased to the operating level over the first 1/2 hour to obtain a coherent base deposit and increase the ferric iron concentration to 2-3 g/l before copper solids are fed to the cell.

Figures 7, 8, and 9 show the basic cell design and layout of a commercial 1.8 metric ton/day unit. Cathodes are constructed from Grade 2 titanium and anodes are made of lead alloyed with 8% antimony and 0.4% arsenic. Recirculation and agitation is accomplished with an axial flow pump which produces large flow rates at low head. A single commercial cell will require 26.5 m^3/min (7000 gpm) at 1.22-1.53 m (4' to 5') head; this represents a pumping power requirement of 0.154 kwh/kg Cu.

Flow from the cells goes to thickening and filtration. Countercurrent washing on a horizontal belt filter is used to keep water from diluting the electrolyte which is returned to the electrowinning cells. The degree of wash can be adjusted to allow various amounts of electrolyte to remain with the residual solids for the purpose of maintaining a certain amount of impurity bleed. Solids are returned to the copper precipitation circuit to supply sulfur for copper sulfide precipitation. If required, solids may be bled from the system through the leach circuit to recover the copper.

The Waste Treatment Section

The waste treatment section, shown in Figure 10, is used to neutralize the waste liquor to a pH suitable for disposal in a containment area. Limestone is used for the initial neutralization, and final pH adjustment is accomplished with slaked lime. Ferrous iron is oxidized to ferric iron with air to facilitate complete iron precipitation. A portion of the precipitated solids is recycled for the purpose of increasing particle size and making the precipitate easier to handle in liquid-solids separation equipment.

This section will require optimization for each particular dust processed. Additional treatment will be required to further insolubilize toxic metals such as arsenic. The typical composition of final liquors after waste treatment is shown in Table 7.

ALTERNATIVE PRECIPITATION METHOD

It is also possible to recover the copper from the leach liquor by precipitation with chalcopyrite concentrate and SO_2. The precipitate is then sent to the slurry electrowinning cell or to the smelter for copper recovery.

When chalcopyrite is used, it takes the place of sulfur in the flowsheet (Figure 4). Sulfur costs approximately $0.41/kgm Cu ($0.0186/lb Cu) at $82.75/m. ton ($75/ton). The use of slurry electrowinning allows the sulfur to be recovered as a by-product and to be subsequently reused in the precipitation; thus the cost of supplying new sulfur is eliminated. However, chalcopyrite can be considered to have no cost since it is already available at the smelter; thus by-product recovery is not necessary to save costs, and the precipitate can be sent to the smelter.

In the precipitation section, leach liquor is mixed with a portion of the chalcopyrite concentrate going to the smelter, and the resulting slurry is sparged with SO_2. The copper sulfate in solution reacts with chalcopyrite to form a variety of copper sulfide compounds. Iron is leached into solution and sulfur dioxide is oxidized to sulfuric acid. Some of the reactions possible are shown in Equations 4, 5, and 6.

$$CuFeS_2 + 3CuSO_4 + 2SO_2 + 4H_2O \rightarrow 2Cu_2S + 4H_2SO_4 + FeSO_4 \qquad (4)$$

$$CuFeS_2 + 2.6CuSO_4 + 1.6SO_2 + 3.2H_2O \rightarrow 2Cu_{1.8}S + 3.2H_2SO_4 + FeSO_4 \qquad (5)$$

$$CuFeS_2 + 1.25CuSO_4 + .25SO_2 + .5H_2O \rightarrow 1.7CuS + .3Cu_{1.8}S + FeSO_4 + .5H_2SO_4 \qquad (6)$$

The amount of co-precipitation of other elements with this method of copper recovery is still being studied. The results depend on the particular concentrate, pH, $Cu^{+2}/CuFeS_2$ mole ratio, and particle size of the concentrate. Research is also continuing in this area because it has good potential for making hydrometallurgical processing of chalcopyrite less difficult in a sulfate system.

SUMMARY

Laboratory and semi-continuous pilot plant investigations have shown that the ElectroSlurry™ process can be incorporated into existing smelter operations for the purpose of recovering copper from flue dusts and tankhouse bleed electrolytes. The process offers the additional advantage of providing a means for purging impurities which degrade the final copper products.

The ElectroSlurry™ cell has the ability to produce good quality copper at high current density with reasonably low power consumption (2.86 kwh/kg Cu at 1076 A/m^2). The process has simple chemistry and achieves high copper recoveries without the use of high pressures or temperatures above ambient boiling points. The sulfate system minimizes corrosion problems associated with chloride based hydrometallurgical processes.

Chalcopyrite has been used as an alternative precipitation reagent and shows a potential for making hydrometallurgical processing of chalcopyrite in a sulfate system less difficult. This technique is under investigation as applicable to the ElectroSlurry™ processing of chalcopyrite.

Work on the additional stabilization of toxic metal precipitates produced in waste treatment is required and will depend on the means of final disposal and local regulations. Methods of recovering other metals from the flue dusts will depend on the characteristics of the particular dust and costs involved.

ACKNOWLEDGEMENTS

The authors would like to thank Alex Jueschke, Jack Otto, Larry Tuttle, Bob Cook, and Jean Kees of Eimco PMD for their important contributions to this effort.

REFERENCES

1. Emmett, R. C., Baczek, F. A., and Wojcik, B. C., "The ElectroSlurry™ Process in Direct Electrowinning of Copper from Concentrate", 1976 SME-AIME Fall Meeting, September 1-3, 1976, Denver, Colorado.
2. Kunda, W., and Mackiw, V. N., "Sulphur and Sulfur Dioxide for Separating Copper from Nickel-Cobalt-Copper Sulfate Solutions", The Canadian Journal of Chemical Engineering, Vol. 39, Dec. 1961, pp. 260-264.
3. Kunda, V., Pearce, R. F., and Mackiw, V. N., "Process for the Precipitation of Metal Values from Solution", U. S. Patent 3,218,161,1965.
4. Hall, R. A., Jones, W. E., O'Neill, C. E., and Illis, A., "Separation of Copper from Arsenic, Antimony or Bismuth in Aqueous Solutions and Slurries", Canadian Patent 1,020,363.
5. Clitheroe, J. B., and Lacy, F. H., "Method for Processing Copper Values", U. S. Patent 3,728,430,1973.

Table 1. Composition of the Electrolyte

Component	Concentration Range
Copper	30 - 40 g/l
Total Iron	30 - 40 g/l
Ferric Iron	1.5 - 3.0 g/l
Sulfuric Acid	100 - 120 g/l
Chloride	150 - 300 mg/l

Table 2. Average Operating Conditions and Copper Recovery Achieved for the Three-Stage, Countercurrent Leaching of Copper Smelting Flue Dusts

	Stage 1	Stage 2	Stage 3
Initial H_2SO_4 Conc., g/l	20-70	85-120	200-150
Final H_2SO_4 Conc., g/l	10-15	25-75	90-125
Initial Fe^{+3} Conc., g/l	∿0	3-6	15-40
Final Fe^{+3} Conc., g/l	∿0	∿0	4-7
Retention Time, HR	0.5-1.0	0.75-2.0	3.0
Temperature, °C	80	80	80
Dust Solubilized, % of Feed	53-40	55-58	71-78
Cu Leached, % of Feed	25-67	65-78	97-98

Table 3. Percent Leach of Various Components For Three Typical Flue Dusts

	Flue Dust A		Flue Dust B		Flue Dust C	
Element	% In Dust	% Leached	% In Dust	% Leached	% In Dust	% Leached
Cu	25.1	96.9	18.4	97.9	16.0	98.0
Fe	25.6	59.4	11.3	83.0	6.4	53.2
As	0.07	75.0	6.3	98.4	10.1	96.0
Zn	2.1	95.0	4.8	92.1	3.7	97.8
Bi	0.11	16.5	0.23	86.1	1.9	20.6
Sb	0.11	80.2	0.23	21.7	0.2	20.0
Mo	0.48	71.4	0.24	68.7	1.8	92.8

Table 4. Analysis of Leach Residue for Three Typical Flue Dusts

Element	Flue Dust A %	Flue Dust B %	Flue Dust C %
Cu	3.9	1.3	0.85
Fe	52.0	6.6	8.9
As	0.089	0.34	1.2
Pb	3.8	11.5	45.2
Zn	0.5	1.3	0.24
Bi	0.48	0.11	4.4
Sb	0.109	0.61	0.43
Mo	0.69	0.26	0.39

Table 5. Percent Precipitation of Leach Liquor Components

Component	Flue Dust A: Conc. In Leach Liquor, g/l	Flue Dust A: % Ppt'd.	Flue Dust B: Conc. In Leach Liquor, g/l	Flue Dust B: % Ppt'd.	Flue Dust C: Conc. In Leach Liquor, g/l	Flue Dust C: % Ppt'd.
Cu	45.0	99.0	16.0	99.3	30.1	98.0
Fe	28.5	0.008	8.3	0.01	7.1	0.02
As	0.097	0.1	5.6	0.015	20.3	0.008
Zn	3.7	0.065	3.9	0.022	3.4	0.047
Bi	0.026	5.0	0.18	0.46	0.39	0.39
Sb	0.163	12.0	0.04	19.8	0.08	18.0
Mo	0.634	5.0	0.4	7.3	3.6	1.5
H_2SO_4	13.0	-	10.0	-	11.1	-

Table 6. Analysis of Slurry Electrowon Cathodes

Cu %	Se ppm	Ni ppm	Fe ppm	Te ppm	As ppm
99.75-99.95	<2	<1	2-6	<2	<1

Sb ppm	Au ppm	Ag ppm	Sn ppm	Bi ppm	S ppm
3.2-<1	<1	5-<1	2.5-<2	.3-.79	10-1600

O_2 ppm	Pb ppm
95-590	13-21.8

Table 7. Typical Composition of Final Liquor After Waste Treatment

Element	Conc. in Feed Liquor to Waste Treatment, mg/l	Conc. in Final Liquor After Waste Treatment, mg/l
Cu	148.0	0.16
Fe	10550.0	0.07
As	6207.0	1.3
Zn	2443.0	0.36
Bi	5.3	0.24
Sb	7.6	<0.2
Mo	823.0	86.5
Pb	9.5	0.51

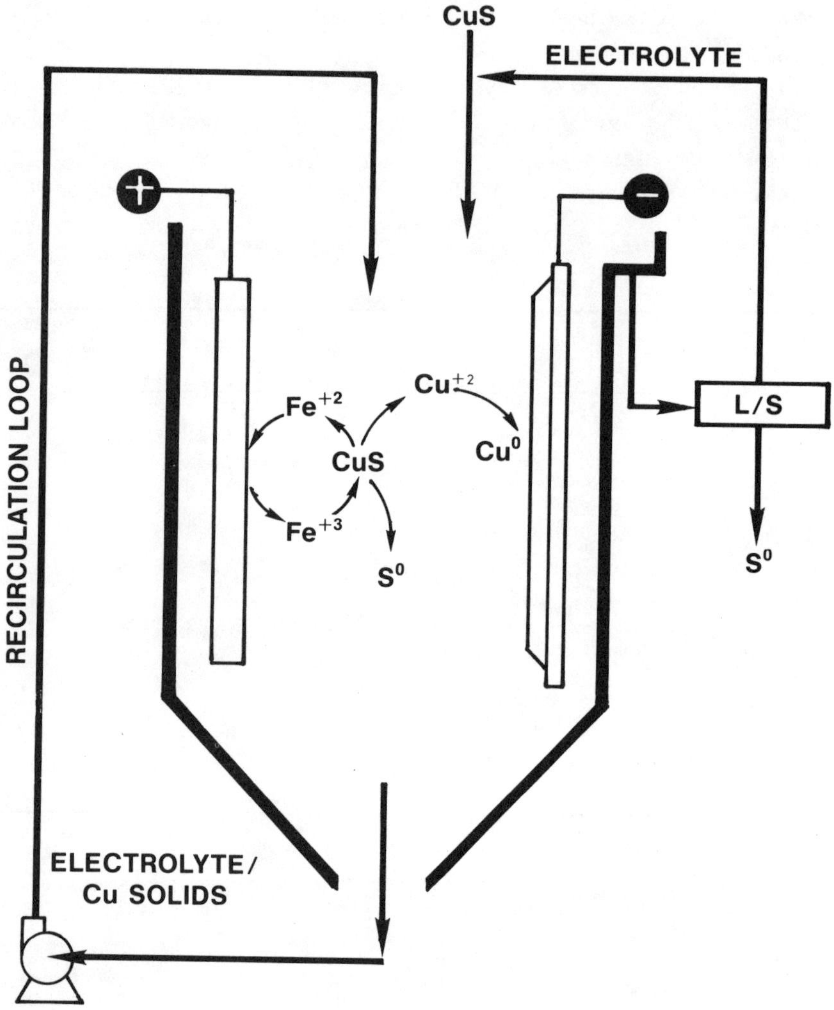

Figure 1. SLURRY ELECTROWINNING OF COVELLITE

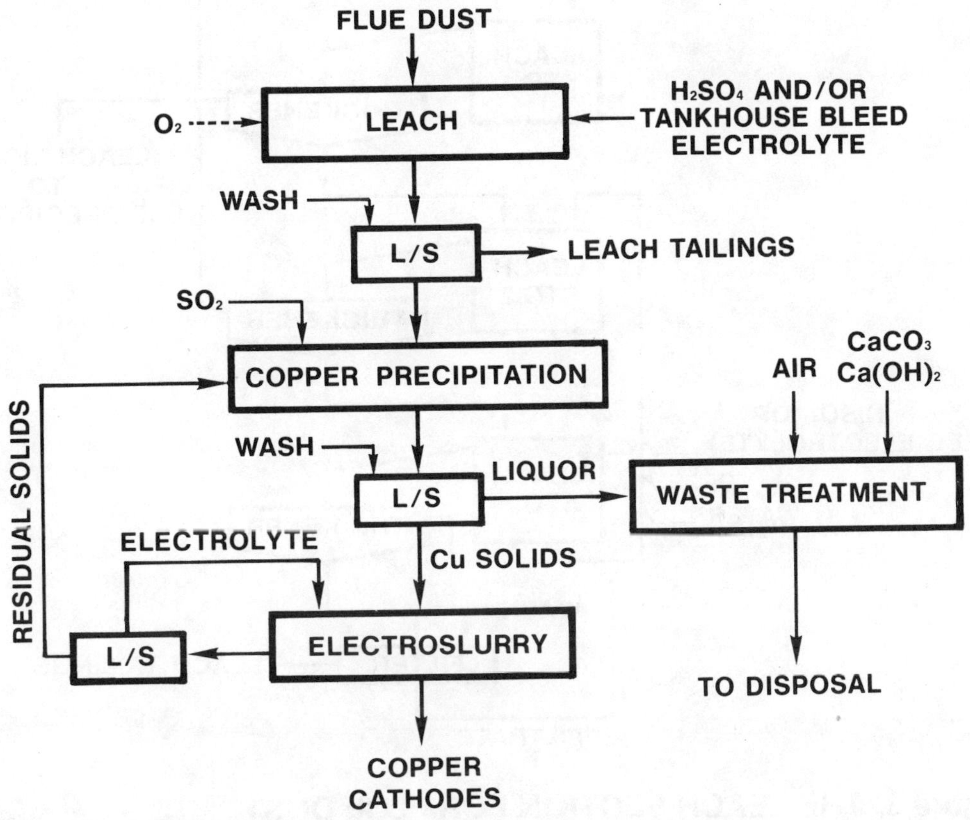

Figure 2. FLUE DUST FLOW SHEET

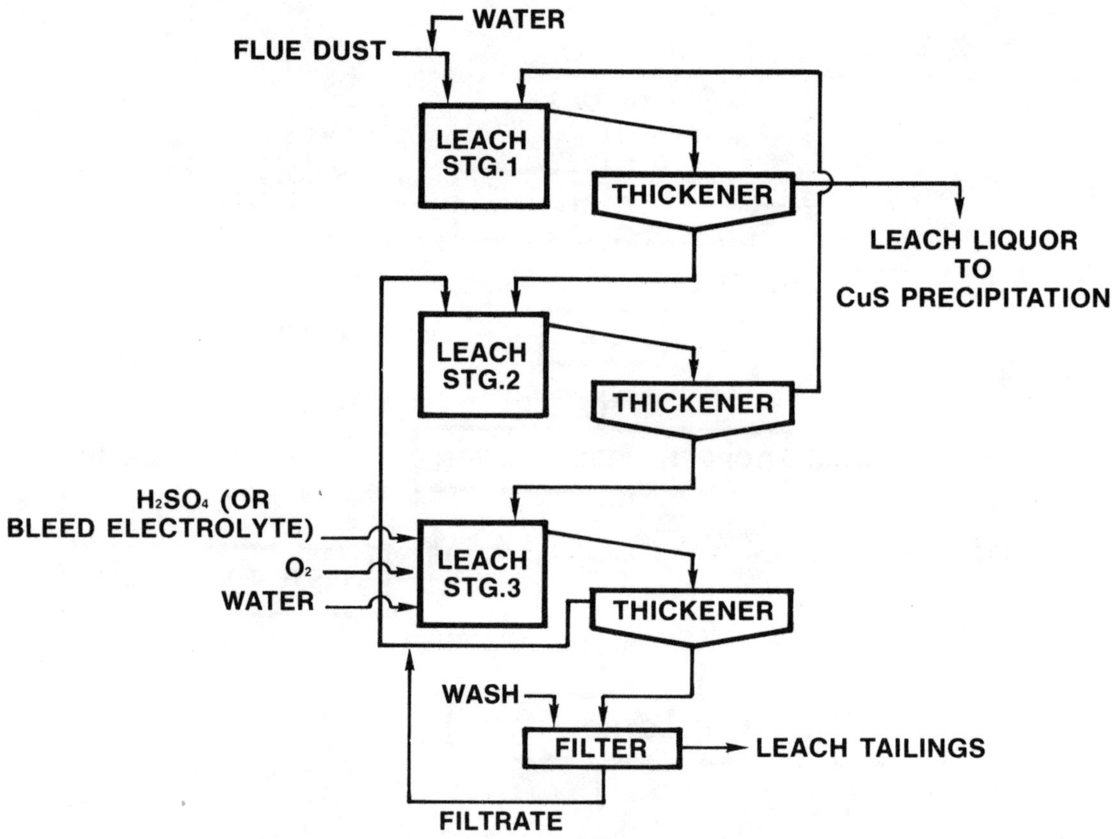

Figure 3. THE LEACH SECTION FOR FLUE DUST

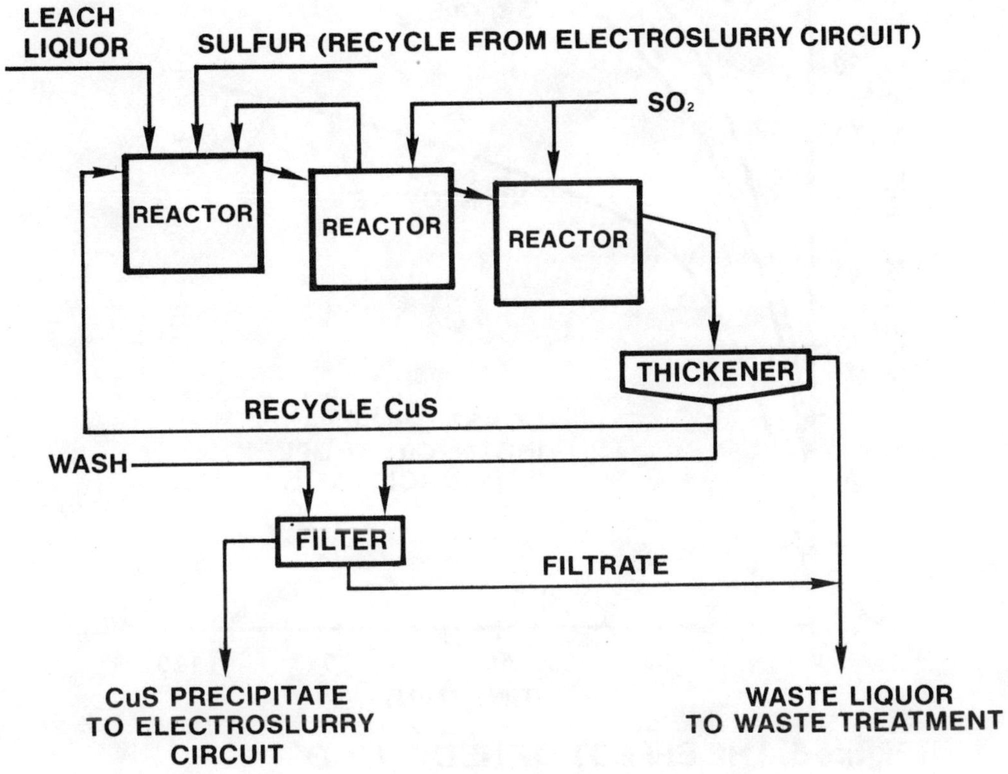

Figure 4. THE CuS PRECIPITATION SECTION FOR FLUE DUST

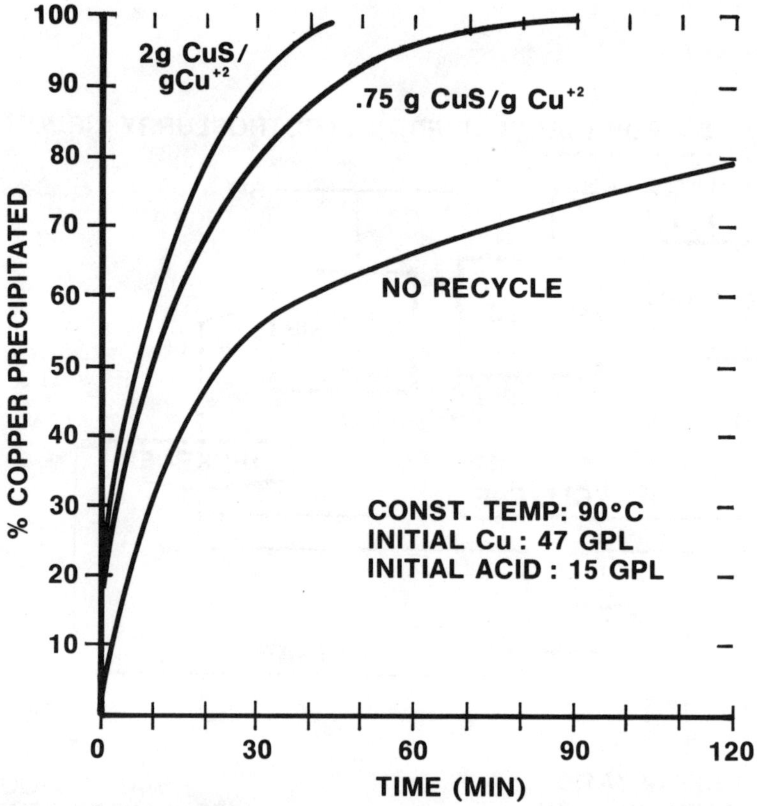

Figure 5. THE EFFECT OF RECYCLING PRECIPITATED CuS ON THE RATE OF Cu^{+2} PRECIPITATION

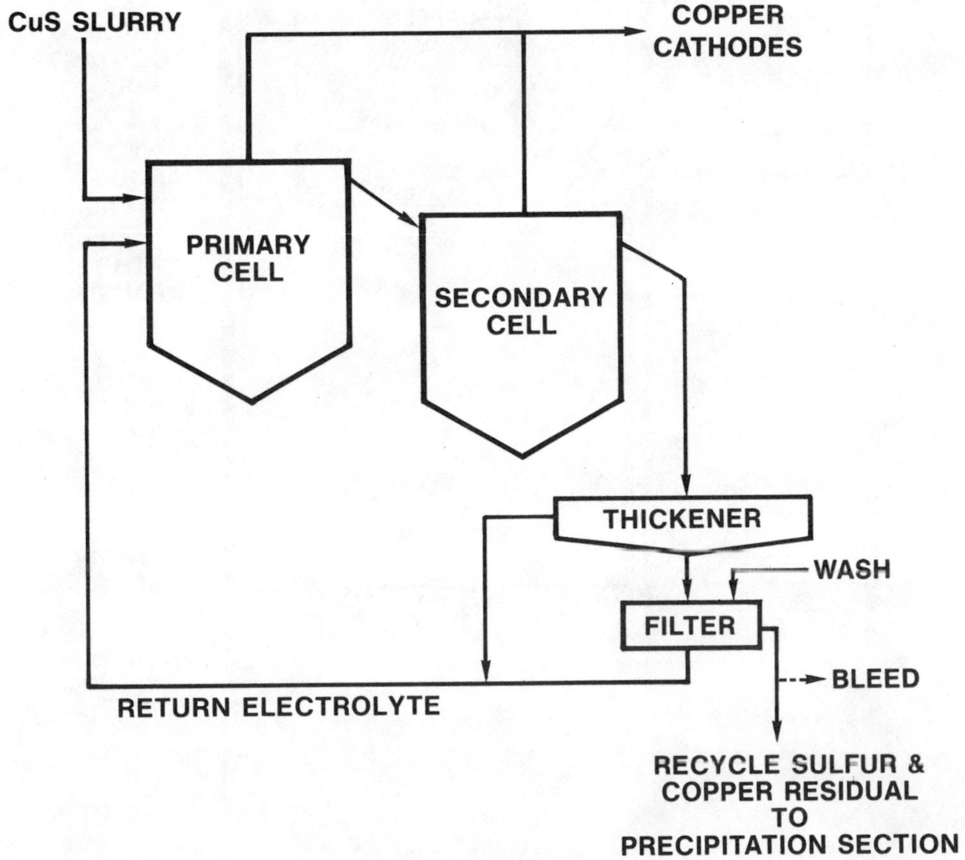

Figure 6. THE ELECTROSLURRY SECTION FOR FLUE DUST

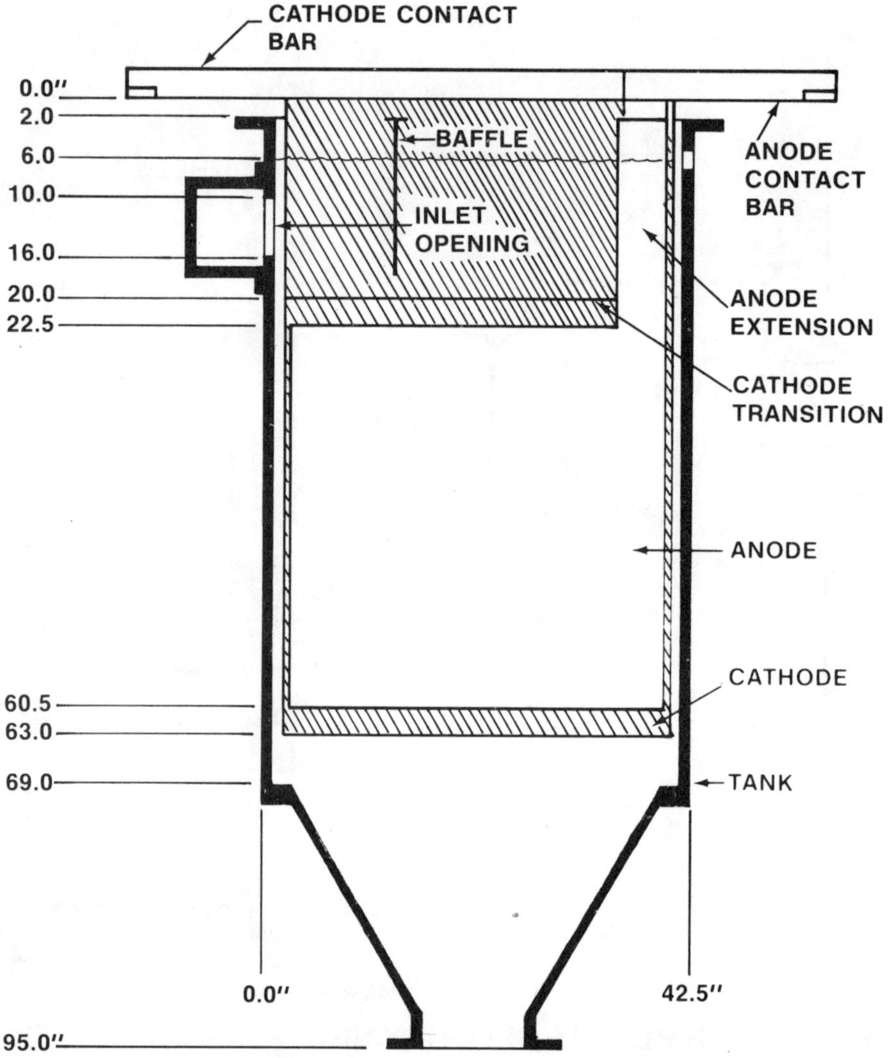

Figure 7. PROPOSED EQUIPMENT ARRANGEMENT FOR A COMMERCIAL CELL

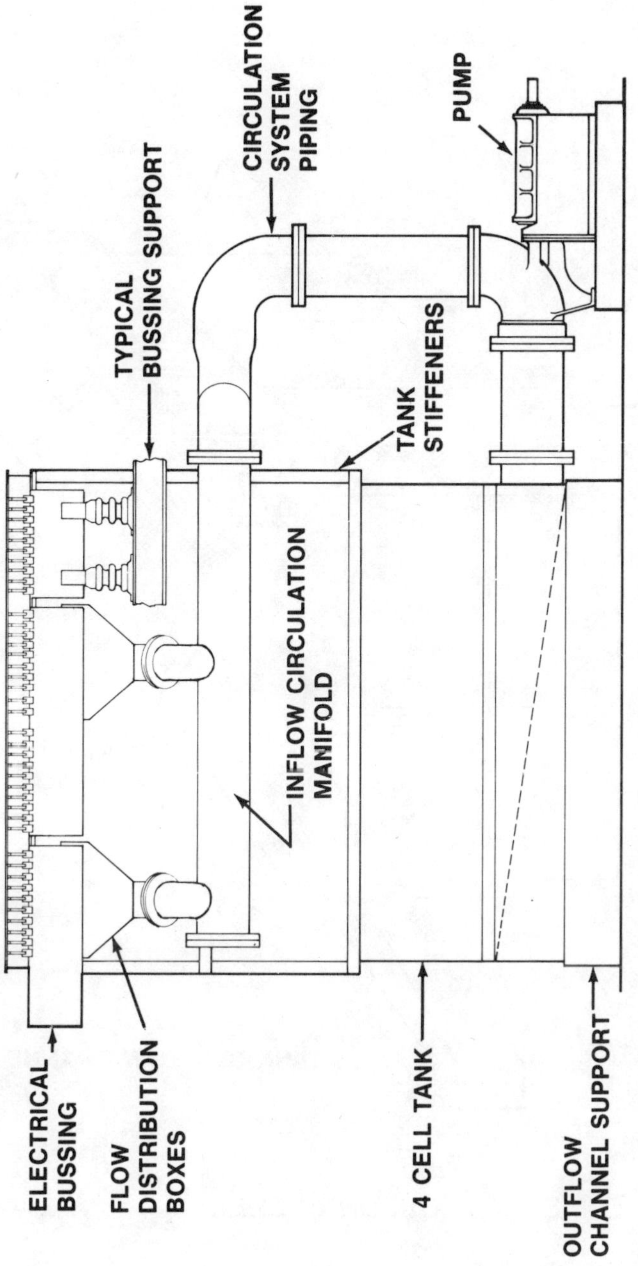

Figure 8. THE ELECTROSLURRY CELL

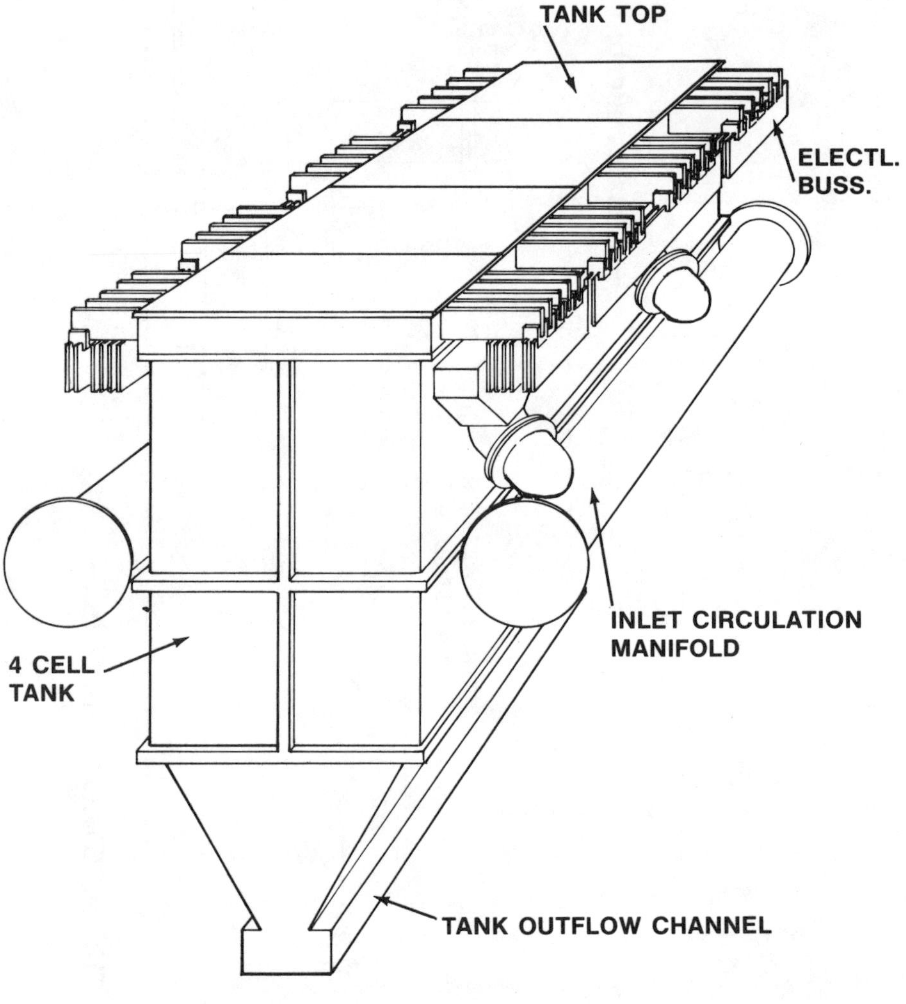

Figure 9. THE ELECTROSLURRY CELL

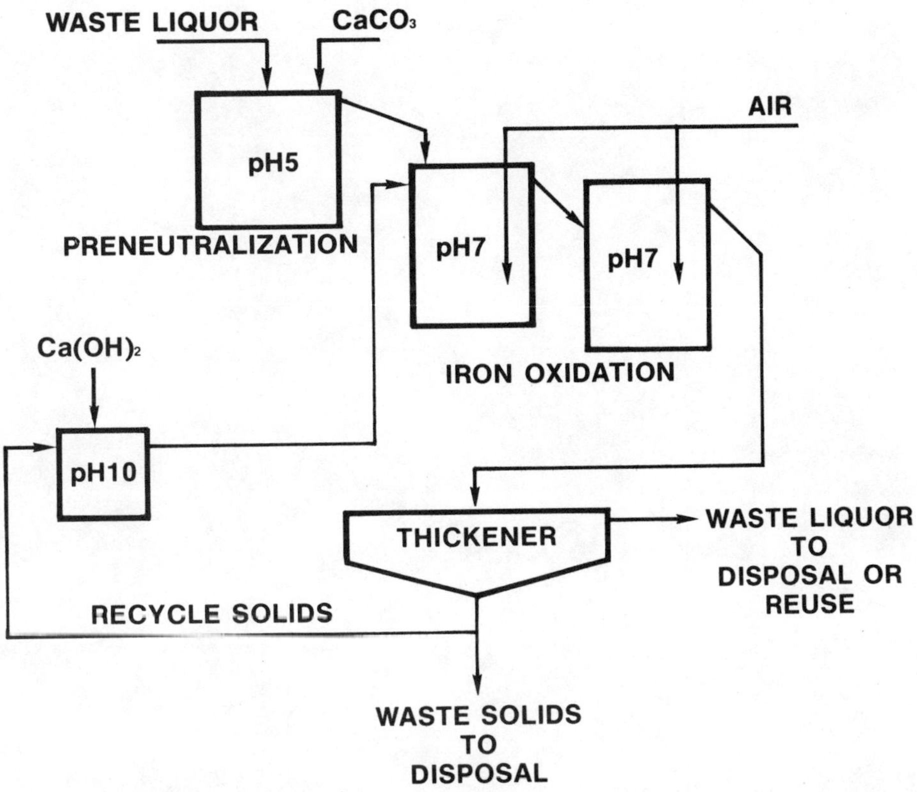

Figure 10. THE WASTE TREATMENT SECTION FOR FLUE DUST

Chapter 13

KENNECOTT PROCESS FOR TREATING COPPER SMELTER FLUE DUSTS

John D. Prater, Manager, Special Projects

W. Joseph Schlitt, Manager, Hydrometallurgy

Kenneth J. Richards, Vice President, Process Technology

Kennecott Minerals Company
Salt Lake City, Utah

ABSTRACT

Kennecott Minerals Company has developed a leach-hydrolysis route for treating flue dusts from copper smelters. The flue dusts are acid leached in an oxygen atmosphere to solubilize copper and fix the arsenic, antimony, and bismuth as stable compounds. Copper is recovered by cementation with iron and the barren solution is recycled. Molybdenum and other values may be recovered by modification of the basic flow sheet. Results of bench scale and pilot plant tests are discussed in terms of the effect of operating parameters.

KENNECOTT PROCESS FOR TREATING COPPER SMELTER FLUE DUSTS

John D. Prater, Manager, Special Projects
W. Joseph Schlitt, Manager, Hydrometallurgy
Kenneth J. Richards, Vice President, Process Technology

Kennecott Minerals Company
Salt Lake City, Utah

INTRODUCTION

The mineral industry is faced with the challenge of controlling the emission of sulfur oxides, particulates and fume from smelting operations to meet increasingly stringent emission standards. The relatively high concentrations of noxious volatile elements including arsenic, antimony, lead, and bismuth in smelter feed compound the problem of effluent treatment.

Kennecott Minerals Company is no exception; the recently constructed copper smelter at Utah Copper Division combines effluent gases from the Noranda-type smelting vessel and converters as feed to the sulfuric acid plants. Cleaning of the gases includes settling chambers and cyclones for recovery of coarse particulates which are recycled to the smelting vessel. Fine particulate and fume are recovered in dry electrostatic precipitators. Since a large fraction of the arsenic, antimony, and bismuth is recovered in the electrostatic precipitator dust, this dust is removed from the circuit to ensure the production of cathodes meeting market specifications. Currently, this dust is stockpiled in concrete lined ponds to ensure complete containment, including any salts which may be solubilized. This is a satisfactory temporary expedient but is not a long range solution for dust disposal.

Several potential leaching processes were explored by Kennecott's Research group; however, none were found to be completely satisfactory for treatment of dusts generated at the UCD smelter. Thus, a research program was initiated resulting in the development of a novel process which fixes the arsenic, antimony, lead, and bismuth as insoluble compounds for discard or further processing and recovers the copper values for recycle to the smelting circuit. Molybdenum which is found in the UCD electrostatic precipitator dust is also recovered. These processes are covered by patents assigned to Kennecott.(1,2) Treatment of the residue and solutions to recover other values is the subject of continuing research. This paper summarizes results of laboratory and pilot plant tests leading to the development of the present process flow sheet.

DESCRIPTION OF THE PROCESS

Dust Composition

The composition of electrostatic precipitator dusts studied in the program varied with their origin. However, except for minor variations in extraction, all responded similarly to the process as developed. Table 1 shows the composition of the dust which is considered typical and which was used in most of the studies.

TABLE 1

COMPOSITION OF FLUE DUSTS

Element	Analysis, %
Cu	13.8
As	11.2
Bi	1.9
Sb	0.36
Pb	17.0
Zn	1.5
Mo	1.6
Fe	3.8
S	10.2

Process Flowsheet

The flowsheet of the Kennecott process is shown in Figure 1 and consists of four principal operations.

An Oxidizing Leach at Elevated Temperature and Pressure. Immediately upon discharge from the electrostatic precipitator, the dust is slurried with recycled leach solution to minimize any environmental problems caused by dusting of the extremely fine material. After adjustment of the pulp density and acidity, the slurry is pumped to the pressure vessel where it is leached with the ferrous sulfate sulfuric acid solution under oxygen pressure to extract the copper and molybdenum and to precipitate the arsenic, antimony, lead, and bismuth as stable insoluble compounds in the leach residue.

Molybdenum Recovery by Solvent Extraction. Following liquid solids separation, molybdenum is recovered from the leach liquor by a Kennecott patented solvent extraction technique.(2) Molybdenum is stripped from the organic solution with a dilute sodium hydroxide solution. Calcium molybdate then is precipitated by addition of milk of lime to recover a molybdenum product and regenerate sodium hydroxide for recycle to the stripping circuit.

Copper Recovery from the Molybdenum Recovery Raffinate by Cementation. Copper is recovered from the molybdenum extraction raffinate in a particulate iron precipitation cone.(3,4) The solution is diluted to an appropriate copper concentration by recycle of copper precipitation cone tailing. Not only is the copper recovered but the dissolved iron provides the necessary iron to fix the arsenic when the solution is recycled to the autoclave leach circuit.

Bleed Stream Treatment to Control Impurity Buildup. The buildup of excess iron, other metallic impurities and sulfate in the leach circuit is controlled by a small bleed stream following copper cementation. This solution is neutralized with milk of lime to precipitate various impurities before discarding to the tailings pond.

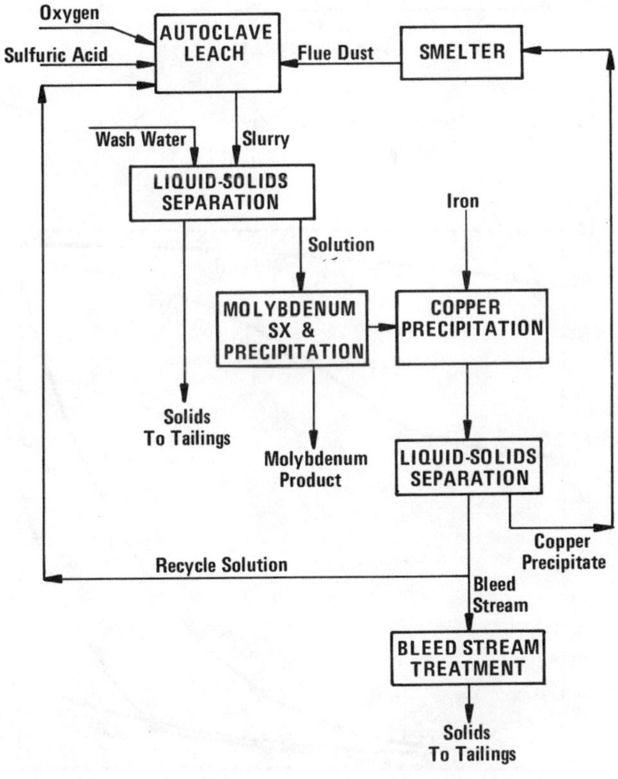

Figure 1. Simplified Process Flow Sheet

OPTIMIZATION OF PROCESS VARIABLES

Autoclave Leach

The primary objectives of the leach are to optimize solubilization of copper in the leach solution and to maximize rejection of the arsenic, antimony, lead, and bismuth in the leach residue. The dust as discharged from the electrostatic precipitator is highly sulfated. Significant solubilization of several components of the dust including copper, molybdenum, and arsenic is noted when the dust is slurried with recycled solution at ambient temperature and pressure. In the oxidizing leach at elevated temperature and pressure additional reactions occur as summarized below:

$$CuS + \tfrac{1}{2}\, O_2 + H_2SO_4 \rightarrow CuSO_4 + S\downarrow + H_2O \qquad (1)$$

$$H_2As_2O_4 + O_2 \rightarrow H_2As_2O_6 \qquad (2)$$

$$2\ FeSO_4 + \tfrac{1}{2}\, O_2 + H_2SO_4 \rightarrow Fe_2\ (SO_4)_3 + H_2O \qquad (3)$$

$$Fe_2(SO_4)_3 + H_2As_2O_6 + 6\ H_2O \rightarrow 2\ Fe\ AsO_4 \cdot 2H_2O\downarrow + 3H_2SO_4 \qquad (4)$$

$$3\ Fe_2(SO_4)_3 + 12H_2O \rightarrow 2HFe_3(SO_4)_2(OH)_6\downarrow + 5H_2SO_4 \qquad (5)$$

Preliminary tests indicated that the reactions were dependent upon the variables of time, temperature, sulfuric acid concentration, oxygen pressure, and iron concentration. Agitation rate was not a significant variable provided that the solids were kept in suspension. Likewise, variations in pulp density from 10 to 40% by weight solids were not significant provided that the solids were maintained in suspension and that iron concentration was in excess of stoichiometric quantity needed to precipitate the soluble arsenic.

The effect of the several significant variables as determined in laboratory tests is shown in Figures 2 through 6. Except for the variable under study, all tests were run for 1 hr at 140°C with a solution containing 2.5% by volume sulfuric acid and 18 g/l Fe as ferrous sulfate at an oxygen overpressure of 345 kPa.

As shown in Figure 2, most copper solubilization occurred during the slurrying operation. Further extraction during leaching was minimal. Arsenic also dissolved during slurrying but it precipitated rapidly during the first 30 min at the test conditions. Molybdenum solubilization decreased slightly and at a constant rate during the autoclave leach.

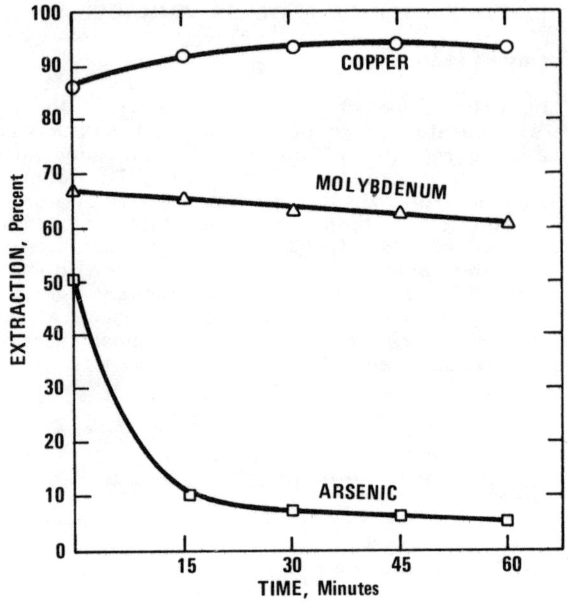

Figure 2. Metal Extraction from Flue Dust as a Function of Time

Data presented in Figure 3 indicate that the temperature markedly affected the molybdenum and arsenic in solution at equilibrium. Temperatures above 100°C were required to precipitate the arsenic as ferric arsenate. Molybdenum extraction also decreased with increasing temperature, presumably due to precipitation of ferric molybdate.

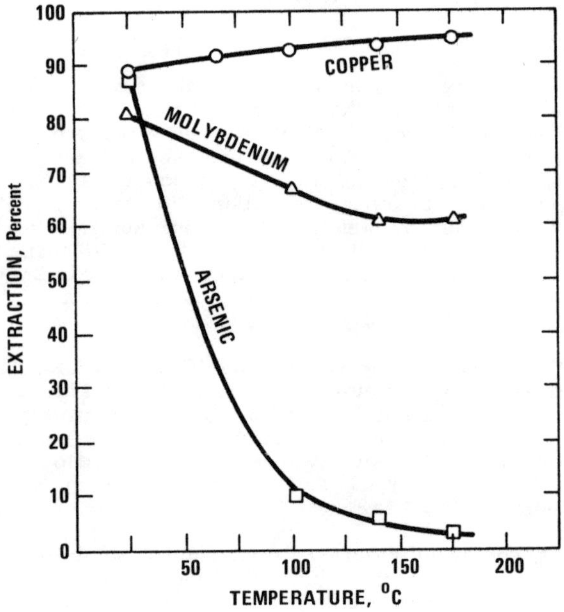

Figure 3. Metal Extraction from Flue Dust as a Function of Temperature

As indicated in Figure 4, arsenic precipitation was incomplete when iron was deficient to precipitate ferric arsenate. An additional test indicated that arsenic precipitation was increased only incrementally by increasing the iron concentration above the stoichiometric requirements. Copper and molybdenum extraction decreased slightly with increasing iron concentration.

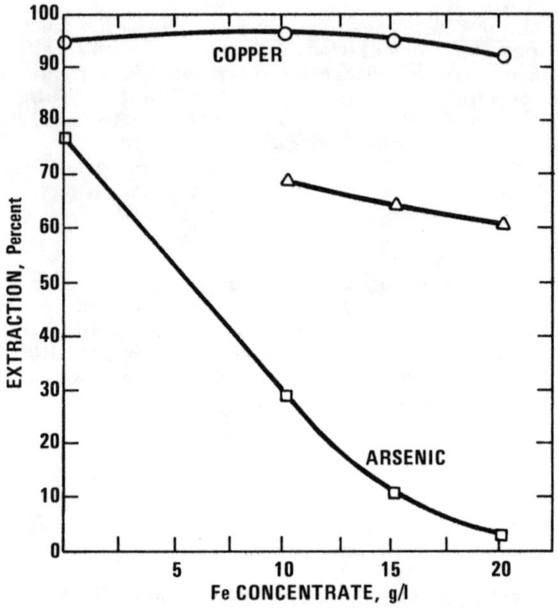

Figure 4. Metal Extraction from Flue Dust as a Function of Fe Concentration

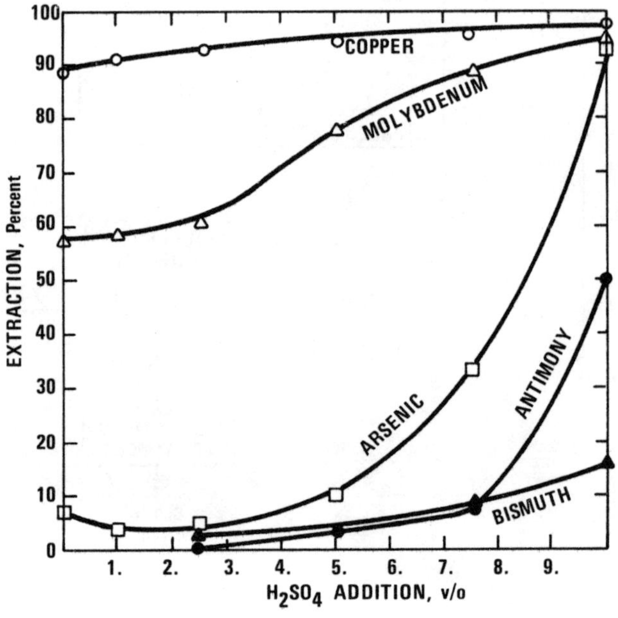

Figure 5. Metal Extraction from Flue Dust as a Function of Acid Concentration

Control of acid concentration is important to achieve the desired results. As shown in Figure 5 solubilization of arsenic and of molybdenum increase significantly as sulfuric acid concentration is increased above 2.5% by volume. Of equal importance from the standpoint of impurity recycle to the smelter is the high solubility of antimony and bismuth noted with increasing acidity. With all other variables studied antimony and bismuth extractions remained low.

Data from the laboratory tests shown in Figure 6 demonstrate that arsenic was effectively precipitated with an oxygen overpressure as low as 200 kPa. Copper and molybdenum extractions were virtually independent of oxygen pressure above 35 kPa.

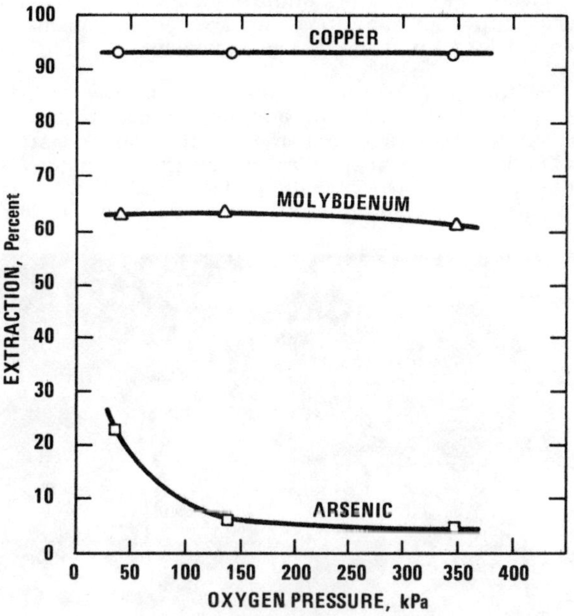

Figure 6. Metal Extraction from Flue Dust as a Function of Oxygen Overpressure

Molybdenum Recovery

Previous test work at KMC-RC had demonstrated that the copper extractant LIX-63 modified with nonylphenol was an effective reagent for recovery of molybdenum from strongly acidic solution. Because of the higher molybdenum concentration in the solutions generated by leaching of UCD electrostatic precipitator dusts the concentrations as defined in the patent assigned to Kennecott(2) were increased to 20% by volume LIX-63 and 6.7% nonylphenol. Equilibrium isotherms demonstrated that raffinates as low as 0.01 g/l Mo from a heading solution containing 2.9 g/l could be attained. The organic solution was loaded to 6 to 7 g/l Mo under these conditions.

The reagent combination will extract some copper from the high copper bearing solutions generated from leaching flue dusts. However, with increased molybdenum loading, copper is displaced from the solvent as shown in Figure 7. A loading of 6 g/l Mo and 0.2 g/l Cu on the organic is attained at high molybdenum recovery, The copper may be reduced to 0.05 g/l Cu prior to stripping molybdenum by washing the organic with a dilute sulfuric acid low copper solution.

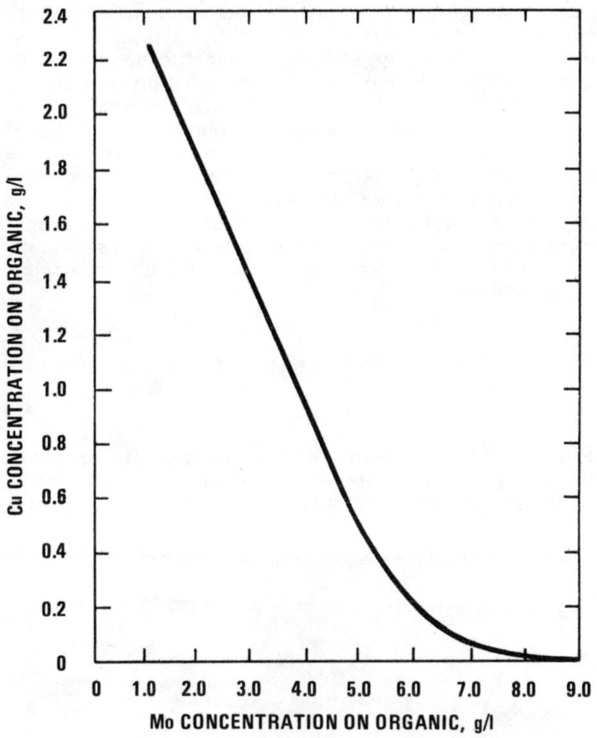

Figure 7. Copper Loading of 20% LIX-63--6.7% Nonylphenol--73.3% Kermac 470B as a Function of Molybdenum Loading

The kinetics of molybdenum extraction with this reagent combination are relatively slow. Typical data developed in continuous tests with varying retention time and organic to aqueous ratios are summarized in Table 2. A mixing time greater than 4 min is required to approach equilibrium.

TABLE 2

EFFECT OF RETENTION TIME AND O/A(1) RATIOS UPON MOLYBDENUM EXTRACTION FROM SOLUTION CONTAINING 1.15 g/l Mo

O/A Ratio	Retention Time, min	Molybdenum Extraction, %
1/3	2.6	84.0
	4.4	97.0
1/4	2.6	72.5
	2.9	78.0
	4.1	91.0
	4.6	95.5
1/5	4.1	87.0
	4.3	92.0

(1) Organic-to-aqueous volume ratio.

The organic solution was effectively stripped with either ammonium hydroxide or sodium hydroxide solution. The choice of stripping reagent is dependent upon the molybdenum product desired. Principal development work at KMC-RC included stripping with sodium hydroxide and precipitating calcium molybdate from the caustic solution with milk of lime. Molybdenum was completely stripped from the organic with a 50 g/l NaOH solution generating solution containing 40 to 50 g/l Mo for molybdenum precipitation. Although molybdenum was not completely precipitated from the caustic solution with milk of lime the regenerated sodium hydroxide with the residual molybdenum was an effective stripping solution. However, if ammonium molybdate were the desired end product, stripping with ammonium hydroxide and crystallizing ammonium molybdate from solution could be an attractive alternative.

PILOT PLANT TESTING

Plant Description

Pilot plant equipment was assembled to develop engineering design criteria and operating data for a full scale operating facility. A view of the pilot plant is shown in Figure 8. Basically, it includes equipment for the following unit operations:

(1) Preslurrying of dust and leach solution

(2) Pressure leach-hydrolysis

(3) Liquid/solid separation

(4) Copper precipitation by cementation with iron

(5) Copper cement recovery

(6) Bleed stream treatment

In addition to the principal pilot plant components shown in Figure 8, a bleed stream was removed from the circuit after the liquid/solid separation and before copper precipitation and treated in continuous mixer-settlers for molybdenum recovery.

The function of the preslurry tank was to prepare a slurry of constant and known composition as feed to the pressure leach-hydrolysis unit operation. Two tanks in parallel were provided for batchwise preparation of the heading leach slurry. Dry electrostatic precipitator dust was received in 209 l

Figure 8. Pilot Plant for Treating Copper Smelter Flue Dusts

(55-gal.) drums from the UCD smelter. Dust was wetted with recycled solution and transferred as a slurry with high pulp density (approximately 40% solids) to one of the preslurry tanks where the slurry was adjusted to 20% solids and 44 g/l (2.5% by volume) sulfuric acid.

The pressure leach-hydrolysis equipment consisted of a Pfaudler 190 l (50-gal.) titanium tank clamped-lid reactor. For the pilot plant program, a loose-fitting, horizontal titanium plate with two 5.1 cm (2 in.) diametrically opposite openings was installed as a baffle to divide the autoclave into two approximately equal compartments. Agitation was provided by a single 3-blade impeller in each compartment driven by a variable-speed, 2 hp motor. Impeller diameters were 25.4 cm and 50.8 cm (10 in. and 20 in.) in the upper and lower compartments, respectively. The reactor was electrically heated with a Trent heating jacket consisting of six parallel, horizontal resistance circuits located between the mantle and the reactor body. The temperature was controlled by an iron-constantan thermocouple in a well submerged in the leach slurry and connected to automatic temperature controlling and recording equipment. Oxygen was introduced through a 0.64 cm (1/4 in.) line discharging in the lower compartment near the outer edge of the lower impeller. A pressure-indicating gauge and safety release valve were installed in the reactor head. Samples could be taken from a sample port extending into the upper compartment of the reactor. Slurry was fed into the top compartment and discharged from the bottom. When operating in the continuous mode a capacitance probe maintained a near-constant level in the reactor by actuating a solenoid and air-actuated valve on the discharge line.

The slurry from the autoclave was discharged directly into an agitated pressure let-down tank which was cooled by spraying with water. In the pilot plant campaigns no attempt was made to recover latent heat from the reactor discharge slurry.

Liquid-solid separation of the autoclave discharge was made by thickening and filtering the thickener underflow. The thickener underflow discharged intermittently by gravity flow directly to the filter tank through a timed, air-actuated valve on the bottom of the thickener. Filtration of the thickener underflow was made on a 90 cm (3 ft) diameter by 15.2 cm (6 in.) face Eimco-belt filter with agitator and motor and variable-speed filter drive. Ancillary equipment included a vacuum pump, two filtrate receivers and filtrate pumps. Two water spray nozzles were provided for washing the filter cake.

A schematic drawing of the cone used for copper precipitation in the pilot plant campaigns is shown in Figure 9. Solution was pumped into the bottom of the cone and flowed upward discharging into a top overflow launder. Particulate iron was continuously fed into the inner cone; the top of the wall of this cone extended approximately 2.54 cm (1 in.) above the solution level. The inner cone was agitated with a laboratory mechanical mixer to ensure wetting of the particulate iron. The wetted iron dropped into the precipitation reaction zone. Sample ports at the indicated positions were installed to follow the rate of precipitation. Operation of the cone was interrupted periodically to remove the precipitated copper from the cone. In the pilot campaign the copper was allowed to settle, the solution decanted, and the solids filtered on a Star pressure filter.

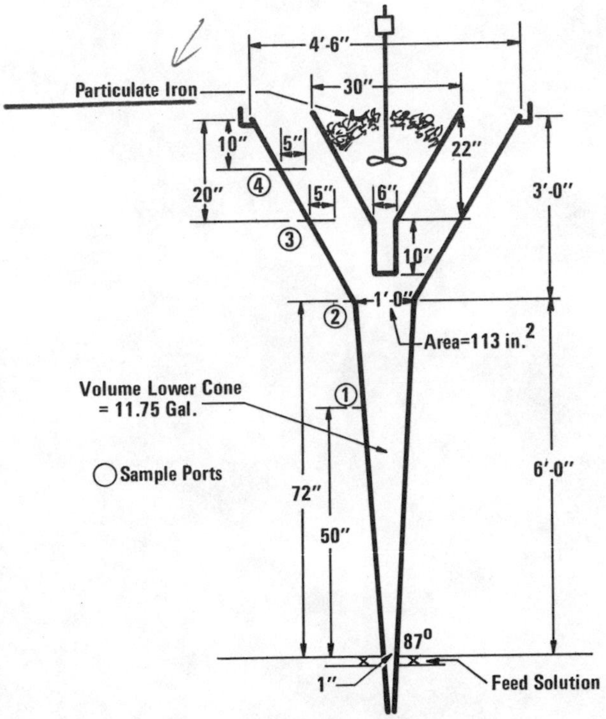

Figure 9. Copper Precipitation Cone

Ancillary equipment included solution and slurry transfer pumps and surge tanks to ensure a fully-integrated flow sheet.

The small scale pilot plant for molybdenum solvent extraction is shown in Figure 10. The mixer-settlers were identical for the extraction, washing, and stripping stages. The mixers were 7.4 cm diameter by 7.8 cm deep (335 ml volume). Three-blade flat turbine impellers, 5 cm in diameter by 1 cm deep, driven at 650 rpm provided mixing. Settlers had a settling area of 43 cm^2.

Pilot Plant Results

The pilot plant was operated in batch and continuous feeding modes. The horizontal baffle was ineffective in converting the vessel to a two-stage reactor. Tracer tests indicated that backmixing occurred and the unit essentially was single-stage. Batch test data from the pilot plant confirmed laboratory data with the notable exception of oxygen overpressure. As indicated previously arsenic was nearly completely precipitated in the laboratory scale tests with an overpressure of 200 kPa; however, in the pilot plant equipment an oxygen overpressure of 345 kPa was required to achieve the same results. The variation is attributed to improved oxygen diffusion and absorption in solution in the laboratory equipment. Analysis of the system indicated that oxygen absorption in solution in the pilot tests was

Figure 10. Molybdenum Recovery Solvent Extraction Equipment

principally obtained from solution of oxygen at the solution-gas interface instead of within the body of the solution. Oxygen consumption averaged 0.05 kg/kg of dust.

Several campaigns were completed in which the reactor was operated in the continuous mode with feed rates of 29 to 59 kg/hr (65 to 130 lb/hr). Although arsenic precipitation was not complete because of short circuiting of solution, results agreed favorably with those calculated for a single stage reactor. Successful operation in a continuous mode was demonstrated so that a properly designed staged reactor should achieve metallurgical results comparable with batch operation.

TABLE 3

SUMMARY OF PILOT PLANT LIQUID SOLIDS SEPARATION DATA

Feed	19% Solids
Flocculant dosage	0.4 kg/t
Thickener Area	0.7 m^2/t/24 hr
Thickener Discharge	40% Solids
Filtration Rate	2670 kg/m^2/24 hr
Wash water	2 kg/kg solids
Washing Efficiency	98%

Separation of liquid and solids is relatively difficult. Addition of flocculants is required to achieve reasonable setting rates. Pilot data for thickening and filtering are summarized in Table 3.

Performance of the cone for copper cementation exceeded projections. Two samples of particulate iron, one a relatively low grade product assaying 65% metallic iron and the second analyzing 86% Fe^0 were used as precipitants. Each performed equally well. Typical results are shown in Table 4. The copper was completely precipitated from solutions with copper concentration as high as 10 g/l at flow rates up to 11.4 l/min with iron consumption only slightly greater than stoichiometric (0.88 g Fe/g Cu). With more dilute copper solutions, copper precipitation was complete; however, iron was consumed by side reactions and the iron required per unit of copper recovered increased. At a higher flow rate of 19 l/min results were inconsistent; copper in the effluent solution varied from 0.05 to 1.9 g/l during the course of the test.

TABLE 4

PERFORMANCE OF PARTICULATE IRON CONE FOR COPPER PRECIPITATION

Feed Rate l/min	Heading Cu, g/l	Tailing Cu, g/l	Fe^{o} Consumption kg/kg Cu
11.4	2.3	0.01	1.54
11.4	10.0	0.01	0.88
19.0	8.7	0.05-1.9	1.04
11.4	27.6	3.8	1.03

Any arsenic, antimony, and bismuth in the leach liquor are partially precipitated with the cement copper. Data indicate that 40 to 50% of the arsenic, 65% of the bismuth, and 65% of the antimony are precipitated. Careful monitoring of the cone during cementation of the copper did not detect that any arsine was liberated. During the continuous feed leach campaigns, arsenic in solution to the precipitation cone was higher than that which will report to the solution under equilibrium operating conditions and thus represented conditions which would favor arsine liberation if any were generated.

Some particulate copper overflowed the cone under all test conditions. The amount of particulate overflow could not be correlated with flow rate, copper concentration or other operating parameters. The copper appeared to be carried out of the cone by a flotation process in which the particles attached to hydrogen gas bubbles generated from dissolution of iron by the acid in solution. The particulate copper settled rapidly when the solution was discharged to the barren solution surge tank.

After the autoclave leach and initial liquid-solid separation, a bleed stream from the filtrate was treated for molybdenum recovery. Metallurgical results as optimized are summarized in Figure 11. At equilibrium 98% of the molybdenum was extracted in four stages. The resultant organic solution assayed 6.2 g/l Mo, 0.028 g/l As and 0.35 g/l Cu. A dilute acid wash of the organic solution removed 84% of the copper and 25% of the arsenic. Molybdenum was then recovered from the organic solvent by stripping with 50 g/l sodium hydroxide. During the continuous campaign some sludge formed in the extraction stage organic-aqueous interface. However, this did not present a serious problem during the operation.

The molybdenum in the strip solution was precipitated batchwise as calcium molybdate with a 20% calcium hydroxide slurry. Addition of lime equivalent to 110% of the stoichiometric for the reaction

$$Na_2MoO_4 + Ca(OH)_2 \longrightarrow CaMoO_4 + 2NaOH$$

produced a barren solution containing 2 to 4 g/l Mo. After adjusting the sodium hydroxide concentration to 50 g/l, the solution was satisfactory for recycle to the stripping cycle. Actual reagent consumption in the pilot plant campaign was 0.96 kg $Ca(OH)_2$ and 1.0 kg NaOH per kilogram of molybdenum recovered.

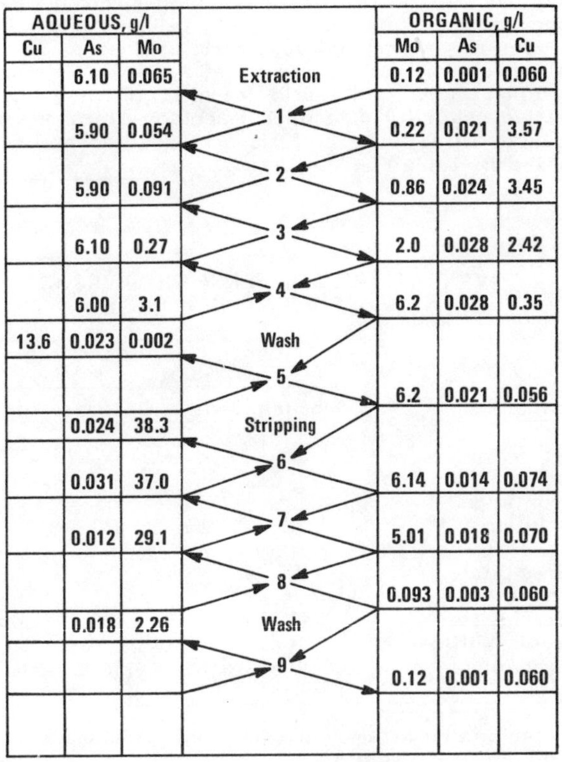

Figure 11. Molybdenum Circuit Profile, 4 Extraction Stages O/A: 1/2.7

Chemical composition of the calcium molybdate recovered in successive lots is shown in Table 5. The data indicate the product should be suitable for molybdenum alloying in the steel industry.

TABLE 5

ANALYSIS OF CALCIUM MOLYBDATE PRODUCTS

Sample	Wet Analysis, %				Spectrographic Analysis, %	
	Mo	Ca	S	As	Cu	Pb
SX-2	42.6	21.4	0.012	0.025	0.05-0.1	<0.001
SX-3	42.8	21.4	0.022	0.020	0.03-0.06	<0.001
SX-4	43.6	20.6	0.039	0.020	0.03-0.06	<0.001
SX-5	43.5	20.8	0.039	0.020	0.05-0.1	<0.001

ENVIRONMENTAL IMPACT

Two waste streams are generated by the process: namely, the leach residue and a gypsum metal hydroxide precipitate from the bleed stream treatment. The two streams are compatible for discharge to a common tailings pond or may be kept separate for possible recovery of additional values from the leach residue. This is a continuing subject for research. The arsenic precipitated in the pressure leach has been identified by scanning electron microscope photographs and energy dispersive x-ray analyses as well crystallized ferric arsenate. This is an extremely stable compound and is virtually insoluble under mildly acid leach conditions. Comparative solubility data when leaching several arsenic products with dilute sulfuric acid solution are summarized in Table 6.

TABLE 6

COMPARATIVE SOLUBILITY OF SEVERAL ARSENIC MATERIALS

	As in Solution, mg/l			
Product	5 hours	3 days	24 days	pH
Leach Residue	<1	<1	0.53	3.2
Leach Residue, no iron added to solution	24	43	60	2.6
Iron Arsenate Precipitate(1)	130			3.5
Iron Arsenate Precipitate(2)	250			3.5
Arsenic Sulfide Precipitate	20	52	150	1.9

(1) Precipitated at ambient conditions with lime @ final pH 10. Fe:As = 5:1

(2) Precipitated at ambient conditions with lime @ final pH 10. Fe:As = 3:1.

Special emphasis was placed upon detection of potential environmental problems during the pilot plant campaign. All personnel were continuously monitored for exposure to lead and arsenic and data interpreted by Kennecott's Environmental Hygiene Group. Arsenic was detected only during short periods of time when the dry dust was transferred from drums to the preslurry tank. At that time arsenic was detected in the 5 $\mu g/m^3$ to 10 $\mu g/m^3$ level. No other dust or toxic metal concentrations approached action or permissible exposure limits. The copper precipitation cone was completely hooded and the gases continuously monitored for arsine. Arsine was not detected in the atmosphere above the solution nor in the ventilation duct leading from the hood at any time even when solutions generated in the continuous mode of operation and containing higher than normal arsenic concentration were processed.

CONCLUSIONS

As a result of the laboratory and pilot plant tests a process has been developed and operating parameters defined for the treatment of typical smelter dusts. The process meets the primary objectives of recovering most of the copper as a product for recycle to the smelting circuit and of removing undesirable impurities as stable compounds which can be discarded under controlled conditions. When molybdenum is present it may also be recovered. A flowsheet of the process is shown in Figure 1. Leaching at 140°C with an oxygen overpressure of 345 kPa extracts 92% of the copper, and 60% of the molybdenum and rejects 94% of the arsenic and more than 99% of the arsenic, lead, and bismuth. Molybdenum first is extracted by solvent extraction and precipitated from the stripping solution with milk of lime as a calcium molybdate product suitable as an additon reagent for the steel industry. Copper is recovered by cementation with iron. The iron reporting to the solution provides the required iron for fixing the arsenic in the leach step. A bleed stream is neutralized with lime to control impurity buildup in the circuit.

The pilot plant demonstrated that the process was technologically viable in both batch and continuous feed modes of operation. The operating parameters were easily controlled within prescribed limits. The process is not extremely sensitive to changes in operating parameters and minor variations in temperature or oxygen overpressure has little impact on the distribution of the elements in solution and in the residue. Iron must be present in at least stoichiometric ratio to arsenic for precipitation of ferric arsenate. Excess has no adverse effect on the process and will be precipitated and discarded with the residue. Although temperatures and pressures required for leaching are above ambient, and the system somewhat corrosive, materials of construction do not appear to be a problem. The leach operation is the most corrosive unit process. Titanium proved to be an adequate material of construction in the pilot plant.

ACKNOWLEDGEMENTS

The authors wish to express sincere thanks to all the personnel at the Kennecott Minerals Company Research Center for their contributions which helped make this paper possible. We wish to especially thank B. A. Wells who performed much of the experimental work and D. R. Clark for his contributions in the pilot plant program.

REFERENCES

1. Prater, J. D., and Wells, B. A., "Recovery of Copper from Arsenic Containing Metallurgical Waste Materials," U.S. Patent 4149880, April 17, 1979.

2. Wells, B. A., and Clark, D. R., "Selective Extraction of Molybdenum from Acidic Leach Liquors," U. S. Patent 4026988, May 31, 1977.

3. Back, A. E., "Use of Particulate Iron in the Precipitation of Copper from Dilute Solutions," Trans. Soc. Min. Eng. Vol. 238, No. 1, 1967, pp. 12-16.

4. Back, A. E., Fisher, K. E., and Kocherhans, J., "Process and Apparatus for the Precipitation of Copper from Dilute Acid Solutions," U.S. Patent 3154411, October 27, 1964.

Chapter 14

RECOVERY OF ZINC FROM METALLURGICAL DUSTS AND FUMES

D. Pearson

Metals Extraction Division
Warren Spring Laboratory
Stevenage, Herts, UK

INTRODUCTION

In 1975 278×10^3 tonnes of zinc was consumed in the United Kingdom, of which 69.3×10^3 tonnes was obtained from secondary sources. Of this secondary zinc 37.7×10^3 tonnes was used in brass making; the remainder was used in non metallic applications - eg oxides and salts.

Many forms of zinc-bearing wastes arise, for example as galvaniser's drosses and residues, fumes from ferrous and non-ferrous metallurgical processes, and filter cakes from rayon manufacture and metal finishing. In the case of zinc-bearing dusts and fumes, the quantities available for recycle are at present restricted by the failure to collect them efficiently, but environmental factors are leading to improvements in gas cleaning, with the result that quantities collected will increase and then exceed those required for the production of chemicals, so that some dusts may either have to be dumped or processed to metal.

The recovery of zinc from steelmaking dusts is practised in W. Germany (1) and Japan (2), employing processes based on reduction in rotary kilns in order to produce sponge iron and a zinc- rich oxide suitable for feeding to the zinc-lead blast furnace. Although the zinc-lead blast furnace is the only zinc-making process used in the UK, no similar treatment of steelmaking dusts is carried out; a pilot scheme was tested by the British Steel Corporation, but development was abandoned for economic reasons. It appears to be cheaper to pelletise suitable dusts and export them for treatment elsewhere.

Against this background the Government of the UK supported a programme to study possible hydrometallurgical methods for recovering zinc from metallurgical dusts and fumes. The major part of the world's zinc is produced by acidic leaching of zinc oxide calcine followed by electrowinning, and the application of a similar process to oxide-products from secondary sources appeared merely to require development in detail and adaptation. A novel variant of the acid-leach process was proposed by MX Processor Reinhardt and Co AB (3), wherein the normal procedure for the purification of electrolytes was replaced by a solvent extraction procedure in which the zinc was extracted with an organic solvent. Alkaline leaching processes have also been described, in particular the ammonium carbonate route of Wendt (4), which although designed for primary feeds might utilise the inherent selectivity of the reagent advantageously in relation to processing secondary materials; in addition a simple caustic-leach-electrowinning process has been invented by workers (5) at University College, Cardiff, Wales.

All these processes, with the exception of caustic leaching, were examined at the Warren Spring Laboratory. The results obtained, process flowsheets, and a commentary on the economics form the subject of this presentation.

PROCESS TECHNOLOGY

Acid-based Processes

Published information on the electrolytic production of zinc is concerned almost exclusively with its production from primary sources. The first stage is an oxidative roast to convert sulphides to oxides; this has the added advantage of removing chlorides and fluorides by volatilisation. The calcine is leached with sulphuric acid or return electrolyte, and impurities are removed from the solution to the extent needed to ensure high electrochemical efficiency in the subsequent electrowinning operation. Roasting is obviously not needed with the waste materials under examination so that removal of halides is not achieved in this way, in fact since the materials are themselves the products of volatilisation and condensation, the concentrations of halides are often enhanced. The effect of halogen impurities has been reported (6) as increasing anode corrosion and causing cathode-sticking, and their removal must form part of the electrolyte purification scheme.

Regarding the effects of metallic impurities on electrolysis, there is a lot of contradictory data on allowable limits, suggesting that some of their effects may be synergistic. Van Niekerk and Allen (7) reported the findings of various authors concerning iron, quoting maxima ranging from $12\text{-}70 \times 10^{-3}$ kg m^{-3}. However, Kerby and Ingraham (8) reported that the presence of 15×10^{-3} kg m^{-3} reduced the current efficiency by 10 per cent.

Non-ferrous metal impurities have an adverse effect on current efficiency; the electro- deposition of zinc (electrode potential 0.76 V) is only possible because of the cathodic hydrogen overvoltage. Non-ferrous metallic impurities such as Co, Ni, Ge etc reduce this overvoltage so that hydrogen is evolved. Quantitative statements of the permissible threshold levels are contradictory, suggested nickel and cobalt concentrations have ranged from 1×10^{-5} kg m^{-3} to 12×10^{-3} kg m^{-3} (8) with generally accepted maxima of 3×10^{-4} kg m^{-3} for Co and 1×10^{-4} kg m^{-3} for Ni. Fukubayashi et al (9) made a systematic study by spiking pure zinc sulphate solutions with nickel, and showed that the effect was temperature dependent and 1×10^{-3} kg m^{-3} was tolerable at temperatures below 40°C. It seems obvious from the foregoing that the permissible levels of impurities can only really be determined in any particular case by experiments with the materials under consideration.

Methods of removing the impurities have invariably consisted of some form of basicity precipitation for iron, and cementation- reduction for the non-ferrous metals. The problem of halide removal does not usually arise with primary materials but Callaway and Koepel (10) have described the precipitation of chloride as cuprous chloride using cement-copper, in connection with copper production. Whilst basicity methods are used to remove iron, granular or crystalline precipitates are necessary in order to minimise loss of zinc, and the jarosite (11) or goethite (12) processes are now used; the former, by allowing precipitation at low pH, also conserves acid, which would otherwise be wasted by neutralisation. The removal of copper, cadmium and germanium by precipitation with zinc dust is relatively easy, but precipitation of nickel and cobalt is more difficult and is carried out hot in the presence of activators such as As_2O_3, Sb_2O_3, Sb/Pb or $CuSO_4$ (12).

A completely different method of purification is described by Reinhardt (3). Instead of removing the impurities from the zinc-bearing solution, the zinc is extracted from the solution by DEHPA, employing the known selectivity of this reagent (13) for zinc over other non-ferrous metals. Acid is generated in the raffinate by transfer of H^+ counter ions, so that the raffinate may be used for leaching. Although extraction of zinc ceases when the pH of the raffinate falls below 2, high salt concentrations have a buffering effect so that 30-40 kg m^{-3} of zinc can be extracted at each pass. Halides are not extracted and hence cannot enter the electrolytic cell. The extractant is stripped by the H^+ present in spent electrolytes, and zinc tenors range from 30-70 kg m^{-3} in the electrowinning-stripping circuit.

Alkali-based Processes

Zinc oxide dissolves in ammonium carbonate solutions, forming a series of complex cations of formula $Zn(NH_3)_n^{++}$ where n is 1-6. For the most efficient leaching of oxide materials Wendt (4) recommended NH_3:CO_2 ratios of 3-3.5:1. Other metals also dissolve in ammoniacal ammonium carbonate solutions. These include Cu, Ni, Co, and in reducing systems Fe. Illis (14) describes the solubility of $FeCO_3(NH_4)_2CO_3$, which on oxidation precipitates iron as $Fe(OH)_3$. In applications for recovering copper, by leaching with ammoniacal solutions, the phenomenon of iron dissolution is seldom observed, because of enhanced oxidation due to the presence of Cu^{2+} ions, but with some steelmaking dusts, which are reducing because metallic iron is present, ammonium carbonate can dissolve iron as well as zinc. Also because of the high cost of the ammoniacal reagents, the lixiviants must be recycled after recovering the zinc. Wendt (4) proposed decomposition of the leach solution by boiling, with precipitation of $ZnCO_3$ and recovery of NH_3 and CO_2 from the off-gases. A more practicable method may be to recover the zinc by solvent extraction, using a chelating reagent. The solvent can be stripped with sulphuric acid and the zinc electrowon from the acid zinc sulphate. A process of this type, developed at the warren Spring Laboratory, is described briefly in this paper and is the subject of a recent UK patent application (15).

The recovery of zinc from oxidised ores, using caustic soda for leaching was described by Merrill and Lang (16); the zinc was recovered from solution by dilution, carbonation, sulphidation or electrolysis. A similar process using caustic soda for treating steelmaking dusts is proposed by University College, Cardiff (5). The dusts are leached with 40% caustic soda solution, and zinc is recovered electrolytically using a cell with a rotating cathode. The economics for this process are claimed to be favourable for a wide range of arc furnace and converter dusts.

EXPERIMENTAL

Both acid and alkali-based processes have been studied at the Warren Spring Laboratory, together with a variety of possible feedstocks. The systems examined were

1. Sulphuric acid leaching and electrowinning applied to
   - (a) Fumes from a secondary copper smelter
   - (b) Fumes from brass ingot melting
   - (c) Waelz oxides produced by the reductive roasting of steelmaking dusts
   - (d) High-zinc arc furnace dusts
2. Ammonium carbonate leaching and electro-winning applied to
   - (a) Dusts from a BOS converter. These dusts contained about 6% Zn and the object was to remove the zinc in order to allow return of the dusts to the blast furnace or converter.

Feedstocks

Flowsheets were developed for materials containing 6-78% of zinc. The end-product in each was cathode zinc.

Analyses of the feedstocks are given in Table 1.

Leaching the Dusts

The leachability of the feedstocks (as % of the total zinc content) was tested by contacting them with an excess of acidic and alkaline reagents. The presence of zinc ferrites inhibited extraction unless the terminal pH was very low; in practice this would entail multi- stage leaching. Highly acid leach solutions complicate the subsequent purification, so that countercurrent leaching should be used whereby the excess acid is neutralised in the leaching circuit; however, this can lead to a build up of the recirculating load of iron, and lead to costly solids-liquid separation. With the BOS converter dust the primary objective was not to recover the zinc but to allow the iron values to be returned to the converter or blast furnace. Leaching results are shown in Table 2.

On the basis of these results, acid-leach processes were studied for the arc furnace and smelter dusts and the Waelz oxide, and an ammonium carbonate leach process for the BOS converter dusts.

Acid Leaching. Sulphuric acid was used in all the acid-leaching work. After suitable preliminary tests with pure acid to establish feasibility and conditions, the lixiviants were derived from return electrolytes or other downstream process solutions.

Copper Smelter Fume and Brass Melting Fumes. Dissolution of these materials proceeded with little difficulty, and single-stage leaching with return

Table 1. Composition of the Feedstocks

Material	Composition (%) Zn	Fe	Pb	Cu	Cd	Ni	Co	Sn	Cr	Na	K	Cl	F
High-Zn EAF Dust	36.0	17.9	6.0	0.3		0.02	0.01	0.2	0.13				
Waelz Oxides from S/M Dusts	57.0	7.6	4.7	0.04	0.24	0.03	0.01					1.57	1.06
BOS Converter Dust	6.7	60.9	0.55	0.05				0.1		0.08	0.12		
Copper Smelter Fume	45.0	0.52	14.9	2.26	4.6	0.05	0.02	0.5				1.89	0.67
Brass melting Fume	78.0	0.08	1.1	0.79	0.67			0.02				1.97	

Table 2. Leachability (as% of total zinc content)

Material	Reagent Sulphuric Acid pH 3-4	Sulphuric Acid pH 2	Sodium* Hydroxide	Ammonia-** Ammonium Carbonate
High-Zn EAF Dust	80	85-90	60	
Waelz Oxides	85	95-99		
Cu Smelter fume	95-99			
BOS Converter Dusts			70	76

*25% w/v, ** 1-molar each

electrolyte at 90°C for 1 h consistently extracted more than 97% of the zinc, using a lixiviant containing typically 100 kg m^{-3} H_2SO_4 and 40 kg m^{-3} Zn. A typical leach solution contained (kg m^{-3}) Zn 110; Cu 3.91; Fe 0.39; Ni 0.045; Cl 5.3.

Waelz Oxide from Steelmaking Dusts. Leaching was more difficult; single stage leaching, finishing at pH 3.35, extracted only 67% of the zinc. Multi-stage leaching using the "neutral" and "acid" procedures customary in the primary industry was more successful. Countercurrent leaching using solutions returned from downstream stages appeared the most economical, but was abandoned because of an increasing circulating load of iron between the two operations. The simpler cross-current technique was quite effective, but the excess acidity had to be neutralised for iron to be removed as hydroxide. When the solutions were to be subsequently purified by cementation, alkaline feedstock could not be used for this because of the re- introduction of halides, but when solvent extraction was used to purify the electrolyte, halides were unimportant, and the excess acid could be used to extract the zinc from additions of feedstock, albeit at lower efficiency. Overall extraction was 98-99%, falling to 93% when losses of zinc in the iron cakes were allowed for. Typical leach solutions, passing to purification, contained (in kg m^{-3}) Zn 100; Fe 9.5, Cd 0.5; Cu 0.15, Cl 3; F 1.5; Ni $9x10^{-3}$; Co $1.5x10^{-3}$.

High-Zinc Arc Furnace Dust. The zinc concentration was much lower with this dust than with the other materials subjected to acid leaching. Two stage cross-current leaching, as described for the Waelz oxides was tested but extraction of iron by the "acid leach" was very high, and usually the filtrates contained 20-22 kg m^{-3} Fe. With these materials the acid leach pulps were also very difficult to filter, as were the slurries formed when additional steelmaking dust was added to neutralise the excess acid and precipitate the dissolved iron. The quantity of dust added at this stage was relatively large so that overall recoveries, about 80% of the zinc, were almost the same in the two stage leach with subsequent neutralisation, as with a single stage "neutral leach". It was accepted however that the low value of the feedstock permitted lower recovery of the zinc, so that a single stage leach could be adopted for this material.

Alkali Leaching. Leaching with ammoniacal ammonium carbonate solution was studied in connection with the BOS-converter dusts. The lixiviants were expected to be highly selective for zinc, thus eliminating extensive solution purification. The leaching operation proved unexpectedly difficult, due to co-dissolution and subsequent precipitation of iron.

Ammonium-carbonate-ammonia solutions were prepared from laboratory grade ammonium bicarbonate-carbamate solution, with ammonia solution added to obtain the required $NH_3:CO_2$ ratio. Preliminary tests showed that the target level of 1% Zn in the leached material could be reached at room temperature only with concentrated solutions and long contact times. Increasing the temperature to 60°C, which would necessitate pressure-resistant, sealed equipment allowed more dilute solutions to be used with 4 h contact. However on standing the leach-filtrates were seen to become turbid, with deposition of brown solids. Unlike analogous processes where ammonium carbonate is used to leach metallic copper or nickel, the dissolution of zinc oxide does not require the use of oxidising conditions. Illis (14) states that oxidation of ferrous ammonium carbonate to ferric hydroxide occurs at E_H-0.15 V (SCE); measurements on these leach solutions gave 0.6 V immediately following filtration, and slow air-oxidation caused precipitation. The solutions were therefore unstable and not suitable for treatment by solvent extraction. Oxidation during filtration also tended to blind the filter medium, making filtration difficult. There appeared to be two methods of overcoming the problem, (a) by preventing the formation of $FeCO_3(NH_4)_2CO_3$ or (b) by oxidising the leach-pulp before filtration. Attempts to prevent the formation of the iron complex, by restricting the concentration of carbonate, were not successful, as the reduced dissolution of iron was accompanied by lowered extraction of zinc. Oxidative leaching was therefore studied, using an overpressure of pure oxygen, and agitation to promote its absorption.

Oxidative Leaching. The study of oxidative leaching techniques involved the use of varying concentrations of reagents, oxygen overpressure, methods of addition of oxygen and leaching temperatures and times. The work eventually resulted in a method whereby 85-90% of the zinc was extracted, but the solutions contained only 0.02-0.10 kg m^{-3} of iron. Details of the work cannot be divulged at the present time but will eventually be published as the patent application (15).

Purification of the Leach Solutions

The requirements regarding the purity of electrolytes for efficient electrowinning of the zinc have been mentioned earlier. Because of the large amount of published information, and the commercial success of cementation processes for acidic zinc electrolytes, these were considered at first to be the preferred route. However, when it was shown that cementation failed to remove nickel and cobalt from some of the solutions, and that organic precipitants were necessary, solvent extraction of the zinc was studied as an alternative process.

In the case of alkaline leach solutions, the proposed process included solvent extraction of the zinc by a chelating extractant, followed by stripping with acid, and purification of the acid-strip solution as required.

Purification of Acid-Leach Solutions. Solutions resulting from the acidic leaching of all the feedstocks tested contained the following impurities - halides (Cl, F), Fe, Cu, Cd, Ni, Co and traces of other metallic elements which would automatically be removed along with the major impurities. A stagewise purification procedure was therefore adopted, employing the techniques used in the primary industry.

ELIMINATION OF CHLORIDE: Two methods are available for removing most of the chloride

(a) Washing the feedstock with water prior to leaching
(b) Precipitation of chloride, as cuprous chloride, from the leach solution.

Exploratory tests in washing the feedstocks generally showed insufficient removal of chloride. Such a process would also make more difficult the water balances in the process, as a nett removal of water in filter cakes is necessary in order to be able to use reasonably small bleed-off volumes.

Chemical precipitation by the following reaction was therefore studied.

$$2Cl^- + Cu^{++} + Cu^o = 2CuCl$$

It is obviously necessary to have an equivalent amount of copper in solution, and it was usually necessary to add copper sulphate. Reactive cement copper was also needed; this was produced either by direct cementation from copper sulphate with zinc dust, or by dissolving the cuprous chloride from previous tests in hot ferrous chloride solution, and cementing out the copper with iron powder. In practice the removal of chloride was found to be most efficient at near boiling temperature and a large excess (at least 100%) of cement copper was required. The technique worked well at the ambient pH of the leach solutions. Some typical results are given in Table 3.

ELIMINATION OF IRON: Iron can be removed by oxidation and hydrolytic precipitation. Thus the ferrous iron in the solution following chloride removal was oxidised by adding manganese dioxide or potassium permanganate and the pH was adjusted to 4.5-5 with lime or caustic soda. When the all-cementation purification route was used it was not possible to use additional feedstock for pH adjustment as this would re-introduce halides; when solvent extraction was used however the presence of chloride was immaterial and further dust could be used for neutralisation.

The iron precipitated as a crystalline goethite and filtration was generally easy. The filter cakes were washed with one displacement volume of hot water; some zinc was held in the cake, but losses were usually less than 5%. Some typical results are shown in Table 4.

Zinc remaining in the filter cake could also include that unleached from added dust; when dusts were used for neutralisation the filter cakes contained 10-17% Zn, indicating that about 25% of the zinc in the added dust had not dissolved, and losses were equivalent to 7-8% of the zinc in the process.

REMOVAL OF NON-FERROUS METAL IMPURITIES BY CEMENTATION: The usual practice in primary zinc production is to cement out metallic impurities in two stages by additions of zinc dust. The first stage, carried out at ambient or moderate temperature precipitates the more electronegative elements; viz, precious metals, copper, cadmium etc; whilst the second stage, carried out at boiling point in the presence of "activators", precipitates principally nickel and cobalt.

Tests on the first cementation were carried out on solutions at ambient temperature (50°C) after filtering off the "iron cake". In general little difficulty was encountered, and the cadmium content of the precipitated solids, particularly those derived from copper smelter fumes appeared to be high enough to justify a credit. Typical results of these tests are presented in Table 5.

Table 3. Removal of Chloride by Cementation at 90°C

Feedstock	Cl in leach soln kg m^{-3}	pH	$CuSO_4 5H_2O$ added kg m^{-3}	time h	Final pH	Final Cl conc kg m^{-3}	Cl removed %
Cu Smelter Fume(a)	7.03	3.75	14.0	1*	4.2	1.79	74.5
	"	2.5	"	1	4.5	1.31	81.4
	"	1.5	"	"	4.0	0.59	91.6
Waelz Oxide(b)	2.70	2.4	12.5	2	4.0	0.250	90.7
	2.30	1.8	10.7	3.5	3.6	0.210	90.9
	2.91	1.2	4.0	1.75	1.5	0.284	90.2

(a) Solution contained (kg m^{-3}) Zn 110, Cu 3.9, Fe 0.4   * at 20°C,
(b) " " " Zn 100, Cu 0.15, Fe 9.5   rest at 90°C

Table 4. Removal of Iron by Precipitation

Feedstock	Initial Fe conc kg m^{-3}	Oxidant	Final pH	Alkali used	Final Fe conc kgx10^{-3} m^{-3}	Zn held up %
Copper Smelter Fume	0.39	$KMnO_4$	5.3	NaOH	0.25	2.2
" " "	"	"	"	$Ca(OH)_2$	1.0	1.4
Waelz Oxide	2.8	"	"	"	0.5	4.9
" "	9.2	"	4.6	NaOH	4.6	5.2
" "	9.2	MnO_2	4.9	"	2.4	5.2
" "	9.2	"	5.1	$Ca(OH)_2$	0.9	9.7

Table 5. Cementation: Copper and Cadmium removal

Feedstock	Initial Conc kg m^{-3}		Zn dust kg m^{-3}	Final Conc kg m^{-3}		Solid Product	
	Cu	Cd		Cu	Cd	Cu %	Cd %
Copper Smelter Fume	1	6	6.5	0.04	0.04	11.8	36
Waelz Oxide	0.12	0.41	1.0	0.0003	0.0015	2.04	14.7

Attempts to reproduce results obtained in the primary industry in the second cementation stage were, however, generally unsuccessful. It was found impossible to guarantee the production of electrolytes containing less than 10^{-3} kg m^{-3} of nickel and cobalt by cementation alone, and, in the pilot scale production of electrolyte from copper smelter fumes, it was necessary to add dimethylglyoxime and 1.nitroso 2 naphthol and filter off the precipitates before successful electrowinning tests were performed. An intensive study of the literature pertaining to the second stage cementation operation provided no clue as to the reasons for failure. Activators Sb_2O_3, As_2O_3 and $CuSO_4$ were used, and cementation was carried out at pH values up to the maximum for stable solutions (5.5) both in open vessels and under anaerobic conditions, with various types of stirring. The feedstocks for the tests were copper-smelter fumes and Waelz oxide; the brass-melting fumes contained no nickel or cobalt. Cementation was abandoned before the work on arc furnace dusts commenced.

A noteworthy feature concerning the removal of nickel and cobalt was that cementation became less efficient as solutions were recycled through the leaching, purification and electrowinning stages, suggesting the build up of some deleterious material in the process. Table 6 presents data concerning pilot scale tests (0.25 m^3 batches of electrolyte containing about 120 kg m^{-3} Zn) with successive additions of zinc dusts. It can be seen that in the virgin batch, nickel and cobalt concentrations did not greatly exceed the target. By the third recycle, nickel concentration was $3 x 10^{-3}$ kg m^{-3} and current efficiency on electrowinning had fallen to 50%.

Samples of solution from batch 4 were subjected to further treatment to remove more of the nickel and cobalt. More electropositive reducing agents (Mg 0.5 kg m^{-3}, Al 1.0 kg m^{-3}) were added but these left the Ni and Co concentrations virtually unchanged. However, addition of the organic precipitants dimethylglyoxime (Na), 1 nitroso 2 naphthol (0.3 kg m^{-3} each) lowered the concentrations to the target and allowed the electrowinning of zinc at an efficiency of 85%.

Solutions prepared from the Waelz oxide also failed to respond to cementation. Results of tests in Table 7 (mainly open vessel tests) show that repeated additions of zinc dust failed to remove the nickel and cobalt.

In some tests the nickel and cobalt appeared to redissolve with prolonged treatment and the presence of a complexing ion was suspected. Of the possible contaminants, fluoride appeared most likely to be the complexant, particularly as the Waelz oxide solutions, from a feedstock with a higher fluorine

Table 6. Cementation: Solutions derived from smelter fumes

Batch	Zn dust added (cumulative) kg m^{-3}	Sb_2O_3 added (cumulative) kgx10^{-3} m^{-3}	concentrations achieved Ni kgx10^{-3} m^{-3}	Co kgx10^{-3} m^{-3}
1 (virgin)	4.0	1.0	3.44	2.10
	6.0	1.8	1.20	1.90
2 (1st recycle)	5.0	1.0	7.25	3.5
	6.5	2.0	5.0	ND
	6.9	3.0	5.0	ND
3 (2nd recycle)	1.0	1.0	9.3	4.4
	7.0	1.0	3.1	3.5
	8.0	2.0	1.15	3.3
4 (3rd recycle)	2.0	nil	3.7	3.35
	4.0	"	3.5	4.1
	7.0	"	3.1	4.0

ND not determined

Table 7. Cementation: Solutions derived from Waelz Oxide

Test	Zn kgm^{-3}	pH	Zn added (cumulative) kg m^{-3}	Temp °C	Time h	Analysis of treated soln (kgx10^{-3} m^{-3}) Ni	Co
1	135	5.3	4.0	90	1	8.6	1.1
			8.0	"	1	8.4	ND
2	92.9	4.5	4.0	"	0.5	12.9	ND
			8.0	"	1.0	7.0	ND
		3.5	9.0	"	"	7.0	ND
			12.5	"	(s)	3.9(a)	
			18.5	"	(s)	4.7(a)	
3	92.2	5.2	4.0	50	1	11.6	1.54
		3.5	8.0	90	0.5	3.0	0.85
		"	8.0	"	3.0	8.0	1.4
		2.3	11.5	"	0.75	5.7	ND

(s) cementation carried out in a sealed vessel
(a) 0.7x10^{-3} kg m^{-3} $CuSO_4$ added
ND not determined

content, showed the worst effects.

A brief study of the effect of fluoride iron was carried out by spiking solutions of zinc sulphate (94 kg m^{-3} Zn) and nickel nitrate (0.01 kg m^{-3} Ni) with sodium fluoride (0, 1 or 2 kg m^3 F). To samples of these solutions were added zinc dust (2 kg m^{-3}) and copper sulphate (0.2 kg m^{-3}). The solutions were heated to 90°C, stirred, and samples were withdrawn at intervals to follow the reduction in the concentration of Ni. The results are given in Fig 1 and show that fluoride ion retards the precipitation of nickel, especially at low pH, although re-dissolution was not apparent. Curve 6 shows that some improvement was obtainable by precipitating the fluoride as calcium fluoride. This effect of fluoride ion does not appear to have been reported in connection with the recovery of primary zinc, so a synergistic mechanism may be present with these solutions. Because most of the metallurgical dusts and fumes contain fluorides, a different method of purification was needed, thus solvent extraction was examined.

REMOVAL OF NON-FERROUS IMPURITIES BY SOLVENT EXTRACTION: The extraction of zinc by DEHPA at low pH was described by D'Arcy et al (13), and processes involving its use have been proposed by Reinhardt (3) and Barthel et al (17). Work at the Warren Spring Laboratory aimed at removing zinc from nickel-plating solutions had shown separation factors ($\frac{Zn}{Ni}$) of $2.2x10^4$ and as the raffinates can be returned to leaching, and complete extraction of zinc is not necessary, the extraction of nickel may be further inhibited by "squeezing out". The first stage in the presently reported work was to produce an isotherm for the extraction and Fig 2 shows the isotherm obtained using a 50 vol% solution of DEHPA in MSB 210 (aliphatic kerosene) with 10% of isodecanol as third-phase inhibitor. A McCabe Thiele construction showed that 4 countercurrent extraction stages would be needed. The solvent was stripped with return electrolyte which would contain about 50 kg m^{-3} Zn and 150-200 kg m^{-3} H_2SO_4; past experience suggested three countercurrent stages for this. After preliminary countercurrent tests which confirmed the data, a complete scheme was devised which included scrubbing, and the recovery of zinc from that part of the raffinate which was bled off to maintain water- balances. The scheme is shown in Fig 3, with 11 stages in all, two being used for scrubbing (with return electrolyte) and two for zinc recovery from the raffinate-bleed. The circuit resembles in some respects that proposed by Reinhardt (3) but incorporates the necessary bleeds of spent electrolyte and raffinate in single circuit, and minimises losses of zinc. For the testwork, 11 mixer settlers of 0.2 dm^3 mixer capacity were used, and the incorporation of a small electro- winning cell allowed complete simulation of the full size circuit.

Before passing to solvent extraction, the solutions required the following prior treatment using techniques similar to those already described viz:

(a) Removal of iron by precipitation so that the solvent would not be blocked by extraction of Fe^{III}.
(b) Removal of cadmium and copper by cementation. Although this could have been effected by solvent extraction it would have entailed further elaboration of the circuit, and, as a cadmium-rich cementation product should have a ready sale, cementation was chosen on grounds of simplicity and economics.

The solutions fed to solvent extraction contained (kg m^{-3}) Zn 67-95, Ni $20\text{-}40x10^{-3}$ Co $1\text{-}4x10^{-3}$, Fe $0.2\text{-}2x10^{-3}$. During operation results were largely as expected, but some additions to the circuit were found to be necessary. The aqueous phase in the scrubbing stages was recycled from settlers to mixers to give a phase ratio of about unity in the mixers, and a filter of active carbon was used ahead of the electrolytic cell to eliminate "solvent burn" and maintain a good quality zinc deposit. Continuous testing showed that raffinates produced from feed solutions containing 67-95 kg m^{-3} Zn contained 40-45 kg m^{-3} Zn. Electrolytes produced by stripping the solvent contained less than $1x10^{-3}$ kg m^{-3} each of Ni and Co, and current efficiencies were in the range 82-87%.

An extraction profile was constructed by analysing the stages at shut down (Fig 4) and this also agreed well with the original extraction isotherm (Fig 2) showing a close approach to equilibrium. Calculated Murphree stage efficiencies were close to 100% in extraction and 50-70% in stripping.

Solvent extraction thus appeared to be an effective way of obtaining high-purity zinc electrolytes from the leach solutions.

Purification of Alkaline Leach Solutions. The concentration of zinc in the ammonium carbonate-based leach solutions was very much lower, typically 6-8 kg m^{-3}, so that concentration as well as purification was needed to obtain solutions for electrowinning. Methods were required which allowed the re-use of the reagents for leaching, since ammonium carbonate is costly.

Three main types of process were available. The solutions could be decomposed by heat, and the ammonia and carbon dioxide recovered by scrubbing the off-gases, whilst the zinc would be precipitated as an impure carbonate, which could be re-dissolved in acid and the solution purified and electrolysed. This type of process was basically unattractive because the problem elements nickel and cobalt would be leached along with the zinc, as would any halides, thus making purification by cementation as difficult as with the acid-based processes. Another method would be to purify the ammoniacal solution and to electrowin the zinc from the purified solution. Such a process would be difficult to operate, as the current efficiency of electro-deposition from ammoniacal solutions is very low unless chemically-reducing conditions can be maintained. The third method would be to use liquid or solid ion exchangers to transfer the zinc to an acidic solution, whilst not transferring the anions which would otherwise interfere with the subsequent purification of the acidic electrolytes.

The last type of process was chosen for development, and liquid ion exchange (solvent extraction) was employed. Extractants were chosen which extracted the zinc with the minimum co-extraction of ammonia, and a process was developed whereby the zinc was stripped with acidic return electrolyte to form the feed for the electrolytic cell. As the interfering anions (Cl^-, F^-) were not

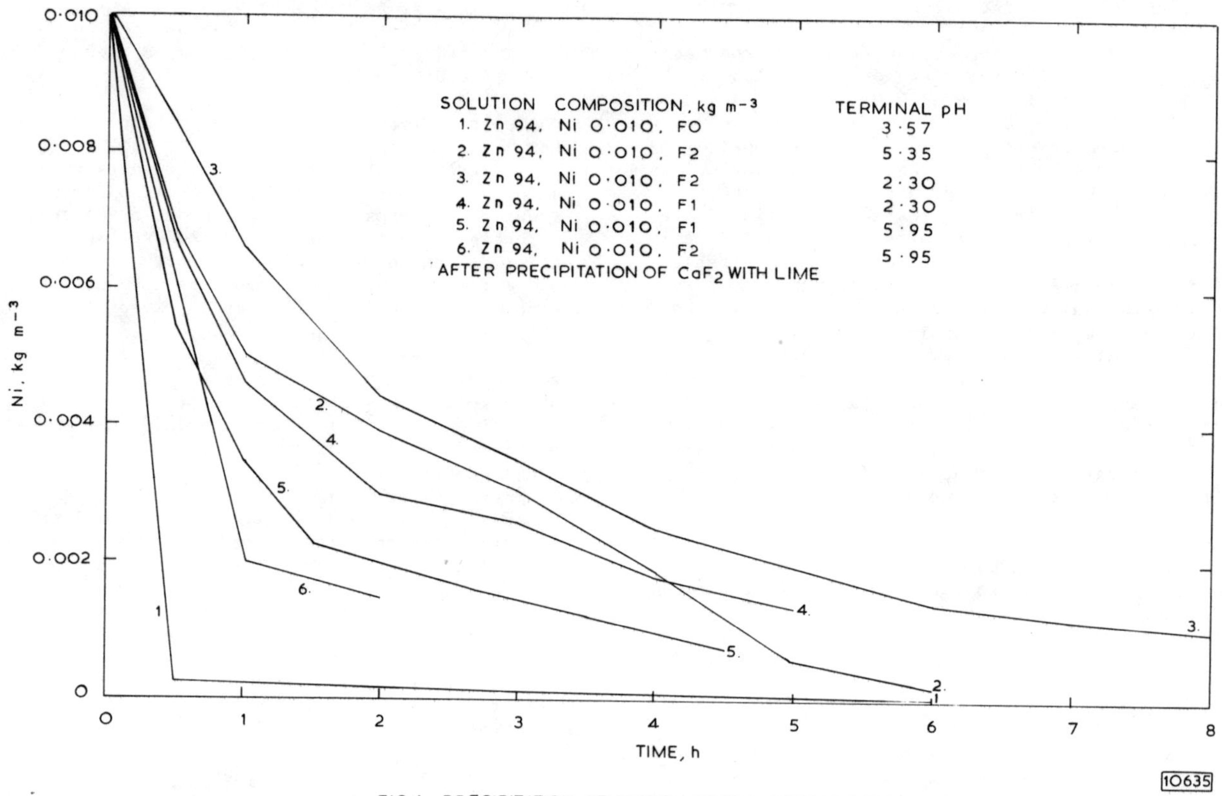

FIG.1 PRECIPITATION OF NICKEL FROM ZINC ELECTROLYTES

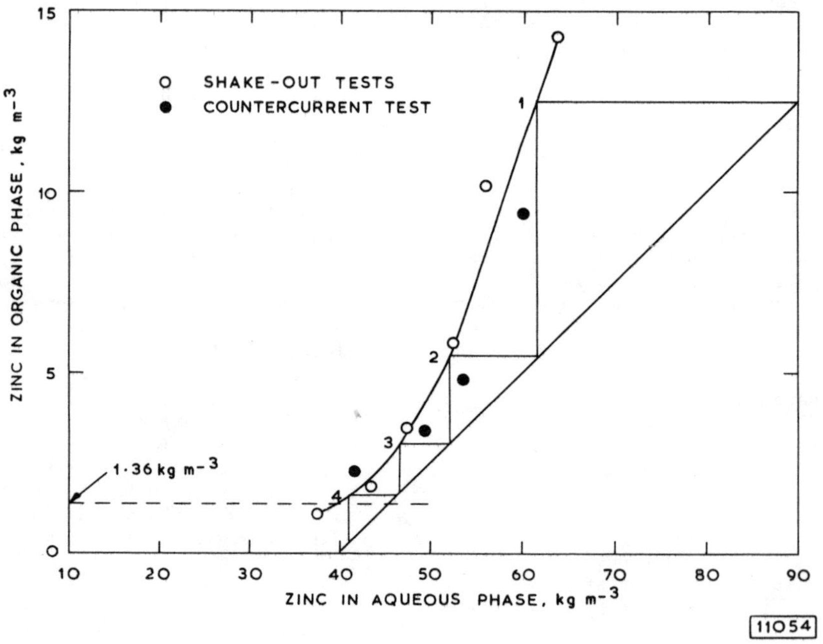

FIG. 2 EXTRACTION ISOTHERM DEHPA - Zn

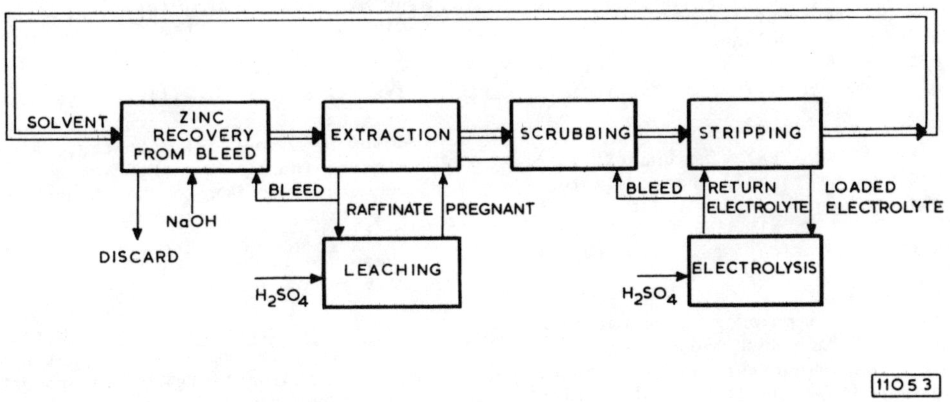

FIG. 3 OUTLINE OF THE PURIFICATION SCHEME FOR WAELZ OXIDE

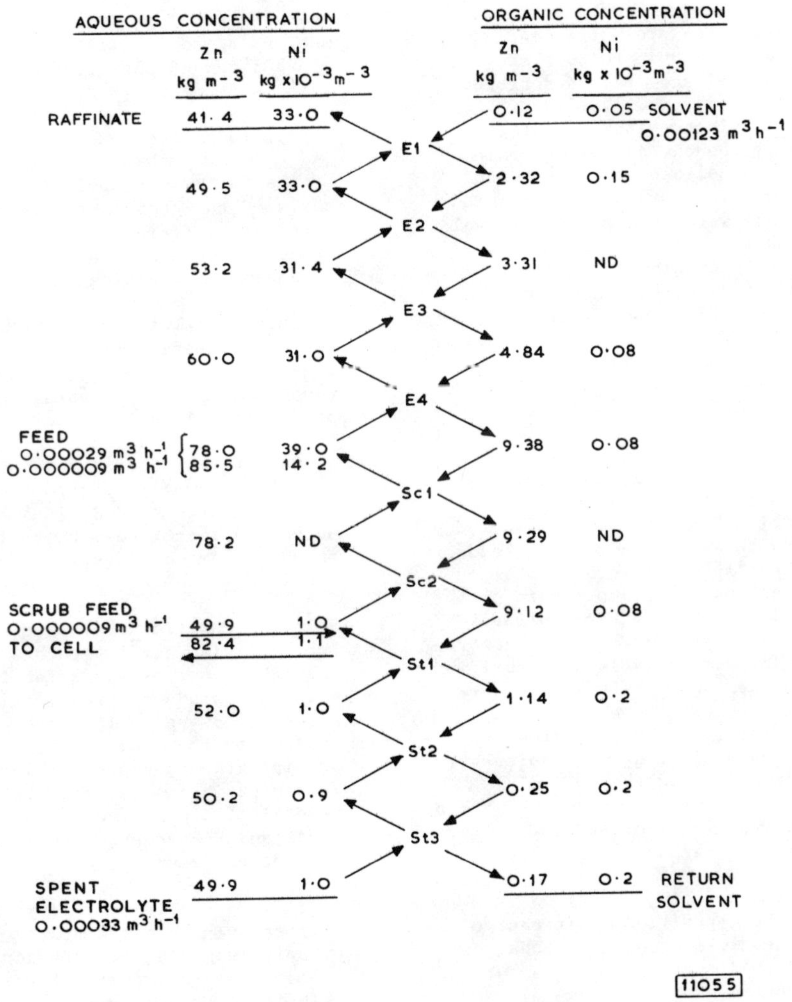

FIG. 4 EXTRACTION PROFILE

extracted, the electrolytes could be freed from copper, nickel and cobalt by cementation, and continuous operation of the solvent extraction showed current efficiencies of 85-90% and recovery of over 98% of the ammonia from the solvent extraction circuit.

PURIFICATION OF ELECTROLYTES: The quantities of iron to be removed were very small, hence simple oxidation (by H_2O_2 or MnO_2) and adjustment of the pH to 4.0 (with ZnO) was all that was required. As the pH of the strip solutions was 2.8, very little alkali was needed.

Nickel and cobalt were removed by cementation at 90°C, using 4 kg m^{-3} of zinc dust and 0.4 kg m^{-3} of cupric sulphate. It is noteworthy this operation succeeded easily because of the absence of complexing anions. Some typical compositions of electrolytes, before and after treatment, are given in Table 8.

Electrowinning

Zinc is electrowon from acid solutions preferably containing at least 80 kg m^{-3} of zinc using lead anodes and aluminium or stainless steel cathodes. Providing that the electrolytes are sufficiently pure, current efficiencies exceeding 90% are possible. The effects of impurities have already been described.

Acid-leach Electrolytes. Electrowinning tests were carried out on laboratory, bench, and pilot scale. For the bench-scale tests, cells containing 5 anodes and 4 cathodes were used. The current was measured directly, and the cathodes were weighed at intervals. A noteworthy feature was the delayed effect of nickel/cobalt impurities in continuous electrolysis. When nickel/cobalt impurities were present, current efficiency could fall from 90% to 60% after 6-12 h operation at densities of 300-400 Am^{-2}, but, provided the concentrations of Ni or Co did not exceed about 10^{-3} kg m^{-3}, efficiency was maintained at 80-90%, producing spent electrolytes containing 30-50 kg m^{-3} of zinc.

Pilot scale tests were carried out on solutions produced from copper-smelter fumes. The cell contained two cathodes of 0.81m x 0.29m constructed from aluminium sheet and three anodes of 99/1 lead/silver. The anode-cathode spacings were maintained at 0.04 m by plastic spacers. The temperature of the electrolyte was kept below 25°C by water-cooling grids; higher temperatures appeared to exacerbate the effect of impurities. The electrolytes were purified by cementation only. Nickel and cobalt levels were above the prescribed limits, but the electrolytes nevertheless gave good deposits at high current efficiency in static bench scale tests. Results with one such solution, which contained (kgx10^{-3} m^{-3}) Ni 3.1, Co 3.3, were satisfactory for about 80 h then efficiency deteriorated rapidly with visible evolution of hydrogen at the cathode (Table 9).

Further bench-scale tests, using solutions from Waelz oxides, which had been purified by intensive cementation to reduce the nickel concentrations to around 10^{-3} kg m^{-3} confirmed that the fall in current efficiency was due to contamination (Table 10).

Electrowinning tests were carried out concurrently with purification by solvent extraction. Current efficiencies of about 85% corresponded to nickel concentrations of about 10^{-3} kg m^{-3}, but the delayed effect of impurities in continuous operation did not allow any test-by-test relationship to be established.

Alkali-leach Electrolytes. These electrolytes originating from the processing of BOS converter dusts by ammoniacal leaching, solvent extraction and electrowinning were purified by precipitation and cementation. Due to the absence of complexing anions, nickel and cobalt were removed satisfactorily (Table 8). Bench scale electro-winning tests on these electrolytes showed high current efficiency. Results are shown in Table 11.

Quality of Electrowon Zinc. Metallic zinc was stripped from the cathodes in the various electrowinning tests. Each product was melted and cast into ingots. Samples were taken by drilling. Table 12 shows the compositions of these samples together with the ASTM specifications for various grades of zinc.

The products meet ASTM high grade (HG) specification except for the lead content in three specimens. The lead had certainly originated from the anodes and it is probable that with prolonged use, Pb/Ag anodes will develop an impervious coating of PbO_2 and HG specifications will then always be able to be met.

Conclusions Based on the Experimental Studies

The following conclusions were reached on the basis of the above experimental studies and these conclusions formed the basis for the construction of process flow sheets and estimation of material balances for treating the various feedstocks.

Leaching. All the zinc-rich materials could be leached satisfactorily with sulphuric acid using single stage or two stage cross-current techniques. Countercurrent techniques were not suitable because of increasing cyclic loads of iron.

Selectivity for zinc is shown in leaching with ammoniacal ammonium carbonate solutions, provided that the correct conditions are chosen, and the method is therefore inherently suitable for the removal of minor amounts of zinc whilst conserving the iron values.

Solution Purification. The conventional treatment for purifying acidic leach solutions from primary calcine is not always effective for treating acidic solutions produced from metallurgical dust and fume products, owing to the higher concentrations of halides in these materials. Although chloride may be precipitated as cuprous chloride, the fluoride remaining in the solution retards or prevents the removal of nickel and cobalt by cementation with zinc dust, and subsequent electrolysis shows falling current efficiency.

It is possible to purify the acidic solution further by precipitating the nickel and cobalt with organic reagents, but the cost of these reagents will probably be too great. Nevertheless, the cementation technique would be suitable for feedstocks which do not contain nickel or cobalt, eg fumes from brass ingot making.

The solvent extraction route appeared to be robust and effective for processing acidic leach solutions,

Table 8. Purification of Electrolytes (from Alkaline SX)

Test		Zn kg m^{-3}	Fe kgx10^{-3} m^{-3}	Ni kgx10^{-3} m^{-3}	Co kgx10^{-3} m^{-3}
1	Acid strip solution	67.3	25.3	11.2	0.4
	Purified electrolyte	68.9	1.1	0.5	0.4
2	Acid strip solution	67.0	15.0	16.2	1.0
	Purified electrolyte	68.0	1.6	0.9	0.9
3	Acid strip solution	67.1	22.0	5.6	0.3
	Purified electrolyte	68.1	0.6	1.0	0.2

Table 9. Pilot Scale Electrolysis of Zinc solution produced by acid leaching

Time (cumulative) h	Current (during period) Ah	Current Density A m^{-2}	Zinc deposited kg	Current Efficiency %	Zinc deposited (cumulative) kg	Average Current Efficiency %
25.17	4454.5	361	4.537	83.5	4.537	83.5
48.54	4213	368	4.118	80.15	8.655	81.9
77.32	5131.5	364	4.601	73.5	13.256	78.8
84.62	1362.9	381	0.773	46.5	14.029	75.9

Table 10. Electrolysis of Zinc solutions produced by acid leaching then purified by intensive cementation

Zinc dust added kg m^{-3}	Zinc concentration kg m^{-3}	Ni concentration kgx10^{-3} m^{-3}	Current Density A m^{-2}	Zn concentration in spent electrolyte	Current Efficiency %
15.3	111	1.05	314	39.9	93.25
24.8	97.9	0.90	330	28.7	88.9
12.9	110	0.89	336	43.9	87.6

Table 11. Electrolysis following SX from alkaline leach solutions

Zinc conc kg m^{-3}	Nickel conc kgx10^{-3} m^{-3}	Current Density A m^{-2}	Duration of Test h	Zinc conc spent electrolyte kg m^{-3}	Current Efficiency %
65.5	0.20	416	30	28.0	94.6
68.3	0.55	375	29	23.6	91.2
68.9	0.52	416	33	31.5	90.0
68.0	0.85	416	34	29.5	88.5

Table 12. Analysis of the electrowon zinc samples

Origin of electrolyte	Anode material	Pb %	Cd %	Fe %	Cu %	Ni %	Co %	Mn %
Copper smelter fume Acid leach-cementation lab scale	Pb	0.0986	0.0030	0.0014	0.0031	0.0011	0.0006	0.0002
Copper smelter fume Acid leach-cementation lab scale	Pb/Ag	0.0207	0.0027	0.0012	0.0006	0.006	nil	nil
Copper smelter fume Acid leach-cementation pilot scale	Pb/Ag	0.110	0.016	0.002	0.002	nil	nil	nil
Brass ingot fume Acid leach-cementation lab scale	Pb/Ag	0.0214	0.0001	0.0015	0.0008	0.0006	0.0006	0.00003
Waelz oxide Acid-leach SX lab scale	Pb/Ag	0.049	-	<0.001	0.003	0.001	<0.001	<0.001
BOS Converter dust Alkali leach SX lab scale	Pb/Ag	0.10	-	0.05	0.05	-	-	0.001
Specifications (ASTM)								
SHG		0.003	0.03	0.003				
HG		0.07	0.03	0.02				
I		0.20	0.50	0.03				

provided the solution was freed from iron, copper and cadmium before SX. The complete isolation of the leaching and elecrowinning circuits, except for controlled recycle, is attractive since it lessens the risk of electrolyte contamination by complexing anions.

The alkaline leach solutions required concentration as well as purification before electrolysis, hence solvent extraction was an obvious choice. Purification of the acid strip solutions following solvent extraction was easily accomplished by cementation with zinc dust.

Electrowinning. The results of the electrowinning tests confirmed the need to reduce nickel and cobalt concentrations to 10^{-3} kg m^{-3} or less. Their effects were usually delayed for 20-30 h, but then a rapid fall in current efficiency occurred. The effect of impurities was exacerbated by high temperatures, and electrolytes had to be maintained below 25°C, using cooling grids or coils in the cells.

FLOWSHEETS

Flowsheets for processing various feedstocks are given in Figs 5 and 6.

(1) Copper-smelter fumes using acid-leaching and purification by precipitation and cementation.
(2) Waelz oxides by acid leaching and purification by solvent extraction.
(3) The flowsheet for the ammoniacal leaching of BOS dusts cannot be presented at this time, but some comments are added on the economics.

Quantities refer to 1 tonne of zinc cathode and are presented in kg or m^3 for (1) and (2)

Reagent and Energy Requirements

Acid Leach-Cementation (Flowsheet 1 (Fig 5) for copper smelter fume)

Reagent	Quantity kg	Unit Cost £kg^{-1}	Cost £t^{-1}Zn	Total Costs £t^{-1}Zn
Sulphuric Acid	114	0.033	3.76	
Zinc Dust	97.7	0.733	71.61	
Copper Sulphate	119.7	0.300	35.91	
Antimony Trioxide	0.027	7.52	0.20	
Manganese Dioxide	0.725	0.100	0.07	
Dimethyl Glyoxime (Na)	4.172	5.60	23.36	
1.nitroso 2 naphthol	4.172	19.00	79.27	214.18
Energy	kWh	£kWh^{-1}		
Electrowinning	3530	0.026	91.78	
Power and heat(equivalent)	1700	0.015*	25.50	
			117.28	117.28
				331.46

*average cost for electricity and fossil fuel

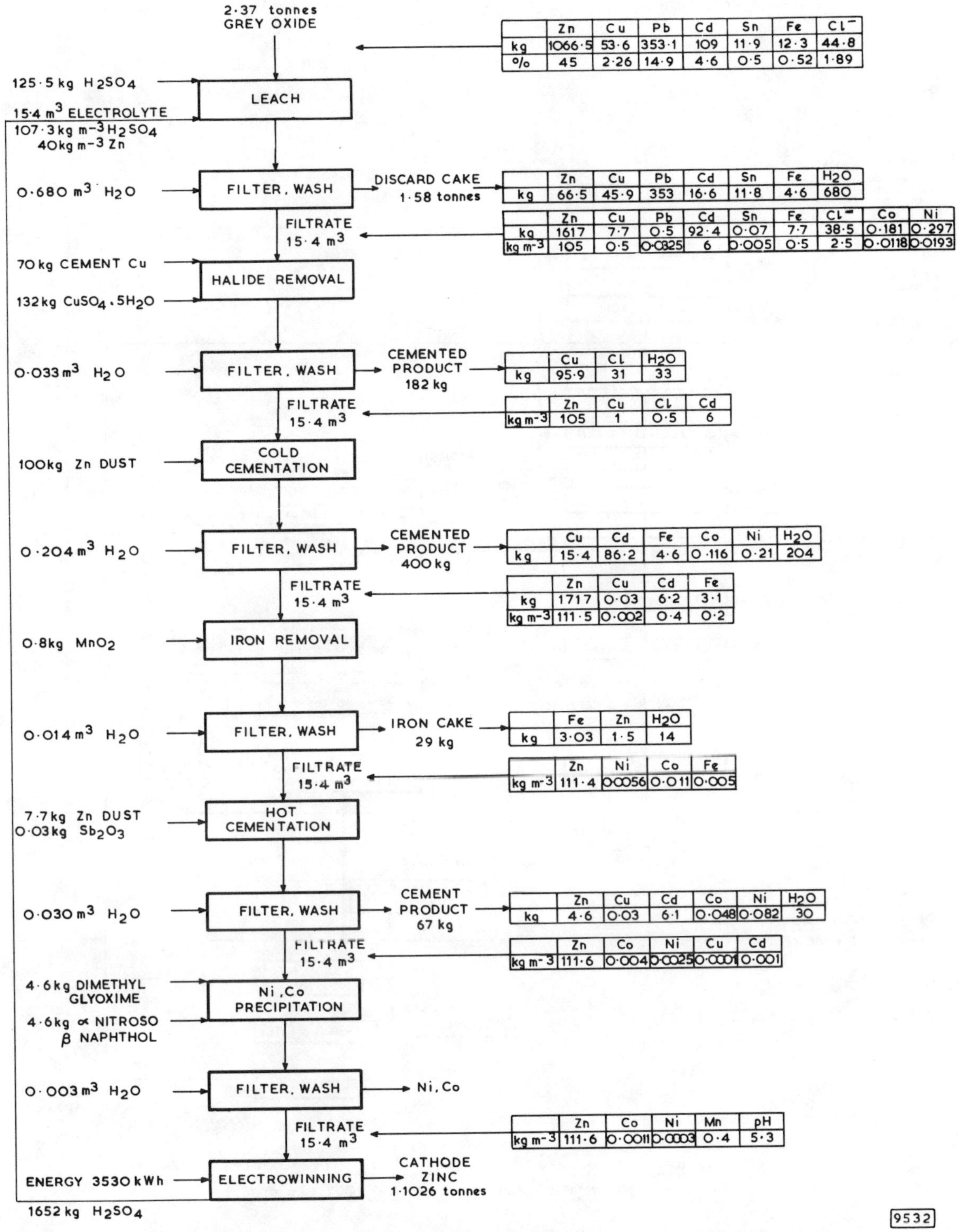

FIG. 5 PROCESS FLOW SHEET FOR COPPER SMELTER FUME (FOR 1 tonne Zn)

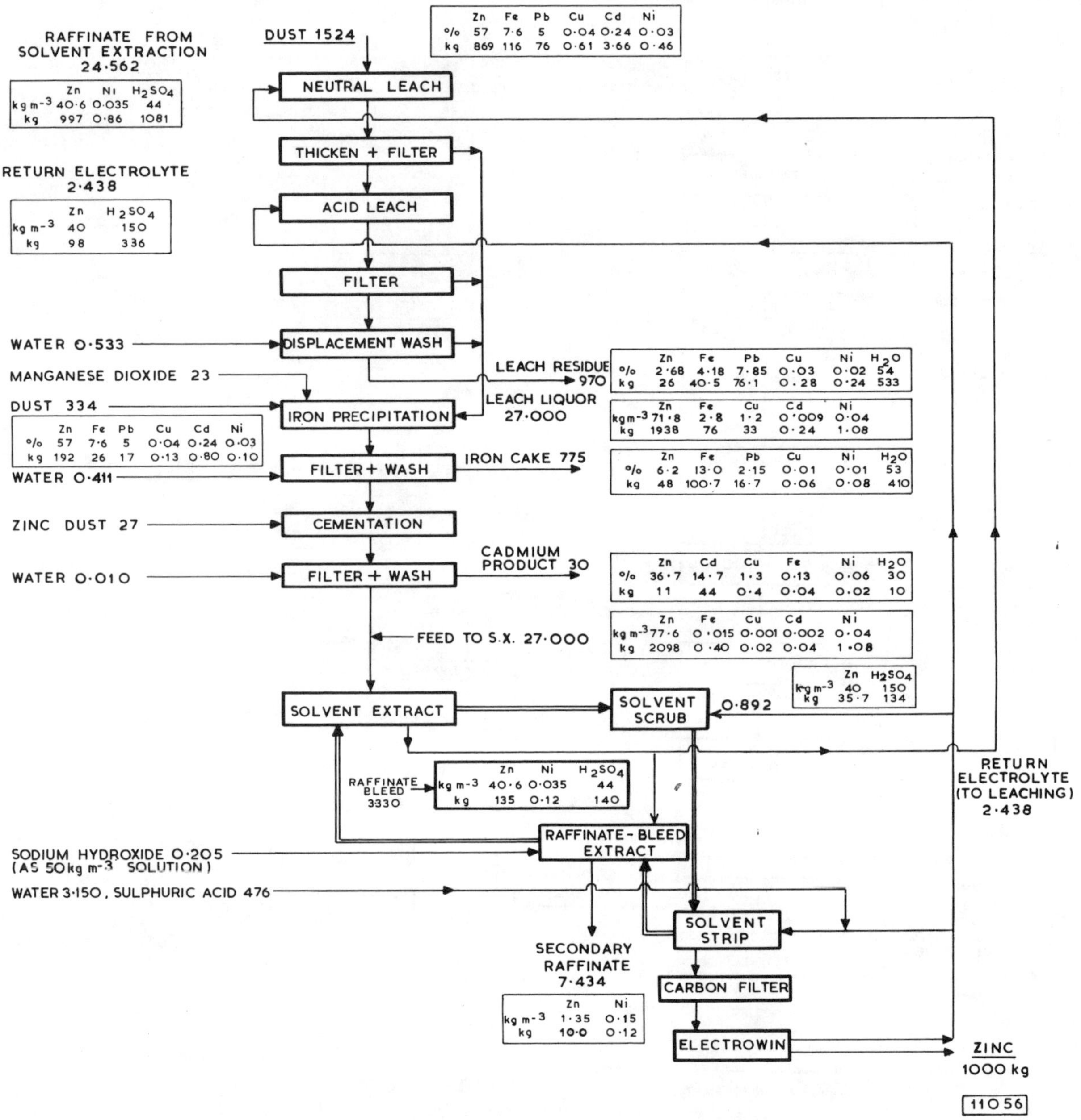

FIG. 6 FLOWSHEET FOR ZINC FROM WAELZ OXIDE
[QUANTITIES FOR 1t ZINC CATHODE (kg or m³)]

Acid Leach-Solvent Extraction (Flowsheet 2 (Fig 6) for Waelz Oxide)

Reagent	Quantity kg	Unit Cost £kg^{-1}	Cost £t^{-1}Zn	Total Costs £t^{-1}Zn
Sulphuric Acid	476	0.033	15.7	
Sodium Hydroxide	205	0.101	20.71	
Manganese Dioxide	23	0.100	2.30	
Zinc Dust	27	0.733	19.79	
DEHPA	1.0	4.00	4.00	
			62.5	62.5
Energy	kWh	£kWh^{-1}		
Electrowinning	3222	0.026	83.77	
Oil	0.053 m^3	160 m^{-3}	8.48	
			92.25	92.25
				154.75

Acid Leach-Solvent extraction for Arc Furnace Dust (Flowsheet 2 (Fig. 6))

Reagent	Quantity kg	Unit Cost £kg^{-1}	Cost £t^{-1}Zn	Total Costs £t^{-1}Zn
Sulphuric Acid	1147	0.033	37.85	
Sodium Hydroxide	209	0.101	21.11	
Zinc Dust	30.3	733	22.21	
DEHPA	1.0	4.0	4.0	
			85.17	85.17
Energy	kWh	£kWh^{-1}		
Electrowinning	3114	0.026	80.97	
Oil	0.06 m^3	160 m^{-3}	9.60	
			90.57	90.57
				175.74

Economics

Predictions of profit and loss are given for the acid based processes 1, 2, 2a.

(1) Throughput related factors: (£t^{-1} Zn)

		Unit value (£ kg^{-1})	(1)	(2)	(2a)
costs	(Feedstocks	(1) 0.0675 (2) 0.0775	160.0	144.0	-
	(Reagents and Energy		331.46	154.75	175.74
income	(Products - zinc	0.35	350	350	350
	(cadmium	2.00	172.4	88.0	5.5
	Value (products - inputs)		30.94	139.25	179.76

Notes (1) Copper smelter fumes by acid leach-cementation
(2) Waelz Oxide " " " SX
(2a) Arc furnace dust " " " "

Prices for feedstocks are estimated current selling prices, where appropriate.

Estimated Capital Charges* £x10^3

Flowsheet No	(1)	(2)	(2a)
Leaching	450	700	496
Purification	464	1833	1833
Electrowinning	2800	2800	2800
Effluent treatment	400	400	400
Working Capital**	800	800	800
Total Cost	4914	6533	6329
Capital Charge £t^{-1} Zn Based on amortization at12% over 10 y	72.8	101.4	97.9
Interest on working capital (12%)	9.6	9.6	9.6
	82.4	110.0	107.5
Overheads:			
Maintenance (5% of capital)	20.6	28.7	27.6
Labour (£180,000 y^{-1})	18	18	18
	38.6	46.7	45.6

Notes * Capital charges for plant producing 10,000 ty^{-1} Zn in (1) (2) (2a).
** based on 3 months stocks

Overall Economic Viability: (£ t^{-1} Zn)

Flowsheet No	(1)	(2)	(2a)
Capital and Interest	82.4	110	107.5
Overheads (labour and maintenance)	38.6	46.7	45.6
	121.0	156.7	153.1
Value of products less reagents and energy	30.94	139.25	179.76
Overall Profit*	-90.06	-17.45	+26.66

* (-) denotes loss

CONCLUSIONS

The technical feasibility of processing various zinc-bearing wastes was proved on a laboratory, and in one case, on a pilot plant- scale. The economics however are discouraging, only the processes for Waelz oxides and EAF dusts approach "break even" (allowing for weekly fluctuations in zinc prices).

If pyrometallurgical processes are not economic in the United Kingdom, hydrometallurgical processes are probably less so. There are several factors which must be considered:-

(1) The scale of operations, 10,000 t y^{-1}, is just too small for a base non-ferrous metal plant incorporating several unit processes.
(2) The lead values in the dusts are not recovered by the hydrometallurgical processes. Exploratory tests have been carried out on leach residues, using ammoniacal ammonium sulphate or sodium chloride as lixiviants, but estimated costs were greater than the value of the lead recovered.
(3) The very high costs of the organic precipitants used with the copper smelter fume are offset adventitiously by the high revenue from cadmium. The economics of an acid leach-solvent extraction process would resemble those for the Waelz oxide, ie about break even.
(4) The copper smelter fume and Waelz oxide already command high prices because of their value for chemicals production, and the costs incurred in producing the Waelz oxide. Both these materials can be treated in the zinc- lead blast furnace.
(5) The process for the BOS converter dusts was designed for a specific purpose - the alleviation of problems concerning waste disposal in the vicinity of the plant. Details of the economics cannot be divulged at this time, but a loss of about £25 t^{-1} dust is forecast, a figure which could be lower than the increased disposal charges if long-haul transportation were necessary.

REFERENCES

1. Serbent, H., Maczek, H, and Rellermeyer, H. Large scale test for the treatment of BF sludge and BOS dusts according to the Waelz Process. Proc. Ironmaking Conf. 1975, 34, pp. 194-205.

2. Shirimatsu, J., and Kasuharu, Y. Effective use of in-plant generated dusts using the SL/RN Process. Proc. Ironmaking Conf. 1975, 36, pp. 384-397.

3. MX Processor Reinhardt and Co. AB. Procédé de récuperation par voie humidé de zinc à partir d'une matiere contenant du zinc et fer. Fr Demande 2 338 997 Jan 1977.

4. Wendt, W.J. Ammonia-Ammonium Carbonate leaching of Low Grade Zinc Ores, Eng. Min. J. 154, (9), pp. 84-90.

5. Pooley, F.D. UK Pat. Appl. 28448/75.

6. Mantell, C.L. Electrochemical Engineering pp 210-224. McGraw-Hill Book Co., New York 1960.

7. Van Niekerk, C.J., and Allen, D.R. The electrolytic extraction of zinc at the Zinc Corporation of South Africa Ltd. J.S. Afr. Inst. Min. Metall. Feb. 1977, pp. 146-149.

8. Kerby, R.C., and Ingraham, T.R. The Effect of impurities on the current efficiency of zinc electrodeposition. Research Report R243. Mines Branch, Ottawa, Canada.

9. Fukubayashi, H.H., O'Keefe, T.J., and Clinton, W.C. Effect of Impurities and Additives on the electrowinning of zinc. Report RI7966. US Bureau of Mines, Washington DC.

10. Callaway, L.A., and Koepel, F.N. Metallurgical Plant of the Andes Copper Mining Company, Trans. AIME, 106, 1933, p. 678-729.

11. Haigh, C.J., and Pickering, R.W. Treatment of Zinc plant residue at the Risdon Works of the Electrolytic Zinc Co of Australia. Proc AIME World Conf. Mining Met. Lead & Zinc, 1970, pp. 423-428.

12. Van den Neste, E. Metallurgie-Hoboken- Overpelts Zinc Electrowinning Plant, CIM Bulletin, Aug 1977, pp. 173-185.

13. D'Arcy, R.S., Churchward, P.E., and Borrowman, S.R. Preliminary technical description of a process for separating copper and zinc by solvent extraction and for chemical refining of base metal concentrates. US Bureau of Mines, Salt Lake City.

14. Illis, A., Nowlan, G.C., and Koehler, H.J. Production of nickel oxide from ammoniacal process streams, CIM Bull. March 1970, pp. 352-61.

15. Pearson, D., and Wilson, J.C. Zinc Recovery from Metallurgical Fumes, UK Pat. Appl. 79 05194, 14.3.79.

16. Merrill, C.C., and Lang, R.S. Experimental caustic leaching of oxidised zinc ores and minerals and the recovery of zinc from leach solutions, United States Bureau of Mines Report RI 6576, 1965.

17. Barthel, G., Fischer, H., and Scheffler, U. Separation and recovery of copper and zinc from sulphuric acid solutions of metal salts by solvent extraction, Proc. Int. Sym. Complex Metall., 1978, pp. 121-133.

ACKNOWLEDGEMENT

The programme on acidic leaching was supported by the Chemicals and Minerals Requirements Board of the Department of Industry (UK).

Chapter 15

RECOVERY OF ALUMINUM FROM FLY ASH BY THE CALSINTER PROCESS

B. Zane Egan
Forest G. Seeley
A. Donald Kelmers

Chemical Technology Division
Oak Ridge National Laboratory
Oak Ridge, Tennessee 37830

INTRODUCTION

Coal ash, presently the fifth most abundant of the solid minerals produced in the United States, is expected to become the fourth most abundant by the end of this decade (1). The production of fly ash from the combustion of pulverized coal amounted to a total of 44 million metric tons (Mt) in 1977 and is expected to increase to 91 Mt by 1985. Coal ash, which results from the inorganic clay and mineral content of the coal, contains several elements and thus can be considered as a mined and crushed mineral resource rather than a waste material. Furthermore, potential problems associated with fly-ash disposal ponds and the leaching of potentially hazardous elements through groundwater infiltration and/or rainfall provide additional incentive for processing fly ash for environmental control as well as for the recovery of certain metal values.

Fly ash could provide an abundant domestic source of aluminum and several other metals. Aluminum is now recovered from bauxite, and over 90% of this ore is imported since domestic sources are limited and cannot meet the U.S. demand. There is a close relationship between the aluminum content of fly ash and the domestic consumption of this metal (Table 1).

Table 1. Relationship between the aluminum and titanium contents of fly ash and U.S. needs in 1977[a]

	Aluminum (Mt)[b]	Titanium (Mt)[b]
Domestic production	0.422	0.214
Imports	5.779	0.285
U.S. demand	5.637	0.456
Available from fly ash	5.062	0.392

[a]Data derived from ref. (2).

[b]Mt = million metric tons.

Based on an estimated average aluminum content of 11.5%, the 44 Mt of fly ash produced in 1977 contained 5.06 Mt of aluminum, while the U.S. demand for that year was 5.64 Mt. Similarly, the titanium content of the fly ash also showed close agreement with the U.S. demand.

Methods of recovering aluminum, other metals, and by-products from coal wastes have been studied for many years, and work is currently under way at several laboratories (3-10). These methods include variations of the Bayer process, leaching with hydrochloric or sulfuric acid, lime sintering, lime-soda sintering, reductive chlorination, and the Calsinter process, which was recently developed at the Oak Ridge National Laboratory. Other nonbauxite aluminum resources in the United States include clay, shale, alunite, dawsonite, and anorthosite; and methods for the recovery of aluminum from these resources have been extensively studied in the past (11-13).

The Calsinter process (Fig. 1) provides a method for solubilizing the aluminum and several other metals in fly ash (14,15). In this process, fly ash is blended with a source of calcium oxide such as limestone, gypsum, and/or flue gas desulfurization scrubber sludge, and the mixture is pelletized and sintered at 1000 to 1250°C. The sintered material is subsequently ground and leached with dilute sulfuric acid. The solids are then removed by filtration, and the metal values are recovered from the leach solution. Aluminum can be recovered as ammonium alum or aluminum sulfate following a solvent extraction step to remove many other metals, which can be subsequently recovered from the solvent extraction regeneration stream. An attractive feature of Calsinter is the fact that two power generation wastes, fly ash and flue gas desulfurization scrubber sludge, can be utilized in a single process to recover valuable resources.

This paper describes the chemical and mineralogical characterization of fly ash and the chemistry of the sinter and leach steps in the Calsinter process. The effects of process parameters on the rate and amount of aluminum solubilization are discussed.

METHODS AND MATERIALS

Sintering and Leaching Steps

In experiments involving short sintering times (≤20 min), the feed materials were ground and sized to -325 mesh, blended, and pressed to form small pellets weighing 0.15 to 1.0 g. The tube furnace used for sintering the pellets was preheated to the desired temperature (900 to 1200°C); the pellets were then placed in a platinum boat, which was pushed into the furnace, held for the desired time, and then removed from the furnace and allowed to cool in air. The heatup and cooldown rates were rapid. In experiments involving larger quantities of material, pellets weighing 10 to 30 g were pressed at 68.9 MPa (10,000 psi) and sintered in a platinum-lined silica tray in a box furnace. The pellets were placed in the furnace at room temperature, and an appropriate temperature increase was programmed to reach the desired sintering temperature (1000 to 1250°C) over a 5-h period. This was necessary to prevent the large pellets from bursting due to rapid evolution of water and CO_2. After being sintered at the maximum temperature for the desired time, the pellets were removed and cooled in air. Thus, with large pellets, the actual heating time was considerably longer than the reported time at sintering temperature.

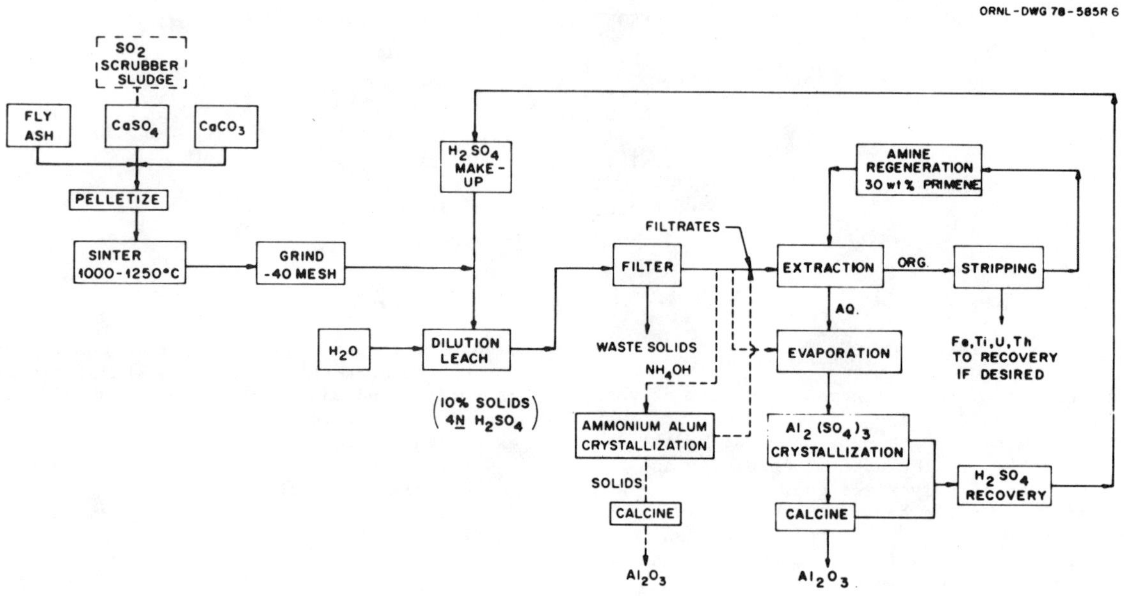

FIGURE 1. Generalized flow chart for the Calsinter process.

The following procedure was used in measuring the maximum aluminum solubilization. Sintered material was ground to -40 mesh, and 10 to 20 g was added to 200 mL of 1 to 4 N H_2SO_4 at room temperature. The slurry was stirred with a Teflon-coated magnetic stirrer and heated to reflux temperature (~110°C) for time periods ranging from 5 to 20 h. The slurry was filtered while hot, and the filter cake was washed with approximately one bed volume of hot, dilute sulfuric acid. For leaching rate studies, the sintered material was ground to -50 + 100 mesh and added to hot 4 N H_2SO_4 in the reflux flask to give a pulp density of 5%. At timed intervals, samples of the stirred slurry were removed via a syringe. The samples were then filtered through a disposable Millipore filter attached to the syringe. The sampling and filtering operation required about 20 s.

Thermogravimetric Analyses

A DuPont Model 990 Thermal Analyzer-Model 951 Thermogravimetric Analyzer was employed for these studies. The components of the sinter mix were sieved, and the -325 mesh fractions were used to prepare pellets weighing about 150 mg. Water or a 1% polyvinyl alcohol solution was added as a binder for pellet preparation. Samples were placed in a platinum boat, and the change in mass was measured in air or helium over the temperature range 25 to 1200°C. Programmed heating rates were varied from 2 to 10°C/min.

Reagents

Much of the work was done with one sample of fly ash recovered from an electrostatic precipitator following the combustion of pulverized coal at approximately 1750°C. The power plant, located in eastern Tennessee, obtained the coal from the Southern Appalachian supply region. Fly-ash samples from nine other locations in the United States were used in some of the tests.

Scrubber sludge generated by a limestone slurry flue gas desulfurization unit at a power plant in the southeastern United States was received as discharged to the waste pond. It was dried at 110°C prior to use as a sintering reagent.

The gypsum ($CaSO_4 \cdot 2H_2O$) and the limestone ($CaCO_3$) were reagent-grade chemicals.

Analyses

Elemental compositions of the various materials and solutions were determined by a number of methods. Many of the solutions were analyzed by atomic absorption (AA) and inductively coupled plasma (ICP) spectrophotometric techniques. Solids were examined directly by neutron activation and X-ray fluorescence methods; some were also solubilized and analyzed by AA and ICP. Crystalline phases were identified by X-ray diffraction (XRD) and scanning electron microscopy (SEM) techniques.

RESULTS AND DISCUSSION

Characterization of Fly Ashes

Fly ashes derived from coals that were obtained from different supply regions showed wide variations in chemical and mineralogical composition. In general, the chemical composition was similar to that of clays and shales and the major chemical elements were silicon and oxygen, accompanied by significant amounts of aluminum, iron, calcium, and a number of other elements, including titanium (Table 2). Many additional elements were present at lower levels. Table 3 shows a typical minor-element analysis for a sample of an eastern Tennessee fly ash.

Both amorphous silicate glass and crystalline phases, such as mullite, spinel, and quartz, have been identified in fly ashes (16) (Table 4). The amorphous-glass phase contains alkali metals and alkaline-earth elements as well as silicon. The spinel phase contains iron and some of the aluminum, along with some first-row transition elements; the composition shown

Table 2. Major constituents of ten fly ashes from power plants throughout the United States

Location of power plant	Element, expressed as oxide (wt %)				
	SiO_2	Al_2O_3	Fe_2O_3	CaO	TiO_2
West Virginia	53.5	26.2	3.6	0.59	1.88
Mississippi	43.2	20.2	18.6	2.73	1.22
Colorado	39.7	19.4	5.4	24.8	1.48
South Dakota	38.8	9.6	5.8	23.1	0.57
North Dakota	44.6	20.8	3.8	12.9	0.88
Washington	44.5	22.3	6.7	7.5	4.45
Tennessee	48.6	28.5	9.0	1.3	1.31
Oklahoma	44.3	16.9	19.6	7.4	0.62
Alabama	55.4	25.5	8.6	1.6	1.16
Tennessee	49.4	28.1	10.8	1.51	1.32

Table 3. Typical analysis showing minor constituents of a fly ash derived from Southern Appalachian supply region coal

Element	Concentration (mg/kg)
Potassium	23,400
Sodium	10,800
Calcium	2,310
Barium	1,200
Strontium	970
Vanadium	311
Zirconium	270
Arsenic	257
Zinc	200
Manganese	184
Vanadium	180
Rubidium	171
Cerium	155
Chromium	155
Copper	130
Lead	121
Gallium	80
Nickel	58
Molybdenum	46
Cobalt	46
Scandium	33
Tin	25
Thorium	24
Uranium	15
Selenium	12
Cadmium	2
Mercury	0.2

($Fe_{2.3}Al_{0.7}O_4$) was an average of the iron and aluminum contents of individual spinel crystals, as measured by SEM-energy dispersive X-ray fluorescence techniques (16). The mullite phase contained some silicon and the balance of the aluminum not present in the spinel phase plus some trivalent and quadrivalent elements, including chromium, titanium, and vanadium. The quartz phase, of course, was primarily SiO_2. Some preliminary work with western lignite ashes, which contained higher CaO contents, indicated a different mineralogical character, suggesting that larger quantities of crystalline calcium aluminosilicates and smaller amounts of mullite were present.

Fly ash from low-lime coal sources generally contained a large fraction of spherical particles, both solid and hollow (cenospheres), with diameters ranging from less than 1 μm to greater than 200 μm. Analysis of the eastern Tennessee fly-ash sample showed that greater than 90% of the material had a particle size of less than 44 μm. The wide variation in particle size, the spherical character, and an example of a cenosphere partially filled with smaller spheres can be seen in Fig. 2. The smooth, rounded surface of the spheres was composed of the amorphous silicate

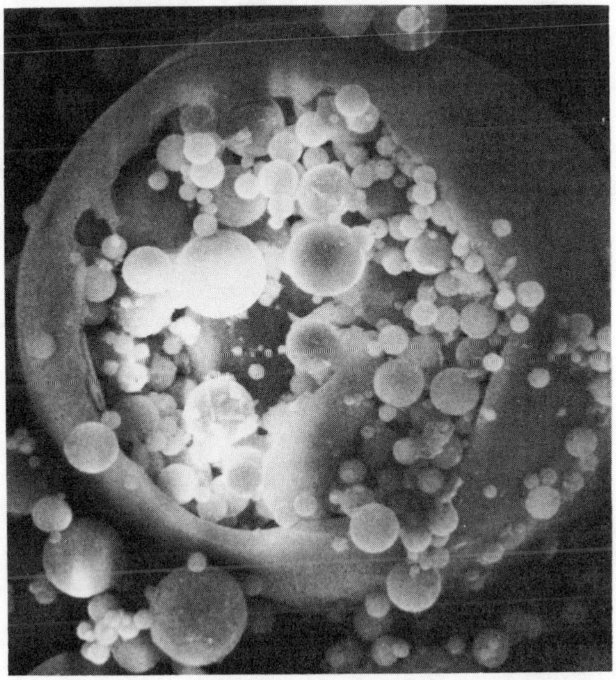

FIGURE 2. Photomicrograph of eastern Tennessee fly-ash particles.

Table 4. Characterization of the amorphous and crystalline phases of fly ash derived from Southern Appalachian supply region coal

	Amorphous phase	Crystalline phase		
		Spinel	Mullite	Quartz
Composition	Silicate glass	$Fe_{2.3}Al_{0.7}O_4$	$3Al_2O_3 \cdot 2SiO_2$	SiO_2
Associated metals	Na, K, Mg, Ca, U	V, Cr, Mn, Co, Cu, Ni, Zn	Ti, V, Cr, Fe, Ga, Zn	
Acid solubility	Soluble	Soluble	Insoluble	Insoluble

glass. This can be removed by etching with 1% hydrofluoric acid, leaving skeletons of mullite and quartz crystallites (16).

Fly ash is converted from glassy spheroids to a crystalline matrix by the Calsinter process (Fig. 3). The composition of the resulting aluminum-containing product depends on the ratio of reactants. Both anorthite and gehlenite products have been found by X-ray diffraction analysis.

Characterization of Scrubber Sludge

Analysis of the dried flue gas desulfurization scrubber sludge (Table 5) used in some tests showed that it is primarily a mixture of calcium sulfite and calcium sulfate containing unreacted calcium carbonate. Based on the iron, aluminum, and silica present, the sludge sample was estimated to contain about 20% fly ash.

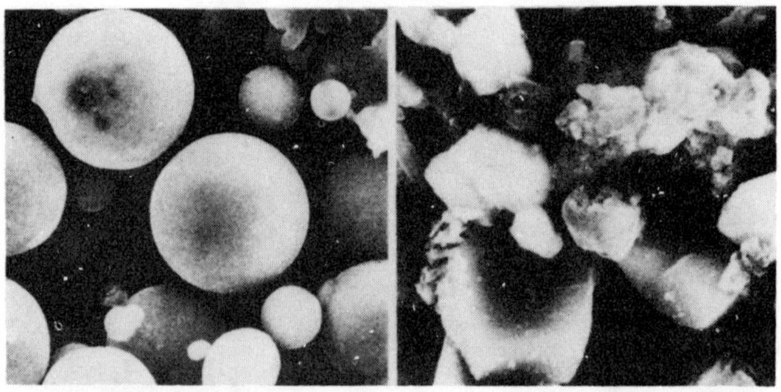

FIGURE 3. Eastern Tennessee fly ash converted from glassy spheroids to a crystalline matrix by the Calsinter process.

Table 5. Analysis of scrubber sludge

Constituent	Wt %
Calcium	21.6
Aluminum	4.0
Iron	4.2
Silica	20.4
Sulfate	4.9
Sulfite	15.7
Carbonate	28.6

Comparison of Calsinter and Direct Acid Leach Processes

One of the advantages of the Calsinter process, as compared with direct acid leach methods, is its applicability to a variety of fly ashes. A mean aluminum dissolution of 98.8% was obtained for ten fly ashes of widely varying composition from different areas of the United States (Table 6). In these tests, samples of the fly ashes were individually sintered with a mixture of gypsum and limestone (1:1:1 weight ratio) for 2 h at 1200°C. The sintered material was then ground to -40 mesh and leached at 10% solids in two stages with 2 N H_2SO_4 under reflux for 20 h. For comparison, duplicate samples of the fly ashes

Table 6. Comparison of the Calsinter and direct acid leach methods for solubilization of aluminum from ten fly ashes

Location of power plant	Aluminum solubilized (%)	
	Calsinter	Direct H_2SO_4 leach
West Virginia	99.2	42
Mississippi	99.1	48
Colorado	99.7	83
South Dakota	99.4	74
North Dakota	99.7	73
Washington	99.3	48
Tennessee	96.0	55
Oklahoma	98.8	62
Alabama	97.8	44
Tennessee	98.6	61

were also leached directly by refluxing in 8 N H_2SO_4 for 20 h. Approximately 96 to 99% of the aluminum was solubilized by the Calsinter method, regardless of the fly ash used, while the values for the direct acid leach method ranged from 42 to 83%. While 83% of the aluminum was recovered from the Colorado ash by the direct acid leach, the recoveries from other ashes were lower. Ashes that showed high aluminum solubility by the direct acid leach generally contained high concentrations of calcium oxide (Table 2). This probably reflects the formation of acid-soluble calcium aluminosilicates during the coal combustion step.

Effects of Sintering Time and Temperature on Aluminum Solubilization

Aluminum solubilization was increased from the 55% obtained by direct acid leaching to greater than about 85% by sintering for 10 min or longer at temperatures of 1000°C or higher (Fig. 4). A substantial increase was noted even at sintering times as short as 2 min. In these tests, mixtures of eastern Tennessee fly ash, gypsum, and limestone (1:1:1 weight ratio) were sintered as indicated (Fig. 4), and the sintered material was leached under reflux with 4 N H_2SO_4 for 20 h. Sintering at 900°C did not appreciably improve the solubilization, compared with direct acid leaching.

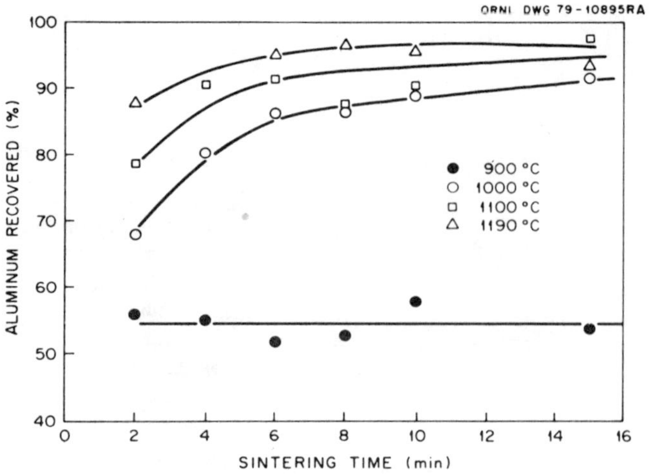

FIGURE 4. Effect of sintering time and temperature on aluminum solubilization.

For process application, sintering at 1000 to 1200°C for 15 min would probably be sufficient to achieve greater than 90% aluminum solubilization.

Effects of Acid Strength and Leach Time on Aluminum Solubilization

The total aluminum dissolved from a given sinter product was highly dependent on the concentration of the sulfuric acid in the leach step. Figure 5 summarizes the results obtained when eastern Tennessee fly ash, gypsum, and limestone (1:1:1 wight ratio)

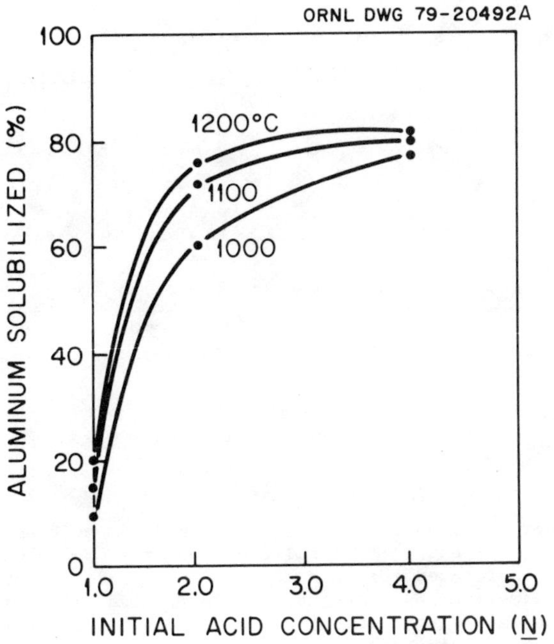

FIGURE 5. Effect of acid strength on aluminum leachability after sintering at 1000, 1100, and 1200°C.

were sintered at 1000 to 1200°C and then leached at 10% pulp density under reflux conditions with 1 to 4 N H_2SO_4 for 20 h. In these single-stage leach tests, good solubilization in all cases was obtained with 4 N H_2SO_4. In other tests using two-stage leaching techniques, 96 to 98% solubilization (Table 6) was obtained with 2 N H_2SO_4. Countercurrent leaching techniques, which likely would be applied in process applications, should achieve optimum aluminum solubilization while the acid strength and leach pulp density could be varied to achieve the desired aluminum concentration and free-acid content of the leach solution.

In tests designed to measure the leach rate, a mixture of fly ash, limestone, and gypsum (1:1:1 weight ratio) that had been sintered at 1200°C for 20 min was leached at 5% pulp density with 4 N H_2SO_4 under reflux conditions, and the leachate was sampled at 2-min intervals. Maximum aluminum solubilization was achieved within 30 min; about half of the aluminum was solubilized in the first 2 min.

Effect of Sinter Composition on Aluminum Solubilization

The compounds formed during sintering, as well as the resulting aluminum solubilization, were found to be dependent on the sinter mix composition. Various mixtures of eastern Tennessee fly ash, limestone, and gypsum were sintered at 1200°C for 20 min and leached with 4 N H_2SO_4 for 5 h under reflux conditions. The primary sinter products were identified by X-ray diffraction analysis, and the leachate was analyzed to determine the percent of aluminum solubilized (Table 7).

When this fly ash was heated without addition of any calcium source, the subsequent acid solubilization of aluminum decreased from about 55% to about 3%. The major component identified in the sintered material was mullite, which remained in the leach residue. The decreased solubility is probably due to conversion of the aluminum in the amorphous silicate glass and spinel phases (Table 4) to mullite during heating. The conversion of spinel in clays to mullite on heating has been reported (17), and the ratio of calcium, aluminum, and silicon for this eastern Tennessee fly ash lies in the mullite region of the ternary phase diagram for $CaO-Al_2O_3-SiO_2$ (18).

When small amounts of the CaO source were added to the sinter mixture, anorthite ($CaO \cdot Al_2O_3 \cdot 2SiO_2$) was formed in the sinter product; with larger additions, gehlenite ($2CaO \cdot Al_2O_3 \cdot SiO_2$) was formed (Table 7). The same compounds were formed when limestone, gypsum, or a limestone-gypsum combination was used as the CaO source. However, the decomposition of $CaSO_4$ to yield

Table 7. Effect of sinter composition on aluminum solubilization and sinter reaction products

Sinter composition (weight ratio)			Aluminum-containing sinter product[a]	Aluminum solubilized[b] (%)
Fly ash	Limestone	Gypsum		
1	0	0	Mullite	3
1	0.25	0	Anorthite	68
1	1	0	Gehlenite	90
1	1.5	0	Gehlenite	95
1	0	1	Anorthite	79
1	1	1	Gehlenite	89

[a]Pellet sintered at 1200°C for 20 min.
Anorthite = $CaO \cdot Al_2O_3 \cdot 2SiO_2$, or $CaAl_2Si_2O_8$.
Gehlenite = $2CaO \cdot Al_2O_3 \cdot SiO_2$, or $Ca_2Al_2SiO_7$.
Mullite = $3Al_2O_3 \cdot 2SiO_2$, or $Al_6Si_2O_{13}$.

[b]Leached with 4 N H_2SO_4 for 5 h at reflux temperature.

CaO was slower than that of $CaCO_3$, and higher temperatures and/or longer times may be required to achieve high aluminum solubilization under process conditions. The formation of anorthite and gehlenite is consistent with the ternary phase diagram (18). The calcium, aluminum, and silicon composition of these sinter mixtures lies in the region of the CaO-Al_2O_3-SiO_2 ternary phase diagram containing anorthite, gehlenite, pseudowollastonite ($CaSiO_3$), and calcium orthosilicate (Ca_2SiO_4). In some tests, β-Ca_2SiO_4 was identified in the sinter products. When large amounts of limestone and gypsum were used, $Ca_5(SiO_4)_2SO_4$ was also formed. Reactions with other metal constituents of the fly ash also occurred since other metals were solubilized by the sinter-leach steps; however, these products were not identified.

The formation of calcium aluminum sulfate ($4CaO \cdot 3Al_2O_3 \cdot SO_3$) has been suggested as a reaction product that is formed from sintering calcium sulfate and coal wastes (19). This compound was not detected in Calsinter products. Other investigators (20,21) indicated that calcium aluminum sulfate was unstable in the presence of free SiO_2. Since the Calsinter process utilizes an acid leach, there is no particular advantage or requirement for producing $4CaO \cdot 3Al_2O_3 \cdot SO_3$.

Aluminum solubilization increased with increasing calcium addition and, under the leach conditions used in these studies reached a value of 90 to 95%. The aluminum that is solubilized may be considered a measure of the aluminum included in the gehlenite and anorthite crystalline phases, while insoluble aluminum indicates the aluminum fraction in mullite.

Thermogravimetric data confirm the decomposition of $CaCO_3$ at ~800°C and of $CaSO_4$ at ~1100°C in these sinter mixtures. The $CaCO_3$ content was varied to give eastern Tennessee fly ash–gypsum–limestone mixtures with the following weight ratios: (a) 1:1:0, (b) 1:1:0.2, (c) 1:1:0.5, and (d) 1:1:1. With no $CaCO_3$ present [curve (a) in Fig. 6], mass losses below 900°C can be attributed primarily to the loss of water of hydration from $CaSO_4 \cdot 2H_2O$ and losses of pellet binder, carbon, and adsorbed water from the fly ash. At higher temperatures, 900 to 1200°C, the mass loss was due to the decomposition of sulfate (22). As $CaCO_3$ was added to the fly ash–gypsum mixture [curves (b)–(d)], the decomposition of the limestone was observed at 700 to 860°C and the higher-temperature decomposition of the sulfate was suppressed. When the fly ash–gypsum–limestone composition reached a weight ratio of 1:1:1 [curve (d)], weight loss due to decomposition of the sulfate was no longer observed at temperatures near 1200°C.

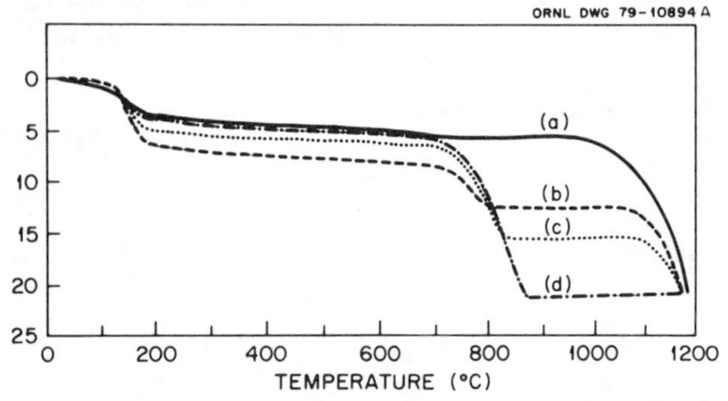

FIGURE 6. Thermograms of fly ash–$CaSO_4 \cdot 2H_2O$–$CaCO_3$ mixtures.

Sample: 150- to 160-mg pellets.
Atmosphere: air; flow rate, 50 mL/min.
Heating rate: curve (a), 2°C/min;
curves (b)–(d), 10°C/min.
Fly ash–$CaSO_4 \cdot 2H_2O$–$CaCO_3$ composition (by weight):
(a) 1:1:0, (b) 1:1:0.2, (c) 1:1:0.5, (d) 1:1:1.

Aluminum solubilization can be correlated with the CaO/Al_2O_3 mole ratio in the sinter mixture (Fig. 7). The amount solubilized was found to be independent of the CaO source (limestone and/or gypsum or lime).

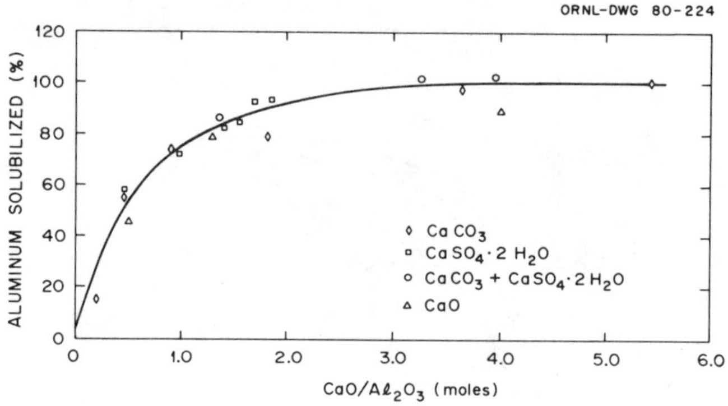

FIGURE 7. Aluminum solubilization from fly ash using different sintering reagents.

Under the conditions selected, the decomposition of $CaSO_4$ did not go to completion, and the CaO values were corrected for unreacted $CaSO_4$, based on thermogravimetric measurements. These results (Fig. 7) show that either of these CaO sources can be utilized in the Calsinter process and that the addition of about 1 mole of calcium per mole of aluminum (2 moles of calcium oxide per mole of alumina) in this eastern Tennessee fly ash would be required to achieve satisfactory aluminum solubilization.

The chemical reactions occurring during the sinter and leach steps of the Calsinter process can be empirically described by a series of equations. This means of representing the reactions assumes that decomposition to calcium oxide occurs before significant conversion of mullite to anorthite and/or gehlenite. Such is the case for limestone since the thermogravimetric results showed that $CaCO_3$ decomposed in these mixtures at temperatures below which any increased aluminum solubilization was observed. With $CaSO_4$, however, the decomposition and reaction with mullite occurred at about the same temperature and thus may proceed simultaneously. The presence of silica has been reported to decrease the $CaSO_4$ decomposition temperature (22).

Examination of Calsinter leach residues by X-ray diffraction techniques showed the presence of anhydrite ($CaSO_4$) rather than gypsum ($CaSO_4 \cdot 2H_2O$); thus $CaSO_4$ is shown in Eqs. (6) and (7). Acid consumption would be greater than that shown in Eqs. (6) and (7) since the dissolution of other metals would consume H_2SO_4 as would any excess lime present. The leaching reactions generate insoluble silica and calcium sulfate, resulting in an increase in viscosity during leaching.

Reagent Decomposition:

$$CaCO_3 \rightarrow CaO + CO_2 \quad (1)$$

$$CaSO_4 \cdot 2H_2O \rightarrow CaO + SO_3 + 2H_2O \quad (2)$$

Sintering:

$$2CaO + SiO_2 \rightarrow Ca_2SiO_4 \quad (3)$$

$$3CaO + 4SiO_2 + \underset{\text{(mullite)}}{[3Al_2O_3 \cdot 2SiO_2]} \rightarrow \underset{\text{(anorthite)}}{3[CaO \cdot Al_2O_3 \cdot 2SiO_2]} \quad (4)$$

$$6CaO + SiO_2 + \underset{\text{(mullite)}}{[3Al_2O_3 \cdot 2SiO_2]} \rightarrow \underset{\text{(gehlenite)}}{3[2CaO \cdot Al_2O_3 \cdot SiO_2]} \quad (5)$$

Dissolution:

$$\underset{\text{(anorthite)}}{[CaO \cdot Al_2O_3 \cdot 2SiO_2]} + 4H_2SO_4 \rightarrow Al_2(SO_4)_3 + CaSO_4 + 2SiO_2 + 4H_2O \quad (6)$$

$$\underset{\text{(gehlenite)}}{[2CaO \cdot Al_2O_3 \cdot SiO_2]} + 5H_2SO_4 \rightarrow Al_2(SO_4)_3 + 2CaSO_4 + SiO_2 + 5H_2O \quad (7)$$

Utilization of Flue Gas Desulfurization Scrubber Sludge as the Calcium Oxide Source

The lime-limestone scrubbing of power plant flue gas for desulfurization generates a volume of scrubber sludge nearly comparable to the fly ash produced from typical coals. A sample of scrubber sludge (Table 5) was obtained and tested as a calcium oxide source for the Calsinter process. Scrubber sludge was found to be effective as a substitute for gypsum in tests involving limestone-gypsum additions for aluminum solubilization (Fig. 8). The aluminum solubilization was essentially the same, becoming greater than 95% after either mixture had been sintered at 1200 to 1250°C for 2 h and leached in 4 $\underline{N}$ H_2SO_4 at reflux temperature for 20 h. In other tests, eastern Tennessee fly ash and scrubber sludge were blended at various ratios, sintered at 1200°C, and leached with 4 $\underline{N}$ H_2SO_4 at reflux for 20 h. Greater than 90% aluminum solubilization was achieved as long as the scrubber sludge comprised 50% or more of the sinter mixture weight. Thus, under process conditions, a power plant generating both fly ash and scrubber sludge could combine these as feed for the Calsinter process rather than having two waste streams for disposal. Although the volume of the solid waste would not be significantly reduced, the waste would be more innocuous. The SO_3 produced in the sintering step could be recovered as sulfuric acid and used in the leaching step.

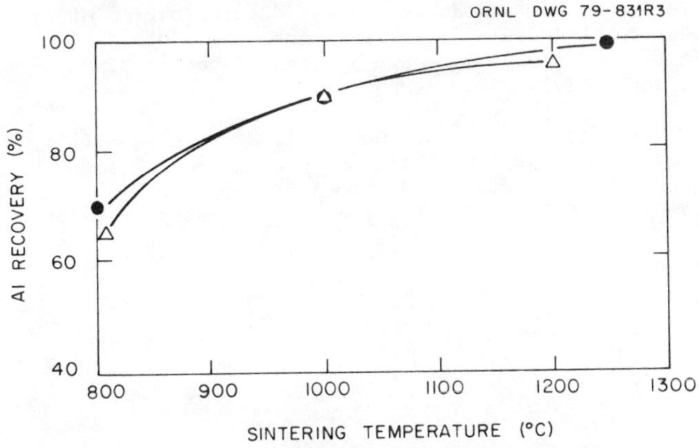

FIGURE 8. Effect on aluminum solubilization of substituting scrubber sludge for gypsum.

● Fly ash–$CaSO_4$–$CaCO_3$

△ Fly ash–scrubber sludge–$CaCO_3$

SUMMARY OF OPTIMUM CALSINTER PROCESS CONDITIONS

Based on these laboratory studies, the following process steps and parameters have been defined for the Calsinter process:

1) Fly ash is blended with a CaO source so that the final mixture contains about 1 mole of calcium per mole of aluminum (2 moles of CaO per mole of Al_2O_3). Limestone, gypsum, or flue gas desulfurization scrubber sludge can be used as the CaO source. In the case of high-lime fly ashes, the calcium content of the fly ash would represent part of the calcium requirements.

2) The blended mixture is pelletized to ensure good solid-solid contact in the subsequent sintering step.

3) The pellets are sintered at 1000 to 1250°C. Reaction is incomplete at lower temperatures, while melting and slagging may occur at higher temperatures. Reaction times of 15 to 30 min are sufficient to yield maximum solubilization of aluminum and several other metals.

4) The sintered pellets are ground to −40 mesh to yield a higher surface area and ensure rapid solubilization in the leaching step.

5) In batch operation, a two-stage leach with 2 $\underline{N}$ H_2SO_4 under reflux conditions (∿110°C) for 15 to 30 min yields essentially complete (96 to 98%) aluminum solubilization. Countercurrent leaching could be employed to optimize acid consumption and aluminum sulfate concentration in the leach solution.

6) Leach solutions are clarified by filtration.

Additional process steps are under development. A solvent extraction method has been tested for the removal of iron and other metals from the sulfuric acid leach solution (23). Precipitation of the aluminum as ammonium alum, basic aluminum sulfate, aluminum sulfate, or aluminum chloride is under investigation. These precipitated solids would then be roasted to form the alumina product, and by-product gases would be recycled.

ACKNOWLEDGMENTS

Excellent laboratory assistance was provided by G. D. Campbell, J. Coombs, L. K. Felker, C. Ling, and J. Renfro.

The authors thank L. D. Hulett, R. L. Sherman, and H. W. Dunn, of the ORNL Analytical Chemistry Division, for their X-ray diffraction and scanning electron microscopy work. Appreciation is also expressed to other members of the Analytical Chemistry Division for analytical support during the study.

This research was sponsored by the Office of Energy Research, U.S. Department of Energy under contract W-7405-eng-26 with the Union Carbide Corporation.

REFERENCES

1. Coal Ash Utilization — Fly Ash, Bottom Ash and Slag, Torrey, S., ed., Noyes Data Corp., Park Ridge, N. J., 1978.

2. U.S. Department of Interior, Bureau of Mines, Minerals in the U.S. Economy: Ten-Year Supply-Demand Profiles for Mineral and Fuel Commodities, 1968-1977, May 1979.

3. Hamer, C. A., "Alumina from Fluidized Bed Combustion Ash of Hat Creek, B.C., Coaly Wastes — Part I: Acid Extraction," Report MRP/MSL 79-21, 1979, Mineral Sciences Laboratories, Canada.

4. Winer, A., and Tiggetts, T. E., "Coal Associated Materials as Potential Nonbauxite Sources of Alumina," Report 78-31, Canada Centre for Mineral and Energy Technology.

5. Nowak, Z., "Iron and Alumina Extraction from Powerplant Fly Ash in Poland," U.S. Bureau of Mines Information Circular 8640, 1974, Washington, D.C.

6. Murtha, M. J., and Burnet, G., "Recovery of Alumina from Coal Ash by High Temperature Chlorination," Proceedings of the Iowa Academy of Science, Vol. 83, 1976, pp. 125-129.

7. Murtha, M. J., and Burnet, G., "New Developments in the Lime-Soda Sinter Process for Recovery of Alumina from Fly Ash," Report IS-M-177, Iowa State University, Ames, Iowa.

8. Chou, K-S., "A Study of the Extractability of Alumina from Fly Ash by the Lime-Sinter Process," Report IS-T-716, 1976, Iowa State University, Ames, Iowa.

9. Chou, K-S., Klemm, W. A., Murtha, M. J., and Burnet, G., "The Lime-Sinter Process for Production of Alumina from Fly Ash," Proceedings of the Fourth International Ash Utilization Symposium, St. Louis, 1976.

10. Cavin, D. C., "A Study of Iron and Aluminum Recovery from Power Plant Fly Ash," Report IS-T-611, 1974, Iowa State University, Ames, Iowa.

11. Hamer, C. A., "Acid Extraction Processes for Non-bauxite Alumina Materials," Report 77-54, 1977, Canada Centre for Mineral and Energy Technology.

12. Walthall, J. H., Miller, P., and Striplin, M. M., Jr., "Development of a Sulfuric Acid Process for Production of Alumina from Clay," Transactions of the American Institute of Chemical Engineers, Vol. 41, 1945, pp. 55-140.

13. Kammermeyer, K., and White, A. H., "The Extraction of Alumina from Kaolin and Other Silicates," Transactions of the American Institute of Chemical Engineers, Vol. 36, 1940, pp. 683-700.

14. Seeley, F. G., Canon, R. M., and McDowell, W. J., "Chemical Development of New Processes for the Recovery of Resource Materials from Coal Ash," Proceedings of the Fifth International Ash Utilization Symposium, Atlanta, 1979.

15. Canon, R. M., Kelmers, A. D., McDowell, W. J., Seeley, F. G., and Watson, J. S., "Metal Removal from Coal Ashes and Wastes," in Solid Waste Research and Development Needs for Emerging Coal Technologies, American Society of Civil Engineers (1979), pp. 120-134.

16. Hulett, L. D., Weinberger, A. J., Ferguson, N. M., Northcutt, K. J., and Lyon, W. S., "Chemical Speciation Studies of Fly Ash — Part I: Characterization of Solids," Report RP-1061, 1979, Electric Power Research Institute, Palo Alto, California.

17. Brindley, G. W., and Nakahira, M., "The Kaolinite-Mullite Reaction Series: III. The High Temperature Phases," American Ceramic Society Journal, Vol. 42, 1959, pp. 319-324.

18. Levin, E. M., Robbins, C. R., and McMurdie, H. F., Phase Diagrams for Ceramists, American Chemical Society, 1964, Figure 630, p. 219.

19. Goodboy, K. P., "Investigation of a Sinter Process for Extraction of Al_2O_3 from Coal Wastes," Metallurgical Transactions, Vol. 78, 1976, pp. 716-718.

20. Fukuda, N., and Kiyoku, N., "Effects of Silica on the Formation of Anhydrous Sulfoaluminate," Semento Gijutsu Nempo, Vol. 20, 1966, pp. 46-48.

21. Didamony, E., "Effect of Silicon Dioxide on the Formation of Calcium Sulfoaluminate," Tonind.-Ztg., Vol. 102, 1978, pp. 583-587.

22. Swift, W. M., Panek, A. F., Smith, G. W., Vogel, G. J., and Jonke, A. A., "Decomposition of Calcium Sulfate: A Review of the Literature," ANL-76-122, 1976, Argonne National Laboratory, Argonne, Illinois.

23. Seeley, F. G., and McDowell, W. J., "Iron(III) Extraction Equilibria in the System: $Fe_2(SO_4)_3$--$(NH_4)_2SO_4$ vs Primene JM-T Sulfate--Toluene," Journal of Inorganic and Nuclear Chemistry, in press.

Chapter 16

RECOVERY OF HIGH PURITY SILVER FROM SULFATED COPPER REFINERY SLIMES

Peter D. Parker and John A. Bonucci

AMAX Base Metals Research & Development, Inc.
Carteret, New Jersey

James E. Hoffmann

Minerals Division
Exxon Research & Engineering Company
Florham Park, New Jersey

ABSTRACT

A hydrometallurgical process has been developed for the recovery of high-purity silver from sulfated residues. The process eliminates the high temperature operations ordinarily associated with silver recovery and refining. The crux of the process is a metathetical leach which maintains the sulfate ion concentration at a level which both allows silver nitrate to be freely soluble and simultaneously depresses the solubility of lead nitrate. The subsequent hydrolytic purification of the silver nitrate solution, the recovery of high-purity silver by electrowinning in a diaphragm cell, and the final product purity are discussed.

RECOVERY OF HIGH PURITY SILVER FROM SULFATED COPPER REFINERY SLIMES

Peter D. Parker and John A. Bonucci

AMAX Base Metals Research & Development, Inc.
Carteret, New Jersey

James E. Hoffmann

Minerals Division
Exxon Research & Engineering Co.
Florham Park, New Jersey

Background

The novel silver recovery process described in this paper was developed at AMAX Base Metals R&D, Inc. as part of an overall process for extracting precious metals from the anode slimes generated during the electrorefining of copper[1]. The overall approach involves sulfation roasting the slimes to improve selenium extraction and recovery followed by the hydrometallurgical flow sheet for recovery of copper, nickel, and silver which is the subject of this paper. Finally, gold and platinum group metals are isolated by chlorination technology.

Anode slimes are the insoluble constituents in a copper electrorefining anode which fall to the bottom of the electrodeposition tank when the anode is being electrolytically dissolved. A typical composition range of copper refinery slimes is shown in Table I.

Table I

Compositional Range of Copper Refinery Slimes (percent)

Se	4-21	Sn	2-5
Te	1-3	Pb	12-29
As	0.2-2	Ag	15-19
Sb	0.1-5	Au	0.4-0.8
Cu	3-5	Pd	0.02-0.2
Ni	0.03-2	Pt	0.002-0.02

After their removal from electrodeposition tanks, anode slimes are commonly treated pyrometallurgically in a doré furnace, - a form of small oil- or gas-fired reverberatory furnace. The slimes are charged to this furnace, melted, and subjected to a series of fluxing and slagging operations in order to produce an impure silver bullion which is further refined by electrolytic parting. There are a number of disadvantages associated with this procedure. These include loss of some metal values contained in the slimes, air pollution, high energy consumption, high operating temperatures, slag recycles and a long processing time. The more volatile metals are a primary concern with respect to loss. Of these selenium is the most important and, at its present market value of \$22-\$33/kg, there is considerable incentive to recover as much of it as possible. Selenium is present in slimes both in elemental form as well as combined with silver and base metals as selenides.

Many slimes treatment processes have been studied with the aim of improving selenium recovery from anode slimes. One of the best known is sulfation roasting in which slimes are mixed with concentrated sulfuric acid and heated to temperatures at which the selenium is oxidized and volatilized as selenium dioxide. This oxide is then recovered from the vapor phase by scrubbing with dilute sulfuric acid. The overall sulfation reactions for the major components of the slimes are represented in Table II.

Table II

Typical Slimes Sulfation Reactions

Type 1 $Se + 2H_2SO_4 \rightarrow SeO_2 + SO_2 + 2H_2O$

Type 2 $Cu + 2H_2SO_4 \rightarrow CuSO_4 + 2H_2O + SO_2$

Type 3 $NiO + H_2SO_4 \rightarrow NiSO_4 + H_2O$

Type 4 $Ag_2Se + 4H_2SO_4 \rightarrow Ag_2SO_4 + SeO_2 + 3SO_2 + 4H_2O$

Origin of the Process

The process which is the subject of this paper was developed to treat sulfation roasted slimes for silver recovery after selenium removal by sulfation roasting has been accomplished. Under the strongly oxidizing conditions which prevail during sulfation roasting, lead, copper, nickel, and silver are all sulfated while the selenium is being oxidized. The heart of the problem lies in then recovering the sulfated silver. In any hydrometallurgical process the aqueous solubility of silver sulfate is too great to be ignored, yet on the other hand it is too low to realistically consider water leaching for recovery of silver. Even at boiling, less than 10 g/liter silver will dissolve. Furthermore, the presence of silver sulfate in slimes charged to a doré furnace would complicate its normal fire refining cycle and could result in excessive losses of silver to the various slags. Hence, conventional furnacing treatment of sulfation roasted slimes was considered inappropriate and there was need to develop a hydrometallurgical process to recover the silver content of the treated material. With this objective in mind, we sought a lixiviant which would solubilize silver and possess all of the following characteristics:

1. Allow the attainment of high concentrations of soluble silver.

2. Depress the concentration of sulfate ions which would otherwise limit silver solubility.

3. Be inexpensive and readily available or easily synthesized.

4. Produce an aqueous silver compound whose characteristics are well known.

Because silver nitrate is one of the most soluble silver salts and the physical and chemical characteristics of its solutions are well known, we sought a means of producing a solution of this salt. The amount of silver sulfate which can be solubilized in a nitrate medium is inversely related to the concentration of sulfate ion in this solution. As this is also true for lead which would introduce contamination problems, we found the most feasible way to selectively leach silver while depressing lead in a nitrate system was to maintain a very small amount of sulfate ion in solution. Hence, we needed a cation which would form a sparingly soluble sulfate salt and thus depress the equilibrium solubility of sulfate ions to a low level. This pointed to the well-known insoluble sulfate compounds formed by calcium, barium, strontium, and lead. Obviously, from a cost and availability point of view, calcium was the most desirable of this cation group.

One can make a preliminary estimate of the effect of $Ca(NO_3)_2$ leaching on silver and lead solubility by assuming that, when an excess of sulfated slimes residue is slurried in a concentrated calcium nitrate solution, the following metathetical exchanges will take place:

$$Ag_2SO_4(c) + Ca(NO_3)_2(aq) + 2H_2O = 2AgNO_3(aq) + CaSO_4 \cdot 2H_2O(c) \quad (1)$$

$$PbSO_4(c) + Ca(NO_3)_2(aq) + 2H_2O = Pb(NO_3)_2(aq) + CaSO_4 \cdot 2H_2O(c) \quad (2)$$

These expressions show that at equilibrium, the concentrations of calcium, silver and lead in solution would be limited fundamentally by the solubilities of their respective sulfate compounds. That is, since there is excess $PbSO_4$, $CaSO_4 \cdot 2H_2O$, and Ag_2SO_4 in the solid phase and a finite concentration of $SO_4^=$ ion in solution, the controlling equilibria are:

$$Ca^{++} + SO_4^= + 2H_2O = CaSO_4 \cdot 2H_2O \quad (3)$$

$$2Ag^+ + SO_4^= = Ag_2SO_4 \quad (4)$$

$$Pb^{++} + SO_4^= = PbSO_4 \quad (5)$$

Hence, the final solution is simultaneously saturated in these three metal sulfates. This system can be defined by five equations shown below. The indicated solubility products were evaluated at 298°K from available thermodynamic data[2].

$$K_{sp_{CaSO_4 \cdot 2H_2O}} = \gamma_\pm^2 m_{Ca} m_{SO_4} = 2.45 \times 10^{-5} * \quad (6)$$

$$K_{sp_{PbSO_4}} = \gamma_\pm^2 m_{Pb} m_{SO_4} = 1.34 \times 10^{-8} \quad (7)$$

$$K_{sp_{Ag_2SO_4}} = \gamma_\pm^3 m_{Ag}^2 m_{SO_4} = 1.25 \times 10^{-5} \quad (8)$$

Charge Balance:

$$m_{Ag} + 2m_{Pb} + 2m_{Ca} = m_{NO_3} + 2m_{SO_4} \quad (9)$$

Ionic Strength:

$$\mu = 0.5(4m_{Ca} + 4m_{Pb} + 4m_{SO_4} + m_{Ag} + m_{NO_3}) \quad (10)$$

We have calculated the theoretical solubilities of Ag^+, Ca^{++}, Pb^{++}, and $SO_4^=$ for a model of the process in which the initial lixiviant was a pure, 1.36 molal solution of $Ca(NO_3)_2$. Since published values for the activity coefficients for $PbSO_4$ and $CaSO_4$ in concentrated solutions are limited, we estimated these data by equating them, respectively, to solutions of $CdSO_4$ and $MnSO_4$ of equivalent ionic strength[3]. Activity coefficients for Ag_2SO_4 were estimated by comparison to data calculated for Rb_2SO_4 using an equation provided by Lewis and Randal[4]:

$$\log \gamma_\pm = \log \gamma_{\pm_{CaCl_2}} + m_{Rb_2SO_4} \cdot {}^4/_3 \, \Delta B*$$

We used a procedure in which we assumed a value for μ, calculated the appropriate $\gamma_\pm$ and solved for the equilibrium ion concentrations which were used to recalculate μ. This iteration was repeated until convergence was obtained. Some additional calculations were performed for the cases in which $Sr(NO_3)_2$, $Pb(NO_3)_2$ or $Ba(NO_3)_2$ were used as the lixiviant instead of $Ca(NO_3)_2$. The calculated solubilities are presented below in Table III.

It is apparent from the results of these calculations that leaching with barium, lead or strontium nitrates results in lower equilibrium sulfate levels than leaching with $Ca(NO_3)_2$. However, while these materials would allow higher final silver nitrate concentrations in solution, barium, lead and strontium nitrates are more expensive and, in addition to the cost consideration, their use would result in increased lead solubilization. Use of a calcium nitrate system, by comparison, achieves a very significant reduction in the solubility of lead. $Ca(NO_3)_2$ provides nitrate for the silver and forms $CaSO_4$ and thereby enhances the solubility of $AgNO_3$ because sulfate ion is removed from solution. At the same time, $CaSO_4$ is just soluble enough to depress lead to acceptable levels and produce the highest silver-to-lead ratio in the final solution.

---

* $\gamma_\pm$ = mean activity coefficient

m = molality

${}^4/_3 \, \Delta B$ = a molality-dependent deviation constant = -0.336 for Rb_2SO_4

Table III

Theoretical Solubilities of Ions in a Nitrate Solution
$(NO_3^-) = 2.72m$

Ion	Solubility in $Ca(NO_3)_2$ System (molal)	Solubility in $Sr(NO_3)_2$ System (molal)	Solubility in $Ba(NO_3)_2$ System (molal)	Solubility in $Pb(NO_3)_2$ System (molal)
Ca^{++}	1.13			
Sr^{++}		0.581		
Ba^{++}			3.52×10^{-3}	
Ag^{+}	0.362	1.40	2.51	2.52
Pb^{++}	7.08×10^{-4}	1.05×10^{-2}	3.26×10^{-2}	3.28×10^{-2}
$SO_4^{=}$	1.12×10^{-2}	6.56×10^{-4}	1.70×10^{-4}	1.69×10^{-4}

Experimental Work

It was immediately apparent on leaching the sulfated slimes, as is, with $Ca(NO_3)_2$ that a variety of other soluble sulfates would dissolve as nitrates along with the silver. Typically, copper, nickel, and iron also formed aqueous nitrates. Some tellurium and residual selenium also dissolved. Consequently, these impurities were removed by a preliminary water leach. The complete flow sheet which we chose to investigate is shown in Figure 1.

Water Leach - Although Ag_2SO_4 has a limited solubility in water, about 10 g/liter, we wished to minimize the amount of silver dissolved in this step and therefore chose the highest solids-to-liquids ratio for leaching consistent with good extraction of the impurities mentioned above. The amount of silver that actually was solubilized in the water leach turned out to be about 5% of the silver content of the slimes.

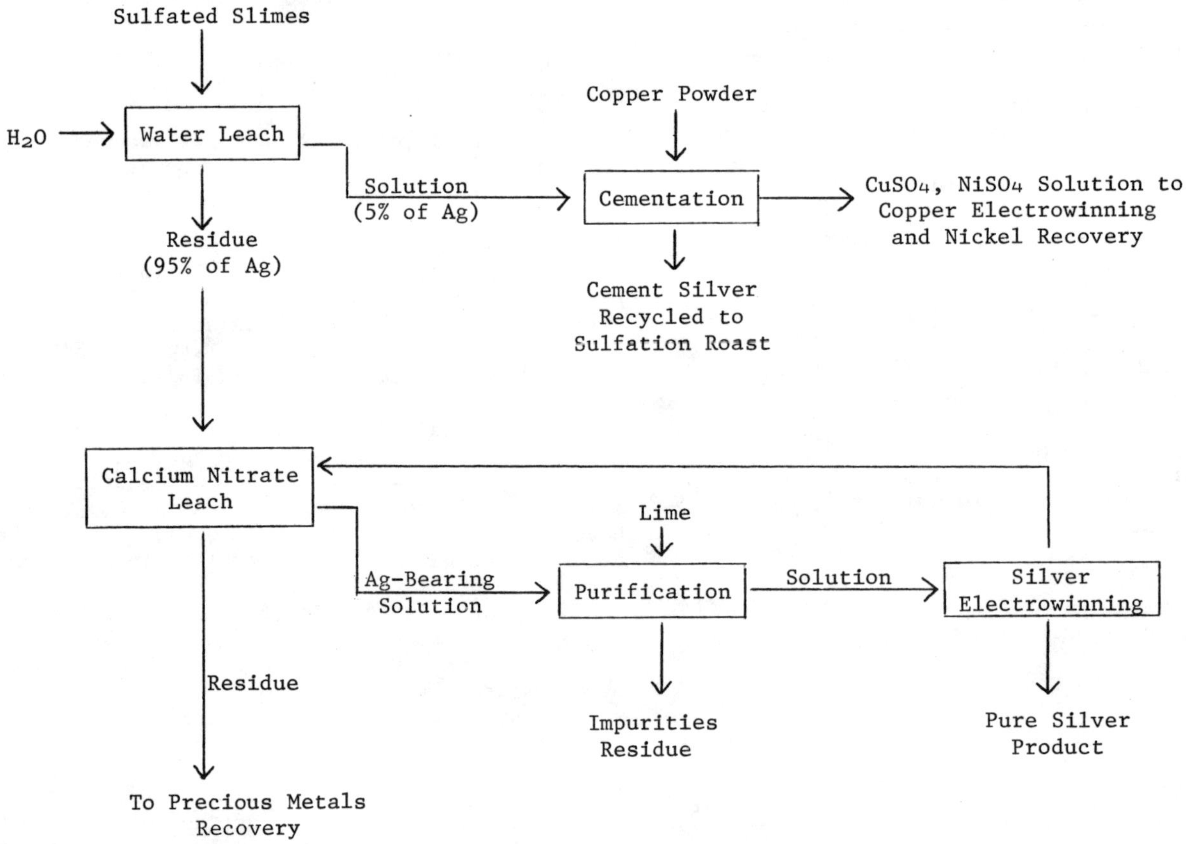

Figure 1

Block Flow Diagram for Silver Recovery Process

Certain of the impurities in the water leach liquor including tellurium, iron, and antimony were immediately rejected to the leach residue by an in-situ pH adjustment to about 3.5 prior to filtration. The solution, containing the dissolved silver was then treated by addition of copper powder to cement out the silver and any other substances reducible by copper. This silver cement product was returned to the sulfation roasting operation. The remaining solution of $CuSO_4$ and $NiSO_4$ was of sufficient purity to be treated for copper recovery by electrowinning and $NiSO_4$ recovery by evaporation of a bleed stream. The composition of the slimes residue after the water leach is shown in Table IV where it is compared to the slimes before any treatment.

Table IV

Typical Sulfation Roast Residues

	As Received %	Water Leached %
Se	0.05	0.02
Te	1.44	1.1
Cu	4.84	0.03
Ni	1.34	0.07
Ag	20.0	22.6

$Ca(NO_3)_2$ Leach - After removing the soluble sulfates as described above, we leached the wet residue with a $Ca(NO_3)_2$ solution to solubilize silver. Typical leaching conditions are noted in Table V. The leach was performed hot, at a solids-to-liquids ratio adjusted to maximize both silver extraction and silver concentration.

Table V

Calcium Nitrate Leaching Conditions

Solids/Liquids Ratio	150 g/liter
Temperature	~100°C
Time	1 hour
Agitation	Vigorous
Lixiviant Composition	$Ca(NO_3)_2$ - 1.0 molar Ag^+ - 0.6 molar

A moderate excess of $Ca(NO_3)_2$ over the stoichiometric requirements for precipitation of sulfates and generation of nitrate ion for the silver is desirable. The beneficial effect of excess calcium ion is to suppress sulfate concentration and thereby increase the solubility of silver in the system. The final silver concentrations were generally about 0.9 molar. The 0.6M Ag concentration in the lixiviant shown in Table V reflects the silver returned to leaching in the spent electrolyte from the silver electrowinning cells.

When we analyzed the leach liquors, we found their silver and lead contents to be higher and their calcium concentration lower than predicted by theory. This differential between the actual and predicted solubilities is shown below in Table VI. The ion concentrations in the theoretical leach liquor were converted from molal to molar units using an estimated solution density which was based on the weighted average of actual $Ca(NO_3)_2$ and $AgNO_3$ solution densities.

Table VI

Solubilities in the $Ca(NO_3)_2$ System
Ion Concentrations (moles/liter)

	Actual	Predicted
Ag^+	0.871	0.337
Ca^{++}	0.781	1.05
Pb^{++}	8.16×10^{-3}	6.60×10^{-4}
$SO_4^=$	2.13×10^{-2}	1.04×10^{-2}
NO_3^-	2.54	2.54

Because of the failure of the theoretical model to predict, even approximately, the actual solubilities of silver, calcium and lead, it was evident that a more accurate description of this system would require consideration of equilibria involving several ion pairs. As a result, in addition to the equilibria described by equations 3, 4 and 5 above, we examined equilibria involving the complexes $CaSO_4(aq)$, $AgSO_4^-$, and $AgNO_3(aq)$ as follows:

$$Ca^{++} + SO_4^{=} = CaSO_4(aq) \qquad (11)$$

$$Ag^{+} + SO_4^{=} = AgSO_4^{-} \qquad (12)$$

$$Ag^{+} + NO_3^{-} = AgNO_3(aq) \qquad (13)$$

The system was now defined by eight equations (14-21):

$$K_{sp_{CaSO_4 \cdot 2H_2O}} = \gamma_{Ca^{++}} \gamma_{SO_4^=} m_{Ca} m_{SO_4} a^2_{H_2O} = 2.45 \times 10^{-5} \qquad (14)$$

$$K_{sp_{PbSO_4}} = \gamma^2_{\pm} m_{Pb} m_{SO_4} = 1.34 \times 10^{-8} \qquad (15)$$

$$K_{sp_{Ag_2SO_4}} = \gamma^2_{Ag^+} \gamma_{SO_4^=} m^2_{Ag} m_{SO_4} = 1.25 \times 10^{-5} \qquad (16)$$

$$K_{assoc_{CaSO_4}} = \frac{\gamma_{CaSO_4(aq)}}{\gamma_{Ca^{++}} \gamma_{SO_4^=}} \cdot \frac{m_{CaSO_4}}{m_{Ca} m_{SO_4}} = 204 \qquad (17)$$

$$K_{assoc_{AgSO_4^-}} = \frac{\gamma_{AgSO_4^-}}{\gamma_{Ag^+} \gamma_{SO_4^=}} \cdot \frac{m_{AgSO_4^-}}{m_{Ag} m_{SO_4}} = 20.0 \qquad (18)$$

$$K_{assoc_{AgNO_3}} = \frac{\gamma_{AgNO_3(aq)}}{\gamma_{Ag^+} \gamma_{NO_3^-}} \cdot \frac{m_{AgNO_3}}{m_{Ag} m_{NO_3}} = 0.51 \qquad (19)$$

Charge Balance:

$$m_{Ag}+2m_{Ca}+2m_{Pb}+m_{x} = m_{NO_3}+2m_{SO_4}+m_{AgSO_4^-} \quad (20)$$

Ionic Strength:

$$\mu = 0.5(4m_{Ca}+4m_{Pb}+4m_{SO_4}+m_{Ag}+m_{NO_3}+m_{AgSO_4^-}+m_{y}) \quad (21)$$

In the above relations, K is the indicated constant, γ is the activity coefficient of the subscripted species and m refers to the molality of the subscripted species. The terms m_x and m_y are correction factors included to account for the presence of small amounts of other cations in solution. The activity of water was included in equation (14) as a further refinement of the system's definition.

Values of association constants for $CaSO_4$(aq), $AgSO_4^-$, and $AgNO_3$(aq) were obtained from the literature[5, 6]. All individual ion activity coefficients were determined by the mean salt method, using reported activity coefficients for equivalent ionic strength solutions of KCl and other salts[3, 4]. Since information on the activity coefficients of neutral ion pairs such as $CaSO_4$ and $AgNO_3$ is not readily available, activity coefficients of unity were assigned to these species. The activity coefficient of $AgSO_4^-$ was approximated by using the activity coefficient of HCO_3^- in solutions of $KHCO_3$ of equivalent ionic strength[5]. Estimation of the activity of water was accomplished by comparison to KCl solutions of equivalent total solute concentration[3].

The values for the concentrations of the various species were determined by the same iterative procedure described earlier. The calculated solubilities, along with the actual values, are shown below in Table VII with the calculated concentrations converted from molal to molar units as explained for the data given in Table VI.

Table VII

Solubilities in the $Ca(NO_3)_2$ System

	Predicted (molal)	Predicted (M)	Actual (M)
Total Ag	0.906	0.841	0.871
Total Ca	0.846	0.785	0.781
Pb^{++}	1.99×10^{-3}	1.85×10^{-3}	8.16×10^{-3}
Total SO_4	1.04×10^{-2}	9.65×10^{-3}	2.13×10^{-2}
Total NO_3	2.72	2.53	2.54

By comparing the results presented in Table VII to those in Table VI, it can be seen that taking into consideration the ion pairs $CaSO_4$(aq), $AgSO_4^-$, and $AgNO_3$(aq) in solution resulted in a great improvement in predicting the solubilities of silver and calcium, even though the values obtained for lead and sulfate were not substantially improved. This is perhaps due to existence of other complexes involving lead and sulfate ions which we have not taken into account.

Purification - We found it advisable to purify the pregnant calcium nitrate-silver nitrate leach liquor prior to recovering its silver content. An advantage of working in the silver nitrate system lies in the fact that silver is very soluble up to a comparatively high pH. The wide stability range of the aqueous silver ion allows the pregnant leach liquor to be readily purified from most other elements by hydrolysis to a pH of about 6. This is illustrated in Table VIII which shows the solubility of the pertinent elements as a function of pH.

Table VIII

Solubility vs. pH for Constituent Metal Ions in Leach Liquor

Metal Ion Activity	pH: Ag	Cu	Ni	Pb	Fe
1	6.2	4.7	6.1	6.2	1.7
10^{-1}	7.2	5.2	6.6	6.7	2.0
10^{-2}	8.2	5.7	7.1	7.2	2.3
10^{-3}	9.2	6.2	7.6	7.7	2.8
10^{-4}	10.2	6.7	8.1	8.2	3.2

Iron and copper are effectively controlled at pH 6. However, complete removal of nickel and lead is impossible. On the other hand, nickel will have been largely removed in the preliminary water leach and consequently represents little problem at this point. Since the system is saturated with $CaSO_4$, lead solubility is limited by the residual sulfate ion concentration. Other impurities, notably antimony, arsenic, selenium, and tellurium, are quantitatively removed by compound formation with ferric iron during the water leach. Summarizing the purification:

1. Cu is removed by hydrolysis at pH 6.

2. Ni is removed by pre-leaching.

3. Fe is removed by hydrolysis.

4. As, Sb, Se and Te are removed by compound formation with trivalent iron.

5. Soluble Pb is controlled by the calcium sulfate equilibrium solubility.

Electrowinning - Despite the general effectiveness of the purification steps noted above, some impurities occasionally survived the hydrolytic purification step and did cause problems. For these reasons, we explored the feasibility of recovering silver electrolytically. The advantage of electrowinning silver from the leach solution is apparent when one compares the deposition potential for silver with those of the various impurities which might pass through the purification step. The comparison is provided in

Table IX. These data show that a favorable condition exists for electrolytic recovery of high purity silver from solution, particularly in light of the fact that impurity elements are present at far lower activities than unity.

Table IX

Deposition Potentials for Various Metals in Nitrate Media

	Unit Activity (mv)	Actual Conc. (mv)
Ag	800	723
Se	740	666
Te	630	565
Cu	340	220
Sb	145	45
Pb	-126	-198
Ni	-230	-281
Fe	-409	-499

The experimental conditions chosen for electrowinning silver are shown in Table X. Theoretically, one could have stripped the electrolyte of all its silver content, but in practice this was not achievable. The electrolyte feed, at pH 6, was essentially free of nitric acid and the only acidity in the bulk electrolyte resulted from the discharge of oxygen at the anode. Figure 2 shows the current efficiency of the process as a function of the rising nitric acid concentration. The reason for the decreasing current efficiency with rising acidity became apparent when the cell was observed in operation. Finely divided, crystalline, silver dendrites which fell to the bottom of the cell were simply being redissolved in the bulk of the acidic electrolyte. Nitric oxide fumes were observed over the bath. However, since the rising acidity moved in step with high silver removal from solution, it was necessary to limit silver removal to that amount which would generate no more than 0.4 moles/liter HNO_3. This permitted operation at a current efficiency of 90% and stripping 0.4 mole/liter of silver satisfied the constraint.

Table X

Electrowinning Conditions for Silver Recovery

Cell Voltage	4 volts
Cathode Current Density	150 ma/cm^2
Anode Current Density	215 ma/cm^2
Anode Material	Duriron®
Cathode Material	Stainless Steel
Electrolyte:	
Temperature	40-45°C
Inlet Composition	$Ca(NO_3)_2$ - 1M $AgNO_3$ - 1M
Exit Composition	$Ca(NO_3)_2$ - 1M $AgNO_3$ - 0.6M HNO_3 - 0.4M

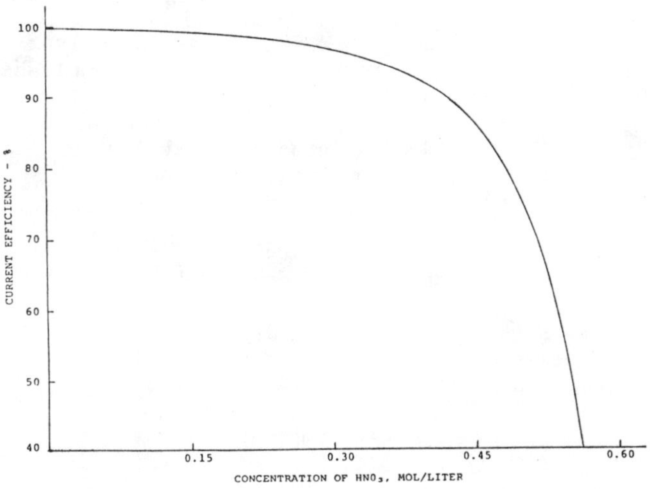

Figure 2

Current Efficiency vs. HNO_3 Concentration

As an alternative, we also investigated a means of neutralizing the acid generated. Since acid generation takes place at the anode, it was a comparatively simple matter to bag the anode and introduce a slurry of calcium hydroxide into the bag to neutralize the acid as it was formed. Testing showed that this measure allowed us to strip the electrolyte to any desired silver concentration and since the spent electrolyte returned to the leaching circuit would be much lower in silver, this allowed us to increase the solid/liquid ratio in the $Ca(NO_3)_2$ leach.

A series of ten locked-cycle bench-scale tests were performed to evaluate electrowinning from the standpoint of product purity and current efficiency. The average current efficiency obtained was 96% in an electrowinning operation that produced approximately 1 lb silver/hour. Overall recovery of silver from the sulfated slimes exceeded 95%. Build-up of impurities in the electrolyte was minimal. The silver electrodeposits were melted and analyzed spectrographically. Average purity of our silver product, as noted in Table XI, exceeded the ASTM specification easily.

Table XI

Purity of Electrowon Silver (Average of 10 Cycles) Wt %

Cu	0.0014	Te	0.0003
Pb	0.0006	Au	<0.0002
Bi	<0.0001	Pd	<0.0004
Fe	0.0044	Ag	>99.99

SUMMARY

The process described for recovering silver from sulfated copper refinery slimes accomplished the following objectives:

1. Eliminated doré furnacing with its attendant air pollution and energy consumption.
2. Reduced the processing time for the silver content from days to hours.
3. Eliminated many recycle problems associated with silver-bearing slags generated by the doré furnace process.
4. Reduced process temperature from 1300 C to 100 C.
5. Evidenced a capability for producing silver of a grade better than four-nines purity.

REFERENCES

1. Hoffman, J.E.; Parker, P.D. and Sabo, A.C., "Extraction and Purification of Silver From Sulfates," U. S. Patent 3,996,046, Dec. 7, 1976.
2. Rossini, F.D., et al, "Selected Values of Chemical Thermodynamic Properties," National Bureau of Standards Circular No. 500, Washington, D.C., 1952.
3. Conway, B.E., Electrochemical Data, Elsevier Publishing Co., New York, 1952, pp 77, 80, 90.
4. Lewis, G.N. and Randal, M., "Data for Aqueous Electrolyte Solutions," Appendix 4, Thermodynamics, 2nd Ed. (Revised by Pitzer and Brewer), McGraw-Hill, New York, 1961, pp 643-647, 653-657.
5. Garrels, R. M. and Christ, C. L., Solutions, Minerals and Equilibria, Freeman, Cooper & Co. San Francisco, 1965, pp 96, 104.
6. Helgeson, H. C., "Complexing & Hydrothermal Ore Deposition," International Series of Monographs on Earth Sciences, Vol 17, The Macmillan Co., New York, 1964, pp 72-73.

Section IV

Fundamental Aspects of Solvent Extraction Phenomena

L.L. Tavlarides and Jan D. Miller,
Section Editors

Chapter 17

DEVELOPMENT OF CHEMICALLY BASED SINGLE AND MULTICOMPONENT METAL LIQUID-LIQUID EXTRACTION MODELS

Ying-Chu Hoh* and Renato G. Bautista

Ames Laboratory USDOE and Department of Chemical Engineering
Iowa State University, Ames, Iowa 50011

ABSTRACT

The chemically based models previously developed to describe and predict the distribution coefficients and separation factors for several different single and multicomponent metal extraction systems are presented in a more general form in this paper. The degree of formation of the aqueous phase inorganic metal complexes, the extraction mechanism of the complexes, and the equilibrium constant for the extraction reaction are used in the development of the models. The concept has been successfully applied to extraction systems where the organic solvent extracts an ionic species and where the organic solvent extracts only the neutral species. In addition, a model to describe a system where the organic solvent is extracting several extractable species has been derived on the same basis.

*Institute of Nuclear Energy Research, Lung-Tan, Taiwan, Republic of China

DEVELOPMENT OF CHEMICALLY BASED SINGLE AND MULTICOMPONENT METAL LIQUID-LIQUID EXTRACTION MODELS

Ying-Chu Hoh* and Renato G. Bautista
Ames Laboratory USDOE and Department of Chemical Engineering
Iowa State University, Ames, Iowa 50011

INTRODUCTION

The growing importance of liquid-liquid extraction in the separation and concentration of metals, either from the primary ore, low grade ore, or from recycle materials has accelerated the research in understanding the role of extractants, diluents, solvents and chelating reagents on the chemistry of the extraction process. For some metal extraction systems, enough data are available in the literature to describe the chemical process taking place. The modeling of these processes using the chemical solution properties as the basis rather than using purely mathematical and/or statistical correlation may make possible the development of a more general model applicable to other systems. There have been many studies reported in the literature[1-17] using chemical solution properties as part of the model developed. In most of these cases, a large amount of data must first be made available, resulting in a more complicated model that is not very useful since the experimental data are already available. In other cases, the models developed are not readily applicable to other systems and to multicomponent metal systems. In this paper, the approach used successfully in developing chemically based models with several metal extraction systems containing different complexing ligands, chelating reagents, diluents, and solvents[11-13 16-17] are presented in a more general form applicable to both single and multicomponent metal systems. In addition, the derivation of a model for the case of an organic solvent extracting several extractable species is presented without confirming experimental data.

MODEL DEVELOPMENT

The chemically based thermodynamic models for liquid-liquid extraction systems presented in this paper have been divided into three categories: (1) Organic solvent extracting an ionic species, such as lanthanides extracted by di-(2-ethylhexyl) phosphoric acid (HDEHP) in kerosene, (2) organic solvent extracting a neutral species such as actinides extracted by tri-n-butyl phosphate (TBP), and (3)) organic solvent extracting several extractable species. These models are developed by making use of the inorganic metal complex stoichiometric stability constants, their liquid-liquid extraction mechanism and the equilibrium constant for the extraction reaction.

Type I: Organic Solvent Extracting an Ionic Species

The extraction mechanism for this type can be written as

$$(M^{+p})_A + p((HG)_q)_0 \rightleftharpoons (M(G_q)_p)_0 + p(H^+)_A \qquad (1)$$

where M^{+D} is the metallic species with a charge of +p, $(HG)_q$ is the organic solvent with a polymer number q and the subscripts A and 0 represent the aqueous and organic phases, respectively. For the case of lanthanides extracted by di-(2-ethylhexyl) phosphoric acid, p = 3, and q = 2.

The equilibrium constant K_e for the extraction reaction given in Equation (1) can be written as:

$$K_e = \frac{[(M(G_q))_p]_0\ [H^+]_A^p}{[M^{+p}]_A\ [(HG)_q]_0^p} \qquad (2)$$

where the brackets represent the activities. By using the relation between the activity and activity coefficient, Equation (2) can be rewritten as:

$$K_e = \frac{(M(G_q)_p)_0\ (H^+)_A^p}{(M^{+p})_A\ ((HG)_q)_0^p} \cdot \frac{\gamma_{(M(G_q)_p)}\ \gamma_{(H^+)}^p}{\gamma_{(M^{+p})}\ \gamma_{(HG)_q}^p} \qquad (3)$$

where the parentheses represent the concentrations. The total metal concentration in the organic phase can be represented by:

$$(M)_0^T = (M(G_q)_p)_0 = K_e \cdot \frac{(M^{+p})_A\ ((HG)_q)_0^p}{(H^+)_A^p} \cdot \frac{\gamma_{(M^{+p})}\ \gamma_{(HG)_q}^p}{\gamma_{(M(G_q)_p)}\ \gamma_{(H^+)}^p} \qquad (4)$$

It is generally assumed that the inorganic metal complexes present in aqueous phase are M^{+3}, MA^{+2},

*Institute of Nuclear Energy Research, Lung-Tan, Taiwan, Republic of China

MA_2^{+1} ... MA_p (1,2,3,...), where A represents a ligand. The formation of the metal inorganic complexes with ligand A can be represented by the following reactions:

$$M^{+p} + A^- \rightleftharpoons MA^{p-1} \quad (5)$$

$$M^{+p} + 2A^- \rightleftharpoons MA_2^{p-2} \quad (6)$$

$$\vdots$$

$$M^{+p} + pA^- \rightleftharpoons MA_p \quad (7)$$

The stoichiometric stability constants $\beta_1, \beta_2 \ldots \beta_p$ for Equations (5), (6), and (7), respectively, can be written as:

$$\beta_1 = \frac{(MA^{p-1})}{(M^{+p})(A^-)} \quad (8)$$

$$\beta_2 = \frac{(MA_2^{p-2})}{(M^{+p})(A^-)^2} \quad (9)$$

$$\vdots$$

$$\beta_p = \frac{(MA_p)}{(M^{+p})(A^-)^p} \quad (10)$$

where the parentheses represent the concentrations of the species. The total metal concentration in the aqueous phase can be represented by:

$$(M)_A^T = (M^{+p}) + (MA^{p-1}) + (MA_2^{p-2}) + \text{---} + (MA_p) \quad (11)$$

Substitution of Equations (8), (9), and (10) into Equation (11) yields:

$$(M)_A^T = (M^{+p})\ [1 + \beta_1(A^-) + \beta_2(A^-)^2 + \text{---} + \beta_p(A^-)^p]$$

$$= (M^{+p})\ [1 + \sum_i^p \beta_i(A^-)^i] \quad (12)$$

The distribution coefficient K_d is defined for a single component system by the ratio of the total concentration of the metallic specie in the organic phase to the total concentration of the metallic specie in the aqueous phase:

$$K_d = (M)_0^T/(M)_A^T \quad (13)$$

Substitution of Equations (4) and (12) into Equation (13) yields:

$$K_d = K_e \cdot \frac{\gamma_{(M^{+p})}\gamma^p_{(HG)_q}}{\gamma_{(M(G_q)_p)}\gamma^p_{(H^+)}} \cdot \frac{1}{1+\sum_i^p \beta_i(A^-)^i} \cdot \frac{((HG)_q)_0^p}{(H^+)_A^p} \quad (14)$$

In 1915, Bjerrum (as cited by 18) introduced the use of a variable by defining the degree of formation of an inorganic metal complex MA_i in systems such as those given by Equations (5), (6) and (7) as follows:

$$\alpha_i = \frac{(MA_i)}{(M)_A^T} = \frac{(MA_i)}{(M^{+p}) + (MA^{p-1}) + (MA_2^{p-2}) + \text{---} + (MA_p)} \quad (15)$$

By using Equations (8), (9) and (10) ..., Equation (15) can be rewritten as:

$$\alpha_i = \frac{\beta_i\ (A^-)^i}{1 + \sum_i^p \beta_i (A^-)^i} \qquad i = 1, 2, \ldots p \quad (16)$$

Therefore, we can define

$$\alpha_o = \frac{1}{1 + \sum_i^p \beta_i(A^-)^i} \quad (17)$$

At this point, the assumption is made that the ratio of the activity coefficients of the molecules in Equation (14) varies slightly with concentration. Should this assumption hold, then the quotient of activity coefficients in Equation (14) can be regarded as a constant as given by:

$$K_\gamma = \frac{\gamma_{(M(G_q)_p)}\gamma^p_{(H^+)}}{\gamma_{(M+p)}\ \gamma^p(HG)_q} \quad (18)$$

Combining this activity coefficient ratio K_γ with the liquid-liquid extraction equilibrium constant K_e gives an effective equilibrium constant

$$K_1 = K_e/K_\gamma \quad (19)$$

for the extraction equilibria. Substitution of Equations (17) and (19) into Equation (14) results in

$$K_d = K_1\alpha_o \frac{((HG)_q)_0^p}{(H^+)_A^p} \quad (20)$$

By defining

$$\phi = \frac{\text{organic concentration}}{\text{aqueous acidity}} = \frac{((HG)_q)_0}{(H^+)_A} \quad (21)$$

Equation (20) becomes

$$K_d = K_1 \alpha_0 \phi^p \quad (22)$$

Equation (22) is the predictive model equation for a single component system in this case. The distribution coefficient K_d is directly proportional to the organic concentration raised to the power p and to its degree of formation, and is inversely proportional to the acidity of the aqueous phase raise to the power p.

By definition, the separation factor for a binary component system of two different metals M and N with the same charge number +p can be written as:

$$S_{M,N} = \frac{(K_d)_M}{(K_d)_N} = \frac{K_{1,M}\, \alpha_{0,M}\, \phi_M^p}{K_{1,N}\, \alpha_{0,N}\, \phi_N^p} \quad (23)$$

Since for a binary component system,

$$\phi_M = \phi_N \quad (24)$$

Equation (23) reduces to

$$S_{M,N} = \frac{K_{1,M}\, \alpha_{0,M}}{K_{1,N}\, \alpha_{0,N}} \quad (25)$$

Defining

$$K_1' = \frac{K_{1,M}}{K_{1,N}} \quad (26)$$

Equation (25) can be rewritten as:

$$S_{M,N} = K_1' \frac{\alpha_{0,M}}{\alpha_{0,N}} \quad (27)$$

Equation (27) is the model equation for a binary component system which shows that the separation factor is proportional to the ratio of the degree of formation for the two components only.

Type II: Organic Solvent Extracting a Neutral Species

The extraction mechanism for this type of reaction can be written as:

$$(M^{+p})_A + p(A^-)_A + q(TBP)_0 = (MA_p \cdot qTBP)_0 \quad (28)$$

where M^{+p} is the metal with a charge of +p, A^- is a ligand, and TBP is tri-n-butyl phosphate. Plutonium and neptunium extracted by TBP are the examples for this category.[17] The equilibrium constant for this type of extraction mechanism can be written as:

$$K_e = \frac{[MA_p \cdot qTBP]_0}{[M^{+p}]_A\, [A^-]_A^p\, [TBP]_0^q} \quad (29)$$

where the brackets denote the activities. As in Equation (3), Equation (29) can be rewritten as:

$$K_e = \frac{(MA_p \cdot qTBP)_0}{(M^{+p})_A\, (A^-)_A^p\, (TBP)_0^q} \cdot \frac{\gamma_{(MA_p \cdot qTBP)}}{\gamma_{(M^{+p})}\, \gamma_{(TBP)}^q\, \gamma_{(A^-)}^p} \quad (30)$$

where γ's are the activity coefficients.

The total metal concentration in the organic phase is equal to

$$(M)_0^T = (MA_p \cdot qTBP)_0 = K_e (A^-)_A^p\, (M^{+p})_A$$

$$(TBP)_0^q \frac{\gamma_{(M^{+p})}\, \gamma_{(TBP)}^q\, \gamma_{(A^-)}^p}{\gamma_{(MA_p \cdot qTBP)}} \quad (31)$$

The total metal concentration in the aqueous phase can be represented by Equation (12). Substitution of Equations (31) and (12) into Equation (13) yields

$$K_d = K_e\, (A^-)_A^p (TBP)_0^q \cdot \frac{1}{1 + \sum_i \beta_i (A^-)^i} \cdot \frac{\gamma_{(M^{+p})}\, \gamma_{(TBP)}^q\, \gamma_{(A^-)}^p}{\gamma_{(MA_p \cdot qTBP)}} \quad (32)$$

By using the same assumption made in the case of type I in Equations (17) and (19), Equation (32) can be rewritten as

$$K_d = K_1 \alpha_0 (A^-)_A^p (TBP)_0^q \quad (33)$$

Equation (33) is the model equation for tye type 2 extraction mechanism. The distribution coefficient K_d is directly proportional to the ligand concentration raised to the power p, to the organic concentration raised to the power q and to its degree of formation.

For a binary component system, the separation factor between metals M and N with the same charge number +p can also be shown as:

$$S_{M,N} = \frac{K_{1,M}\, \alpha_{0,M}\, (A^-)_{A,M}^p\, (TBP)_{0,M}^q}{K_{1,N}\, \alpha_{0,N}\, (A^-)_{A,N}^p\, (TBP)_{0,N}^q} \quad (34)$$

because $(TBP)_{0,M} = (TBP)_{0,N}$ and $(A^-)_{A,M} = (A^-)_{A,N}$. Therefore Equation (34) reduces to

$$S_{M,N} = K_1' \frac{\alpha_{0,M}}{\alpha_{0,N}} \tag{35}$$

where K_1' has already been defined in Equation (26). This model equation has the same form as Equation (27). Therefore, it can be concluded that the model equation for binary metal component systems is the same regardless of the differences in the extraction mechanisms so long as the metals to be separated have the same charge number.

Type III: Organic Solvent Extracting Several Extractable Species

When more than one extractable species are being extracted, the reaction mechanisms can be represented by the following reactions:

$$(M^{+p})A + p(HG)_{q})_0 \rightleftharpoons (M(G_q)_p)_0 + p(H^+)_A \tag{36}$$

$$(MA^{p-1})_A + ((p-1)(HG)_q)_0 \rightleftharpoons (MA(G_q)_{p-1})_0 + (p-1)(H^+)_A \tag{37}$$

$$\vdots$$

$$(MA_p)_A + ((HG)_q)_0 \rightleftharpoons (MA_p(G_q))_0 + (H^+)_A \tag{38}$$

The equilibrium constants for Equations (36), (37), and (38) can be written as

$$K_{e,1} = \frac{[H^+]_A^p[M(G_q)_p]_0}{[M^{+p}]_A[(HG)_q]_0^p} \tag{39}$$

$$K_{e,2} = \frac{[(MA(G_q)_{p-1}]_0[H^+]_A^{p-1}}{[MA^{p-1}]_A[(HG)_q]_0^{p-1}} \tag{40}$$

$$\vdots$$

$$K_{e,3} = \frac{[MA_p(G_q)]_0[H^+]_A}{[MA_p][(HG)_q]_0}, \tag{41}$$

respectively. The brackets represent activities for each species. By using Equations (4) and (18), Equations (39), (40), and (41) can be rewritten as

$$K_{e,1} = \frac{(M(G_q)_p)_0\ (H^+)_A^p}{(M^{+p})_A((HG)_q)_0^p} \cdot K_{\gamma,1} \tag{42}$$

$$K_{e,2} = \frac{(MA(G_q)_{p-1})_0(H^+)_A^{p-1}}{(MA^{p-1})_A((HG)_q)_0^{p-1}} \cdot K_{\gamma,2} \tag{43}$$

$$\vdots$$

$$K_{e,p} = \frac{(MA_p(G_q))_0(H^+)_A}{(MA_p)((HG)_q)_0} \cdot K_{\gamma,p} \tag{44}$$

The parentheses denote the concentrations. The total metal concentrations in the organic phase can be represented by

$$(M)_0^T = (M(G_q)_p)_0 + (MA(G_q)_{p-1})_0 + \text{---} + (MA_p(G_q))_0 \tag{45}$$

Substitution of Equations (42), (43), and (44) into Equation (45) yields

$$(M)_0^T = \frac{K_{e,1}}{K_1}\ \frac{(M^{+p})_A((HG)_q)_0^p}{(H^+)_A^p} + \frac{K_{e,2}}{K_{\gamma,2}}\ \frac{(MA^{p-1})_A((HG)_q)_0^{p-1}}{(H^+)_A^{p-1}}$$

$$+ \text{---} + \frac{K_{e,p}}{K_{\gamma,p}} \cdot \frac{(MA_p)_A((HG)_q)_0}{(H^+)_A} \tag{46}$$

By using Equations (8), (9), (10), (19), and (21), Equation (46) can be rewritten as

$$(M)_0^T = (M^{+p})[K_1\phi^p + K_2\beta_1(A^-)\phi^{p-1} + K_3\beta_2(A^-)^2\phi^{p-2}$$

$$+ \text{---} + K_p\beta_{p-1}(A^-)^{p-1}\phi] \tag{47}$$

The total metal concentration in the aqueous phase can be represented by Equation (12). Substitution of Equations (12), (16), (17), and (47) into Equation (13) yields

$$K_d = K_1\alpha_0\phi^p + K_2\alpha_1\phi^{p-1} + K_3\alpha_2\phi^{p-2} + \text{---} + K_p\alpha_{p-1}\phi \tag{48}$$

Equation (48) is the thermodynamic model equation for the case when more than one extractable species are being extracted.

Data for Model Calculations

The ligand and hydrogen ion concentrations of the equilibrated aqueous phase can be determined experimentally. The organic concentration $((HG)_q)_0$ can be calculated by material balance as in the following equation:

$$((HG)_q)_0^I - p(M^{+p})_0^T + (A^-)_0^T = ((HG)_q)_0^E \tag{49}$$

where $(HG)_q^I$ and $(HG)_q^E$ are the initial and equilibrium organic concentrations with a polymer number q, (M^{+p}) denotes a metal ion concentration with a +p charge, and (A^-) represents the free ligand concentration. Subscript 0 denotes the organic phase and superscripts E, I, and T represent the equilibrium, initial and total concentrations, respectively.

The inorganic metal complex stoichiometric stability constants can be obtained from the literature or experimentally by the method described by Rossotti.[18] The degree of formation α_i can be calculated by Equations (16) and (17) together with the ligand concentrations of the inorganic metal complex and stoichiometric stability constants.

The distribution coefficient K_d at a specified condition can be determined by analysis of the metal concentration in the organic and the aqueous phase. When all of the above information are available, the liquid-liquid extraction effective equilibrium constants K_1, K_2, ... K_p can then be estimated by linear regression analysis.

After the effective equilibrium constants are established, the distribution coefficients at other experimental conditions can be calculated at specified ligand, hydrogen ion, and organic concentrations together with the total ionic strength of the aqueous phase.

The separation factor for a binary metal component system can be determined from the ratio of the distribution coefficients of the two components of interest in the system. The ratio of the effective equilibrium constant K_1' in Equation (27) and (35) can be obtained by linear regression analysis of the separation factors and the degree of formation of the two components. After the value of K_1' is established, the separation factors at other experimental conditions can be calculated from the available information on the degrees of formation of the inorganic metal complexes as well as the total ionic strength of the aqueous phase.

SUMMARY AND CONCLUSIONS

The generalized form of the chemically based models presented in this paper for the metal extraction systems where the extractable species are either ionic or neutral have been experimentally proven for several different systems.[11-13, 16-17] It is anticipated that other liquid-liquid metal extraction systems especially those with similar chemical and physical properties would behave in similar fashion, and thus allow the use of the above technique to describe and predict their distribution coefficients and separation factors. The chemical solution properties required are kept to a minimum and the resulting model developed is kept simple. The multicomponent model is based on the model developed for the single metal component system. These models make use of the aqueous phase inorganic metal complex stoichiometric stability constants expressed as its degree of formation, their extraction mechanism of the complexes, and the equilibrium constant for the extraction reaction.

The model equation for the organic solvent which extracts an ionic species can be represented by the equation

$$K_d = K_1\alpha_0\phi^p$$

where K_1 is the effective equilibrium constant, α_0 is the degree of formation and ϕ is the ratio of the organic solvent concentration to the aqueous acidity and p is the charge of the metal being extracted.

For the organic solvent which extracts only a neutral species, the model equation can be represented by the equation

$$K_d = K_1\alpha_0(A^-)^p_A(TBP)^q_0$$

where (A^-) is the ligand concentration in the aqueous phase, p is the charge of the metal being extracted and q is the stoichiometric number of the chelating reagent in the reaction.

For binary component systems, the model equation for the separation factor is the same for all cases and can be written as

$$S_{M,N} = K_1' \frac{\alpha_{0,M}}{\alpha_{0,N}}$$

where K_1' is the ratio of the effective equilibrium constants for the binary metal system and $\alpha_{0,M}$ and $\alpha_{0,N}$ are the degree of formation of metals M and N, respectively. This equation is limited to systems where the chemical and physical properties of the metals being separated are similar.

In the case where the organic solvent is extracting several extractable species, the model equation takes the form

$$K_d = K_1\alpha_0\phi^p + K_2\alpha_1\phi^{p-1} + K_3\alpha_2\phi^{p-2} + \text{---} K_p\alpha_{p-1}\phi$$

where K_1, K_2, ... K_p are the effective equilibrium constants for each of the extractable species.

ACKNOWLEDGEMENT

This research was supported in part through the Graduate Fellowship Award to one of the authors (Y.C.H.) by the Institute of Nuclear Energy Research, Chinese Atomic Energy Council, Taiwan, Republic of China, and the U.S. Department of Energy, Director of Energy Research, Office of Basic Energy Sciences, WPAS-KC-03-02-02.

REFERENCES

1. Glueckauf, E., 1958, "The Physical Chemistry of Solvent Extraction of Electrolytes", Belgische Chemische Industrie, Vol. 23, No. 11, Nov., pp. 1215-1230.

2. Lloyd, P. J. and Oertel, M. K., 1963, "A Theoretical Basis for the Evaluation of Long Chain Amine Extractants and Its Application in Desing Calculations for Separation Efficiencies and the Number of Theoretical Stages", Unit Processes in Hydrometallurgy, Wadsworth, M. E., and David, F., eds., Gordon and Breach Science Publishers, New York, pp. 453-477.

3. Oshsenfeld, W., Baumgarel, G., and Schmieder, H., 1969, "Empirical Functions for the Calculations of Simultaneous Pu(IV) and U(VI) Distribution in the System Metal Nitrates-Nitric Acid/Tributyl Phosphate-N-Dodecane", Solvent Extraction Research, Kertes, A. S. and Marcus, Y., eds.,

Wiley, New York, pp. 373-381.

4. Horner, D. E., 1970, "A Mathematical Model and A Computer Program for Estimating Distribution Coefficients for Plutonium, Uranium, and Nitric Acid in Extractions with Tri-n-Butyl Phosphate", Oak Ridge National Laboratory ORNL-TM-2711.

5. Ionannou, T. K., Bautista, R. G., and Smutz, M., 1971, "Correlating Multicomponent Equilibrium Data for the Extraction of Lanthanides with Di-(2-ethylhexyl)-Phosphoric Acid as the Solvent", Proceedings International Solvent Extraction Conference, Gregory, J. G., Evans, B., and Weston, P. C., eds., Society of Chemical Industry, London, Vol. II, pp. 957-965.

6. O'Brien, W. G., and Bautista, R. G., 1974, "A Thermodynamic Activity Model for a Single Lanthanide Nitrate Liquid-Liquid Extraction System", Proceedings International Solvent Extraction Conference, Jeffreys, G. F., ed., Society of Chemical Industry, London, Vol. 2, pp. 1399-1408.

7. Forrest, C. and Hughes, M. A., 1975, "The Modelling of Equilibrium Data for the Liquid-Liquid Extraction of Metals, Part II. Models for the Copper/LIX 64N and Chromate/Aliquat 336 Systems", Hydrometallurgy, Vol. 1, pp. 139-154.

8. Hughes, M. A., Andersson, S., and Forrest, C., 1975, "Distribution Surfaces in Solvent Extraction Systems", International Journal of Mineral Processing, Vol. 2, pp. 267-276.

9. Bauer, G., and Chapman, T. W., 1976, "Measurement and Correlation of Solvent Extraction Equilibria. The Extraction of Copper by KELEX 100", Metallurgical Transactions B, Vol. 7B, Dec., pp. 519-527.

10. Cognet, M. C., and Renon, H., 1976/1977, "Influence of Aqueous Phase Composition Upon Copper Extraction by Cationic Extractants: A Thermodynamic Interpretation", Hydrometallurgy, Vol. 2, pp. 305-314.

11. Nevarez, M., and Bautista, R. G., 1978, "A Thermodynamic Equilibrium Model to Predict the Cobalt Distribution Coefficient in the Cobaltous Chloride-Hydrochloric Acid-Water-Tributyl Phosphate Liquid-Liquid Extraction System", American Institute of Chemical Engineers Symposium Series No. 173, Vol. 74, pp. 97-106.

12. Hoh, Y. C., Nevarez, M., and Bautista, R. G., 1978, "A Predictive Thermodynamic Model for the Distribution Coefficients of Neodymium in the $Nd(NO)_3)_3$-HNO_3-H_2O-1M HDEHP-Amsco Liquid-Liquid Extraction System", Industrial and Engineering Chemistry Process Design & Development, Vol. 17, pp. 88-91.

13. Hoh, Y. C., and Bautista, R. G., 1978, "Chemically Based Model to Predict Distribution Coefficients in the Cu-LIX 65N and Cu-KELEX 100 Systems", Metallurgical Transactions B, Vol. 9B, Mar., pp. 69-75.

14. Whewell, R. J., and Hughes, M. A., 1979, The Modelling of Equilibrium Data for the Liquid-Liquid Extraction of Metals. Part III. An Improved Chemical Model for the Copper/LIX 64N System", Hydrometallurgy, Vol. 4, pp. 109-124.

15. Whewell, R. J., Hughes, M. A., and Middlebrook, P. D., 1979, "The Modelling of Equilibrium Data for the Liquid-Liquid Extraction of Metals. Part IV. The Effect of the Diluent on the Copper/LIX 64N System", Hydrometallurgy, Vol. 4, pp. 125-133.

16. Hoh, Y. C., and Bautista, R. G., 1979, "Liquid-Liquid Extraction Equilibrium Models for Binary and Ternary Lanthanides-HDEHP Systems", Industrial and Engineering Chemistry Process Design & Development, Vol. 18, July, pp. 446-453.

17. Hoh, Y. C., and Bautista, R. G., 1979, "Chemically Based Models to Predict Distribution Coefficients in the Pu(IV) and Np(IV) Nitrate-TBP Systems", Proceedings of the International Solvent Extraction Conference, Lucas, B. H., Ritcey, G. M., and Smith, H. W., eds., Canadian Institute of Mining and Metallurgy, CIM Special Volume 21, Vol. 1, pp. 273-278.

18. Rossotti, F. J. C. and Rossotti, H., 1961, The Determination of Stability Constants, McGraw-Hill, London.

Chapter 18

EXTRACTION OF NICKEL AND COBALT FROM ACIDIC SOLUTIONS USING LIX63-DNNS MIXTURES

K. Osseo-Asare

Department of Materials Science and Engineering
The Pennsylvania State University
University Park, PA 16802

Helen S. Leaver

Research Metallurgist
AMAX Extractive Research & Development, Inc.
Golden, CO 80401

John M. Laferty

Manager, Applied Technology
AMAX, Inc.,
Denver, CO. 80212

ABSTRACT

LIX63 in combination with DNNS selectively extracts nickel and/or cobalt from acidic leach liquors with pH values as low as 1.0. Nickel/cobalt and cobalt/nickel selectivities are achieved by varying the relative concentrations of LIX 63 and DNNS. Increasing LIX63 concentration retards both extraction and stripping rates while DNNS has the opposite effect. Slope analyses of equilibrium data indicate that both LIX63 and DNNS are present in the extracted complexes. Examples of possible flowsheets are presented.

EXTRACTION OF NICKEL AND COBALT FROM ACIDIC SOLUTIONS USING LIX63-DNNS MIXTURES

K. Osseo-Asare
Associate Professor of Metallurgy

The Pennsylvania State University
University Park, PA 16802

Helen S. Leaver
Research Metallurgist

AMAX Extractive Research & Development, Inc.
Golden, CO 80401

John M. Laferty
Manager, Applied Technology

AMAX Inc., Denver CO 80212

ABSTRACT

LIX63 in combination with DNNS selectively extracts nickel and/or cobalt from acidic leach liquors with pH values as low as 1.0. Nickel/cobalt and cobalt/nickel selectivities are achieved by varying the relative concentrations of LIX63 and DNNS. Increasing LIX63 concentration retards both extraction and stripping rates while DNNS has the opposite effect. Slope analyses of equilibrium data indicate that both LIX63 and DNNS are present in the extracted complexes. Examples of possible flowsheets are presented.

INTRODUCTION

Most materials from which nickel is produced - ores, concentrates and metallurgical intermediates - also contain cobalt. Since both elements can be marketed in a relatively pure form only, their separation becomes mandatory. In the case of hydrometallurgical treatment of nickel-bearing materials, additional problems arise from the presence of a number of impurities such as iron, aluminum, magnesium and manganese. Traditionally these various elements were separated by selective precipitation (1). Apart from the inherent chemical complexity, this approach has the disadvantage that it involves a number of costly unit operations, e.g., for liquid-solid separations.

The application of solvent-extraction could, in principle, effectively overcome these drawbacks associated with precipitation, in much the same way as it has opened the possibility of economically treating impure copper solutions. However, in contrast to copper, suitable selective extractants for nickel and cobalt were not available when AMAX embarked upon a research program aimed at the recovery of nickel and cobalt from low-grade oxide ores. Consequently the development of applicable solvent extraction techniques was made part of this program. These techniques would have to provide means for isolating nickel and cobalt from impure solutions as well as for separating the two elements. Since it had been established that nickel-cobalt separation could be effectively achieved in chloride media (2), the initial phase of the investigation was directed at isolation of nickel and cobalt from dilute, impure sulfate solutions. Once this objective was achieved, the elements could, if necessary be converted to a chloride medium and separated by known means.

In 1960, Magner reported (2) on results of test work using a solution of dinonylnaphthalene sulfonic acid (DNNS) in kerosene to extract nickel and cobalt from synthetic sulfate leach solutions. However, it was readily established that this approach was not practical for treatment of real, impure, leach solutions because of excessive co-extraction of impurities such as iron, aluminum and magnesium. AMAX research (3) established that the desired selectivity could be achieved by using a mixed extractant consisting of LIX63 and DNNS. It was further discovered that this selectivity was not limited to the separation of nickel and cobalt from impurities but could be extended to the separation of nickel and cobalt by judicious variation of the relative concentrations of LIX63 and DNNS in the organic phase.

This paper presents a quantitative treatment of the manner in which changes in the relative amounts of the two extractants affect the overall solvent extraction-stripping performance. In addition, the paper attempts to explain the mechanism of the synergism of the mixed-extractant system and it suggests means to apply it to nickel-cobalt hydrometallurgical systems.

EXPERIMENTAL

Organic Reagents

LIX63 was used as received from General Mills (now Henkel Corp.). DNNS was prepared by repeatedly contacting the commercial product NaSUL-AS (trade name for the ammonium salt of dinonylnaphthalene sulfonic acid, marketed by R. T. Vanderbilt) with 600 gpl H_2SO_4 at O/A = 1/1 until titration showed that conversion to the acid form was complete. To mixtures of the two extractants, Solvesso 150, a product of Exxon, was added as diluent.

Mixture Design

The experimental design and data analysis utilized the COMPU-SERV Response Surface Methodology (RSM) computer program (4). The loading, selectivity and

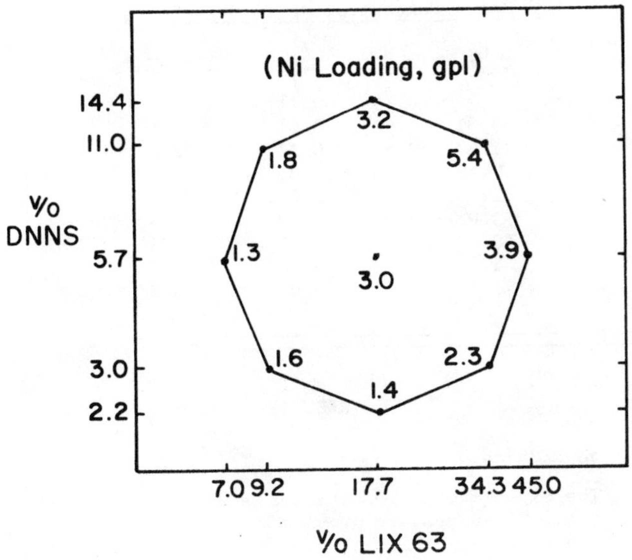

Figure 1a. Effect of variation in LIX63 and DNNS concentrations on gpl Ni loading from sulfate solutions (O/A = 1/5, initial aqueous nickel concentration = 4.2 gpl, feed pH = 1.1, 60 minute contact, 25°C).

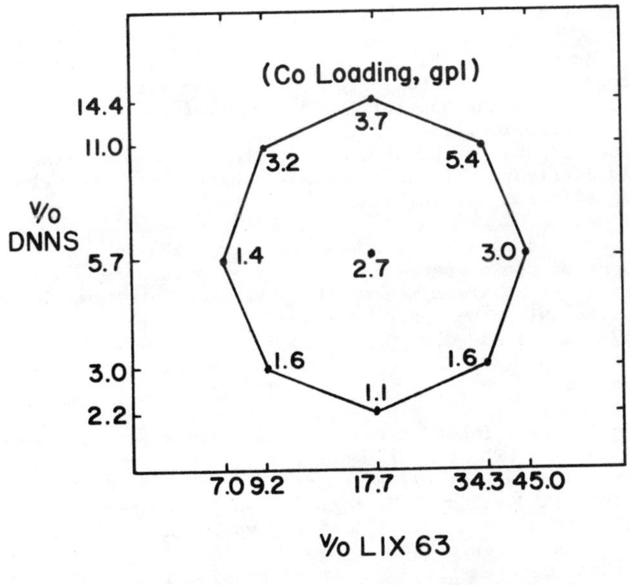

Figure 1b. Effect of variation in LIX63 and DNNS concentrations on gpl Co loading from sulfate solutions (O/A = 1/5, initial aqueous cobalt concentration = 4.2 gpl, feed pH = 1.1, 10 minute contact, 25°C).

stripping tests were performed with 9 different mixes limited to the region 7 v/o ≤ LIX63 ≤ 45 v/o and 2.2 v/o ≤ DNNS ≤ 14.4 v/o (v/o is used here to represent volume percent). The extractant concentration ranges for the interfacial tension studies were 2 v/o ≤ LIX63 ≤ 58 v/o and 1.5 v/o ≤ DNNS ≤ 20 v/o. Five replicate tests were performed using the center point solvent mixes.

The nickel loading tests were performed with an aqueous solution (4.2 gpl Ni, pH 1.1) obtained by sulfuric acid leaching of laterite ore. Using conventional shakeouts, each organic mixture was loaded at O/A = 1/5 for 60 minutes. The long contact time and high A/O ratio were used to ensure that each mixture was loaded to capacity. Cobalt loading tests were performed with a synthetic cobalt sulfate solution (4.2 gpl Co, pH 1.1). Ten minute shakeouts were used at O/A = 1/5. Portions of the loaded organics resulting from the extraction tests were saved for the stripping experiments. Using 450 gpl H_2SO_4, the loaded organic mixtures were contacted at O/A = 1/1 for ten minutes. For the selectivity tests, laterite leach liquor (4.2 gpl Ni, pH 1.1) was spiked with Co and Fe to give 4.2 gpl each. Ten minute shakeouts were used at O/A = 1/5.

Interfacial tension measurements were carried out with a Fisher ring tensiometer. Laterite leach liquor (4.2 gpl Ni, pH 1.7) and the organic mixtures were shaken at O/A = 1/5 for 60 minutes. After phase separation, equal volumes of the aqueous and organic phases were placed together in a measuring dish and the interfacial tension recorded. All experiments were performed at room temperature.

Mechanism Studies

Since the method of slope analysis requires that the extractants be maintained essentially in the uncomplexed state, very dilute nickel sulfate solutions (5×10^{-3} M Ni) were used for the equilibrium studies; 5×10^{-1} M Na_2SO_4 being used as supporting electrolyte in order to keep the ionic strength constant. Three sets of experiments were performed to study the effects of pH, DNNS concentration and LIX63 concentration on nickel extraction. The first set used three solvent mixes containing 7 v/o DNNS and 1, 4 and 12 v/o LIX63 respectively; shakeouts were performed with aqueous solutions in the pH range 0.5 ≤ pH ≤ 1.5. For the second set of experiments, the equilibrium pH was fixed at about pH 1. Five solvent mixes were used, all of which contained 30 v/o LIX63; DNNS was varied in the range 0.3 v/o ≤ DNNS ≤ 2.3 v/o. An equilibrium pH of about 1 was also used for the third set of experiments. DNNS concentration was 7 v/o and LIX63 was varied in the range 3 v/o ≤ LIX63 ≤ 8 v/o. Preliminary experiments indicated that 4 hour contact times were required to achieve equilibrium.

The rate studies used solvent mixes containing 30 v/o LIX63 and 5, 7.8, 13 and 15.5 v/o DNNS. Laterite leach liquor (4.3 gpl Ni, pH 2) was utilized for the extraction tests; stripping solution was 450 gpl H_2SO_4. Both the equilibrium and rate studies were performed with O/A = 1/1 at room temperature (~25°C).

Table 1. Co-Extraction of Co, Ni, Fe with LIX63-DNNS Mixtures

Composition of Solvent Mixture		Metal Concentration in Organic, gpl			
v/o LIX63	v/o DNNS	Ni	Co	Fe	Ni+Co+Fe
7.0	5.7	1.2	0.19	0.11	1.4
9.2	2.9	0.81	0.62	0.01	1.4
9.2	11.0	1.6	0.52	0.22	2.3
17.7	2.2	0.56	0.62	0.01	1.2
17.7	5.7	1.2	1.6	0.02	2.8
17.7	14.4	3.0	0.52	0.30	3.8
34.3	2.9	0.90	0.74	0.01	1.6
34.3	11.0	1.9	3.6	0.04	5.5
45.0	5.7	1.0	2.0	0.02	3.0

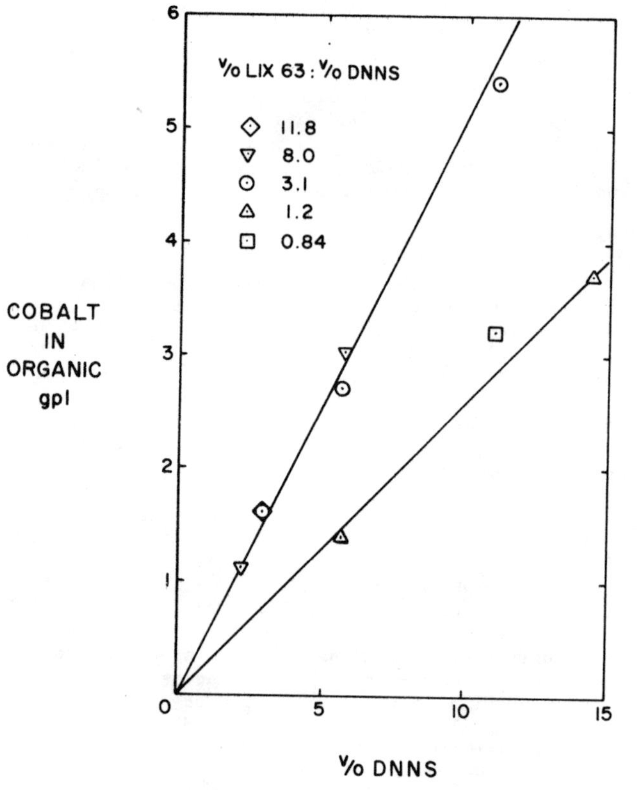

Figure 2. Effect of LIX63/DNNS ratio on cobalt extraction from sulfate solutions (O/A = 1/5, initial aqueous cobalt concentration = 4.2 gpl, feed pH = 1.1, 10 minute contact, 25°C).

RESULTS

Mixture Design

The experimental results are summarized in Figures 1 to 4. These results provide a vivid illustration of the manner in which changes in LIX63 and DNNS content can affect the overall performance of the LIX63-DNNS solvent extraction system.

A comparison of the extraction data in Figures 1a and 1b indicates that cobalt loading from cobalt-rich solutions is very similar to nickel loading from nickel-rich solutions. Loading is highest when LIX63 and DNNS concentrations are increased simultaneously. However, as can be seen from Figure 2, at a given DNNS concentration, loading increases with increase in LIX63 concentration until the ratio of v/o LIX63 to v/o DNNS reaches 3; beyond this ratio further increase in LIX63 concentration achieves no further improvement in loading.

According to Figure 3a, the best nickel stripping conditions (over 90%) are found in the narrow region located in the top left hand corner of the experimental region, i.e., high DNNS and low to intermediate LIX63. On the other hand, the cobalt stripping data (Figure 3b) indicate that cobalt strips readily (≥98%) from all the mixtures.

The selectivity results are summarized in Table 1. None of the mixtures is selective towards Fe. However, there are mixtures that are selective towards either Ni or Co. For example, 17.7 v/o LIX63-14.4 v/o DNNS loads 3.0 gpl Ni and 0.52 gpl Co, while 34.3 v/o LIX63-11 v/o DNNS loads 1.9 gpl Ni and 3.6 gpl Co.

The interfacial tension results are given in Figure 4. The mixtures which gave the largest interfacial tensions, ≥17 dynes/cm, are located in the high LIX63 region. With the exception of these three, all the other mixtures tested gave interfacial tension values below 3 dynes/cm. The extent to which LIX63 improves the interfacial tension is strongly influenced, however, by the DNNS level. Thus, when the DNNS content is greater than 6 v/o, the presence of LIX63 appears to have little effect on interfacial tension.

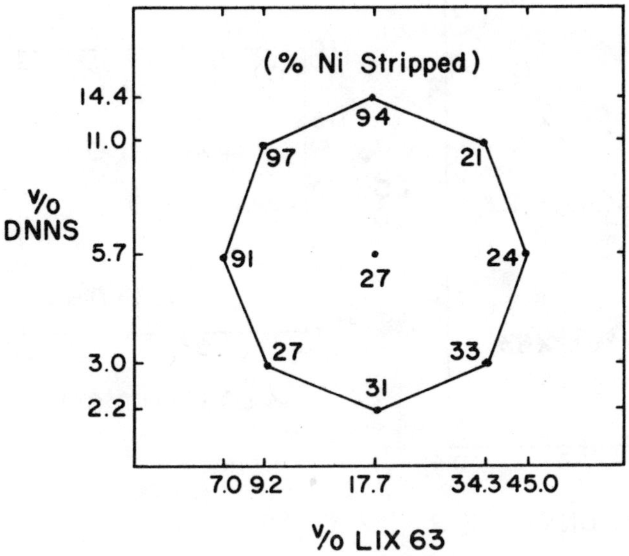

Figure 3a. Effect of variation in LIX63 and DNNS concentrations on percent Ni stripped using 450 gpl H_2SO_4. (O/A = 1/1, 10 minute contact, 25°C).

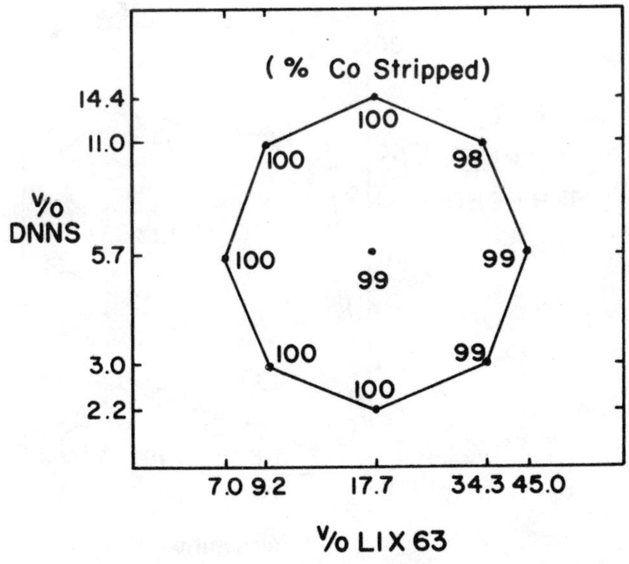

Figure 3b. Effect of variation in LIX63 and DNNS concentrations on percent Co stripped using 450 gpl H_2SO_4. (O/A = 1/1, 10 minute contact, 25°C).

Mechanism Studies

Figures 5a, 5b and 5c show, respectively, the effects of pH, DNNS and LIX63 on Ni extraction in terms of log D versus pH, log D versus log(v/o DNNS), and log D versus log(v/o LIX63). The distribution coefficient, D, is the ratio of nickel concentration in the organic phase to nickel concentration in the aqueous phase. The extraction and stripping rate results are given in Figures 6a and 6b, respectively.

DISCUSSION

Mixture Design

The experimental results were used with the RSM program to derive contour maps for the characteristics studied. The contours were obtained by fitting a second order equation to the experimental data:

$$Y = B_o + B_1X_1 + B_2X_2 + B_{11}X_1^2 + B_{21}X_2X_1 + B_{22}X_2^2 \quad (1)$$

where Y = characteristic value
B_o, B_1, etc = regression coefficients
X_1, X_2 = log (v/o LIX63), log (v/o DNNS)

The following prediction equations were obtained for nickel loading, nickel stripping, interfacial tension, Ni/Co and Ni/Fe selectivity:

Nickel loading (NL), gpl, as response

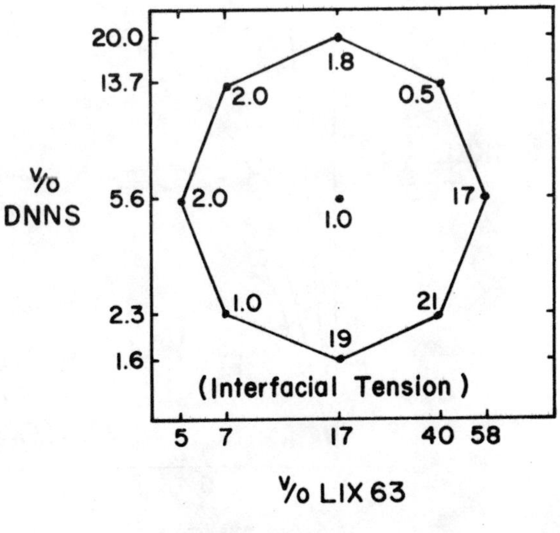

Figure 4. Effect of variations in LIX63 and DNNS concentrations on organic/aqueous interfacial tension (dynes/cm), following nickel extraction from laterite leach liquors (4.2 gpl Ni, pH = 1.7, O/A = 1/5, 60 minute contact, 25°C).

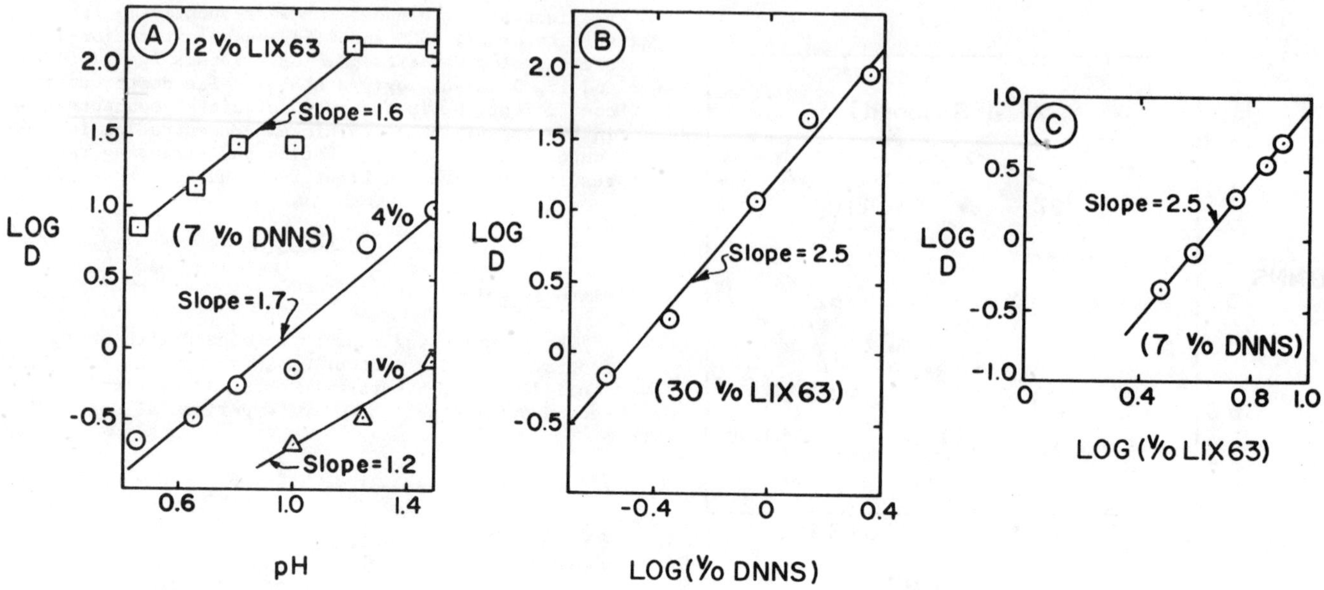

Figure 5. Slope analysis of nickel extraction from sulfate solutions (5×10^{-3} $\underline{M}$ Ni, 5×10^{-1} $\underline{M}$ Na_2SO_4, O/A = 1/1, 4 hour contact, 25°C), (A) effect of pH (7 v/o DNNS); (B) effect of DNNS (30 v/o LIX63, pH = 1); (C) effect of LIX63 (7 v/o DNNS, pH = 1).

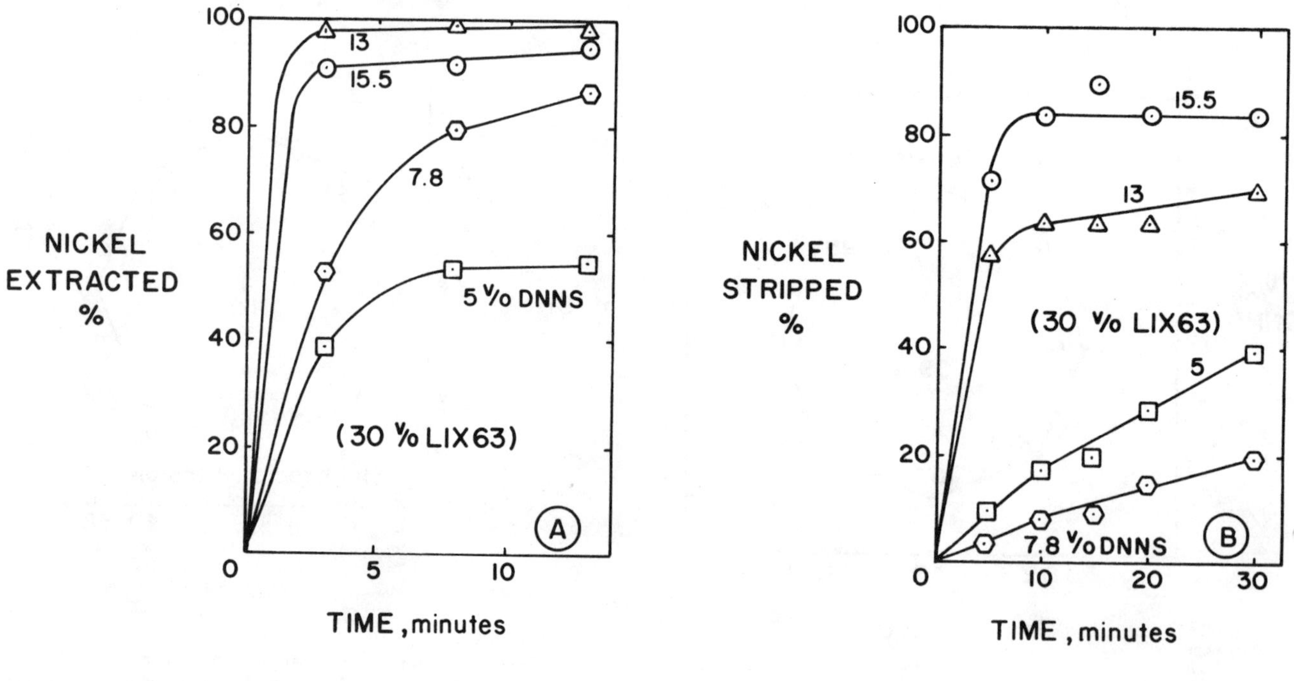

Figure 6. (A) Effect of DNNS on rate of nickel extraction (30 v/o LIX63; feed conditions: 4.3 gpl Ni, pH2; O/A = 1/1; 25°C); (B) Effect of DNNS on nickel stripping (30 v/o LIX63, 450 gpl H_2SO_4; O/A = 1/1; 25°C).

$$\log (NL) = -0.924 + 1.26X_1 + 0.233X_2 - 0.568X_1^2 + 0.986X_2X_1 - 0.708X_2^2 \quad (2)$$

Percent nickel stripped (NS) as response

$$\log (NS) = 2.17 - 1.54X_1 + 1.15X_2 + 1.04X_1^2 - 2.21X_2X_1 + 1.38X_2 \quad (3)$$

Interfacial tension (IT), dynes/cm, as response

$$IT = 11.8 - 18.1X_1 - 7.1X_2 + 23.9X_1^2 - 36.4X_2X_1 + 25.1X_2^2 \quad (4)$$

Ni/Co selectivity (D_{Ni}/D_{Co}) as response

$$\log (D_{Ni}/D_{Co}) = 2.38 - 3.24X_1 + 0.059X_2 + 1.58X_1^2 - 2.36X_2X_1 + 2.29X_2^2 \quad (5)$$

Ni/Fe selectivity (D_{Ni}/D_{Fe}) as response

$$\log (D_{Ni}/D_{Fe}) = 0.334 + 5.20X_1 - 2.79X_2 - 2.64X_1^2 + 2.78X_2X_1 - 1.23X_2^2 \quad (6)$$

The contour plots provided by the computer have been reproduced in Figure 7. Figure 7a shows that high nickel loading is achieved by increasing both LIX63 and DNNS (i.e., top right hand corner of the region of interest). On the other hand, as can be seen from Figure 7b, this region is hard to strip. According to Figure 7b, the best stripping conditions are attained by decreasing LIX63 while keeping DNNS at a high level (see top left hand corner). Thus, in order to obtain improved nickel stripping while at the same time providing acceptable nickel extraction, a compromise in LIX63 concentration must be found. Accordingly, RSM was requested to solve the following optimization problem:

Determine the maximum percent nickel stripped (NS) subject to the constraints

$$75\% \le NS \le 100\%$$
$$3 \text{ gpl Ni} \le NL \le 6 \text{ gpl Ni} \quad (7)$$

The optimization chose a solvent mixture of composition 16.3 v/o LIX63-14.4 v/o DNNS giving 80% nickel stripping and 3 gpl nickel loading.

The "optimum" solvent mixture 16.3-14.4 v/o given by the RSM program, is based on prediction equations which are approximations of the true response surface. Therefore in order to discover the true optimum, various solvent mixtures in the neighborhood of the 16.3-14.4 v/o mixture were explored. Stripping isotherms were determined for six extractant mixtures in which DNNS was kept constant at 15 v/o and LIX63 was varied from 30 v/o to 14 v/o. The results shown in Figure 8 confirmed the computer predictions. Nickel stripping improves dramatically as the LIX63 concentration is lowered; below 18 v/o LIX63 however, there is no further advantage.

Figures 7c, 7d and 7e have as responses, nickel selectivities over cobalt and iron respectively, and cobalt selectivity over iron. (As the measure of selectivity, the ratio of distribution coefficients has been used). It can be seen that raising DNNS concentration increases Ni/Co selectivity but lowers Ni/Fe selectivity. These results suggest that variation in the relative amounts of LIX63 and DNNS can produce a cobalt-selective or a nickel-selective solvent. For example, the solvent 34.3 v/o LIX63-11 v/o DNNS is more selective towards cobalt. (3.6 gpl Co and 1.9 gpl Ni are coextracted), while the solvent 17.7 v/o LIX63-14.4 v/o DNNS is more selective towards nickel (0.52 gpl Co and 3.0 gpl Ni are coextracted).

Cobalt loading from cobalt-rich solutions is comparable with nickel loading from nickel-rich solutions (see Figure 1). However when the two metals are present together, they behave differently. At a given DNNS concentration, the organic affinity for Co increases at the expense of Ni as the LIX63 concentration increases; this is illustrated in Figure 9 with data taken from Table I (open data points) and Figure 1b (solid data points). On the other hand, it can be seen from Figure 10 that as the DNNS concentration increases, nickel loading increases relative to that of cobalt. The solid data points in Figure 10 represent cobalt loading in the absence of nickel; unlike the mixed nickel-cobalt system (open data points), cobalt extraction increases with DNNS concentration. This behavior of cobalt must be attributed to differences in the kinetics of cobalt and nickel extraction as well as in the stability of the extracted species. It will be shown below that increase in DNNS concentration enhances the rate of nickel extraction. Thus, as DNNS concentration is raised, nickel extraction should improve. It would seem that cobalt extraction rate does not depend as strongly on DNNS concentration. This difference in kinetics may also explain why, in sharp contrast with nickel, cobalt stripped readily from all the mixtures investigated.

The response surface for interfacial tension is shown in Figure 7e. High interfacial tensions are given by mixtures which contain high LIX63 and low DNNS concentrations (bottom right hand corner). According to Figure 7b, such mixtures strip poorly. Conditions of low interfacial tension (top left hand corner, Figure 7e), however, correlate strongly with improved stripping (top left hand corner, Figure 7b).

The mixed LIX63-DNNS solvent extraction system can be used to process nickel-cobalt streams of widely varying nickel/cobalt ratios simply by altering the mixture composition. The following examples are presented to illustrate possible configurations for practical solvent extraction processes. These examples by no means exhaust the possibilities. In the illustrations shown in Figures 11, 12 and 13, the blocks represent mixers-settlers. In Figure 11, nickel and cobalt are coextracted; cobalt is then stripped first, followed by nickel. In Figure 12, cobalt is extracted first; the cobalt-barren raffinate is then sent for nickel extraction. In Figure 13, nickel is extracted first, the nickel barren

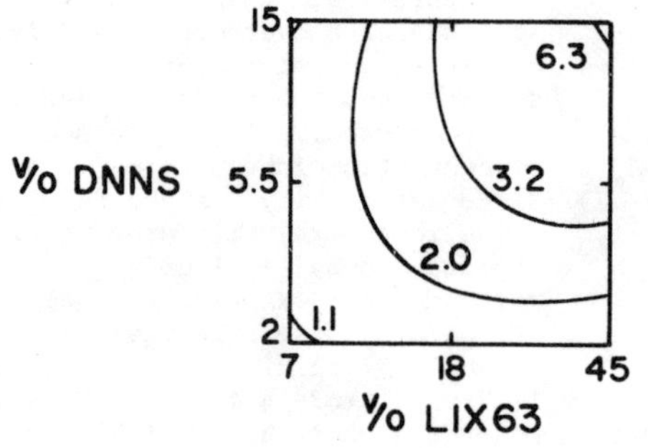

(a) Nickel loading (g/l)

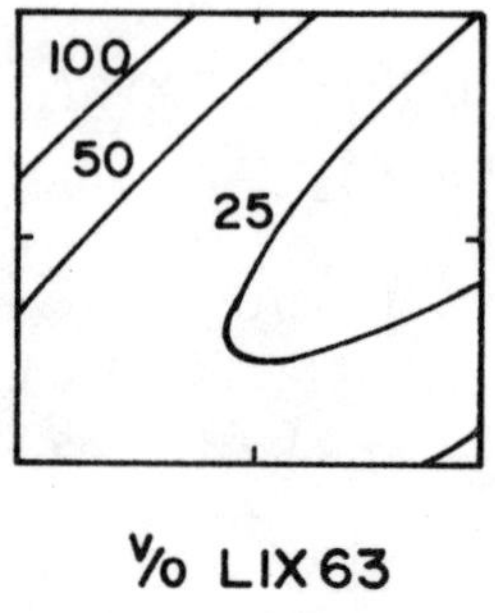

(b) Nickel Stripping (%)

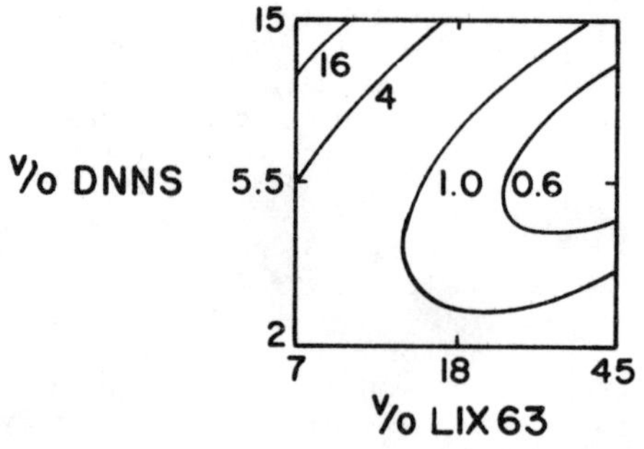

(c) Ni/Co selectivity (D_{Ni}/D_{Co})

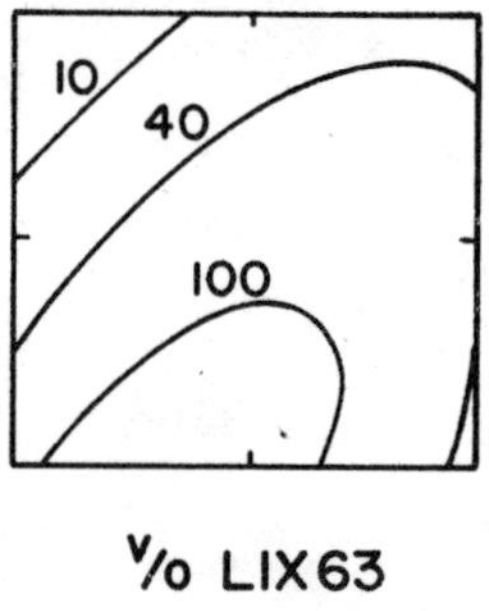

(d) Ni/Fe selectivity (D_{Ni}/D_{Fe})

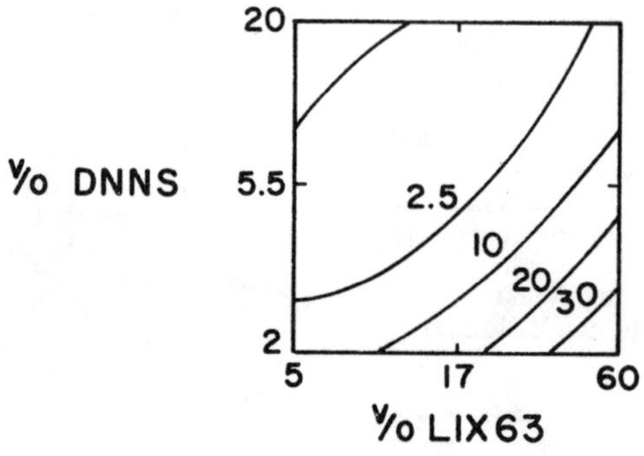

(e) Interfacial tension (dynes/cm)

Figure 7. Response surfaces generated by RSM.

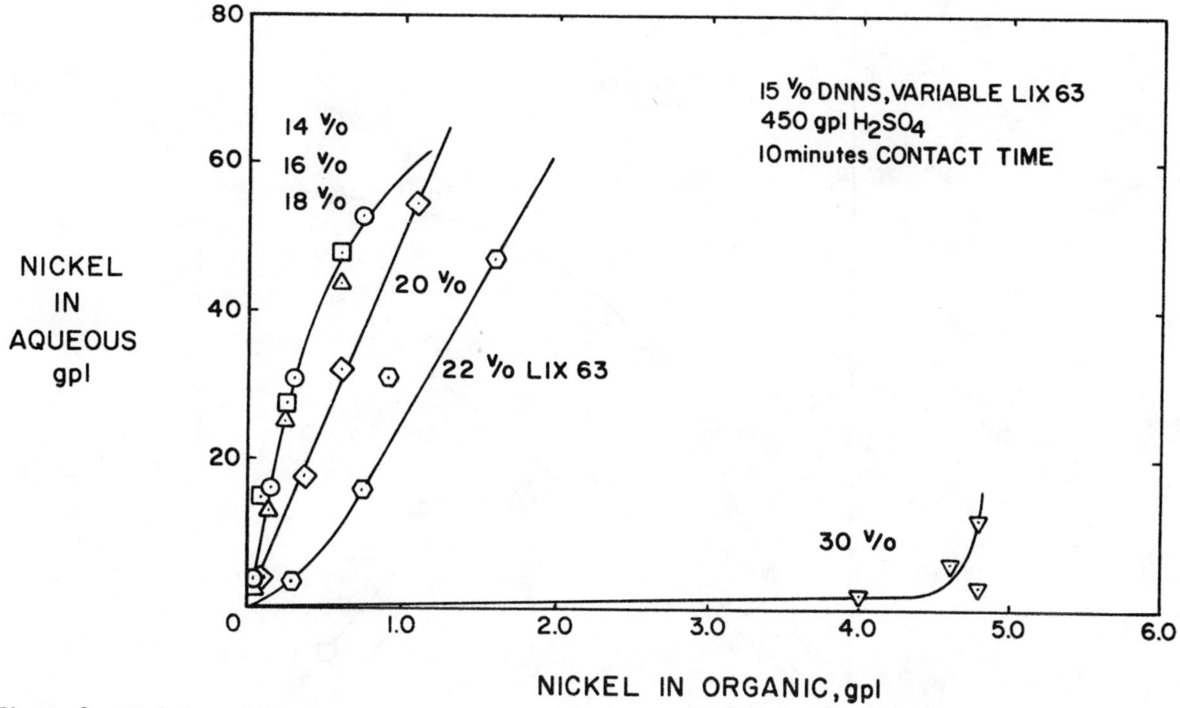

Figure 8. Nickel stripping isotherms for solvent mixes of composition 15 v/o DNNS and variable LIX63.

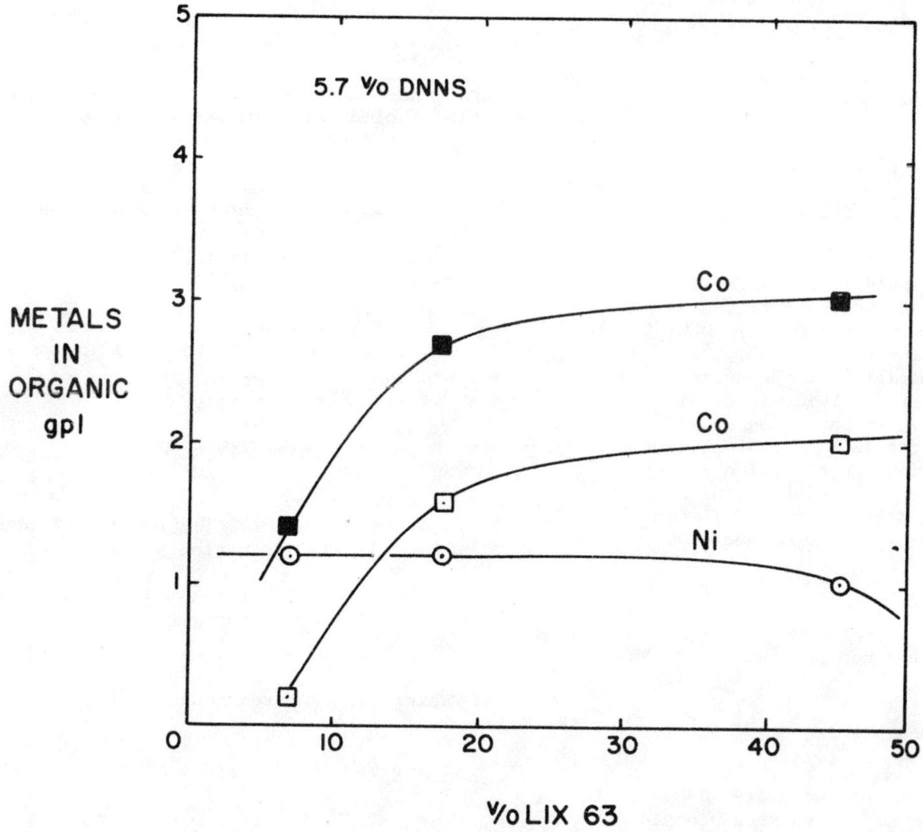

Figure 9. Effect of LIX63 concentration on nickel and cobalt extraction using synthetic cobalt solution (solid data points) and cobalt-spiked laterite leach liquor (open data points).

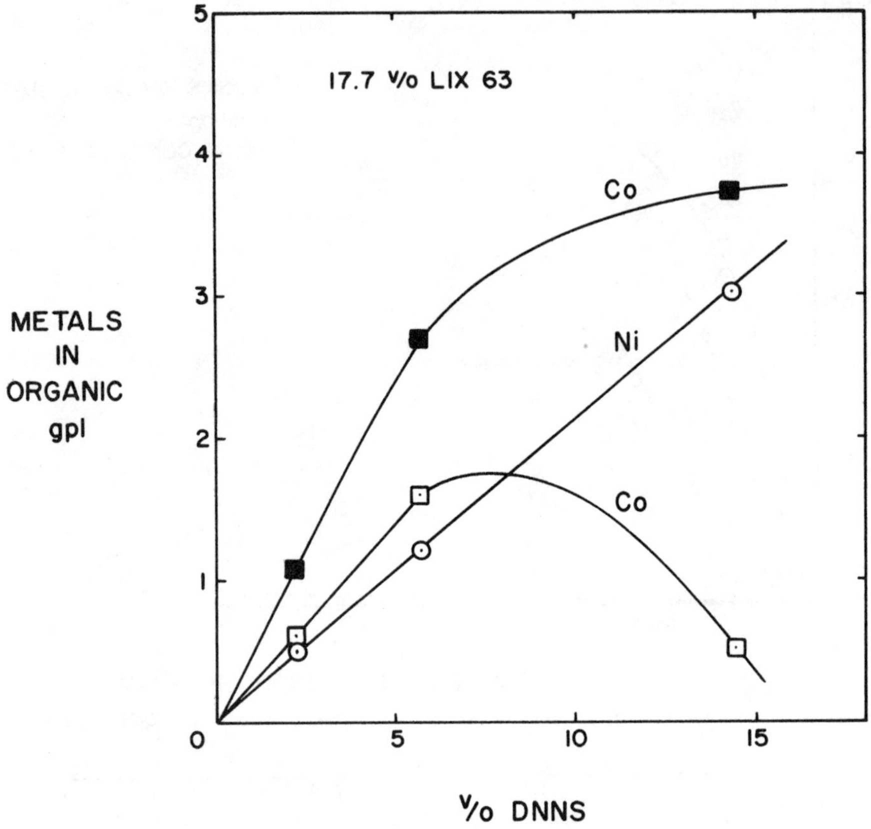

Figure 10. Effect of DNNS concentration on nickel and cobalt extraction using synthetic cobalt solution (solid data points) and cobalt-spiked laterite leach liquor (open data points).

raffinate becoming the feed for a cobalt extraction circuit.

Referring back to Figure 1 and Table I, it can be seen that it would be advantageous to use an organic mixture such as 17.7 v/o LIX63-14.4 v/o DNNS for the flowsheets given by Figures 11 and 13. Since cobalt strips more readily than nickel, the coextracted cobalt can be stripped first followed by nickel stripping with a higher concentration of mineral acid. The flowsheet illustrated in Figure 12 is best used when cobalt concentration (in the Ni+Co process solution) is much higher than nickel concentration. The mixture 34.3 v/o LIX63-11 v/o DNNS is cobalt-selective and can be applied advantageously.

Mechanism Studies

The extraction process can be visualized thus:

$$Ni^{++} + mHD + nHL \rightleftarrows NiD_mL_nH_{m+n-x} + xH^+ \qquad (8)$$

where HD and HL represent respectively, DNNS and LIX63, and $NiD_mL_nH_{m+n-x}$ is the extracted metal complex. Once x, m, and n are known, the form of the extracted complex can be determined:

$$NiD_mL_nH_{m+n-x}$$

For example, if n=zero, it would mean that LIX63 is not involved in the complex. The standard method of determining x, m and n is the technique of slope analysis. (5).

Writing the extracted complex as $\overline{Ni}$ and the aqueous nickel ion as Ni, the equilibrium constant for reaction 8 becomes:

$$K = \frac{[\overline{Ni}][H^+]^x}{[Ni][HD]^m[HL]^n} \qquad (9)$$

By means of the distribution coefficient, $D = [\overline{Ni}]/[Ni]$. Equation 9 can be transformed to

$$K = \frac{D\ [H^+]^x}{[HD]^m\ [HL]^n} \qquad (10)$$

Taking logs and rearranging gives

$$\log D = \log K + xpH + m\log\,[HD] + n\log\,[HL] \qquad (11)$$

According to Equation 11,

a. If LIX63 and DNNS concentrations are held constant, a plot of log D versus pH should give a line with slope x.

b. If LIX63 and the pH are held constant, a plot of log D versus log [HD] should give a line with slope m.

c. If DNNS and pH are held constant, a plot of log D versus log [HL] should give a line with slope n.

Examination of Figure 5a shows that under the conditions of the present experiments, x may assume values between 1 and 2. The values of both m and n are given as 2.5 by Figures 5b and 5c. Since both LIX63 and DNNS were used as commercial grade materials instead of purified organic compounds, the values of x, m and n must be taken as estimates. Nevertheless, the results of the equilibrium studies are very significant in that they show that both LIX63 and DNNS are incorporated into the extracted nickel complex.

The results of the rate studies indicate that both the rate of extraction and the rate of stripping are enhanced by increase in DNNS concentration. In Table II, (d%Ni/dt), the initial rates of nickel extraction and stripping are presented.

The results of the mechanism studies indicate that while both LIX63 and DNNS are present in the extracted complex these two extractants play different roles in the extraction process. DNNS provides both equilibrium and kinetic synergism; that is, DNNS influences the extent of extraction (shown by the equilibrium studies) as well as the rate of extraction and stripping (shown by the rate studies). The effect of LIX63 concentration on extraction and stripping rates was not investigated extensively. Nevertheless, preliminary results show that increasing LIX63 concentration retards extraction rate. For example, a 1 v/o LIX63-7 v/o DNNS solvent attains equilibrium in fifteen minutes; increasing LIX63 concentration to 12 v/o and 70 v/o prolongs equilibrium time to thirty minutes and four hours, respectively. The difference in LIX63 and DNNS kinetic behavior must be attributed to the ability of DNNS to decrease the organic aqueous interfacial tension (6).

CONCLUSIONS

1. Nickel loading from nickel-rich solutions is very similar to cobalt loading from cobalt-rich solutions; loading is highest when LIX63 and DNNS concentrations are increased simultaneously. At a given DNNS concentration, increasing LIX63 content enhances loading up to a point beyond which no significant increment is gained. Fixing LIX63 concentration and varying DNNS concentration similarly results in a plateau in extraction. Thus, at 17.7 v/o LIX63, nickel loading more than doubles from 1.4 gpl to 3.0 gpl when DNNS concentration is increased from 2.2 v/o to 5.7 v/o. Raising DNNS concentration further to 14.4 v/o yields only 0.2 gpl additional loading.

2. Cobalt strips readily (≥98%) from all the mixtures investigated (7 v/o ≤ LIX63 ≤ 45 v/o, 2.2 v/o ≤ DNNS ≤14.4 v/o). The solvent mixture of composition 34.3 v/o LIX63-11 v/o DNNS gave 98% Co stripping; in comparison, nickel was found to be hardest to strip from this mixture (21% stripping). In general, relative to cobalt, nickel strips with difficulty; however, increased DNNS concentration dramatically improves stripping. Thus, while the two mixtures 17.7-5.7 v/o and 17.7-14 v/o both load approximately 3 gpl Ni, the 5.7 v/o DNNS mixture strips only 30% of its nickel while the 14.4 v/o DNNS mixture achieves 94% stripping at O/A = 1/1 (450 gpl H_2SO_4 strip acid). Mixtures in the neighborhood of 16 v/o LIX63-15 v/o DNNS give the most desirable nickel extraction-stripping characteristics.

3. Variation in the relative amounts of LIX63 and DNNS can produce a cobalt-selective or a nickel-selective extractant system. In general, for a given LIX63 concentration, Ni loads at the expense of Co as the DNNS concentration increases. By manipulating mixture composition, the mixed extractants can therefore be used to process solutions of widely varying nickel/cobalt ratios.

4. Both LIX63 and DNNS are present in the extracted metal complexes; in the range of LIX63 and DNNS compositions investigated, the extracted species contain Ni, LIX63 and DNNS in the mole ratios Ni/LIX63/DNNS: 1/(2-3)/(2-3).

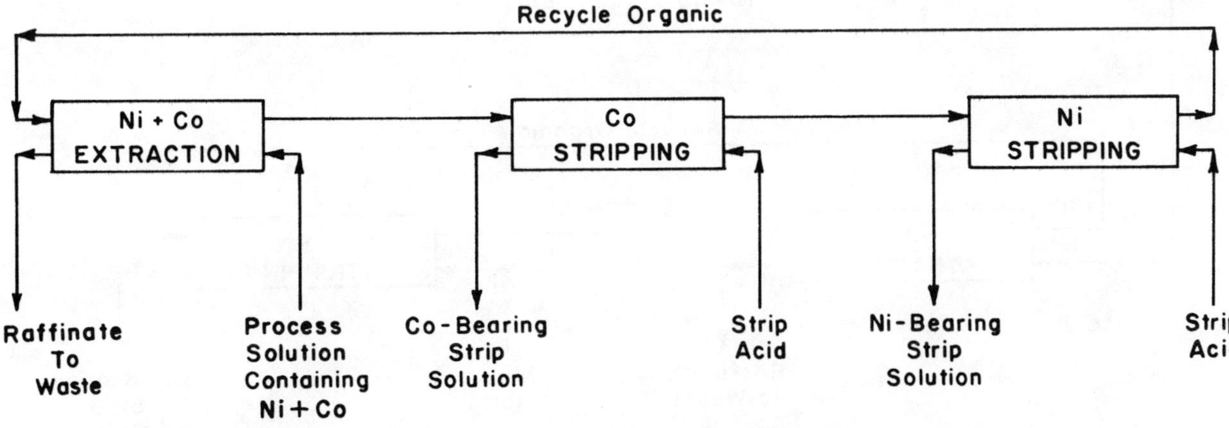

Figure 11. Proposed flowsheet for nickel and cobalt coextraction.

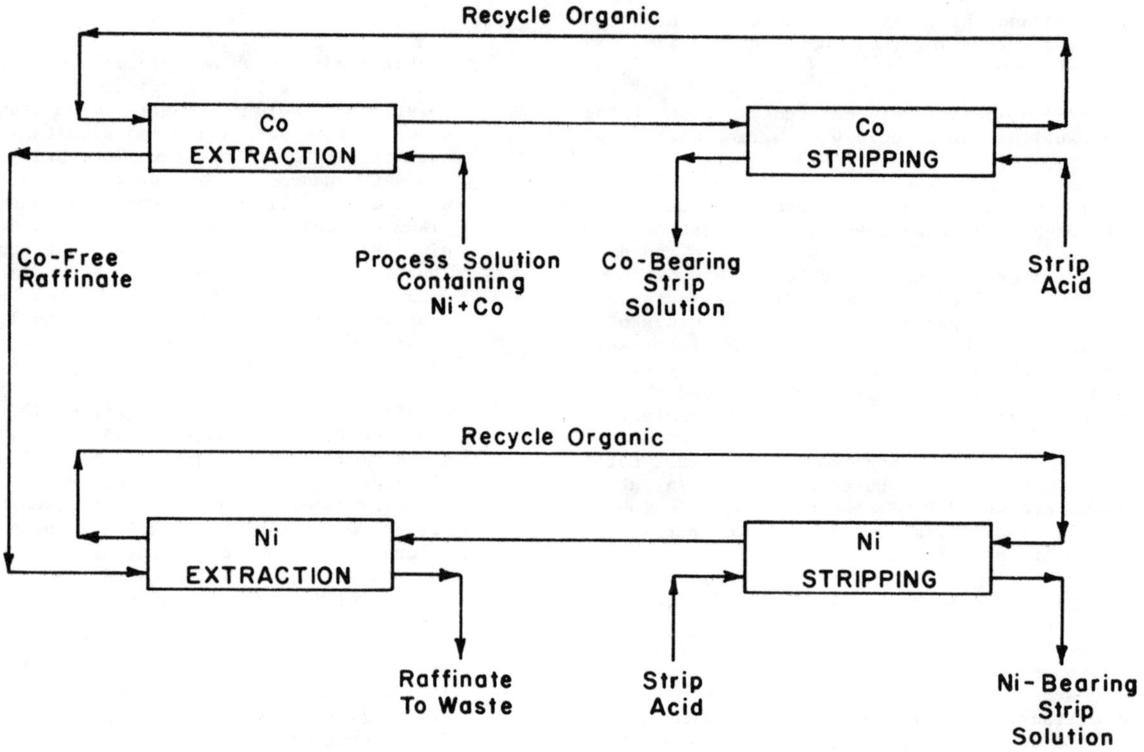

Figure 12. Proposed flowsheet for selective cobalt extraction.

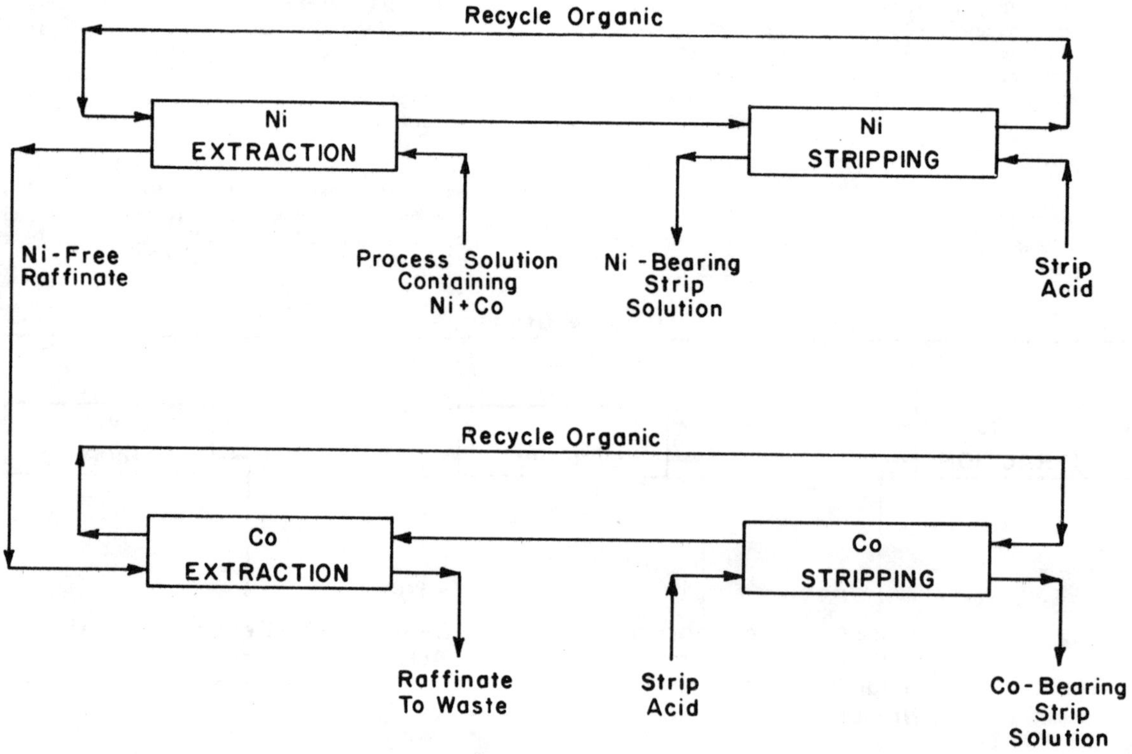

Figure 13. Proposed flowsheet for selective nickel extraction.

Table 2. Effect of DNNS Concentration on the Initial Rates of Nickel Extraction and Stripping

Solvent Mixture		d(%Ni)/dt (% per minute)	
v/o LIX63	v/o DNNS	Extraction	Stripping
30	15.5	30	14.4
30	13	33	12.0
30	7.8	18	0.8
30	5	13	2.0

5. Raising DNNS concentration increases both the rate of extraction and stripping, the effect on stripping being more pronounced.

ACKNOWLEDGEMENT

The authors thank Dr. J. Roorda for his helpful comments.

REFERENCES

1. Boldt, J. R. and Queneau, P., The Winning of Nickel, Van Nostrand, Princeton, 1966.

2. Magner, J. E., "Cobalt-Nickel Separation by Solvent Extraction", in P. Queneau, Ed., Extractive Metallurgy of Copper, Nickel and Cobalt, Interscience, New York, 1961.

3. Fekete, S. O., Meyer, G. A. and Wicker, G. R., "The Selective Extraction of Nickel and Cobalt from Acid Leach Solutions Using a Mixed Solvent System", TMS AIME Preprint No. A77-95.

4. COMPU-SERV Network Inc., Response Surface Methodology User's Manual. (5000 Arlington Centre Blvd., Columbus, Ohio 43220)

5. Flett, D. S., Okuhara, D. N. and Spink, D. R., "Solvent Extraction of Copper by Hydroxy Oximes", J. Inorg. Nucl. Chem., Vol. 35, pp. 2471-2487 (1973).

6. Osseo-Asare, K. and Keeney, M. E., "Aspects of the Interfacial Chemistry of Nickel Extraction with LIX63-HDNNS Mixtures", Met. Trans. B., Vol. 11B, pp. 63-67 (1980).

Chapter 19

MASS TRANSFER AND REACTION RATES IN THE SOLVENT EXTRACTION OF METALS

Samuel W.-S. Tse, E. S. Vargas, and T. W. Chapman

Chemical Engineering Department
University of Wisconsin
Madison, Wisconsin 53706

ABSTRACT

Models for interphase mass transfer rates in the solvent extraction of metals are developed for zinc and copper chlorides being extracted by triisooctylamine and for copper extraction by LIX 64N from ammonia solution. The approach is to adopt a stagnant film model in which film thicknesses are known and to derive the interfacial metal fluxes for the case where all reversible reactions are fast and locally equilibrated.

A growing drop mass transfer cell is used to measure extraction rates. It is found that for the cases studied the transfer rates are mass-transfer limited. On the other hand, copper extraction by LIX from acidic solution exhibits slow reaction kinetics, and earlier kinetics models are confirmed. Interactions that cause varying selectivity in the simultaneous extraction of copper and zinc chlorides are demonstrated.

MASS TRANSFER AND REACTION RATES IN THE SOLVENT EXTRACTION OF METALS

Samuel W.-S. Tse, E. S. Vargas, and T. W. Chapman

Chemical Engineering Department
University of Wisconsin
Madison, Wisconsin 53706

Although the kinetics of reaction-rate-limited metal extraction have received some attention in recent years, there has been less explicit consideration of the rate of extraction in mass-transfer-limited cases. In an earlier paper (1) we discussed a two-film mass-transfer model for metal extraction and showed how nonlinearities arise from the reaction stoichiometry, even when the interfacial reaction is fast. Furthermore, it was shown that interactions can arise between two metals being extracted simultaneously so that selectivity can depend on mass transfer conditions as well as the equilibrium properties of the system.

In this paper we develop mass transfer models for the extraction rates in several systems of practical interest. We also describe an apparatus for measuring rates on a known interfacial area and compare some observed rates with those predicted by the models. The extraction systems we have studied are zinc chloride-triisooctylamine (TIOA), cupric chloride-TIOA, mixed zinc and copper chlorides-TIOA, and copper sulfate-ammonia-LIX 64N.

Amine Extraction Systems

The extraction of metal chlorides by a tertiary amine such as TIOA may be described by the following reversible reactions:

(a) metal complex formation in the aqueous phase

$$M^{+2} + nCl^- \rightleftharpoons MCl_n^{2-n} \qquad (1)$$

with stability constants

$$\beta_n = \frac{[MCl_n^{2-n}]}{[M^{+2}][Cl^-]^n}, \quad n = 1,2,3,4\ ; \qquad (2)$$

(b) amine salt formation at the interface

$$R_3N + H^+ + Cl^- \rightleftharpoons R_3NHCl \qquad (3)$$

with equilibrium constant

$$K_s = \frac{[R_3NHCl]}{[R_3N][H^+][Cl^-]}\ ; \qquad (4)$$

(c) metal complex extraction at the interface

$$2R_3NHCl + MCl_4^{=} \rightleftharpoons (R_3NHCl)_2MCl_2 + 2Cl^- \qquad (5)$$

with equilibrium constant

$$K_M = \frac{[(R_3NHCl)_2MCl_2][Cl^-]^2}{[MCl_4^{=}][R_3NHCl]^2}\ , \qquad (6)$$

or

$$2R_3NHCl + MCl_2 \rightleftharpoons (R_3NHCl)_2MCl_2 \qquad (7)$$

with

$$K_M' = \frac{[(R_3NHCl)_2MCl_2]}{[MCl_2][R_3NHCl]^2}\ ; \qquad (8)$$

and

(d) amine salt dimer formation in the organic phase

$$2R_3NHCl \rightleftharpoons (R_3NHCl)_2 \qquad (9)$$

with equilibrium constant

$$K_D = \frac{[(R_3NHCl)_2]}{[R_3NHCl]^2}\ . \qquad (10)$$

For copper and zinc chlorides in acidic 4M NaCl we have fit metal distribution equilibrium data by the parameter values shown in Table 1 (2,3). This model is fairly accurate in predicting the individual extraction equilibria but only moderately successful in predicting the simultaneous extraction equilibria. Because no better model is yet available, we have adopted this equilibrium model as the description of the chemistry needed to derive expressions for mass-transfer controlled extraction rates. This or some other, more accurate equilibrium correlation must be used in order for the mass transfer model to be consistent with the extraction chemistry.

TABLE 1

Parameters for the Zinc and Copper Chloride-TIOA Equilibrium Models

	Copper	Zinc	TIOA-HCl
β_1 (M^{-1})	1.3	5.3	
β_2 (M^{-2})	0.3	3.1	
β_3 (M^{-3})	-	0.64	
β_4 (M^{-4})	-	1.5	
K_s (M^{-2})			1.51×10^4
K_M	-	1.56×10^5	
K_M' (M^{-2})	230	-	
K_D (M^{-1})			8

One simplification that is made here is to assume that the amine salt formation is complete under extraction conditions so that interfacial transfer of HCl need not be considered.

Transport Model for Single Metal Extraction

For single metal extraction following Eq. 5, such as extraction of zinc, the mass transfer situation is shown schematically in Figure 1. Extraction occurs as the metal is transported from the bulk aqueous solution through the diffusion layer to the interface as M^{2+}, MCl^+, MCl_2, MCl_3^- and $MCl_4^=$. The metal is extracted into the immiscible organic phase according to Eq. 5. The metal-amine complex is transported through the organic diffusion layer to the bulk organic solution. TIOA is transported to the interface in the form of salt and dimer. The chloride ions generated by the exchange reaction diffuse to the bulk aqueous phase. By stoichiometry, the total flux of chloride is equal to twice the total flux of metal.

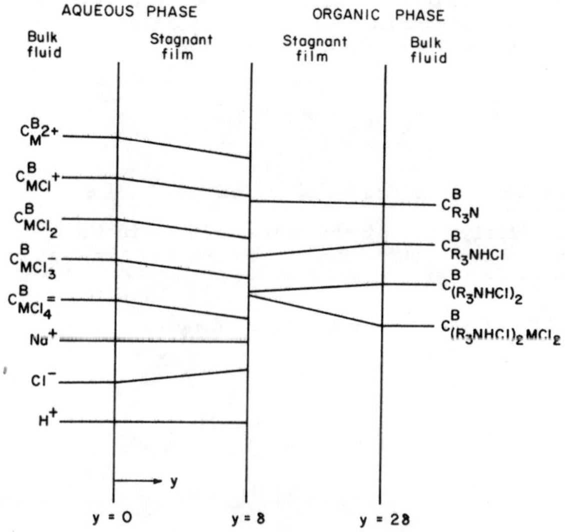

FIGURE 1. The model system for single metal extraction by TIOA

Model assumptions

The following simplifying assumptions are employed in the mass transfer model:

(1) The aqueous and organic species are completely immiscible.

(2) The exchange reaction is equilibrated at the interface; that is, the exchange reaction is fast.

(3) All the homogeneous reactions are fast and equilibrated locally.

(4) All the reaction equilibria can be described by the law of mass action based on concentrations rather than activities.

(5) The film thicknesses for the diffusion layers in the two phases are equal and designated as δ.

(6) All the solute species have the same diffusivity designated as D.

Intraphase transport equations

Each chemical species in the model is identified by a subscript index:

$1 = M^{2+}$	$6 = Cl^-$
$2 = MCl^+$	$7 = R_3NHCl$
$3 = MCl_2$	$8 = (R_3NHCl)_2$
$4 = MCl_3^-$	$9 = (R_3NHCl)_2MCl_2$
$5 = MCl_4^=$	

The governing equations for the steady-state film model are:

(1) Continuity of metal

$$\frac{d}{dy}(N_1 + N_2 + N_3 + N_4 + N_5) = 0 \tag{11}$$

(2) Continuity of chloride

$$\frac{d}{dy}(N_2 + 2N_3 + 3N_4 + 4N_5 + N_6) = 0 \tag{12}$$

(3) Continuity of amine-metal complex

$$\frac{d}{dy} N_9 = 0 \tag{13}$$

(4) Continuity of total amine

$$\frac{d}{dy}(N_7 + 2N_8 + 2N_9) = 0 \tag{14}$$

By use of the reaction stoichiometry, the integrated form of the continuity equations may be written in terms of the interfacial flux:

(1) Total flux of metal

$$N_1 + N_2 + N_3 + N_4 + N_5 = N_M \tag{15}$$

(2) Total flux of chloride

$$N_2 + 2N_3 + 3N_4 + 4N_5 + N_6 = 2N_M \tag{16}$$

(3) Flux of amine-metal complex

$$N_9 = N_M \tag{17}$$

(4) Total flux of amine

$$N_7 + 2N_8 + 2N_9 = 0 \tag{18}$$

In these equations N_M is flux of metal across the interface.

Nondimensionalization

The problem can be made dimensionless by defining the following dimensionless quantities:

flux $$\bar{N}_i = \frac{N_i \delta}{D c_M} \quad (19)$$

where c_M = total metal concentration in the bulk aqueous phase.

concentration $$\bar{c}_i = c_i / c_M \quad (20)$$

distance $$\eta = y/\delta \quad (21)$$

The stability constants and equilibrium constants are normalized using the total bulk metal concentration, and they are identified as β_1^*, β_2^*, β_3^*, β_4^*, K_M^*, K_D^*, and K_s^*. The definitions of these dimensionless parameters are given as

$$\beta_n^* = \beta_n c_M^n \; ; \; n = 1,2,3,4 \, , \quad (22)$$

$$K_M^* = K_M \, , \quad (23)$$

$$K_D^* = K_D c_M \quad (24)$$

and $$K_s^* = K_s c_M^2 \quad (25)$$

The total flux of metal is normalized in the same way as that for each individual species, namely

$$\bar{N}_M = \frac{N_M \delta}{D c_M} = \frac{N_M}{k c_M} \quad (26)$$

where k is the aqueous phase film mass transfer coefficient. Since the flux of metal may be expressed as the product of the film mass transfer coefficient and the concentration driving force, the dimensionless flux $\bar{N}_M$ may also be expressed as

$$\bar{N}_M = \Delta c_M / c_M \quad (27)$$

where Δc_M is the effective concentration driving force in the system.

It should be noted that the dimensionless flux may be viewed as a correction factor to the mass-transfer-limited flux of aqueous metal (kc_M). Alternatively, it may be viewed as a correction factor to the maximum driving force. The effective driving force is proportional to the correction factor according to Eq. 27. The main purpose of the two-film mass transfer model is to estimate this correction factor at given concentrations in the bulk aqueous and organic phases so that the driving force can be formulated properly. For extraction conditions, the correction factor ranges between 0 and 1; for stripping conditions, it takes on negative values. $\bar{N}_M$ equals +1 for maximum metal flux from aqueous to organic phase and equals 0 when the system is at equilibrium.

With the dimensionless quantities defined in this section, the flux relations can be written in the dimensionless form as follows:

(1) Dimensionless total flux of metal

$$\bar{N}_1 + \bar{N}_2 + \bar{N}_3 + \bar{N}_4 + \bar{N}_5 = \bar{N}_M \quad (28)$$

(2) Dimensionless total flux of chloride

$$\bar{N}_2 + 2\bar{N}_3 + 3\bar{N}_4 + 4\bar{N}_5 + \bar{N}_6 = 2\bar{N}_M \quad (29)$$

(3) Dimensionless flux of amine-metal complex

$$\bar{N}_9 = \bar{N}_M \quad (30)$$

(4) Dimensionless total flux of amine

$$\bar{N}_7 + 2\bar{N}_8 + 2\bar{N}_9 = 0 \quad (31)$$

Equations describing aqueous phase

The dimensionless flux equations (Eqs. 28 and 29) may be rewritten by the use of Fick's Law,

(1) Transport of metal

$$-\frac{d}{d\eta}(\bar{c}_1 + \bar{c}_2 + \bar{c}_3 + \bar{c}_4 + \bar{c}_5) = \bar{N}_M \quad (32)$$

(2) Transport of chloride

$$-\frac{d}{d\eta}(\bar{c}_2 + 2\bar{c}_3 + 3\bar{c}_4 + 4\bar{c}_5 + \bar{c}_6) = 2\bar{N}_M \quad (33)$$

In addition to the transport equations, there are homogeneous equilibria that have to be considered. In dimensionless form, the stability relations for the chlorocomplexes are written as:

$$\beta_n^* = \frac{\bar{c}_{n+1}}{\bar{c}_1 \bar{c}_6^{-n}} \; ; \; n = 1,2,3,4 \quad (34)$$

Equations describing organic phase

Similarly, the dimensionless fluxes in the organic phase (Eqs. 30 and 31) may be rewritten in the form of Fick's Law,

(1) Transport of amine-metal complex

$$-\frac{d}{d\eta}\bar{c}_9 = \bar{N}_M \quad (35)$$

(2) Transport of total amine

$$\frac{d}{d\eta}(\bar{c}_7 + 2\bar{c}_8) = 2\bar{N}_M \quad (36)$$

In dimensionless form, the dimerization equilibrium of the amine salt is given by

$$K_D^* = \bar{c}_8 / (\bar{c}_7)^2 \quad (37)$$

Boundary conditions at the interface

At the interface, the equilibrium of the metal exchange reaction is represented in dimensionless form by

$$K_M^* = \frac{\bar{c}_9^{\circ} \, \bar{c}_6^{\circ 2}}{\bar{c}_7^{\circ 2} \, \bar{c}_5^{\circ}} \quad (38)$$

where the superscript '°' indicates the interface. At the interface, there is also continuity such that the flux of metal in the aqueous side is equal to that in the organic phase. This relationship has already been used in obtaining Eq. 17.

Boundary conditions outside the diffusion layer

At the outer edge of the diffusion layer, the concentration of each species is equal to that in the bulk. These bulk species concentrations can easily be computed using total analytical concentrations and the homogeneous equilibrium constants at given metal chloride, sodium chloride, hydrogen chloride, total amine and metal-amine concentrations.

The two-film mass transport model thus consists of four first order ordinary differential equations (Eqs. 32, 33, 35, 36) and six non-linear algebraic equations (Eqs. 34, 37, 38). The unknowns are the concentration profiles of species 1 through 9, and the flux of metal. The model can be solved using the bulk concentrations and the film thickness as prescribed conditions.

Due to the coupling between the flux and the profiles, and because of the non-linear nature of the algebraic equations, a special mathematical technique is required. A mixed collocation technique has been used to solve the transport equations. An introduction to the application of this method has been presented by Stewart (4).

First, the transport equations describing the organic phase (Eqs. 35 and 36) are solved analytically to express the concentration profiles of the organic species in terms of the metal flux. Hence, Eq. 35 becomes

$$\bar{c}_9 = \bar{c}_9^B + (2 - \eta)\, \bar{N}_M \tag{39}$$

and Eq. 36 becomes

$$\bar{c}_7 + 2\bar{c}_8 = (\bar{c}_7^B + 2\bar{c}_8^B) - 2(2 - \eta)\, \bar{N}_M \tag{40}$$

The superscript B denotes a bulk condition. With the dimer formation equilibrium, Eq. 40 can be written as

$$2K_D^* \,\bar{c}_7^2 + \bar{c}_7 = (\bar{c}_7^B + 2\bar{c}_8^B) - 2(2 - \eta)\, \bar{N}_M \tag{41}$$

Equations 39 and 41 can be written at the interface, where η equals one, to provide the following two relations for the interfacial organic concentrations.

$$\bar{c}_9^{\circ} = \bar{c}_9^B + \bar{N}_M \tag{42}$$

$$2K_D^* \,\bar{c}_7^{\circ 2} + \bar{c}_7^{\circ} = (\bar{c}_7^B + 2\bar{c}_8^B) - 2\bar{N}_M \tag{43}$$

The profiles of the aqueous species and the metal flux are estimated by using collocation at two interior points by the method of coefficients. The profiles of the metal species being extracted (species 5) and the free chloride (species 6) are approximated by a first order shifted Legendre polynomial as shown:

$$(\bar{c}_i - \bar{c}_i^B) = \eta \sum_{j=0}^{1} c_{ij} P_j(\eta) \tag{44}$$

The collocation points are the roots of the second order term in the shifted Legendre polynomial ($P_2(\eta)$), and the values are 0.211325 and 0.788675. Equation 44 can be expressed in a simpler quadratic form by defining new unknowns a_i and b_i instead of c_{i1} and c_{i2}, and Eq. 44 is rewritten as

$$\bar{c}_i = \bar{c}_i^B + a_i\eta + b_i\eta^2 \tag{45}$$

Hence, the profiles for species 5 and 6 are approximated by the following equations:

$$\bar{c}_5 = \bar{c}_5^B + a_5\eta + b_5\eta^2 \tag{46}$$

$$\bar{c}_6 = \bar{c}_6^B + a_6\eta + b_6\eta^2 \tag{47}$$

The profiles of other aqueous species can be expressed in terms of species 5 and 6 and are written as

$$\bar{c}_1 = \frac{\bar{c}_5}{\beta_4^* \,\bar{c}_6^4} \tag{48}$$

$$\bar{c}_2 = \frac{\beta_1^* \,\bar{c}_5}{\beta_4^* \,\bar{c}_6^3} \tag{49}$$

$$\bar{c}_3 = \frac{\beta_2^* \,\bar{c}_5}{\beta_4^* \,\bar{c}_6^2} \tag{50}$$

$$\bar{c}_4 = \frac{\beta_3^* \,\bar{c}_5}{\beta_4^* \,\bar{c}_6} \tag{51}$$

At this point, the two transport equations (Eqs. 32 and 33) are written at the two interior collocation points to generate four algebraic equations. The interfacial equilibrium equation may be used by setting η equal to one in Eqs. 46 and 47 for $\bar{c}_5$ and $\bar{c}_6$. Equation 42 may be substituted for $\bar{c}_9^{\circ}$. These five equations along with Eq. 43 constitute a set of non-linear algebraic equations in the unknowns a_5, a_6, b_5, b_6, $\bar{N}_M$, and $\bar{c}_7^{\circ}$. These unknowns can be computed using a standard algorithm such as Newton's method.

For metals such as copper and cobalt where the extraction is better modelled by the exchange of the MCl_2 species, the development of the corresponding mass transport model is completely analogous to the previous case except that a different exchange mechanism is used and the presence of MCl_3^- and $MCl_4^=$ species is neglected in the diffusion layer.

<u>Transport model for simultaneous extraction of two metals</u>

The development of the two film mass transport model for the simultaneous extraction of two metals by TIOA follows exactly the same pattern as the single metal case. The governing equations to be considered are:

(1) Continuity of the first metal
(2) Continuity of the second metal
(3) Continuity of chloride
(4) Continuity of the first metal-amine complex
(5) Continuity of the second metal-amine complex
(6) Continuity of total amine

As with one metal, these differential equations, the homogeneous equilibria, and the two interfacial exchange reaction equilibria are solved together to obtain the concentration profiles and the fluxes for both metals. The solution method is again a two-point collocation in the aqueous diffusion layer.

<u>Typical flux results</u>

The two-film model for TIOA extraction is capable of predicting the dimensionless fluxes for given conditions in the bulk. In mathematical form, the flux for single metal A extraction is expressed as

$$N_A = N_A(c_M, Cl, c_{org}, A_{org}; K_A, K_D, \underset{\sim}{\beta}_A, D, \delta) \quad (52)$$

where

A_{org} = concentration of metal A in the bulk organic phase

Cl = total chloride concentration in the bulk aqueous phase

c_{org} = total amine concentration

$\underset{\sim}{\beta}_A$ = vector of metal A chlorocomplex stability constants

By use of the following dimensionless parameters,

$$X_1 = Cl/c_M \quad (53)$$

$$X_2 = c_{org}/c_M \quad (54)$$

and $$X_3 = A_{org}/c_{org} \quad (55)$$

Equation 52 may be rewritten as

$$\bar{N}_A = \bar{N}_A(X_1, X_2, X_3; K_A^*, K_D^*, \underset{\sim}{\beta}_A^*) \quad (56)$$

To illustrate the parametric effects, we can investigate the effects of these concentration parameters upon the flux for the copper-TIOA system. Curve 1 in Figure 2 shows the behavior of Eq. 56 with respect to X_3 for a specific combination of c_{Cu}, X_1, and X_2. Note that a value of X_3 equal to 0.5 corresponds to complete amine loading and a negative flux indicates stripping.

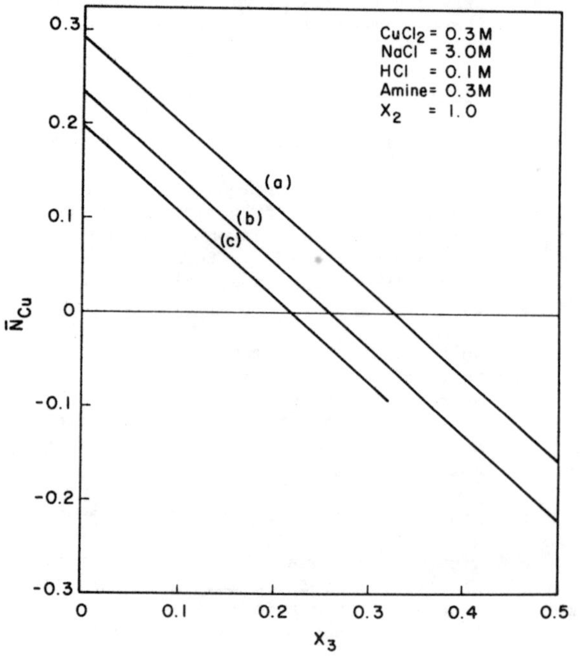

FIGURE 2. Effect of the presence of zinc upon the flux of copper. Curve a: no zinc. Curve b: 0.05 M Zn-TIOA in the organic phase. Curve c: 0.03 M $ZnCl_2$ in the aqueous phase.

Figure 2 also shows the model's prediction of the effect of the presence of zinc upon the flux of copper. Curve b corresponds to the conditions in curve a but with the additional presence of 0.05 molar zinc in the organic phase. The flux of copper into the organic phase is suppressed due to the decrease in free amine salt. Curve c corresponds to the effects of 0.03 molar zinc chloride in the aqueous phase in addition to the conditions in curve a. Again, the flux of copper into the organic phase is found to be suppressed due to the competition for amine salt by zinc extraction.

It should be cautioned that although these curves appear to be linear in copper loading, the application of a linear rate expression to predict the mass transfer rate in a real process is limited. This is because all the bulk concentrations change with conversion within a contactor, and the flux depends on these concentrations in a non-linear manner. Consequently, it is necessary to use the complete two-film mass transfer model to estimate the flux at any given conversion inside the contactor.

In order to design contacting equipment for solvent extraction processes, it is generally necessary to integrate the local flux over reactor volume or residence time. Such integration requires knowledge of both the interfacial transfer rate and the flow conditions inside the contactor. The model that we have developed here to estimate the flux of metal should be useful for equipment design calculation, especially because they can be done so efficiently. In the following sections, the validity of the two-film model will be verified by comparing the model's prediction with experimental data obtained from a growing drop mass-transfer cell.

Growing drop mass-transfer cell

Bauer (5) developed a growing drop cell to study the transfer rate of the copper-Kelex 100 extraction system. He has discussed the construction, operation and mathematical analysis of this equipment in detail. A diagram of the growing drop cell is shown in Figure 3. Basically, the geometry is characterized by the continuous formation of fresh interfacial area and penetration-type mass transfer. An aqueous drop grows uniformly at the tip of a capillary into the continuous organic phase and subsequently detaches and coalesces with the aqueous solution at the receiver neck. One advantage of the growing drop cell is that it continuously generates clean interface. This is vital for interfacial rate studies as the build up of surface active impurities at the interface can directly affect the rate of the transport process. Also, the growing drop cell provides a known interfacial area across which mass transfer takes place, and the convective mass transfer can be analyzed. The principle is essentially the same as that of the dropping mercury electrode used in polarography. This equipment is particularly useful to confirm whether the extraction rate in TIOA system is limited by mass transfer or reaction kinetics and whether our model is adequate. Tunison has discussed the use of growing drop rate data for reactor analysis (6).

The mathematical models to describe mass transfer at the growing drop interface and at the neck are reviewed in the following section.

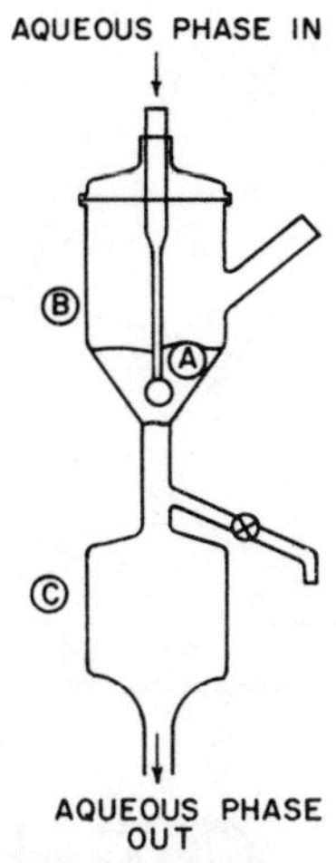

A. DROP FORMATION CAPILLARY
B. ORGANIC PHASE CHAMBER
C. AQUEOUS PHASE CHAMBER

FIGURE 3. Diagram of Growing Drop Mass Transfer Cell

Growing drop mass transfer

Ilkovic (11) was the first to develop a penetration model for drop-formation transfer based on uniform stretching of the surface of a spherical drop. The result of his analysis is

$$M_\tau = 1.31\,(\Delta c)\,4\pi\left(\frac{3q\tau}{4\pi}\right)^{2/3}\left(\frac{D\tau}{\pi}\right)^{1/2} \tag{57}$$

where

M_τ = total mass transferred in time τ

Δc = concentration driving force between the bulk and the interface

q = volumetric flow rate

τ = drop life

The average drop area is given by

$$\bar{A} = \frac{12}{5}\,\pi\left(\frac{3q\tau}{4\pi}\right)^{2/3} \tag{58}$$

Putting Eq. 58 into Eq. 57, we get

$$M_\tau = \left(\frac{5}{3}\right)(1.31)(\Delta c)\bar{A}\left(\frac{D\tau}{\pi}\right)^{0.5} \tag{59}$$

The mass transfer rate may also be expressed in the form of the film theory as

$$M_\tau = (\Delta c)\,\bar{A}\,\frac{D}{\delta_d}\,\tau \tag{60}$$

Consequently, an effective film thickness in the growing drop may be identified as

$$\delta_d = \left(\frac{1}{1.31}\right)\left(\frac{3}{5}\right)(D\pi\tau)^{0.5} \tag{61}$$

Flat surface mass transfer

For unsteady diffusion to a plane interface in a stagnant medium the penetration model is

$$M_\tau = (\Delta c)\,A_n(4D\tau/\pi)^{0.5} \tag{62}$$

where A_n is the interfacial area. The corresponding finite-stagnant-film relation is

$$M_\tau = (\Delta c)\,A_n(D/\delta_n)\tau \tag{63}$$

so that comparison of these two equations yields the effective film thickness at the neck δ_n to be

$$\delta_n = 1/2\,(D\pi\tau)^{0.5} \tag{64}$$

The complete mass transfer picture in the growing drop cell is a combination of the above two transport processes. The observable rate of extraction of metal A is given by an unsteady mass balance on the organic phase in the cell:

$$V\frac{dc_A}{dt} = N_{A,d}\,\bar{A}_d + N_{A,n}\,A_n \tag{65}$$

$$= \bar{N}_A c_M D(\bar{A}_d/\delta_d + A_n/\delta_n) \tag{66}$$

where V = organic phase volume

t = time

c_A = organic metal A concentration

Let us define $\bar{c}_A = c_A/c_M$

and
$$\theta = \frac{tD}{V}\left(\frac{\bar{A}_d}{\delta_d} + \frac{A_n}{\delta_n}\right) \tag{67}$$

Then in dimensionless form, Eq. 66 can be written as

$$\frac{d\bar{c}_A}{d\theta} = \bar{N}_A \tag{68}$$

which relates the dimensionless flux to the rate of change in organic metal concentration. Now we can use the mass transfer model, which provides the dimensionless flux as a function of bulk concentrations and the equilibrium constants, to predict the extraction rate in the growing drop cell. Conversely, we can use the growing drop experimental data to confirm our mass-transport model for TIOA extraction.

Growing drop experiments

Growing-drop TIOA-extraction measurements were performed for copper chloride, for zinc chloride, and simultaneously for these two metals. The experimental procedure is briefly outlined below:

(1) Five milliliters of TIOA salt solution is charged into the organic chamber.
(2) The aqueous flow rate is set at the desired flow rate.
(3) Mechanical mixing of the organic reservoir is induced periodically with a medicine dropper.
(4) Drop life is estimated periodically by counting the number of drops formed in a certain time interval.
(5) All of the organic phase is removed at the

end of each run and analyzed for metal concentration.

In order to reduce statistical interdependence of the results within each set of runs, fresh organic solution was used for the run at each time duration.

Two problems were encountered in the experiments. First, the heavier metal-amine complex accumulated at the bottom of the organic chamber in the vicinity of the drop and neck. We solved the mixing problem by reducing the organic volume to 5 ml and induced mechanical mixing. Secondly, we had a coalescence problem during the zinc extraction and the simultaneous two metal extraction. Drop coalescence was not fast, and there were persistently three to four drops held up at the bottom of the organic chamber. In these cases the receiver area available for mass transfer was corrected by taking the area to be four times the actual area of cross section in the neck. With this correction the mass transferred at the bottom of the cell amounted to 67 percent of the total metal exchanged at most.

The total metal content of the organic sample was determined by stripping and EDTA titration with PAN indicator. The copper content in the organic copper-zinc mixture was determined with a Perkin-Elmer Model 402 Ultraviolet Visible Spectrophotometer. We determined the copper concentration by comparing the sample absorbance to the absorbances of standard TIOA-Cu-xylene solutions at this particular wavelength. The zinc content in the organic phase was calculated by difference from the total metal titration. The precision of the spectrophotometric technique based on repeated analyses of the same sample was better than 4%. Since the absolute value of copper concentration was generally an order of magnitude smaller than that of zinc, the error in the determination of zinc from the two metal mixture should therefore be the error in titration, approximately 1%.

The results for the growing drop experiments and the corresponding theoretical predictions are presented in Figures 4, 5 and 6 which present the fractional loading of metal versus dimensionless time θ. Each point is a single experimental run. Figure 4 shows the copper extraction in the growing drop cell for 0.1 M aqueous copper, 2 M NaCl, 0.1 M HCl and 0.2 M TIOA. The aqueous flow rate was 2.357 ml/min, and the average drop life was 0.814 second. Figure 5 shows the zinc extraction data. The aqueous zinc concentration was 0.0513 M; NaCl, HCl, and amine were 2 M, 0.04 M and 0.1 M, respectively. The flow rate was 2.06 ml/min, and the average drop life was 1.253 second. For single metal extraction, the model predicts satisfactorily the rate of the extraction process.

Figure 6 presents the simultaneous extraction of copper and zinc. The concentrations of $CuCl_2$, $ZnCl_2$, NaCl, HCl and amine were 0.09862 M, 0.02478 M, 2 M, 0.05 M and 0.1 M, respectively. The aqueous flow rate was 2.35 ml/min, and the average drop life was 1 second. The zinc data are in reasonable agreement with the model's prediction, but copper extraction efficiencies seem to be smaller. However, the experimental copper extraction does show a maximum in Figure 6 as predicted by the model. Note that the copper loading at finite contact time exceeds its apparent equilibrium level by a factor of two.

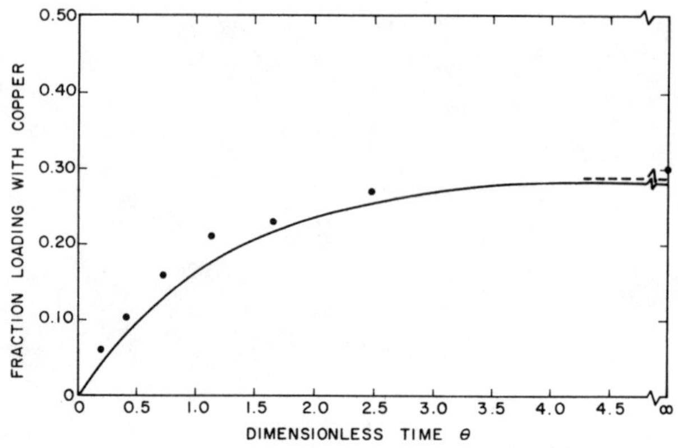

FIGURE 4. Comparison between experimental and theoretical extraction of copper chloride by TIOA in a growing drop cell. (Aqueous phase: 0.1 M $CuCl_2$, 2 M NaCl, 0.1 M HCl; organic phase: 0.2 M TIOA; average flow rate = 2.357 ml/min, average drop life = 0.814 sec.)

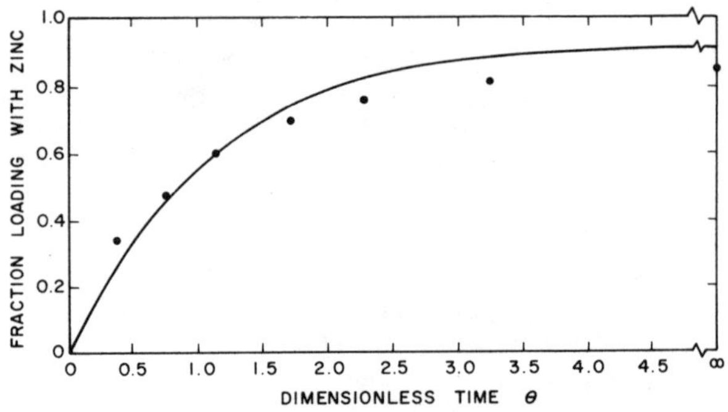

FIGURE 5. Comparison between experimental and theoretical extraction of zinc chloride by TIOA in a growing drop cell. (Aqueous phase: 0.0513 M $ZnCl_2$, 2 M NaCl, 0.04 M HCl; organic phase: 0.1 M TIOA; average flow rate = 2.06 ml/min, average drop life = 1.25 sec.)

The Copper-LIX 64N System

Measurements of the rate of copper extraction by LIX 64N we also made, with and without the addition of ammonia to copper sulfate solutions. Conditions for several runs in the growing drop cell are shown in Table 2. The question at hand is whether or when this system is controlled by reaction kinetics or mass transfer.

Earlier workers (7,8,9) have shown that copper extraction by LIX from acidic solution is slow and limited by the reaction kinetics. From their work a composite rate expression for the extraction reaction can be constructed,

TABLE 2

Summary of the Concentration Levels for Copper Extraction in the Growing-Drop Cell
(Initial LIX 64N Concentration = 0.404 M)

	Aqueous Phase				Organic Phase	
Run	Total Ammonia (M)	Total Sulfate (M)	Total Copper (M)	pH	Copper (M) (at equilibrium)	Initial Ratio (aqueous copper / free extractant)
R-1	0.0	0.081	0.081	4.26	--	0.2
R-2	0.0	0.236	0.236	3.82	0.188	0.584
R-3	0.0	0.364	0.236	0.86	0.148	0.584
R-4	5.9	0.369	0.241	11.09	0.201	0.596
R-5	3.52	0.366	0.236	10.69	0.199	0.584
R-6	3.46	0.364	0.078	10.41	0.198	0.193

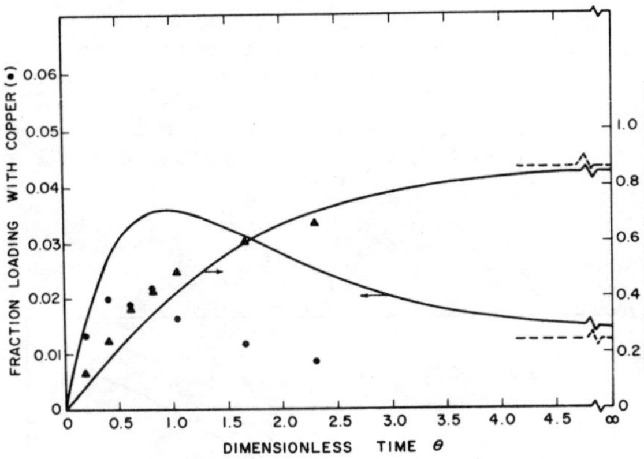

FIGURE 6. Comparison between experimental and theoretical simultaneous extraction of copper and zinc chlorides by TIOA in a growing drop cell. (Aqueous phase: 0.09862 M $CuCl_2$, 0.02478 M $ZnCl_2$, 2 M NaCl, 0.05 M HCl; organic phase: 0.1 M TIOA; average flow rate = 2.35 ml/min, average drop life = 1 sec.)

$$N_{\mathcal{M}} = k_f (Cu^{+2})(R^{63}H)^{1/2}(R^{65N}H)^{3/2} \quad (69)$$

where $k_f = 3.88 \times 10^{-4}$ (cm/sec)/(mol/cm^3)^2 and $R^{63}H$ and $(R^{65N}H)$ are the concentrations of LIX 63 and LIX 65N, the active components in LIX 64N. As a comparison the two-film mass transfer model presented earlier for metal-acid liquid ion exchange should apply to the copper-LIX system with the equilibrium constant K = 0.05 (10).

We have attempted to develop a mass transfer model for copper extraction from ammonia solutions that is analogous to that given for metal chlorides above. Unfortunately, numerical convergence difficulties have prevented computation of species profiles and fluxes in the case where ammonia complexing is extensive. On the other hand, the cause of the numerical difficulties, the near irreversibility of the extraction reaction at high pH, allows a considerable simplification of the mass transfer model. That is, under mass-transfer control the interfacial concentration of the limiting reactant, copper or LIX, should be close to zero, at least during the initial stages of the extraction process.

In Table 2, the ratio of bulk copper concentration to initial bulk LIX concentration is given. If this ratio is less than 0.5, copper should be the limiting reactant; otherwise, LIX should be limiting if the film mass transfer coefficients in the two phases are comparable. Thus, the expected initial extraction rates for fast reaction kinetics may be estimated by

$$N_{\mathcal{M}} = k\, c_{Cu^{+2},\infty} \quad (70)$$

or

$$N_{\mathcal{M}} = -k\, c_{HR,\infty} \quad (71)$$

Experimental Results

Figure 7 shows the experimental results from Runs R-2 (acidic solution) and R-5 (ammonia solution)

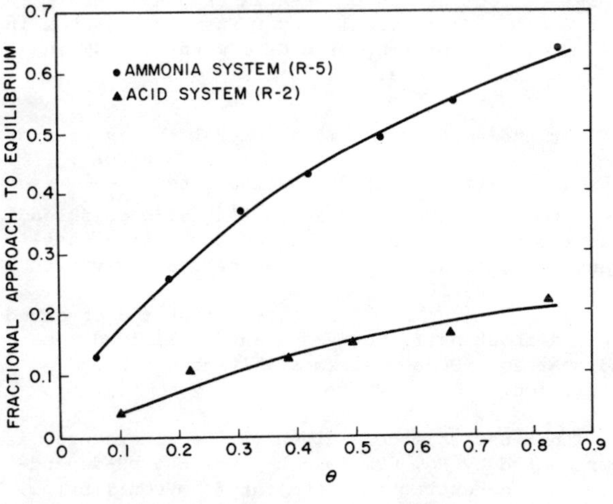

FIGURE 7. Comparison Between Copper Extraction From Acid and Ammonia Solutions by LIX 64N in a Growing-Drop Cell

as the fractional approach of the organic copper to its ultimate equilibrium value as a function of the dimensionless contact time θ. It is seen that the rate for the ammonia system is comparable to that observed in the amine systems, but the acid solution displays a considerably slower extraction rate.

From the initial rate observed in Run R-5 and our theoretical estimate of the mass transfer coefficients in the cell we compute an interfacial aqueous copper concentration of zero, which is consistent with Eq. 70 and indicates a fast, virtually irreversible interfacial reaction.

Figure 8 presents the rate data for Runs R-4 and R-5 in terms of the remaining free reagent plotted as a logarithmic function of θ. Also shown is a theoretical curve based on Eq. 71, reagent-limited mass transfer. The data indicate an extraction rate even greater than this estimated mass-transfer limit.

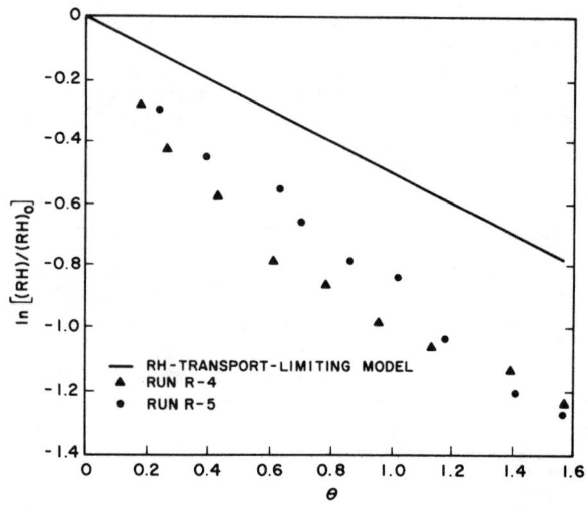

FIGURE 8. Comparison Between Experimental and Theoretical Copper Extraction Rates in a Growing-Drop Cell with LIX 64N and Ammonia.

For the acidic system in Run R-1 the extraction data were differentiated, and the rate $\frac{d(CuR_2)}{dt}$ is presented in Figure 9 as a function of copper loading. Also shown are calculated curves based on a mass-transfer model and on the kinetics model given in Eq. 69.

It is seen that the observed rate is about half the mass-transfer-limited rate and that Eq. 69 does a remarkably good job of prediction.

The kinetics models given in the literature, as represented by Eq. 69, do not predict a pH-dependence for the extraction rate far from equilibrium. Figure 10 compares the results of Runs R-2 and R-3. It is seen that the extraction at the lower pH is considerably slower.

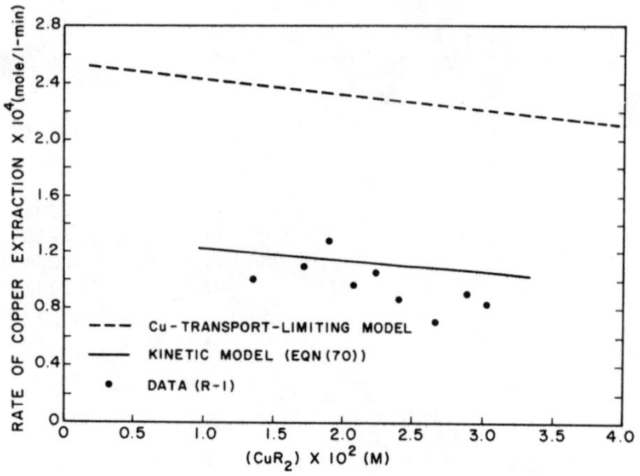

FIGURE 9. Comparison Between Experimental and Theoretical Initial Rates of Copper Extraction from an Acid System by LIX 64N in a Growing-Drop Cell (Run R-1)

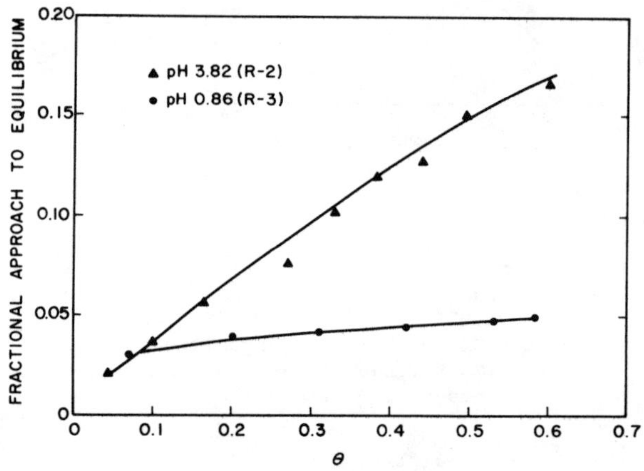

FIGURE 10. Effect of pH on Rate of Copper Extraction from an Acid Solution by LIX 64N in a Growing-Drop Cell

Discussion

A two-film mass transfer model for the extraction of metals by TIOA has been developed, and an efficient solution scheme demonstrated. This model accounts in a simple but essential manner for the multicomponent nature of this solvent extraction system arising from the multiple chemical reactions. Its validity has been tested by running the extraction in a growing drop cell. Agreement between the data and predictions demonstrates that the TIOA extraction is limited by mass transfer and the model is essentially correct.

For the case of simultaneous two-metal extraction, the zinc extraction data are in reasonable agreement with the prediction. The experimental degree of copper extraction was observed to be much smaller than the theoretical prediction. However, the data do show a predicted peak. The deviation of the copper data from the model may possibly be explained by the following factors:

(1) Because of the uncertainty in the value of the copper extraction equilibrium constant, the proper value might be considerably smaller than 230 $(mole/l)^{-2}$, which has been used in the present calculations. Also, incorrect stability constant values or dimerization equilibrium constant might have been chosen.

(2) The solutions are generally non-ideal, perhaps moreso in the copper and zinc mixtures. The activities of the identified species may be quite different from their concentrations. In particular, some important association reactions may have been overlooked. For example, there might be formation of a Cu-Zn complex in the aqueous or organic phase or higher order amine salts might have been formed.

(3) The samples might have changed due to a photochemical redox reaction, $Cu^{+2} \rightarrow Cu^{+1}$ in the organic phase. Since the analytical method determined only the cupric-TIOA complex concentration, the actual copper concentration in the organic phase may be larger than the measured value.

In any case it is clear that the equilibrium model is not adequate in the two-metal case. It is seen that the copper loading at large time is quite a bit lower than the value predicted for system equilibrium. Thiel (2) also experienced difficulty in predicting the simultaneous equilibria so that we must recognize that our thermodynamic model is not yet satisfactory.

In summary, the two-film mass transport model is capable of predicting the extraction rates in mass-transfer-controlled TIOA solvent extraction systems. However a better thermodynamic model is required to improve the model's predictions of the fluxes and loading of metals in the multiple metal extraction.

In the case of copper extraction from ammonia solution by LIX, it has been found that the process is mass-transfer limited in contrast to the situation in acidic solutions. The simplified irreversible reaction model provides a fair estimate of the rate far from equilibrium. Problems remain, however, in predicting the approach to equilibrium adequately and in characterizing the rates during simultaneous extraction or stripping with a second metal such as nickel. Work on these problems is continuing.

References

1. Chapman, T. W., Caban, R., and Tunison, M. E., 1975, "Rates of Liquid-Liquid Ion Exchange in Metal Extraction Processes", AIChE Symp. Ser., 71, No. 152, p. 128.

2. Thiel, M., August 1974, "Separation of Copper and Zinc by Liquid-Liquid Extraction with Tri-isooctyl Amine", M.S. Thesis, University of Wisconsin, Madison.

3. Tse, S. W.-S., November 1978, "Mass Transfer Rates in the Solvent Extraction of Metal Chlorides by Tri-isooctyl Amine", M.S. Thesis, University of Wisconsin, Madison.

4. Stewart, W. E., "Solution of Transport Problems by Collocation Methods", in Lectures in Transport Phenomena, AIChE Continuing Education.

5. Bauer, G. L., 1974, "Solvent Extraction of Copper: Kinetics and Equilibrium Studies", Ph.D. Thesis, University of Wisconsin, Madison.

6. Tunison, M. E. and Chapman, T. W., 1978, "Characterization of the Kenics Mixer as a Liquid-Liquid Extractive Reactor", AIChE Symp. Ser., 74, No. 173, 112.

7. Miller, J. D. and Atwood, R. L., 1975, J. Inorg. Nucl. Chem., 37, 2539.

8. Atwood, R. L., Thatcher, D. N., and Miller, J. D., 1975, Met. Trans., 6B, 465.

9. Flett, D., Okuhara, D. N., and Spink, D. R., 1973, J. Inorg. Nucl. Chem., 35, 1967.

10. Bautista, R. G. and Hoh, Y.-C., 1978, Met. Trans., 9B, 69.

Chapter 20

EQUILIBRIUM AND KINETICS OF COPPER EXTRACTION FROM AMMONIACAL SOLUTIONS BY HYDROXOXIMES WITH PARTICULAR EMPHASIS ON TRANSPORT PHENOMENA

J.D. Miller, H.H. Haung and E.F. Pereira

Department of Metallurgy and Metallurgical Engineering
University of Utah
Salt Lake City, Utah 84112

ABSTRACT

The chemistry of copper extraction from ammoniacal solutions by hydroxyoxime extractants was studied. Equilibrium measurements were made by shakeout experiments with subsequent analyasis of the aqueous and/or organic phases. The kinetic experiments were carried out in a single drop reaction cell in order to establish the details of the intrinsic reaction kinetics for this system.

Equilibrium measurements indicated that the extraction involves complete displacement of ammonia ligands from the coordination sphere of the cupric ion by the extractant molecules. The extraction reaction has the following stoichiometry:

$$Cu(NH_3)^{2+}_{i(aq)} + 2HR_{(o)} \rightleftarrows CuR_{2(o)} + iNH_{3(aq)} + 2H^+_{(aq)}$$

and as might be expected the extraction coefficient was found to be sensitive to the ammonia concentration and the pH of the system. These results were interpreted in terms of the mass action expression and explained by taking into consideration the stability of the various cupric ammonia complexes,

$$\log E = \log K - \log (1 + \sum_{i=1}^{6} \beta_i [NH_3]^i) + 2 \log [HR]_{(o)} + 2pH$$

From single drop experiments, it appears that the kinetics of extraction are controlled by reactant transfer in the dispersed phase, a conclusion which is supported by a low apparent activation energy of 5.3 kcal/mole. Based on first principles, the reaction rate was explained to be due to molecular diffusion for small drop sizes which behaved as stagnant spheres, while convective transport contributed significantly to the reaction rate for larger drop sizes and other circumstances which result in internal circulation. Rate control by reactant transport in dispered phase is valid under normal conditions regardless of which phase is the dispersed phase. It was predicted and demonstrated that only when the ratio of the reactant concentration in the continuous phase to the reactant concentration in the dispersed phase is less than approximately 0.05 will resistance due to reactant transport in the continuous phase become significant. The rate of copper extraction from ammoniacal solutions is at least one order of magnitude faster than the rate of extraction from acidic solutions. This was to be expected in as much as extraction from acidic solution has been reported to be controlled by a complex surface reaction having an apparent activation energy from 7 to 15 kcal/mole.

EQUILIBRIUM AND KINETICS OF COPPER EXTRACTION FROM AMMONIACAL SOLUTIONS BY HYDROXYOXIMES WITH PARTICULAR EMPHASIS ON TRANSPORT PHENOMENA

J.D. Miller, H.H. Haung and E.F. Pereira

Department of Metallurgy and Metallurgical Engineering
University of Utah
Salt Lake City, Utah 84112

The chemistry of copper extraction from ammoniacal solutions by hydroxyoxime extractants was studied. Equilibrium measurements were made by shakeout experiments with subsequent analysis of the aqueous and/or organic phases. The kinetic experiments were carried out in a single drop reaction cell in orer to establish the details of the intrinsic reaction kinetics for this system.

Equilibrium measurements indicated that the extraction involves complete displacement of ammonia ligands from the coordination sphere of the cupric ion by the extractant molecules. The extraction reaction has the following stoichiometry:

$$Cu(NH_3)^{2+}_{i(aq)} + 2HR_{(o)} \rightleftarrows CuR_{2(o)} + iNH_{3(aq)} + 2H^+_{(aq)}$$

and as might be expected the extraction coefficient was found to be sensitive to the ammonia concentration and the pH of the system. These results were interpreted in terms of the mass action expression and explained by taking into consideration the stability of the various cupric ammonia complexes,

$$\log E = \log K - \log (1 + \sum_{i=1}^{6} \beta_i [NH_3]^i) + 2 \log [HR]_{(o)} + 2pH$$

From single drop experiments, it appears that the kinetics of extraction are controlled by reactant transfer in the dispersed phase, a conclusion which is supported by a low apparent activation energy of 5.3 kcal/mole. Based on first principles, the reaction rate was explained to be due to molecular diffusion for small drop sizes which behaved as stagnant spheres, while convective transport contributed significantly to the reaction rate for larger drop sizes and other circumstances which result in internal circulation. Rate control by reactant transport in the dispered phase is valid under normal conditions regardless of which phase is the dispersed phase. It was predicted and demonstrated that only when the ratio of the reactant concentration in the continuous phase to the reactant concentration in the dispersed phase is less than approximately 0.05 will resistance due to reactant transport in the continuous phase become significant. The rate of copper extraction from ammoniacal solutions is at least one order of magnitude faster than the rate of extraction from acidic solutions. This was to be expected in as much as extraction from acidic solution has been reported to be controlled by a complex surface reaction having an apparent activation energy from 7 to 15 kcal/mole.

INTRODUCTION

A significant research effort has been devoted to develop ammonia oxidative leaching processes for copper sulfide concentrates (1,2,3) and low grade native copper ore (4). This has lead to process development studies and finally to construction of plants to treat copper sulfide concentrates and copper scrap (5,6). When compared with conventional sulfuric acid leaching processes, the major advantage of the ammoniacal system is its ability to reject iron and many acid consuming gangue minerals.

An important unit operation in any hydrometallurgical process is the concentration and purification of the leach liquor for metal recovery, which frequently is accomplished by solvent extraction. A major class of extractants for the recovery of copper from leach liquors has been hydroxyoximes; chelating extractants such as LIX reagents developed by General Mills. Currently other companies such as Shell Chemical Ltd. and Acorga Ltd. produce similar extractants. The most widely used extractant, however, is LIX 64N a mixture of hydroxyoximes consisting of two components, the aliphatic hydroxyoxime (LIX 63) and the aromatic β hydroxyoxime (LIX 65N). The β hydroxyoxime is capable of existing as either the syn- or anti-isomer. Detailed discussion on the chemistry and the structure of LIX 64N has been reported in other literature (7).

A considerable amount of research has been done on the extraction of copper from acidic solutions both with regard to equilibrium and kinetic behavior (7-10). Very little is known about important factors which govern the equilibrium and kinetic behavior of copper extraction from ammoniacal solutions. A research program involving equilibrium and kinetic measurements of copper extraction from ammoniacal solution by LIX reagents has been completed. In the equilibrium study, the results have been analyzed from solution thermodynamic considerations. In the kinetic study, the results have been analyzed in terms of heterogeneous reaction kinetics especially taking into consideration transport processes in both the dispersed and continuous phases.

Equilibrium in Copper Extraction

The chemistry of copper extraction from acidic solution by hydroxyoxime reagents has been extensively investigated by several researchers (7,11). Spectrophotometric evidence (7) indicates that a copper β hydroxyoxime chelation complex is the extracted product and that the stoichiometry is given by the following reaction:

$$Cu^{+2}_{(aq)} + 2HR_{(o)} \rightleftarrows CuR_{2(o)} + 2H^+_{(aq)} \qquad (1)$$

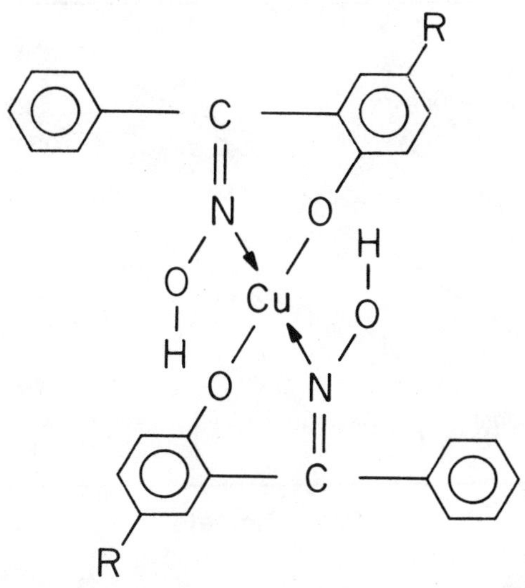

EXTRACTED COMPLEX, $R_2^{65}Cu$

The extraction coefficient (copper concentration in the organic phase divided by copper concentration in aqueous phase) can be calculated from the equilibrium constant, K:

$$\log E = \log K + 2 \log [HR]_{(o)} + 2pH \qquad (2)$$

where,

E = extraction coefficient
 $= [CuR_2]_{(o)}/[Cu^{+2}]_{total(aq)}$,
K = equilibrium constant for reaction presented in Equation 1 including activity coefficients,
$[HR]_{(o)}$ = concentration of extractant in organic phase

In the ammoniacal system this analysis must be altered due to the presence of free ammonia. To begin with, acid neutralization must be considered,

$$H^+_{(aq)} + NH_{3\ free\ (aq)} \rightleftarrows NH^+_{4\ (aq)} \qquad (3)$$

which has a positive effect on the extraction; and secondly the stability of the cupric-ammine complexes must be considered,

$$Cu^{+2}_{(aq)} + iNH_{3\ free\ (aq)} \rightleftarrows Cu(NH_3)^{2+}_{i(aq)} \qquad (4)$$

which has a negative effect on the extraction.

Copper extraction from these solutions can be represented in the following way;

$$Cu(NH_3)^{2+}_{i(aq)} + 2HR_{(o)} \rightleftarrows$$

$$CuR_{2(o)} + 2NH^+_{4(aq)} + (i-2)NH_{3(aq)} \qquad (5)$$

Thermodynamically, however, the copper extraction reaction written for acidic solutions, Equation 1, can still be applied to the ammonia system, but the activity of the cupric ion will be much lower than the total copper concentration. In ammoniacal solutions, the cupric ion concentration can be calculated from the stability constants for the formation of cupric-ammine complexes according to reactions as given in Equation 4,

$$[Cu^{+2}]_{aq} = \frac{[Cu(NH_3)_i^{2+}]_{(aq)}}{\beta_i [NH_3]^i_{free(aq)}} \qquad (6)$$

where

$[Cu(NH_3)_i^{2+}]_{(aq)}$ = concentration of the ith cupric-ammine complex in the aqueous solution,
$[Cu^{2+}]_{(aq)}$ = concentration of simple hydrated cupric ion in the aqueous solution,
$[NH_3]_{free(aq)}$ = concentration of free ammonia in the aqueous solution
β_i = overall stability constant of the ith cupric-ammine complex

The values of β_i for the cupric-ammine complexes are listed below(12),

$$\log \beta_1 = 4.15 \quad \log \beta_2 = 7.67 \quad \log \beta_3 = 10.54$$

$$\log \beta_4 = 12.67 \quad \log \beta_5 = 12.17 \quad \log \beta_6 = 9.67 \qquad (7)$$

If metal hydrolysis can be ignored, then the fraction of cupric ion to total cupric concentration in the aqueous phase can be expressed as,

$$\frac{[Cu^{2+}]_{(aq)}}{[Cu]_{total\ (aq)}} = \frac{1}{1 + \sum_{i=1}^{6} [NH_3]^i_{free(aq)}} \qquad (8)$$

Consequently, an additional term will contribute to the expression for the copper extraction coefficient,

$$\log E = \log K + \log \left(1 + \sum_{i=1}^{6} \beta_i [NH_3]^i_{free\ (aq)}\right)$$

$$+ 2 \log [HR]_o + 2pH \qquad (9)$$

From examination of the stability constants for the cupric-ammine complexes, it is seen that the tetra-ammine cupric ion and the penta-ammine cupric ion have the highest β values. This suggests that for normal ammoniacal leaching systems, in which the concentration of free ammonia is about 16 gpl the average number of ammine ligands coordinated to a cupric ion will be between 4 and 5.

Kinetics of Copper Extraction

Several fundamental kinetic studies have been reported on the extraction of copper by LIX reagents in acidic systems (10,11,13). These studies have employed a variety of experimental techniques. The results indicate that the copper extraction reaction by LIX

65N/LIX 63 reagents is heterogeneous and that the rate is controlled by a surface chemical reaction at the aqueous/organic interface. Results also show that the addition of LIX 63 to LIX 65N enhances the reaction kinetics without disturbing the final equilibrium, LIX 63 being characterized as a catalyst in the system. Flett (11) indicated that the activation energy for LIX 65N is about twice as large as that for a mixture of LIX 65N and LIX 63, a result which supports this catalytic role played by LIX 63 in acidic extraction systems.

The proposed surface reaction mechanisms by several investigators differ significantly. Results from experiments with AKUFVE system by Flett (11) and the Lewis cell by Fleming (13) indicate that the rate limiting step is the addition of an extractant molecule, the second ligand for the cupric ion, at the interface. However, the results from single drop experiments (10) indicate that the rate of extraction is independent of hydrogen ion concentration at relatively high pH values (2.21 - 4.98) and the rate controlling step was suggested to be the addition of the first extractant molecule.

Hydrodynamics of Liquid Spheres Moving Through an Immiscible Liquid Field

Among the three techniques that have been used for the study of extraction kinetics; the AKUFVE system, the Lewis cell and Single Drop Reaction Cell, the single drop technique seems to have two unique features; a well defined interfacial area and fairly well established hydrodynamic behavior for liquid spheres.

Liquid spheres differ from solid spheres in that the dispersed liquid can be set in motion by the viscous drag from the continuous liquid through which the drop moves. Additionally, drops can be subjected to deformation because of the differences in pressure acting on various parts of the surface. As a result of these considerations, the terminal velocity of a liquid drop may differ significantly from the corresponding velocity for a solid sphere. Figure 1 shows the terminal velocity of a liquid drop as a function of the drop diameter. Three distinct regions can be identified (14-16). For small drops in region A, the liquid boundary layer inside the drop remains rigid and the liquid drop will be spherical in shape and behave like a solid sphere. The terminal velocity will generally be that of a solid sphere of equal size and density moving in the same continuous phase. In region B, the boundary layer inside the drop will no longer remain rigid, and the momentum from the continuous phase will be transferred to the drop which develops internal circulation. Under these circumstances the terminal velocity will be greater than that of the non-circulating drop, the solid sphere. As the drop size is increased, region C, the larger drop will travel at about the same speed, independent of size. Oscillation of the drop can be observed in this region and the internal flow pattern is a combination of eddy flow and circulation.

Viscosity, interfacial surface tension and density difference will also influence circulation of the liquid sphere. For example, fully developed circulation in a drop of specified size can be damped out if the viscocity of the dispersed phase is increased (17).

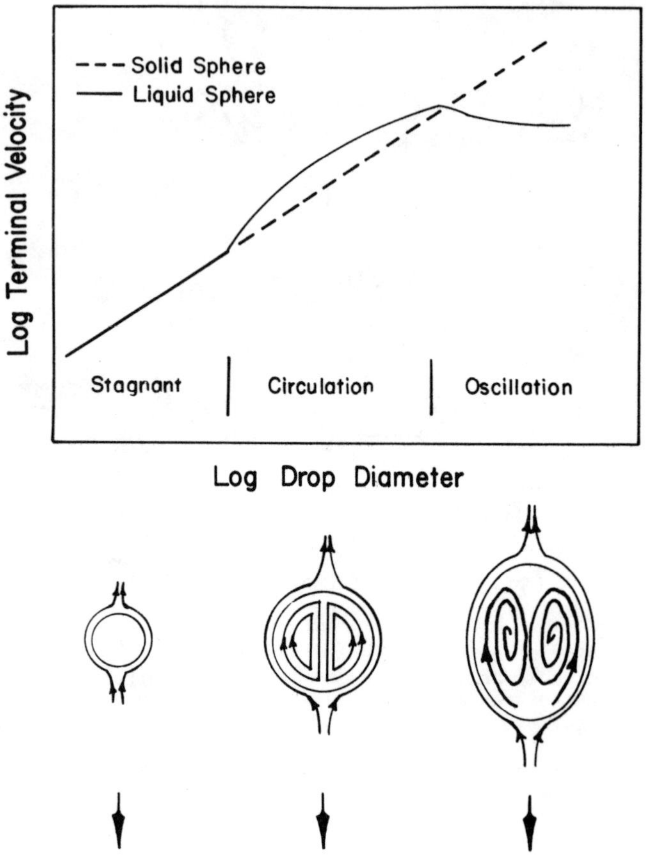

Figure 1. Terminal velocity and internal flow pattern for a liquid sphere as a function of drop diameter.

The surface tension reduces the circulation in such a way that the higher surface tension the more energy it takes to create the new surface at the front of the drop and hence a higher surface tension reduces the extent of circulation (18).

Mass Transfer in Single Drop Reaction Cells

Mass Transfer in the Dispersed Phase

Due to the change in the internal flow pattern of the moving drop, mass transfer in the dispersed phase must also be evaluated in three different regions; the stagnant region, the circulation region and the oscillation region.

The mass transfer equation for a stagnant drop can be derived from the continuity equation based on Fick's 2nd law for which the transport depends only on molecular diffusion. The integrated expression for mass transfer in the stagnant drop is found to be (19),

$$\frac{C_o - \bar{C}_t}{C_o - C_s} = 1 - \frac{6}{\pi^2} \sum_{n=1}^{\infty} \frac{1}{n^2} \exp\left[- Dn^2\pi^2 t/a^2\right] \qquad (10)$$

where

D = diffusion coefficient

a = drop radius

t = reaction time

C_o = Initial concentration in dispersed phase (uniform distribution of concentration is assumed)

$\bar{C}_t$ = average concentration in dispersed phase at time t

C_s = surface concentration at time t

If internal mass transfer is the rate controlling step, surface concentration, C_s, will be equal to the equilibrium concentration at the interface. As will be shown from the equilibrium study, for the typical experimental conditions that were used in the kinetic study ($[NH_3]_{total}$ = 0.5<u>M</u>, pH = 10, Extractant 1 to 10 vol.%) the extent of extraction will be close to 100% as long as the copper concentration is greater than 0.05 gpl. This means that in most kinetic experiments the surface concentration of reactant would be close to zero. However for higher free ammonia concentrations this may not be true and the equilibrium concentration at the interface would have to be considered.

For short reaction times, the integrated rate equation can be approximately by the following parabolic relationship; (19)

$$\frac{C_o - \bar{C}_t}{C_o} = \frac{\Delta\bar{C}_t}{C_o} = 3.38 \left(\frac{Dt}{a^2}\right)^{1/2} \qquad (11)$$

and the rate of mass flow can be written as

$$I = \frac{4}{3}\pi a^3 \frac{d\bar{C}_t}{dt} = -7.09\ C_o\ a^2 \left(\frac{D}{t}\right)^{1/2} \qquad (12)$$

The mass transfer equation for a circulating drop is more complicated in that both molecular diffusion and convect ve transport must be considered. Kronig and Brink(20) have derived a mass transfer equation using streamline functions together with an expression for molecular diffusion from one streamline to another. The integrated mass transfer equation is, (21,22)

$$\frac{C_o - \bar{C}_t}{C_o} = 1 - \frac{3}{8} \sum_{n=1}^{\infty} B_n^2 \exp\left[-\lambda_n \frac{16\ Dt}{a^2}\right] \qquad (13)$$

where

λ_n = eigenvalues

B_n = function of eigenfunction

For short reaction times (t_{max} < 25 sec) and considering the values for λ_n and B_n(23), the equation can be simplified as,

$$\frac{C_o - \bar{C}_t}{C_o} = 4.65 \left(\frac{Dt}{a^2}\right)^{1/2} \qquad (14)$$

and the rate of mass flow can be derived as,

$$I = -\ 9.73\ C_o a^2 \left(\frac{D}{t}\right)^{1/2} \qquad (15)$$

A comparison of Equation 15 with Equation 12 indicates that for short reaction times, the rate of transport in a circulating drop is increased by almost 40 percent due to the convective contribution to the transport process.

Because of the complexity of the internal flow pattern of the oscillating drop, there is no satisfactory model for internal mass transfer derived from first principles but some emphirical equations exist (24). Drop oscillation occurs when the diameter of the drop is greater than 5 mm, a size which generally is not of interest in solvent extraction systems. As a result, internal mass transfer for the oscillating drop will not be considered.

Mass Transfer in the Continuous Phase

When the reaction rate is controlled by mass transfer in the continuous phase, the reactant species has to pass through the mass transfer boundary layer adjacent to the drop surface. Levich (25) has derived a theoretical expression for mass transfer in the continuous phase. The rate of mass flow can be determined by the following equation,

$$I = 5.79 \left(\frac{\mu_c}{\mu_c + \mu_d}\right)^{1/2} \left(\frac{DU}{a}\right)^{1/2} a^2\ \Delta C_c \qquad (16)$$

where,

μ_c, μ_d = viscosities of the continuous phase and dispersed phase, respectively

U = terminal velocity

ΔC_c = concentration difference between the bulk and the surface concentrations in the continuous phase

Other investigators have used dimensional analysis to obtain empirical correlations; Skelland and Cornish (26) for stagnant drops and Garner and Tayeban (27) for circulating and oscillating drops.

Relative Significance of Mass Transfer in the Dispersed and Continuous Phases

In order to assess the relative significance of mass transfer in the dispersed and continuous phases, a comparison was made, as shown in Figure 2, where the modified rate of mass flow (mass flow rate divided by the initial concentration of the reactant in the dispersed phase) is plotted as a function of the drop travel time. Figure 2 was constructed by using the stagnant model for mass transfer in the dispersed phase (Equation 12) and the Levich equation for mass transfer in the continuous phase (Equation 16). A family of curves for several initial concentration ratios, $C_{c,o}/C_{d,o}$, is shown where $C_{c,o}$ is the initial concentration of the continuous phase reactant and $C_{d,o}$ is the initial concentration of

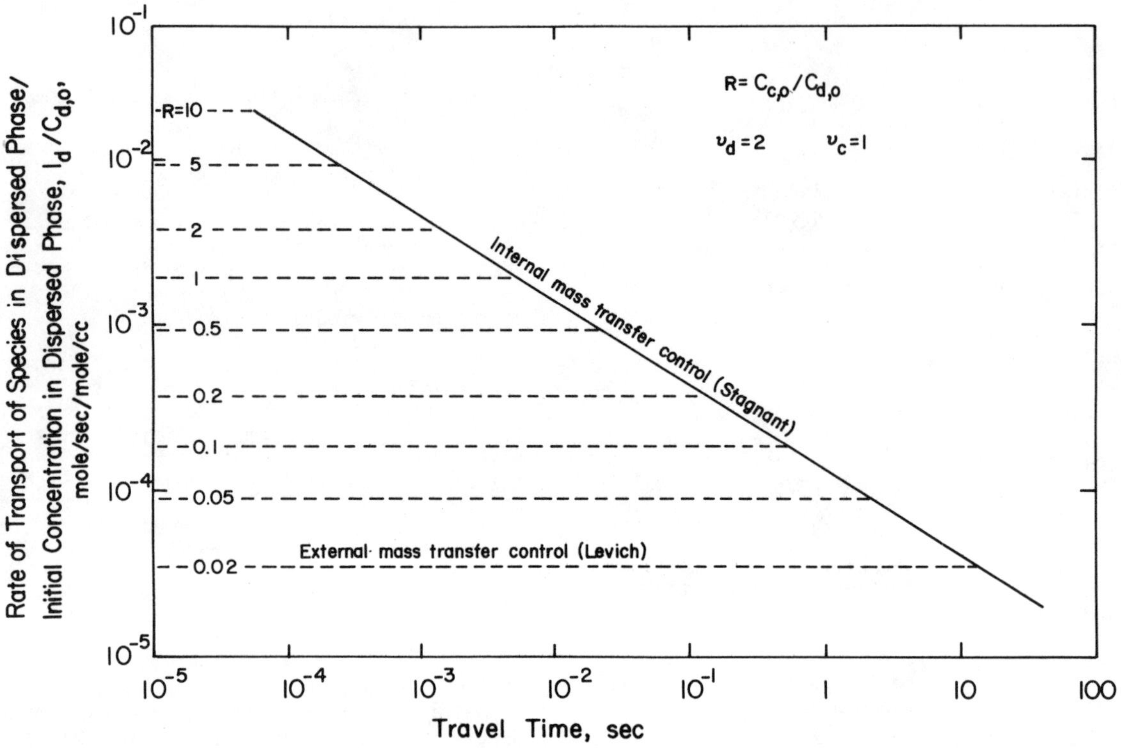

Figure 2. Representation of the relative importance of transport processes in the continuous and in the dispersed phase for various reactant concentration ratios. Intersection of horizontal dashed line with diagonal line indicates the critical time (t_c) during which the rate is limited by mass transfer in the continuous phase.

Conditions: $D_c = 7.5\times10^{-6}$ cm^2/sec, $D_d = 4\times10^{-6}$ cm^2/sec,

$a = 0.1$ cm; $\mu_c = 1$ cp, $\mu_d = 1.68$ cp,

$U = 10$ cm/sec; $\nu_d = 2$, $\nu_c = 1$ (ν = stoichiometry coefficients)

dispersed phase reactant. The mathematical analysis indicates that the rate is controlled initially by mass transfer in the continuous phase and after some critical time resistance due to mass transfer in the dispersed phase becomes predominant. This critical time, t_c, depends on the ratio of the reactants' initial concentrations. Normally, the time interval during which the reaction rate would be limited by mass transfer in the continuous phase is too small to be significant unless the initial concentration ratio $C_{c,o}/C_{d,o}$, is smaller than 0.05. This suggests that in most solvent extraction systems mass transfer in the continuous phase will be faster than mass transfer in the dispersed phase so that generally for a fast chemical reaction mass transfer in the dispersed phase will be rate limiting.

End Effects in Single Drop Reaction Cells

In the single drop experiments mass transfer can occur not only during the travel time but also during the drop formation and coalescence period. As a result, the extent of reaction may be greater than the values predicted by a particular model. A much used technique to correct for these effects has been to plot the overall extent of reaction as a function of the drop travel time, and then extrapolate to the ordinate at zero time to determine the extent of reaction contributed by the end effects (23,28-29). Atwood (10) also tried to correct for end effects by extrapolating the rate curve to the abcissa value for no reaction to determine a lag time which represents the time that the drop is not traveling but during which reaction occurs.

Two different mechanisms for internal mass transfer to a growing drop have been proposed. The surface stretch model (30) assumes that all elements at the surface remain and stretch as the drop grows. The penetration model (31) assumes that the growth of a drop is preceded by the addition of fresh elements which penetrate to the surface. Popovich (31) showed that the above models could be reduced to a general form,

$$N = 4\pi G \, \Delta C \, \left(\frac{3qt}{4\pi}\right)^{2/3} \left(\frac{Dt}{\pi}\right)^{1/2} \tag{17}$$

where

N = total amount of mass (moles) transferred in time t (sec).

q = volume rate of flow into the drop, cc/sec.

ΔC = difference between bulk concentration and surface concentration, moles/cc.

G = constant, 1.31 for surface stretch model, 1.52 for penetration model.

According to Equation 17, the fraction extracted by transport processes in the dispersed phase during drop formation for the typical conditions specified in this kinetic study will be less than 1% of the total extraction, taking G = 1.31.

Most of the studies on mass transfer during the drop coalescence period have been with respect to the diffusion from the newly settled drop phase to the coalescence interface (23,32). In this kinetic study on copper extraction however, the mass transfer from the continuous phase to the coalescence interface can also play a significant role. For both cases important parameters to determine the rate of reaction during the drop coalescence will be the coalescence interface, diffusion coefficients, frequency of drop coalescence and the volume of the drop. Experimental results indicate that for a small coalescence area, reaction during drop coalescence is negligible.

EXPERIMENTAL

The research effort on copper extraction from ammoniacal solution has been divided into two major sections. The equilibrium study and the kinetic study. The experimental approach was to identify the important variables in each section in order to provide fundamental information for further application.

Chemicals

The primary extractant for this study was LIX 64N (Lot no. 3K19304 and GH19245C). In a few experiments other extractants, LIX 65N (lot no. 6A19205) and LIX 54, a β-diketone extractant (33) (Lot no. 7A15808) were used. All extractants were provided by General Mills and used without further purifications. The organic diluent was commercial Escaid 200 which has less than 1% aromatic content.

Extraction Equilibrium

The chemistry of copper extraction from ammoniacal solution was investigated by using a shake-out test procedure with separatory funnels. The organic solution, LIX 64N in Escaid 200, was shaken in a separatory funnel with an aqueous solution containing copper and ammonia from which the extraction was to be made. After about 30 minutes of vigorous shaking on the wrist shaker the phases were allowed to separate. Each phase was analyzed for copper by atomic absorption; for the aqueous phase the amount of free ammonia had to be neutralized by sulfuric acid, and for the organic phase the copper was stripped out with an aqueous sulfuric acid solution. The results were used to calculate the distribution coefficient and the amount of copper loaded in the organic phase.

Extraction Kinetics

The kinetic study was carried out in a single drop reaction cell. The apparatus consisted of a reaction cell, orifice for drop formation, provision for drop removal and temperature controller. The schematic diagram of the apparatus for a short cell length is shown in Figure 3. A similar type of construction was

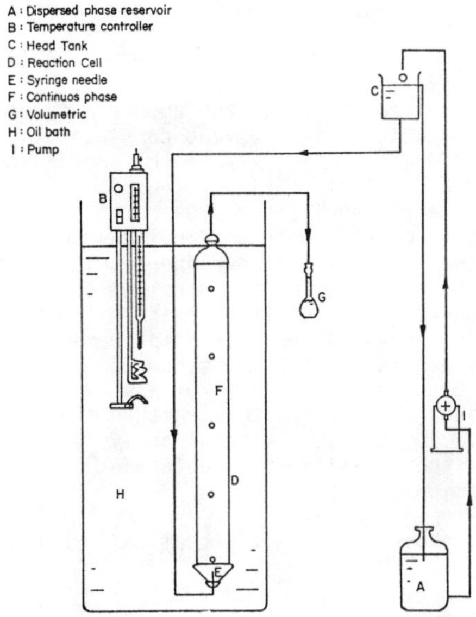

Figure 3. Schematic diagram of Single Drop Reaction Cell.

used for longer cell lengths as well. Depending upon the density difference of the dispersed phase and continuous phase, the data can be collected either for a falling drop or a rising drop. After the dispersed phase was selected, the solution was introduced from the head tank to the reaction cell by hydrostatic pressure for the short cell lengths and by pressure from a nitrogen gas cylinder for the long cell lengths. For organic dispersed experiments, the organic drops were formed at the bottom of the cell from the syringe needle. After rising through the cell, the drop coalesced at the top of the aqueous/organic interface, and discharged through a tube, being collected outside the cell. For aqueous dispersed experiments, the system was reversed, and the aqueous drops were allowed to fall through the cell. The size of the drop was varied by using stainless steel needles with different size openings which ranged from gauge numbers 24 to 33.

By controlling the head pressure, a constant frequency of droplet generation could be obtained which was usually on the order of 100 drops per minute. Under this drop frequency the end effect was found to be insignificant. The average drop size was calculated from the number of drops required to collect a specific volume and the contact time during the steady state travel was measured by timing the travel of a drop from the tip of the needle to the coalescence interface. The travel time was varied by changing the length of the cell which ranged from 13 to 160 cm for the short cell and from 47 to 390 cm for the long cell. The amount of copper extracted or amount of LIX reagent consumed for each experiment was determined by measuring the copper concentration with an atomic absorption spectrophotometer (the same method used in the equilibrium study).

Experiments were designed so that the kinetics of extraction could be investigated for both rising drops (organic dispersed) and falling drops (aqueous dispersed). The rate data were analyzed in terms of the theoretical considerations associated with mass transfer phenomena to determine the effect of copper concentration in the aqueous solution, LIX 64N concentration in organic phase, drop size and temperature.

RESULTS AND DISCUSSION

Extraction Equilibrium

It has been reported that the copper extraction by LIX 64N in acidic media involves complete displacement of coordinated water molecules of the cupric ion by the LIX 64N extractant. In this study an infrared spectrophotometric analysis was used to compare the copper complex extracted from ammoniacal solution with that from the acidic solution. The results indicate that the extracted complex is identical for both systems and suggest that copper extraction from ammonia solution also involves complete displacement of coordinated ammine ligands. The absence of a primary absorption band near 3300 cm^{-1} for any ammonia molecules retained by the chelate structure confirms this conclusion. In addition, Kjedahl analysis for ammonia showed that the organic phase contained a negligible amount of ammonia.

From the above findings the chemical reaction can be written as,

$$Cu(NH_3)^{2+}_{i(aq)} + 2HR_{(o)} \rightleftarrows CuR_{2(o)} + 2H^+ + i(NH_3)_{(aq)} \qquad (5)$$

and the extraction coefficient can be presented as,

$$\log E = \log K - \log\left(1 + \sum_{i=1}^{6} \beta_i [NH_3]^i\right) + 2 \log [HR]_o + 2pH \qquad (9)$$

Effect of Free Ammonia

Figure 4 shows the effect of free ammonia concentration on copper extraction at a constant pH value (10.2 to 10.5), maintained by keeping the ratio of $[NH_3]_{free}/[NH_4^+]$ constant at 2:1. The system had an excess amount of extractant so that the concentration of the extractant can be considered to be constant, and the amount of coordinated ammonia can be neglected. As expected from the extraction coefficient equation, Equation 9, the results (Figure 4) show that the copper extraction decreases as the ammonia concentration increases for both LIX 64N and LIX 54. The results indicate that the LIX 64N extractant forms a more stable copper complex than LIX 54 in ammoniacal solution.

According to Equation 8, a plot of log E and log $[NH_3]_{free}$ should be indicative of the number of coordinated ammonia molecules, Figure 5 is such a plot and the average slopes for both extractants is seen to be 4.5. This suggests that the cupric ion in this ammoniacal solution is coordinated by an average of 4 to 5 ammonia molecules which is in good agreement with the value that would be predicted from the stability constants for copper-ammine complexes.

Effect of pH

The effect of pH on copper extraction was determined by keeping the total amount of ammonia, $[NH_3]_{free}$ plus $[NH_4^+]$, constant at 0.5 M. In this case, the extraction of copper is more complicated due to the fact that the pH value has a positive effect on extraction while free ammonia concentration has a negative effect, shown by Equation 9. As a result Figure 6

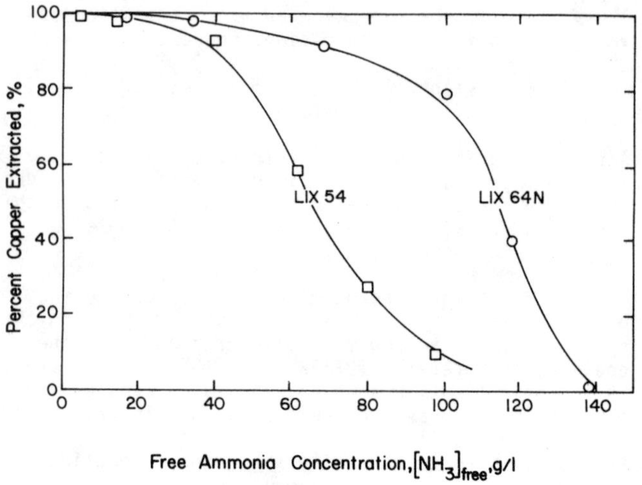

Figure 4. Effect of ammonia concentration on copper extraction at constant pH (10.2-10.5).

LIX 64N: 0.004M Copper; 40% LIX 64N in Escaid 200; A/O 1:1; NH_3/NH_4^+ 2:1

LIX 54: 0.004M Copper; 38% LIX 54 in Cyclosol 63, A/O 1:1; NH_3/NH_4^+ 2:1

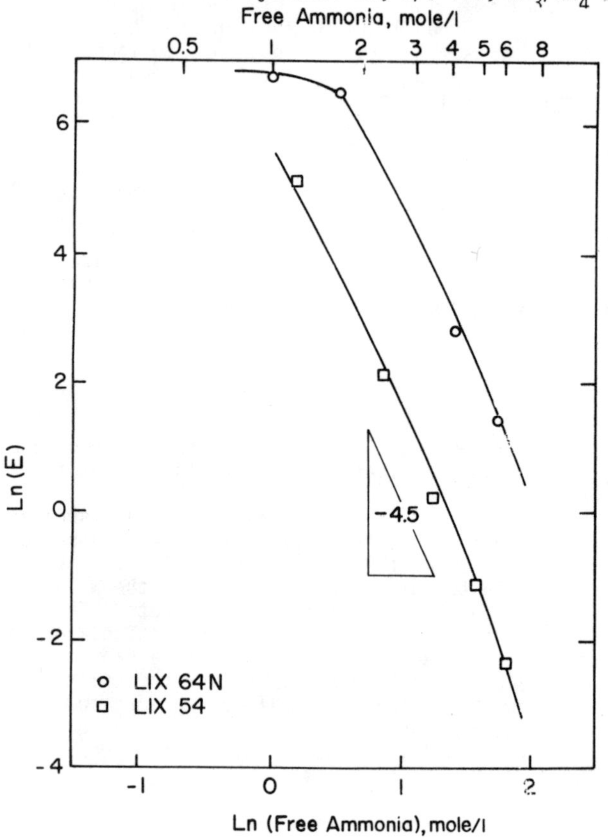

Figure 5. Slope Analysis plot indicative of the number of ammonia molecules coordinated by the cupric ion under the conditions specified in Figure 4.

$$\log E = \log K + \log\left(1 + \sum_{i=1}^{6} \beta_i [NH_3]^i\right) + 2pH + 2 \log [HR]_{(o)}$$

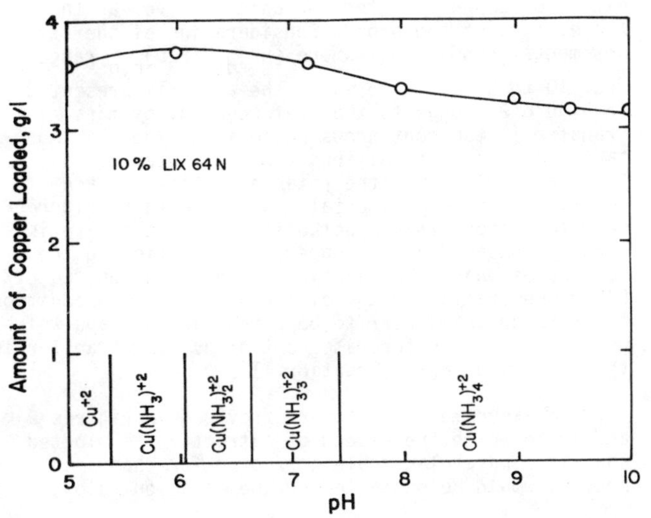

Figure 6. Effect of pH on copper extraction at a constant total ammonia concentration of 0.5 M.

shows a parabolic-type of curve for a plot of copper extraction as a function of pH. For low pH values, copper extraction increases as the pH increases, which may be due to the pH effect only, because the concentration of free ammonia is negligibly small at these low pH values. However, at high pH values the concentration of the ammonia and the number of ammonia ligands will be increased significantly. In this region the ammonia effect will overcome the pH effect and the copper extraction will decline. Predominate copper species are indicated in Figure 6 for various pH regions. Similar effects have been found in the extraction of nickel from ammoniacal solution (34).

Comparison between Acidic and Ammoniacal Systems

According to the equilibrium expression, Equation 9, the thermodynamics for copper extraction seems to be more favorable in the ammoniacal system than in the acid system. The reason is due to high pH values in the ammoniacal solution overwhelming the negative effect of the free ammonia if the concentration of free ammonia is not extremely high. This analysis is supported by the experimental results of the equilibrium extraction isotherms for both acidic and ammoniacal systems (shown in Figure 7). It can be seen that the isotherm for the ammoniacal system is steeper than that for the acid system suggesting that the amount of copper loaded into the organic phase will be higher for the same concentration of copper in the aqueous phase. The pH values listed in Figure 7 were initial values; no attempt was made to buffer the pH.

Other information also can be found from the extraction isotherms. First, due to the steeper isotherm the number of stages required in the extraction from the ammoniacal solution is normally less than that required for extraction from acidic solution. (35)

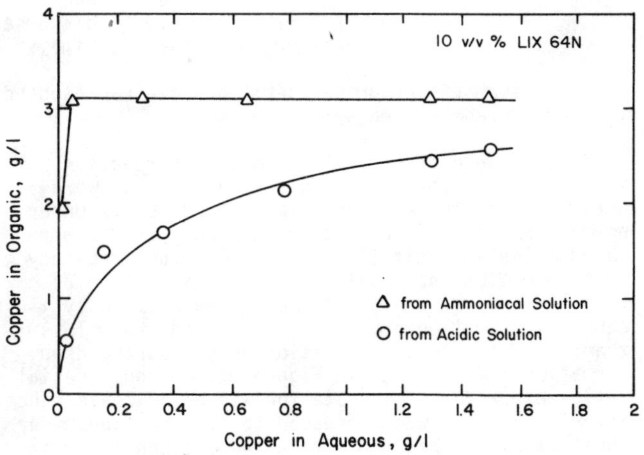

Figure 7. Equilibrium extraction isotherms of copper extraction by 10 V/V % LIX 64N from ammoniacal solution (0.5 M NH_3, 0.5 M NH_4^+, pH = 9.70) and from acidic solution (pH = 1.79).

Second, at relative low cupric concentrations in ammoniacal solutions, the isotherm has already reached maximum loading which indicates that the equilibrium concentration of the extractant required is quite small. This observation simplifies subsequent analysis of the rate data as explained in the introduction during the derivation of transport equations.

Extraction Kinetics

The kinetics were studied in both rising drop and falling drop experiments. The results were used to analyze the effect of drop size, temperature and reagent concentrations both in the dispersed phase and in the continuous phase. By considering all these effects, the rate controlling step was identified and a comparison between experimental results and theoretical predictions was made. Normally, the organic phase consisted of the diluent, Escaid 200, which contained from 1 to 50 volume percent of extractant (LIX 64N). The copper concentration in the aqueous phase ranged from 0.5 to 4.3 gpl copper, as cupric sulfate, with 1 M total ammonia (0.5 M free ammonia) at a pH of 9.75. The drop size varied from 0.84 to 1.52 mm in radius.

Rising Drop Experiments

In this series of experiments the continuous phase was always the ammoniacal solution of copper sulfate, and the dispersed phase was the organic solution (LIX 64N in Escaid 200). Normally the initial copper concentration in the organic was zero. The amount of copper extracted into the organic solution is related to the change in the LIX concentration by the stoichiometry of the reaction,

$$-\Delta[HR]_{d,t} = 2\Delta[CuR_2]_{d,t} \qquad (18)$$

where

$\Delta[HR]_{d,t}$, $\Delta[CuR_2]_{d,t}$ = change in concentration of extractant and copper, respectively, in dispersed organic phase at time, t

All data from rising drop experiments are represented in the figures by open symbols.

Significance of end effects and back reaction kinetics. End effects in this kinetic study were determined by the extrapolation method (23). Under normal experimental conditions the end effects were found to be insignificant due to the high frequency of drop formation and small coalescence area of 0.64 cm^2 (see Figures 9 and 16). As mentioned previously it is expected from theoretical considerations that less than 1% of the total extraction occurs during drop formation. Also shown in Figure 16, for one special set of experiments, is data for which the coalescence interfacial area was increased to 1.54 cm^2 and under these circumstances end effects were found to be important.

The signficance of the back reaction in the range of variables evaluated in this study was assessed by initially loading the organic phase with copper at the 10% and 30% level. No decrease in the extraction rate could be observed when the results were compared to the rate of extraction when the initial copper concentration in the dispersed organic phase was zero. From these results the back reaction kinetics were not considered to contribute significantly to the overall reaction rate.

Effect of LIX 64N and copper concentration. Experimental results for two drop sizes with several concentrations of LIX 64N in the dispersed organic phase and several concentrations of copper in the continuous aqueous phase are presented in Figure 8. Because the reaction is directly proportional to the interfacial area, Figure 8 was plotted in terms of extent of copper reacted per unit area versus the travel time of the drop. Consideration of the experimental conditions, where $[HR]_{d,o}/[Cu]_{c,o}$ ranges from 10 to 0.22, in terms of the analysis presented in Figure 2, suggests that rate control by mass transfer in the continuous phase should not be important under these conditions and that internal mass transfer would limit the rate for a fast surface reaction. The experimental results shown in Figure 8, seem to support this hypothesis in that the rate is largely independent of copper concentration in the continuous phase, but distinctly dependent on the LIX 64N concentration in the dispersed phase. In addition the rate curves appear to be parabolic in shape which would be expected for rate control by mass transfer in the dispersed phase (Equations 11 and 14).

More experimental data are provided in Figures 9 and 10 in which the extent of extraction is plotted against a parabolic scale corrected for change in area as would be suggested by the equation below,

$$[CuR_2]_{d,t} = k'[HR]_{d,o}D^{1/2}(t/a^2)^{1/2} \quad (19)$$

where the value of k' depends on whether the drop is circulating or stagnant. A linear relationship can be found for all concentrations of LIX 64N. Figures 9 and 10 also indicate, as mentioned earlier, that the correction for end effects during drop formation and drop coalescence is unnecessary.

The reaction order calculation based on experimental results indicates that the rate has a first order dependence on the extractant concentration and zero order dependence on the copper concentration. These results suggest that the rate of extraction is limited by mass transfer in the dispersed phase.

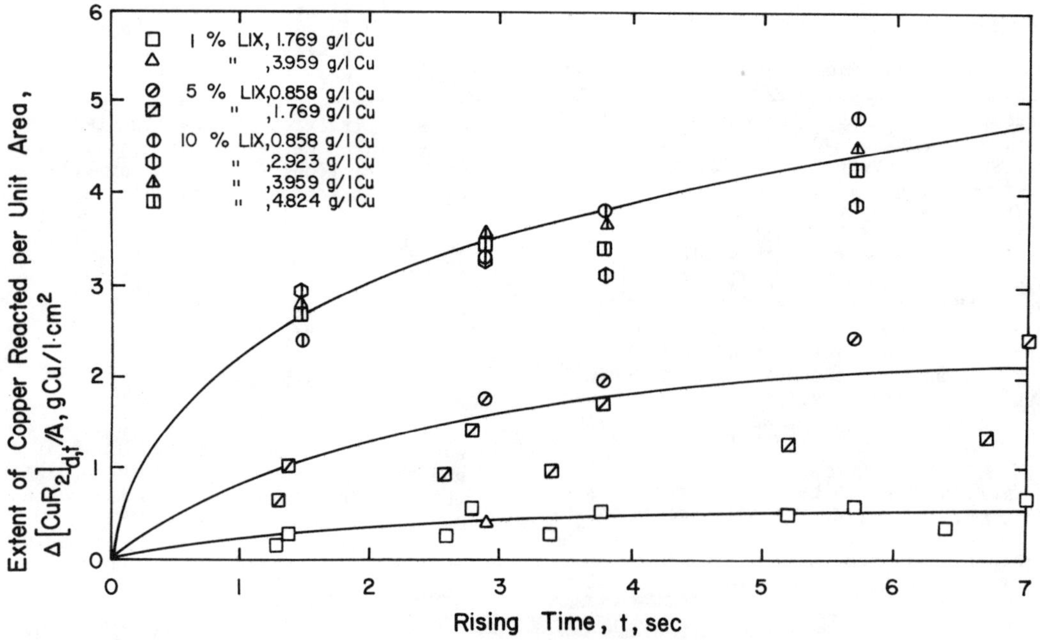

Figure 8. Copper extraction into rising drop for various dispersed (organic) and continuous (aqueous) phase reactant concentrations

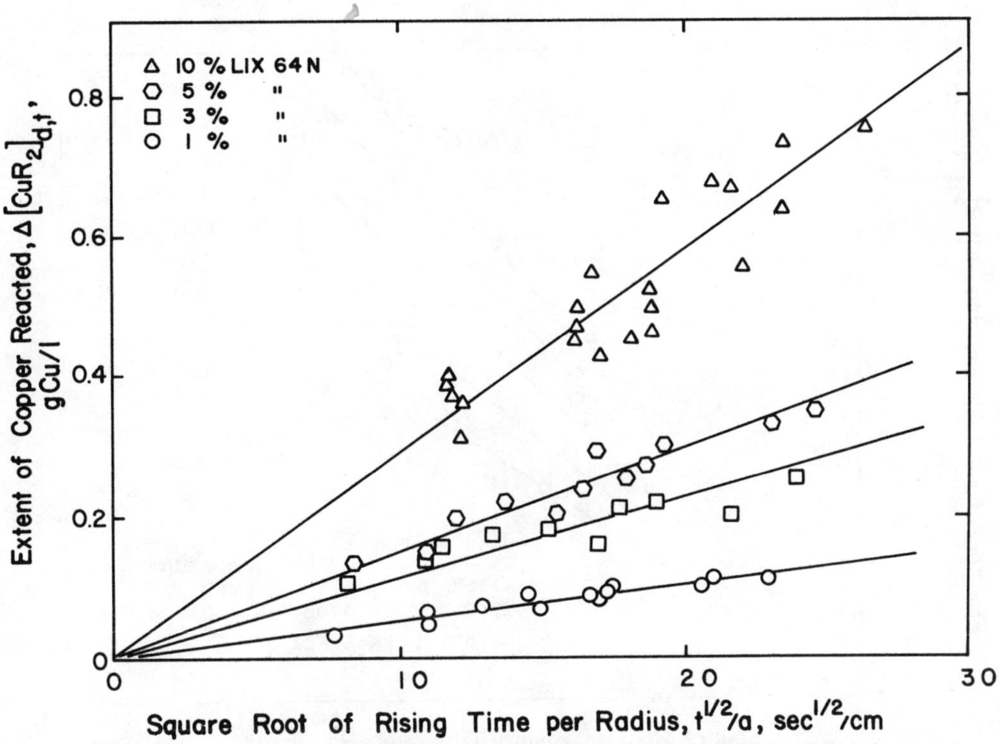

Figure 9. Parabolic plot of rate data illustrating that the kinetics conform to the linear relationship required for rate control by mass transfer in the dispersed phase. Various copper concentrations.

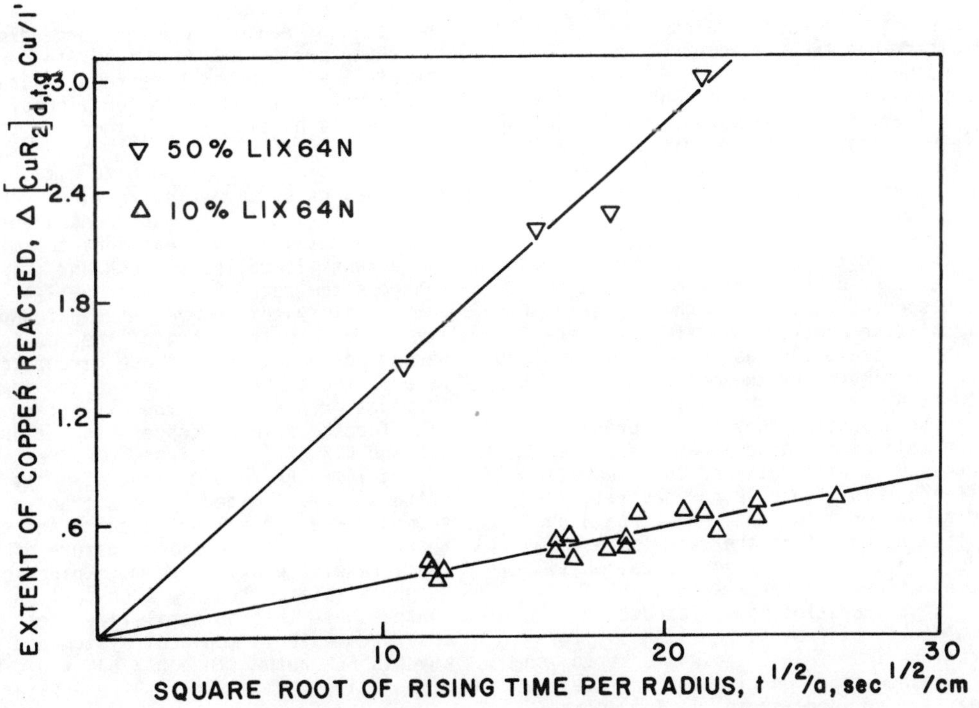

Figure 10. Parabolic plot of rate data illustrating that the kinetics conform to the linear relationship required for rate control by mass transfer in the dispersed phase. Various copper concentrations.

Effect of temperature. The hypothesis of rate control by mass transfer in the dispersed phase also is supported by the low temperature coefficient observed from the data presented in Figure 11. This figure shows that the rate of copper extraction increases as the temperature increases. It also shows

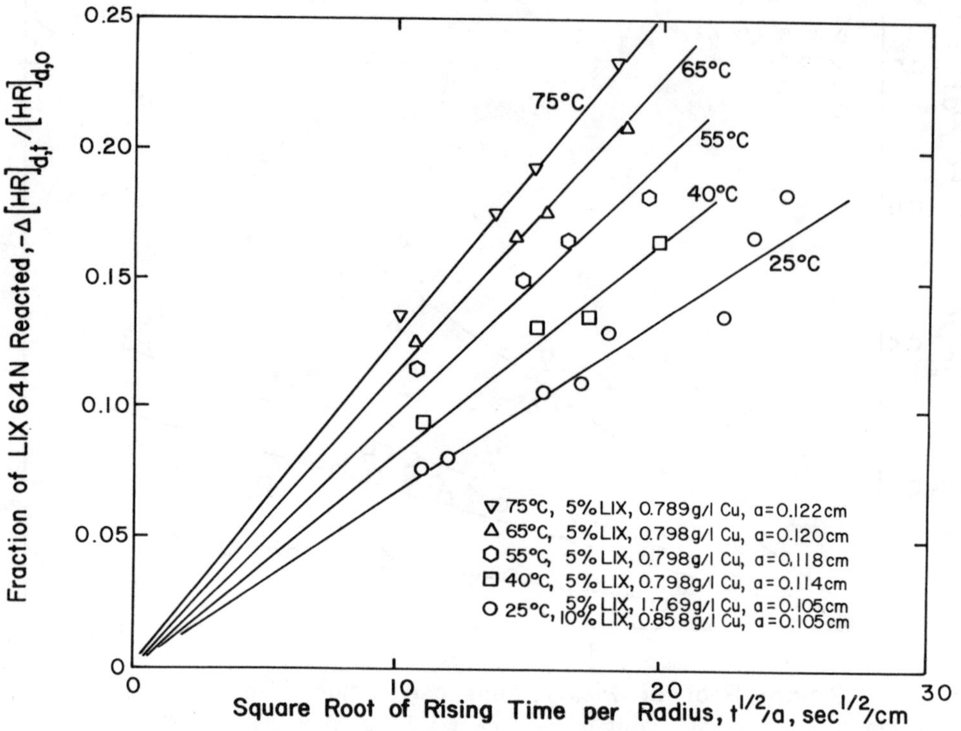

Figure 11. Parabolic plot of rate data for extraction into rising drops at various temperatures.

that the rate curves for all temperatures are parabolic which indicates that the rate controlling step does not change in the temperature range of this study. For this parabolic plot, the dependent variable has been normalized with respect to the initial concentration of the extractant in the dispersed phase. The corresponding Arrhenius calculation reveals a low apparent activation energy of 5.0 kcal/mole.

Comparison of theory with experimental results. In order to theoretically predict the rate of reactant transport in the dispersed phase for both the stagnant model and the circulation model, the diffusion coefficient of LIX 64N in Escaid 200 has to be known and, in this regard, was estimated by the method of Wilke and Chang (36) to be 4.34×10^{-6} cm^2/sec. In subsequent Figures in this section theoretical predictions are given by solid and dashed lines. A close agreement between experimental data and the theoretical predictions is shown in Figure 12 for several concentrations of LIX 64N for both small and large drop sizes. The results indicate that the rate follows the stagnant model (molecular diffusion, Equation 11) for LIX 64N concentrations greater than 5%, and that the rate follows the circulation model (convective transport plus molecular diffusion, Equation 14) for LIX 64N concentrations less than 5%. This observation which indicates a change from the circulation model to the stagnant model for an increase in LIX concentration has not been satisfactorily explained on the basis of the change in the physical properties of the system.

The effect of the drop size on the kinetics of the extraction is shown in Figure 13. For small drops (a = 0.108 cm), the experimental data seems to be in the stagnant region; and for large drops (a = 0.135 cm) the experimental data seems to be close to the circulation region. These results are consistent with the expected hydrodynamic behavior of larger drops which would have a greater tendency to circulate.

Kinetics of extraction with different LIX reagents. The kinetics of copper extraction by LIX 65N from acidic solution is slow and consequently LIX 63 is added to catalyze the reaction, the mixture of the extractants is called LIX 64N and this combination enhances the rate significantly (7-10). In the ammoniacal system however, the rate appears to be limited by the transport of the reactant in the dispersed phase and under these circumstances should not be expected to be dependent on the composition and the chemical properties of the extractant. Such is the case for the rate of copper extraction by LIX 64N, LIX 65N and LIX 54 (a β-diketone extractant) as shown in Figure 14. The results indicate that the rate of extraction is limited by mass transfer in the dispersed phase for all extractants and the rates are insensitive to the chemical nature of the extractant. These results provide one more piece of evidence which supports rate control by transport in the dispersed phase. Note that no catalytic effect from LIX 63 is observed in the ammoniacal system and it is clear why LIX 65N rather than LIX 64N would be suitable in ammoniacal systems such as the Arbiter process.

Change in rate control from mass transfer in the dispersed phase to mass transfer in the continuous phase. As mentioned previously, mass transfer in the dispersed phase is normally slower than mass transfer in the continuous phase. However, the latter step can become rate limiting if the initial concentration ratio $[Cu]_{c,o}/[HR]_{d,o}$ is less than 0.05 (see Figure 2).

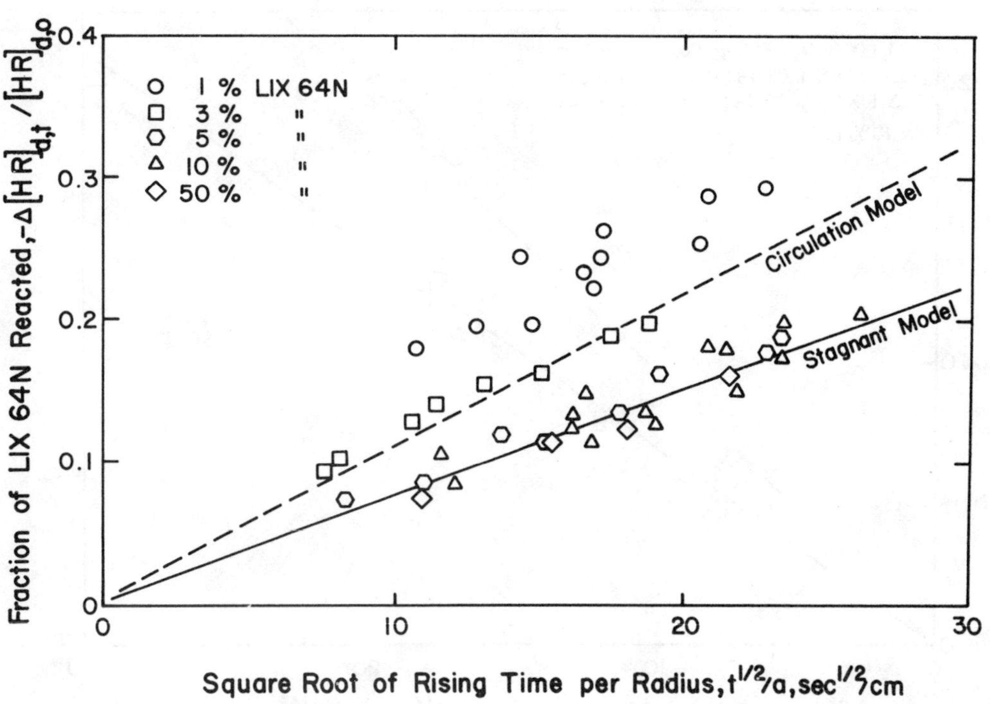

Figure 12. Comparison of experimental results with theoretical predictions for rate control by mass transfer in the dispersed phase (See equations 11 and 14) showing the concentration effect. Various copper concentrations.

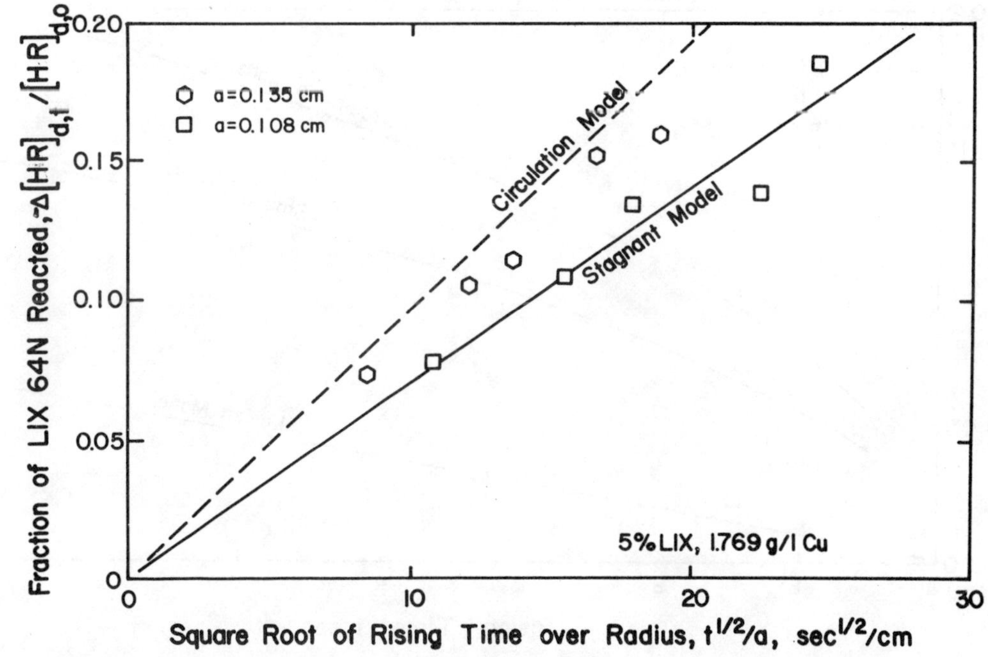

Figure 13. Comparison of experimental results with theoretical mass transfer showing the drop size effect.

Experiments were run to test this hypothesis by selecting concentration ratios of 0.14 and 0.027. When the concentration ratio was 0.14, the experimental results, as expected, followed the stagnant model of internal mass transfer with results similar to those presented previously. However when the concentration ratio was lowered to 0.027, the rate data did not fit the dispersed phase transport model, but was found to fit the Levich model for external mass transfer as shown in Figure 15.

Long cell experimental results. In order to con-

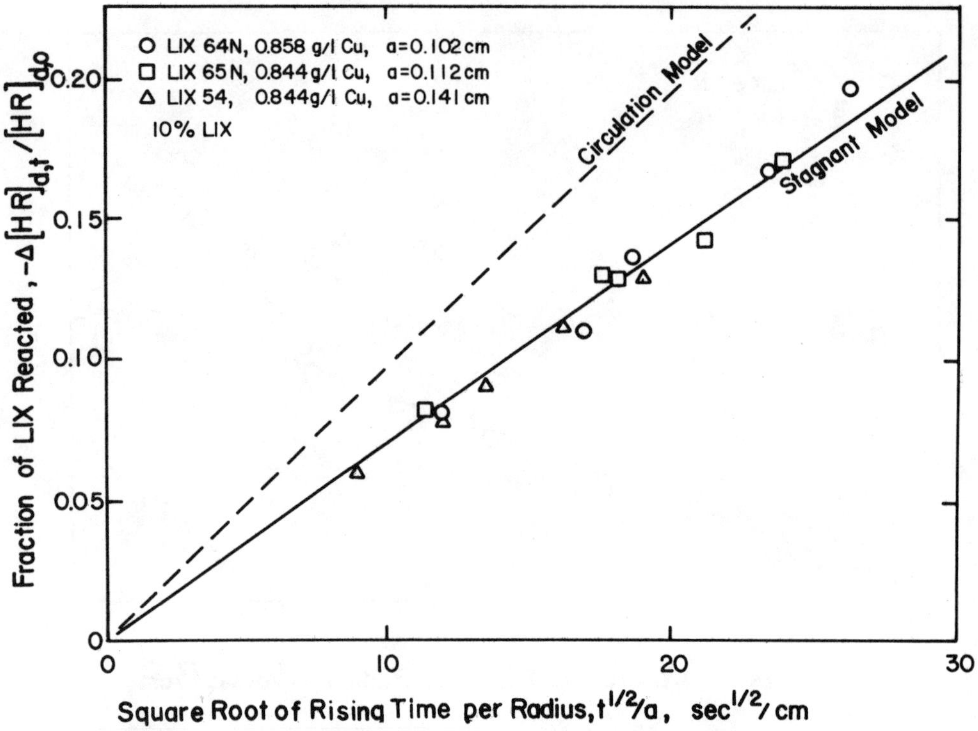

Figure 14. Parabolic plot of rate data for different chelating extractants. Note that the kinetics are essentially independent of the extractant and that 'no' catalytic effect from LIX 63 is observed.

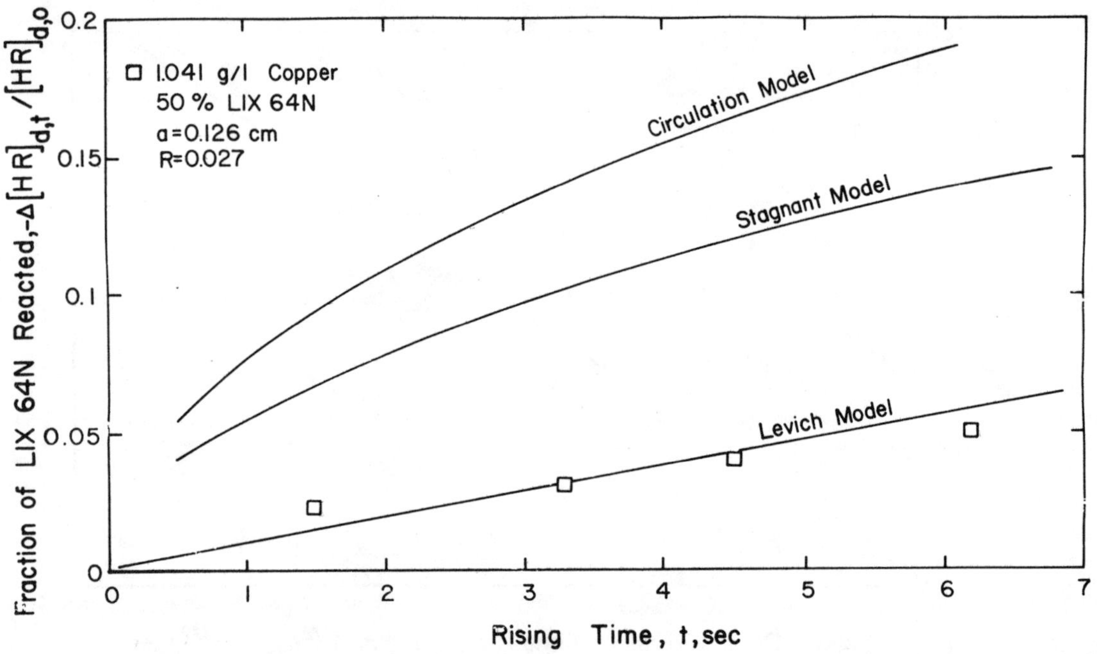

Figure 15. Demonstration of change in rate control from dispersed phase mass transfer limiting to continuous phase mass transfer limiting for small values of the parameter R defined as $[Cu]_{c,o}/[HR]_{d,o}$. For $R < 0.05$ reaction rates should be limited by reactant transport in the continuous phase. See Figure 2.

firm the results obtained with short cells, experiments were performed in which the main objective was to achieve a greater extent of reaction by longer contact time. Figure 16 shows the results obtained compared with those results from shorter cell experiments under similar conditions. These results show

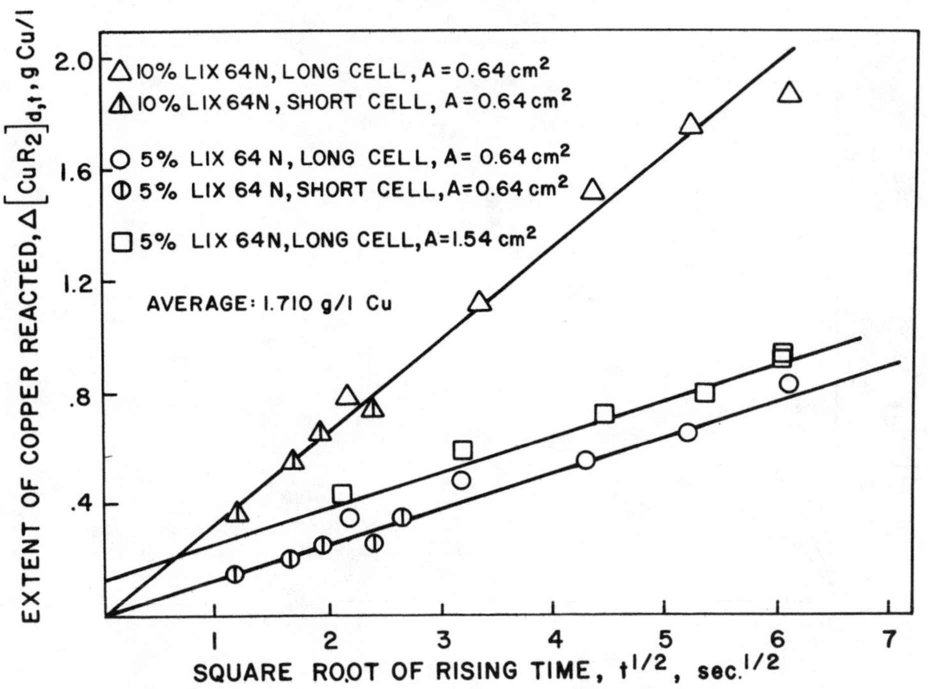

Figure 16. Comparison of experimental results obtained from long and short cells for similar experimental conditions. The effect of the aqueous/organic interface area is shown. Note that extrapolation of the curve with larger interfacial area to the ordinate at time zero determines the extent of reaction which can be attributed to the end effects.

that the rate curves are parabolic, consistent with the results of the shorter cell experiment, indicating that the rate controlling step does not change even for longer contact times. In another set of experiments the coalscence interfacial area was increased from 0.64 cm^2 to 1.54 cm^2 and under these circumstances a signifcant end effect can be detected.

Falling Drop Experiments

In this series of experiments the aqueous phase and the organic phase were reversed. The aqueous phase became the dispersed drops and the organic solution became the continuous phase. Under these circumstances, of course, the aqueous drops fell through the continuous organic phase. All data from falling drop experiments are represented by closed symbols.

Effect of copper and LIX 64N concentration. If the reaction were still controlled by mass transfer in the dispersed phase as predicted from arguments associated with the construction of Figure 2, the rate should be dependent on the dispersed aqueous phase copper concentration and independent of the continuous organic phase LIX 64N concentration. The falling drop results, shown in Figures 17 and 18 indicate the rate increases as the copper concentration increases (Figure 17) and that the rate does not change as the LIX 64N concentration increases (Figure 18). Again, as for the rising drop experiments, the rate follows parabolic kinetics for both the small drop ($\bar{a}$ = 0.087 cm) experiments, results of which are shown in Figure 19, and the large drop ($\bar{a}$ = 0.128 cm) experiments, the results of which are shown in Figure 20. Analysis of the rate constants shows that the kinetics are first order with respect to the copper concentration for both drop sizes and suggests that the rate is still limited by reactant transport in the dispersed phase even when the phases are inverted.

Comparison of theory with experimental results. The mass transfer equations for the stagnant model and circulation model can still be applied for the falling drop. However, the diffusing species is no longer the hydroxyoxime molecule but the copper-ammonia complex. The diffusion coefficient for the hydrated cupric ion has been reported in the literature to be 7.5 x 10^{-6} cm^2/sec (37). If the diffusion coefficient of copper ammonia complex is assumed to be the same as the hydrated cupric ion, the reaction rate can be predicted from first principles. Using Equations 11 and 14 the comparison of experimental data with theoretical predictions can be made and is shown in Figures 21 and 22 where close agreement between theory and experiment is observed. Figure 21 shows that as the copper concentration decreases the drops tend to circulate more. A similar effect was seen for rising drops when the hydroxyoxime concentration was reduced. However, for the same experimental conditions, the falling aqueous drops tend to circulate more than the rising drops. Compare Figure 21 with Figure 12. This may be due to a viscosity effect, more circulation for lower viscosity drops. The viscosity of the aqueous drop for these conditions is about 1.0 cp while the viscosity of the organic drop is about 1.7 cp.

The effect of the drop size on the kinetics of extraction in the falling drop experiments is shown in Figure 23. Again the results appear to follow the anticipated hydrodynamic behavior in which the larger the drop the greater the circulation.

The combined effect of both concentration and drop

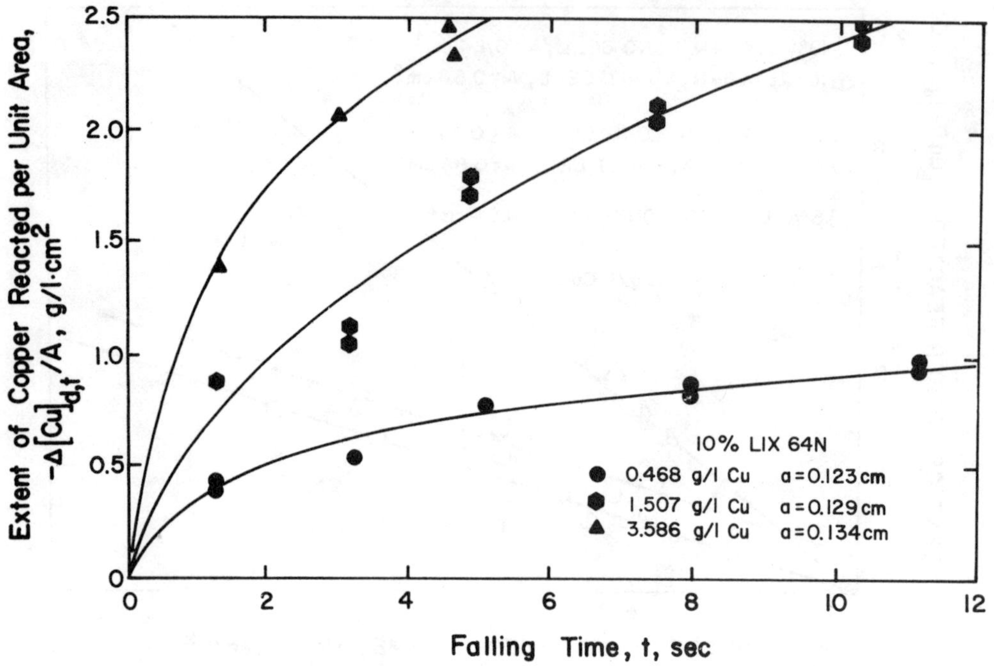

Figure 17. Copper extraction from large falling drops for various initial copper concentrations.

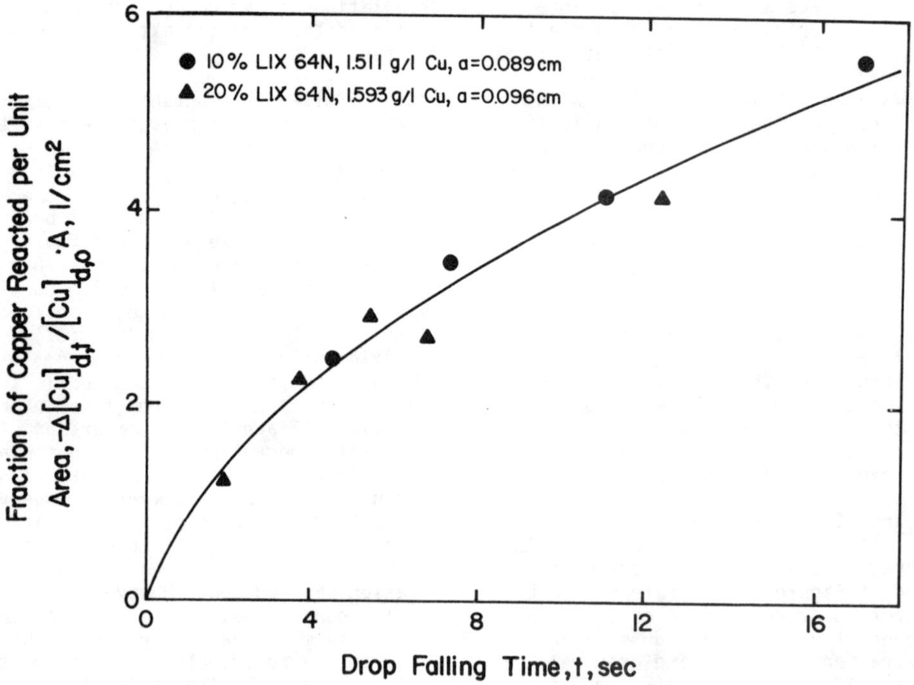

Figure 18. Copper extraction from falling drops for various initial LIX 64N concentrations.

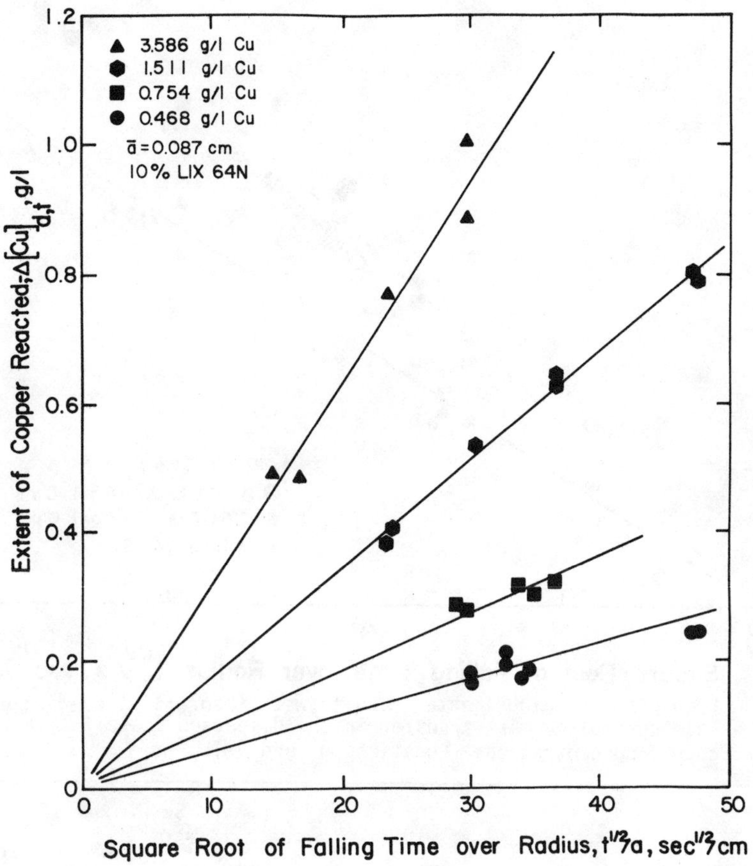

Figure 19. Parabolic plot of rate data for small falling drops illustrating that the kinetics conform to the linear relationship required for rate control by mass transfer in the dispersed phase.

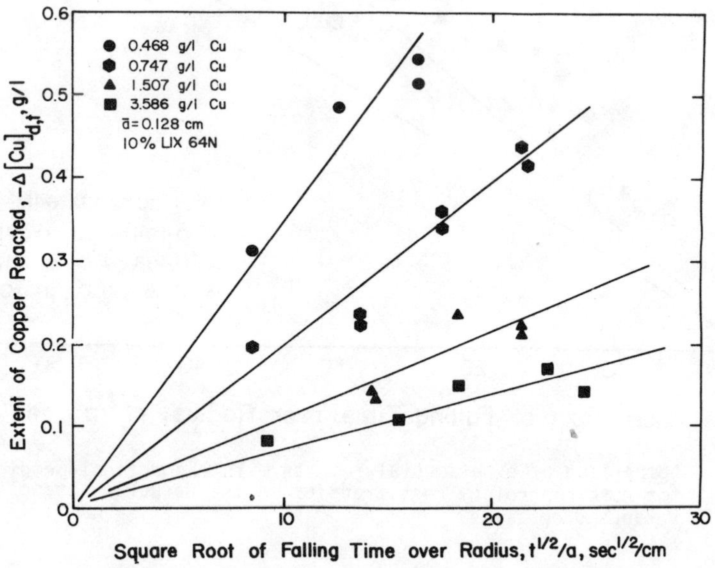

Figure 20. Parabolic plot of rate data for large falling drops illustrating that the kinetics conform to the linear relationship required for rate control by mass transfer in the dispersed phase.

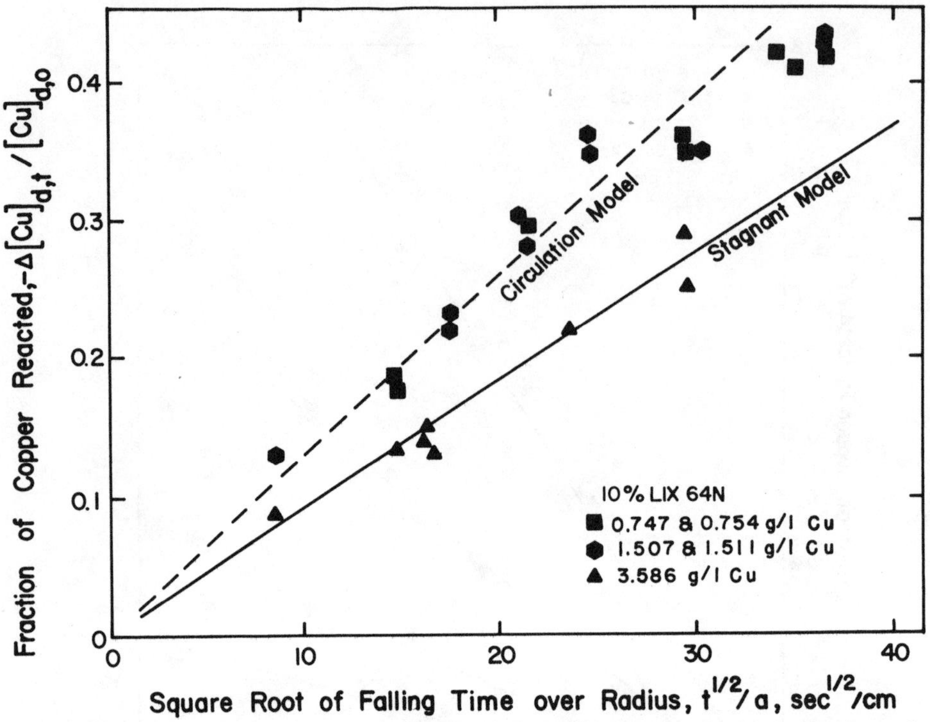

Figure 21. Comparison of experimental results with theoretical predictions for rate control by mass transfer in the dispersed phase for various copper concentrations (Equations 11 and 14).

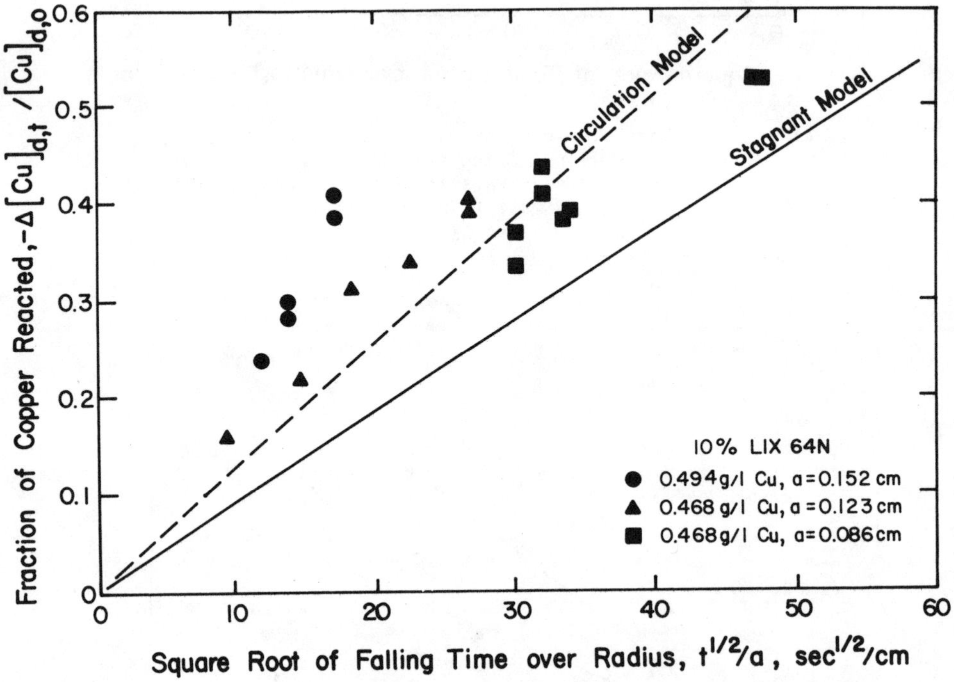

Figure 22. Comparison of experimental results with theoretical predictions for rate control by mass transfer in the dispersed phase for various drop sizes.

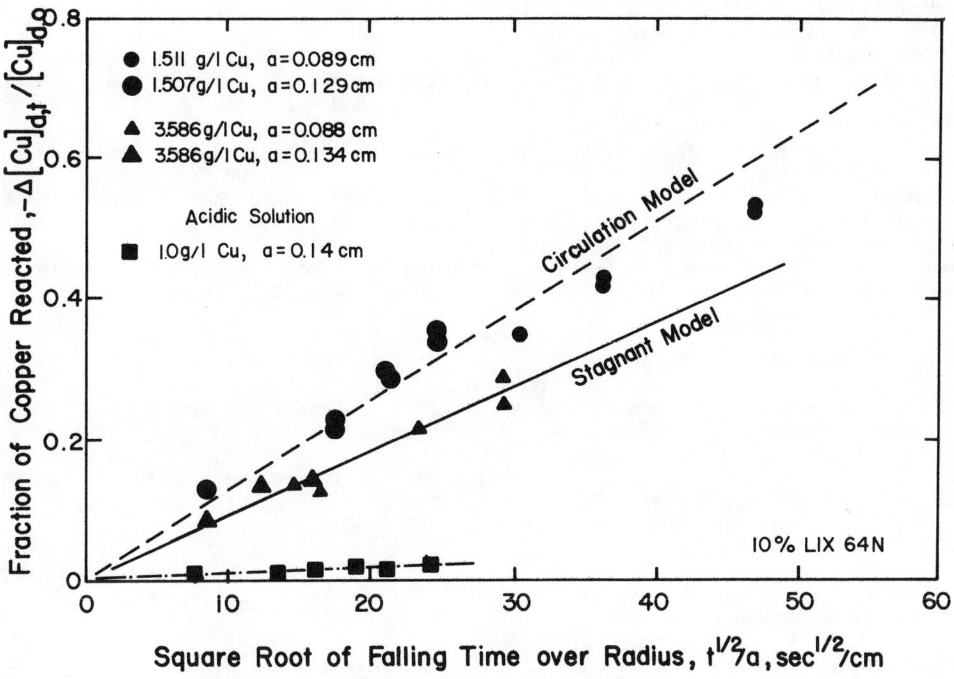

Figure 23. Combination effect of drop size and concentration on extraction kinetics compared with theoretical prediction for dispersed phase reactant mass transfer control. The kinetics of extraction in acidic solution is also shown which illustrate the rate is slower and not controlled by mass transfer processes.

size on the kinetics of extraction is shown in Figure 23. As the concentration decreases or drop size increases the amount of circulation increases.

All the experimental results provide strong evidence that extraction of copper by LIX 64N from ammoniacal solution normally will be controlled by transport of the reactant in the dispersed phase. This result is quite different from the kinetics of copper extraction from acidic solution, a reaction which has been shown to be limited by a surface chemical reaction (9,10). The difference in the reaction rate between extraction from ammoniacal solution and from acidic solution for similar experimental conditions is also shown in Figure 23 and the rate of copper extraction from ammoniacal solution is about one order of magnitude greater than extraction from acidic solution. Because the extraction rate in acid systems is much slower than the slowest type of mass transport, as predicted from the stagnant model, as well as other strong evidence, such as reaction orders, etc., it is clear that the extraction rate in acid systems is limited by a surface chemical reaction. The reason for the slow surface chemical reaction in acidic media is undoubtedly due to the contribution from the back reaction kinetics which is not a factor in ammoniacal systems. In addition, the high polarity of coordinated water molecules in the hydration sheath of the cupric ion may inhibit the extraction rate, in acid systems via a hydrogen bond linkage with the aqueous phase. On the other hand, displacement of the much less polar, coordinated ammonia molecules may represent a significantly lower resistance which facilitates the rate of reaction to such an extent that the rate becomes limited by reactant transport.

SUMMARY AND CONCLUSIONS

The equilibrium state and the kinetics involved in copper extraction from ammoniacal solutions by hydroxyoxime extractants have been studied. The equilibrium study involved shake-out tests, infrared analysis, as well as chemical analysis. The kinetic experiments were accomplished in a single drop reaction cell of variable length. The experimental results and data analysis suggest the following conclusions:

1. Copper extraction from ammoniacal solution by hydroxyoxime extractants involves a complete displacement of ammonia ligands from the coordination sphere of the cupric ion. The chemical reaction can be written as,

$$Cu(NH_3)^{+2}_{i(aq)} + 2HR_{(o)} = CuR_{2(o)} + iNH_{3(aq)} + 2H^+_{(aq)}$$

2. The extraction coefficient is sensitive to the ammonia concentration and the pH of the system, which various cupric ammine complexes,

$$\log E = \log K - \log(1 + \sum_{i=1}^{6}[NH_3]^i_{(aq)}) + 2\log[HR]_{(o)} + 2pH$$

3. Theoretical analysis and the experimental results from single drop experiments indicate that the kinetics of extraction appear to be controlled by reactant mass transfer in the dispersed phase (irrespective of which phase is dispersed), a conclusion which is supported by a low apparent activation energy of 5.3 kcal/mole.

4. The mass transfer process is explained to be

due to molecular diffusion for small drop sizes, while convective transport contributes to the reaction rate for larger drop sizes provided the viscosity of the dispersed phase is not too high.

5. As a result, rate of copper extraction from ammoniacal solutions is significantly faster than from acidic solutions in which the rate is known to be controlled by a chemical reaction at the liquid/liquid interface. The reason for the low extraction kinetics in the acidic system is the significant contribution of back reaction kinetics to the overall rate of reaction which is not significant in ammoniacal systems.

REFERENCES

1. M.B. Shirts, P.A. Bloom and W.A. McKinney, "Double Roast-Leach Electrowinning Process for Chalcopyrite Concentrates," USBM R.I. 7006 (1975).

2. L.W. Beckstead and J.D. Miller, "Ammonia, Oxidation Leaching of Chalcopyrite-Reaction Kinetics," Met. Trans. AIME, 8B, 19-29 March (1977).

3. "In Clean-Air Production Aribiter Process is First Off the Mark," EM/J, 174, No. 2, 74 (1973).

4. R.D. Groves, T.H. Jeffers and G.M. Potter, "Leaching Coarse Native Copper Ore with Dilute Ammonium Carbonate Solution," Solution Mining Symposium, AIME, 381 (1975).

5. M.C. Kuhn, N. Arbiter and H. Kling, "Anaconda's Arbiter Process for Copper," CIM Bulletin, February 62-69 (1974).

6. W. Kunda, H. Veltman and D.J.I. Evans, "Production of Copper from the Ammine Carbonate System," Copper Metallurgy, ed. by P. Ehrlich, AIME Symposium on Copper Metallurgy, Denver, Colorado, Feb. 27 (1970).

7. R.L. Atwood and J.D. Miller, "Structure and Composition of Commercial Copper Chelate Extractants," SME/AIME, Trans., 254, 319 (1973).

8. J.D. Miller and R.L. Atwood, "Discussion of the Kinetics of Copper Solvent Extraction with Hydroxy Oximes," J. Inorg. Nucl. Chem., 37, 2539-2542 (1975).

9. D.R. Spink and D.N. Okuhara, "Comparative Equilibrium and Kinetics of KELEX 100/120 and LIX 63/65N/64N Systems for Extraction of Copper," International Symposium on Hydrometallurgy, AIME, 497 (1973).

10. R.L. Atwood, D.N. Thatcher and J.D. Miller, "Kinetics of Copper Extraction from Nitrate Solutions by LIX 64N," TMS/AIME Trans., 6B, 465 (1975).

11. D.S. Flett, D.N. Okuhara and D.R. Prink, "Solvent Extraction of Copper by Hydroxy Oximes," J. Inorg. Nucl. Chem., 35, 2471 (1973).

12. L.G. Sillen and A.E. Martell, Stability Constants of Metal-Ion Complexes, Chemical Society, London, Special Publication No. 17, 152 (1964).

13. C.A. Fleming, "The Kinetics and Mechanism of Solvent Extraction of Copper by LIX 64N and KELEX 100," N.I.M. Report Johannesburg, S. Africa, No. 1793 (1976).

14. S. Hu and R.C. Kinter, "The Fall of Single Liquid Drops Through Water," AIChE, 1, No. 1, 42 (1955).

15. R.C. Kinter, "Drop Phenomena Affecting Liquid Extraction," Advances in Chemical Engineering, Drew ed., Academic Press, 4, 51 (1963).

16. A.I. Johnson and L. Braida, "The Velocity of Fall of Circulating and Oscillating Liquid Drops Through Quiescent Liquid Phase," Can. J. Chem. Engr., Dec., 165 (1957).

17. H. Lamb, Hydrodynamics, Sixth ed., Cambridge Univ. Press, 450 (1932).

18. W.N. Bond and D.A. Newton, "Bubbles, Drops and Stokes' Law," Phil. Mag. 6, 151, 794 (1927).

19. J. Crank, The Mathematics of Diffusion, Clarendon Press, Oxford, 2 ed., 89 (1975).

20. R. Kronig and E.R. Brink, "On the Theory of Extraction from Falling Droplet," App. Sci. Res. A2, 143 (1950).

21. P.W. Berg and J.L. McGrego, Elementary Partial Differential Equation, Holden-Day, San Francisco, (1969).

22. H. Margenau and G.M. Murphy, The Mathematics of Physics and Chemistry, D. Van Nostrand, New York, 2nd ed., 253 (1943).

23. A.I. Johnson and A.E. Hamielec, "Mass Transfer from Large Inside Drops," AIChE J., 6, No. 1, 145, (1960).

24. D.M. Rose and R.L. Kintner, "Mass Transfer from Large Oscillating Drops", AIChE, J., 12, 530 (1966).

25. V.G. Levich, Physicochemical Hydrodynamics, English Translation, Prentice-Hall, 395 (1962).

26. A.H.P. Skelland and A.R.H. Cornish, "Mass Transfer from Spheroids to an Air Stream," AIChE, J., 9, 73 (1963).

27. F.H. Garner and M. Tayeban, "The Importance of the Wake in Mass Transfer from both Continuous and Dispersed Phase System I., "Anales Real Sco. Espan. Espan. Fis. Quim. Sev. B-Quim, 56B, 479 (1960).

28. F.B. West, P.A. Robinson, A.C. Morgenthaler, T.R. Beck and D.K. McGregor, "Liquid-Liquid Extraction from Single Drops," Ind. Eng. Chem., 43, 243 (1951).

29. W. Light and W.F. Pansing, "Solute Transfer from Single Drops in Liquid-Liquid Extraction," Ind. Eng. Chem. 45, 1885 (1953).

30. H. Groothius and H. Kramers, "Gas Absorption by Single Drops During Formation," Chem. Engr. Sci., 4, 17 (1955).

31. A.T. Popovich, R.E. Jervis and O. Tross, "Mass Transfer During Single Drop Formation," Chem. Engr. Sci., 19, 357 (1964).

32. P.M. Heertjes, W.A. Holve and H. Talsma "Mass Transfer between Isobutanol and Water in Spray-Column" Chem. Eng. Sci., 3, 122 (1954).

33. Preliminary Report, "LIX 54 - A New Reagent for Metal Extractions from Ammoniacal Solutions," General

Mill's Chemicals, Inc., Mineral Industries, 2030 East Broadway, Tucson, Arizona 85719.

34. N.M. Rice, M. Nedved and G.M. Ritcey, "The Extraction of Nickel from Ammoniacal Media and its Separation from Copper, Cobalt and Zinc Using Hydroxyoxime Extractants, "Hydrometallurgy, 3, No. 1, 35 (1978).

35. J.B. Scuffham and G.A. Rowdon, "Solvent Extraction of Metals from Ammoniacal Solutions," Mining Engineering, SME/AIME, 25, No. 12, p. 33, Dec. (1973).

36. C.R. Wilke and P. Chang, "Correlation of Diffusion Coefficients in Dilute Solutions, "AIChe, J., 1, No. 2, 264 (1955).

37. Y.H. Li and S. Gregory, "Diffusion of Ions in Sea Water and in Deep-Sea Sediments," Geochimica et Cosmochimica Acta, 38, 703 (1974).

Chapter 21

EXPERIMENTAL METHODS IN SOLVENT EXTRACTION KINETICS: USE OF THE LIQUID JET RECYCLE REACTOR FOR LIX65N-COPPER (II) CHELATION KINETICS

R. W. Freeman

Department of Chemical and Biochemical Engineering
Rutgers-The State University of New Jersey
New Brunswick, New Jersey 08903

L. L. Tavlarides

Department of Chemical Engineering
Illinois Institute of Technology
Chicago, Illinois 60616

ABSTRACT

The experimental methods used in obtaining kinetic data for solvent extraction systems can affect the validity of the results. The method of contacting the two liquids should allow for accurate determination of the surface area, hydrodynamics, and diffusional processes. The appropriateness of the Liquid Jet Recycle Reactor is discussed and the modelling of the mass transfer and chelation reaction is presented. The LJRR is used to investigate the chelation of copper by LIX65N in toluene from aqueous copper sulfate sulfuric acid solutions. The ranges of experimental variables are: Copper, 1.8-13.8 g-dm^{-3}, sulfuric acid 0-1.6 g-dm^{-3}, LIX65N, 0.1-1.0 mol-dm^{-3}. Data on the initial rate of copper chelation at the interface is generated. Nonlinear regression analysis shows that the following expression adequately describes the data within experimental error.

$$\text{rate} = \frac{5.69\times10^{-6}[\text{Cu}^{++}][\text{LIX65N}]}{1 + 1.43\times10^{2}[\text{H}^{+}]^{\frac{1}{2}}[\text{LIX65N}]^{\frac{1}{2}}}$$

where $[Cu^{++}]$ and $[H^{+}]$ are the concentrations of the ionic species. The above rate expression is derived as an intrinsic interfacial rate expression, free of diffusional or surface area contributions.

EXPERIMENTAL METHODS IN SOLVENT EXTRACTION KINETICS: USE OF THE LIQUID JET RECYCLE REACTOR FOR LIX65N-COPPER (II) CHELATION KINETICS

R. W. Freeman

Department of Chemical and Biochemical Engineering
Rutgers-The State University of New Jersey
New Brunswick, New Jersey 08903

L. L. Tavlarides

Department of Chemical Engineering
Illinois Institute of Technology
Chicago, Illinois 60616

INTRODUCTION

The intrinsic kinetics of the solvent extraction of metals can play a significant role in the modelling, design and evaluation of hydrometallurgical solvent extraction operations. In order to be useful, the experimental determination of intrinsic kinetic parameters must be able to distinguish true chemical reaction effects from diffusional resistances, surface area dependencies, convective contributions and the effect of interfacial impurities. A wide variety of liquid-liquid contacting devices have been used to study solvent extraction processes, as illustrated in Table 1. The liquid jet technique has been employed with some success in the study of the liquid diffusitivites of gases (Chaing 1954, Scriven 1959) and for some liquid-liquid systems (Quinn 1961, Fosberg 1967) the version used in this study, the liquid jet recycle reactor (LJRR), is particularly suited to the study of interfacial reactions as might occur in the chelation or complexation of metal ions during hydrometallurgical extractions. The recycle feature allows small amounts of extractant to contact an aqueous solution through a clean, well defined interface for up to several hours. This allows acquisition of meaningful data even under conditions where mass transfer is an order of magnitude slower than diffusional processes, a condition which can occur for some extractants and/or experimental conditions. The inherent advantages of the liquid jet technique are retained in this configuration and allow the rate of interfacial transport to be determined independent of surface area, diffusional and convective contributions.

The LJRR is used here to study the kinetics of the solvent extraction of copper by the commercially important oxime extractant LIX65N*. Copper is extracted from aqueous solutions via a chelation reaction in which two molecules of LIX65N complex the metal ion, making it soluble in the organic phase. Although this system has received considerable attention in the literature (Spink 1974, Flett 1973, Atwood 1975, Fleming 1978) the experimental methods employed have some doubt as to whether the rate data obtained were free from the effects of diffusion, convection or interfacial blocking by stagnant films. The LJRR provides a means of obtaining data on the intrinsic rate of the copper chelation reaction.

*Product of General Mills Chemicals

TABLE 1. Experimental Contactors for Liquid-Liquid Kinetic Data

Contactor	Description	Comments	References
AKUFVE	Stirred Cell	Surface area and mixing ill defined, hydrodynamics unknown	Spink 1974 Flett 1973
Lewis Cell	Fluids layered in cell, each layer stirred separately	Accumulation of surfactants, surface waves complicate hydrodynamics, data scatter $\pm$ 30%	Olander 1964 Bulicka 1976 Fleming 1978
Diaphragm Cell	Fluids partitioned by diaphragm	Surface area unknown in porous membrane	Onda 1975
Single drop	Drop falls or rises in liquid	Three hydrodynamic regimes too complicated for simplified analysis, surfactant contamination	Atwood 1975 Whewell 1975
Growing drop	Drop formation in liquid	Complex hydrodynamics, surface contamination	Bauer 1974
Liquid jet	Jet of one fluid passes through second fluid	Short contact times, well defined interfacial area	Quinn 1961 Fosberg 1967 Freeman 1980

APPARATUS AND EXPERIMENTAL PROCEDURE

The liquid jet apparatus as employed here operates with a downward flowing jet of aqueous solution surrounded by a cocurrently flowing stream of the organic phase extractant. Figure 1 is a photograph of the jet in operation. The aqueous phase contains the

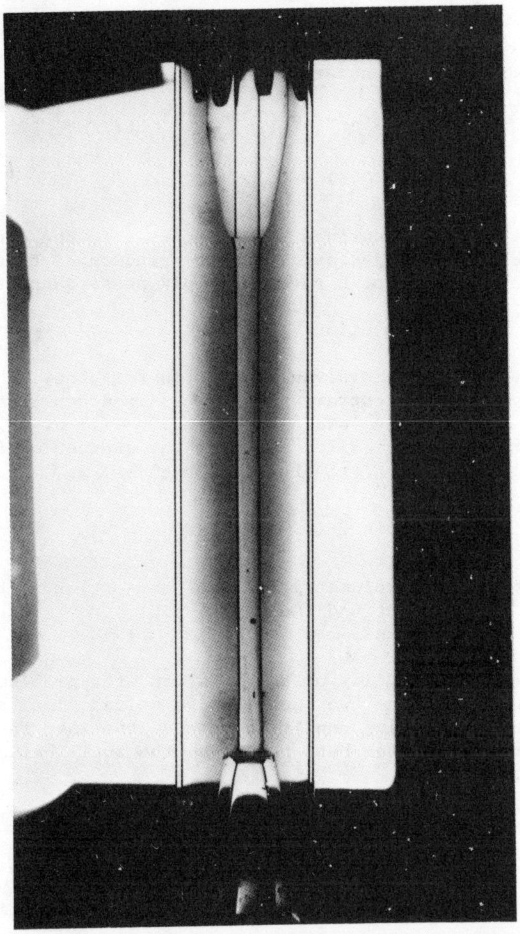

Figure 1. Photograph of jet in operation. Aqueous flow 85 cm^3-min^{-1}.

metal ions which are extracted across the interface between the jet and the outer fluid.

Figure 2 is a schematic of the Liquid Jet Recycle Reactor apparatus as employed in this work. The recycle path for the organic extractant is readily apparent from the illustration and contains approximately 30 mℓ of fluid. The aqueous phase is fed from the reservoir by gravity down through the constant head device, rotameter, and the heat exchanger (to maintain constant temperature operation) and into the jet chamber. The organic phase is pumped in a loop through a heat exchanger to the jet chamber where it flows cocurrent to the jet. Leaving the jet chamber, the organic phase passes through a recording UV-visible spectrophotometer equipped with a flow cell and back to the pump. The aqueous phase leaves the jet chamber and goes to a back pressure regulating device. The constant head vessel, jet chamber and the water bath in which it is immersed are isolated from vibration.

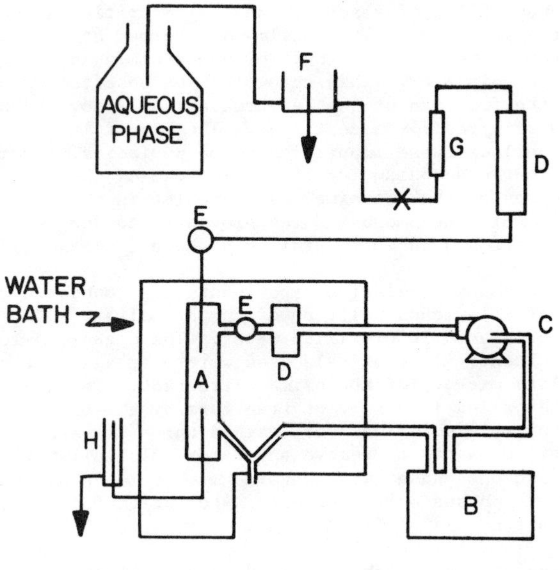

Figure 2. Schematic of the Liquid Jet Recycle Reactor. A) jet chamber, B) UV-Visible spectrophotometer, C) recirculating pump, D) constant head device, G) rotameter, H) variable head backpressure regulator.

The jet chamber is glass tubing with various inlet and outlet tubes fused onto it. The nozzle consists of a length of precision bore glass tubing with the end tapered. The nozzle tip is free from irregularities when observed under 20X magnification. The jet receiver is constructed from glass tubing flame tapered and countersunk at a 45° angle. To avoid contamination of the fluids, all wetted surfaces, including the gear pump for the organic phase are either teflon or glass.

An experiment is conducted by first establishing the jet flow and then pumping the organic fluid into the jet chamber while adjusting the back pressure regulator device to maintain the jet. After the annulus is filled, any aqueous phase which overflowed the receiver during the start up is drained out through the drain. The outlet of the jet chamber is then connected to the spectrophotometer, closing the loop path, and the rate of recycle adjusted to the desired value. The accumulation of the organic phase copper complex is monitored spectrophotometrically, using the visible light absorption peak at 666 nm.

Accurate determination of the interfacial area available for mass transfer is accomplished by taking photographs from a rigid isolated camera mount so that focal distance will remain unchanged. To eliminate measurement errors of the actual jet dimension due to curvature effects, a stainless steel rod machined to various diameters is inserted in the position of the jet and photographed under optically identical conditions. Comparison of these known diameters with photographed diameters of the jet gives the actual jet dimensions. The total organic phase volume is determined by calibration of the apparatus with known volumes of liquid. The flux through the interface, N, is simply related to the rate of accumulation of the transfering species by the material balance equation

$$N = \frac{\text{volume organic}}{\text{area of jet}} \frac{dC_j}{dt} \tag{1}$$

The LJRR has the ability to deal with systems where stagnant films of reactants and/or impurities gather at the interface. These films are formed at the bottom of the jet and their presence causes a change in the jet diameter. This change can be observed visually and the fraction of the jet area actually covered by this film readily ascertained. The liquid jet technique also allows these impurities to be periodically removed. A slight underbalancing of the jet causes the film to rip away and be entrained as a droplet in the aqueous jet. This phenomena and the procedure to deal with it are described in more detail elsewhere (Freeman 1977).

The copper chelation experiments are conducted using a synthetic leach solution of copper sulfate and sulfuric acid in distilled water. The organic diluent is toluene. All chemicals used were of reagent grade quality except for the oxime extractant. The LIX65N was supplied in a solvent free form by the manufacturer. The organic solutions were washed three times with distilled water to remove any water soluble impurities. The aqueous phases were satured with toluene and the organic phases saturated with water prior to experimentation.

The values for the independent concentration variables are varied in accordance with a central composite experimental design. The copper concentration is varied from 1.75 g-dm^{-3} to 13.75 g-dm^{-3}, the acid from zero to 1.55 g-dm^{-3} and the extractant from a nominal 0.1 mol-dm^{-3} to 1.0 mol-dm^{-3}.

ANALYSIS

Analysis of mass transfer and mass transfer with interfacial reaction for the liquid jet configuration can yield simple, useful expressions through the use of appropriate modelling.

Hydrodynamics

Rigorous solution of the hydrodynamics of liquid-liquid jets presents formidable difficulties (see Yu 1975). However, for the limiting case of a completely relaxed jet with cocurrent flow, an analytical solution can be obtained (Freeman 1977). This solution relates the surface velocity to the average jet velocity through

$$\beta = \text{surface velocity/avg. jet velocity}$$

$$= \frac{a - bQ^o/Q^a}{a-c} \qquad (2)$$

where $a = 2\mu^a[M^4 - 2M^2 + 1 - M^4\ell n(M) + \ell n(M)]$; $b = \mu^0[M^2 - 1 - 2\ell n(M)]$; $c = \frac{1}{2}\mu^0[M4 - 4M2 + 3 + 4\ell n(M)]$. M is the ratio of the chamber radius to the jet radius, μ the viscosity. Inherent in the assumption of a relaxed jet are the assumptions of a non-accelerating flowfield and constant jet diameter.

Penetration Theory Formulation

Assumption of a relaxed jet greatly simplifies the hydrodynamics and allows the penetration theory approach to be used in analysis of the mass transfer. For short contact times as are found in the liquid jet device, the depth of penetration is small, allowing planar geometry to be adopted and the effect of velocity gradients to be neglected. Assuming uncoupled diffusion perpendicular to the jet axis, the conservation equations for the jth species can be written

Aqueous Phase

$$U_s \frac{\partial C_i^a}{\partial x} = \mathcal{D}_j^a \frac{\partial^2 C_j^a}{\partial y^2} \qquad (3)$$

Organic Phase

$$U_s \frac{\partial C_i^o}{\partial x} = \mathcal{D}_j^o \frac{\partial^2 C_j^o}{\partial y^2} \qquad (4)$$

subject to the boundary conditions

$$C_j(x, \pm\infty) = C_j^b \qquad (5)$$

$$C_j(0,0) = C_j^b \qquad (6)$$

where U_s is the surface velocity, x is in the direction parallel to the jet axis, y is perpendicular to the jet axis. Superscript b refers to bulk phase quantities.

Mass Transfer Solution

In a process involving mass transfer alone, the interfacial concentration can be assumed constant. This assumption allows either equation (3) or (4) to be solved analytically, yielding the expression for W, the molar rate of transfer from the jet, as

$$W = -4\sqrt{\mathcal{D}\beta QL}\ (C^b - C^s) \qquad (7)$$

where Q is the volumetric flow of the jet, L is the length of the jet and the superscript s represents the interfacial values.

Complete solution of the mass transfer problem requires that the two equations be solved simultaneously. The usual coupling condition relates the interfacial concentrations for both phases by the equilibrium condition

$$[C^{os}/C^{as}] = m$$

where m is the distribution coefficient. The solution for the coupled equations is

$$W = \frac{4\sqrt{\mathcal{D}\mathcal{D}^a\beta QL}}{\sqrt{\mathcal{D}^a} + m\sqrt{\mathcal{D}^o}}\ (C^{ba} - mC^{bo}) \qquad (8)$$

The applicability of the above analysis to the LJRR has been verified experimentally (Freeman 1980). Experiments were conducted using systems of known characteristics and the analysis outlined above was able to predict the diffusivities of toluene and benzene in water to within 10% of the literature values.

Mass Transfer with Interfacial Reaction

The chelation of copper by the extractant molecules is considered to occur at the interface. The penetration theory formulation of equations (3) through (6) is still valid. The components diffusing through the organic phase are HR and CuR_2; through the aqueous phase, Cu^{++} and H^+. Equations (3) and (4) must now be coupled by the interfacial reaction rate

$$\mathcal{D}_j \left.\frac{\partial C_j}{\partial y}\right|_{y=0} \equiv N_j \equiv \text{rate of reaction at the interface} \qquad (9)$$

In general, the nonlinearity of the reaction rate expression precludes an analytical solution. However, for reaction rates slow as compared to diffusion, an approximate solution can be obtained.

The molar rate of mass transfer from the jet is given by

$$W = \int_0^L 2\pi r\, R(C_j^s(x))dx \qquad (10)$$

where $R(C_j^s(x))$ is the rate of surface reaction evaluated at each point x down the jet. Note that the interfacial concentration of the jth species, C_j^s is changing down the length of the jet. For reactions slow as compared to the diffusion, the interfacial concentrations will not change rapidly from the initial value. A logical first approximation under these conditions is to represent each C_j^s by an approximate constant value, say C_j^*, so that

$$W = R(C_j^*)\ \text{Area}$$

Equation (7) can then be used to provide a first estimate of C_j^*

$$C_j^* \approx C_j^b - W_j/4\sqrt{\mathcal{D}_j \beta Q L} \qquad (11)$$

This value for C_j^* does not, of course, make equation (11) exact. However, this value of C_j^b must lie between the extremes of $C_j^s(x)$ and so is a reasonable choice for a first approximation. Thus the solution to diffusion with interfacial reaction can be approximated in a simple way to yield first estimates of reaction rate parameters.

RESULTS

The kinetics of the copper chelation reaction are analyzed through the functionality of the flux across the interface. The flux for each experiment is calculated from the observed organic phase copper concentration versus time curve by use of equation (1). Figure 3 is an example of the data obtained and shows that the rate of accumulation is independent of the copper complex concentration over the range studied, as is expected for initial rate data. The maximum percent loading observed in these experiments is about 14%.

The effects of diffusional resistances are accounted for by analyzing the rate as a function of interfacial concentrations as estimated by use of equation (11). The difference between the estimated interfacial concentration and the bulk phase composition is at most 1.7% for the extractant, 2% for the copper and 8.3% for the hydrogen ion. Representative percent differences are 0.75%, 1.5% and 6.5% for extractant, copper, and hydrogen ion. These results indicate that the reaction rate is controlling and that the previous assumptions are valid. The bulk phase concentrations of the cupric and hydrogen ions in the aqueous phase are calculated in accordance with the copper sulfate and bisulfate ionic equilibria (Freeman, to be published). The organic phase extractant concentration is determined by the method of ultimate loading.

The overall dependence of the reaction rate on the three independent concentration variables is readily apparent from the data. The rate is approximately

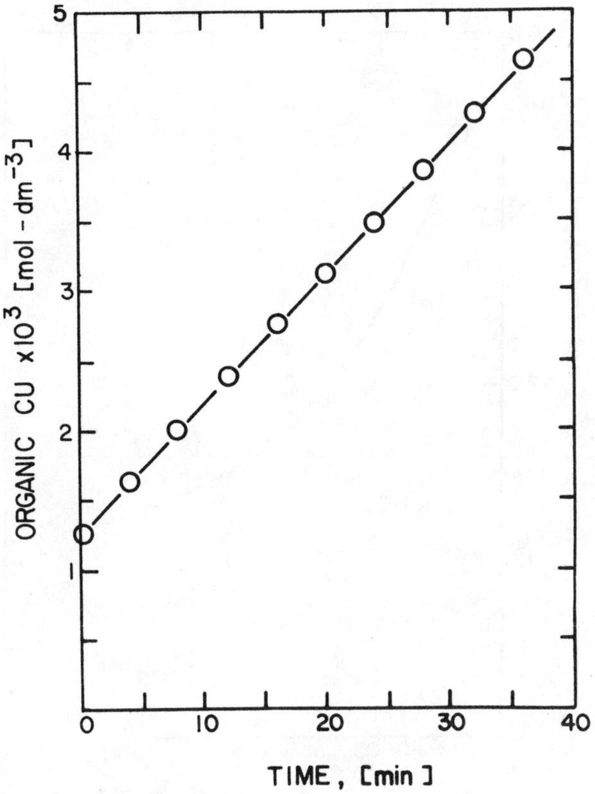

Figure 3. Accumulation of copper in the organic phase as a function of time.

first order with respect to copper ion concentration, somewhat less than first order with respect to the extractant concentration, and an inverse order with respect to the hydrogen ion concentration. Figure 4 illustrates the dependence on the hydrogen ion concentration. Detailed analysis of the experimental data is accomplished using least squares nonlinear regression techniques.

Reproducibility

It is necessary to have an adequate knowledge of the errors inherent in the experimental observations themselves in order to determine adequacy of fit for a particular rate model. To this end, a number of replicate experiments were performed at the central point in the experimental grid. For repeat runs using the same liquid solutions and flow conditions, the reproducibility of the LJRR reactor is excellent, about ±3% as illustrated by the results in Table 2. Experimental errors in phase composition result is an overall error of about ±6%.

Evaluation of Models

Evaluation of various models is performed using standard analysis of variance techniques. The number of observations and the inherent experimental error as discussed above indicate that an F test ratio of 2.9 or greater is sufficient for the rejection of a model at the 90% confidence level.

The models which have been proposed in the literature have been of the general power law form

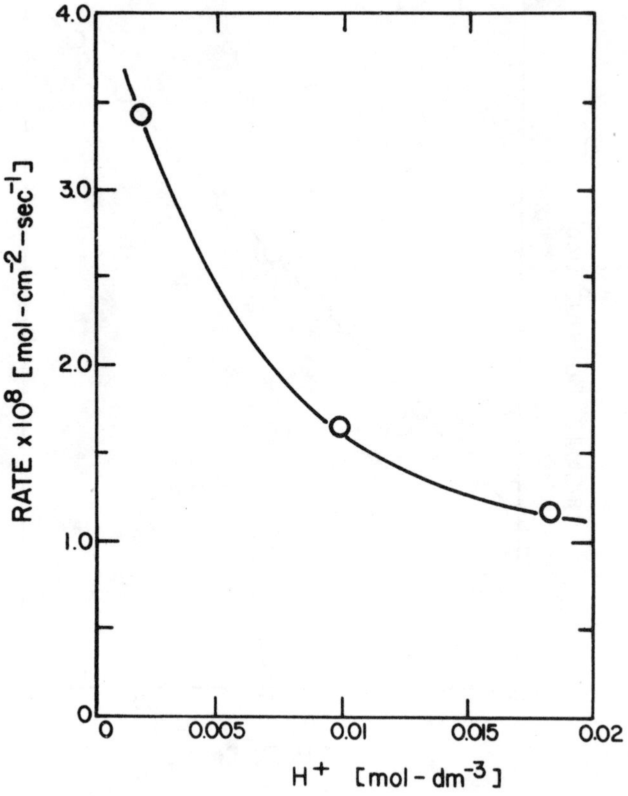

Figure 4. Rate of extraction as a function of interfacial hydrogen ion concentration $[Cu^{++}] = 0.125$, $[HR] = 0.55$.

TABLE 2. Reproducibility of Liquid Jet Recycle Reactor, Constant Liquid Compositions, Set 1; [LIX65N] = 0.57, $[CuSO_4] = 0.126$, $[H_2SO_4] = 0.0078$, Set 2; [LIX65N] = 0.52, $[CuSO_4] = 0.127$, $[H_2SO_4] = 0.0076$

No.	Rate × 10^8, Set 1 mol-cm^{-2}-sec^{-1}	Rate × 10^8, Set 2 mol-cm^{-2}-sec^{-1}
1	1.61	1.51
2	1.54	1.55
3	1.58	1.58
4	1.57	1.59
Average	1.58	1.56
Standard Deviation	0.03	0.04

$$r = k[Cu]^a[HR]^b/[H^+]^c \tag{12}$$

Spink (1974) proposed values of a = 1, b = 1 and c = 1. This model was tested and the F test ratio found to be 80.4, more than sufficient to reject the model. Fleming (1976) reported apparent orders of reaction such that a = 1, b = 1.1, c = .6. An improvement in fit for this model was noted, but the F test ratio of 16.1 indicated excessive lack of fit. The best fit power law expression was found to be

$$\text{rate} = 5.4 \times 10^{-8} \frac{[Cu]^{.962}[HR]^{.623}}{[H^+]^{.414}} \tag{13}$$

This model yields an F test ratio of 3.3, still sufficient for rejection.

Best Fit Model

The best fit model obtained at the present stage of the analysis is given below

$$\text{rate} = \frac{5.69\ 10^{-6}[Cu^{++}][LIX65N]}{1 + 1.43\times10^{2}[H^+]^{\frac{1}{2}}[LIX65N]^{\frac{1}{2}}} \tag{14}$$

The F test value for this model is F = 2.9, just the value necessary to preclude rejection at the 90% confidence level. This model was obtained by an empirical fit to the observed data, and the expression is not the result of any obvious mechanistic pathway. Further work to interpret the data in terms of mechanistic models is currently in progress.

CONCLUSIONS

The liquid jet recycle reactor is shown to be a useful tool in the study of metal extraction processes. The readily determined interfacial area, ability to simplify the hydrodynamics, the ability to observe and control the accumulation of surface impurities, and the short contact times combine to provide an effective experimental technique.

The method of obtaining intrinsic diffusion free rate data has been demonstrated and the data utilized in the evaluation of several models. The appropriateness of the analysis is enhanced by the short contact times inherent in the liquid jet technique. The calculated correction for the interfacial concentration is only a small fraction of the experimentally accessible bulk concentrations. The contact times of other contacting systems may well be one or two orders of magnitude greater than the 0.05 seconds of the liquid jet. For these larger times the interfacial concentrations could be substantially different from the bulk concentrations, placing greater demands on the diffusion modelling and casting serious doubt on the observed rate dependencies.

Interpretation of the LIX65N reaction rate data obtained with the LJRR indicates that the power law models are inadequate to explain the data. The best fit model for the rate of copper extraction given by equation (14) is not derived from any obvious mechanistic pathway. The models which assume the rate controlling step occurs during a dissociative or non-dissociative surface reaction with negligible sorption effects also do not appear to be able to satisfy the statistical requirements. Additional experimental work and analysis is in progress.

NOMENCLATURE

a,b,c	constants defined in equation (2)
C	molar concentration
CuR_2	copper-extractant complex
$\mathcal{D}$	diffusivity
HR	extractant molecule
M	ratio of jet chamber radius to jet radius
m	distribution ratio
N	flux [mol-cm^{-2}sec^{-1}]
Q	volumetric flow rate

R	rate of surface reaction
r	radius of jet
U	velocity in x direction
W	molar rate of transfer [mol-sec^{-1}]
x	direction parallel to jet axis
y	direction perpendicular to jet axis

Greek

β	ratio of surface velocity to average velocity
μ	viscosity

Superscripts

a	aqueous phase
b	bulk phase
o	organic phase
s	surface

Subscripts

j	component index

REFERENCES

Atwood, R.L., Thatcher, D.N., Miller, J.D., 1975, "Kinetics of Copper Extraction from Nitrate Solutions by LIX64N," Met. Trans. B., Vol. 6B, Sept., pp. 465-473.

Bauer, G., 1974, "Solvent Extraction of Copper: Kinetic and Equilibrium Studies," Ph.D. Thesis, University of Wisconsin-Madison, Madison, Wisconsin.

Bulicka, J., Prochazka, J., 1976, "Mass Transfer Between Two Turbulent Liquid Phases," Chem. Eng. Sci., Vol. 31, pp. 137-146.

Chaing, S.H., Toor, H.L., 1959, "Interfacial Resistance in the Absorption of O_2 by H_2O," AIChE J., Vol. 5, No. 2, pp. 165-168.

Fleming, C.A., 1976, Report No. 1793, National Institute for Metallurgy, Johannesburg, S. Africa.

Fleming, C.A., et al, 1978, "The Kinetics of the Extraction of Copper by LIX65N and the Catalytic Role of LIX63N in this Reaction," J. Appl. Chem. Biotechnol., Vol. 28, pp. 443-452.

Flett, D.S., Okuhara, D.N., Spink, D.R., 1973, "Solvent Extraction of Copper by Hydroxy Oximes," J. Inorg. Nucl. Chem., Vol. 35, pp. 2471-2487.

Fosberg, T.M., Heideger, W.I., 1967, "Interphase Mass Transfer in Binary Liquid Systems: Laminar Liquid Jets," Can. J. of Chem. Eng., Vol. 45, pp. 82-89.

Freeman, R.W., 1977, "Mass Transfer from a Liquid-Liquid Jet," M.S. Thesis, Illinois Institute of Technology, Chicago, Ill., 101 pp.

Freeman, R.W., Tavlarides, L.L., 1980, "Study of Interfacial Kinetics for Liquid-Liquid Systems I: The Liquid Jet Recycle Reactor," Chem. Eng. Sci., Vol. 35, pp. 559-566.

Freeman, R.W., to be published, "Kinetics of Liquid-Liquid Interfacial Reactions: Chelation of Copper by Acid Chelating Reagents," Ph.D. Thesis, Illinois Institute of Technology, Chicago, Ill.

Olander, D.R., Reddy, L.B., 1964, "The Effect of Driving Force on Liquid-Liquid Mass Transfer," Chem. Eng. Sci., Vol. 19, pp. 67-73.

Onda, K., et al, 1975, "Study of Mass Transfer Between Phases by a Diaphragm Cell, Physical Absorption of CO_2 into Water or Aqueous Electrolyte Solutions," J. Chem. Eng. Japan, Vol. 8, pp. 25-29.

Quinn, J.A., Jeannin, P.G., 1961, "Interfacial Resistance: Diffusion into a Laminar Liquid-Liquid Set," Chem. Eng. Sci., Vol. 15, pp. 243-250.

Scriven, L.E., Pigford, R.L., 1959, "Fluid Dynamics and Diffusion Calculations for Laminar Liquid Jets," AIChE J., Vol. 5, pp. 397-402.

Spink, D.R., Okuhara, D.N., 1974, "Comparative Equilibrium and Kinetics of an Alkylated Hydroxy Quinoline and a β-Hydroxy Oxime for the Extraction of Copper," Metall. Trans., Vol. 5, pp. 1935-1942.

Whewell, R.J., Hughes, M.A., Hanson, C., 1975, "Kinetics of the Solvent Extraction of Copper II within LIX Reagents: Single Drop Experiments," J. Inorg. Nucl. Chem., Vol. 37, pp. 2303-2307.

Yu, H., Scheele, G.F., 1975, "Laminar Jet Contraction and Velocity Distribution in Immiscible Liquid-Liquid Systems," Int. J. Multiphase Flow, Vol. 2, pp. 153-169.

INTERFACIAL FLOW ON A STATIC LIQUID DROP WITH FORCED INTERNAL CIRCULATION. PART I-BASIC FLOW.

S. B. Joo and Lawrence Burkhart

Ames Laboratory, U.S. DOE and Department of
Chemical Engineering, Iowa State University
Ames, Iowa

ABSTRACT

A hemispherical, constant volume drop with forced internal circulation was considered as a model approximating a real drop at a similar stage of formation. Axially symmetric steady flow within a drop contaminated by surfactants was described by solving the full Navier-Stokes equations written in terms of a meridian plane stream function. Rotational flow occurred inside the drop above a computed Reynolds number of 5.8, and the change in interfacial flow caused by imposing a surface-tension gradient at the surface had a significant effect on the internal flow.

Although the work is quite fundamental in nature, the rotational flow described in the paper is responsible for much of the bulk transport of solute to the interface in hydrometallurgical processes such as solvent extraction.

INTERFACIAL FLOW ON A STATIC LIQUID DROP WITH FORCED INTERNAL CIRCULATION. PART I-BASIC FLOW.

S. B. Joo and Lawrence Burkhart

Ames Laboratory, U.S. DOE and Department of Chemical Engineering, Iowa State University
Ames, Iowa

INTRODUCTION

Fluid motion within forming liquid drops has received considerable attention in single drop studies because of its pronounced effect on the rate of mass transfer during the period of drop formation. A qualitative description of this phenomenon was first reported by Garner and Skelland (1951), who found that flow patterns within forming drops are similar to those observed in drops falling or rising in a second fluid. However, early investigators were plagued by experimental difficulties in observation and measurement during such a brief event as drop formation. In an attempt to avoid these problems, Constan and Calvert (1963) developed a forced flow technique in which a drop of constant volume was formed by adding fluid through a center nozzle and removing it at the same rate through an annulus located at the outer edge of the nozzle. In a careful photographic study of the flow patterns within forming drops, Burkhart, et al. (1976) found that the flow patterns inside of a forming drop were indeed very similar to those observed in drops with forced flow created by the method of Constan and Calvert (1963).

This technique made it possible to observe motion of the type observed in forming drops and revealed a somewhat unexpected phenomenon on the surface of the drop. Ordered circulation patterns were found above certain Reynolds numbers. These axially unsymmetric circulation patterns were considered as a manifestation of hydrodynamic instability. The objective of the present study was to construct a mathematical description for the basic state of the stability analysis, i.e., a mathematical description of the state of the drop before any perturbations were introduced. In Part II the emergence of axially unsymmetric circulatory motion is predicted from a non-circulatory flow. Thus, the basic state is an axially symmetric steady flow.

Flow patterns of the basic state are strongly dependent upon the Reynolds number, which requires that both inertial and viscous terms be considered in solving the Navier-Stokes equations. Then, the governing equations for this problem become non-linear partial differential equations. In addition, it is necessary that the interface be contaminated by surfactants in order that the interfacial circulation may occur. As a result, the flow patterns must be computed and analyzed for different interfacial properties such as the surface-tension gradient and the surface viscosity.

STATEMENT OF THE PROBLEM

A liquid drop of constant volume is assumed to have been formed on concentric nozzles. Flow within the drop is created by continuously injecting liquid through the center nozzle and withdrawing it at an equal rate from the outer annulus. The flow field is assumed to be axially symmetric so that there is no circumferential component of the velocity. Under these conditions the flow patterns can be described in terms of a stream function in the meridian plane as shown in Figure 1. With this treatment for a drop of Newtonian fluid with constant density and viscosity, the continuity equation is automatically satisfied and the Navier-Stokes equation can be written as

$$\frac{1}{r\sin\theta}\frac{\partial(\psi, E^2\psi)}{\partial(r,\theta)} - \frac{2E^2\psi}{r^2\sin^2\theta}\left(\frac{\partial\psi}{\partial r}\cos\theta - \frac{1}{r}\frac{\partial\psi}{\partial\theta}\sin\theta\right)$$

$$= \frac{1}{Re}E^4\psi \tag{1}$$

where the Jacobian and the differential operator, E^2, are defined by

$$\frac{\partial(\psi, E^2\psi)}{(r,\theta)} = \frac{\partial\psi}{\partial r}\frac{\partial(E^2\psi)}{\partial\theta} - \frac{\partial\psi}{\partial\theta}\frac{\partial(E^2\psi)}{\partial r}$$

$$E^2 = \frac{\partial^2}{\partial r^2} + \frac{\sin\theta}{r}\frac{\partial}{\partial\theta}\left(\frac{1}{\sin\theta}\frac{\partial}{\partial\theta}\right) \tag{2}$$

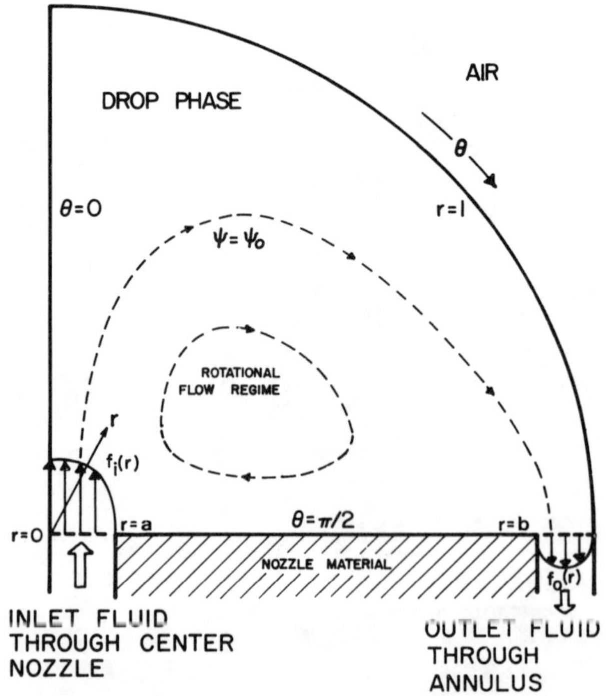

FIGURE 1. Meridian plane of a hemispherical drop.

The velocity components, determined by differentiating the stream function, are

$$U_r = -\frac{1}{r^2 \sin\theta}\frac{\partial\psi}{\partial\theta} \quad ; \quad U_\theta = +\frac{1}{r\sin\theta}\frac{\partial\psi}{\partial r} \tag{3}$$

The equations have been written in terms of dimensionless quantities by using the drop radius R and the maximum velocity of the incoming fluid V as characteristic parameters. Thus, the Reynolds number in Equation (1) is given as

$$Re = RV\rho/\mu \tag{4}$$

The non-linear terms in Equation (1) are usually neglected for falling drops because the internal flow patterns are independent of Reynolds number for a wide range of Reynolds numbers. However, for flow within forming drops, which is strongly dependent upon Reynolds number (Burkhart, et al., 1976), both inertial and viscous terms must be considered.

In order to complete the problem statement, it is necessary to furnish boundary conditions along the periphery of the meridian plane. First, the axial symmetry condition requires

$$\frac{\partial U_r}{\partial\theta} = 0 \; ; \quad U_\theta = 0 \quad \text{at } \theta = 0 \tag{5}$$

The boundary at $\theta = \pi/2$ is composed of three parts: the fluid inlet at the center nozzle, the surface of nozzle material, and the fluid outlet at the annulus. By assuming that fluid enters and leaves in a direction normal to the base of the drop, the velocity components at $\theta = \pi/2$ are

$$U_r = 0 \; ; \quad U_\theta = f(r) \quad \text{at } \theta = \frac{\pi}{2} \tag{6}$$

in which f(r) is defined by

$$f(r) = \begin{cases} f_i(r) & \text{at } o \leq r \leq a \\ o & \text{at } a \leq r \leq b \\ f_o(r) & \text{at } b \leq r \leq 1 \end{cases} \tag{7}$$

The function $f_i(r)$ is the velocity distribution at the inlet of the center nozzle, and $f_o(r)$ designates the velocity at the outlet of the annulus.

Next, the boundary conditions with respect to r are considered. Since the velocity must be finite everywhere, it is necessary to say

$$U_r \text{ and } U_\theta \text{ are bounded at } r=0. \tag{8}$$

The boundary conditions at r=1 concern the balance of the normal and tangential forces at the interface. The normal component includes forces exerted by internal flow and surface tension and determines the shape of the interface since the force exerted in the direction normal to the interface by surface tension is directly proportional to the curvature of the interface. Therefore, when the interface is given an arbitrary curvature, the normal component of interfacial force balance will not be satisfied in general. Thus the r-component of the interfacial force balance is neglected, assuming that the interface is spherical, and a velocity condition which is true for both the true and spherical interface imposed. That is, the normal component of velocity vanishes at the interface,

$$U_r = 0 \quad \text{at } r = 1 \tag{9}$$

When the interface is contaminated by surfactants, they will be swept downstream of the bulk flow along the interface and will produce a variation in the surface tension. This variation will affect the normal component of interfacial force balance and, in turn, the curvature of the interface. For example, studying the effects of surfactants on flow behavior of a falling drop, Schechter and Farley (1963) found that only a sinusoidal distribution of the surface-tension gradient is compatible with a spherical interface. However, when the magnitude of the variation is small this effect will be of minor importance. In this case, the variation in surface tension can be ascribed to an arbitrary distribution and its effect is considered through the tangential component of the interfacial force balance. For a gaseous continuous phase, the shear force transmitted from it to the drop can be neglected and the tangential component of interfacial force balance derived by Scriven (1969) becomes

$$\frac{\partial}{\partial r}\left(\frac{U_\theta}{r}\right) = -S_I\frac{\partial\gamma}{\partial\theta} + S_v\left[\frac{\partial}{\partial\theta}\left\{\frac{1}{\sin\theta}\frac{\partial(U_\theta\sin\theta)}{\partial\theta}\right\} + 2U_\theta\right] \quad \text{at } r = 1 \tag{10}$$

Here the dimensionless groups, S_I and S_v, and dimensionless surface tension γ are defined as

Dimensionless surface tension;

$$\gamma = \frac{\sigma_o - \sigma}{\sigma_o - \sigma_{\pi/2}} \tag{11}$$

Surface-tension gradient groups;

$$S_I = Re\cdot S_o = \frac{RV\rho}{\mu}\cdot\frac{(\sigma_o - \sigma_{\pi/2})/R}{\rho V^2} \tag{12}$$

Surface viscosity group;

$$S_v = \frac{\mu_s/R}{\mu} \tag{13}$$

where σ_o and $\sigma_{\pi/2}$ designate surface tensions at $\theta = 0$ and $\theta = \pi/2$ respectively, and μ_s is the surface viscosity.

Equations (5) through (10) provide boundary conditions for Equation (1) through the relations between the stream function and velocity components given in Equation (3).

MATHEMATICAL SOLUTION

The solution of Equation (1) is difficult because it is non-linear and it is a partial differential equation. The latter difficulty is resolved by reducing the equation to an equivalent set of ordinary differential equations by means of an appropriate Legendre polynomial series representation of the θ-dependence of the stream function, $\psi(r,\theta)$. It is then required that the expansion series satisfy all boundary conditions with respect to θ. For the problem posed here, the appropriate stream function is

$$\psi(r,\theta) = \sin^2\theta \left[\sum_{\substack{\ell=2 \\ \text{even}}}^{\infty} g_\ell(r)\{P_\ell(\cos\theta) - A_\ell\} + \int_0^r r f(r)\,dr \right] \tag{14}$$

in which A_ℓ is a shorthand notation for $P_\ell(\cos \pi/2)$, $P_\ell(\cos \pi/2)$ is the ℓth order Legendre polynomial, and $g_\ell(r)$ is the corresponding coefficient. The Legendre polynomials are restricted to be even orders in order to satisfy the conditions along the axis of the drop. This also makes the subsequent calculation more convenient since the Legendre polynomials of even order are orthogonal in the range of $0 \le \theta \le \pi/2$ whereas Legendre polynomials are orthogonal in the range $-\pi/2 \le \theta \le \pi/2$.

For practical calculation, it is necessary to truncate the infinite series after an appropriate number of terms. That is, the series is approximated as

$$\psi(r,\theta) = \sin^2\theta \left[\sum_{\substack{\ell=2 \\ \text{even}}}^{N_t} g_\ell(r)\{P_\ell(\cos\theta) - A_\ell\} + \int_0^r r f(r)\,dr \right] \tag{15}$$

This expression can then be substituted into Equation (1) and the resulting equation divided by sinθ and integrated from θ = 0 to θ = π/2. After considerable algebra, it is possible to write the result in the following compact form

$$\frac{1}{2n+1} L_n^4 g_n + \sum_\ell d_{\ell n}(D_\ell^4 g_\ell)$$
$$= -\frac{Re}{r^2}\Big[\sum_\ell \sum_m \Big(C^1_{\ell mn} g_\ell''' g_m + C^2_{\ell mn} g_\ell'' g_m' + C^3_{\ell mn}\frac{g_\ell'' g_\ell}{r}$$
$$+ C^4_{\ell mn}\frac{g_\ell' g_m}{r^2} + C^5_{\ell mn}\frac{g_\ell g_m}{r^3}\Big) + \sum_\ell \big(G^1_{\ell n}(r) g_\ell''' + G^2_{\ell n}(r) g_\ell''$$
$$+ G^3_{\ell n}(r) g_\ell' + G^4_{\ell n}(r) g_\ell\big) + 2a_n F(r)\tilde{D}(L_o^2 F(r))\Big]$$

$$\text{for } n = 2, 4, 6, \cdots, Nt \tag{16}$$

where the coefficient functions, $G^i_{\ell n}(r)$ (i = 1,2,3,4) are given by

$$G^1_{\ell n}(r) = 2(a_{\ell n} - A_\ell a_n)F$$

$$G^2_{\ell n}(r) = -4(a_{\ell n} - A_\ell a_n)\frac{F}{r} - b_{\ell n}F'$$

$$G^3_{\ell n}(r) = [-\{2a_{\ell n}\{2+\ell(\ell+1)\} + 4(c_{\ell n} + A_\ell a_n)]\frac{F}{r^2}$$

$$G^4_{\ell n}(r) = \{2(a_{\ell n} - A_\ell a_n) + b_{\ell n}\}\tilde{D}(L_o^2 F) + \{2\ell(\ell+1)a_{\ell n}$$
$$+ (4+\ell(\ell+1))b_{\ell n}$$
$$+ 4c_{\ell n}\frac{F'}{r^2} + 8\{\ell(\ell+1)a_{\ell n} + 2(a_{\ell n} - A_\ell a_n)$$
$$- 2c_{\ell n}\}\frac{F}{r^3} \tag{17}$$

in which F and F(r) designate $\int_0^r r f(r)\,dr$ and the differential operators are defined as

$$L_\ell^2 = \frac{d^2}{dr^2} - \{2 + \ell(\ell+1)\}\frac{1}{r^2}$$

$$L_\ell^4 = \frac{d^4}{dr^4} - 2\{2 + \ell(\ell+1)\}\frac{1}{r^2}\frac{d^2}{dr^2} + 4\{2 + \ell(\ell+1)\}$$
$$\frac{1}{r^3}\frac{d}{dr} + (\ell^2+\ell-2)(\ell^2+\ell+4)\frac{1}{r^4}$$

$$D_\ell^4 = 4\left[\frac{1}{r^2}\frac{d^2}{dr^2} - 2\frac{1}{r^3}\frac{d}{dr} + \{2 - \ell(\ell+1)\}\frac{1}{r^4}\right]$$

$$\tilde{D} = \frac{d}{dr} - \frac{2}{r} \tag{18}$$

The coefficient constants $C^i_{\ell mn}$ (i = 1,2,3,4,5) are defined in terms of constants, A_n, $A_{\ell n}$, $b_{\ell n}$, $c_{\ell n}$, $d_{\ell n}$, $A_{\ell mn}$, $b_{\ell mn}$, $c_{\ell mn}$, $d_{\ell mn}$, which are integrals of various products of single, double, and triple products of Legendre polynomials and their derivatives according to the number of subscripts. Their detailed expressions and values are given in Joo (1979).

It is then necessary to write the boundary conditions with respect to r by using the coefficient function $g_\ell(r)$. To do this, the velocity components are expressed in terms of the series representation,

$$U_r = -\frac{1}{r^2}\left[2\cos\theta\left\{\sum_{\substack{\ell=2 \\ \text{even}}}^{Nt}(P_\ell(\cos\theta) - A_\ell)g_\ell + F\right\}\right.$$
$$\left. + \sin\theta\left\{\sum_{\substack{\ell=2 \\ \text{even}}}^{Nt}\frac{\partial P_\ell(\cos\theta)}{\partial\theta}g_\ell\right]\right.$$

$$U_\theta = \frac{\sin\theta}{r}\left[\sum_{\substack{\ell=2 \\ \text{even}}}^{Nt}\{P_\ell(\cos\theta) - A_\ell\}g_\ell' + F'\right] \tag{19}$$

Use of these expressions reduces the boundary conditions as follows:

$$g_n = 0 \; ; \quad g_n' = 0 \quad \text{at } r = 0 \tag{20}$$

$$g_n = 0 \; ; \quad \text{at } r = 1$$

$$\frac{1}{2n+1}(g_n'' - 2g_n') = -S_I \Gamma_n + \sum_{\substack{\ell=2 \\ \text{even}}}^{Nt} S_{\ell n}\, g_\ell'$$

$$\text{for } n = 2,4,\cdots, Nt \tag{21}$$

where Γ_n and $S_{\ell n}$ are defined as

$$\Gamma_n = \int_o^{\frac{\pi}{2}} \frac{\partial r}{\partial \theta} P_n(\cos\theta)\, d\theta$$

$$S_{\ell n} = S_v \left[-\frac{\ell(\ell+1)}{2n+1}\delta_{\ell n} + \int_o^{\frac{\pi}{2}} \cos\theta \frac{\partial P_\ell(\cos\theta)}{\partial\theta} P_n(\cos\theta)\, d\theta\right] \tag{22}$$

and $\delta_{\ell n}$ is the Kronecker delta. Equation (16) with boundary conditions in Equations (20) through (22) completely defines a two-point boundary value problem for a set of non-linear ordinary differential equations. There are Nt/2 fourth-order differential equations with 2Nt boundary conditions.

The method of quasilinearization, an application of the Newton-Raphson-Kantorovich approximation in function space, was employed to obtain a numerical solution. The non-linear problem was reduced to an iterative linear problem where the non-linear terms in Equation (16) were replaced by the linear terms of a Taylor's series. This technique was used successfully for Reynolds numbers up to 110. For higher Reynolds numbers than this, convergence became difficult.

The amount of computation increased very markedly with number of terms retained in the series. In addition, increasing the number of terms made the convergence of successive iterations much slower. However, the series converged so rapidly that a reasonable approximation could be obtained by retaining only a few terms. A typical ratio between the magnitude of g_2 and g_4 was approximately 10, and g_6 decreased in a similar manner for the Reynolds number range discussed in this paper. As a result, there was no observable difference between flow patterns obtained by retaining two and three terms in the series.

RESULTS AND DISCUSSION

In order to compare the computed flow patterns with the experimental observations reported by Burkhart, et al. (1976), the geometry of the drop was chosen as a=0.9 and b=0.1, and the Reynolds number was redefined in terms of the inner nozzle diameter and the average velocity at the center nozzle. Thus, for a parabolic velocity profile at the center nozzle, the Reynolds number given by Equation (4) becomes ten times greater than the Reynolds number defined in this manner.

Initial calculations were carried out for a system without surfactant and the values of the various surface groups were set equal to zero. Later, the effects of the presence of surfactants on flow behavior were demonstrated by the change in flow patterns when finite values were assumed for the surface groups.

Figure 2 shows the development of rotational flow patterns computed for S_o=0 and S_v=0 (surfactant free system). The streamlines were almost straight for low Reynolds numbers and became curved more and more with increasing Reynolds numbers until a rotational flow appeared at a Reynolds number of 5.8. The region of rotational flow grew with increasing Reynolds numbers and a fully developed rotational was achieved at a Reynolds number of 10.0. For Reynolds numbers greater than 11.0, it became difficult to obtain convergence in solving the boundary value problem of coefficient functions in the Legendre polynomial series representation. However, it is also difficult, experimentally, to maintain a stable drop at higher Reynolds numbers due to deformation of the interface by increased jet action from the center nozzle.

In the flow computations discussed so far, the velocity distributions at the inlet and outlet of the concentric nozzles were assumed to be parabolic and uniform, respectively. A different assumption for the velocity distribution at the outlet of the annulus did not change the flow patterns appreciably. However, a change in the velocity distribution at the center nozzle had a significant effect on flow

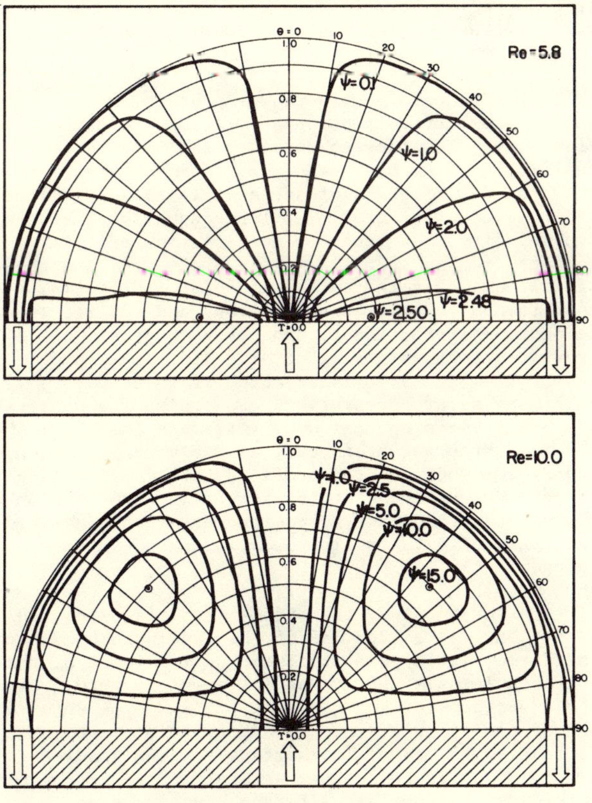

FIGURE 2. Growing rotational flow region with increasing Reynolds number.

behavior. Not unexpectedly, when plug flow instead of parabolic distribution was assumed, the Reynolds number at which rotational flow appeared decreased from 5.8 to 3.5.

Flow through the center nozzle must have a parabolic distribution at distances far from the nozzle mouth. However, the velocity profile at the mouth will be modified by an end effect. A qualitative effect of this change may be deduced from the development of the velocity profile by the efflux of a jet emerging from a long slit and mixing with the surrounding fluid discussed by Schlichting (1968). It is expected that a parabolic profile would be slightly flattened. This suggests that a rotational flow appear at a Reynolds number somewhat lower than 5.8.

The computed Reynolds number criterion of 5.8 for the emergence of rotational flow is lower than the value of 9.7 experimentally observed by Burkhart, et al. (1976). There are several differences between the present mathematical model and the system used in their experiment, however. The transmission of a shear force from the continuous phase and the effects of surfactants were not considered in the computations. If they had been considered, a value greater than 5.8 for the computed Reynolds number criterion would have resulted. This would decrease the difference between theory and experiment. But, as will be shown later, changing the interfacial conditions does not affect the flow significantly at distances far from the interface. Since the rotational flow emerges near the center, the differences in interfacial conditions do not have a pronounced effect on the Reynolds number criterion. On the other hand, the discrepancy may be partly attributed to experimental difficulty in determining accurately the exact point at which a rotational flow region actually begins. However, the similarity between computed and experimentally observed flow patterns is obvious.

The presence of surfactant at the interface will modify the surface tension as well as the surface viscosity. In order to consider the effects of these modifications on flow behavior, it is necessary to know the distribution of surfactants along the interface. Since the material balance equation includes the velocity distribution, the solution of this problem requires an iterative procedure in which a first approximation of the surfactant distribution is used to obtain a velocity distribution, and the resulting velocity distribution is then used to obtain an improved surfactant distribution. If this treatment of the problem is to be manageable, it is necessary to make many severe assumptions concerning the physical properties of a contaminated interface and the mechanism of surfactant transport, as was done by Levich (1962) for much simpler problems involving falling drops. Hence, the distribution of surface tension and the values of surface viscosity were simply assumed and their effects on flow behavior were considered independently.

In order to study the effect of surface-tension variation, a distribution of the type used by Schechter and Farley (1963) was utilized and the value of the surface viscosity group was set equal to zero. Imposing a surface-tension gradient of this type retarded interfacial flow and the retardation increased with increasing values of the surface-tension gradient group, S_o. Figure 3 shows values of S_o which make the interface stagnant at different Reynolds numbers. These are the maximum values allowable for S_o since a further increase in S_o results in a flow in the reverse direction.

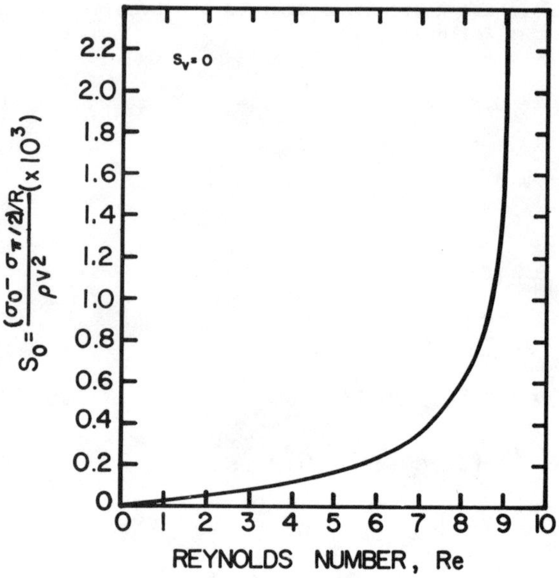

FIGURE 3. Values of the surface-tension gradient group S_o at which the interface becomes stagnant.

The value of S_o which makes the interface stagnant at a Reynolds of 5.0 is 1.70×10^{-4}. For a water drop at room temperature with a diameter of 1.0 cm this value of S_o corresponds to a drop in surface tension $\sigma_o - \sigma_{\pi/2}$, of 1.23×10^{-7} N/m. This is very small compared to the surface tension of pure water, 7.275×10^{-2} N/m. Therefore, interfacial flow can be reduced to zero without deforming the shape of the drop significantly.

It is also of interest to compare this value with a surface-tension change observed on the surface of a bubble rising at a terminal velocity through water containing isoamyl alcohol as a surfactant. For this system, Levan and Newman (1976) found that the surface tension change was 4.0×10^{-8} N/m and that the surfactant formed a gaseous monolayer. Since the radius of the bubble was 1.1×10^{-5} m, this change in surface tension corresponds to a change of 9.1×10^{-6} N/m in the present system. Although the manner of surfactant accumulation must be different for interfaces of different dimensions and also for drops and bubbles, this comparison suggests that the interface of a drop with forced internal circulation can become stagnant easily at moderate Reynolds numbers due to the effect of a surface-tension gradient caused by gaseous monolayers.

The change in interfacial flow caused by a surface-tension gradient affects the internal flow in turn. Figure 4 shows two examples of the change in internal flow in which the interfacial flow was reduced to zero (stagnation). For the upper figure the conditions were $S_o = 2.39 \times 10^{-4}$, $S_v = 0$, and Re = 6.0. For the lower figure they were $S_o = 3.53 \times 10^{-4}$, $S_v = 0$, and Re = 7.0. As was expected, the change is not significant near the center of the drop.

The change in internal flow can also be considered by comparing the θ-components of the internal velocities with and without a surface-tension gradient, as shown in Figure 5. This comparison revealed that the

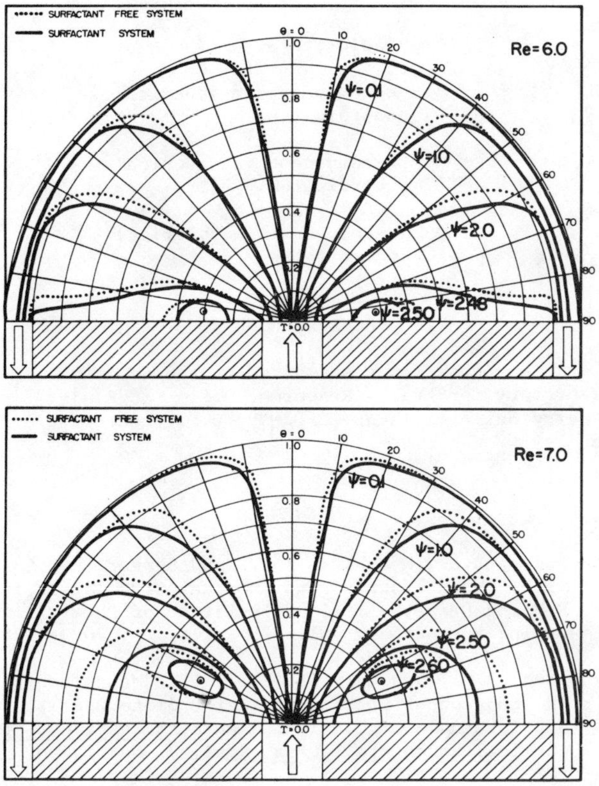

FIGURE 4. Change in internal flow patterns due to a surface-tension gradient caused by the presence of surfactants.

effect of surfactants on internal flow within a drop with forced internal circulation is quite different from the surfactant effect for a moving drop. The presence of surfactants retards internal flow everywhere within a moving drop since the driving force for internal flow is supplied by the retarded interfacial flow. On the other hand, for the case of a drop with forced internal circulation, the flow is retarded in the interfacial region, but is accelerated in other regions. For the particular case shown in Figure 5, the fluid at $r < 0.8$ is accelerated in the θ-direction when the interfacial flow is retarded by the presence of surfactants. This happens because the decreased amount of flow in the interfacial region must be compensated in some other part of the drop when a constant amount of liquid is injected and withdrawn through the concentric nozzles. This must be considered in explaining the difference in the effects of the presence of surfactants on mass transfer in forming and in moving drops.

A unique feature of this study is that the effect of surface viscosity was considered in studying the effects of surfactants on the flow behavior of drops. This effect is represented by the surface viscosity group, S_v. Figure 4 shows a change in interfacial flow for different values of surface viscosity. The interfacial flow is retarded more for increased surface viscosity group. The expression for S_v in Equation (13) indicates that the smaller the drop, the greater the effect of retardation. Actually, Davis and Acrivos (1966) observed that the surface of a small bubble became stagnant more easily than that of a large bubble. This may partly be attributed to the

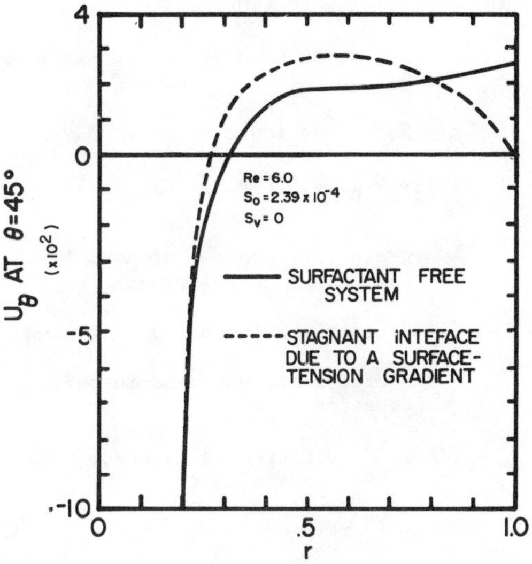

FIGURE 5. Comparison of θ-components of velocities with and without surfactants.

radial dependence of the retardation effect of surface viscosity.

The values of S_v for gaseous monolayers on a bulk fluid of viscosity of 1.0×10^{-3} N·s/m² (kg/m·s) range from 1.5×10^{-3} to 1.5×10^{-2}, since gaseous monolayers have surface viscosities ranging from 10^{-8} to 10^{-7} kg/s. It may thus be noted from Figure 6 that the change in interfacial velocity for the case considered in this figure ($r=1.0$ and $\theta=\pi b$) is less than 1% for gaseous monolayers. This provides quantitative evidence for the neglect of the effect of surface viscosity, assumed by most investigators in the study of the effect of surfactants on interfacial flow.

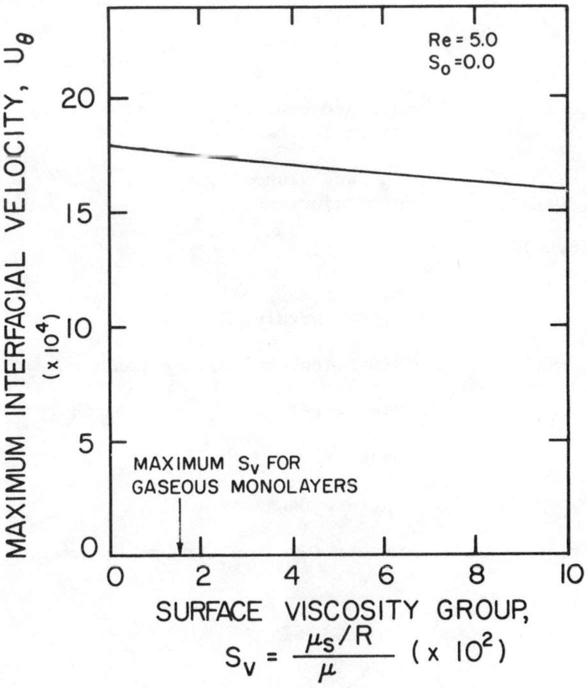

FIGURE 6. Effect of surface viscosity on interfacial flow.

NOMENCLATURE

A_n — Values of Legendre polynomials of n-th order at $\theta = 0$.

a — Radius of inner nozzle.

a_n, $a_{\ell n}$, $b_{\ell n}$, $c_{\ell n}$, $d_{\ell n}$, $a_{\ell mn}$, $b_{\ell mn}$, $c_{\ell mn}$, $d_{\ell mn}$ — Constants defined in terms of integrals of Legendre polynomials in the basic flow calculation.

b — Distance to inside edge of annulus.

$C^i_{\ell mn}$ — Coefficient constants in basic flow equations.

$\tilde{D}$, D^4_ℓ, E^2, L^2, L^2_ℓ, L^4_ℓ — Differential operators with respect to r.

$F(r)$ — $\equiv \int_0^r rf(r)dr$

$f_i(r)$, $f_o(r)$, $f(r)$ — Velocity distribution of basic flow at $\theta = \pi/2$.

$G^i_{\ell n}(r)$ — Coefficient functions in basic flow equations.

$g_\ell(r)$ — Coefficient functions for basic flow series expansion.

ITR — Order of disturbance flow spherical harmonic series truncation.

Nt — Order of basic flow Legendre polynomial series truncation.

$P_\ell(\theta)$ — ℓth order Legendre polynomial.

R — Drop radius.

Re — Reynolds number.

r — Dimensionless radial coordinate.

S_o, S_I — Surface-tension gradient groups, dimensionless.

S_v — Surface viscosity group, dimensionless.

$\bar{U}(r, \theta)$ — Basic flow.

V — Maximum basic flow velocity at the center nozzle.

γ — Dimensionless surface tension.

μ — Viscosity.

ρ — Density.

σ — Surface tension.

$\psi(r, \theta)$ — Stream function for basic flow.

$( )'$ — $\equiv \frac{d}{dr}$ or $\frac{d}{dx}$.

$(\dot{})$ — $\equiv \frac{d}{d\theta}$.

Subscripts

b — Basic state.

θ — θ-component in spherical coordinates.

ϕ — ϕ-component in spherical coordinates.

r — r-component in spherical coordinates.

ACKNOWLEDGMENT

This research was supported by the Assistant Secretary for Energy Research, Office of Basic Energy Sciences, WPAS-KC-02-03, U.S. Department of Energy.

REFERENCES

Burkhart, L.E., Weathers, P.W., and Sharer, P.C., 1976, "Mass Transfer and Internal Circulation in Forming Drops," American Institute of Chemical Engineers Journal, Vol. 22, No. 6, pp. 1090-1096.

Constan, J.L., and Calvert, S., 1963, "Mass Transfer in Drops under Conditions that Promote Oscillation and Internal Circulation," American Institute of Chemical Engineers Journal, Vol. 9, No. 1, pp. 109-115.

Davis, R.E., and Acrivos, A., 1966, "The Influence of Surfactants on the Creeping Motion of Bubbles," Chemical Engineering Science, Vol. 21, No. 8, Aug., pp. 681-685.

Garner, F.H., and Skelland, A.H.P., 1951, "Liquid-Liquid Mixing as Affected by the Internal Circulation within Droplets," Transactions, Institution of Chemical Engineers (London), Vol. 29, pp. 315-321.

Joo, S.B., Ph.D. Thesis, 1979, Iowa State University.

Levan, M.D., and Newman, J., 1976, "The Effect of Surfactant on the Terminal and Internal Velocities of a Bubble or Drop," American Institute of Chemical Engineers Journal, Vol. 22, No. 4, pp. 695-701.

Levich, V.G., 1962, Physicochemical Hydrodynamics, Prentice-Hall, Englewood Cliffs, N.J.

Schechter, R.S., and Farley, R.W., 1963, "Interfacial Tension Gradients and Droplet Behavior," Canadian Journal of Chemical Engineering, Vol. 41, No. 3, June, pp. 103-107.

Schlichting, H., 1968, Boundary-Layer Theory, McGraw-Hill, New York.

INTERFACIAL FLOW ON A STATIC LIQUID DROP WITH FORCED INTERNAL CIRCULATION: PART II - STABILITY ANALYSIS.

S. B. Joo and Lawrence Burkhart

Ames Laboratory, U. S. DOE and Department
of Chemical Engineering, Iowa State University
Ames, Iowa

ABSTRACT

Axially unsymmetric ordered circulation patterns on the surface of a hemispherical constant volume drop with forced internal motion were considered as a manifestation of hydrodynamic instability, which was affected by internal flow as well as by the properties of the interface. A simplified mathematical model was analyzed in order to study the mechanism of onset of interfacial circulation from a non-circulatory basic state. Values of the critical Reynolds numbers were obtained when the basic state became a fully developed rotational flow and the values changed for different interfacial properties.

The work describes a heretofore unreported surface flow pattern on drop surfaces. From a practical point of view, any motion in the interface of a droplet can have a significant effect of mass transfer rates in solvent extraction processes. Although surface motion in droplets actual processing equipment is very complex, continued study of the fundamental nature of such phenomena should ultimately lead to more effective equipment design procedures.

INTERFACIAL FLOW ON A STATIC LIQUID DROP WITH FORCED INTERNAL CIRCULATION: PART II - STABILITY ANALYSIS.

S. B. Joo and Lawrence Burkhart

Ames Laboratory, U. S. DOE and Department
of Chemical Engineering, Iowa State University
Ames, Iowa

INTRODUCTION

The rate at which mass is transferred through a fluid interface is significantly affected by the motion of fluid near the interface. The conventional picture of interfacial flow on a forming drop is the axially symmetric flow as discussed in Part I. However, careful observation of the interface of a drop in which there is forced internal circulation shows that axially unsymmetric interfacial circulation occurs above a certain Reynolds number, based upon internal flow conditions, and that there are several definite circulation patterns which change in number and shape according to the Reynolds number. Poonawalla (1976) observed this type of motion on the surface of a forced circulation drop formed on Teflon concentric nozzles submerged in water. The drop fluid was a mixture of mineral oil and varsol. High-speed motion pictures of the movements of tiny hollow glass beads on the surface of the drop showed that there was no circulatory motion for Reynolds numbers lower than 10, but that two, four, and six circulation patterns occurred in the Reynolds number ranges of 10 to 34, 60 to 150, and 220 to 260, respectively. The Reynolds numbers were based upon inlet velocity, nozzle diameter, and drop viscosity and density. Figure 1 shows a schematic diagram of these circulation patterns. A similar interfacial movement was observed by Bakker, et al. (1966) on the surface of a drop forming at a single nozzle tip but no ordered circulation patterns were identified, possibly because of the short time available for observation.

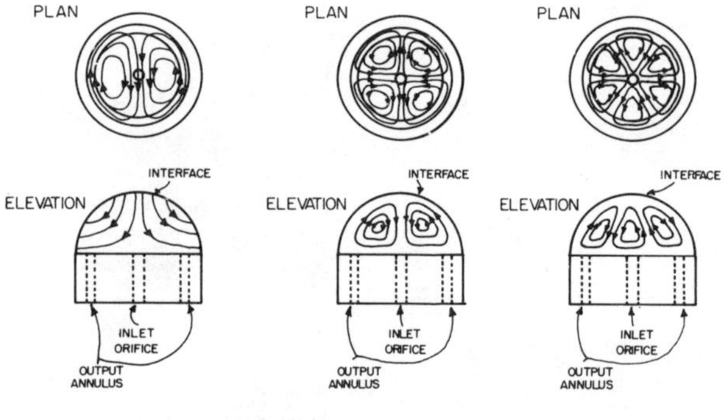

FIGURE 1. Flow patterns of interfacial circulation on a drop with forced internal circulation.

Although surfactants were not added intentionally in Poonawalla's experiments, their presence was necessary because pure water cannot support the large velocity gradient at the interface where the surface flow is radially inward. This can be seen experimentally by using a submerged jet of water impinging against either a flat air-water or organic-water interface surrounded by a circular boundary. Similar circulation patterns will occur at such an interface but removal of the interfacial film destroys the circulatory motion. This experiment also suggests that the effect of the surrounding medium on the existence of these interfacial flow patterns is negligible. The purpose of this investigation was to develop, from hydrodynamic stability theory, a mathematical model for explaining this motion.

The stability analysis presented is somewhat different from that usually found, in that the system of non-stationary basic states, like flows in rotating spheres and cylinders, have closed boundaries which are easier to describe. On the other hand, many problems for systems containing an interface have a stationary basic state and predict small scale convection cells as a disturbance flow, as occurs in Benard or Marangoni instability problems. Because of these well known problems, one may have an erroneous notion that every disturbance flow at an interface should appear as small convection cells. However, when there is a two dimensional motion in the basic state, a disturbance flow will be stretched in the direction of the motion, as noted by Jeffreys (1928). The large scale circulatory motion to be predicted as a disturbance flow in this analysis may be understood from this point of view.

The analysis is based on linear stability theory. In order to overcome the difficulty caused by two-dimensional coordinate dependence of the basic flow, the disturbance flow is represented by a toroidal-poloidal field and the defining scalars are expanded by means of spherical harmonics. The critical Reynolds numbers are determined by solving the resulting eigenvalue problem such that the criterion gives a sufficient condition for instability. Aside from the desire to know the critical Reynolds number, it is also of interest to determine the most unstable disturbance flow pattern. However, the analysis will predict an unstable regime of flow only from a non-circulatory basic state. Explanation of the transition between different circulation patterns was not attempted.

Stability was also affected by interfacial properties. Their effects were demonstrated by the change in critical Reynolds number according to different interfacial properties or the change in initial disturbance growth rate at a given Reynolds number.

There are some difficulties in devising a mathematical model for this phenomenon. First is the question of how to introduce motion in the circumferential direction since there is no driving force for flow in this direction. The system described here is different from those involving the surface of an open channel (Merson and Quinn, 1965) or a hemispherical air bubble on a wall (Kenning and Cooper, 1966) in that a non-uniform flow causes a driving force in the direction of interfacial circulation in the latter cases. Second, the interfacial circulation appears above a certain Reynolds number, and the number of circulation patterns takes only even integer numbers as a function of continuously increasing Reynolds number. Thus, it is necessary to devise a way to account for a discrete change from a continuous variable.

For these reasons, hydrodynamic stability theory was employed. The basic premise was that the flow system of forced internal circulation discussed in Part I becomes unstable above a certain Reynolds number. Motion in the circumferential direction was introduced as a disturbance and stability analysis used to show whether the disturbance grew or damped out. It is also conceivable that the stability is affected by a surface tension gradient caused by a non-uniform distribution of surfactant swept downstream of the bulk flow along the interface as well as by interfacial properties, which also affect the growth of disturbances at the interface.

FORMULATION OF THE LINEAR STABILITY PROBLEM

One quadrant of a hemispherical drop with forced internal circulation is illustrated in Figure 2. The basic state (i.e. the initial condition of the drop before any disturbance is introduced) is an axially symmetric steady flow in which there is no ϕ-component of velocity. This steady flow was perturbed and hydrodynamic stability theory was used to determine the growth rate of the disturbance. For linear analysis, where the disturbances are infinitesimal and the nonlinear terms with respect to the disturbance are neglected, the governing equations for an incompressible fluid can be written as:

$$\frac{\partial \bar{u}}{\partial t} + \bar{u} \cdot \nabla \bar{U} + \bar{U} \cdot \nabla \bar{u} = -\nabla P + \frac{1}{Re} \nabla^2 \bar{u} \qquad (1)$$

$$\nabla \cdot \bar{u} = o \qquad (2)$$

The Reynolds number, Re, is defined in terms of drop radius R, maximum velocity at the center nozzle V, drop density ρ and drop viscosity μ. $\bar{u}$ is the disturbance flow and $\bar{U}$ is the basic flow obtained in Part I as:

$$\bar{U} = \begin{Bmatrix} -\left[\sum_{s}^{Nt} (\sin\theta\, P_s + 2\cos\theta\, P_s) \frac{g_s}{r^2} + 2\cos\theta \frac{\phi}{r^2}\right] \\ \sum_{s}^{Nt} (\sin\theta\, P_s \frac{g_s'}{r}) + \sin\theta \frac{\phi'}{r} \\ 0 \end{Bmatrix} \qquad (3)$$

where $\phi(r)$ is defined by

$$\phi(r) = F(r) - \sum_{\substack{b=2 \\ \text{even}}}^{ITR} A_b g_b(r) \qquad (4)$$

The function F(r) is the forcing function and $g_b(r)$ is the coefficient function in the Legendre polynomial series representation for the basic flow. Since the latter was given as numerical data, it was necessary

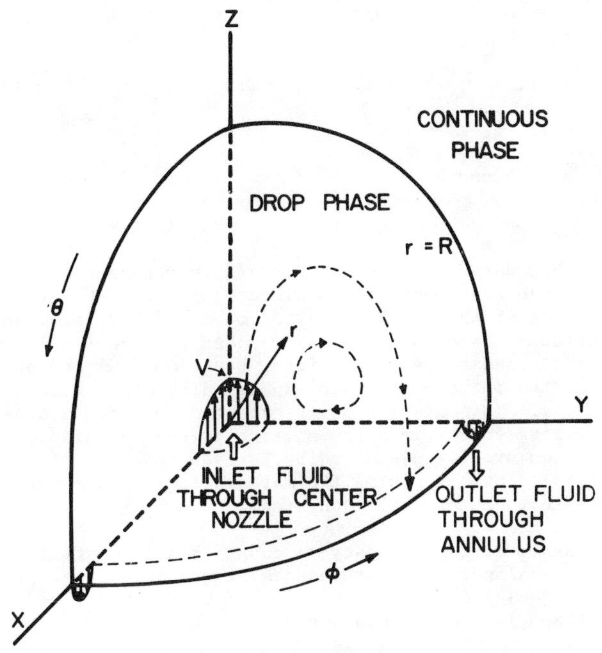

FIGURE 2. One quadrant of a hemispherical drop with forced internal circulation.

to employ a numerical technique to solve Equation (1).

The pressure term, ∇P, in Equation (1) was eliminated by taking the curl of the equation, and the independent variable, t, was eliminated by assuming an exponential time dependence expressed as:

$$\bar{u}\,(r, \theta, \phi, t) = \bar{u}\,(r, \theta, \phi)\, e^{-ct} \qquad (5)$$

where c is a complex number, $c = c_r + ic_i$, the real part representing the disturbance growth rate. Under these conditions, Equation (1) becomes:

$$\begin{aligned} &\nabla \times [\nabla \times (\nabla \times \bar{u})] - \lambda(\nabla \times \bar{u}) \,| \\ &Re\,[\nabla \times (\bar{u} \cdot \nabla\bar{U} + \bar{U} \cdot \nabla\bar{u})] = \bar{o} \end{aligned} \qquad (6)$$

in which λ is defined as $\lambda = Re \cdot c = \lambda_r + i\lambda_i$. At this point, no restrictions other than incompressibility have been imposed. Thus, Equation (6) may be used as a starting point for any linear stability analysis involving an incompressible fluid.

The value of λ_r serves as a criterion for stability. Disturbances for which $\lambda_r < 0$ are damped while those for which $\lambda_r > 0$ are amplified and instability sets in. When λ_r vanishes for negative λ_i, the transition to instability takes place following a stationary pattern of motion, i.e. the principle of exchange of stabilities holds. This simplifies the mathematical problem since only the stationary regime of instability needs to be considered. However, the validity of the principle has been proven for only a few problems (Chandrasekhar, 1961) and simply assumed in many other problems involving a fluid interface (Scriven and Sternling, 1964, Berg and Acrivos, 1965). It was assumed but not proven that the principle can be applied to this model.

The boundary conditions for the disturbance flow are:

$$\bar{u}\ (r,\ \theta,\ \phi) = \bar{u}\ (r,\ \theta,\ \phi + 2\pi) \tag{7}$$

$$\bar{u}\ (r,\ \tfrac{\pi}{2},\ \phi) = \bar{o} \qquad \text{at } \theta = \tfrac{\pi}{2} \tag{8}$$

$$\bar{u} \text{ is bounded} \qquad \text{at } r = 0 \tag{9}$$

$$\bar{u}_r = 0 \qquad \text{at } r = 1 \tag{10}$$

The boundary condition at $\theta = \pi/2$ expresses the assumption that there is no disturbance at either the surface of the nozzle material or the fluid inlet and outlet. The boundary conditions at r=1 are the interfacial conditions. Therefore, Equation (10) indicates that the interface remains spherical under disturbances. This assumption is expected to offer a stabilizing effect in a way similar to the cellular convection problem discussed by Scriven and Sternling (1964), thereby making the mathematical system more stable than the actual physical system.

Next, it is necessary to consider a tangential force balance for disturbances at the interface. The interfacial force balance derived by Scriven (1960) is written in vector notation as

$$\frac{\partial}{\partial r}\left(\frac{\bar{u}_s}{r}\right) = -S_I \nabla_s \gamma' \tag{11}$$

The dimensionless surface tension and the dimensionless group, S_I, are defined by

$$\gamma' = \frac{\sigma_o - \sigma}{\sigma_o - \sigma_{\frac{\pi}{2}}} \tag{12}$$

$$S_I = R_e \cdot S_o = \frac{RV\rho}{\mu}\ \frac{(\sigma_o - \sigma_{\frac{\pi}{2}})/R}{\rho V^2} \tag{13}$$

The effect of surface viscosity has been neglected in the force balance.

For later usage, Equation (11) can be further manipulated by applying the scalar product of the surface gradient operator on the unit sphere and the continuity equation to give

$$\left[\frac{\partial}{\partial r}\left\{\frac{1}{r^2}\frac{\partial}{\partial r}(r^2 u_r)\right\}\right]_{r=1} = -S_I \nabla_s^2 \gamma' \tag{14}$$

Use of this equation requires that the disturbance in surfactant concentration at the interface be known since the disturbance in the surface tension arises from its dependence on the local concentration of surfactants, which has been disturbed as a result of flow disturbances.

The starting point for this problem is the surfactant material balance for the basic state, which can be written as

$$j_{nb} = \nabla_s \cdot (\Gamma_b \bar{U}_s) - D_s \nabla_s^2 \Gamma_b \tag{15}$$

$\Gamma_b \bar{U}_s$ represents the flux transferred along the interface by the tangential component of the flow velocity. The flux of surfactants from the bulk fluid to the interface is determined by two processes: adsorption or desorption, and diffusion through the bulk fluid. The former being the rate determining step, the concentration of surfactants may be assumed to have a constant value of c_o throughout the bulk phase. This is a reasonable assumption for a drop with forced internal circulation since the bulk fluid is continuously renewed by the forced convection. It is also assumed that the surfactant concentration along the interface deviates very slightly from a concentration of Γ_e which is at equilibrium with the bulk surfactant concentration c_o, as was assumed by Levich (1962) for contaminated falling drops. In this case, the surfactant material balance can be approximated as

$$j_{nb} = \Gamma_e \nabla_s \cdot \bar{U}_s - D_s \nabla_s^2 \Gamma_{bd} \tag{16}$$

where Γ_{bd} is defined by $\Gamma_b = \Gamma_e + \Gamma_{bd}$ with $\Gamma_{bd} \ll \Gamma_e$.

Introducing disturbances into Equation (16) and subtracting the material balance of the basic state gives

$$j_n' = \Gamma_e \nabla_s \cdot \bar{u}_s - D_s \nabla_s^2 \Gamma_{bd}' \tag{17}$$

The relation can be simplified further by assuming that surfactants are so slowly adsorbed or desorbed as to be essentially insoluble in the bulk fluid during the period of development of instability. Then, use of the relation $\Gamma_{bd}' = \Gamma'$ and the continuity equation reduces Equation (17) to

$$\left[\frac{\partial}{\partial r}(r^2 u_r)\right]_{r=1} = \frac{D_s}{RV\Gamma_e} \nabla_s^2 \Gamma' \tag{18}$$

Finally, combining Equations (14) and (18) gives

$$\left[\frac{\partial}{\partial r}\left\{\frac{1}{r^2}\frac{\partial}{\partial r}(r^2 u_r)\right\}\right]_{r=1} = -\frac{S_I}{\alpha}\left[\frac{\partial}{\partial r}(r^2 u_r)\right]_{r=1} \tag{19}$$

Here the dimensionless number, α, will be called a stability group and is defined by

$$\alpha = \frac{D_s(\sigma_o - \sigma_{\pi/2})/R}{k\Gamma_e V} \tag{20}$$

in which k is a proportionality constant between surface tension and surfactant concentration. This linear relation is a customary assumption for gaseous monolayers (Griffith, 1962; Berg and Acrivos, 1965).

Equation (6) and the boundary conditions, Equations (7) through (10) and (19), define the governing system for the stability analysis. The system will not be solved for arbitrary values of λ in general. Rather, solutions for only certain values of λ are of interest, i.e., it is an eigenvalue problem in λ.

SOLUTION OF THE STABILITY EQUATIONS

When the basic flow is a function of one independent variable as in Couette or Poiseuille flow, the partial differential equations governing a disturbance flow can be reduced to ordinary differential equations by analyzing the disturbance flow in terms of two-dimensional periodic waves. This customary introduction of normal modes was not possible for the present case because the basic flow depended on two spatial coordinates. In order to overcome this difficulty, the disturbance flows were represented in terms of a toroidal-poloidal field and the defining scalars were expanded by an appropriate spherical harmonic series. Munson and Joseph (1971) employed a similar approach for flow between concentric rotating spheres.

Following Chandrasekhar (1961), any solenoidal vector field can be represented in terms of two vector components as

$$\bar{u} = \bar{T} + \bar{S} \tag{21}$$

in which $\bar{T}$ is a toroidal component and $\bar{S}$ is a poloidal component. These components are defined by their defining scalars, Ψ and Φ, as

$$\bar{T} = \nabla x(\frac{\Psi}{r}\bar{r}) \quad ; \quad \bar{S} = \nabla x[\nabla x(\frac{\Phi}{r}\bar{r})] \tag{22}$$

With this treatment, the continuity equation is satisfied automatically and the components of velocity are

$$\bar{u} = \begin{bmatrix} -\frac{1}{r^2}\frac{1}{\sin\theta}\frac{\partial}{\partial\theta}(\sin\theta\frac{\partial\Phi}{\partial\theta}) + \frac{1}{\sin^2\theta}\frac{\partial^2\Phi}{\partial\theta^2} \\ \frac{1}{r\sin\theta}\frac{\partial\Psi}{\partial\phi} + \frac{1}{r}\frac{\partial^2\Phi}{\partial r\partial\theta} \\ -\frac{1}{r}\frac{\partial\Psi}{\partial\theta} + \frac{1}{r\sin\theta}\frac{\partial^2\Phi}{\partial r\partial\phi} \end{bmatrix} \tag{23}$$

When the defining scalars are expanded in spherical harmonics with coefficients which are allowed to be functions of r, it is necessary to make the expansion series satisfy the boundary conditions with respect to θ and ϕ. Therefore, they were chosen as

$$\Psi(\bar{r}) = \sum_{\ell=o}^{\infty}\sum_{m=-\ell}^{+\ell}\cos\theta\, T_\ell^m(r)Y_\ell^m(\theta,\phi)$$

$$\Phi(\bar{r}) = \sum_{\ell=o}^{\infty}\sum_{m=-\ell}^{+\ell}\cos\theta\, S_\ell^m(r)Y_\ell^m(\theta,\phi) \tag{24}$$

for $\ell - m$ = odd

in which Y_ℓ^m is the spherical harmonic defined as

$$Y_\ell^m(\theta,\phi) = e^{im\phi}P_\ell^{|m|}(\cos\theta)$$

and $P_\ell^{|m|}(\cos\theta)$ is the associated Legendre polynomial.

The value m is the axial wave number and the eigenvalue problem is solved over integer values of m. Thus, the summation over m in Equation (24) may be eliminated. On the other hand, it is necessary to truncate the infinite series at some appropriate number of terms for practical calculation. Hence, Equation (24) was written as

$$\Psi = \sum_{\ell}^{ITR}\cos\theta\, T_\ell Y_\ell^m$$

$$\Phi = \sum_{\ell}^{ITR}\cos\theta\, S_\ell Y_\ell^m \tag{25}$$

where $\ell = (m+1), (m+3), \cdots$, ITR, and the summation limit, ITR, denotes the order of truncation.

The velocity components of the disturbance flow, written in terms of these series and obtained from Equation (23) are,

$$\bar{u} = \sum_{\ell}^{ITR}\begin{bmatrix} \cos\theta\,\frac{\ell(\ell+1)+2}{r^2}S_\ell Y_\ell^m \\ \cos\theta\,(\frac{1}{r}S'_\ell Y_\ell^m + \frac{im}{r\sin\theta}T_\ell Y_\ell^m) - \sin\theta\,\frac{1}{r}S'_\ell Y_\ell^m \\ \cos\theta\,(\frac{im}{r\sin\theta}S'_\ell Y_\ell^m - \frac{1}{r}T_\ell\frac{\partial Y_\ell^m}{\partial\theta}) + \sin\theta\,\frac{1}{r}T_\ell Y_\ell^m \end{bmatrix} \tag{26}$$

The components of this expression are complex. It is understood that the real parts represent physical quantities. For integer m, the periodicity condition with respect to ϕ is satisfied and all components vanish at $\theta = \pi/2$. It can also be shown that the disturbance flows represented by m=1, m=2, and m=3 will grow to be two, four, and six circulation patterns respectively, as shown in Figure 1 if the system is unstable.

Substituting Equations (3) and (26) into Equation (1) gives a vector equation in terms of Legendre polynomials, spherical harmonics, and basic and disturbance flow component functions. This equation, multiplied by a unit poloidal, $\bar{S}_n^{-m}$, (n = m+1, m+3, $\cdots$, ITR) and subsequent integration gives a set of ordinary differential equations. By following the same procedure with a series of unit toroidals, $\bar{T}_n^{-m}$ (n = m+1, m+3, $\cdots$, ITR), instead of poloidals, a second set of differential equations is obtained. The final result of this tedious procedure is

$$\sum_{\ell}^{ITR}[\{C_{\ell n}^1 S_\ell'''' + (C_{\ell n}^2 + \lambda C_{\ell n}^3 r^2)\frac{S_\ell''}{r^2} + C_{\ell n}^4\frac{S_\ell'}{r^3} + (C_{\ell n}^5 + \lambda C_{\ell n}^6 r^2)\frac{S_\ell}{r^4} + iM\{H_{\ell n}^1(r)T_\ell'' + H_{\ell n}^2(r)\frac{T_\ell'}{r} + H_{\ell n}^3(r)\frac{T_\ell}{r^2}\}] = 0$$

$$\sum_{\ell}^{ITR}[\{D_{\ell n}^1 T_\ell'' + (D_{\ell n}^2 + \lambda D_{\ell n}^3 r^2)\frac{T_\ell}{r^2}\} + iM\{K_{\ell n}^1(r)S_\ell'' + K_{\ell n}^2(r)\frac{S_\ell'}{r} + K_{\ell n}^3(r)\frac{S_\ell}{r^2}\}] = 0 \tag{27}$$

for ℓ, n = (m+1), (m+3), ··· , ITR

where M is defined as $m \cdot R_e$ and the coefficient functions, $H^j_{\ell n}$ and $K^j_{\ell n}$ (j=1,2,3), are defined by

$$H^1_{\ell n}(r) = h^1_{o\ell n}\frac{\phi}{r^2} + \sum_b^{ITR} h^2_{b\ell n}\frac{g_b}{r^2}$$

$$H^2_{\ell n}(r) = h^3_{o\ell n}\frac{\phi'}{r} + h^4_{o\ell n}\frac{\phi}{r^2} + \sum_b^{ITR}(h^5_{b\ell n}\frac{g'_b}{r} + h^6_{b\ell n}\frac{g_b}{r^2})$$

$$H^3_{\ell n}(r) = h^7_{o\ell n}\phi'' + h^8_{o\ell n}\frac{\phi'}{r} + h^9_{o\ell n}\frac{\phi}{r^2} + \sum_b^{ITR}(h^{10}_{b\ell n}g''_b + h^{11}_{b\ell n}\frac{g'_b}{r} + h^{12}_{b\ell n}\frac{g_b}{r^2})$$

$$K^1_{\ell n}(r) = k^1_{o\ell n}\frac{\phi}{r^2} + \sum_b^{ITR} k^2_{b\ell n}\frac{g_b}{r^2}$$

$$K^2_{\ell n}(r) = k^3_{o\ell n}\frac{\phi'}{r} + k^4_{o\ell n}\frac{\phi}{r^2} + \sum_b^{ITR}(k^5_{b\ell n}\frac{g'_b}{r} + k^6_{b\ell n}\frac{g_b}{r})$$

$$K^3_{\ell n}(r) = k^7_{o\ell n}\phi'' + \sum_b^{ITR} k^8_{b\ell n}g''_b \tag{28}$$

The coefficient constants, $C^j_{\ell n}$ (j=1~6) and $D^j_{\ell n}$ (j=1~3) in Equation (27) and $h^j_{b\ell n}$ (j=1~12) and $k^j_{b\ell n}$ (j=1~8) in Equation (28), are derived in terms of integrals of various products of Legendre polynomials and associated Legendre polynomials over a unit sphere. Their expressions and values are given in Joo (1979). For m≠0, Equation (27) becomes complex. That is, the governing equations for disturbance flows of two, four, and six circulation patterns are complex.

The boundary conditions of Equations (9), (10), and (19) written in terms of T_ℓ and S_ℓ are

$$\left.\begin{aligned} T_n(o) &= 0 \\ S_n(o) &= S'_n(o) = 0 \end{aligned}\right\} \quad \text{at } r=0 \tag{29}$$

$$S_n(1) = 0 \qquad \text{at } r=1 \tag{30}$$

$$S''_n(1) = (2 - \frac{S_I}{\alpha})\, S'_n(1) \tag{31}$$

These equations provide four boundary conditions needed on S_n for each n since the system of Equation (27) is fourth order with respect to S_n. However, only one boundary condition on T_n for each n has been obtained, whereas the equation is second order with respect to T_n. This one more necessary condition is found by using the property of a toroidal vector field that $\nabla_s \cdot \bar{T}_n = 0$. Since for any vector field, $\bar{T}$

$$\nabla_s \cdot T = \frac{1}{\sin\theta}\frac{\partial}{\partial\theta}(T_\theta \sin\theta) + \frac{1}{\sin\theta}\frac{\partial T_\phi}{\partial\phi}$$

it follows that $\nabla_s \cdot T = 0$ when $\bar{T}$ has the form

$$\bar{T} = (0,\ (r\sin\phi)^{-1}\,\partial\Psi/\partial\phi,\ -r^{-1}\,\partial\Psi/\partial\theta)\ .$$

However $\nabla\cdot\bar{T} = 0$ when $\bar{T}$ is toroidal and therefore $\nabla_r\cdot\bar{T} \equiv \nabla\cdot\bar{T} - \nabla_s\cdot\bar{T} = r^{-2}\,\partial/\partial r\,(r^2 T_r) = 0$ at r=1. Use of (24) in this last yields (32). Therefore, one has

$$T'_\ell(1) - 2\,T_\ell(1) = 0 \qquad \text{at } r=1 \tag{32}$$

When the governing equations are complex, these conditions must also be written in terms of real and imaginary parts.

The equations written in terms of real and imaginary parts will double the size of the governing system. However, examining the two sets of equations obtained for real and imaginary parts reveals that they define identical eigenvalue problems when appropriate conditions are imposed on real and imaginary parts of T_ℓ and S_ℓ. As a result, it is only necessary to use one of the two sets,

$$\sum_\ell^{ITR}[D^1_{\ell n}T''_\ell + (D^2_{\ell n} + \lambda D^3_{\ell n}r^2)\frac{T_\ell}{r^2} - M\{K^1_{\ell n}(r)S'' + K^2_{\ell n}(r)\frac{S'_\ell}{r} + K^3_{\ell n}(r)\frac{S_\ell}{r}\}] = 0$$

$$\sum_\ell^{ITR}[C^1_{\ell n}S''''_\ell + (C^2_{\ell n} + \lambda C^3_{\ell n}r^2)\frac{S''_\ell}{r^2} + C^4_{\ell n}\frac{S'_\ell}{r^3} + (C^5_{\ell n} + C^6_{\ell n}r^2)\frac{S_\ell}{r^4} + M\{H^1_{\ell n}(r)T''_\ell + H^2_{\ell n}(r)\frac{T'_\ell}{r} + H^3_{\ell n}(r)\frac{T_\ell}{r^2}\}] = 0 \tag{33}$$

for ℓ, n = (m+1), (m+3), ···, ITR

with boundary conditions

$$S'_\ell(0) = S_\ell(0) = T_\ell(0) = 0 \qquad \text{at } r=0$$

$$S''_\ell(1) = (2 - \frac{S_I}{\alpha})S'_\ell(1)$$

$$S_\ell(1) = 0 \qquad \text{at } r=1$$

$$T'_\ell(1) = 2T_\ell(1) \tag{34}$$

This eigenvalue problem was solved as an initial value problem using fourth-order Runge-Kutta integration. In order to start the integration, estimates were made for the unknown boundary conditions at one boundary and, since the system was linear, these guesses were combined such that the required conditions at the other boundary were satisfied. When the estimates are made orthogonal and the required conditions are made homogeneously, the determinant of a matrix whose components are the integrated results for the required conditions must be zero in order that there exists a non-trivial solution. Thus, the determinants are obtained as a function of λ and the values of λ which make the determinant zero can be found as eigenvalues. There occurs a spectrum of eigenvalues. These eigenvalues are computed as a function of Reynolds number and the critical Reynolds number is the value at which the smallest eigenvalue becomes zero. It should be mentioned here that the basic flow becomes different according to different Reynolds numbers, unlike Poiseuille or Couette flow in which only one velocity profile needs to be considered for all Reynolds numbers.

The amount of computation increases rapidly with the number of terms retained in the spherical harmonic series. Fortunately, the series converges very rapidly and can be approximated with reasonable accuracy by using only a few terms. Convergence at a Reynolds number of 100 is only slightly slower than at a Reynolds number of 10.

RESULTS AND DISCUSSION

Since this analysis considered the stability of the system discussed in Part I, the geometry of the drop and the Reynolds numbers in this section are the same as those in the corresponding discussion in Part I.

In order to study the stability as a function of Reynolds number, it was necessary to know how the surface-tension distribution changed with Reynolds number so that the corresponding basic state could be determined. In the basic flow calculation, the distribution was assumed to be always sinusoidal, and the effect of a variation in surface-tension gradient was observed by changing the decrease in surface tension independently of the Reynolds number. Accordingly, the stability calculation was carried out for an arbitrarily chosen decrease in surface tension of 9.1×10^{-6} N/m, the value observed by Levan and Newman (1976) on the surface of bubble rising at a terminal velocity. The effect of a change in the basic state produced by different surface-tension gradients was considered separately.

It is not possible to observe a negative initial disturbance growth rate since a disturbance of infinitesimal size dies out with time. However, the growth rate may be used as a measure of the instability of the system. Thus, it may be seen from Figure 3 that the instability of the system increases with increasing Reynolds numbers, and that the critical Reynolds numbers appear when the basic flow becomes a fully developed rotational flow. The critical Reynolds numbers are 10.4, 10.7, 11.3, and 12.2 for two, four, six, and eight circulation patterns, respectively. It may also be noted that the instability decreases with increasing wave number. To illustrate this more clearly, the critical Reynolds numbers are shown as a function of wave number in Figure 4. Applying the principle of selective amplification to this result explains the experimental observation that a flow of

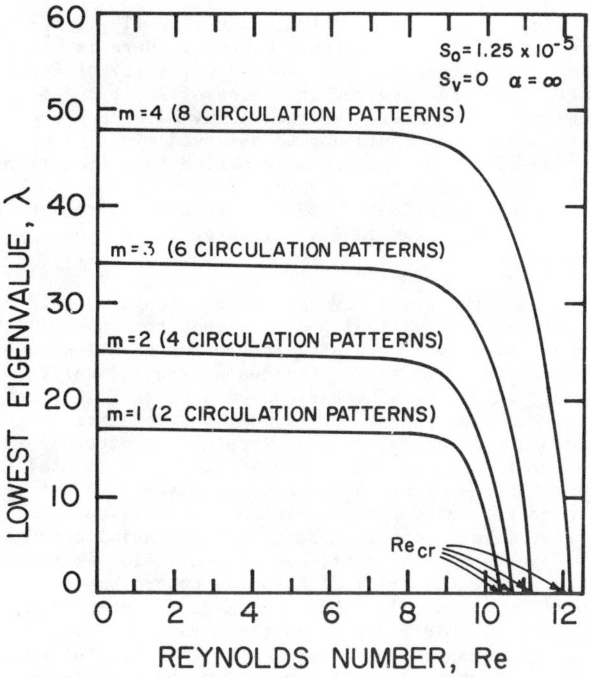

FIGURE 3. Determination of critical Reynolds numbers for different disturbance flow patterns.

two circulation patterns appears first when the system becomes unstable as the Reynolds number increases.

Another aspect, which can be seen in Figures 3 and 4, is that the system is most unstable to the disturbance of two circulation patterns; and the stability increases more rapidly for larger numbers of circulation patterns. It is interesting that the change of stability in this manner coincides with the change in

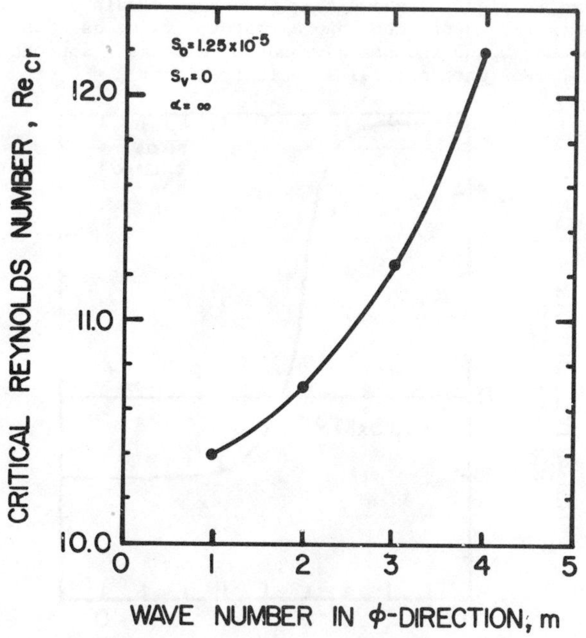

FIGURE 4. Critical Reynolds numbers as a function of wave number in the circumferential direction.

the Reynolds numbers at which Poonawalla (1976) observed circulation patterns. However, there is no fundamental difference between the stability of 8 circulation patterns and the stability of 2, 4, and 6 circulation patterns. The reason why more than 6 circulation patterns could not be observed might have been due to the difficulty of experimentally maintaining a drop at high enough Reynolds numbers due to the impingement of the jet. The occurrence of more than 6 circulation patterns in flat circular interfaces supports this view.

Although the stability characteristics predicted by the present mathematical model -- that the instability of the system increases with increasing Reynolds numbers such that the system becomes hydrodynamically unstable at Reynolds numbers above a certain critical value, and that the system is most unstable for a disturbance resulting in two circulation patterns -- agree well with what was expected from experimental observation, there is a significant difference between the computed and expected critical Reynolds numbers. The experimental observation that interfacial circulation appears when a rotational internal flow emerges suggests that the critical Reynolds number for the present system should occur at a Reynolds number of approximately 6.0 rather than the computed value of 10.4. The simplifications in estimating the interfacial properties and the assumption that the interface remains spherical under a surface-tension disturbance may account for this difference. However, another plausible explanation is the nature of linear stability analysis. Since the disturbance in linear stability analysis is restricted to be of infinitesimal size, the critical Reynolds numbers obtained by the linear method are often much higher than those experimentally observed.

For all previous calculations in which the effect of the stability group α on the stability was not considered, a value of S_I/α was set equal to 0 ($\alpha = \infty$). This choice of $\alpha = \infty$ can be explained as follows. The curve in Figure 5 shows that the negative disturbance growth rate approaches one asymptotic value as α approaches 0 and another asymptotic value as α approaches ∞. For mathematical simplicity either $\alpha = 0$ or $\alpha = \infty$ could have been used. However, $\alpha = 0$ implies that there is no variation in surface tension so $\alpha = \infty$ was used.

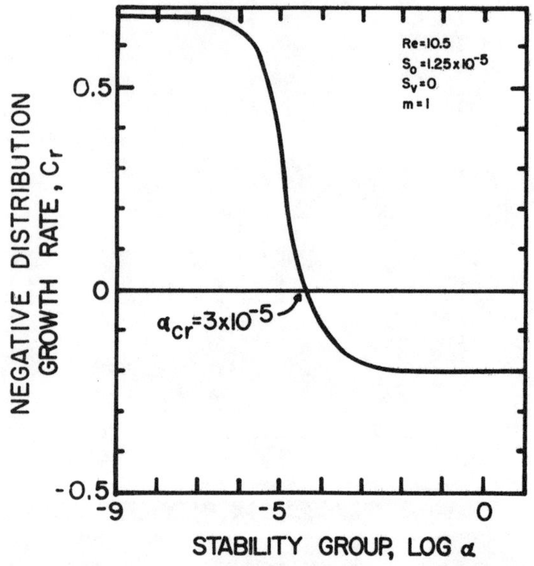

FIGURE 5. Determination of the critical stability group at a Reynolds number of 10.5.

Berg and Acrivos (1965) assumed that $D_s/k\Gamma_e$ was 5×10^{-7} for gaseous monolayers. The corresponding values of α become 1.25×10^{-10} and 2.5×10^{-15}, respectively. Figure 6 indicates that the stability of this system is almost independent of the properties of monolayers. Instability increases with increasing values of the stability group. However, it should be remembered that there were several severe assumptions made in the mathematical model with regard to transport mechanisms of surfactants and properties of monolayers. The model could have been too simplified to describe a real system accurately.

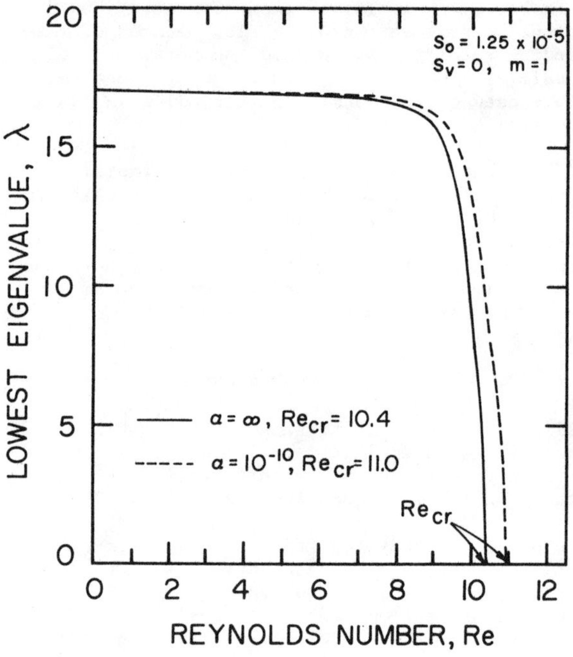

FIGURE 6. Effect of stability group on critical Reynolds number. The critical Reynolds number of 10.4 at $\alpha = \infty$ increases to 11.0 at $\alpha = 10^{-10}$.

For values of the stability group between 10^{-7} and 10^{-2}, the disturbance growth rate at a Reynolds number of 10.5 increases markedly with increasing values of the stability group, which indicates that instability increases with increasing surface-tension gradient and increasing surface diffusivity. As a result of this increase, there appears a critical value of α at which a stable system becomes unstable for a further increase in the stability group. By analogy with the concept of a critical Reynolds number, this may be called a critical stability group. But it may be deduced from the asymptotic behavior, shown in Figure 5, that the critical stability groups can be obtained for only a certain range of Reynolds numbers around 10 whereas critical Reynolds numbers can be obtained for any interfacial conditions. The effect of the stability group on the stability can also be demonstrated by the change in critical Reynolds number as shown in Figure 6.

When the basic flow was retarded to a different degree by different values of the surface-tension gradient group S_o, the result of the stability calculation was changed. Figure 7 shows the change of the initial disturbance growth rate at a Reynolds number of 10 for different basic states corresponding to different values of the surface-tension gradient group. The stability increases when the basic flow is retarded more, although not significantly.

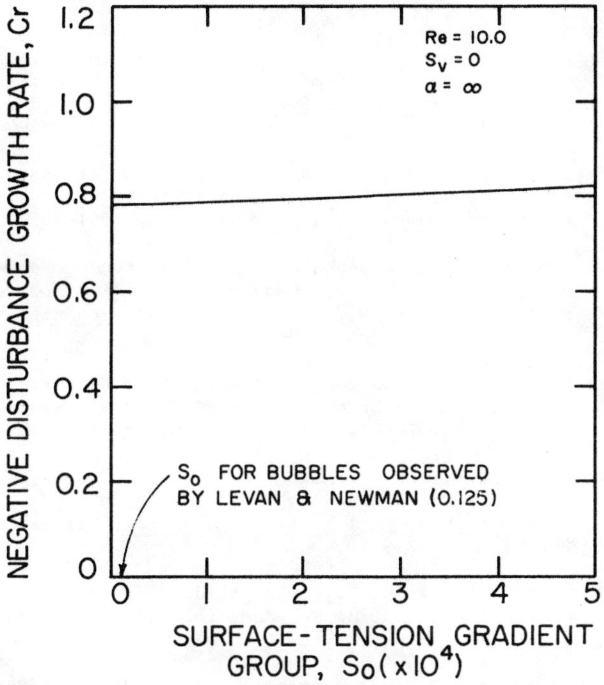

FIGURE 7. Effect of retardation in basic flow on initial disturbance growth rate.

Finally, calculations were made to see how the stability changed when a certain portion of the drop was assumed to have no disturbance. A spherical core at the center of the drop was assumed to have no disturbance. As expected, the stability increased as the radius of the undisturbed core increased, as shown in Figure 8. It may be noted from the curve that the increase in the negative disturbance growth rate (indicating greater stability) becomes more rapid for larger spherical regions, which suggests that the region near the interface plays a more important role than the region near the center of the drop in determining the stability.

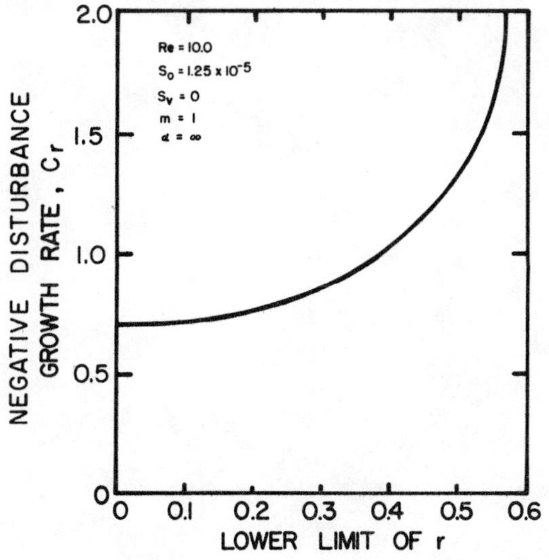

FIGURE 8. Decreasing instability as the region of no disturbance grows.

NOMENCLATURE

A_n	Values of Legendre polynomials of n-th order at $\theta = 0$.
$c = c_r + ic_i$	Initial disturbance growth rate.
c_o	Surfactant concentration in bulk fluid.
D_s	Surface diffusivity.
ITR	Order of disturbance flow spherical harmonic series truncation.
j_n	Flux of surfactants at interface in normal direction.
k	Proportional constant between surface-tension and interfacial surfactant concentration.
m	Wave number in the circumferential direction.
Nt	Order of basic flow Legendre polynomial series truncation.
$P_\ell(\theta)$	ℓth order Legendre polynomial.
$P_\ell^m(\theta)$	Associated Legendre polynomial.
R	Drop radius.
Re	Reynolds number.
Re_{cr}	Critical Reynolds number.
r	Dimensionless radial coordinate.
$\bar{r}$	Position vector.
S_o, S_I	Surface-tension gradient groups.
$\bar{S}(\bar{r})$	Poloidal field for disturbance flow.
$\bar{S}_\ell^m(\bar{r})$	Poloidal field for defining scalar $S_\ell(r)Y_\ell^m(\theta, \phi)$.
$\bar{T}(\bar{r})$	Toroidal field for disturbance flow.
$\bar{T}_\ell^m(\bar{r})$	Toroidal field for defining scalar $T_\ell(r)Y_\ell^m(\theta, \phi)$.
$\bar{U}(r, \theta)$	Basic flow.
$\bar{u}(r, \theta, \phi)$	Disturbance flow.
V	Maximum basic flow velocity at the center nozzle.
$Y_\ell^m(\theta, \phi)$	Spherical harmonic.
α	Stability group.
$\Gamma(\theta)$	Interfacial surfactant concentration.

γ	Dimensionless surface tension.
λ	Eigenvalue for linear stability calculation.
μ	Viscosity.
ρ	Density.
σ	Surface tension.
$\Phi(\bar{r})$	Defining scalar for toroidal field.
$\phi(r)$	$\equiv F(r) + \sum^{Nt}_{\ell} A_{\ell} g_{\ell}(r)$
$\Psi(\bar{r})$	Defining scalar for poloidal field.
$\psi(r, \theta)$	Stream function for basic flow.
∇, ∇^2	Gradient and Laplacian operators.
$(\ )'$	$\equiv \frac{d}{dr}$ or $\frac{d}{dx}$

Subscripts

b	Basic state.
d	Deviation.
e	Equilibrium.
n	Normal direction.
r	Real number or r-component in spherical coordinates.
s	Surface.
θ	θ-component in spherical coordinates.
ϕ	ϕ-component in spherical coordinates.

ACKNOWLEDGMENT

This research was supported by the Assistant Secretary for Energy Research, Office of Basic Energy Sciences, WPAS-KC-02-03, U.S. Department of Energy.

REFERENCES

Bakker, C.A.P., Buytenen, P.M.V., and Beek, W.J., 1966, "Interfacial Phenomena and Mass Transfer," Chemical Engineering Science, Vol. 21, No. 11, Nov., pp. 1039-1046.

Berg, J.C., and Acrivos, A., 1965, "The Effect of Surface Active Agents on Convection Cells Induced by Surface Tension," Chemical Engineering Science, Vol. 20, No. 8, Aug., pp. 737-745.

Chandrasekhar, S., 1961, Hydrodynamic and Hydromagnetic Stability, Oxford University Press, New York.

Griffith, R.M., 1962, "The Effect of Surfactants on the Terminal Velocity of Drops and Bubbles," Chemical Engineering Science, Vol. 17, No. 12, Dec., pp. 1057-1070.

Jeffreys, H., 1928, "Instability in Fluid Motion," Proceeding of the Royal Society (London), Series A, Vol. 118, pp. 195-208.

Joo, S.B., 1979, Ph.D. Thesis, Iowa State University.

Kenning, D.B.R., and Cooper, M.G., 1966, "Interfacial Circulation Due to Surface Active Agents in Steady Two-Phase Flows," Journal of Fluid Mechanics, Vol. 24, Part 2, pp. 293-306.

Levan, M.D., and Newman, J., 1976, "The Effect of Surfactant on the Terminal and Internal Velocities of a Bubble or Drop," American Institute of Chemical Engineers Journal, Vol. 22, No. 4, pp. 695-701.

Merson, R.L., and Quinn, J.A., 1965, "Stagnation in a Fluid Interface," American Institute of Chemical Engineers Journal, Vol. 11, No. 3, pp. 391-395.

Munson, B.R., and Joseph, D.D., 1971, "Viscous Incompressible Flow Between Concentric Rotating Spheres. Part 1: Basic Flow," Journal of Fluid Mechanics, Vol. 49, Part 2, pp. 289-303.

Poonawalla, F.F., 1976, "Interfacial Flow Patterns on a Static Liquid Drop with Forced Internal Circulation," M.Sc. Thesis, Iowa State University.

Scriven, L.E., and Sternling, C.V., 1964, "On Cellular Convection Driven by Surface-Tension Gradients," Journal of Fluid Mechanics, Vol. 19, Part 3, pp. 321-340.

Scriven, L.E., 1960, "Dynamics of a Fluid Interface," Chemical Engineering Science, Vol. 12, No. 2, May, pp. 98-108.

Skelland, A.H.P., and Caenepeel, C.L., 1972, "Effects of Surface Active Agents on Mass Transfer During Droplet Formation, Fall, and Coalescence," American Institute of Chemical Engineers Journal, Vol. 18, No. 6, pp. 1154-1163.

Chapter 23

FACTORS AFFECTING DROPLET SIZE DISTRIBUTIONS PRODUCED IN DISPERSED PHASE MIXERS

R. Mackelprang, J.A. Herbst, and J.D. Miller

Department of Metallurgy and Metallurgical Engineering
University of Utah
Salt Lake City, Utah 84112

ABSTRACT

Droplet size distributions were determined by stabilizing in gelatin and measurement of the distributions by means of computerized image analysis. Several variables that affect droplet sizes were studied and the resulting size distributions were modeled via a population balance equation. Key parameters in the model were related to the physical properties and operating conditions of the systems examined using a multiple linear approach.

FACTORS AFFECTING DROPLET SIZE DISTRIBUTIONS PRODUCED IN DISPERSED PHASE MIXERS

R. Mackelprang, J. A. Herbst and J. D. Miller

Department of Metallurgy and Metallurgical Engineering
University of Utah
Salt Lake City, Utah 84112

Introduction

Dispersed phase mixers are an important reactor type in several major industries and are becoming very popular in the metallurgical industry for the recovery of metal values from low grade aqueous solutions via solvent extraction. A key aspect of optimizing dispersed phase reactions, such as found in solvent extraction systems, is to understand the variables that affect the drop size distribution. The droplet size distribution and interfacial area are critical parameters of most solvent extraction systems since these often control the reaction rate between phases and may also influence significant economic issues such as organic losses, power consumption and capital-cost of equipment. Due to the wide use of mixer(s)/settler(s) in solvent extraction systems an important contribution can be made by quantifying the effect of important operating variables which determine the performance of these systems.

To accomplish this objective simple accurate measurements of the drop size distributions in the mixer must be made. Numerous techniques have been utilized for measuring drop size distributions including: chemical reactions (7); sedimentation techniques (1,9) simulation with hot wax (5,20) ; light transmission (3, 19, 24); dilution/stabilization techniques (17, 21); photographic techniques (4, 6, 8, 12, 14, 16, 20); and encapsulation techniques (14). Each technique has some advantages and limitations and each can be effective in certain systems or cases. Few, however can be used universally. Hence, after careful consideration of all of these techniques and after preliminary testing of many of these techniques, a dilution/stabilization technique known as gelatin embedding (17) was selected as the best possible method for stabilizing droplets in this project.

Finally, any complete description of a drop size distribution should include an accurate mathematical model for describing the size distribution. This model should also be useful for predicting size distributions in terms of the variables that affect it without introducing unnecessary complexity. In this regard a useful mathematical model was developed based on population balance concepts.

Experimental Systems and Techniques

The liquid-liquid system investigated in this project was studied at six different phase ratios and employed six organic and eight aqueous phases of different compositions. The standard organic phase was selected as a 10% by volume LIX 64N in 90% by volume Escaid 200, whereas the standard aqueous phase was selected as a 0.25 molar sodium sulfate solution which is believed to represent the ionic strength of typical industrial systems. A summary of the organic and aqueous phases used in this project is presented in Table 1. Details of these systems and their physical properties can be found elsewhere (13) (18).

The experimental equipment utilized in this project consisted of two agitating units, three different vessels and one impeller for mining. The primary agitation unit used was a Benco Model ELB Agitator which was mounted on a dynamo-meter stand and equipped with an Ametek Model L-2 force gage for power input measurements. This unit was capable of speed settings up to 1100 RPM and torque measurements of up to 20 in-lbs. A secondary agitating unit consisted of a Vickers hydraulic drive transmission assembly powered by a 5.0 horsepower motor. This unit was connected to the impeller shaft by a coupling arrangement which would also permit the insertion of a torque transducer for power measurements. High speeds and power inputs could be attained with this unit using small impellers.

The primary vessel used in this project was a modified, four liter beaker constructed of pyrex glass. This beaker was designed to allow for the mounting of four equally spaced baffles of 1.3 cm and provided with an enlarged spout to facilitate sampling. This vessel had an inside diameter of 15.9 cm and was run at a constant volume of 3.25 liters or a liquid depth of 17.2 cm in a batch mode. A similar vessel to this was utilized for the continuous tests and a somewhat different, totally enclosed vessel of larger volume was used for the high speed tests, details of these are reported elsewhere (13).

The only impeller employed in this project was a single, flat blade turbine impeller with a diameter of 6.35 cm and a blade height of 0.8 cm. The impeller was constructed of 316 stainless steel and came with the ELB agitator kit.

The sampling procedure was essentially the same regardless of the agitating equipment or vessel being used. The organic droplets inside the vessel were sampled by withdrawing a small amount of emulsion from the desired point in the vessel by means a small siphon. These small siphons were constructed from 3 mm diameter glass tubing by heating and bending into a "V" shape. Generally this construction allowed the

Table 1. Aqueous and Organic Phase Composition and Physical Properties

System	Composition	pH	Ionic Strength	Density (g/cc)	Viscosity (poise x 10^2)
Aqueous 1	Distilled Water	6.4	0.000	1.00	1.00
Aqueous 2	.25 M Na_2SO_4	6.8	0.750	1.03	1.18
Aqueous 3	.875 M Na_2SO_4	7.2	2.625	1.10	1.61
Aqueous 4	1.50 M Na_2SO_4	7.7	4.500	1.17	2.16
Aqueous 5	.25 M Na_2SO_4 + .0826 M H_2SO_4	2.0	1.000	1.03	1.26
Aqueous 6	.25 M Na_2SO_4 + .1815 M NH_4OH	11.0	0.930	1.03	1.26
Aqueous 7	.25 M Na_2SO_4 + .0826 M H_2SO_4 + .051 M $CuSO_4$	2.0	1.100	1.03	1.26
Aqueous 8	.225 M $NaHCO_3$ + 0.225 M Na_2CO_3	9.8	1.35	1.04	1.30
Organic 1	Escaid 200			0.751	1.49
Organic 2	Cyclohexane			0.761	0.762
Organic 3	Xylene			0.860	0.610
Organic 4	Isooctane			0.693	0.520
Organic 5	10% LIX in 90% Escaid 200			0.764	1.720
Organic 6	5% Adogen 383 in 95% Xylene at 0.076 M Sebacyol Chloride			0.900	0.650

siphon to be placed in the sample lip of the vessel in a manner such that the outside stem of the siphon was vertical while the inside stem was located at the desired sampling point which was usually at a point mid-way between the liquid surface and the impeller and mid-way between the vessel wall and the impeller shaft.

Once in place, the siphon was initiated and allowed to run briefly after which a couple of drops of emulsion were allowed to fall into a petri dish containing 4 to 6 ml of freshly prepared, 8% by weight pigskin gelatin solution at 28° to 30°C. The emulsion drops were mixed briefly and the gelatin solution manipulated to allow the formation of a thin layer of gelatin with representative drops to cover the bottom of the petri dish. The petri dish was then immediately placed in an ice water bath where it solidified almost instantly after which it was scanned microscopically for quality. Poor quality samples were discarded whereas good samples were refrigerated at 10°C and then analyzed within the next 24 hour period.

Most samples were analyzed using the Quanta Met 720, a computerized image analyzer that counts and sizes the embedded drops. A minimum of 300 drops per sample were sized in this manner, however, as many as 1000 per sample were often counted.

Experimental Results

The major variables affecting drop sizes which were investigated in this project include: (1) the time to reach equilibrium; (2) speed/power input; (3) phase ratio or phase fraction; (4) the presence and absence of an extractant-LIX74N, (5) the behavior of pure diluents; (6) the rate of mass transfer across the interface; (7) aqueous phase pH; and (8) aqueous phase ionic strength. The ability to predict continuous flow distributions at steady state from batch data was also examined.

The approach to equilibrium or steady state was investigated mainly to note the appropriate time for sampling to obtain a true steady state sample. The approach to equilibrium was noted to be a complex function of several variables of which speed/power input and phase fraction are the most significant. Therefore, a high phase fraction system agitating at a low speed was chosen as a limiting case with all subsequent cases to be tested utilizing higher speeds and lower phase fractions and hence shorter times for equilibrium to occur. The test consisted of taking samples at times of; 1.25, 2.5, 5.0, 10.0 20.0, 30.0, 40.0, and 50 minutes after the start of agitation. It was found upon analysis of these respective size distributions that significant changes occurred for times up to 30.0 minutes. Thus all subsequent tests were conducted after allowing a miniumum of thirty minutes for equilibrium to occur.

The experimental design performed on the standard system consisted of four levels of phase ratio; 0.2, 0.3, 0.4, and 0.5 and three levels of speed/power input; 450rpm/2.5×10^{-3} HP, 550rpm/3.7×10^{-3} HP, and 650rpm/5.7×10^{-3} HP. Typical results are illustrated in Figure 1 for speed/power inputs at two levels of phase fraction and Figure 2 where four levels of phase fraction are illustrated at two levels of speed. The results of these tests concur with the general findings of other researchers, i.e., an increasing fineness of droplet size distribution as the speed/power input increases and as the phase fraction decreases.

The effect of mass transfer on the droplet size distribution was studied by adding copper as dissolved cupric sulfate to the standard system at a pH of 2.0 and then sampling the size distribution at times of: 0, 20, 40, 80, 160, 320 and 1800 seconds after addition.

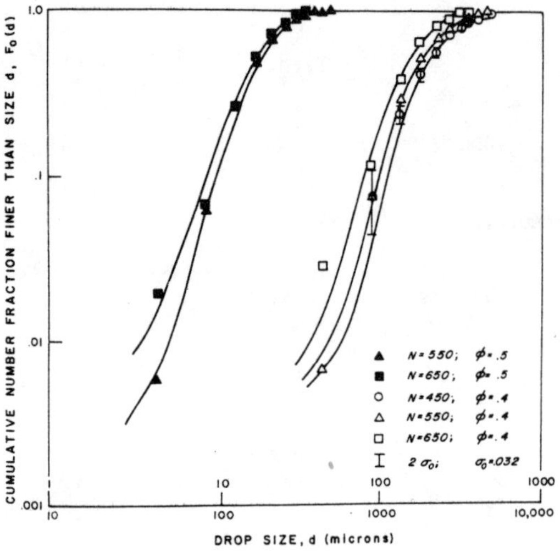

FIGURE 1. The effect of speed/power input at phase fractions of .5 and .4 in the standard system (Upper scale right side plots, lower scale left side plots).

The respective size distributions for these times were noted to differ very little and hence it is concluded that mass transfer in this system has no effect on droplet size distribution which is not too surprising since the physical properties of the system remained relatively constant.

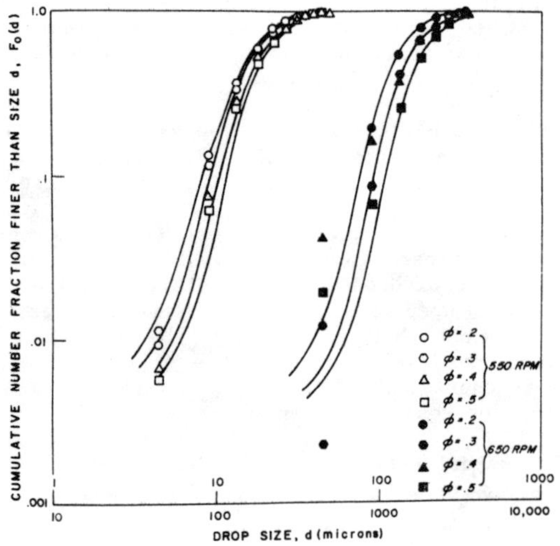

FIGURE 2. The effect of phase fraction at speeds of 550 and 650 rpm in the standard system (Upper scale right side plots, lower scale left side plots).

The effect of the presence of the extractant, LIX 64N, as 10% of the organic volume was evaluated by comparing the size distribution obtained in a pure Escaid 200 system to the size distribution of a system in which the organic phase contained 10% LIX 64N. The results are given in Figure 3 which shows the size distributions for the pure Escaid 200 system to be significantly coarser than the 10% LIX 64N system.

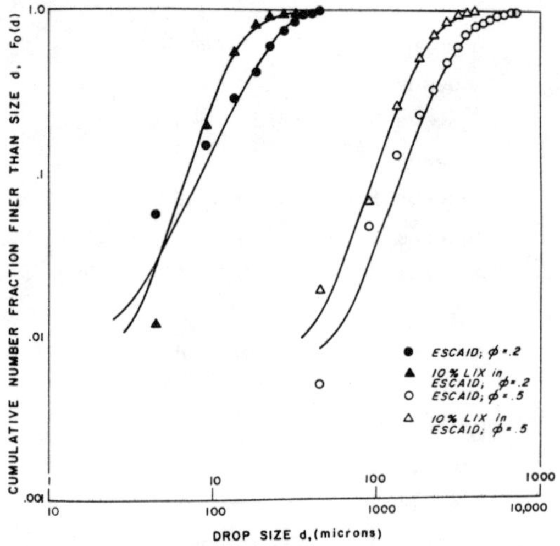

FIGURE 3. The effect of LIX 64N presence in Escaid 200 on the drop size distribution at a speed of 650 rpm in the standard aqueous system (Upper scale right side plots, lower scale left side plots).

This phenomenon can be explained in terms of the effect of the LIX 64N concentration on the liquid/liquid interfacial tension. With LIX 64N present, the interfacial tension was found to be 16.2 dynes/cm as compared to 45.2 dynes/cm for the pure Escaid. The lower interfacial tension results in a lower resistance to breakage forces, hence a finer drop size results. Essentially the same behavior was noted when other diluents with similar interfacial tensions such as isooctane and cyclohexane were used. Pure xylene with a lower interfacial tension than any of the other diluents again displayed a finer size distribution.

The variation in droplet sizes caused by changes in the aqueous phase composition was also investigated. Obviously the most common changes in the aqueous phase are those resulting from dissolved salts or ionic strength and changes in pH.

The effect of dissolved salts or ionic strength level was studied by measuring the change in the size distribution at four levels of sodium sulfate addition. The results are illustrated in Figure 4 and indicated that small additions of salt have little effect on the droplet size distribution while large addition of salt causes the distribution to become progressively finer. The explanation for this phenomenon arises from changes in the physical properties of the aqueous phase, namely density and viscosity which are changing to promote breakage, and also as a result of electrostatic surface forces. In the latter case preferential adsorption of an ion at the organic surface may lead to droplet charging or the formation of a dense double layer which inhibits coalescence and enhances droplet shattering during breakage.

The pH of the aqueous phase in the standard system

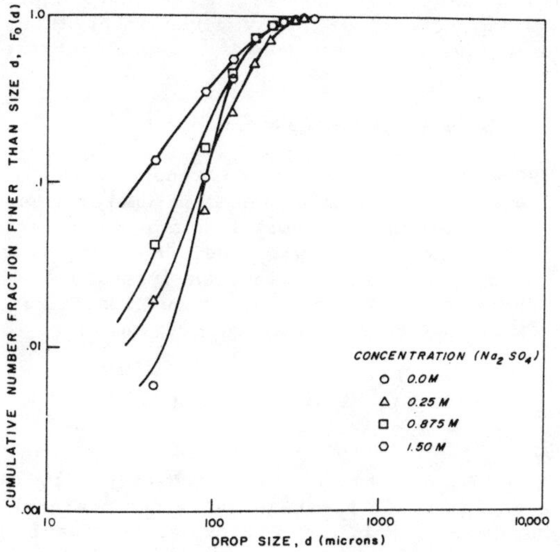

FIGURE 4. The effect of aqueous ionic strength on size distribution in the LIX 64N/Escaid 200 system at a phase fraction of .5 and speed of 650 rpm.

was also noted to affect droplet size distribution appreciably in some cases. Here it was found that size distributions from low and high pH ranges differed from those at moderate pH ranges. The change was most dramatic in the case of high pH's as shown in Figure 5.

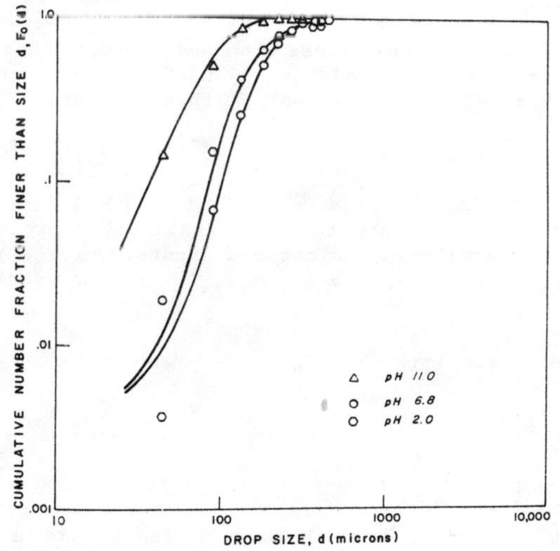

FIGURE 5. The effect of aqueous pH on drop size distribution in the standard organic system at a phase fraction of .5 and a speed of 650 rpm.

The size distribution at a pH of 11 was much finer than at lower pH's. This behavior was again traced to a large decrease in interfacial tension which changes from 16.2 dynes/cm at moderate pH to about 5.2 dynes/cm, at a high pH - a decrease of over two thirds. These results may reflect the conversion of the acid form of the extractant into the sodium salt which could significantly alter the interfacial tension and electrostatic forces thus accounting for the fine drop size distribution.

The less drastic changes in size distributions that occur at low pH are not conclusive and should be checked at very low pH values which may display variations more clearly. Nevertheless, it is thought that lowering the pH may affect the charge on the drops and hence droplet electro-kinetic behavior which in turn may influence breakage and coalescence mechanisms.

The last topic of interest investigated in this project was the appropriateness of using batch tests to simulate continuous tests in the LIX 64N/Escaid 200 -acid/sulfate system. These tests were performed in a slightly different vessel which was modified for continuous flow. The size distributions produced by this vessel in the batch mode were compared to the size distributions generated in the continuous mode with retention times of 2.16 and 0.54 minutes. The results are shown in Figure 6 and indicate that very similar size distributions occur between batch tests and continuous tests with moderate to high retention times.

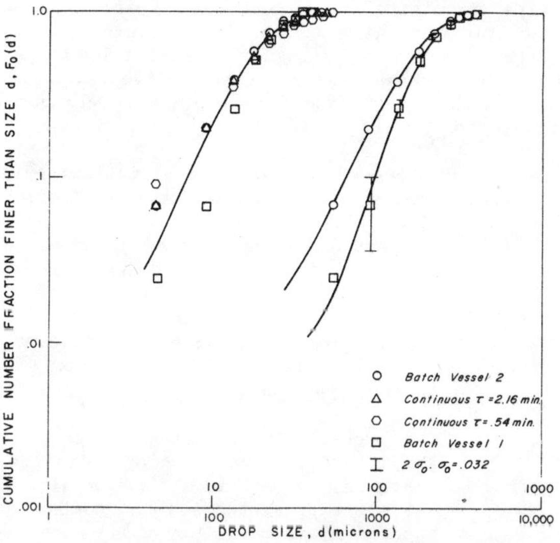

FIGURE 6. Drop size distribution differences produced by vessels one and two and continuous mode operations with the standard system at a phase fraction of .5 and a speed of 650 rpm (Upper scale right side plots, lower scale left side plots).

At short retention times the agreement becomes worse and thus the simulation is worse. However, we can conclude from these results that batch tests can be used to simulate continuous tests run at moderate to high retention times.

Mathematical Model

Mathematical models for describing drop size distributions are numerous and varied. Several models have utilized a population balance approach with good success (2, 5, 23), but have incorporated some questionable assumptions, manipulations and excessive complexity into their development and use. Hence it was desired to examine these models in an effort to improve on their utility.

Since the scope of this topic is rather extensive it is impossible to report all the details of the model development here. Instead a general outline will be presented. The details of the model development are given elsewhere (13, 25).

The model development begins with the macroscopic population balance equation initially proposed by Hulbert and Katz (10):

$$\frac{1}{V}\frac{\partial}{\partial t}\{V\overline{\psi}(\underline{\zeta},t)\} = -\sum_{j=1}^{J}\frac{\partial}{\partial \zeta_j}\{x_j\overline{\psi}(\underline{\zeta},t)\} - \overline{D}(\underline{\zeta},t) + \overline{B}(\underline{\zeta},t) + \frac{1}{V}(Q_{in}\overline{\psi}_{in} - Q_{out}\overline{\psi}_{out}) \quad (1)$$

The term on the left hand side of Equation 1 represents the overall time rate of change in the average population density function for a droplet assembly. Where $\overline{\psi}(\underline{\zeta},t)d\underline{\zeta}$ gives the number of particles in a system per unit volume ($\underline{\zeta}$ represents the transposed vector of properties $\zeta_1, \zeta_2, \zeta_3 \ldots \zeta_n$). The first term on the right hand side represents the change in properties resulting from continuous processes such as mass transfer while the next two terms represent the changes occuring as a result of catastrophic processes such as breakage and coalescence events. The final term on the right hand side represents changes due to flow in and out of a vessel.

We now proceeed to make the following initial assumptions and simplifications for the mixer:

(1) The system being modeled is a batch mixer and hence flow in and out are zero ($Q_{in} = Q_{out} = 0$)

(2) The volume in the mixer is constant, hence the volume terms, V, cancel from each term in the equation

(3) The model is concerned with only one property ζ_1, that being droplet size expressed as volume v, and a size distribution represented as a number density functon by volume, $\psi(v,t)$.

(4) Size changes due to continuous processes such as chemical reactions are negligible, thus $x_1 = d\zeta_1/dt$ is eliminated from the equation.

(5) The system being modeled is isothermal and homogeneous and will approach a state of equilibrium in a finite time. Equation (1) therefore reduces to:

$$\partial/\partial t\,\{\overline{\psi}(v,t)\} = \overline{B}(v,t) - \overline{D}(v,t)$$

$$= \overline{B}_b(v,t) + \overline{B}_c(v,t) - \overline{D}_b(v,t) - \overline{D}_c(v,t) \quad (2)$$

Where the subscripts b, c, refer to the processes of breakage and coalescence respectively, hence $\overline{B}_b$ refers to the birth of drops due to breakage and so forth.

Each term on the right hand side of Equation 2 must now be developed individually and thus we begin with the term $\overline{D}_b(v,t)$ since it is the most basic to the model development. We are interested here in obtaining an expression for the death of droplets due to breakage of drops in any given interval v to v+dv per unit time and volume which we shall represent as:

$$\overline{D}_b(v,t)dv = -S(v,t)\overline{\psi}(v,t)dv \quad (3)$$

Here the term S(v,t) is known as the selection function which is a lumped parameter representing the fraction of droplets of a given size which are breaking per unit time. For simplicity it is assumed that the selection function is independent of time and is proportional to droplet "size" as represented in Equation 4;

$$\overline{D}_b(v,t)dv = -kv^m\,\overline{\psi}(v,t)dv \quad (4)$$

where k is a constant. The exponent m will take on a value of unity for breakage rates proportional to drop volume, two thirds for rates proportional to drop area, one third for rates proportional to drop diameter and zero if the breakage rate is independent of drop size.

Likewise the term representing the birth due to breakage of new drops into the interval v to v+dv from all larger intervals v to v_{max} can be represented as:

$$\overline{B}_b(v,t)dv = \int_v^{v_{max}} b(v,v')\,kv'^m\overline{\psi}(v',t)dv'dv \quad (5)$$

Where b(v,v') is known as the breakage function and represents the number of daughter droplets of size v resulting from a breakage of a droplet of size v'. An assumption now must be made as to the form of b(v,v'). In this case it was assumed that binary breakage occurs producing a uniform distribution of daughter drops. Equation 5 then becomes:

$$\overline{B}_b(v,t)dv = \int_v^{v_{max}} 2kv'^{(m-1)}\,\overline{\psi}\,(v',t)dv'dv \quad (6)$$

Now turning to the contributions of coalescence events we can state that the death of droplets out of this interval via coalescence events over the range of all interval sizes, v_{min} to $v_{max}-v$ is represented as:

$$\overline{D}_c(v,t)dv = -\int_{v_{min}}^{v_{max}-v}\lambda(v,v',t)\,\overline{\psi}(v,t)\,\frac{\overline{\psi}(v',t)}{\overline{n}(t)}\,dv'dv \quad (7)$$

Where the term $\lambda(v,v',t)$ is known as the coalescence function which represents the probability that a drop of size v will coalesce with a drop of size v' in a dt of time. The form of this function has been assumed to be constant, hence independent of droplet size and time. Furthermore, if V_{max} is assumed to be large with respect to dv', the integration of Equation 7 gives:

$$\overline{D}_c(v,t)dv = -\lambda\overline{\psi}(v,t)dv \quad (8)$$

Finally the birth of droplets into the interval v to v+dv from coalescence events occuring among drops from all smaller size intervals from v_{min} to the size v can be represented as:

$$\overline{B}_c(v,t)dv = \frac{\lambda}{2}\frac{1}{\overline{n}(t)} \int_{v_{min}}^{v} \overline{\psi}(v-v',t)\overline{\psi}(v',t)dv'dv \qquad (9)$$

Combining each of these respective expression into the reduce population balance equation and rearranging gives:

$$\frac{\partial}{\partial t}\left(\overline{\psi}(v,t)\right) = -\overline{\psi}(v,t)\{kv^m+\lambda\} + 2k\int_{v}^{v_{max}} v'^{(m-1)}\overline{\psi}(v',t)dv'$$

$$+\ \lambda/2\overline{n}(t) \int_{v_{min}}^{v} \overline{\psi}(v-v',t)\overline{\psi}(v',t)dv' \qquad (10)$$

The form of this equation now requires that the parameters k, λ, m and v_{max} be specified prior to obtaining a solution. However, by making the equation dimensionless we can eliminate the parameters k, λ and thereby attempt solutions by choosing various values for the parameter m (from zero to one) once a reliable estimate of v_{max} is obtained.

The following non dimensional quantities are proposed for making Equation 10 dimensionless:

$$v_o = (\lambda/2k)^{1/m} \qquad (11)$$

$$t_o = 1/\lambda \qquad (12)$$

$$v^* = v/v_o \text{ and } dv^* = dv/v_o \qquad (13)$$

$$t^* = t/t_o \text{ and } dt^* = dt/t_o \qquad (14)$$

$$f_o(v,t) = f_o^*(v^*,t^*)/v_o \qquad (15)$$

$$\overline{n}^*(t^*) = \overline{n}(t)/\overline{n}(0) \qquad (16)$$

$$\overline{\psi}^*(v^*,t^*)dv^* = \overline{n}^*(t)f_o^*(v^*,t^*)dv^* \qquad (17)$$

Upon substituting these expressions into Equation 10 one obtains:

$$\partial/\partial t^*\{\overline{\psi}^*(v^*,t^*)\} = -\overline{\psi}^*(v^*,t^*)(1+\frac{v^{*m}}{2})$$

$$+ \int_{v^*}^{v^*_{max}} v'^{*\,m-1}\ \overline{\psi}^*(v'^*,t^*)dv'^* \qquad (18)$$

$$+\ 1/2\overline{n}^*(t^*)/\overline{n}(0) \int_0^{v^*} \overline{\psi}^*(v^*-v'^*,t^*)\overline{\psi}^*(v'^*,t^*)dv'^*$$

The solutions for the selected values of the parameter m were generally obtained by using a fourth order interpolating polynomial to integrate and a fourth order Adams-Bashforth-Moulton method to differentiate (13). However, it was found that for the case of m=1 an analytical solution was possible (25) in the steady state form of Equation 18 which produces an exponential number density function, $f_o(v)$, with a dimensionless mean size of 1.0, that is:

$$f_o(v) = \frac{1}{v_o} e^{-v/v_o} \qquad (19)$$

Numerical solutions for the cases where the parameter m equals 1, 2/3, 1/3 and 1/6, corresponding to selection functions which are proportional to drop volume, drop area, drop diameter and the square root of the drop diameter, are shown in Figure 7.

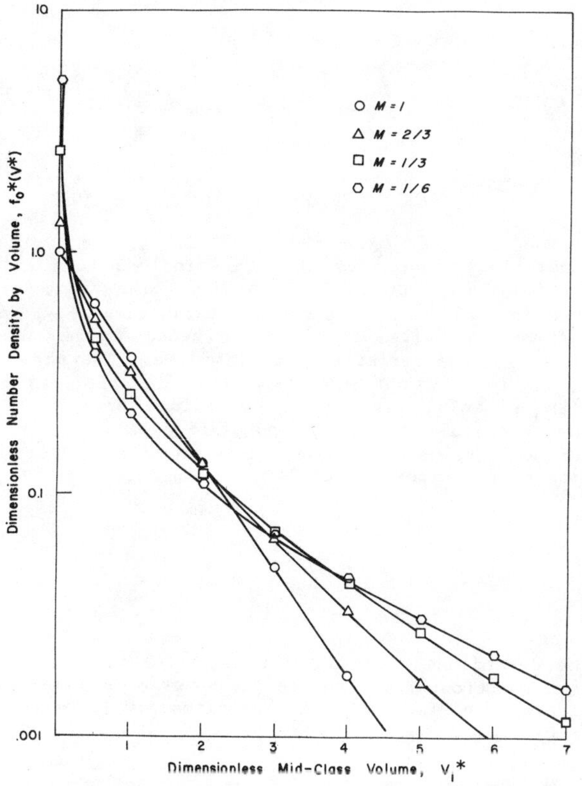

FIGURE 7. Dimensionless drop size distributions generated by the population balance Equation 18 for a batch mixer.

The formulation of the dimensionless population balance equation results in the recognition of the importance of the parameters m and v_o. Therefore, to fully utilize and develop the potential of the model it becomes necessary to relate these parameters to the system properties and operating conditions.

The parameter m is the single most important parameter of the model since it stresses the importance of breakage processes on the droplets as well as the final form of the mathematical solution and, in this regard, can be thought of as a distribution modulus.

The solutions to the model for various cases of m indicates that larger values of m result in coarser size distributions than those with smaller values of m. This behavior suggests that the parameter m can be related to system variables and operating conditions. To test this possibility an empirical equation developed from knowledge of important liquid physicochemical properties and operating conditions was constructed. Important variables were grouped together in dimensionless groups via Buckingham Pi Theory and the coefficients and exponents determined by multiple linear regression on experimental data. Values for the parameter m were selected by plotting each set of experimental data transformed to dimensionless form against the dimensionless distributions generated by the model for several cases of the parameter m and then choosing the value of m that best explained the experimental data. Hypothesis tests were then performed on each term to test its significance in the model. The final accepted equation for m was:

$$m = e^{0.19} (1+\phi)^{1.76} (\nu_d/\nu_c)^{0.37} (H/D)^{0.52} (N'_{we})^{-0.38} \qquad (20)$$

The parameter v_o on the other hand is the basis for converting to or from the dimensionless form and is the basis for the simplified model. The parameter v_o is also a function of the other parameters k, λ, and m and has the units of volume and hence may be thought of as a characteristic size or volume. The parameter v_o is readily obtained from the dimensionless mean size or volume, which is determined both from the numerical solution to the population balance equation for each value of m, and from the dimensional mean size of the experimental data by the relation:

$$v_o = \bar{V}/\bar{V}^* \qquad (21)$$

The dependence of the dimensionless mean size, which is a function of v_o, on the parameter m is shown in Figure 8 and is seen to be approximately linear.

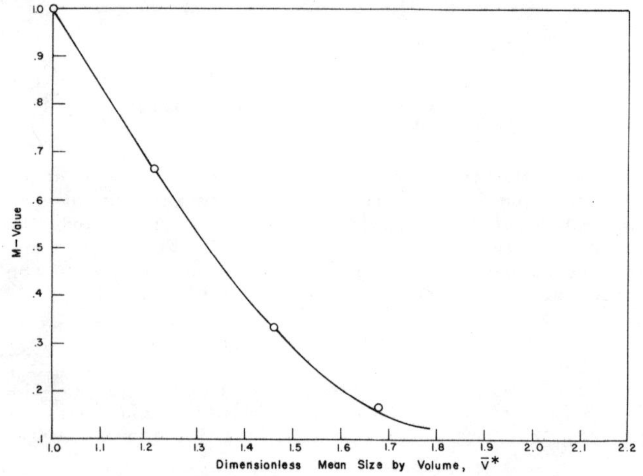

FIGURE 8. The parameter m as a function of the dimensionless mean size by volume.

It may be obvious by now that once the parameter m can be predicted as by Equation 20 and the corresponding dimensionless mean size estimated from Figure 8 all that remains is to predict the dimensional mean volume. At this point all the major parameters of the model are known and the size distribution can be determined.

Since the mean size can be related to the parameter v_o which in turn is a function of the parameter m; it seems logical to assume that the mean size is a function of the same variables that determine the parameter m. (Independent considerations of the factors affecting the mean size will lead to the same conclusion.) Hence developing an analogous expression for the mean size by the same method as that used to develop the expression for the parameter m we get:

$$\bar{V} = D^3 e^{-6.37} (H/D)^{-1.24} (\bar{T})^{0.03} (N'_{we})^{-1.54} (\phi)^{0.89} \qquad (22)$$

The large dependence of this equation on the impeller size, D, is the result of the dimensionless analysis by the Buckingham Pi Theory and unfortunately limits this equation to systems with similar sized impellers to the one used here.

Dimensionless Correlation of Experimental Data and Prediction of Dimensional Size Distributions

The experimental data generated in this project was transformed into dimensionless form utilizing m-values calculated from Equation 20 and v_o values calculated from the data mean sizes with assistance from Figure 8. Examples of experimental data that fit the model predictions for various cases of the parameter m are shown in Figures 9, 10, 11 and 12.

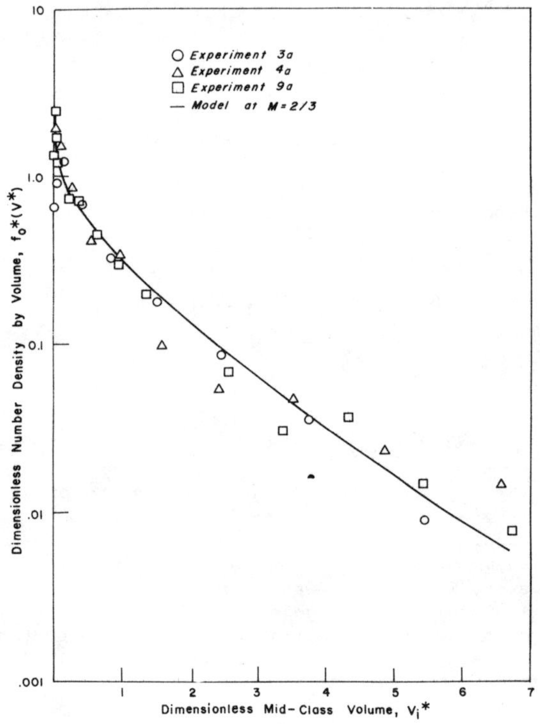

FIGURE 9. Dimensionless drop size distributions with m-values near two-thirds compared to the model at m equal two-thirds.

Values of m for the data obtained in this project ranged from 0.69 to 0.13 with most falling in the neighborhood of m equal to one-half.

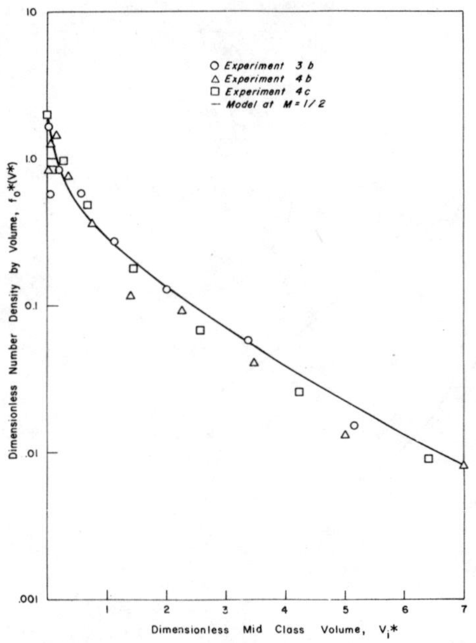

FIGURE 10. Dimensionless drop size distributions with m-values near one-half compared to the model at m equals one-half.

The upper limit of m-values is believed to be in the range of m equal to 0.85 to 1.0 for most conventional solvent extaction systems.

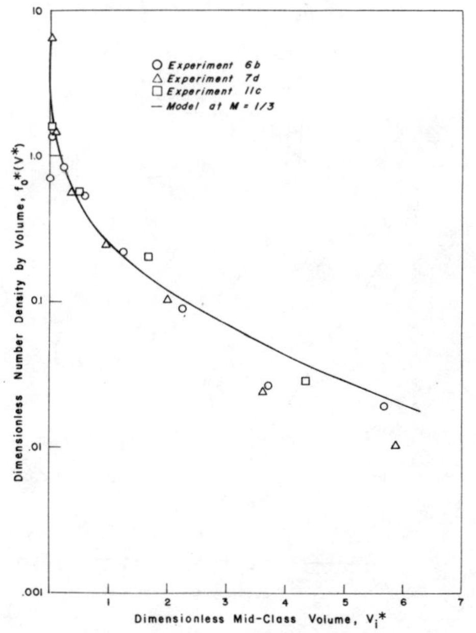

FIGURE 11. Dimensionless drop size distributions for data with m-values near one-third compared to the model prediction at m equal one-third.

Large values of m seem to correspond to low intensity of agitation. Few distributions can be found with m-values larger than one as complete emulsifications of the organic phase cannot occur in these systems. Values of m near zero on the other hand are believed to be the lower limit for the parameter values and these would apply only to very fine, collodial dispersions produced by very intense agitation not commonly used in solvent extaction systems.

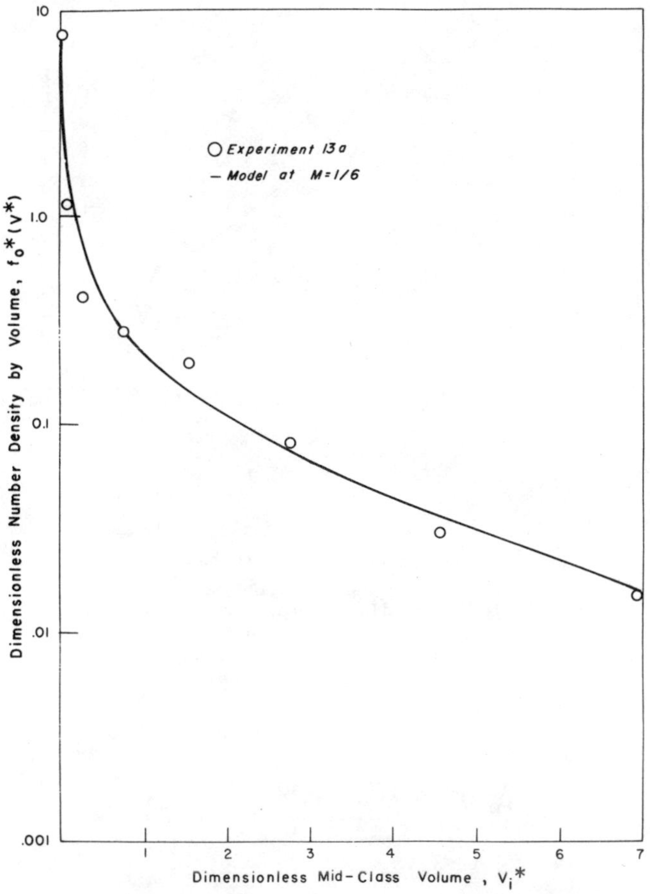

Figure 12. Dimensionless drop size distribution for data with a m-value near one-sixth compared to the model prediction at m equal one-sixth.

It is apparent by now that once the population balance equation has been solved to yield dimensionless size distributions and mean sizes for a reasonable range of m-values, and once equations become availiable for calculating accurate m-values and dimensional mean sizes from the system physical properties and operating conditions, then the entire dimensional size distribution can be predicted.

To demonstrate the validity and accuracy of this procedure the solutions to the population balance equation and Equations 20, 21 and 22 for m and v_o were used to generate the dimensional distributions for several experimental tests conducted during this project as well as experimental tests conducted by other researchers.

Examples of project data versus model predictions are shown in Figures 13 and 14 for cases with m-values ranging from 0.14 to 0.58 (13).

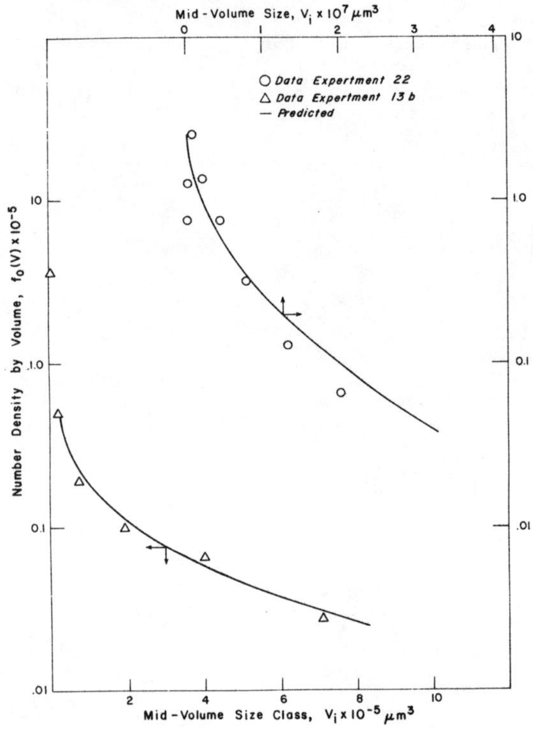

FIGURE 13. Predicted dimensional drop size distributions compared to actual data for experiments 2a and 13b from reference (13).

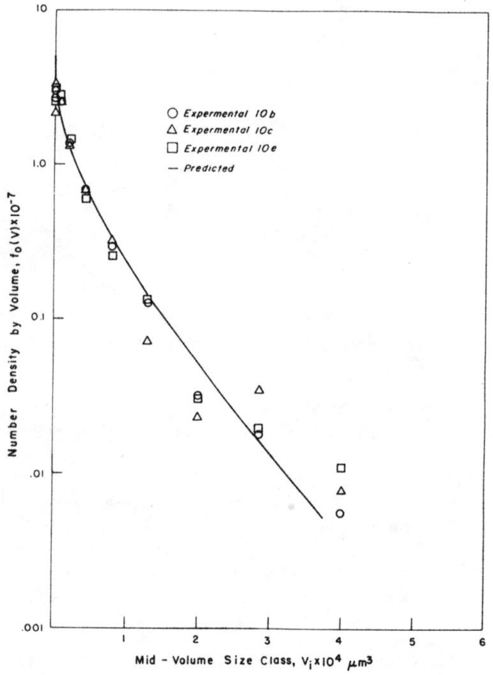

Figure 14. Predicted dimensional drop size distributions compared to actual data for experiments 10b, 10c, and 10e from reference (13).

Examples of data correlations using the results of other researchers are shown in Figures 15 and 16.

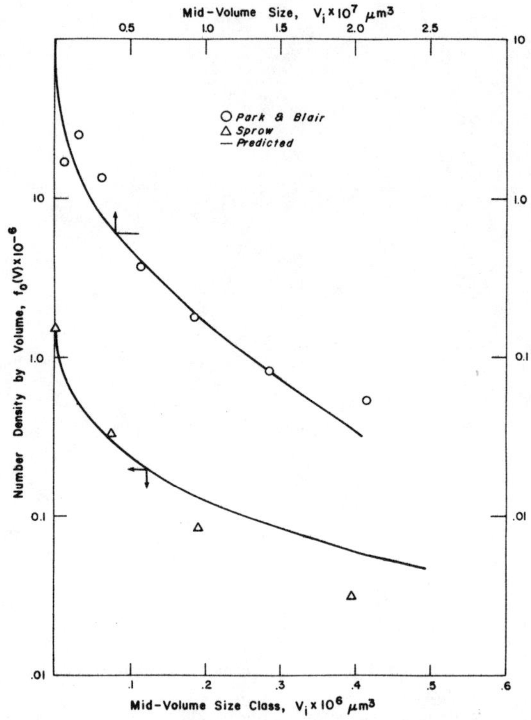

FIGURE 15. Predicted dimensional drop size distributions compared to actual data for the work of Park and Blair (16) and Sprow (22).

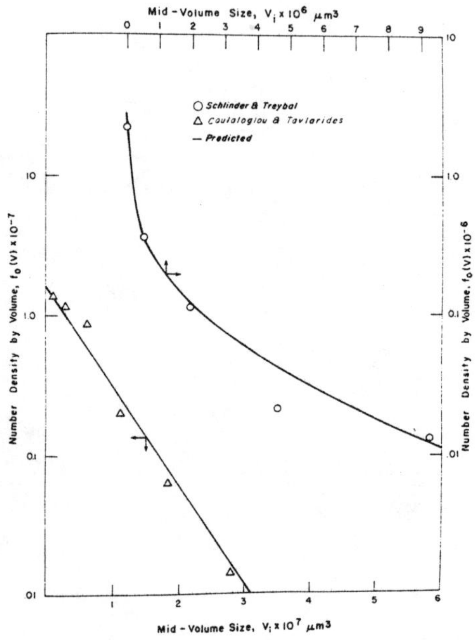

FIGURE 16. Predicted dimensional drop size distribution compared to actual data for the work of Schlinder and Treybal (19) and verification of Equation 19 for the data of Coulaloglou and Tavlarides (6).

The model predictions for the work of Coulaloglou and Tavlarides (5) shown in Figure 16 utilizes the mean size reported by the authors instead of calculating one by Equation 22 since the impeller sizes are quite different.

Summary and Conclusions

Several important variables were found to influence drop size disribution in dispersed phase mixers: (1) agitation time; (2) speed/power input; (3) phase fraction; (4) the presence or absence of an extractant; (5) aqueous pH; and (6) ionic strength of the aqueous. Other factors such as simultaneous chemical reaction and continuous mode operaton at moderate to long retention times seemed to have little effect on size distribution in the system studied.

The model developed here is unique in that it simplifies the mathematics with logical assumptions and reduces the number of adjustable parameters to a minimum via dimensional analysis. This allows one to relate the significant parameters of the model to the system physical properties and operating conditions via simple multiple linear regression expressions. It was demonstrated that this model permits the prediction of dimensional size distributions without experimentation for similar bench scale systems and has considerable potential for future application.

Acknowledgements:

The authors wish to thank The U.S. Bureau of Mines, Salt Lake Metallurgy Center, for partial funding of this research; the Kennecott Copper Corp. for computer time donated to allow numerical solutions of the population balance equations; Mr. Jaime Sepulveda for helpful discussions; and Mr. Jim Smith of the U. of U. Medical Center for assistance in operating the Quanta Met Image Analyzer.

Notation

English

$\overline{B}$	Average change in population due to discrete birth events $((l^{-6}t^{-1}))$
$\overline{B}_b$	Average birth of drops due to breakage events $(l^{-6}t^{-1})$
$\overline{B}_c$	Average birth of drops due to coalescence events $(l^{-6}t^{-1})$
D	Impeller diameter (l)
$\overline{D}$	Average change in population due to discrete death events $(l^{-6}t^{-1})$
$\overline{D}_b$	Average death of drops due to breakage events $(l^{-6}t^{-1})$
$\overline{D}_c$	Average death of drops due to coalescence events $(l^{-6}t^{-1})$
$F_o(d)$	Cumulative number density function by diameter (-)
$f_o(v)$	Number density function for droplet volume (l^{-3})
$f_o^*(v^*)$	Dimensionless number density function by volume (-)
H	Liquid height (l).
k	Selection function constant (variable units)
m	Selection function size dependence parameter
N'_{We}	Modified Weber number (-)
$\overline{n}(t)$	Average number of drops per unit volume (ℓ^{-3})
$\overline{n}^*(t)$	Dimensionless average number of drops per unit volume (-)
Q_{in}	Flow rate into the mixer (l^3t^{-1})
Q_{out}	Flow rate out of the mixer (l^3t^{-1})
S(v,t)	Droplet selection function (t^{-1})
t	Time (t)
t*	Dimensionless time (-)
t_o	Characteristic time-inverse of the coalescence function (t)
V	Total volume of liquid in the mixer
v	Volume of a drop (l^3)
v*	Dimensionless drop volume (-)
$\overline{v}$	Average drop volume (l^3)
v′	Volume of drops in different size classes (l^3)
v_o	Parameter for dimensional analysis (-)
$\overline{v}^*$	Dimensionless mean drop size (-)
v_{max}	Maximum drop volume (l^3)
v_{min}	Minimum drop volume (l^3)
ζ_i	Property of interest.
λ	Coalescence function (t^{-1})
ϕ	Phase fraction (-)
$\overline{\Psi}$	Average population density (l^{-6})
$\overline{\Psi}^*$	Dimensionless population density (-)
ν_c	Continuous phase kinematic viscosity (l^2t^{-1})
ν_d	Disperse phase kinematic viscosity (l^2t^{-1})
I	Ionic strength $I = 1/2\ \Sigma nC_i(Z_i)^2$.
I_o	Standard ionic strength equal to .75.
$\overline{I}$	Dimensionless ionic strength equal to I/I_o.
x_j	Time rate of change of property ζ_i

References

1. Assenov, A. and Slater, M. J., "Measurement of Drop Size in Liquid/Liquid Systems by Sedimentation," Hydrometallurgy, Vol. 2, Elsevier Scientific Publishing Co., Amsterdam, 1976, p. 157.

2. Bajpai, R. K. and Ramkrishna, D., "A Coalescence Redispersion Model for Drop-Size Distributions in an Agitated Vessel," Chem. Eng. Sci., Vol. 31, 1976, p. 913.

3. Calderbank, P. H., "Physical Rate Processes in Industrial Fermentation," Trans. Inst. Chem. Engs., Vol. 36, 1958, p. 443.

4. Chen, H. T. and Middleman, S., "Drop Size Distribution in Agitated Liquid-Liquid Systems," AIChE Journal, Vol. 13, No. 5, Sept. 1967, p.989.

5. Coulaloglou, C. and Tavlarides, L. L., "Drop Size Distributions and Coalescence Frequencies of Liquid-Liquid Dispersions in Flow Vessels," AIChE Journal, Vol. 22, No. 2, March 1976, p. 289.

6. Coulaloglou, C. and Tavlarides, L. L., "Description of Interaction Processes in Agitated Liquid-Liquid Dispersions," Chem. Eng. Sci., Vol. 32, 1977, p. 1289

7. Fernandes, J. B., and Sharma, M. M., "Effective Interfacial area in Agitated Liquid-Liquid Contactors, " Chem. Eng. Sci., Vol. 22, 1967, p. 1267.

8. Giles, J. W., et.al., "Drop Size Distributions in Agaitated Liquid-Liquid Systems with Simultaneous Interface Mass Transfer and Chemical Reaction," Solvent Extraction Proceedings International Solvent Extraction Conference, Edited by J.A. Gregory, Soc. Chem. Ind., London, 1971, p. 94.

9. Grosssman, G., Determination of Droplet Size Distributions in Liquid-Liquid Dispersions," Ind. Eng. Chem. Process Des. and Develop., Vol. 2, No. 4. 1972, p. 537.

10. Hulburt, H. M. and Katz, S., "Some Problems in Particle Technology. A Statistical Mechanical Formulation," Chem. Eng. Sci., Vol. 19, 1964, p. 574.

11. Komasawa, I., T. Sasakura, and T. Otake, "Behavior of Reacting and Coalescing Dispersed Phase in a Stirred Tank Reactor," J. of Chem. Eng. of Japan, Vol. 2 No. 2, 1969, p. 209.

12. Luhning, R. W., and Sawistowski, H., "Phase Inversion in Stirred Liquid-Liquid Systems," Solvent Extraction Proceedings International Solvent Extraction Conference, Edited by J.A. Gregory, Soc. Chem. Ind., London, 1971, p. 873.

13. Mackelprang, R. B., "Mathematical Modeling of Disperse Phase Mixers," M.S. Thesis, University of Utah, August, 1979.

14. Mlynek, V. and Resnick, W., "Drop Sizes in Agitated Liquid-Liquid Systems," AIChE Journal, Vol. 18, No. 1, Jan. 1972, p. 122.

15. Nishikawa, M., "Drop Size Distribution in a Mixing Vessel," Research Paper, Dept. of Chem. Eng. Kyoto University, Japan, 1975.

16. Park, J. Y. and Blair, L. M., "The Effect of Coalescence on Drop Size Distribution in an Agitated Liquid-Liquid Dispersion," Chem. Eng. Sci., Vol. 30, 1975, p. 1057.

17. Prokop, A., Ludvik, M. and L. E. Erickson, "Growth Models of Cultures with Two Liquid Phases," Biotechnol. Bioeng., Vol. 14, 1972, p. 587.

18. Ritcey, G. M. and Ashbrook, A. W., "Solvent Extraction in Process Metallurgy," Short Course Text sponsored by TMS-AIME, Denver, Colorado, Feb. 1978.

19. Schindler, H. D. and Treybal, R. E., "Continuous-Phase Mass Transfer Coefficients for Liquid Extraction in Agitated Vessels," AIChE Journal, Vol. 14, No. 5, Sept. 1968, p. 790.

20. Shinnar, R., "On the Behavior of Liquid Dispersions in Mixing Vessels," J. Fluid Mech., Vol. 10, 1961, p. 259.

21. Sprow, F. B., "Drop Size Disteributions in Strongly Coalescing Agitated Liquid-Liquid Systems," AIChE Journal Vol. 13, No. 5, Sept. 1967, p. 995.

22. Sprow, F. B., "Distribution of Drop Sizes Produced in Turbulent Liquid-Liquid Dispersion," Chem. Eng. Sci., Vol. 22, 1967, p. 435.

23. Valentas, K. J. and Amundson, N. R., "Influence of Droplet Size-Age Distribution on Rate Processes in Dispersed Phase Systems," Ind. and Eng. Chem. Fund., Vol. 7, No. 1, Feb. 1968, p.66.

24. Vermeulen, T., Williams, G. M., and G. E. Zanglois, "Interfacial Area in Liquid-Liquid and Gas-Liquid Agitation," Chem. Eng. Progr., Vol. 51, No. 2, Feb. 1955, p. 85-F.

25. Herbst, J. A. "Rate Processes in Multiparticle Metallurgical Systems" Chapt. 2 in Rate Processes of Extractive Metallurgy, Ed. H. Y. Sohn and M. E. Wadsworth, Plenum Press, 1979, p. 53.

Chapter 24

ANALYSIS OF PHASE TRANSFER CATALYTIC REACTIONS IN LIQUID-LIQUID SYSTEMS

James Lee Hibbard and Doraiswami Ramkrishna

School of Chemical Engineering
Purdue University
West Lafayette, Indiana 47907

ABSTRACT

A population balance model accounting for the transient variations in drop size distributions in a stirred liquid-liquid system is analyzed for its effect on chemical reaction conversion in a stirred, batch reactor. The analysis is applied specifically for phase transfer catalytic reactions, which indicates that transient effects of evolution of drop size distributions may be important in such systems.

ANALYSIS OF PHASE TRANSFER CATALYTIC REACTIONS IN LIQUID-LIQUID SYSTEMS

James Lee Hibbard and Doraiswami Ramkrishna

School of Chemical Engineering
Purdue University
West Lafayette, Indiana 47907

ABSTRACT

A population balance model accounting for the transient variations in drop size distributions in a stirred liquid-liquid system is analyzed for its effect on chemical reaction conversion in a stirred, batch reactor. The analysis is applied specifically for phase transfer catalytic reactions, which indicates that transient effects of evolution of drop size distributions may be important in such systems.

INTRODUCTION

An integral aspect of reaction engineering is elucidation of the role of physical transport processes on conversion rates in chemical reaction systems. This goal is indeed difficult to accomplish in dispersed phase reactors in which the different reactants are distributed between the dispersed and continuous phases. Thus in a liquid-liquid reaction system, in which reaction occurs in the dispersed phase droplets, the overall reaction rate at any instant is the cumulative result of reaction rates in individual droplets. The reaction rate in a single drop will depend on the concentration of the reacting species in the droplet at the instant of consideration. The concentration of reacting species in a droplet varies in several ways. First there is the obvious influence of chemical transformation; second, the droplet is an open system permitting the transfer of species across the interface; a further serious complication is that resulting from breakage and coalescence of droplets which establishes a continuously changing statistical distribution of reaction species among a population of droplets, which is itself changing continually. Indeed the problem of dealing with how concentration in a single droplet changes during the period in which the droplet retains its identity (that is, it neither breaks nor coalesces with another droplet) comes under the usual framework of transport phenomena in single phase systems (Bird, Stewart and Lightfoot, 1960). Although the application of transport equations to single droplets is no simple matter, the complications that arise from the dynamic effects of coalescence and redispersion are even more challenging.

The objective of the present work is to demonstrate the effect of temporal evolution of drop size distributions on reaction conversion in a liquid-liquid system. To accomplish this objective a transient-coalescence redispersion model will be used. The model is useful as a diagnostic tool but is unrealistic from a practical viewpoint. Although there is increasing interest in the experimental measurement of transient size distributions, there is not sufficient information to derive a quantitatively significant transient model for the drop size distributions.

In view of the rising interest in the commercial exploitation of phase transfer catalytic reactions, together with the fact that many of them are rapid reactions, they seemed ideally suited for the demonstrations of this paper.

PHASE TRANSFER CATALYSIS

Phase transfer catalysis (PTC) involves the addition of small amounts of a complexing agent, typically a quaternary ammonium or phosphonium salt, to the liquid-liquid dispersion. The phase-transfer catalyst complexes with the reactant in the aqueous phase to form an ion-pair. The solubility characteristics of this ion-pair are different from that of the original anion. If the proper catalyst has been selected, the ion-pair complex will have preferential solubility in the organic phase. Migration of the complex to the organic phase will bring the two reactants into close proximity and promote the homogeneous reaction. After reaction the phase transfer catalyst is free to return to the aqueous phase to begin a new cycle.

The mechanism of PTC as envisaged by Starks (1971) is presented in the following schematic.

$$\text{aqueous phase} \quad Q^+X^- + A^- \underset{k_2}{\overset{k_1}{\rightleftarrows}} Q^+A^- + X^-$$

$$\text{organic phase} \quad Q^+X^- + AB \overset{k_3}{\leftarrow} Q^+A^- + BX$$

Q^+X^- is the phase transfer catalyst, a quaternary ammonium or phosphonium salt. The catalyst complexes with the nucleophilic reactant A^- to form the ion-pair Q^+A^-, which is transferred to the organic phase, where it reacts with compound BX. The driving force for this movement derives from high solubility of the complex in the organic phase together with facilitation by chemical reaction. The organic product is AB which remains in the organic solvent while the catalyst appears as an ion-pair. The anion in the ion-pair is the same moiety contained by the original catalyst salt, so that transfer of the ion-pair to the aqueous phase regenerates the catalyst and closes the catalyst loop. The partition coefficient, π_{QA} defined by

$$\pi_{QA} = \frac{[QA]_{o,e}}{[QA]_{a,e}}$$

determines the quality of the catalyst, higher values being characteristic of better catalysts.

Kinetic studies of phase transfer catalytic reactions have been carried out in the past by allowing for reaction rates to be limiting by controlling the amount of catalyst and maintaining high stirring rates

(Starks & Owens, 1973). In a commercial reactor, however, the objective would be one of providing high conversion rates so that the foregoing restraint on the amount of catalyst would be less meaningful. The importance of physical rate processes relative to chemical reaction is best assessed by appropriate dimensionless groups which arise from non-dimensionalizing the reactor equations. A convenient way to do this would be to assume an infinite mixing rate between the dispersed phase so that a lumped dispersed phase model would be applicable. We consider this next.

Lumped Dispersed Phase Model

For a well-stirred batch reactor, the following dimensionless equations are readily identified

$$\frac{dU}{d\tau} = - UV \quad (1)$$

$$\frac{dV}{d\tau} = - P_1(V - P_3X) - P_2UV \quad (2)$$

$$\frac{dW}{d\tau} = - P_1(W - P_4Y) + P_2UV \quad (3)$$

$$1 = V + W + X + Y \quad (4)$$

$$P_5(1-Z) = P_2(1-U) + V + X \quad (5)$$

$$KP_5YZ = X[P_2(1-U) + 1 - W - Y] \quad (6)$$

We refer to the Nomenclature for explanation of the symbols. The dimensionless groups P_1-P_5 can be interpreted easily. The magnitude of P_1, which measures the importance of mass transfer rate relative to reaction rate, is the most important to the present study. Parameters P_5 and P_2 are the initial mole ratios of the reactants A and B to total moles of the catalyst. The ratio P_5/P_2 is the initial mole ratio of the reactants and will determine, which reactant will be first consumed.

Hibbard (1978) has investigated the solution of Eqs. (1) - (6) for a range of the parameters P_1 through P_5. The values of the parameters were derived from a reaction studied by Herriott and Picker (1975) which consisted of a chloride displacement on 1-chlorooctane with the nucleophile thiophenoxide ion ($C_6H_5S^-$) to form the product n-octylphenylthio ether ($C_6H_5SC_8H_{12}$). This reaction was carried out by Herriott and Picker in a batch reactor containing water and benzene. From amongst the different catalysts used by them, the selection made here is tetrabutylphosphonium chloride ($(C_4H_5)_4PCl$). This catalyst extracts 81% of the thiophenoxide ion from the aqueous to the organic phase, and has a rate constant of 3.5×10^{-3} liters per mole per second. The correspondence between these chemicals and the symbols used for the PTC mechanism is identified in Table 1.

While the necessary partition coefficients are obtainable from Herriott and Picker (1975) it is difficult to choose a rate constant that will correspond exactly to their results. This is because their kinetic analysis involves a second order reaction with concentrations of the reactants in their parent phases and a rate constant linearly related to the total catalyst concentration. Although good quantitative correlation was obtained the equations of Herriott and Picker are not a true kinetic representation of PTC mechanism. Despite this shortcoming, however, their results are useful for assessing the approximate value of the rate constant k_3 in this work.

For the calculations in this work, the reaction vessel was assumed to be of the type used in the mass transfer studies of Harriott (1962). This helped in a more realistic assessment of the mass transfer effects. The reactor dimensions are summarized in Table 2.

The pertinent physicochemical data used in the calculations are presented in Table 3. The area appended volumetric mass transfer coefficient, K_La was calculated from data obtained by Brian et al (1969). The details are available elsewhere (Hibbard, 1978). Hibbard (1978) has made substantially more calculations than presented in this work; here, we consider the mass transfer limited situations only, which are diagnosed by the behavior of the dimensionless ratio

$$\psi \equiv \frac{t_{0.9}}{t^*_{0.9}}$$

where $t_{0.9}$ is the time required for 90% conversion when mass transfer limitations exist and $t^*_{0.9}$ is the corresponding time when no mass transfer limitations exist. Figure 1 shows a plot of ψ versus stirrer speeds for the chosen reactor at various total catalyst concentrations computed by Hibbard (1978). Quite expectedly, the steep rise in conversion times occurs when stirrer speeds are so small that mass transfer rates are controlling. Note specially that the transition from the mass transfer-controlled to the reaction-controlled situations is unaffected by the amount of catalyst present, i.e. although absolute conversion times may be decreased by augmenting the amount of catalyst, the <u>relative</u> conversion times in the two regimes remain the same. Indeed this behavior may be expected of a PTC system since an increase in the catalyst concentration would increase both the reaction and mass transfer rates.

Dispersed Phase Mixing

The times required for the two rate processes, mass transfer and reaction have been related through a lumped dispersed phase model the basis of which is an infinite mixing rate between the dispersed phase droplets. Alternatively, the coalescence frequency w of the organic phase droplets had been assumed to be infinity. Certainly, the assumption of infinite rate of mixing is not always valid; the effect of a finite rate of mixing is to introduce a concentration distribution of the reactants among the droplet population, the nature of which would also depend on the reaction and mass transfer rates.

Table 4 gives the coalescence frequency and its inverse for various stirring speeds roughly estimated by adapting a correlation due to Komazawa et al (1970). The inverse of the coalescence frequency is the time required on the average for one dispersed phase volume to be mixed. Since $\frac{1}{w}$ depends on the -2.5 power of the stirring speed, it is sensitive to any change in agitation rate. To accomplish the mixing of one dispersed phase volume, the inverse of the coalescence frequency ranges from 4700 s at 50 rpm to 2.63 s at 1000 rpm. It is possible from Table 4 to compare the time scales of mixing and reaction. Letting U(t) be the fraction of unaccomplished reaction in time t, the value of $U(\frac{1}{w})$ would measure the impact of mixing on conversion. For example, at a catalyst concentration of 5×10^{-2} moles per liter, less than 1% of the reaction occurs in the time $\frac{1}{w}$ at any stirring speed 200 rpm or higher. The assumption of infinite mixing for such reaction conditions may be reasonable. High catalyst concentrations would therefore imply that segregational effects would play an important role in determining reaction conversion.

POPULATION BALANCE ANALYSIS

Clearly, incomplete mixing situations can only be dealt with by population balance models. Quantitatively significant population balance models have been limited by lack of adequate information about breakage and coalescence rates of droplets. Curl (1963, 1967) used a single-step coalescence-redispersion model in which drops were assumed to be identical in size and coalescive interaction between droplets was immediately followed by breakage into two equally sized droplets. Bajpai et al (1976) considered a similar coalescence-redispersion model in which however drops were allowed to be of different sizes and the redispersion event produced two drops of arbitrary sizes (of course subject to the obvious constraint that breakage preserves the total volume of the droplets). Based on this model, the stabilized drop size distribution in a batch vessel was found to be exponential in drop volume and corroborated rather well by experimental data.

We present here a transient model for the evolution of drop size distribution based on the work of Bajpai et al (1976). There is as yet no quantitative verification of this model for transient situations but it is a useful implement to inquire into the role of such transient phenomena in reactor behavior because, (i) the model is reasonable near equilibrium and (ii) it is sufficiently simple to obtain without any extensive computation answers in diagnosis of the importance of the phenomena in question. There are some drawbacks of the model, however. These will be discussed while presenting the model.

Population Balance Model

In extending the model of Bajpai et al (1976) to the present situation, it is assumed that coalescence and redispersion events are not affected by solute concentrations, mass transfer or chemical reaction. Briefly the model assumes that the interaction between two droplets involves coalescence of the pair followed by an immediate redispersion into two droplets whose sizes are randomly distributed except for the requirement of volume conservation. The total number of droplets thus remains constant throughout.

The details of the drop interaction model are available in Bajpai et al (1976).

Here we consider the predictions of the foregoing model by first analyzing the transient evolution of drop size distributions without accounting for chemical reaction. This is permissible since the model neglects the effect of the solutes on the coalescence-redispersion phenomena. The dimensionless population balance equation is given by

$$\frac{\partial}{\partial\tau'}f(x,\tau')=\int_x^\infty \frac{dy''}{y''}\int_0^{y''} dy'\ f(y''-y',\tau')f(y',\tau')-f(x,\tau') \tag{7}$$

A complete explanation of the symbols is provided in the Nomenclature. Equation (7) has been solved analytically for a variety of initial conditions by Hibbard (1978). Figure 2 shows the transient evolution of the drop size distributions from a normal initial size distribution.

It is opportune at this stage to reflect on the transient predictions of the drop size distribution of Bajpai et al (1976). Most importantly, it demands that the total number of droplets remain constant. Undoubtedly, the rate or change of droplet number must decrease progressively after the onset of stirring. The data of Tsukiyama et al (1974) indicate a quick drop in droplet number and thereafter only a slight variation so that the number of drops remains materially constant. However, this behavior is more characteristic of higher stirring speeds. At lower speeds, the initial period in which the droplet number changes is more pronounced so that the transient model presented here is not applicable during that period. An even more serious objection to the model here is that it demands that the interfacial area <u>decrease</u> monotonically to the equilibrium value. Thus the initial interfacial area must always be <u>more</u> than that at equilibrium. This implies that the model visualizes an initial spurt of breakage, producing very small drops which grow by coalescence and redispersion. In short, the dispersion is assumed to respond rapidly with an abrupt change at the beginning in the total number of droplets which is followed by a slower period of adjustment of the size distribution. It is difficult to accept this requirement of the model under all circumstances from a practical viewpoint. The merit of the model is its simplicity which makes it a valuable diagnostic tool for assessing the role of drop size distributions in determining physical and chemical rate processes in liquid-liquid dispersions. The essential point is that a furtherance of sophistication in the absence of extensive experimental correlation would lead to wasteful calculations. Whether or not an elaborate experimental undertaking is advisable may be established by such simple diagnostic models. Unfortunately, it is not always that this aspect of modelling is appreciated.

The total interfacial area at time τ' relative to that at equilibrium, denoted by $\xi(\tau')$, is given by

$$\xi(\tau') = \frac{\int_0^\infty x^{2/3}f(x,\tau')dx}{\Gamma(5/3)} \tag{8}$$

For a uniform initial distribution (with the largest drop volume 7 times the smallest) it is found that $\xi(0) = 1.0821$, this implies that the interfacial area changes by about 8.2% during evolution to the equilibrium distribution. For the normal initial distribution with a 25% standard deviation, $\xi(0) = 1.0996$ which corresponds to a 10% reduction in the interfacial area. When a Dirac delta distribution is used $\xi(0) = 1.1077$ i.e. a 10.8% reduction in the interfacial area. In all these cases the change in interfacial area is somewhat limited. This is characteristic of the drop size distribution model on hand because of the constraint of constant total droplet number. The interfacial area would vary much more during the period (one not covered by the model) in which the number of droplets changes. We may expect however that if the present model predicts effects on reaction conversion due to segregation and evolving interfacial area, which are of the same order as the change in interfacial area, then the effects predicted cannot be ignored.

The analytical solution to the drop size distribution model allows us to define a characteristic time representing the time scale of change of interfacial area[†]. Indeed the actual time for $\xi(\tau')$ to reach its asymptotic value of unity is infinity. However, an operational definition is to resort to a time it takes for $\xi(\tau')$ to decay to, say 1.005. Denoting this transient period by t_t, the unaccomplished reaction at the end of this time is presented in Table 5 for different catalyst concentrations. At the slow speed of 50 rpm, it is found that there is less than 10% of unaccomplished reaction at $Q_T=5\times10^{-2}$ whereas approximately 24% of reaction is left over when $Q_T=1\times10^{-2}$. Clearly in both cases, most of the reaction is over during the transient period of evolution of the drop size distribution. Even at 200 rpm about 30% of the reaction is complete during the transient period at the higher catalyst concentration. It should therefore be clear that the period in which drop size distributions evolve must have a considerable effect on overall conversion especially at lower speeds and higher catalyst concentrations. A point of particular interest that deserves emphasis is the following. The lumped dispersed phase analysis showed that for stirring speeds exceeding 200 rpm mass transfer effects are negligible so that variation in interfacial area may be expected to have only a small effect regardless of the catalyst concentration. However, transient analysis of the drop size distribution shows that at higher catalyst concentrations, reaction times compete favorably with such transient times during which drop size distributions evolve. Associated with transient drop size are transient reactant concentration distributions (whether or not mass transfer is controlling) among the drop population on which the overall conversion would depend. The prediction of such effects cannot be made without a population balance analysis, which is considered next.

The dimensionless population balance equation in the multivariate density $f(x,z,\tau')$ is given by

$$\frac{\partial}{\partial\tau'} f(x,\underset{\sim}{z},\tau') + \sum_{i=1}^{3} \frac{\partial}{\partial z_i}[f(x,\underset{\sim}{z},\tau')\dot{\underset{\sim}{z}}]$$
$$= \int_x^\infty \frac{dy''}{y''} \int_0^{y''} dy' \prod_{i=1}^{3} \int_0^{z_i} dz_i' \, f(y''-y',\underset{\sim}{z}-\underset{\sim}{z}',\tau')\, f(y',\underset{\sim}{z}',\tau')$$
$$- f(x,\underset{\sim}{z},\tau') \tag{9}$$

The vector $z \equiv (z_1,z_2,z_3)$ features the dimensionless molar amounts of $BX(z_1)$, $QA(z_2)$ and $QX(z_3)$, all in a single drop of dimensionless volume x. This equation, together with the proper expressions for reaction rates, mass balances, and mass fluxes provide a complete mathematical description of the system. The dimensionless equations for mass transfer and chemical reaction in a single drop of dimensionless volume x are found to be

$$\dot{z}_1 = -\frac{P_6}{x} z_1 z_2 \tag{10}$$

$$\dot{z}_2 = -\frac{K_o^*(x)a^*(x)}{x}(z_2-P_8xX) - \frac{P_6P_2}{x} z_1 z_2 \tag{11}$$

$$\dot{z}_3 = -\frac{K_o^*(x)a^*(x)}{x}(z_3-P_9xY) + \frac{P_6P_2}{x} z_1 z_2 \tag{12}$$

[†]Note that this is also the time scale for the evolution of the drop size distributions. This time although not unrelated to the time scale of mixing $\frac{1}{w}$ considered earlier, is not necessarily the same as the latter, since the former would depend on the initial departure from the equilibrium distribution.

The dimensionless versions of mass balances for A, B and X in the reactor become

$$1 = X + Y + N \int_0^\infty dx \prod_{i=1}^{3} \int_0^\infty dz_i (z_2+z_3)\, f(x,\underset{\sim}{z},\tau') \tag{13}$$

$$P_5(1-Z) = P_2 + X + N \int_0^\infty dx \prod_{i=1}^{3} \int_0^\infty dz_i (z_2-P_2z_1) f(x,\underset{\sim}{z},\tau') \tag{14}$$

$$P_5\, KYZ = X[1 + P_2 - Y - N \int_0^\infty dx \prod_{i=1}^{3} \int dz_i (z_3+P_2z_1)\, f(x,\underset{\sim}{z},\tau')] \tag{15}$$

The dimensionless parameters in (10)-(15) are explained in the Nomenclature.

The solution of Eq. (9) with $\dot{z}$ determined from (10)-(12) coupled with (13)-(15) is a formidable task. Shah, Ramkrishna and Borwanker (1977) have presented a simulation technique, which is a very convenient way to obtain solutions to such problems in population balances. The reader is referred to the above publication for an elaboration of this simulation technique. Hibbard (1978) discusses in detail the application of the technique to the present problem. We briefly summarize the difficulties in a simulation of this type and how they were overcome. If a sample of N droplets is to be simulated, then 3N+3 differential equations must be solved simultaneously so that the reciprocal effects of the dispersed and continuous phases could be accounted for during the interval of quiescence[†] in the simulation. This is somewhat of a tall order. The assumption was therefore made that during the interval of quiescence no change occurs in the continuous phase variables X and Y. The three differential equations for any given droplet can then be solved independently of all other droplets. After the N droplets have been updated (w.r.t. solute concentrations), new values of X and Y can be calculated with the corresponding conservation equations. Hibbard (1978) further discusses the merits of this assumption. Here we merely present the results.

Results and Discussion

Calculations are presented for the situation under which mass transfer controls the reaction. The extreme case of the lowest stirring speed of 50 rpm and a catalyst concentration of 5×10^{-2} moles per liter were chosen for the calculations. At the reaction conditions given above, the lumped dispersed phase analysis has shown that the apparent reaction rate is primarly under the control of mass transfer. We attempt below answers to the following two questions.

(i) When the rate controlling process is mass transfer are there segregational effects on overall conversion?

(ii) What is the effect of the transient period of evolution of drop size distributions on the conversion rate?

[†]See Shah, Ramkrishna, and Borwanker (1977). The interval of quiescence is a period during which no droplet loses its identity.

First, the two effects above will be measured in terms of $t_{0.9}$ the time for 90% conversion.

In regard to (i), the term "segregation" deserves further clarification. The question that arises is with respect to the drop size distribution of the segregated population. In order to make a meaningful comparison between the effects of different orders of mixing (or segregation) we retain the equilibrium exponential distribution. Table 6 presents the values of $t_{0.9}$ for the complete mixing model (lumped dispersed phase analysis), partial mixing model (results of simulation with droplet interactions) and the complete segregation model (results of simulation without droplet interaction). Clearly, the results for partial mixing are much closer to those of the complete mixing model and substantially different from the results for complete segregation. Indeed, the closeness between the complete mixing model and the partial mixing model arises from the time scale for mixing being much smaller than that for mass transfer. (If on the other hand, the time scale for mixing had been much larger than that for mass transfer, then the partial mixing model would have been closer to the complete segregation model). It is reasonable to state that the time scale for mixing is not very different from the time scale for the dynamics of the drop size distribution. Hence in situations in which rate processes such as mass transfer (or reaction) occur with a time scale smaller than t_t (or alternatively $\frac{1}{w}$ segregational effects become very important. In such cases the temporal evolution of the drop size distributions from an initial distribution of droplets must play an important role in determining the transfer rate. Note that the complete segregation model based on an equilibrium dropsize distribution would be somewhat meaningless in these circumstances. If however the mass transfer process is initiated only after the dropsize distribution has stabilized (such as in the situation under study by withholding the catalyst until an equilibrium distribution has been established), then the complete segregation model based on the stabilized drop size distribution would become applicable under the above condition (i.e., the time scale for mass transfer is much smaller than $\frac{1}{w}$).

Next we consider question (ii) which deals with the effect of the temporal evolution of the drop size distribution. Table 7 shows the calculations of $t_{0.9}$ as predicted by the population balance model for the exponential, uniform and normal initial distributions. The exponential initial distribution indicates that the reaction was allowed to take place after the drop size distribution has been allowed to stabilize. The simulation shows a reduction in $t_{0.9}$ when the reaction is allowed to take place right from the initial distribution. This is understandable since initially the mass transfer area is more at the outset in accord with the stipulations of the model. Note that the reduction in $t_{0.9}$ of about 7 to 8% is quite significant considering the fact that the transient model allowed for variation in the interfacial area only to the same extent. Recently, Barnea, Hoffer and Resnick (1978) used a lumped dispersed phase model to analyze the effect of transient interfacial area variations on mass transfer controlled reactions. They allowed considerably more variation in the interfacial area than was done here so that the effects on conversion were more pronounced. The work of Barnea et al presumes that the time scale for mass transfer is smaller than the time scale for the evolution of the interfacial area while the mixing time scale is considered to be essentially zero. This is contrary to the position taken here that the time scales for droplet interaction (mixing) and transient variation in interfacial area cannot be very disparate since after all it is coalescence and redispersion that changes the interfacial area after the initial spurt of breakage.

Curl (1963) has estimated substantial segregational effects for a continuous stirred tank reactor which has also a spread of residence times. Similar results must hold for the reaction system considered here.

One further observation deserves mention. The emphasis here has been on mass-transfer limited situations, almost implying that reaction-controlled situations are not likely to be influenced by droplet phenomena. This implication is not reasonable since in the phase transfer catalytic system, one could envisage situations in which considerably higher amounts of catalyst are used, which would still be reaction-controlled at higher speeds. If it is assumed that t_t (and hence possibly $\frac{1}{w}$) is independent of the catalysis concentration then considerably more reaction may take place during periods in which dropsize distributions evolve. For nonlinear kinetics, reactant concentration distributions among the evolving droplet population could become sufficiently complex as to produce strong transient effects.

CONCLUSIONS

Reactions in liquid-liquid dispersions may display complicated behavior the nature of which depends on the magnitudes of four different time scales. (i) the reaction time scale, (ii) mass transfer time scale, (iii) time scale for droplet interaction (i.e., coalescence and redispersion) and (iv) time scale for transient evolution of dropsize distribution. The conclusions drawn here pertain to phase transfer catalysis.

The relative time scale for reaction and mass transfer is governed by the stirring speed. The transition from the mass transfer controlled to the reaction controlled regimes with increasing agitation is essentially independent of the total amount of catalyst present. This is because increased catalyst amounts augment reaction rate as well as reaction facilitated mass transfer.

The relative time scale for drop interaction and reaction may be changed by varying the amount of catalyst. The more the catalyst the larger the extent of reaction during the state of segregation; the result is a decrease in the time scale for reaction while the time scale for mixing is assumed to remain the same. Under this circumstance if the catalyst is added after the drop size distribution has stabilized then a totally segregated model based on the equilibrium distribution model becomes more applicable than a complete mixing model. For the catalyst concentration used in this work the complete mixing model was significantly better than the totally segregated model. If on the other hand, catalyst is present right from the onset of stirring, with a small relative time scale for reaction and drop interaction, the influence of the transient in the evolution of dropsize distribution on reaction conversion will be considerable.

Finally, the relative time scale for drop interaction and transient evolution of drop size distribution depends on the extent of departure of the initial size distribution from the equilibrium value. Thus the time scale for transients would be smaller than that for interaction if the initial size distribution is close to the equilibrium. If the initial size distribution is far removed then we would expect that the time scale for transients to be larger than the time scale for mixing. From the standpoint of the model here the two time scales in this case are not significantly different. Although the dynamic dropsize distribution model may be somewhat unrealistic its simplicity makes it a useful diagnostic tool for the purposes of this paper.

NOMENCLATURE

$a(v)$ — Surface area of drop of volume v

$a^*(x)$ — $= a(v)/\bar{v}^{2/3}$

$F(x,\tau')$ — Cumulative drop size distribution

$f(x,\tau')$ — Normalized dropsize distribution $\equiv \frac{n(v,t)}{N_o}$

$f(x,z,\tau')$ — Normalized multivariate distribution of drop volume and amounts of reactants $\equiv \bar{V}\, m_Q^2\, m_{BX_i}\, \frac{1}{N_o}\, n(v,\underset{\sim}{m},t)$

k — Rate constant

K — Equilibrium constant

K_La — Mass transfer coefficient x interfacial area per unit volume of dispersion

$K_o(v)$ — Mass transfer coefficient for drop of volume v: $K_o^*(x) \equiv \frac{K_o(v)}{w_o N_o \bar{v}^{2/3}}$

$\underset{\sim}{m}$ — Vector representing total molar amounts of reactants
$m_1 \equiv$ moles of BX, $m_2 \equiv$ moles of QA, $m_3 \equiv$ moles of QX
$m_{BX_i} \equiv$ moles of BX initially present in the system
$m_Q \equiv$ the total moles of catalyst in all forms

$n(v,t)$ — Number density of drop volume v

$n(v,\underset{\sim}{m},t)$ — Multivariate number density of drops of volume v and reactant amounts

N_o — Total number of drops in dispersion

P_i — Dimensionless groups : $P_1 \equiv \frac{K_La}{k_3Q_T}$,

$P_2 \equiv \phi\frac{[BX]_i}{Q_T}$ or $\frac{m_{BX_i}}{m_Q}$, $P_3 \equiv \pi_{QA}\frac{\phi}{(1-\phi)}$

$P_4 \equiv \pi_{QX}\frac{\phi}{(1-\phi)}$, $P_5 \equiv (1-\phi)\frac{[A^-]_i}{Q_T}$ or

$\frac{[A^-]_i V_o}{m_Q}$, $P_6 \equiv \frac{m_Q k_3}{x_o N_o \bar{v}}$, $P_8 \equiv \pi_{QA}\frac{\bar{v}}{V_a}$,

$P_9 \equiv \pi_{QX}\frac{\bar{v}}{V_a}$

Q_T — Total amount of catalyst

t — Time

U — Dimensionless concentration of [BX]: $= \frac{[BX]}{[BX]_i}$

v — Drop volume, $\bar{v} \equiv$ average drop volume

V — Dimensionless concentration of $[QA]_o$: $=\phi\frac{[QA]_o}{Q_T}$; Also volume

V_T — Total volume of dispersion

w — Coalescence frequency w_o, interaction frequency $= \frac{w}{N_o}$

W — Dimensionless concentration of $[QX]_o$: $= \phi\frac{[QX]_o}{Q_T}$

x — Dimensionless drop diameter $v/\bar{v}$

X — Dimensionless concentration of QA: $(1-\phi)\frac{[QA]_a}{Q_T}$

Y — Dimensionless concentration of QX: $(1-\phi)\frac{[QX]_a}{Q_T}$

Z — Dimensionless concentration of A^-: $\frac{[A^-]}{[A^-]_i}$

z — Dimensionless molar amounts of reactants in drop of dimensionless volume x.
$\underset{\sim}{z} \equiv (z_1,z_2,z_3)$
$z_1 \equiv \frac{m_1}{m_{BX_i}}$, $z_2 \equiv \frac{m_2}{m_Q}$, $z_3 \equiv \frac{m_3}{m_Q}$

Greek Symbols

ϕ — Dispersed phase fraction

ψ — Ratio defined by $t_{0.9}/t^*_{0.9}$

μ — Viscosity

ρ — Density

σ — Interfacial tension

τ — Dimensionless time $= k_3Q_Tt/\phi$

τ' — Dimensionless time $= w_oN_ot$

ξ — Dimensionless interfacial area of dispersed phase

Subscripts

a — aqueous phase

o — organic phase (when no explanation is included otherwise)

LITERATURE CITED

Bajpai, R.K., D. Ramkrishna and A. Prokop, "A Coalescence Redispersion Model for Drop Size Distribution in an Agitated Vessel," Chem. Eng. Sci., 31, 913 (1976).

Barnea, D., M.S. Hoffer and W. Resnick, "Dynamic Behavior or an Agitated Two-Phase Reactor with Dynamic Variation in Drop Diameter- I. Numerical Solutions," Chem. Eng. Sci., 33, 205 (1978).

Bird, R.B., W. E. Stewart and E.N. Lightfoot, "Transport Phenomena, Wiley, New York (1960).

Brian, P.L.T., H.R. Hales and T.K. Sherwood, "Transport of Heat & Mass Between Liquid Spherical Particles in an Agitated Tank," A.I.Ch.E.J., 15, 727, (1969).

Brown, D.E., and K. Pitt, "Drop Size Distribution of Stirred Non-Coalescing Liquid-Liquid System," Chem. Eng. Sci., 27, 577 (1972).

Curl, R.L., "Dispersed Phase Mixing: I. Theory and Effects in Simple Reactors," A.I.Ch.E.J., 9, 175 (1963).

Miller, R.S., J.L. Ralph and R.L. Curl, "Dispersed Phase Mixing: II - Measurements in Organic Dispersed Systems," A.I.Ch.E.J., 9, 196 (1963).

Curl, R.L., "Dispersed Phase Mixing Effects on Second Moments in Dominantly First-Order, Backmix Reactors" Chem. Eng. Sci., 22, 343 (1967).

Harriott, P., "Mass Transfer to Particles: Part I. Suspended in Agitated Tanks," A.I.Ch.E.J., 8, 93 (1962).

Herriott, A.W. and D. Picker, "Phase Transfer Catalysis. An Evaluation of Catalysts," J. Am. Chem. Soc., 97, 2345 (1975).

Hibbard, J.L., Population Balance Analysis of Reactions in Liquid-Liquid Dispersions: Phase Transfer Catalysis, M.S. Thesis, Purdue University, 1978.

Jones, R.A., "Applications of Phase-Transfer Catalysis in Organic Synthesis," Aldrichimica Acta, Acta, 9, 35 (1976).

Nishikawa, M., "Drop Size Distribution in a Mixing Vessel," Paper Presented at the Engineering Foundation Mixing Conference, Rindge, New Hampshire (1977)

Komazawa, I., S. Morioka and T. Otake, "Studies of the Interaction Rate of Dispersed Phase and Resulting Chemical Reaction in a Stirred Tank," Kagaku, Kogaku, 34, 538 (1970).

Shah, B.H., D. Ramkrishna and J.D. Borwanker, "Simulation of Particulate Systems Using the Concept of the Interval of Quiescence," A.I.Ch.E.J, 23, 897 (1977).

Starks, C.M., "Phase-Transfer Catalysis I. Heterogeneous Reactions Involving Anion Transfer by Quaternary Ammonium and Phosphonium Salts," J. Am. Chem. Soc., 93, 195 (1971).

Starks, C.M. and R.M. Owens, "Phase Transfer Catalysis II. Kinetic Details of Cyanide Displacement on 1-Halooctanes," J. Am. Chem. Soc., 95, 3613 (1973).

Tsukiyama, S., A. Takamura nd N. Nakura, "The Increasing Rate of Droplets Number of Mechanical Liquid-Liquid Dispersion," Chem.Pharm. Bull., 22, 1902 (1974).

TABLE 1

The Chemicals for the Phase Transfer Catalysis Reaction

Symbol	Chemical Species
A^-	Thiophenoxide ion
AB	n-Octylphenylthio ether
BX	1-Chlorooctane
QX	Tetrabutylphosphonium Chloride
X^-	Chloride ion

TABLE 2

Dimensions of the Reactor

Reactor Diameter	T	20.32 cm
Reactor Height	H	20.32 cm
Impeller Diameter	D	7.62 cm

TABLE 3

$D_a = 0.37 \times 10^{-5}$ cm^2/s

$k_3 = 3.5 \times 10^{-3}$ ℓ/mole-s

$K = 0.5$

$\mu_a = 0.0156$ g/cm-s

$\mu_o = 0.0066$ g/cm-s

$\pi_{QA} = 8.89$

$\pi_{QX} = 8.89$

$\rho_a = 1.08$ g/cm^3

$\rho_o = 0.879$ g/cm^3

$\bar{\rho} = 1.06$ g/cm^3

$\sigma = 35.0$ dynes/cm

$\phi = 0.1$

TABLE 4

Comparison of the Time Scales for Reaction and Dispersed Phase Mixing

RPM	w, s^{-1}	$\frac{1}{w}, s$	$Q_T=5x10^{-2}$ $U(\frac{1}{w})$	$Q_T=1x10^{-2}$ $U(\frac{1}{w})$	$Q_T=5x10^{-3}$ $U(\frac{1}{w})$
50	.000213	4700	.69	.90	.94
75	.000586	1710	.86	.95	.97
100	.00120	831	.93	.97	.98
150	.00332	302	.96	.99	1.00
200	.0681	147	.98	1.00	1.00
300	.0188	53.3	.99	1.00	1.00
400	.0385	26.0	1.00	1.00	1.00

TABLE 5

The Relative Time Scales for the Transient and Reaction

Stirring Speed (rpm)	t_t (s)	$U(t_t)$ $Q_T=5x10^{-2}$	$U(t_t)$ $Q_T=1x10^{-2}$
50	37,620.	0.1	.2374
75	13,650.	0.1	.4681
100	6,650.	.1904	.6327
150	2,413.	.5172	.8198
200	1,176.	.7180	.9034
300	426.6	.8878	.9617

TABLE 6

The Effects of Segregation

Degree of Segregation	$t^{\dagger}_{0.9}$ $sx10^{-4}$	Percentage Increase in $t_{0.9}$ Over that for Perfect Mixing‡
Perfect Mixing	1.62	
Partial Mixing	1.70	5.1
Complete Segregation	2.01	24.3

† $Q_T = 5 \times 10^{-2}$ moles per liter; Agitation = 50 rpm

‡ (The 95% confidence limits on the percent increase are less than 0.05%)

TABLE 7

The Effects of the Transient

Initial Density	$t^{\dagger}_{0.9}$ $sx10^{-4}$	Percent Reduction in $t_{0.9}$ from that for the Exponential Density‡
Initial Density	1.70	
Uniform (B=1/7)	1.58	7.0
Normal (σ=1/4)	1.56	8.0

† $Q_T = 5 \times 10^{-2}$ moles per liter; Agitation = 50 rpm

‡ (The 95% confidence limits on the percent reduction are less than 0.05%)

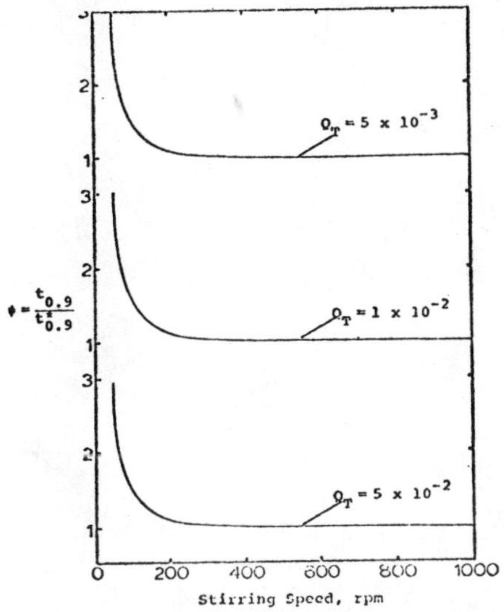

Figure 1. φ as a Function of P_1

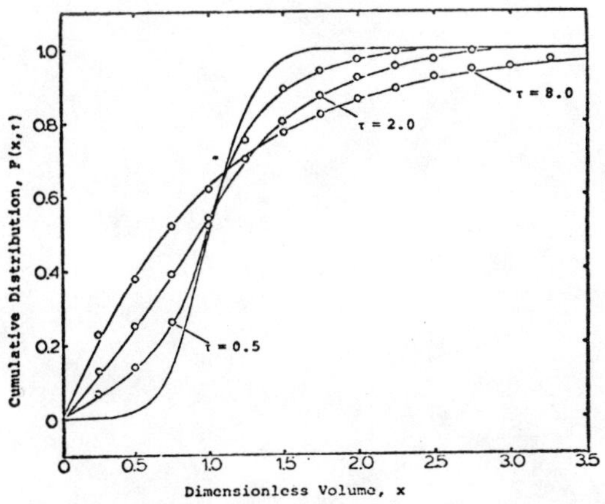

Figure 2. Cumulative Distribution as a Function of Time: Normal Initial Density

Chapter 25

EFFECT OF CONTACTOR TYPE ON METAL SEPARATION BY SOLVENT EXTRACTION*

Ali Altway[1], M. E. Tunison[2], and T. W. Chapman

Chemical Engineering Department
University of Wisconsin
Madison, Wisconsin 53706

ABSTRACT

Simulations of various contactor configurations for simultaneous extraction or stripping of two metals reveal the dependence of conversion and selectivity on operating and design parameters. A two-film mass transfer model is used to provide the rates in countercurrent column and stirred-tank cascade calculations. Even in these mass-transfer-limited cases the stoichiometry of the extraction reactions introduces nonlinearities and interactions between the two metal transfer rates, but these can be represented in a suitable model. One conclusion is that a countercurrent column should be considerably more efficient than a cascade of mixer-settlers of comparable size.

*Paper presented at 72nd Annual Meeting, A.I.Ch.E., San Francisco, California, November 28, 1979

[1]Present Address: Surabaya Institute of Technology, Surabaya, Indonesia

[2]Present Address: Exxon Research and Development Laboratories, Baton Rouge, Louisiana

EFFECT OF CONTACTOR TYPE ON METAL SEPARATION BY SOLVENT EXTRACTION

Ali Altway, M. E. Tunison, and T. W. Chapman

Chemical Engineering Department
University of Wisconsin
Madison, Wisconsin 53706

INTRODUCTION

There are several interesting hydrometallurgical processes that involve the fractionation of a mixture of metal ions by solvent extraction. These include processes proposed for treatment of manganese nodules (1,2), the Falconbridge process for separation of metal chlorides (3), and the Canadian Mines Branch process for recovery of base metals from ammoniacal solution (4). An important characteristic of these processes is that they exploit differential aspects of metal extractability rather than depending upon complete reagent selectivity. As a result, proper process design and control become critical for successful operation on a large scale.

Part of the design problem in metal separations is the selection of a liquid-liquid contactor that exhibits the suitable performance characteristics with regard to both efficiency of extraction as well as metal selectivity. Ritcey and Lucas (5) have reported an interesting pilot-plant study comparing six different contactors for the separation of copper from zinc and nickel by Kelex 100. They found selectivities to vary over a wide range, depending on both contactor type and operating conditions. It is our intention here to provide a framework for interpreting such results.

The purpose of this paper is to present mathematical models for several classes of liquid-liquid contactors that can simulate real contactors and characterize their performance in terms of basic physical and chemical parameters. For purposes of illustration, the models will be presented for idealized situations with regard to the extraction chemistry, but the framework can readily accommodate the complexities introduced by the chemistry of any particular extraction system.

The extraction system we consider here is one we have used before (6,7):

$$A^{+2} + 2HR \underset{}{\overset{K_A}{\rightleftarrows}} AR_2 + 2H^+ \tag{1}$$

$$B^{+2} + 2HR \underset{}{\overset{K_B}{\rightleftarrows}} BR_2 + 2H^+ \tag{2}$$

Metals A and B can both be extracted by reagent HR by a cation-exchange reaction at the liquid-liquid interface. It is assumed that the equilibria may be represented by simple mass-action equilibrium constants:

$$K_A = \frac{c_{AR_2}\, c^2_{H^+}}{c_{A^{+2}}\, c^2_{HR}} \tag{3}$$

$$K_B = \frac{c_{BR_2}\, c^2_{H^+}}{c_{B^{+2}}\, c^2_{HR}} \tag{4}$$

It is also assumed that the reaction is heterogeneous and fast enough to be limited by mass transfer.

In an earlier paper (7) we discussed the analysis of equilibrium stage cascades for such extraction reactions. We also indicated that equipment efficiency can generally alter selectivity from its equilibrium value.

The analysis of real equipment requires consideration of not only mass balances, flow configurations and equilibrium conditions but also local reaction rates. For our model mass-transfer-limited system the interfacial fluxes of A and B are coupled. A two-film mass transfer model yields the fluxes as the solution of two coupled cubic equations or a single fifth-order polynomial. Thus, it is found that local fluxes depend in a nonlinear manner on all six bulk concentrations, the film mass transfer coefficients of all six species, and the two equilibrium constants.

In this paper we shall present calculations for a countercurrent cascade of continuous stirred-tank reactors (CSTR), which should represent real mixer-settler units, and countercurrent and cocurrent plug-flow reactors (PFR), which should represent various columns and static mixers, respectively. Because there are so many parameters and variables that affect the behavior of these operations, it is difficult to express results in a compact form. For purposes of comparison, we shall consider two specific cases which are summarized in Table 1.

TABLE 1

Feed Conditions and System Parameters Employed in Two Case Studies

Parameters: $K_A = 100.0;\ K_B = 10.0$

$k_{H^+} = 3k_{A^{+2}} = 3k_{B^{+2}}$

$k_{HR} = k_{BR_2} = k_{AR_2} = \varepsilon k_{A^{+2}}$

Case	Aqueous Feed	Organic Feed
I	$c_{A^{+2}} = 1.0$ M	$c_{AR_2} = c_{BR_2} = 0.0$
	$c_{B^{+2}} = 1.0$ M	$c_{HR} = 2.0$ M
	$c_{H^+} = 0.0033$ M (pH ≃ 2.5)	

TABLE 1 cont.

Case	Aqueous Feed	Organic Feed
II	$c_{A^{+2}} = 0.1$ M	$c_{AR_2} = 0.2$ M
	$c_{B^{+2}} = 0.2$ M	$c_{BR_2} = 0.8$ M
	$c_{H^+} = 0.0033$ M to 4.0 M	$c_{HR} = 0.2$ M

For the chosen values of the equilibrium constants Case I corresponds to extracting conditions for both metals. Case II covers the range from extracting to stripping conditions depending on pH and phase ratio; B is stripped more readily than A.

The relative values of the film mass transfer coefficients reflect the high mobility of H^+ in water. The parameter ε allows for the possibility of more efficient mass transfer in one phase.

The results of the model calculations will be expressed as stream outlet concentrations as functions of contactor configuration and size, phase ratio, mass transfer coefficients and feed concentrations.

MIXER-SETTLERS

The most common contactor employed for base metal extraction is a countercurrent cascade of mixer-settlers. As a first approximation to describing the performance of such units, one should do equilibrium stage calculations based on mass balances and phase equilibrium between the streams leaving each stage. Real processes then might be characterized by defining stage efficiencies to compare actual performance with the equilibrium case.

One relatively simple but realistic model for analyzing stage efficiencies in mixers is the continuous stirred-tank reactor model (CSTR) in which the composition of each phase in the mixer is taken to be uniform and the amount of metal extraction in the mixer is proportional to the interfacial area in the contacting stage and the intrinsic heterogeneous reaction or transfer rate corresponding to the outlet concentrations.

The framework for such calculations has been given earlier (7). Results will be presented below.

DIFFERENTIAL CONTACTORS

As alternatives to mixer-settlers for liquid-liquid contacting, there are a number of possibilities available. Most of these must be viewed as differential contactors, which exhibit concentrations and therefore transfer rates that vary as continuous functions of position through the volume of the contactor. The principal example is the packed column, but columns with internal agitation, bubble columns, and static in-line mixers must also be considered on a local, differential scale to calculate the overall degree of extraction from intrinsic metal transfer kinetics and the hydrodynamics of the equipment.

To demonstrate the development of appropriate equipment models we present here an analysis of steady countercurrent extraction in a packed column. For simplicity axial dispersion and droplet segregation are neglected. For the transfer rates we use a two-film mass transfer model that incorporates the essential features of the metal extraction chemistry. It is assumed that the flow and bulk stream concentration gradients are one-dimensional, and the film mass transfer coefficients are taken to be uniform and known from conventional solvent extraction correlations.

MODEL OF A COUNTERCURRENT COLUMN

We consider a countercurrent column of height L and a cross-sectional area S into which aqueous feed enters at the top with volumetric flow rate Q_{aq} and organic feed enters the bottom at rate Q_{org} as indicated in Figure 1. Vertical position in the column z is measured from the top.

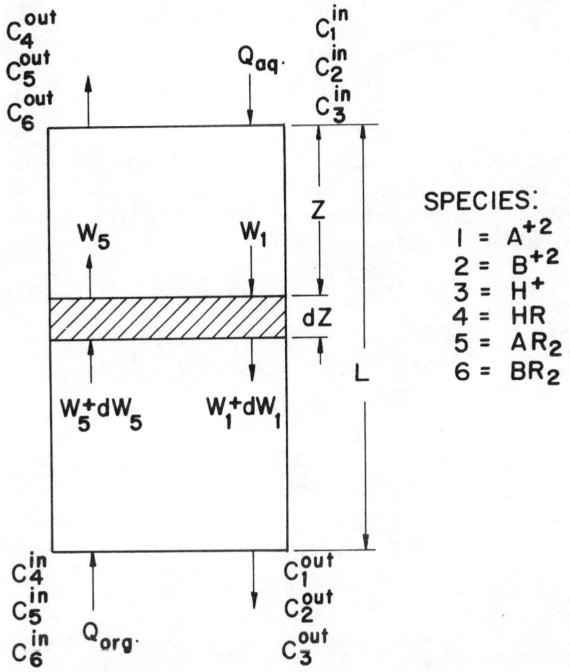

FIGURE 1. Schematic diagram of a differential countercurrent contactor

Mass balances on metals A and B in the aqueous phase may be written in the differential volume Sdz as

$$\frac{dw_A}{dz} = -aN_A S \tag{5}$$

and

$$\frac{dw_B}{dz} = -aN_B S \tag{6}$$

where w_i is the aqueous molar flow rate of metal i, N_i is the interfacial flux of i into the organic phase, and a is the interfacial area per unit volume of column. These differential mass balances may be written in dimensionless form as

$$\frac{d\xi_1}{d\zeta} = \frac{k_1 aSL}{w_A^{in}} c_1^{in} N_A^* \qquad (7)$$

and
$$\frac{d\xi_2}{d\zeta} = \frac{k_2 aSL}{w_B^{in}} c_2^{in} N_B^* \qquad (8)$$

where ξ_i is conversion of species i defined as

$$\xi_i = (c_i^{in} - c_i)/c_i^{in} \qquad (9)$$

Also
$$w_A^{in} = c_1^{in} Q_{aq} \qquad (10)$$

and similarly for B. The dimensionless fluxes are defined by

$$N_A^* = N_A/k_1 c_1^{in} \qquad (11)$$

and
$$N_B^* = N_B/k_2 c_2^{in} \qquad (12)$$

The numerical subscripts 1, 2, ... 6 refer to species A^{+2}, B^{+2}, H^+, HR, AR_2, and BR_2. Accordingly, k_1 and k_2 are the aqueous film mass transfer coefficients for A^{+2} and B^{+2}, respectively.

Although the fluxes are primarily functions of the local stream compositions, we have found it convenient to write Equations (7) and (8) formally as the following integral equations:

$$\xi_1(\zeta) = (1+R)\, k_1 a\tau \int_0^{\zeta} N_1^*\, d\zeta \qquad (13)$$

$$\xi_2(\zeta) = (1+R)\, k_2 a\tau \int_0^{\zeta} N_2^*\, d\zeta \qquad (14)$$

where

$$\zeta = z/L\ , \qquad (15)$$

$$\tau = SL(1-\phi)/Q_{aq} \qquad (16)$$

and ϕ is the volume fraction of organic phase in the fluid mixture. The parameter τ may be viewed as a residence time; if S were corrected for packing volume to be the average cross-sectional area open for flow, τ would become the residence time of the aqueous phase. Thus, the product ($k_1 a\tau$) may be considered as either a dimensionless residence time or dimensionless column height.

For simplicity the hold-up fraction ϕ has been taken to be R/(R+1) where R is the volumetric organic to aqueous feed ratio, Q_{org}/Q_{aq}.

If an appropriate rate model is available to estimate the dimensionless fluxes N_i^* as functions of position or conversion, Equations (13) and (14) may be solved for conversion profiles, and particularly the outlet conversions $\xi_i(1)$, as functions of the various operating parameters. In general, these fluxes are determined by either mass transfer conditions or reaction kinetics and may be expressed as either an analytic model or an empirical correlation of experimental rate data. For the purposes of this illustration we adopt a two-film mass transfer model that is relatively simple to formulate but which also reflects the essential features of the reversible ion exchange chemistry.

MASS TRANSFER MODEL

We have presented the two film model in earlier discussions (6,8). The situation at the interface during extraction is viewed as shown in Figure 2.

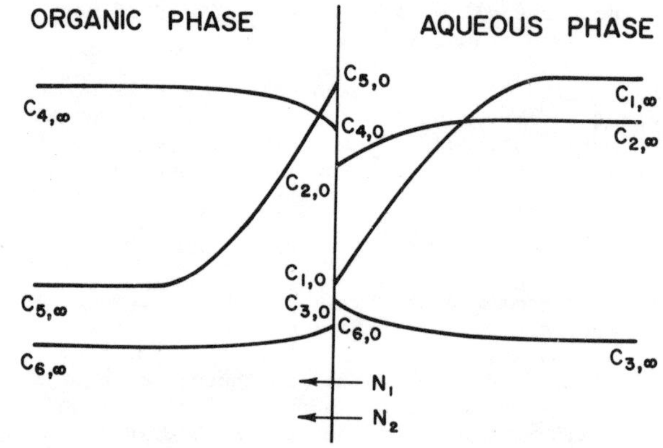

FIGURE 2. Concentration profiles near the interface during simultaneous extraction of two metals

The flux of each species at the interface is written in terms of its film mass transfer coefficient:

$$N_i = k_i(c_{i,\infty} - c_{i,0}) \quad \text{for} \quad i = 1,2,3$$

$$\text{or} \quad N_j = k_j(c_{j,0} - c_{j,\infty}) \quad \text{for} \quad j = 4,5,6. \qquad (17)$$

The stoichiometry of the heterogeneous reactions relates each of these fluxes to the rate of metal transfer:

$$N_1 = N_5 = N_A, \qquad (18)$$

$$N_2 = N_6 = N_B, \qquad (19)$$

$$\text{and} \quad N_3 = N_4 = -2\,(N_A + N_B) \qquad (20)$$

If the reaction kinetics are fast enough to make the process mass-transfer controlled, the interfacial concentrations $c_{i,0}$ should satisfy the equilibrium conditions (3) and (4). Equations (17) may be solved for $c_{i,0}$ in terms of N_A and N_B. When these are substituted into Equations (3) and (4) one obtains

$$\frac{[\bar{c}_5+N_1^*k_1/k_5]\left[\bar{c}_3+2\left(N_1^*k_1/k_3+N_2^*\frac{k_2}{k_3}\frac{c_2^{in}}{c_1^{in}}\right)\right]^2}{[\bar{c}_1-N_1^*]\left[\bar{c}_4-2\left(N_1^*\frac{k_1}{k_4}+N_2^*\frac{k_2}{k_4}\bar{c}_2\right)\right]^2} = K_A \tag{21a}$$

and

$$\frac{\left[\bar{c}_6+N_2^*\frac{k_2}{k_6}\bar{c}_2\right]\left[\bar{c}_3+2\left(N_1^*k_1/k_3+N_2^*\frac{k_2}{k_3}\frac{c_2^{in}}{c_1^{in}}\right)\right]^2}{\left[\bar{c}_2-N_2^*\frac{c_2^{in}}{c_1^{in}}\right]\left[\bar{c}_4-2\left(N_1^*\frac{k_1}{k_4}+N_2^*\frac{k_2}{k_4}\bar{c}_2\right)\right]^2} = K_B \tag{21b}$$

where $\bar{c}_i = c_{i,\infty}/c_1^{in}$ (22)

Equations (21) constitute two simultaneous cubic equations for N_1^* and N_2^* in terms of local bulk concentrations and the film mass transfer coefficients. If the bulk concentrations are known at a point in the column these algebraic equations may be solved to provide values of N_i^* to substitute into Equations (13) and (14). The six bulk concentrations are related to the local conversion values by Equation (9) for c_1 and c_2 and by overall material balances which yield

$$\bar{c}_3 = \frac{c_3^{in}}{c_1^{in}} + 2\left(\xi_1 + \frac{c_2^{in}}{c_1^{in}}\xi_2\right) \tag{23a}$$

$$\bar{c}_4 = \frac{c_4^{in}}{c_1^{in}} - \frac{2}{R}\left[(\xi_{12}-\xi_1) + (\xi_{22}-\xi_2)\frac{c_2^{in}}{c_1^{in}}\right] \tag{23b}$$

$$\bar{c}_5 = \frac{c_5^{in}}{c_1^{in}} + \frac{1}{R}(\xi_{12}-\xi_1) \tag{23c}$$

and

$$\bar{c}_6 = \frac{c_6^{in}}{c_1^{in}} + \frac{1}{R}(\xi_{22}-\xi_2)\frac{c_2^{in}}{c_1^{in}} \tag{23d}$$

where

$$\xi_{i2} = \xi_i(1) \tag{24}$$

the conversion in the aqueous outlet stream.

METHOD OF SOLUTION

It is seen that even this fairly simple rate model yields a nonlinear dependence of the metal fluxes on the species concentrations in both phases. The solution of Equations (7) and (8) or (13) and (14) for the outlet conversions will therefore be a difficult task. Especially in the case of countercurrent streams, simulation of column profiles will ordinarily involve a trial-and-error computation in which outlet stream compositions must be guessed and the profiles integrated repeatedly with the fluxes being evaluated numerous times by perhaps a second trial-and-error calculation. Such a procedure would be tedious and time-consuming.

Our approach to facilitating the necessary calculations has been to replace the integrals in Equations (13) and (14) by an optimal quadrature formula based on assuming that $N_i^*(\zeta)$ may be approximated well by a truncated series of orthogonal polynomials (9). The result is the approximate equations

$$\xi_1^{(n)}(\zeta_j) \doteq (1+R)\,k_1 a\tau \sum_{i=0}^{n+1} H_{ji} N_1^*(\zeta_i);\ j=1,2,\ldots n+1 \tag{25}$$

and

$$\xi_2^{(n)}(\zeta_j) \doteq (1+R)\,k_2 a\tau \sum_{i=0}^{n+1} H_{ji} N_2^*(\zeta_i);\ j=1,2,\ldots n+1 \tag{26}$$

where n is the number of interior quadrature abscissas, ζ_j, used in the approximation to the integral and H_{ji} are the corresponding quadrature weights. The points $\zeta_0 = 0$ and $\zeta_{n+1} = 1$ are also used. Appropriate quadrature points and weights have been given by Guertin (10).

The effectiveness of this method rests on the fact that Equations (25) and (26) are very accurate approximations, even for n as small as 4 or 6. Thus, for a chosen value of n, one has a system of 4×(n+2) simultaneous algebraic equations to solve for the 4×(n+2) unknowns, $\xi_1(\zeta_j)$, $\xi_2(\zeta_j)$, $N_1^*(\zeta_j)$, and $N_2^*(\zeta_j)$. These may be solved by a Newton's-method algorithm in a few seconds on a digital computer.

RESULTS FOR COUNTERCURRENT COLUMNS

Although the computations yield conversion profiles, we are primarily interested in the compositions of the outlet streams. From the terminal conversion values, the outlet concentrations of all species may be obtained from Equations (23).

Some typical results of the computations are shown in Figures 3 through 7. Figures 3 and 4 correspond to extraction conditions, Case I. Figures 5 to 7 demonstrate performance under stripping conditions, Case II. The figures are organized to show the relative effects of several system parameters on conversion and selectivity in countercurrent columns. The parameters of principal interest here are the dimensionless residence time or column height ($k_1a\tau$), the phase ratio R, the relative acid level in the aqueous feed $\bar{c}_{H^+}$, and the ratio of organic to aqueous metal film mass transfer coefficients ε. Note that ε can depend on which phase is dispersed and on the average drop size.

Figure 3 shows the effects of ($k_1a\tau$) and R on the extent of metal extraction. Figure 4 indicates the extraction selectivity in this case.

Figure 5 shows the effectiveness of selective stripping at various stripping acid concentrations. Figures 6 and 7 show the aqueous outlet metal levels under selective stripping conditions.

COMPARISON AMONG REACTOR TYPES

The above results may be compared with those predicted for other types of equipment. For example we may investigate the performance of countercurrent equilibrium stages and cascades of stirred tanks operating with the same feed streams.

Figure 8 shows the conversion in three countercurrent CSTRs under conditions comparable to those

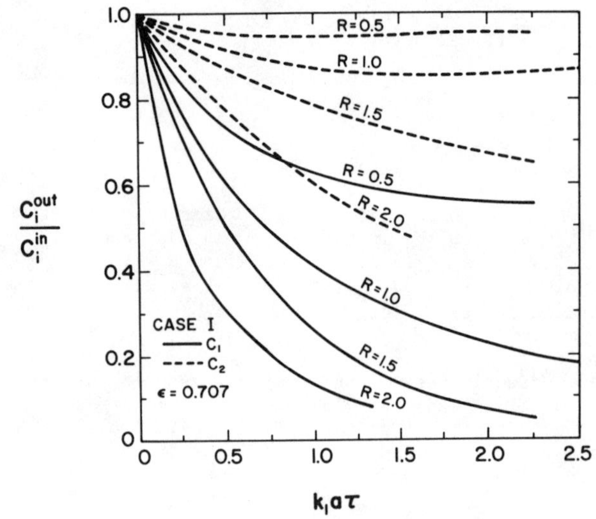

FIGURE 3. Extraction of two metals in a countercurrent column. Effect of phase ratio R.

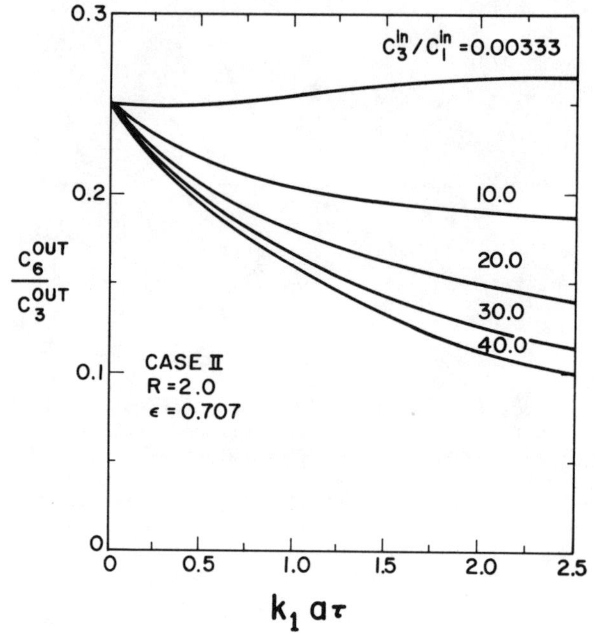

FIGURE 5. Selectivity of stripping. Effect of inlet acid level at R = 2.0.

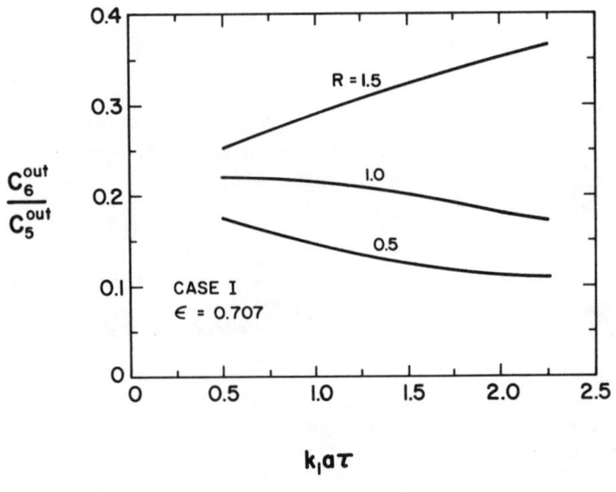

FIGURE 4. Selectivity during extraction of two metals. Effect of phase ratio R.

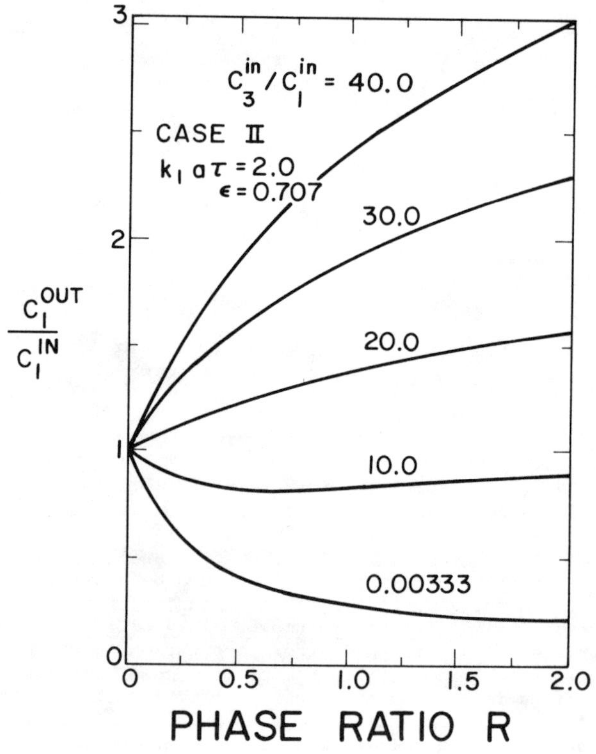

FIGURE 6. Degree of stripping of metal A as a function of phase ratio and acid level.

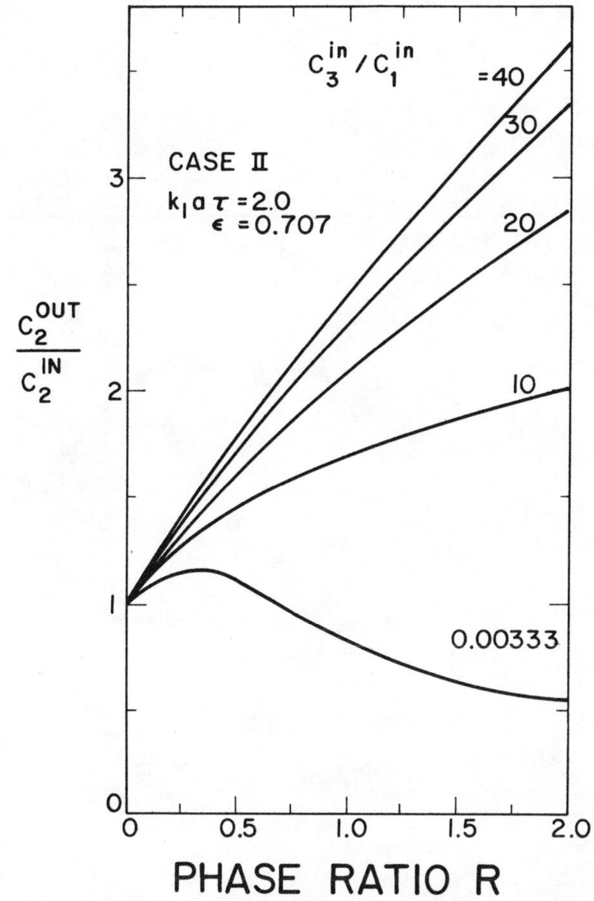

FIGURE 7. Degree of stripping of metal B as a function of phase ratio and acid level.

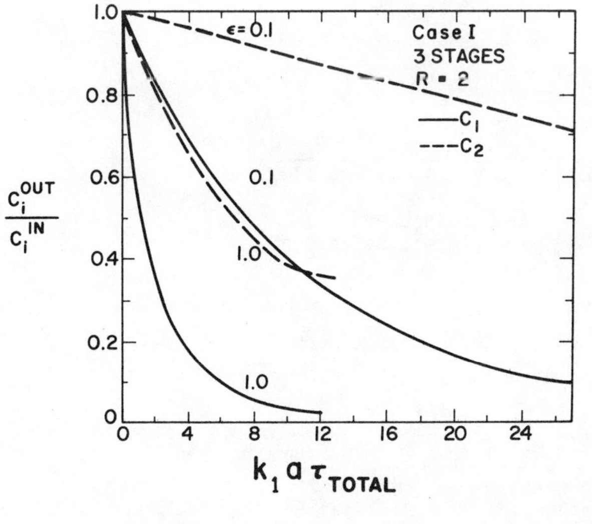

FIGURE 8. Extraction of two metals in a countercurrent cascade of three mixer-settlers as functions of dimensionless residence time. Effect of ε at R = 2.

in Figure 3. Figure 9 gives the selectivity for such a cascade under stripping conditions. In these two figures performance is represented as a function of total aqueous residence time in the three stages; thus, $k_1 a\tau$ is defined analogously to that for a column. Each stage is taken to contribute $(k_1 a\tau)/3$.

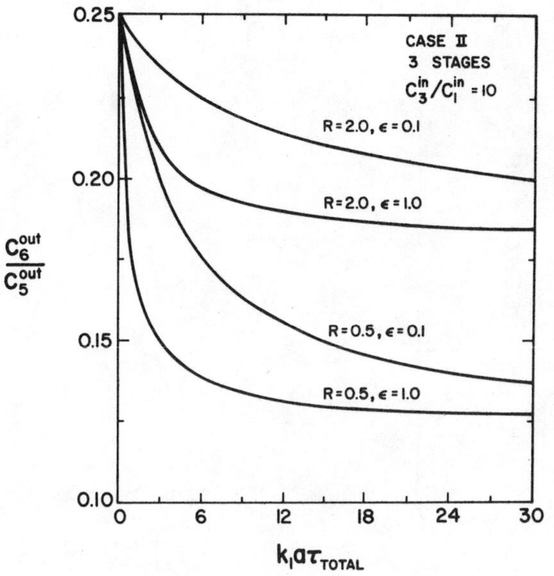

FIGURE 9. Effects of R and ε on selective stripping in a countercurrent cascade of three mixer-settlers.

The corresponding performance of three equilibrium stages for Cases I and II is given in Figures 10 and 11.

A comparison of the conversions for the three contactor models for conditions of Case I is given in Table 2. A similar comparison for selective stripping, Case II, is given in Table 3.

DISCUSSION

Because so many parameters are involved it is difficult to arrive at many general conclusions about the results given here. It is seen that the cascade of three stirred tanks is considerably less efficient in extraction than a countercurrent column with the same total $k_1 a\tau$ value. For example, comparison of Figures 3 and 8 indicates that the former should be about six times larger than the latter, at comparable $k_1 a$ values, to achieve the same degree of extraction. On the other hand, selective stripping of B appears to be more effective in the staged mixers (see Table 3).

Qualitatively the performance of a column is similar to that of an equilibrium-stage cascade with regard to conversion if $k_1 a\tau$ at $\varepsilon = 1$ is comparable in magnitude to the number of ideal stages. Therefore, it may be helpful to view $k_1 a\tau$ as a number of "trans-

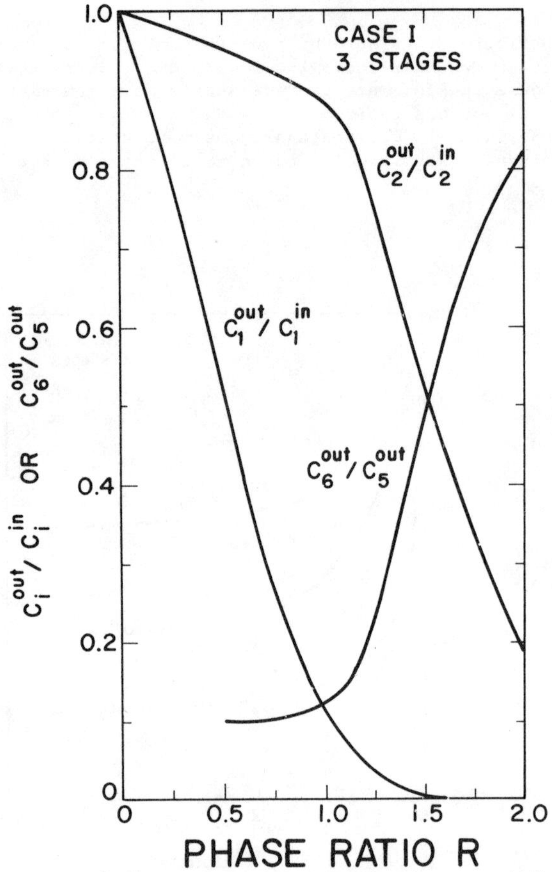

FIGURE 10. Simultaneous extraction in three equilibrium stages as a function of R.

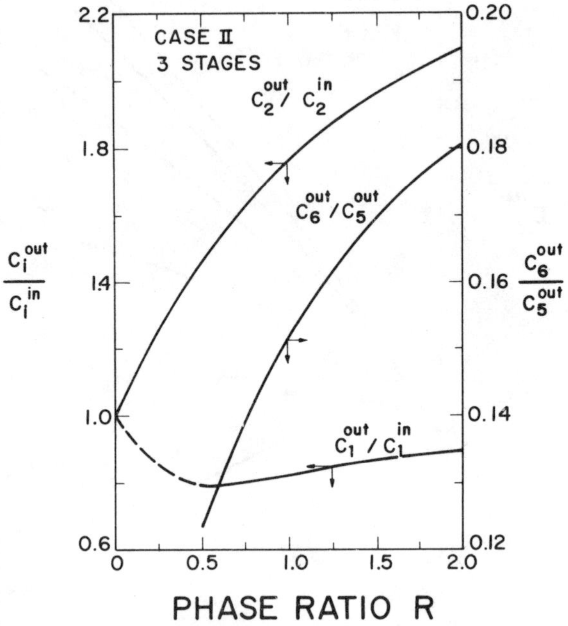

FIGURE 11. Simultaneous stripping in three equilibrium stages as a function of R. c_3^{in}/c_1^{in} = 10.

TABLE 2

Conversions in Simultaneous Extraction of Two Metals. $\bar{c}_2^{in} = 1.0$; $\bar{c}_3^{in} = 0.00333$; $\bar{c}_4^{in} = 2.0$; $\bar{c}_5^{in} = \bar{c}_6^{in} = 0.0$; $K_A = 100$; $K_B = 10$; $\bar{k}_2 = 1.0$; $\bar{k}_3 = 3.0$; $\bar{k}_4 = \bar{k}_5 = \bar{k}_6 = \varepsilon$; $k_1 a\tau = 1.5$ (total) for column and CSTR cascade. The entries are c_5^{out}/c_6^{out}.

	countercurrent plug-flow model		CSTR model (3 stages)		equilibrium stages
R	ε = 1.0	ε = 0.1	ε = 1.0	ε = 0.1	(3 stages)
0.5	0.4710/0.049	0.161/0.018	0.364/0.058	0.110/0.013	0.454/0.046
1.0	0.759/ 0.160	0.228/0.027	0.492/0.110	0.12/ 0.014	0.891/0.106
1.5	0.908/ 0.359	0.289/0.036	0.534/0.133	0.124/0.015	0.997/0.465
2.0	0.963/ 0.578	0.348/0.45	0.553/0.147	0.126/0.015	0.999/0.819

TABLE 3

Conversions in Selective Stripping of Two Metals. $\bar{c}_2^{in} = 1.0$; $\bar{c}_3^{in} = 10.0$; $\bar{c}_4^{in} = 2.0$; $\bar{c}_5^{in} = 8.0$; $\bar{c}_6^{in} = 2.0$; $K_A = 100$; $K_B = 10$; $\bar{k}_2 = 1.0$; $\bar{k}_3 = 3.0$; $\bar{k}_4 = \bar{k}_5 = \bar{k}_6 = \varepsilon$; $k_1 a\tau = 2.0$ (total for column or CSTR cascade). The entries are $\psi_5/\psi_6{}^+$. $({}^+\psi_i = c_i^{out}/c_i^{in})$

	countercurrent plug-flow model		CSTR model (3 stages)		equilibrium stages
R	ε = 1.0	ε = 0.1	ε = 1.0	ε = 0.1	(3 stages)
0.5	1.028/0.675	1.017/0.854	1.042/0.555	1.025/0.789	1.051/0.521
1.0	1.013/0.765	1.009/0.92	1.019/0.654	1.016/0.856	1.022/0.622
1.5	1.008/0.824	1.006/0.945	1.01 /0.71	1.012/0.882	1.011/0.682
2.0	1.007/0.749	1.01/ 0.897	1.006/0.859	1.005/0.958	1.007/0.726

fer units" in the column.

Generally, it appears that there is an inverse relation between conversion and selectivity, particularly with regard to variation of R. When reagent is not fed in excess, competition between A and B enhances selectivity. This effect is seen in the data of Ritcey and Lucas (5).

The effect of ε is also interesting with regard to selectivity. The location of the rate-controlling resistance can shift with the value of ε and alter the relative values of N_1^* and N_2^*. This may be the cause of the changes in extraction and selectivity observed by Ritcey and Lucas (5) in the Mixco column when the identity of the continuous phase was changed.

The important point here is that these model calculations can be done with only moderate effort and expense. These should help considerably in the development of any particular multi-metal extraction process. In conjunction with bench- or pilot-scale data, the models can identify relevant parameter values for optimization, scale-up and process control.

The contactor models can accommodate more realistic, specific rate functions. They can also be generalized to describe more complex flow phenomena such as axial dispersion and non-uniformity of rate parameters. Tunison has already analyzed the effect of drop segregation in a cocurrent differential contactor (11).

REFERENCES

1. Agarwal, J. C., et al., Feb. 22-26, 1976, "Metal Separation by Fluid Ion Exchange in the Processing of Ocean Nodules", paper presented at 105th Annual AIMF Meeting, Las Vegas, Nev.

2. Skarbo, R. R., Dec. 10, 1974, "Selective Stripping Process", U.S. Patent 3,853,725.

3. Wigstol, E. and Froyland, K. E., 1974, "Separation of Copper from Cobalt", U.S. Patent 3,887,679.

4. Ritcey, G. M. and Lucas, B. H., Feb. 1975, Extraction and Separation of Copper, Nickel, Zinc, and Cobalt from Ammoniacal Solution Using Kelex 100", CIM Bulletin.

5. Ritcey, G. M. and Lucas, B. H., 1975, "Evaluation of Contactors for the Extraction and Separation of Copper, Zinc, and Nickel from Ammoniacal Solutions Using Kelex 100", I. Chem. E. Symp. Ser., No. 42, p. 12.1.

6. Chapman, T. W., Caban, R. and Tunison, M. E., 1975, "Rates of Liquid-Liquid Ion Exchange in Metal Extraction Processes", AIChE Symp. Ser., 71, No. 152, 128.

7. Chapman, T. W., 1978, "Equilibrium Stage Calculations for the Solvent Extraction of Metals", AIChE Symp. Ser., 74, No. 173, 120.

8. Tunison, M. E. and Chapman, T. W., 1976, "The Effect of Diffusion Potential on the Rate of Liquid-Liquid Ion Exchange", IEC Fund., 15, 196.

9. Kopal, Z., 1961, Numerical Analysis, John Wiley and Sons, New York.

10. Guertin, E. W., 1972, "Application of Orthogonal Collocation to Axial Integration in the Design of Chemical Reactors", Ph.D. Thesis, University of Wisconsin, Madison.

11. Tunison, M. E., 1976, "Characterization of Plug-Flow, Two-Phase Reactors with Applications to the Solvent Extraction of Metals", Ph.D. Thesis, University of Wisconsin, Madison.

Section V

Electrochemical Reactions and Solution Chemistry

Renato G. Bautista and T.W. Chapman,
Section Editors

Chapter 26

ELECTROCHEMICAL PROCESSES IN THE LEACHING OF METAL SULFIDES AND OXIDES

J. Brent Hiskey and Milton E. Wadsworth

ABSTRACT

The aqueous solution dissolution of metal sulfides and oxides often occurs by electrochemical processes. Galvanic and/or corrosion couples may form if the kinetics of electron or ion transfer at mineral surfaces are slow compared to electron conduction through the solid phases. The kinetics of several metal sulfide and oxide systems are explained by electrochemical reactions related to mixed potentials which form as a result of the coupled anodic and cathodic processes involved.

ELECTROCHEMICAL PROCESSES IN THE LEACHING OF METAL SULFIDES AND OXIDES

J. Brent Hiskey and Milton E. Wadsworth

J. Brent Hiskey is Senior Scientist, Kennecott Minerals Company Process Technology
M.E. Wadsworth is Professor of Metallurgy and Metallurgical Engineering, University of Utah
Salt Lake City, Utah

INTRODUCTION

Fifteen years ago a distinguished electrochemist referred to electrochemistry as an "underdeveloped science." In the article in question, Professor Bockris (1) stated that many areas of natural phenomena and applied technology are not perceived as being electrochemically based. At that time, this was especially true for much of non-ferrous extractive metallurgy and was a particularly accurate description of hydrometallurgy. During the last decade, there has been considerable advancement in applying the theory of electrochemistry to our knowledge of hydrometallurgical processes and in understanding the mechanisms of underlying electrode reactions. Impressive as these advances are, there is still a growing need for electrochemical research of mineral systems and electrochemical engineering to apply this work in practical situations.

In a broad sense electrochemistry is defined as the physical chemistry of ions in solution and of phenomena which occur at the electrified interface. Bockris and Reddy (2) refer to electrochemistry as being divided into two fundamental domains, ionics and electrodics. Ionics is restricted to the behavior of ions in solutions, examples being many familiar redox processes. Electrodics represents the main part of electrochemistry, and is the study of charged interfaces and the conditions governing charge transfer reactions across them. Considering hydrometallurgical applications, electrodics suggests the following relationship for a conducting solid (bulk or surface)

$$\text{Electrodics: Solid} \rightleftarrows \text{Ions in solution} + \text{Electrons in solid}$$

An excellent example of the field of electrodics is the study of metal corrosion in aqueous solution.

The electronic properties of most metal sulfides and of certain oxides are such that corrosion and galvanic couples are readily established in aqueous systems. This is true for p-type, n-type or metallic like compounds for which $t_e \simeq 1$, where t_e is the electronic transport number. As a result of their electronic conductivity, certain minerals can participate in coupled charge transfer processes analogous to a metal corroding in an electrolyte, and the kinetics of leaching can be related to the potential of the solid in contact with the aqueous electrolyte. From a thermodynamic point of view most hydrometallurgical dissolution reactions are strongly favored such that back reaction kinetics normally are unimportant. Build up on ions in solution, however, may markedly affect the kinetics by influencing the potential at the mineral (electrode) solution interface.

Properties of some sulfide and oxide minerals

Several investigators have presented information on the electrochemical behavior of sulfide and oxide minerals. Metal sulfides have received the most attention because they are recognized as valuable sources of non-ferrous metal. Koch (3) has reviewed the electrochemistry of sulfide minerals and Shuey (4) has reviewed the electronic properties of both oxide and sulfide minerals. Included in Table I are resistivities, electronic, and structural information for selected sulfide and oxide minerals. The covalent character of most sulfides provides non-localization of charge resulting in appreciable intrinsic electronic conductivity. The ability of many sulfides to form non-stoichiometric compounds results in increased conduction via electron holes or excess electrons. Oxides in general are more ionic in nature and usually have higher resistivities compared to sulfide minerals.

A fundamental property of semiconducting minerals is the characteristic "rest" potential. For interfacial electrode processes, the "rest" potential corresponds to the equilibrium (no net anodic or cathodic current) electrode potential. Rest potentials for several metal sulfides are shown in Table II. It is important to remember that a mineral electrode system will establish and maintain a certain equilibrium potential that depends not only on the solution composition but also on the composition of the solid phase. Figure 1 illustrates results for the electrochemical oxidation of Cu_2S to CuS in an acid solution (10). The steps in the data have been explained on the basis of the sequential formation of a series of CuS compounds with corresponding rest potentials as follows:

$$Cu_2S \rightarrow Cu_{1.93}S \rightarrow Cu_{1.83}S \rightarrow Cu_{1.67} \rightarrow Cu_{1.4}S \rightarrow Cu\,S$$

From the work of Etienne and Peters (11), the standard free energy of formation of $Cu_{1.93}S$ and $Cu_{1.83}S$ were calculated. At 25°C, the standard free energy of formation values are

$$\Delta G^\circ_{298^\circ K}(Cu_{1.93}S) = -19{,}282 \pm 640 \text{ cal mole}^{-1}$$

TABLE I

Electronic and Structural Properties of Selected Sulfide and Oxide Minerals

Formula	Name	Resistivity OHM-M	Usual Conductor Type	Structure	Ionic Structure
Cu_5FeS_4	Bornite	10^{-3}-10^{-6}	p	Tetragonal	$(Cu^+)_5Fe^{3+}(S^{2-})_4$
Cu_2S	Chalcocite	$4x10^{-2}$-$8x10^{-5}$	p	Orthorhombic	$(Cu^+)_2S^{2-}$
$CuFeS_2$	Chalcopyrite	$2x10^{-4}$-$9x10^{-3}$	n	Tetragonal	$Cu^+Fe^{3+}(S^{2-})_2$
CuS	Corellite	$8x10^{-5}$-$7x10^{-7}$	metallic	Hexagonal	$(Cu^+)_2S_2^{2-}$
PbS	Galena	$1x10^{-5}$-$7x10^{-6}$	n&p	Cubic	$Pb^{2+}S^{2-}$
MoS_2	Molybdenite	7.5-$8x10^{-3}$	n&p	Hexagonal	$Mo^{4+}(S^{2-})$
FeS_2	Pyrite	$3x10^{-2}$-$1x10^{-3}$	n&p	Cubic	$Fe^{2+}S_2^{2-}$
ZnS	Sphalerite	$3x10^{-3}$-$1x10^{-4}$	-	Cubic	$Zn^{2+}S^{2-}$
SnO_2	Cassiterite	10^{+2}-10^{-2}	n	Tetragonal	$Sn^{4+}O_2^{4-}$
Cu_2O	Cuprite	10^{11}-10	p	Cubic	$(Cu^+)_2O^{2-}$
Fe_2O_3	Hematite	$2.5x10^{-1}$-$4x10^{-2}$	n&p	Trigonal	$(Fe^{3+})_2(O^{2-})_3$
Fe_3O_4	Magnetite	$2x10^{-4}$-$4x10^{-5}$	n&p	Cubic	$Fe^{3+}[Fe^{3+}Fe^{2+}](O^{2-})_4$
MnO_2	Pyrolusite	10^{-1}-10^{-3}	n	Tetragonal	$Mn^{4+}(O^{2-})_2$
TiO_2	Rutile	10^4-10	n&p	Tetragonal	$Ti^{4+}(O_2^{2-})_2$
UO_2	Uraninite	20-$4x10^{-1}$		Cubic	$(U^{+4})_{1-x}(O^{2-})_{2+x}$

TABLE II

Mineral	Rest Potential (V vs. SHE)	Conditions Solution	Temp	Ref.
FeS_2	0.63	1.0M H_2SO_4	25°C	5,6
$CuFeS_2$	0.52	1.0M H_2SO_4	20°C	7
Cu_2S	0.44	1.0M H_2SO_4	20°C	8
CuS	0.42	1.0M $HClO_4$	25°C	9
PbS	0.28	1.0M H_2SO_4	20°C	8
ZnS	-0.24	1.0M H_2SO_4	20°C	8
FeS	-0.28	1.0M H_2SO_4	20°C	8

and

$\Delta G_{298°K}(Cu_{1.83}S) = -18,501 \pm 620$ cal mole^{-1}

for the electrode reaction

$$Cu_{1.93}S \rightarrow Cu_{1.83}S + 0.1\ Cu^{2+} + 0.2e \quad (1)$$

The thermodynamically predicted equilibrium (rest) potential for the $Cu_{1.93}S/Cu_{1.83}S$ couple is approximately 0.50 v (vs. SHE) which appears reasonable when considering the uncertainty of the free energy values. In comparison, the data of Mathieu and Rickert (12) for the $Cu_{1.9}S/Cu_{1.8}S$ couple yield an equilibrium potential of about 0.46 v (vs. SHE). This value is in excellent agreement with the rest potentials observed for this region of Cu/S stoichiometry (see Fig. 1).

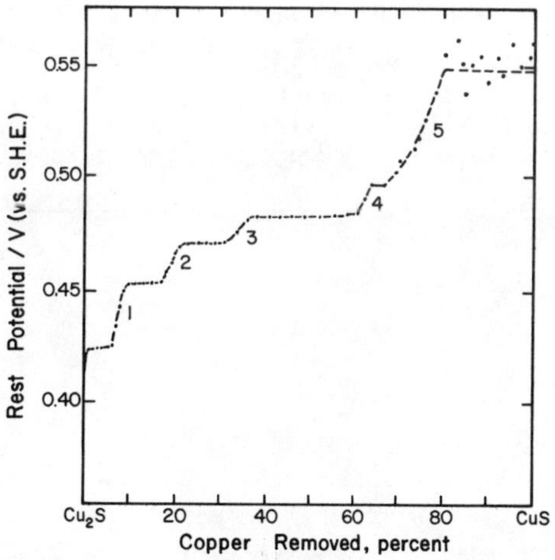

FIGURE 1. Rest potentials measured for successive solid phases formed during the anodic oxidation of Cu_2S to CuS in acid solution. The vertical steps correspond to: 1, $Cu_{1.93}S$; 2, $Cu_{1.83}S$; 3, $Cu_{1.7}S$; 4, $Cu_{1.4}S$; 5, CuS.

Moh (13) reported the existence of blue remaining covellite, designated Type B, formed in the presence of oxidizing agents and having the composition $Cu_{1+x}S$ with variable values of x from approximately 0.09 to

0.43 at room temperature. Another form of blue remaining covellite having a different structure, designated Type A, formed in a pure Cu-S system, was reported to have a solid-solution region with x values from approximately 0.06-0.20. Frenzel (14) reported blue remaining covellite to be a defect structure with excess copper up to $x \approx 0.4$. Clearly confusion still exists over the compositions of intermediate oxidation products. Potter (15) has identified two distinct types of blue-remaining covellite (blaubleiblender) formed using electrochemical cells. Instead of a solid solution the compounds blaubleiblender I, $Cu_{1.1}S$ and blaubleiblender II, $Cu_{1.4}S$ apparently form as distinct compounds. He established a series of Cu-S structures as follows:

$Cu_{1.97 \pm .003}S$	tetragonal djurleite
three metastable phases	proto-djurleite proto-hexagonal chalcocite proto-digenite
$Cu_{1.76}S$ to $Cu_{1.82}S$	low digenite
$Cu_{1.4}S$	blaubleiblender I
$Cu_{1.1}S$	blaubleiblender II

Cavallotti and Salvago (16) reported low digenite (face-centered cubic) to exist in the composition range $Cu_{1.76}S$ to $Cu_{1.79}S$ with a transformation to a high temperature form above 76°C. Brennet et al. (17) investigated the anodic dissolution of Cu_2S and observed the sequence: $Cu_2S \rightarrow Cu_{1.8}S \rightarrow Cu_{1.1}S \rightarrow CuS$. While discrepancies appear concerning the exact compositions it is clear that there are a series of identifiable Cu-S structures, some of which are metastable, with appreciable variation in copper content. Also the presence of copper and iron in the electrolyte influences the character and composition of the solid phases. Based on solid state properties those designated as copper deficient $Cu_{2-x}S$ extend from the high copper region down to approximately $Cu_{1.4}S$. Phases of lower copper (Cu/S < 1.4) content, extending to covellite are viewed as having excess copper, or as $Cu_{1+x}S$. Figure 2 illustrates results

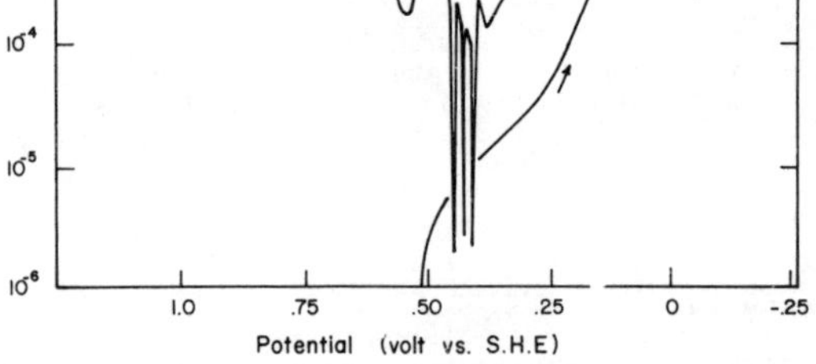

FIGURE 2. Potential-current diagram showing the formation of several intermediate Cu-S layers formed on the surface of chalcopyrite.

obtained by Warren (7) for the oxidation of a thin layer of Cu_2S formed by the cathodic reduction of chalcopyrite ($CuFeS_2$). The rest potential of the $CuFeS_2$ is +0.52 volts. As voltage was decreased the cathodic branch, indicated by the arrow on the rising curve, resulted and at the plateau value ($\sim 6 \times 10^{-4}$ amp) Cu_2S formed in a thin layer. Toward the upper part of the loop at -0.20 volts the voltage was reversed and increased anodically. The sharp peaks indicate a series of Cu-S intermediate solid phases with rest potentials in the region of 0.40 to 0.55, essentially in the same region as those reported by Koch and McIntyre (10).

Thermodynamics

The thermodynamic relationships of various solids in contact with aqueous solutions of specific composition are best illustrated by potential-pH or Pourbaix diagrams. Figure 3 represents sections of the Cu-S (a) and Cu-Fe-S (b) systems under standard conditions (298°K, 1 atm total pressure, and 10^{-3} M total dissolved species). A problem inherent with

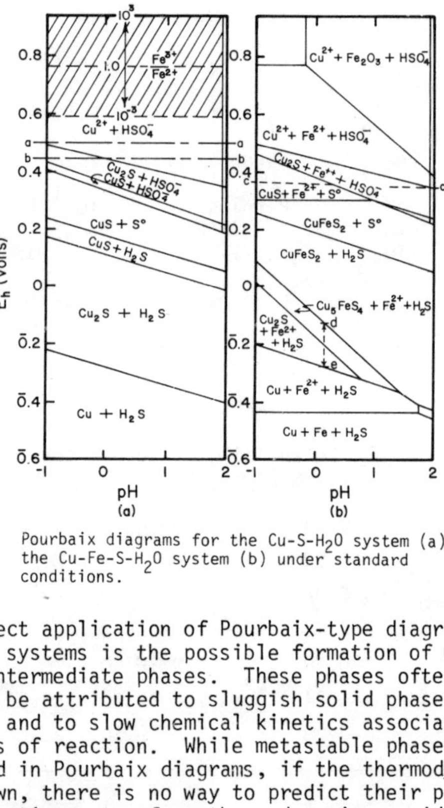

FIGURE 3. Pourbaix diagrams for the $Cu-S-H_2O$ system (a) and the $Cu-Fe-S-H_2O$ system (b) under standard conditions.

the direct application of Pourbaix-type diagrams to kinetic systems is the possible formation of metastable intermediate phases. These phases often result and may be attributed to sluggish solid phase transitions and to slow chemical kinetics associated with products of reaction. While metastable phases can be depicted in Pourbaix diagrams, if the thermodynamics are known, there is no way to predict their presence in a kinetic system from thermodynamic considerations alone. In the Cu-S diagram (Fig. 3a) one can easily predict that Cu_2S cannot coexist with elemental sulfur: i.e., the reaction

$$Cu_2S + HSO_4^- + 7H^+ + 6e^- = 2CuS + 4H_2O \quad (2)$$

$$(E = 0.359 - 0.069\ pH + 0.010 \log a_{HSO_4^-})$$

has a more positive potential than the reaction

$$HSO_4^- + 7H^+ + 6e^- = S° + 4H_2O \quad (3)$$

$(E = 0.338 - 0.069\ pH + 0.010 \log a_{HSO_4^-})$

resulting in a favored net reaction

$$Cu_2S + S = 2CuS \quad (4)$$

$$\Delta G^\circ_{298} = -2{,}900 \text{ cal mole}^{-1}$$

Consequently, in hydrometallurgical systems elemental sulfur is not found in intimate contact with Cu_2S. During the oxidation of Cu_2S with oxygen or Fe^{3+}, elemental sulfur forms outside its range of thermodynamic stability. The shaded portion of Fig. 3(a) indicates the potential of the Fe^{3+}/Fe^{2+} couple for Fe^{3+}/Fe^{2+} activity ratios from 10^{-3} to 10^3. All reactions having half cell potentials less than the Fe^{3+}/Fe^{2+} couple are thermodynamically favored to react anodically in the presence of ferric ion. However, the sequence of reactions products do not necessarily correspond to those having the greatest thermodynamic drive. For example the formation of chalcocite (Cu_2S) is strongly favored during the oxidation of covellite according to the reaction

$$CuS + 3Fe^{3+} + 2H_2O = \tfrac{1}{2}Cu_2S + \tfrac{1}{2}HSO_4^- + 7/2\ H^+ + 3Fe^{2+} \quad (5)$$

$$\Delta G^\circ = -28{,}530 \text{ cal mole}^{-1}$$

Actually covellite is oxidized directly to cupric ion and elemental sulfur

$$CuS + 2Fe^{3+} = Cu^{2+} + 2Fe^{2+} + S^\circ \quad (6)$$

$$\Delta G^\circ = -8{,}330 \text{ cal mole}^{-1}$$

even though the driving force is less favorable. This result indicates a substantial difference in the relative reaction rates. At low temperatures, the oxidation rate of elemental sulfur to bisulfate is slow, therefore Cu^{2+} and S° coexist at conditions where elemental sulfur is metastable. Similarly the oxidation of Cu_2S to cupric ion in solution is highly favored ($\Delta G^\circ = -79{,}320$ cal mole^{-1})

$$Cu_2S + 4H_2O + 10Fe^{3+} = 2Cu^{2+} + HSO_4^- + 7H^+ + 10Fe^{2+} \quad (7)$$

The anodic oxidation of Cu_2S in the presence of ferric ion actually occurs according to the following reaction

$$Cu_2S + 2Fe^{3+} = CuS + Cu^{2+} + 2Fe^{2+} \quad (8)$$

even though the thermodynamic drive ($\Delta G = -11{,}130$ cal mole^{-1}) is less favorable.

While kinetic factors explain the formation of metastable phases, these phases develop potentials which result in the formation of solid intermediates in sequence progressing from states of lowest potential to those of highest potential. The potentials of reactions (6) and (8) are illustrated in Figure 3(a) by the lines b-b and a-a respectively for $a_{Cu^{2+}} = 10^{-3}$. At $a_{Cu^{2+}} = 1$, a-a will move to 0.59 volts and b-b to 0.53 volts. The formation of metastable states in hydrometallurgical systems is of practical importance since the final product may be metastable, as in the case of elemental sulfur. Related side reactions may also be affected. For example the reactions (5) and (7) are acid producing while reactions (6) and (8) are not.

Figure 3(b) illustrates the Cu-Fe-S system and includes the sulfides $CuFeS_2$ (chalcopyrite) and Cu_5FeS_4 (bornite). As in the case of the oxidation of CuS with ferric ion, the formation of bisulfate is favored over the formation of S°. Nevertheless, the principal product in acid ferric sulfate solutions is elemental sulfur ion formed according to the reaction

$$CuFeS_2 + 4Fe^{3+} = Cu^{2+} + 5Fe^{2+} + 2S^\circ$$

$$\Delta G^\circ = -33{,}100 \text{ cal mole}^{-1} \quad (9)$$

The half-cell potential for chalcopyrite in equilibrium with sulfur and its ions is indicated by the line c-c, Figure 3(b), for a concentration typical of that found in copper dump leaching ($a_{Cu^{2+}}\, a_{Fe^{2+}} \approx 10^{-3}$). If the reaction is carried out in acid ferric sulfate solution, essentially stoichiometric quantities of Fe^{2+} and S° are produced according to Eq. (9).

Electrochemical Processes

Electrochemical processes are unique in that the solid assumes a uniform potential throughout, providing ohmic resistance is negligible. Consequently the chemical reaction may be influenced over relatively long distances. The potential may directly affect the kinetics of the reaction, if slow discharge is involved, and may also serve to stabilize intermediate solid phases. Concepts developed in the field of corrosion are translatable to the electron conducting solids in hydrometallurgical systems, interfacial charge-transfer reactions being the fundamental denominator of both. Apart from this important feature both behave as internally short circuited (no external applied voltage) systems, where the solid assumes a mixed potential determined by spatially separated anodic and cathodic processes. At the mixed potential, the sum of all anodic currents, ΣI_a, is equal to the negative sum of all cathodic currents, ΣI_c, such that

$$\Sigma I_a = -\Sigma I_c$$

where $I_a = i_a A_a$ and $I_c = i_c A_c$. The terms A_a and A_c are the respective anodic and cathodic areas, and i_a and i_c are the corresponding current densities. As in corrosion processes, two types of mixed potential regimes may be operative (i.e. local cell and galvanic couples). The local cell type involves a single phase having both anodic and cathodic regions on its surface such that the total surface area of the solid is equal to the sum of anodic and cathodic areas. A galvanic couple is operative where two or more solid phases are in electrical contact. In this union each solid assumes either anodic or cathodic behavior and has its own separate surface area for reaction.

Individual electrode polarization curves are helpful in visualizing the concept of a mixed potential. Figure 4 depicts hypothetical polarization curves for the dissolution of a metal sulfide in the presence of a cationic oxidant N^{n+}. If MS is placed in contact with a solution containing an oxidant N^{n+} (e.g. Fe^{3+} in acid solutions or $Cu(NH_3)_4^{2+}$ in alkaline solutions) having a more positive equilibrium potential, a corrosion cell will result in the anodic oxidation of MS and the cathodic reduction of N^{n+}.

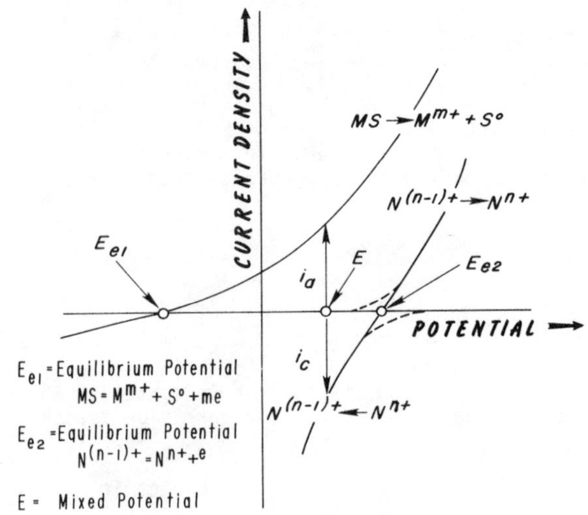

FIGURE 4. Current density potential diagram showing curves for metal sulfide with rest potential E_{e1} and redox ions with rest potential E_{e2} forming mixed potential E.

The half cell reactions are

$$MS = M^{m+} + S^\circ + me^- \qquad (10)$$

and

$$N^{n+} + e = N^{(n-1)+} \qquad (11)$$

The equilibrium potentials E_{e1} and E_{e2} for half-cell reactions (10) and (11) are illustrated in Figure 4. Equilibrium electrode potentials are independent of the properties of the electrode surface, however, electrode surface properties influence to a great extent the shape of the partial current-voltage curves. The value of the mixed potential is established by both the partial equilibrium potentials and the shape of the current-voltage curves. Hence, the mixed potential depends upon the properties of the electrode surface. Potentials more positive than the equilibrium potential drive the half-cell reaction in the net anodic direction and potentials more negative than the equilibrium potential drive the reactions in the net cathodic direction.

The net current density (i) for the system is the sum of the partial current densities i_a (anodic) and i_c (cathodic). The mixed potential (E) occurs where the net current density is zero; the values of the partial current densities, therefore are equal and of the opposite sign ($i_a = -i_c$) as indicated in Figure 4. The anodic overvoltage η_a is defined as $(E-E_{e1})$ and is positive. The cathodic overpotential η_c is defined as $(E-E_{e2})$ and is negative.

The electrochemical bias for charge transfer in simple reversible reactions may be expressed quantitatively by the Butler-Volmer equation. This fundamental equation of electrodics relates the current density at a solid-solution interface to the established overpotential. Applied to the specific example of a metal sulfide undergoing anodic dissolution, Eq. (10), the Butler-Volmer equation, for a single electron transfer process, is

$$i = zF\vec{k}\exp\left[\frac{\beta FE}{RT}\right] - zF\overleftarrow{k}[M^{m+}]\exp\left[\frac{-(1-\beta)FE}{RT}\right] \qquad (12)$$

The first term in the above equation is the partial current density for the anodic (forward) direction of Eq. (10) and the second term the partial current density for the cathodic (reverse) direction, F is the Faraday, $\vec{k}$ and $\overleftarrow{k}$ are the rates for the forward and reverse directions, respectively, in the absence of potential, and β is the transfer coefficient. The rate $\vec{k}$ may contain a concentration term for ad-ions on the mineral surface which may change if the stoichiometry of the mineral varies. At the equilibrium potential E_{e1}, the partial current densities of Eq. (10) are equal $|\vec{i}| = |\overleftarrow{i}|$ and are represented by the exchange current density, i_o. The exchange current density is given by the following relationships

$$i_o = zF\vec{k}\exp\left[\frac{\beta FE_{e1}}{RT}\right] = zF\overleftarrow{k}[M^{m+}]\exp\left[\frac{-(1-\beta)FE_{e1}}{RT}\right] \qquad (13)$$

and it follows directly that

$$i = i_o\left\{\exp\left[\frac{\beta F\eta}{RT}\right] - \exp\left[\frac{-(1-\beta)F\eta}{RT}\right]\right\} \qquad (14)$$

By definition the overvoltage η is equal to the difference between E and E_{e1}. Equation (14) is sometimes referred to as a simplified form of the Butler-Volmer relationship.

If the metal sulfide is allowed to contact an aqueous solution containing an oxidant, N^{n+}, a corrosion couple involving net anodic and net cathodic reactions will result and a steady-state mixed potential will be established. The Butler-Volmer expression for the net anodic, i_a, and net cathodic, i_c, current densities for reactions (10) and (11) are

$$i_a = z_aF\vec{k}_a\exp\left[\frac{\beta_a n_a FE}{RT}\right] - z_aF\overleftarrow{k}_a[M^{m+}]\exp\left[\frac{-(1-\beta_a)n_a FE}{RT}\right] \qquad (15)$$

and

$$i_c = z_cF\vec{k}_c[N^{(n-1)+}]\exp\left[\frac{\beta_c n_c FE}{RT}\right] - z_cF\overleftarrow{k}_c[N^{n+}]\exp\left[\frac{-(1-\beta_c)n_c FE}{RT}\right] \qquad (16)$$

At this stage certain simplifying assumptions help in the development of the rate equations. Commonly single electron transfer processes are favored energetically such that $n_a = n_c = 1$. The number of electrons required for each half-cell reaction according to Eqs. (10) and (11) are $z_a = m$ and $z_c = 1$, respectively. Under certain conditions, the mixed potential may be evaluated in terms of conventional rate parameters. For example, the first of these conditions corresponds to a case where the anodic current due to the anodic oxidation of MS is much greater than the partial anodic current for the anodic oxidation of $N^{(n-1)+}$ (i.e. both i_a and i_c fall within their respective Tafel regions). Hence, the anodic and cathodic currents acquire the following form

$$i_a = mF\vec{k}_a \exp\left[\frac{\beta_a FE}{RT}\right] \quad (17)$$

and

$$i_c = -F\overleftarrow{k}_c [N^{n+}] \exp\left[\frac{-(1-\beta_c)FE}{RT}\right] \quad (18)$$

For the relationship $i_a = -i_c$

$$\exp\left[\frac{\beta_a FE}{RT}\right] = \left[\frac{\overleftarrow{k}_c}{m\vec{k}_a}\right]^{\phi} [N^{n+}]^{\phi} \quad (19)$$

where

$$\phi = \frac{\beta_a}{(1+\beta_a-\beta_c)}$$

The rate of leaching becomes

$$-\frac{dn_{MS}}{dt} = \frac{i_a}{mF} = \vec{k}_a^{(1-\phi)} \left[\frac{\overleftarrow{k}_c}{m}\right]^{\phi} [N^{n+}]^{\phi} \quad (20)$$

where n_{MS} is the number of moles of MS per cm^2. The transfer coefficients β_a and β_c represent the symmetry of the activation barrier for the anodic and cathodic charge transfer reactions, respectively. For symmetrical barriers $\beta = 0.5$. The value of ϕ lies usually in the range 0.4-0.7 and for the special case where $\beta_a = \beta_c = 0.5$, $\phi = 0.5$. The rate of leaching will appear as a fractional order for the oxidant even though cathodic discharge is first order with respect to the oxidant.

A second case assumes that the partial anodic currents for the anodic oxidation of MS and $N^{(n-1)+}$ are of comparable magnitude at the mixed potential. This is equivalent to saying that only the i_a falls within the Tafel region. Therefore the anodic current density (i_a) is again given by Eq. (17) and the cathodic current density (i_c) will be

$$i_c = F\vec{k}_c [N^{(n-1)+}] \exp\left[\frac{\beta_c FE}{RT}\right]$$

$$- F\overleftarrow{k}_c [N^{n+}] \exp\left[\frac{-(1-\beta_c)FE}{RT}\right] \quad (21)$$

and for $i_a = -i_c$

$$\exp\left[\frac{\beta_a FE}{RT}\right] =$$

$$\left[\frac{\overleftarrow{k}_c}{m\vec{k}_a}[N^{n+}] - \frac{\vec{k}_c}{m\vec{k}_a}[N^{(n-1)+}] \exp\left[\frac{FE}{RT}\right]\right]^{\phi} \quad (22)$$

There is no exact analytical solution for the mixed potential term and the apparent reaction order will vary as the mixed potential changes. Assuming $\beta_a = \beta_c = 0.5$, an approximate solution for the mixed potential results

$$\exp\left[\frac{\beta_a FE}{RT}\right] = \left[\frac{\overleftarrow{k}_c [N^{n+}]}{\vec{k}_c [N^{(n-1)+}] + m\vec{k}_a}\right]^{1/2} \quad (23)$$

The leaching rate is now expressed as

$$-\frac{dn_{MS}}{dt} = \vec{k}_a \left[\frac{\overleftarrow{k}_c [N^{n+}]}{\vec{k}_c [N^{(n-1)+}] + m\vec{k}_a}\right]^{1/2} \quad (24)$$

If $k_c [N^{(n-1)+}] >> m\vec{k}_a$, the rate of leaching will be proportional to the $[N^{n+}]/[N^{(n-1)+}]$ ratio to the one half power and will be a unique function of the oxidation potential of the solution.

A third case reflects a situation where the anodic and cathodic partial current densities for the net cathodic process are much greater than those for the net anodic or leaching process. The expression for the mixed potential is

$$\exp\left[\frac{\beta_a FE}{RT}\right] = \left[\frac{\overleftarrow{k}_c [N^{n+}]}{\vec{k}_c [N^{(n-1)+}]}\right]^{\beta_a} \quad (25)$$

and insertion into Eq. (16) yields a rate equation of the following form:

$$-\frac{dn_{MS}}{dt} = \vec{k}_a \left[\frac{\overleftarrow{k}_c [N^{n+}]}{\vec{k}_c [N^{(n-1)+}]}\right]^{\beta_a} \quad (26)$$

Nicol et al. (18) observed all three cases for the dissolution of UO_2 in acid ferric sulfate solutions. These are illustrated in Figure 5 where the log (leaching rate) is plotted versus log (Fe^{2+}) for fixed Fe^{3+} concentrations. As will be discussed later, the dissolution of UO_2 in both acid and alkaline carbonate solution is electrochemically controlled.

When conducting or semi-conducting minerals in electrical contact are subject to leaching simultaneously in the same process, galvanic interactions may occur which can cause an enhanced corrosion of one or more members of the electrochemically coupled mineral regime, and cathodic protection of other mineral members. The mineral with the lower "rest" potential will react anodically, whereas, cathodic behavior (passivation) may be imposed on the mineral processing the higher "rest" potential. As shown in Table II, pyrite possesses the highest "rest" potential of the mineral sulfides listed. If pyrite is brought into electrical contact with another sulfide mineral, the oxidation of the other sulfide mineral will be accelerated under leaching conditions.

An excellent example of this phenomena was recently observed in the leaching of a chalcopyrite bearing material (19). Figure 6 illustrates schematically the polarization curve for chalcopyrite superimposed on that for pyrite. In this example log |i| is plotted versus potential (SHE). Galvanic coupling between

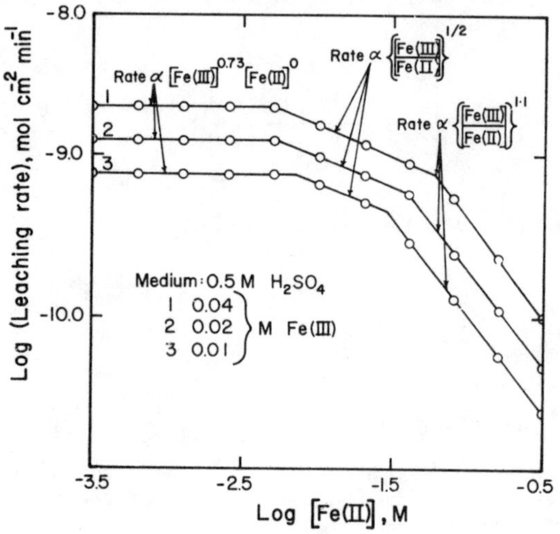

FIGURE 5. Effect of $[Fe^{II}]$ and $[Fe^{III}]$ on the leaching of UO_2 in sulfuric acid at 25°C.

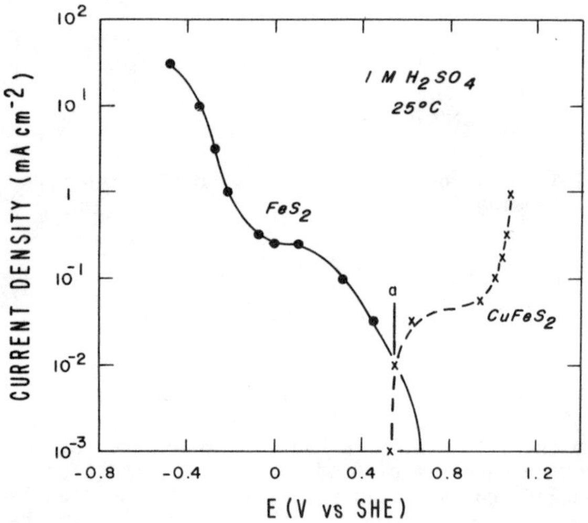

FIGURE 6. Polarization curves for pyrite (FeS_2, cathodic branch) and chalcopyrite ($CuFeS_2$, anodic branch) with mixed potential a.

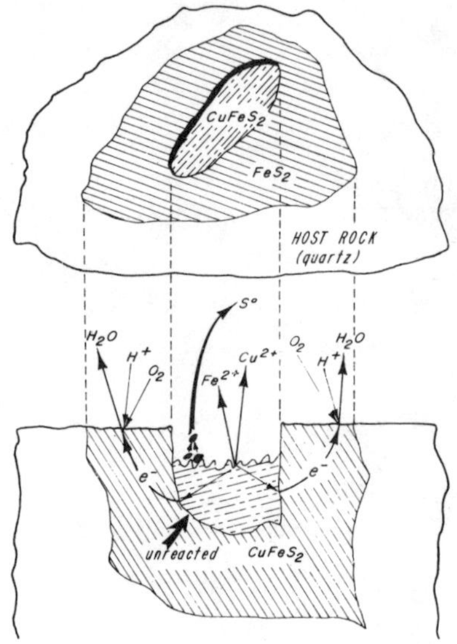

FIGURE 7. Galvanic coupling of pyrite (FeS_2) and chalcopyrite showing flow of electrons from anodic $CuFeS_2$ particles to cathodic FeS_2 particle.

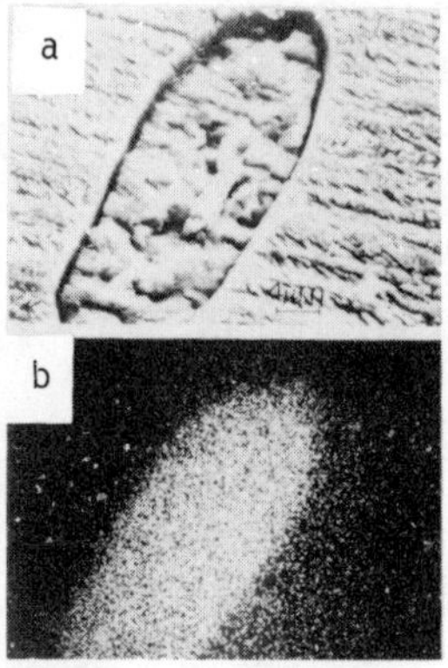

FIGURE 8. a) SEM photograph showing preferentially attacked $CuFeS_2$ in contact with pyrite; b) copper emission scan of area.

pyrite and chalcopyrite within the leaching environment will establish a mixed potential at the voltage represented by line a of Figure 6. At the mixed potential, pyrite is an inert electrode, and oxygen adsorbed on the pyrite surface is discharged cathodically according to Eq. (27),

$$\frac{1}{2}\,O_2 + 2H^+ + 2e = H_2O \tag{27}$$

Chalcopyrite reacts anodically according to the following reaction

$$CuFeS_2 = Cu^{2+} + Fe^{2+} + 2S° + 4e \tag{28}$$

The galvanic leaching mechanism is depicted in Figure 7 showing the flow of electrons from the anodic $CuFeS_2$ particle to the cathodic FeS_2 particle resulting in the discharge of oxygen at the particle surface.

Figure 8 shows typical corrosion of $CuFeS_2$ in intimate (locked) contact with FeS_2 after 6 weeks of leaching. The enhanced corrosion of chalcopyrite is clearly evident with the chalcopyrite attacked more vigorously than the surrounding phase. Peters, et al. (20) described the galvanic interactions between pyrite and galena. The leaching of mixtures of discrete mineral particles revealed an accelerated dissolution of galena and a retarded dissolution of pyrite. These findings are completely predictable from a galvanic leaching model analyzed in terms of electrochemical affects.

ELECTROCHEMICAL/CHEMICAL CONSIDERATIONS AND SELECTED METAL OXIDE/SULFIDE SYSTEMS

Oxidation of uranium dioxide in acid and alkaline carbonate solutions:

Nicol and Needes (18,21) studied the leaching of UO_2 as an electrochemical reaction and developed models describing the leaching process based on a charge transfer mechanism. Detailed electrode kinetic measurements (22,23) were obtained for the anodic oxidation of UO_2 and redox reactions at the UO_2 surface. The electrochemical mechanism correlated well with experimental data for both acid and alkaline systems. The dissolution mechanism was interpreted in terms of an anodic process involving two successive single electron-transfer reactions. The mechanism consistent with the electrochemical observations for acidic solutions is as follows

$$UO_2 + H_2O = UO_2OH + H^+ + e \quad (29)$$

$$UO_2OH = UO_3 + H^+ + e \quad (30)$$

$$UO_3 + 2H^+ = UO_2^{2+} + H_2O \quad (31)$$

For alkaline carbonate solutions the electrochemical data supported the following mechanism

$$UO_2 + HCO_3^- = UO_2HCO_3 + e \quad (32)$$

$$UO_2HCO_3 + OH^- = UO_2CO_3 + H_2O + e \quad (33)$$

$$UO_2CO_3 + 2CO_3^{2-} = UO_2(CO_3)_3^{4-} \quad (34)$$

Recently, Hiskey (24) presented chemical leaching results for the dissolution of UO_2 in ammonium carbonate. These results confirm an anodic dissolution sequence that includes both charge-transfer and chemical-reaction steps. Furthermore, these experiments supported a mechanism involving consecutive one-electron-transfer reactions. At low anodic potentials, the dissolution of surface product films (i.e. UO_3 or UO_2CO_3) is considered fast in comparison to charge transfer steps.

During leaching the anodic steps are coupled with the cathodic reduction of the oxidant. For alkaline carbonate systems oxygen is an important oxidant, the cathodic discharge of oxygen at the UO_2 surface occurs by the reaction

$$O_2 + 2H_2O + 4e = 4OH^- \quad (35)$$

With few exceptions, the leaching of UO_2 has been found to obey a half order dependence on oxygen partial pressure. Independent experiments have demonstrated that oxygen is reduced cathodically at a UO_2 surface with a first-order dependence on oxygen. This fact along with the irreversible nature of oxygen reduction at the UO_2 surface verifies the apparent one-half order with respect to oxygen partial pressure. Such a dependence was previously developed theoretically using the Butler-Volmer equation for the special case where i_a and i_c each fall within the corresponding Tafel region (see Eq. 20).

The leaching of a solid is considered electrochemically controlled when the rate of leaching is equal to the rate of anodic charge-transfer. For instance Figure 9 illustrates how closely the observed chemical leaching rates compare with the predicted electrochemical rates for the alkaline carbonate dissolution of UO_2 in the presence of various oxidants. Similarly, close agreement was found between chemical and electrochemical data for oxidants in acidic solutions. It is interesting to analyze the effectiveness of various oxidants in terms of their equilibrium potentials for several important cathodic half-cell reactions. Table III provides the equilibrium for several important cathodic half-cell reactions. If oxidants were selected solely by thermodynamic (equilibrium potential) criterion, then the predicted rates would reflect the following order

$$H_2O_2 > O_2 > Fe(CN)_6^{3-} > Cu(NH_3)_4^{2+}$$

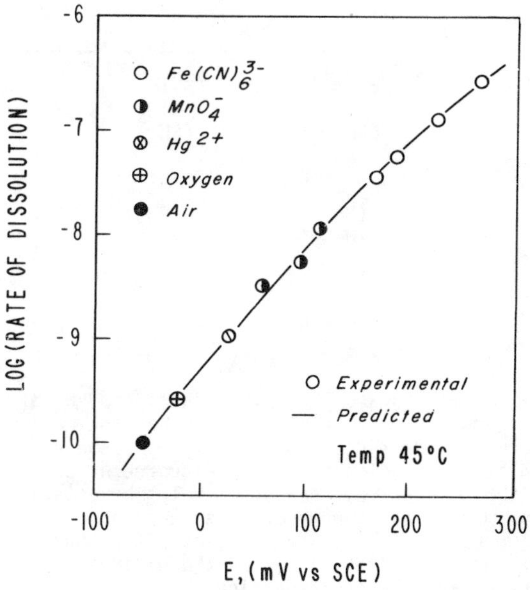

FIGURE 9. Experimental and predicted rates from electrochemical measurements for UO_2 in the presence of several oxidants.

Actually, these oxidants rank in the following order for the leaching of UO_2

$$Fe(CN)_6^{3-} > H_2O_2 > Cu(NH_3)_4^{2+} > O_2$$

Oxygen is a comparatively poor oxidant for the leaching of UO_2 even though it possesses a comparatively high equilibrium potential. This fact is clearly explainable in terms of the electrode kinetics of the half-cell reactions. The typical pattern of current-voltage curves is shown in Figure 10 for the anodic

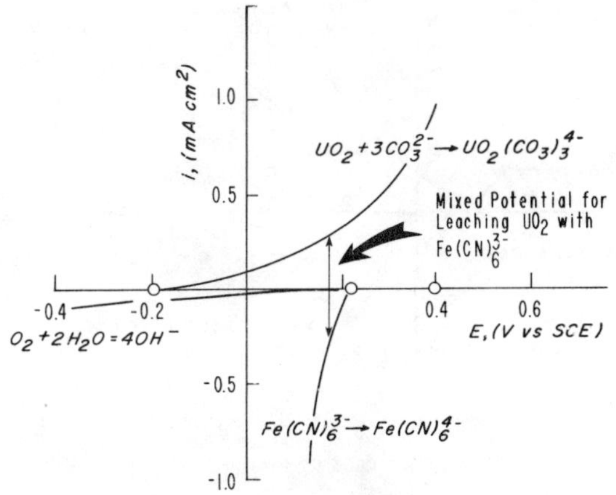

FIGURE 10. Polarization curves showing mixed potential for the leaching of UO_2 with $Fe(CN)_6^{3-}$.

TABLE III

Calculated Equilibrium Potentials for Several Cathodic Half-Cell Reactions, 25 C

Reaction	Conditions	E_c (V vs SCE)
$O_2 + H_2O + 4e = 4OH^-$	P_{O_2} = 1 atm, pH = 10	0.40
$Fe(CN)_6^{3-} + e = Fe(CN)_6^{4-}$	$[Fe(CN)_6^{3-}] = [Fe(CN)_6^{4-}]$ pH 12	0.22
$Cu(NH_3)_4^{2+} + e = Cu(NH_3)_2^{+} + 2NH_3$	$[Cu(NH_3)_4^{2+}] = [Cu(NH_3)_2^{+}]$ $[NH_3] = 10^{-2}$	0.07
$H_2O_2 + 2e = 2OH^-$	$[H_2O_2] = 10^{-2}$, pH = 10	0.88

oxidation of UO_2 and cathodic reduction of O_2 and $Fe(CN)_6^{3-}$. It must be emphasized that the most effective oxidant develops the highest anodic mixed potential during leaching. To a degree a large equilibrium electrode potential helps by establishing a high overvoltage, however, the shape of the current-voltage curves (electrode kinetics) is probably more important since it establishes the ultimate mixed potential and the corresponding leaching current. From Figure 10, it is clearly evident that the slope of the cathodic reduction of oxygen is very low resulting in an extremely small leaching current corresponding to mixed potential position. The extremely reversible $Fe(CN)_6^{3-}/Fe(CN)_6^{4-}$ couple, unlike the irreversible discharge of oxygen, is characterized by a steep polarization curve which yields a mixed potential at E_M' with the magnitude of i_a' equivalently greater.

The dissolution of UO_2 in ammoniacal carbonate solution with hydrogen peroxide as an oxidant deserves special consideration. In comparison to oxygen, the H_2O_2 system exhibits a measurable increase in the rate of UO_2 leaching (21,25). This enhancement in the leaching kinetics can be explained by electrochemistry concepts which take into account both anodic and cathodic effects. Figure 11 compares the polarization curves for both oxygen and hydrogen peroxide systems.

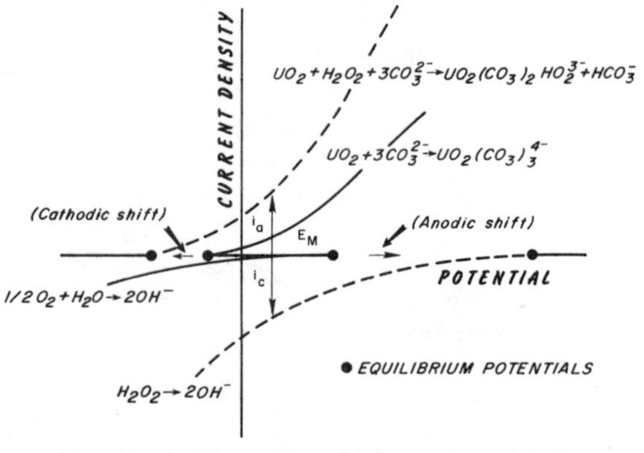

FIGURE 11. Polarization curves of UO_2, oxygen and hydrogen peroxide illustrating enhanced current for leaching in the case of hydrogen peroxide.

In general hydrogen peroxide has the ability to shift the mixed potential to more anodic values, thereby increasing the anodic current (i_a). The higher equilibrium potential for the H_2O_2 half-cell reaction (see Table III) favors an anodic mixed potential shift. In addition, H_2O_2 can react with $UO_2(CO_3)_3^{4-}$ to form a $UO_2(CO_3)_2HO_2^{3-}$ species with an associated cathodic shift from

$$UO_2(CO_3)_3^{4-} + 2e = UO_2 + 3CO_3^{2-}$$

$$E = -0.22 - 0.03 \log \frac{[CO_3^{2-}]^3}{[UO_2(CO_3)_3^{4-}]} \qquad (36)$$

to

$$UO_2(CO_3)_2HO_2^{3-} + HCO_3^- + 2e = UO_2 + 3CO_3^{2-} + H_2O_2$$

$$E = -0.38 - 0.03 \log \frac{[CO^{2-}]^3\,[H_2O_2]}{[UO_2(CO_3)_2HO_2^{3-}][HCO_2^-]} \qquad (37)$$

The cathodic shift in potential is equal to

$$\Delta E_c = E_{37} - E_{36}$$

$$= -0.16 - 0.03 \log \frac{[UO_2(CO_3)_3^{4-}][H_2O_2]}{[UO_2(CO_3)_2HO_2^{3-}][HCO_3^-]}$$

where the logarithmic term is equivalent to the inverse equilibrium constant for the reaction

$$UO_2(CO_3)_3^{4-} + H_2O_2 = UO_2(CO_3)_2HO_2^{3-} + HCO_3^- \qquad (38)$$

Komarov (26) reports an equilibrium constant of 1.6 x 10^2 for this reaction at infinite dilution. Therefore, the cathodic shift in the pH region where H_2O_2 is dominant is approximately -0.09 v. The dashed lines in Figure 11 are the respective polarization curves for H_2O_2 reduction and UO_2 oxidation in the H_2O_2 system. The mixed potential for these conditions is noted by E_M and the increase in current density is clearly apparent.

Electrochemical dissolution of galena:

The electrochemical dissolution of galena in

aqueous solutions has received considerable attention. The electrical conductivity of galena is unusually high, and galena is recognized as having both n- and p- type conductivity. Eadington and Prosser (27) have demonstrated that the oxidation of galena is noticeably influenced by the semi-conductor type. For non-stoichiometric PbS, the rate of oxidation was much higher for sulfur-rich (p-type semi-conductor) specimens than for lead-rich (n-type) specimens. In aqueous solution lead forms a large array of sparingly soluble or insoluble compounds; these materials can form surface film which mask the original electronic properties of the solid. Paul et al. (28) addressed the role of insoluble films as they determine the transition between active-passive behavior during the electrochemical dissolution of galena. They also developed dissolution mechanisms for various aqueous solutions.

The simplified Pourbaix diagram shown in Figure 12 illustrates that PbS can coexist with either sulfur or lead according to the following reactions

$$PbS = Pb^{2+} + S^\circ + 2e \qquad (39)$$

or

$$PbS + 2H^+ + 2e = Pb + H_2S \qquad (40)$$

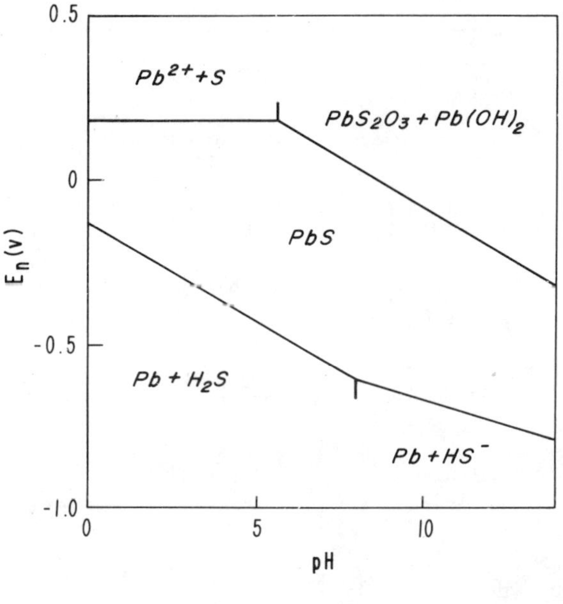

FIGURE 12. Simplified E_h-pH diagram for the Pb-S system.

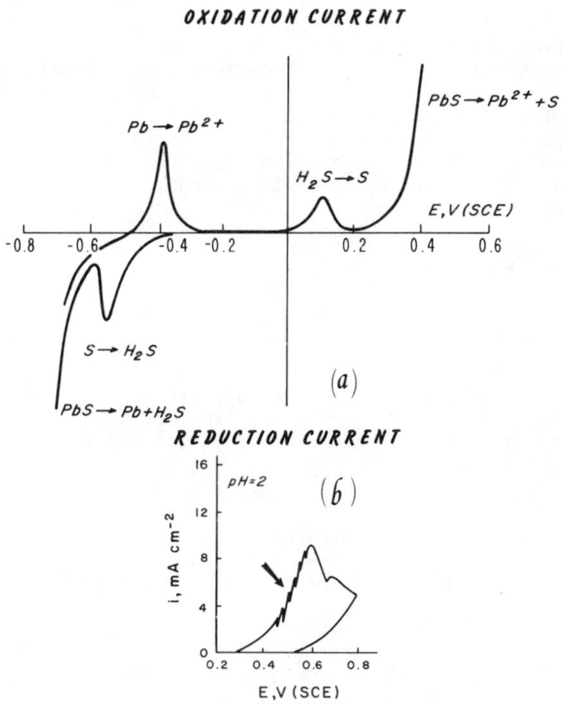

FIGURE 13. a) Polarization and reduction for the cyclic-voltammetric reaction of galena in a perchlorate solution; and b) anodic dissolution of galena showing sequential passivation and oxidation.

The galena electrode is acknowledged as one of the most reversible among sulfide minerals. The voltammogram (28) appearing in Figure 13a illustrates the complex nature of the electrode kinetics. The anodic and cathodic sweeps in perchlorate solutions reveal several interesting features. During the anodic sweep there is evidence of an oxidation peak at 0.1 v which corresponds to the oxidation of H_2S to S°. The anodic dissolution of galena according to the reaction

$$PbS = Pb^{2+} + S^\circ + 2e \qquad (39)$$

at pH 2 proceeds without loss of activity until approximately 0.5 v. The electrode then experiences passivation-reactivation fluctuations in the curve as denoted by the arrow in Figure 13b. This region is a result of the sulfur layer that inhibits the diffusion of Pb^{2+} from the surface. As the concentration of Pb^{2+} increases at the galena surface there is an attendant increase in pH. The higher pH promotes the oxidation of sulfur to a higher oxidation state, and the electrode surface will be quickly passivated either by the precipitation of lead thiosulfate or the direct formation of lead thiosulfate according to

$$PbS + 3H_2O = PbS_2O_3 + Pb^{2+} + 6H^+ + 8e \qquad (41)$$

Reactivation occurs as the PbS_2O_3 dissolves as a result of the continued diffusion of lead ions from the surface through the pores of the sulfur deposit. There is a well defined passivation of the electrode surface beginning at about 0.6 v caused by the precipitation of a basic lead sulfate ($PbO \cdot PbSO_4$ or $3PbO \cdot PbSO_4$) on the galena surface. On the cathodic side of the sweep, a peak which reflects the reduction of S° to H_2S appears near -0.55 v. The peak corresponding to the oxidation of Pb to Pb^{2+} can only occur following the reduction of galena to lead and hydrogen sulfide. It appears that the formation of lead on the galena surface does not reduce its cathodic acitivity. Similar cyclic-voltammetric behavior was observed for chloride and sulfate solutions. However, these systems exhibited a pronounced influence from both

$PbCl_2$ and $PbSO_4$ formed from the respective solutions.

From the study described above Paul et al. (28) suggested a mechanism for the dissolution of galena in acidic perchlorate solutions. The overall anodic reaction is given in Eq. (39). Tafel slopes fail to support this simple two electron-transfer reaction. Therefore, the oxidation of galena is assumed to take place according to distinguishable one-electron-transfer steps,

$$PbS \rightarrow Pb^{2+} + S^- + e \quad (42)$$

$$S^- \rightarrow S^\circ + e \quad (43)$$

The "S^-" sulfur intermediate has been proposed elsewhere (23). The Butler-Volmer treatement for steady state conditions ($d[S^-]/dt = 0$) yields the following equation for the net anodic current

$$i_a = \frac{2F\vec{k}\ k'\ \exp[(2-\beta_a-\beta_a')FE/RT]}{\overleftarrow{k}\ [Pb^{2+}]_s \exp[-\beta_a FE/RT] + k' \exp[(1-\beta_a)FE/RT]} \quad (44)$$

where $[Pb^{2+}]_s$ is the concentration of lead ions at the electrode surface, $\vec{k}$ and $\overleftarrow{k}$ are the respective forward and reverse rate constants for reaction (42), k' is the specific rate of reaction (43), and the other terms are analogous to those previously defined. At sufficiently positive values of E or at low $[Pb^{2+}]$, since the first term of the denominator of Eq. (44) is small compared to the second term, the expression for i_a reduces to

$$i_a = 2F\vec{k} \exp\ (1-\beta_a)FE/RT) \quad (45)$$

Conversely, when the $[Pb^{2+}]$ is high and E is low the second term in the denominator can be neglected and

$$i_a = \frac{2F\vec{k}\ k'}{\overleftarrow{k}\ [Pb^{2+}]} \exp\ [(2-\beta_a')FE/RT] \quad (46)$$

Both cases cited above along with an additional condition when $[Pb^{2+}]_s$ approaches zero have been verified experimentally. Therefore, the proposed mechanism involving consecutive one-electron-transfer reactions is an accurate description of the anodic dissolution of galena.

<u>Oxidation of chalcopyrite in ammoniacal solutions with oxygen</u>

Beckstead and Miller (42) have explained the kinetics of the oxidation of chalcopyrite in oxygenated ammoniacal solution on the basis of an electrochemical model. The stoichiometry of the reactions involved were established by Forward and Mackiw (43) and more recently Kuhn et al. (44) have described the Arbiter process based upon the ammoniacal dissolution of copper sulfides using oxygen. The overall reaction for chalcopyrite is

$$CuFeS_2 + 4NH_3 + \frac{17}{4}O_2 + 2OH^- =$$

$$Cu(NH_4)^{2+} + \tfrac{1}{2}Fe_2O_3 + 2SO_4^= + H_2O \quad (47)$$

Thiosulfate is formed as an intermediate and is subsequently oxidized to the sulfate (43). Beckstead and Miller (42) found the kinetics to be approximately half order with respect to hydroxyl and cupric ions and oxygen at low pressures. At high pressures the rate became zero order with respect to oxygen. Sigmoidal reaction curves were explained on the basis of the autocatalytic effect of cupric ion. Two parallel reactions, one of which involves cupric ion, were postulated. Figure 14 illustrates typical rate data obtained showing the effect of cupric ion on the rate of reaction. The kinetic data were well correlated

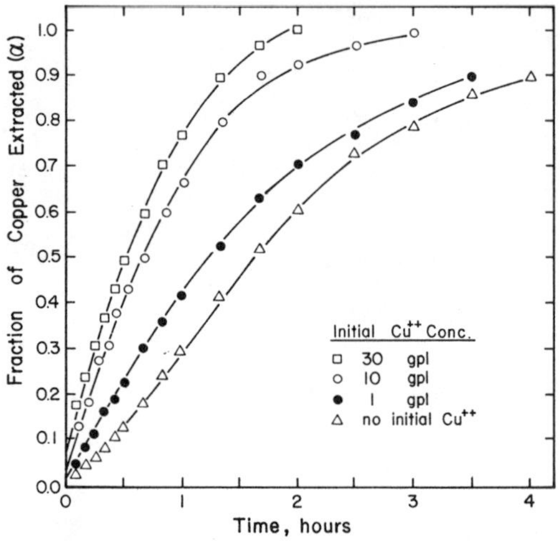

FIGURE 14. Fraction reacted versus time plots for different initial cupric ion concentrations at 90°C, 1% solids, 1.8 M ($NH_4^+ + NH_3$), pH = 1.02, 325 x 400 mesh particles, 2.18 atmospheres O_2 pressure.

over broad ranges of temperature, pressure, and cupric ion concentration by the rate equations

Low Pressure ($P_{O_2} <$ 1 atmosphere)

$$\frac{d\alpha}{dt} = \frac{k_\ell}{d_o} [OH^-]^{\frac{1}{2}} [O_2]^{\frac{1}{2}} (k_1 + k_2 [Cu^{2+}])^{\frac{1}{2}} (1-\alpha)^{2/3} \quad (48)$$

High Pressure ($P_{O_2} >$ 1 atmosphere)

$$\frac{d\alpha}{dt} = \frac{k_h}{d_o} [OH^-]^{\frac{1}{2}} (k_1 + k_2) [Cu^{2+}])^{\frac{1}{2}} (1-\alpha)^{2/3} \quad (49)$$

where k_1, k_2, k_ℓ, and k_h are emirical rate constants, d_o is the initial particle diameter and α is the fraction of $CuFeS_2$ reacted. The approximate half-power orders are characteristic of electrochemical control in leaching. The half cell reactions proposed include the anodic dissolution of $CuFeS_2$

$$CuFeS_2 + 19OH^- =$$

$$Cu^{2+} + \tfrac{1}{2}Fe_2O_3 + 2SO_4^{2-} + \frac{19}{2}H_2O + 17e^- \quad (50)$$

and two parallel cathodic reactions involving the

discharge of oxygen without and with the cupric ion autocatalytic effect

$$O_2 + 2H_2O + 4e^- = 4OH^- \qquad (35)$$

$$O_2 + 2H_2O + Cu^{2+} + 4e^- = 4OH^- + Cu^{2+} \qquad (51)$$

The Butler-Volmer current density equations corresponding to the above half cell reactions were proposed to be

$$i_a = 17F[OH^-]\,k_a \exp\left[\frac{\beta_a FE}{RT}\right] - 17F[Cu^{2+}][SO_4^{2-}]\,k_a' \exp\left[\frac{-(1-\beta_a)FE}{RT}\right] \qquad (52)$$

$$i_{c1} = 4F[OH^-]\,k_{c1} \exp\left[\frac{\beta_{c1}FE}{RT}\right] - 4F[O_2]\,k_{c1}' \exp\left[\frac{-(1-\beta_{c1})FE}{RT}\right] \qquad (53)$$

$$i_{c2} = 4F[OH^-][Cu^{2+}]\,k_{c2} \exp\left[\frac{\beta_{c2}FE}{RT}\right] - 4F[Cu^{2+}][O_2]\,k_{c2}' \exp\left[\frac{-(1-\beta_{c2})FE}{RT}\right] \qquad (54)$$

The kinetic results were consistent with the assumption that the anodic and cathodic partial current densities were large and within the Tafel region for each of the above reactions. Accordingly, using the relationship $i_a = (i_{c1} + i_{c2})$, and assuming $\beta_{c1} = \beta_{c2}$

$$\exp\left[\frac{\beta_a FE}{RT}\right] = \left[\frac{4[O_2](k_{c1}' + [Cu^{2+}]k_{c2}')}{17[OH^-]k_a}\right]^m \qquad (55)$$

where $m = \beta_a/(1+\beta_a-\beta_c)$. Expressed in terms of α, the fraction of $CuFeS_2$ reacted

$$\frac{d\alpha}{dt} = -\frac{1}{n_o}\frac{dn}{dt} = \frac{i_a A}{17Fn} \qquad (56)$$

where n = moles $CuFeS_2$ remaining at time t
n_o = original number of moles of $CuFeS_2 = w_o/M_c$
M_c = molecular weight, $CuFeS_2$
A = area at time t, $= \frac{6w_o f}{\rho_c d_o}(1-\alpha)^{2/3}$
f = geometric shape factor
w_o = original weight, $CuFeS_2$
ρ_c = density, $CuFeS_2$
d_o = original particle diameter

The concentration of copper in solution is related to the solution volume, V, the fraction, α, reacted and the original concentration, $[Cu^{2+}]_o$ by

$$[Cu^{2+}] = [Cu^{2+}]_o + \frac{w_o}{M_c V}\alpha \qquad (57)$$

Experimental results indicate $m \simeq 0.5$. If this value is used, the rate expression for the anodic dissolution of $CuFeS_2$ in oxygenated ammoniacal solution is given by

$$\frac{d\alpha}{dt} = K_o\,(k_1 + k_2[Cu^{2+}]_o + k_2'\alpha)(1-\alpha)^{1/2} \qquad (58)$$

where

$$K_o = \frac{127f}{d_o}[OH^-]^{1/2}[O_2]^{1/2} \qquad (59)$$

and $k_1 = k_{c1}' k_a^{1/2}$; $k_2 = k_{c2}' k_a^{1/2}$; and $k_2' = k_2 w_o/VM_c$. The zero order kinetics at larger oxygen pressure was explained on the basis of Langmuir adsorption of oxygen where $[O_2]$ refers to the surface concentration and is given by

$$[O_2] = \frac{k_o K P_{O_2}}{1 + K P_{O_2}} \qquad (60)$$

where k_o is the total concentration of surface sites and K is the equilibrium constant for adsorption. Eqs. (58), (59) and (60) explain the experimental results as depicted by Eqs. (48) and (49) for low and high pressures respectively. Experimentally K_o may be held constant. Figure 15 illustrates the theoretical correlation of Eq. (58) for three sizes of $CuFeS_2$ at 90°C, 1% solids, 1.8 M $(NH_4^+ + NH_3)$, pH = 10.2, and P_{O_2} = 2.18 atm. The slope is equal to K_o and varies with d_o^{-1} as required by Eq. (59).

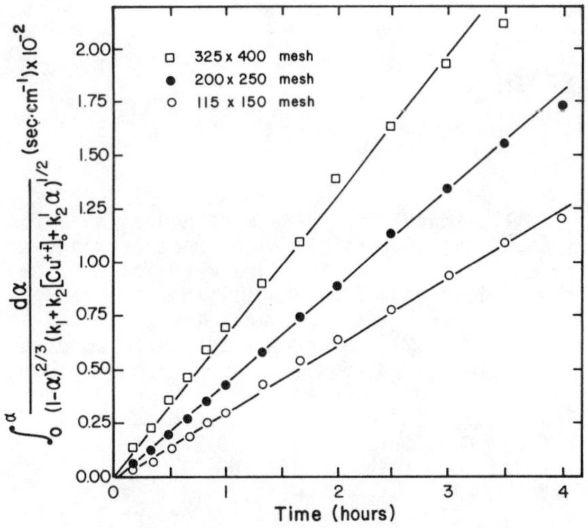

FIGURE 15. Plot of data for three sizes of $CuFeS_2$. Solid lines are theoretical, based upon the electrochemical model.

OXIDATION OF CHALCOCITE AND CHALCOPYRITE IN ACIDIC FERRIC SULFATE SOLUTIONS

Marcantionio (29), in an analysis of the acid ferric sulfate leaching of natural Cu_2S, has identified a sequence of reactions in which the solid phases observed were $Cu_2S \rightarrow Cu_{1.8}S \rightarrow Cu_{1.2}S \rightarrow CuS$. Digenite, $Cu_{1.8}S$, was found to form rapidly. At 30°C, 48x65 mesh Cu_2S particles were completely converted to $Cu_{1.8}S$ in 3.5 minutes in 0.03 M Fe III solutions, indicative of the rapid diffusion of Cu^+ through the sulfur lattice. The final product at 30°C, and low Fe III concentrations, was blue-remaining covellite or blaubleiblender of the composition $Cu_{1.2}S$. Normal covellite, CuS, was never observed as a product of reaction, in agreement with the findings of Thomas, et al. (30). A second stage of leaching in which $Cu_{1.2}S$ reacted to form elemental sulfur progressed much more slowly requiring elevated temperatures to proceed to completion. Sullivan (31) as early as 1930 identified two stages of leaching of Cu_2S. These are usually depicted as

$$Cu_2S + 2Fe^{3+} = Cu^{2+} + CuS + 2Fe^{2+} \tag{61}$$

and

$$CuS + 2Fe^{3+} = Cu^{2+} + S^\circ + 2Fe^{2+} \tag{62}$$

Marcantonio has identified two stages but has provided greater detail regarding the kinetics of each stage. The sequential formation of compounds can only result if the mineral assumes rest potentials proceeding from lower to higher values. The following half-cell reactions are consistent with the observed results

Stage Ia:

$$5Cu_2S = 5Cu_{1.8}S + Cu^{2+} + 2e^- \tag{63}$$

Stage Ib:

$$\frac{5}{3}Cu_{1.8}S = \frac{5}{3}Cu_{1.2}S + Cu^{2+} + 2e^- \tag{64}$$

Stage II:

$$\frac{5}{6}Cu_{1.2}S = \frac{5}{6}S^\circ + Cu^{2+} + 2e^- \tag{65}$$

Figure 16 illustrates rate data obtained at 30°C for various Cu_2S initial particle sizes under strong stirring conditions. The kinetics were found to be first order with respect to [Fe III] up to approximately $\alpha = 0.27$ ($Cu_{1.46}S$), and mass transport controlled. For $\alpha < .27$, measured rates correspond to calculated rates based upon transport through a liquid limiting boundary film. Also initial mass transport control is consistent with the measured acivation energy of 2.8 kcal/mole. For $\alpha > 0.27$ the rate diminished rapidly leveling off at $\alpha = 0.40$. For the range $.27 > \alpha > 0.40$ the kinetics were directly related to [Fe III] and the activity of mobile cuprous ions in the Cu_{1+x} lattice, designated here as (Cu_ℓ^+). This latter range of compositions is essentially that reported for blue-remaining covellite (13). The dependence of the rate on [Cu_ℓ^+] is consistent with a

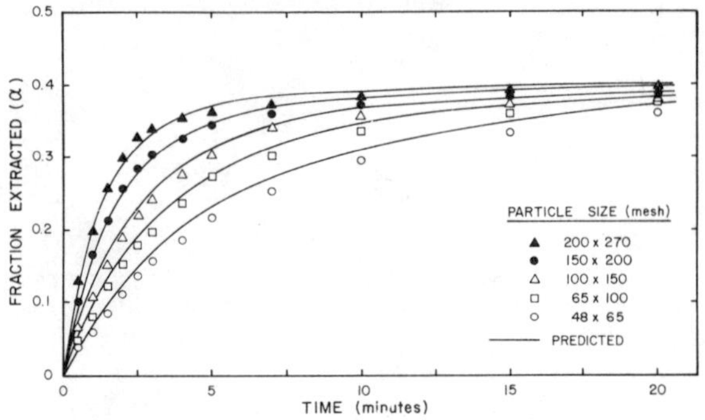

FIGURE 16. Effect of ferric ion concentration on the first stage dissolution of 48 x 65 mesh Cu_2S at 30°C.

series of compounds formed in a sequence of increasing rest potentials or a broad solid solution region for blue-remaining covellite. The formation of digenite or other $Cu_{2-x}S$ compounds, is masked kinetically since initial kinetics are mass transport controlled. The proposed kinetic model is

$$i_c = \frac{FD}{\delta}[Fe^{III}]_s - \frac{FD}{\delta}[Fe^{III}] \tag{66}$$

also, at steady state,

$$i_c = -Fk_s[Fe^{III}]_s[Cu_\ell^+] \tag{67}$$

where D = average diffusion coefficient for the Fe III species

δ = thickness of the diffusion boundary layer

$[Fe^{III}], [Fe^{III}]_s$ = bulk and surface Fe III concentrations

k_s = surface rate constant.

Consistent with a heterogeneous process the kinetics depend upon the initial surface area, A_o; and $A_o = 3w_of/r_o\rho_c$ where w_o is the original weight of Cu_2S, ρ_c is the density of Cu_2S, f is a geometric sphericity factor, and r_o is the particle radius. The maximum extraction for stage I, is $\alpha_o = 0.40$. If α = fraction of total copper extracted

$$[Fe^{III}] = [Fe^{III}]_o\ (1-b\alpha) \tag{68}$$

and

$$[Cu_\ell^+] = \frac{2\rho_c}{M_c}(1-\alpha/\alpha_o) \tag{69}$$

where $[Fe^{III}]_o$ refers to the initial concentration, $b = 2/\phi$, ϕ is the molar Fe III/Cu ratio and M_c is the molecular weight of Cu_2S.

The fraction reacted, α, is related to i_c by

$$\frac{d\alpha}{dt} = -\frac{1}{n_o}\frac{dn}{dt} = -\frac{i_cA_o}{2Fn_o} \tag{70}$$

where n is the moles of Cu remaining in time t and n_o is the original number of moles of Cu present as Cu_2S. Under steady state conditions, Eq. (66) and (67) result in the rate equation

$$\frac{d\alpha}{dt} = \frac{k'}{r_o} \frac{\gamma\pm[Fe^{III}][Cu_\ell^+]}{([Cu_\ell^+] + \frac{D}{\delta k_s})} \quad (71)$$

where $k' = 3w_o fD/2\rho_c n_o \delta$. Equation (71), with Eqs. (68) and (69), may be integrated numerically resulting in the solid curves of Figure 16. At high Fe III concentration the rate diminished due to the decrease in $\gamma\pm$, the activity coefficient of the ferric ion.

The apparent influence of (Cu_ℓ^+) may reflect a variation in voltage due to the change in $[S_\ell^{2-}]$ which is related stoichiometrically. In CuS (4), cuprous ion is present co-valently bonded as $(Cu^+)_2S_2^{2-}$. Consistent with the solid state properties of $Cu_{1+x}S$ the reaction in Stage I may be viewed as

$$Cu^+(\text{lattice}) \rightarrow Cu^{2+}(aq) + e^- \quad (72)$$

$$S^{2-}(\text{lattice}) \rightarrow \tfrac{1}{2}S_2^{2-}\ (\text{lattice}) + e^- \quad (73)$$

$$Cu^+(\text{lattice}) + \tfrac{1}{2}S_2^{2-}\ (\text{lattice}) \rightarrow \tfrac{1}{2}(Cu^+)_2S_2^{2-} \quad (74)$$

For each two moles of Fe III reduced, one mole of Cu_ℓ^+ appears in solution as Cu^{2+} and one mole of Cu_ℓ^+ becomes fixed due to the strong co-valent character of the $(Cu^+)_2S_2^{2-}$ bond.

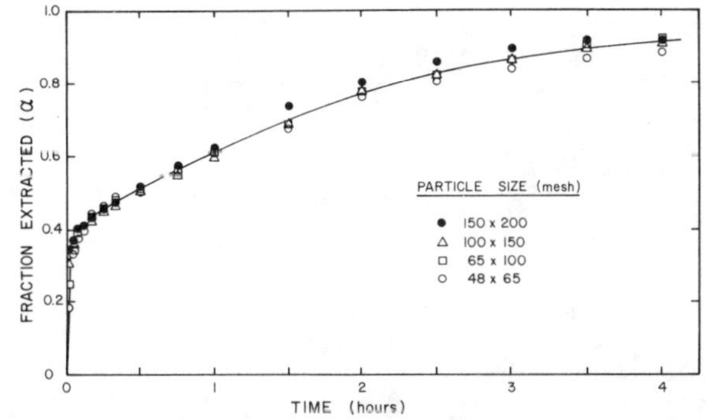

FIGURE 17. Effect of particle size on the second stage dissolution of Cu_2S at 75°C and 1.12 M Fe III (Fe III/Cu = 3/1).

Figure 17 illustrates typical results obtained for Stage II leaching, at 75°C, for various initial Cu_2S particle sizes. The rate is independent of the initial Cu_2S particle size. The blue-remaining covellite formed in Stage I is porous, consisting of uniform columnar crystals. The rate is sufficiently slow that diffusion into the columnar array does not influence the kinetics. The rate was found to vary with [Fe III] with an apparent half-power order and the activation energy was 18.0 ± 1 kcal/mole. Consistent with the physical appearance of the crystals the kinetics were correlated by cylindrical crystalline geometry and a corrosion couple such that $A_a + A_c = A$. The known solid state properties suggest the following steps for the Stage II reaction

$$(Cu^+)_2S_2^{2-}\ (\text{lattice})$$

$$\rightarrow Cu^+S_2^{2-}\ (\text{lattice}) + Cu^{2+}\ (aq) + e^- \quad (75)$$

$$Cu^+S_2^{2-}\ (\text{lattice}) \rightarrow \tfrac{1}{2}(Cu^+)_2S_2^{2-}\ (\text{lattice}) + S^\circ + e^- \quad (76)$$

The net anodic and cathodic processes may be expressed by

$$i_a = 2Fk_a \exp\left[\frac{\beta_a FE}{RT}\right] - [Cu^{2+}]\ k'_a \exp\left[\frac{-(1-\beta_a)FE}{RT}\right] \quad (77)$$

$$i_c = F[Fe^{II}]\ k_c \exp\left[\frac{\beta_a FE}{RT}\right] - Fk'_c\ [Fe^{III}] \exp\left[\frac{-(1-\beta_c)FE}{RT}\right] \quad (78)$$

The cathodic partial current density of Eq. (77) and the anodic partial current density of Eq. (78) are small as evidenced by the fact that large ferrous and cuprous ion concentrations were required to affect the kinetics appreciably. Assuming the net anodic and cathodic reactions are each in the Tafel region the mixed potential may be evaluated. Using shrinking core cylindrical geometry the crystallite area is given by

$$A = k\ A_o\ [(1-\alpha)/(1-\alpha_o)]^{\frac{1}{2}} \quad (79)$$

where k contains crystallite density and appropriate geometric factors, and A_o is original area of the crystallite at the beginning of Stage II. Using the relationship $d\alpha/dt = i_a A/2Fn_o$, and the calculated mixed potential

$$\frac{d\alpha}{dt} = \frac{k\ A_o}{n_o}\left[\frac{k'_c\gamma\pm[Fe^{III}]_o}{2k_a}\right]^m \left[\frac{1-\alpha}{1-\alpha_o}\right]^{\frac{1}{2}} (1-b\alpha)^m \quad (80)$$

where $m = \beta_a/(1+\beta_a-\beta_c)$. The solid curves of Figure 18 were calculated by numerical integration of Eq. (80) with m = 0.5.

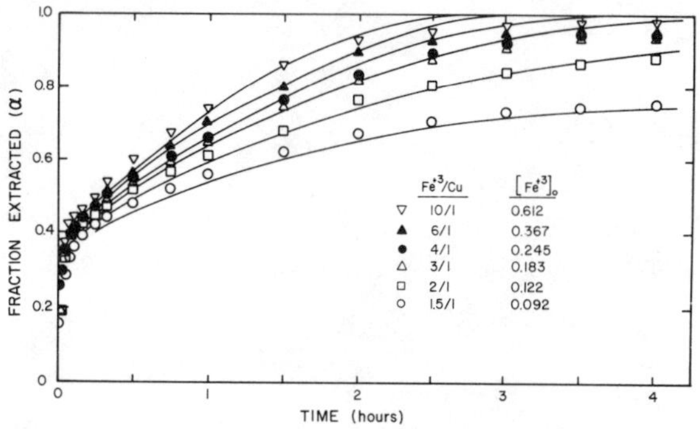

FIGURE 18. Effect of ferric ion concentration on the second stage dissolution of 48 x 65 mesh Cu_2S at 75°C.

These results indicate the slow kinetics of Stage II leaching do not involve diffusion through sulfur layers but are due to slow electron transfer at the mineral sulfur interface.

The anodic dissolution of chalcopyrite in acidic ferric sulfate solutions occurs according to the half-cell reaction (28) forming elemental sulfur. Initially a defect chalcopyrite structure forms followed by passivation by a layer of elemental sulfur. Also, sulfate ion forms but in relatively smaller proportion than elemental sulfur. Initially ferrous ion is released much more rapidly than cupric ions (7,32,33) forming an intermediate structure according to the reaction

$$CuFeS_2 \rightarrow Cu_{1-x}Fe_{1-y}S_2 + xCu^{2+} + yFe^{2+} + 2(x+y)e^- \quad (81)$$

where $y > x$ initially. As the reaction proceeds x approaches y. Elemental sulfur forms according to the reaction

$$Cu_{1-x}Fe_{1-y}S_2 =$$

$$(1-x)Cu^{2+} + (1-y)Fe^{2+} + 2S^\circ + 2(2-x-y)e \quad (82)$$

In the first step of the reaction the intermediate product serves as a passivation layer by limiting the reaction to ionic transport through the layer. This is well illustrated in Figure 19 where anodic polarization curves are shown for various ferrous ion levels. The plateau region, indicating diffusional control, moves up regularly with increasing $[Fe^{2+}]$. The electrode is not passivated by elemental sulfur at this point as noted by Jones and Peters (33) since sulfur, being an electronic insulator, would not respond to the ferrous ion present. The increased plateau current results from the oxidation of ferrous iron to ferric iron at the electrode surface. Although Warren (7) found sulfur in this region it did not appear as a dense continuous layer but occurred as patches and around etched regions of the surface. These results are in contrast to those observed by Munoz et al. (34) who found massive layers of sulfur on fine chalcopyrite leached in acid ferric sulfate solutions. Under these conditions the rate of reaction was not influenced by the addition of ferrous or ferric ions to the solution. It is thus apparent that protective sulfur layers do form after sufficient reaction has occurred. Figure 20 shows anodic polarization curves obtained by Jones (35) in sulfuric acid for various temperatures. For temperatures above 66°C a new passivation region at higher potentials is evident with lower current carrying capability than at the lower temperatures. In this region the rapid increase in current for the low temperatures is caused by discharge of water resulting from transfer of electrons through electron conducting surface layers. At higher temperatures discharge of water does not occur, a result which is consistent with the formation of dense sulfur layers of low electronic conductivity.

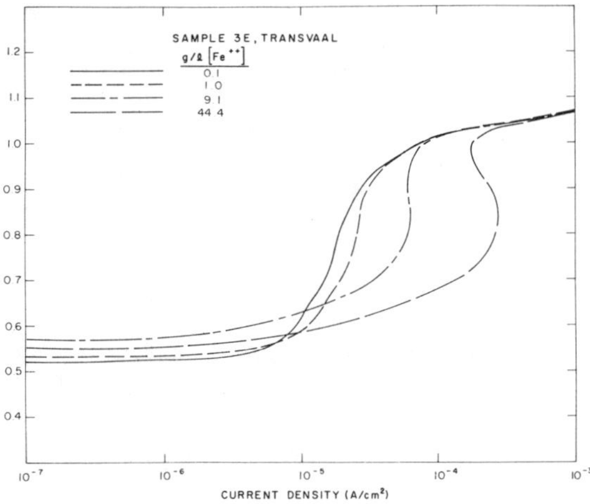

FIGURE 19. Effect of ferrous ion concentration on the anodic polarization curve of transvaal $CuFeS_2$ in 1 M H_2SO_4, 40 mv/min, 25 C. $[Cu^{++}]$ was held constant at 0.90 g/l. (Warren[34]).

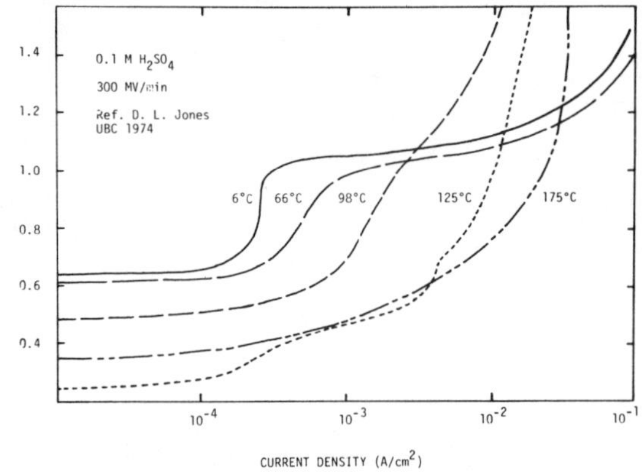

FIGURE 20. Polarization curves for Craigmont chalcopyrite at various temperatures showing passivation at higher temperatures.

Munoz et al. (35) carried out an extensive study of the leaching of fine chalcopyrite and as noted earlier showed that the kinetics were controlled by diffusion through protective layers of sulfur. The principal overall reaction is

$$CuFeS_2 + 4Fe^{3+} \rightarrow Cu^{2+} + 5Fe^{2+} + 2S^\circ \quad (9)$$

with the formation of some sulfate, particularly at high anodic over voltages. Munoz et al. (34) explained the kinetics of leaching of monosized fractions of chalcopyrite by a model in which the main rate limiting step is electronic transport, accompanied by the outward diffusion of coupled Cu^{2+} plus Fe^{2+} ions, through the sulfur layer. Equally plausible is the outward diffusion of coupled Fe^{3+} and Cu^+. Figure 21 illustrates the transport processes proposed. The activation energy was observed to be 83.7 kJ/mol (20 kcal/mole) which compares favorably with the activation energy (23 kcal/mol) for the conductivity of electrons in elemental sulfur. Figure 22 presents the rate data for three monosized fractions of chalcopyrite. The solid lines of Figure 22 were calculated using the diffusion equation for spheres, for diffusion of a reactant through the product layer

$$1-(2/3)\alpha - (1-\alpha)^{2/3} = \frac{k_p'}{d_o} t \quad (83)$$

where α is the fraction reacted and d_o is the initial particle diameter. The rate constant k_p' is given by

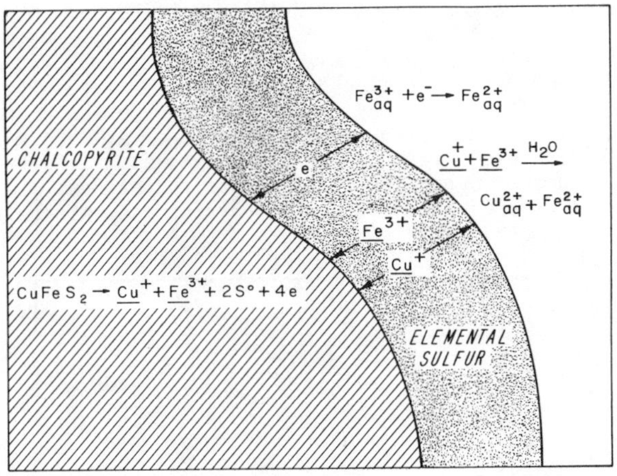

FIGURE 21. Schematic illustrating transport processes in the acid ferric sulfate leaching of chalcopyrite.

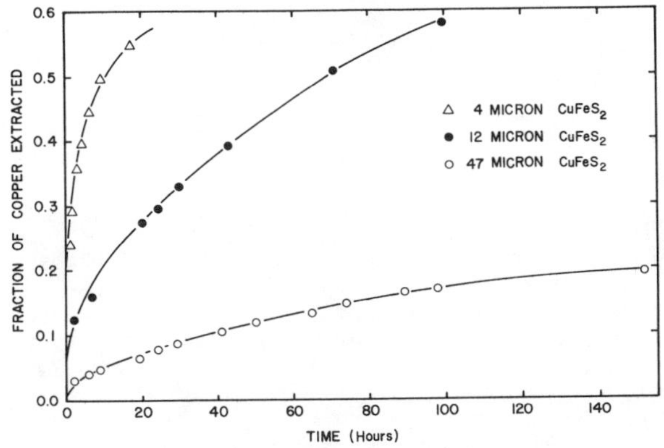

FIGURE 22. A plot of fraction of copper extracted from monosize chalcopyrite particles as a function of time for 1.0 M H_2SO_4, 0.25 M $Fe_2(SO_4)_3$, 90°C. 0.5 percent solids and 1200 rpm.

$$k'_p = \frac{8\chi}{\rho}\left(\frac{D_i\varepsilon}{\tau}\right)a_i$$

where: $\frac{D_i\varepsilon}{\tau}$ = effective diffusivity, ε is the porosity and τ is the tortuosity

a_i = activity of reacting species (electrons) at the $CuFeS_2$-S° boundary

χ = stoichiometric factor, moles $CuFeS_2$ per mole of electrons

ρ = molar density of chalcopyrite

The experimental value of k'_p was determined to be 2.24×10^{-13} cm^2/s.

The rate of leaching was derived using Wagner's theory of oxidation which relates the intrinsic mass transport or electrical conductivity of products of reaction to rates of reaction. The flux of a charged diffusing anion, cation or electrons (mol cm^{-2} sec^{-1}) is given by

$$J_i = \frac{300}{96{,}500}\frac{t_i\sigma}{|z|F}\left[\frac{d\mu_i}{dx} - z_iF\frac{d\phi}{dx}\right] \quad (84)$$

for modest voltage gradients, where

$d\mu_i/dx$ = chemical potential gradient

$d\phi/dx$ = electric field strength

F = Faraday (23,060 cal/equiv. volt)

t_i, z_i = transport number and charge for i th species

σ = electrical conductivity

Electrical neutrality requires the condition

$$J_{Cu^+} + 3J_{Fe^{3+}} = J_e \quad (85)$$

The coupling equation consistent with the observed stoichiometry of the reaction is

$$J_{Cu^+} = J_{Fe^{3+}} = \frac{1}{A}\frac{dn(CuFeS_2)}{dt} \quad (86)$$

If transport of electrons through the sulfur layer is rate controlling then the transport number t_e for electrons is expected to be much smaller than the transport number $t_{Fe^{3+}}$ for Fe^{3+}. Combination of equations (84), (85), and (86) for the condition $t_e \ll t_{Fe^{3+}}$, results in the general equation

$$\frac{dn}{dt}(CuFeS_2) = -A\frac{300}{96{,}500}\frac{t_e\sigma}{4F}\frac{d\mu_e}{dx} \quad (87)$$

For spherical geometry equation (87) may be integrated, assuming quasi steady state, resulting in the equation

$$\frac{d\alpha}{dt} = \frac{3x300}{96{,}500}\frac{M}{\rho}\frac{t_e\sigma}{Fd_o^2}\frac{\Delta G(1-\alpha)^{1/3}}{1-(1-\alpha)^{1/3}} \quad (88)$$

where $\Delta G = \mu_e^o - \mu_e^i$,

μ_e^o, μ_e^i = the chemical potentials for the outer and inner boundary of the sulfur layer

ΔG = the calculated Gibbs molar free energy for the overall reaction

M, ρ = molecular weight and density of $CuFeS_2$

From k'_p and calculated values for ΔG at temperature, the solid lines of Figure 22 were calculated using a best fit value of 7.6×10^{-13} ($ohm^{-1}cm^{-1}$) for $t_e\sigma$. This value compares to a measured value of $\sim 10^{-13}$ for pure sulfur. The correlation was considered satisfactory since estimate values of the activity coefficients of $Cu^{2+}(aq)$, $Fe^{2+}(aq)$ and $Fe^{3+}(aq)$ were used to calculate ΔG. Also, the sulfur produced contained impurities which have been noted to increase the electronic conductivity of sulfur. The proposed model also explains the modest response of the systems to the solution concentration of ferric ion.

CATHODIC CONVERSION OF CHALCOPYRITE

Shirts et al. (36) have demonstrated that chalcopyrite will react cathodically by bringing it into physical contact with a metal having a lower potential. Three metals, Cu, Fe, and Pb. were used. The reactions observed were

$$CuFeS_2 + Cu + 2H^+ \rightarrow Cu_2S + Fe^{2+} + H_2S \quad (89)$$

$$2CuFeS_2 + Fe + 4H^+ \rightarrow Cu + 2Fe^{2+} + 2H_2S \quad (90)$$

$$2CuFeS_2 + Pb + 6H^+ \rightarrow Cu_2S + 2Fe^{2+} + Pb^{2+} + 3H_2S \quad (91)$$

$$CuFeS_2 + Pb + 4H^+ \rightarrow Cu + Fe^{2+} + Pb^{2+} + 2H_2S \quad (92)$$

Galvanic couples are readily established because of the excellent electronic conductivity of all phases. The principal cathodic reaction is

$$2CuFeS_2 + 6H^+ + 2e^- = Cu_2S + 2Fe^{2+} + 3H_2S \quad (93)$$

It is evident from Figure 3(b), line d-e, that there is a large driving force for the cathodic conversion of $CuFeS_2$ to Cu_2S by copper. The half-cell reaction involving the copper is

$$Cu_2S + 2H^+ + 2e^- = 2Cu + H_2S \quad (94)$$

Reaction (94), having a more negative potential than reaction (93), will react anodically. It is also evident from Figure 3(b) that metallic iron will reduce $CuFeS_2$ to metallic copper as will metallic Pb in acid chloride solutions.

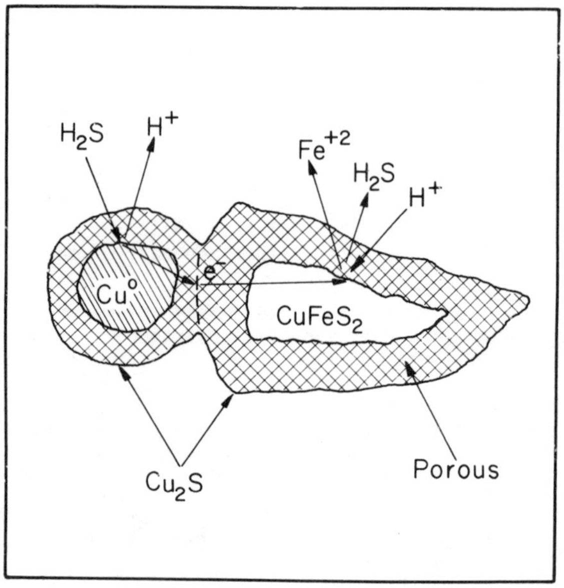

FIGURE 23. Schematic representation of the $CuFeS_2$-Cuo galvanic couple in sulfuric acid.

Hiskey and Wadsworth (37) measured the kinetics of the galvanic conversion of $CuFeS_2$ using copper. Figure 23 illustrates the galvanic $CuFeS_2$-Cu couple. Both the cathodic and anodic reactions produce Cu_2S which welds and establishes good electrical contact. It was found that at lower temperatures, stirring disrupted the bond and retarded the reaction. At 90°C stirring had less effect on the overall kinetics. Figure 24 illustrates results obtained for an intimate mixture of copper powder and chalcopyrite concentrates.

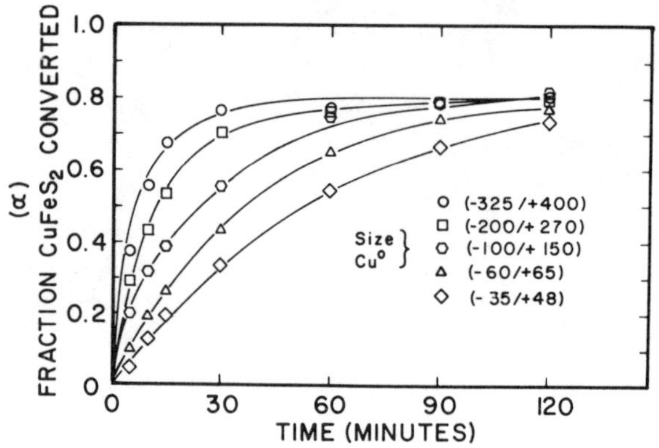

FIGURE 24. Fraction reacted versus time for cathodic reduction of chalcopyrite by various sizes of metallic copper.

Several sizes of copper were used with 325/400 mesh chalcopyrite with excess copper, molar $Cu/CuFeS_2$ ratio of 1.5. The extent of the reaction was measured by the fraction α, indicated by the amount of iron released to the solution. The curves of Figure 24 illustrate the importance of the copper, anodic surface area. Initially the chalcopyrite surface area is larger than that of the copper and the rate is predominantly anodically controlled. As the chalcopyrite reacts, its area becomes smaller and the kinetics shift from predominantly anodic control to predominantly cathodic control. The levelling off below $\alpha = 1.0$ was attributed to lack of perfect mixing of the initial solid phases. Careful blending increased the extent of reaction. In a galvanic system it is necessary to normalize the data in terms of the anodic (A_a) and cathodic (A_c) surface areas by the relationships $I_a = i_a A_a$ and $I_c = i_c A_c$ where I_a and I_c refer to the total anodic and cathodic currents respectively. Additional tests showed the rate of reaction to be essentially independent of [Fe II] in solution. The experimental activation energy was approximately 11.5 kcal/mole. Figure 25 depicts the initial rate versus initial hydrogen ion concentration. The rate order for hydrogen ion is in the range 0.48-0.52 or essentially one-half. Assuming single charge transfer in the activated state, the net cathodic and anodic electrode processes may be represented by the equations

$$I_c = A_c k_c 2F \exp[\frac{\beta_c F E_c}{RT}] - A_c k_c' a_{H^+} 2F \exp[\frac{-(1-\beta_c) F E_c}{RT}] \quad (95)$$

and

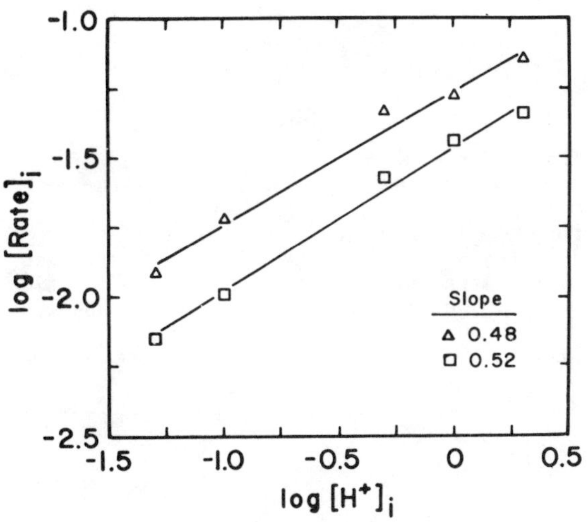

FIGURE 25. Log rate versus log $[H^+]$ showing half-power dependence on hydrogen ion concentration.

$$I_a = A_a k_a 2F \exp\left[\frac{\beta_a F E_a}{RT}\right] - A_a k_a' a_{H^+} 2F \exp\left[\frac{-(1-\beta_a) F E_a}{RT}\right] \quad (96)$$

The rates k_c and k_a may contain the activity of H_2S. In this case a_{H_2S} is constant since H_2S continually evolves at atmospheric pressure. Also k_c may conceivably contain [Fe II] although experimental results showed little dependence on [Fe II]. When steady-state currents are established E is the mixed potential such that $E_a = E_c = E$ and $I_c = -I_a$. An explicit solution for the mixed potential is not possible unless the Tafel slopes are essentially equal, or $\beta_a = \beta_c = \beta$. Using this approximation

$$\exp\left[\frac{-(1-\beta)\ E}{RT}\right] = a_{H^+}^{-(1-\beta)} \left[\frac{A_a k_a + A_c k_c}{A_a k_a' + A_c k_c'}\right]^{1-\beta} \quad (97)$$

Substitution in Eq. (95) results in the overall equation

$$\frac{d\alpha}{dt} = -\frac{I_c}{2mF} = \frac{a_{H^+}^{\beta_c} A_c A_a (k_c' k_a - k_c k_a')}{(A_a k_a + A_c k_c)^{\beta} (A_a k_a' + A_c k_c')^{1-\beta}} \quad (98)$$

The constant value of the activation energy with temperature indicates $k_c k_a' << k_c' k_a$. Also the hydrogen ion rate dependence suggests $\beta \simeq 0.5$. Using this value the rate becomes

$$\frac{d\alpha}{dt} = \frac{a_{H^+}^{1/2} A_c A_a k_c' k_a}{(A_a k_a + A_c k_c)^{1/2} (A_a k_a' + A_c k_c')^{1/2}} \quad (99)$$

Equation (99) provides for the observed anodic control for the initial condition $A_c k_c >> A_a k_a$ and $A_c k_c' >> A_a k_a'$ observed experimentally. Similarly cathodic control results when $A_a > A_c$ for large values of α.

The relationship between A_a, A_c and α are, assuming spherical geometry

$$A_a = \frac{3\phi m M_a}{\rho_a r_{oa}} (1-\alpha/\phi)^{2/3} \quad (100)$$

and

$$A_c = \frac{3\alpha_m m M_c}{\rho_c r_{oc}} (1-\alpha/\alpha_m)^{2/3} \quad (101)$$

The maximum conversion is designated as α_m and is related to the completeness of mixing of the Cu and $CuFeS_2$ phases initially. The other terms are:

α = fraction of $CuFeS_2$ converted
ϕ = mole Cu/moles $CuFeS_2$
M_a, M_c = molecular weights Cu and $CuFeS_2$
ρ_a, ρ_c = densities of Cu and $CuFeS_2$
r_{oa}, r_{oc} = initial particle radius of Cu and $CuFeS_2$
m = moles $CuFeS_2$ in the system

Initially for the case $A_c >> A_a$ for the Eqs. (99) and (100) may be combined. The integrated form for constant a_{H^+} becomes

$$1-(1-\alpha/\phi)^{1/3} = \frac{k_{oa}}{r_{oa}} t \quad (102)$$

where

$$k_a = a_{H^+}^{1/2} \left(\frac{k_c'}{k_c}\right)^{1/2} k_a \frac{mM_a}{\rho_a} \quad (103)$$

and k_a is the linear velocity of the chalcopyrite interface. Figure 26 illustrates the correlation of initial data points, for which $a_{H^+} \simeq$ constant, according to Eq. (102) for three sizes of copper shot. The inverse r_{oa} dependence predicted by Eq. (102) was observed with an average k_a of 5×10^{-5} cm/min. Using direct electrochemical measurements Nicol (38) has confirmed the galvanic character of $CuFeS_2$ reduction by Cu, Fe and Pb. The kinetics of $CuFeS_2$ reduction were found to involve a single electron transfer process. Also the rate was found to be proportional to hydrogen ion to a fractional order and directly proportional to the sulfide concentration.

Chae (39) investigated the kinetics of reduction of $CuFeS_2$ using metallic lead in acidified chloride solutions. The rate was relatively insensitive to particle size. Figure 27 illustrates results for several sizes of particulate lead used to reduce 270x325 mesh chalcopyrite. Also, it was noted that results obtained by pressing a lead pellet against a polished $CuFeS_2$ electrode resulted in an increased rate of reaction, attributed to a decrease in the contact resistance due to the applied pressure. In contrast to the $CuFeS_2$-Cu reaction, a common product capable of fusing the particles does not form in the case of the $CuFeS_2$-Pb system. The related insensitivity of the rate to

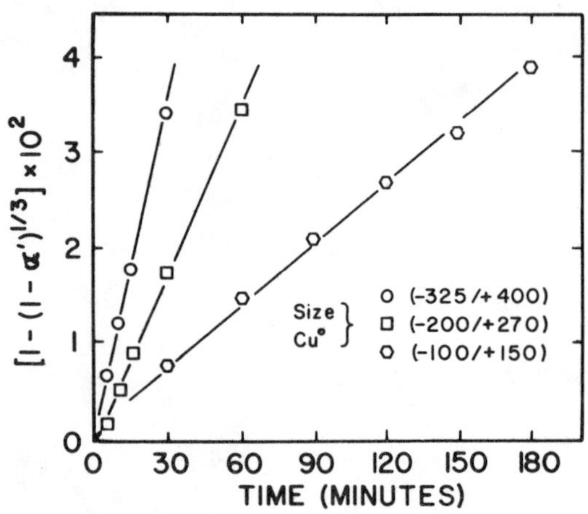

FIGURE 26. Correlation of initial kinetics for the cathodic conversion of chalcopyrite for various sizes of copper.

particle size has been attributed to the probability of particle contact coupled with particle-particle contact areas which provide opposing influences on the geometry of the system with a resultant nullifying effect. The reactions depicted in equations (91) and (92) resulted with the exception that the defect chalcocite $Cu_{1.96}S$ (djurleite) formed instead of Cu_2S followed by reduction to a high surface area metallic copper as a final product. The rate was sensitive to chloride ion concentration and the apparent activation energy was 6.7 kcal/mole.

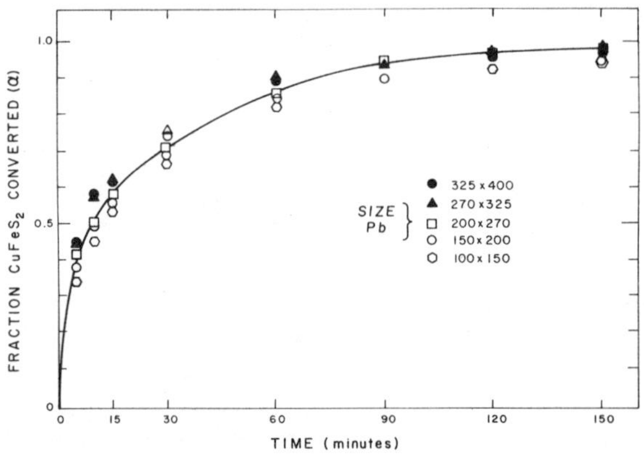

FIGURE 27. The conversion of 270 x 325 mesh chalcopyrite with various sizes of lead at 89°C and 350 rpm. 1 M HCl.

Sohn (40,41) has investigated the reduction of $CuFeS_2$ in cupric sulfate solutions using SO_2 as the reductant. The important half-cell reactions involved are:

cathodic:

$$2CuFeS_2 + 3Cu^{2+} + 4e^- = Cu_5FeS_4 + Fe^{2+} \quad (104)$$

$$Cu_5FeS_4 + 3Cu^{2+} + 4e^- = 4Cu_2S^* + Fe^{2+} \quad (105)$$

anodic:

$$SO_2 + 2H_2O = HSO_4^- + 3H^+ + 2e^- \quad (106)$$

In the sequence of reactions initially a defect structure was observed with a main lattice parameter near that of bornite but slightly greater. As the reaction proceeded bornite formed followed by djurleite, $Cu_{1.96}S$. Figure 28 illustrates fraction of contained iron released to solution as a function of time for three monosized fractions of $CuFeS_2$.

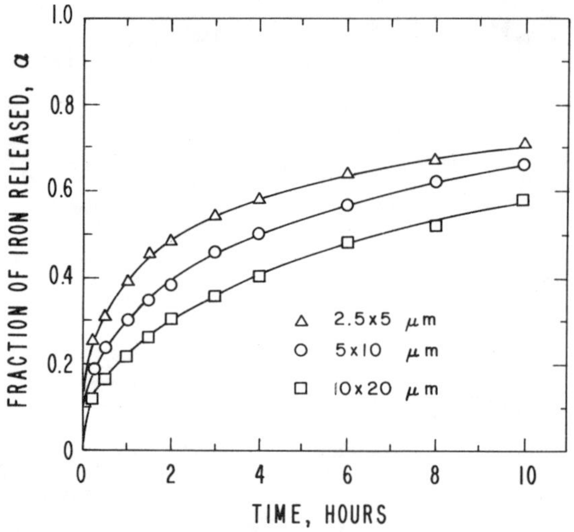

FIGURE 28. A plot of iron extracted from three monosize chalcopyrite fractions as a function of time for 90°C, 0.5% solids, molar ratio of $Cu^{++}/CuFeS_2$ = 7.5 and 1200 rpm.

The rate of conversion increased with copper concentration but was only slightly affected by the partial pressure of SO_2 gas. The most striking effect was that of ferrous ion. Figure 29 illustrates results for the 2.5x5 μm fraction with and without added Fe^{2+} to solution. A striking increase in the rate of reaction was noted. When the iron was added after 45 minutes a sharp increase in rate was noted. Warren (7) noticed a similar effect when reducing $CuFeS_2$ cathodically in the absence of cupric ion. In this case H_2S gas was evolved. Sohn (40) carried out similar cathodic reduction experiments in the presence of Cu^{2+}. The catalytic effect has been attributed to the lowered activation energy for the surface charge transfer process involving both iron and copper according to the following reactions at the electrode surface:

$$Fe^{2+}(aq) + Cu^{2+}(aq) = Cu^+(aq) + Fe^{3+}(aq) \quad (107)$$

$$Cu^+(aq) = Cu^+ \text{ (lattice)} \quad (108)$$

* The defect structure $Cu_{1.96}S$ (djurleite) formed in solutions of modest acidity. Solutions containing one molar or greater H_2SO_4 formed digenite, $Cu_{1.8}S$.

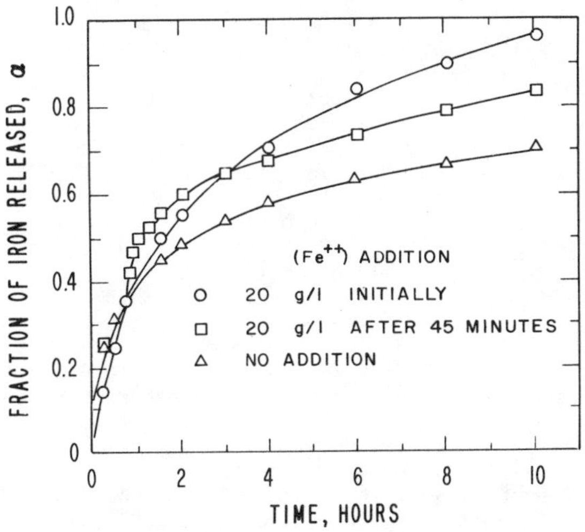

FIGURE 29. Fraction of 2.5 x 5 μm chalcopyrite reduced as a function of time, showing catalytic effect of Fe^{2+}

$$Fe^{3+}(aq) + e^{-} = Fe^{2+}(aq) \tag{109}$$

Figure 30 illustrates the cathodic polarization curves obtained for various Fe^{2+} solution concentrations. The sharp passivation peak formed just above 0.4 volts was attributed to the formation of an initial defect structure. Appreciable enhancement of the current in this region resulted when the ferrous iron concentration was increased. At more negative potentials again the ferrous ion concentration increased the current although in this region only a slight increase occurred above 1 gpl Fe^{2+}. In this region $Cu_{1.96}S$ forms.

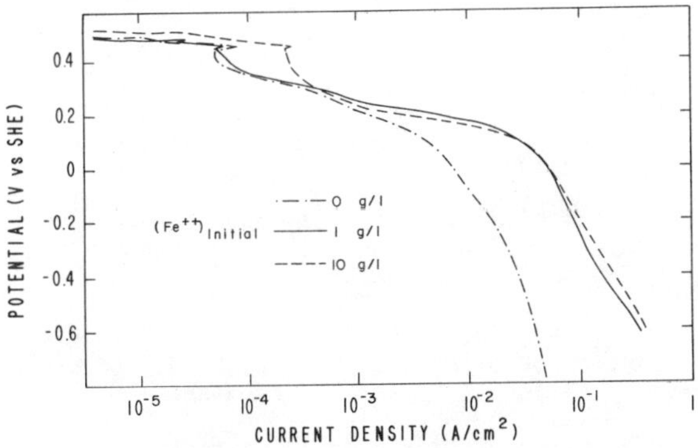

FIGURE 30. Cathodic polarization curves for transvaal chalcopyrite for various ion concentrations for 10 g/l of Cu^{++}, pH = 1.8 and 25°C.

The kinetics of the process surprisingly follow logarithmic kinetics, usually observed for thin tarnish films in which high voltage gradients occur. In this system thick layers of $Cu_{1.96}S$ form with a visible band of bornite which moves through the particle. Microscopic examination of a partially reacted particle clearly showed a narrow band of bornite (less than 0.5 μm in thickness) adjacent to the unreacted core of $Cu_{1.96}S$. Figure 31 illustrates the data of Figure 28 plotted according to the

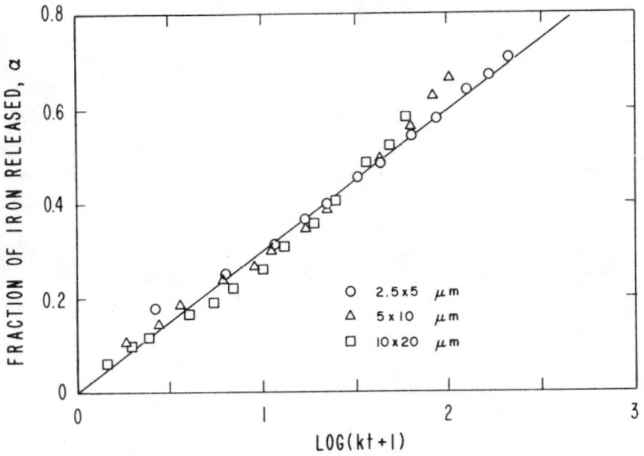

FIGURE 31. A plot of the logarithmic law for the data shown in Figure 28.

logarithmic rate law.

The electrochemical nature of the process was demonstrated (40) by using a diagram cell with graphite electrodes. The cathodic cell contained $CuFeS_2$ and graphite powder in a solution of acidified cupric sulfate. The anode compartment was fed SO_2 gas and contained a suspension of graphite powder. Each compartment had a solid graphite electrode. The effect of ferrous iron addition was examined with separate additions to each compartment. Ferrous iron added to the catholyte increased the current markedly while addition to the anolyte decreased the current slightly.

REFERENCES

1. J.O'M. Bockris, "Electrochemistry: The Underdeveloped Science," J. Electroanal. Chem., 1965, vol. 9, pp. 408-434.

2. J.O'M. Bockris and A.K.N. Reddy, Modern Electrochemistry, vol. 1, p. 3, Plenum Publishing Corp., New York, 1973.

3. G.H. Koch, Modern Aspects of Electrochemistry, No. 10, pp. 211-237, Plenum Press, New York, 1975.

4. R.T. Shuey, Semiconducting Ore Minerals, Elsevier Publishing Company, New York, 1975.

5. T. Biegler and D.A. Swift, "Anodic Behavior of Pyrite in Acid Solutions," Electrochimica Acta, 1979, vol. 24, pp. 415-420.

6. N. Masuko and Y. Hisamatsu, "On the Electrochemical Behavior of Pyrite," Denki-Kagaku, 1963, vol. 31, pp. 907-12.

7. G.W. Warren, "The Electrochemical Oxidation of $CuFeS_2$," Ph.D. Dissertation, Univ. of Utah, 1978.

8. D.M. Chizhikov and V.N. Kovylina, "Investigation of Potentials and Anodic Polarization of the Sulfides of Metals and Their Alloys," Trudy Chetvertogo Soveshchaniya po Electrokhimii, 1956, pp. 715-719.

9. H. Majima, "How Oxidation Effects Selective Flotation of Complex Sulphide Ores," Can. Met. Quart., 1969, vol. 8, pp. 269-273.

10. D.F.A. Koch and R.J. McIntyre, "The Application of Reflectance Spectroscopy to a Study of the Anodic Oxidation of Cuprous Sulfide," J. Electroanal. Chem., 1976, vol. 71, pp. 285-296.

11. A. Ettienne and E. Peters, "Thermodynamic Measurements in the Cu-S System in the Temperature Range 40-80°C," IMM Trans., Sect. C., 1972, vol. 81, pp. 176-181.

12. H.J. Mathieu and H. Rickert, "Electrochemical-Thermodynamic Studies on the Copper-Sulfur System at 15-90°," J. Phys. Chem., 1972, vol. 79, pp. 315-30.

13. G.H. Moh, Mineral Soc. Japan, Spec. Paper 1, 1971, pp. 226-232.

14. G. Frenzel, N. Jb. Miner, 1959, Abh 93, pp. 87-132.

15. R.W. Potter II, "An Electrochemical Investigation of the System Copper-Sulfur," Econ. Geol., 1977, vol. 72, pp. 1524-1542.

16. P. Cavallotti and G. Salvago, Electrochim Metal, 1969 vol. IV, no. 3, pp. 181-210.

17. P. Brennet, S. Jafferalli, J.M. Vanseveren, J. Vereecken, and R. Winand, Met. Trans., 1974, vol. 5, pp. 127-134.

18. M.J. Nicol, C.R.S. Needes, and N.P. Finkelstein, "Electrochemical Model for the Leaching of Uranium Dioxide: 1-Acid Media," In Leaching and Reduction in Hydrometallurgy, A.R. Burkin, ed., pp. 1-11, IMM, London, 1975.

19. V.K. Berry, L.E. Murr, and J.B. Hiskey, "Galvanic Interaction between Chalcopyrite and Pyrite during Bacterial Leaching of Low-Grade Waste," Hydrometallurgy, 1978, vol. 3, pp. 209-326.

20. E. Peters, I.H. Warren, and H. Veltman, First Tutorial Symposium on Hydrometallurgy, The University of Denver, 1972.

21. C.R.S. Needes, M.J. Nicol, N.P. Finkelstein, "Electrochemical Model for the Leaching of Uranium Dioxide: 2-Alkaline Carbonate Media," In Leach and Reduction in Hydrometallurgy, A.R. Burkin, ed., pp. 12-19, IMM, London, 1975.

22. M.J. Nicol, C.R.S. Needes, and N.P. Finkelstein, "The Mechanism of the Anodic Dissolution of UO_2 in Aqueous Solutions," Natn. Inst. Metall., Report PIN 171, Oct. 1973, 37 p.

23. C.R.S. Needes and M.J. Nicol, "The Study of Some Redox Reactions at a UO_2 Surface," Natn. Inst. Metall., Report PIN 165, Sept. 1973, 60 p.

24. J.B. Hiskey, "Kinetics of Uranium Dioxide Dissolution in Ammonium Carbonate," IMM Trans., Sect. C. 1979, vol. 88, pp. 145-152.

25. J.B. Hiskey, "Hydrogen Peroxide Leaching of Uranium in Carbonate Solutions," Submitted for publication.

26. E.V. Komarov, "Complex Formation in the $UO_2(NO_2)_2$-$KHCO_3$-H_2O_2-H_2O System," Russ J. Inorg. Chem. 1959, vol. 4, pp. 591-596.

27. P. Eadington and A. Prosser, "Oxidation of Lead Sulphide in Aqueous Suspensions," IMM Trans., Sect. C, 1969, vol. 78, pp. 74-82.

28. R.L. Paul, M.J. Nicol, J.W. Diggle and A.P. Sanders," The Electrochemical Dissolution of Galena in Aqueous Solutions," Report No. 1878, National Institute for Metallurgy, May 1977.

29. P. Marcantonio, "Chalcocite Dissolution in Acidic Ferric Sulfate Solutions," Ph.D. thesis, Department of Mining, Metallurgical and Fuels Engineering, University of Utah, 1976.

30. G. Thomas, T.R. Ingraham, and R.J.C. MacDonald, "Kinetics of Dissolution of Synthetic Digenite and Chalcocite in Aqueous Acidic Ferric Sulphate Solution," Can. Met. Quart., 1967, vol. 6, pp. 281-292.

31. J.D. Sullivan, "Chemistry of Leaching Chalcocite," U.S. Bureau of Mines, Tech. Paper 473, 1930.

32. J.P. Baur, H.L. Gibbs, and M.E. Wadsworth, "Initial-Stage Sulfuric Acid Leaching of Chalcopyrite Using Radiochemical Techniques," U.S. Dept. of Interior, Bureau of Mines, RI 7823, 1974.

33. D.L. Jones and E. Peters, "The Leaching of Chalcopyrite with Ferric Sulfate and Ferric Chloride," Vol. II, Extractive Metallurgy of Copper, J.C. Yannopoulos and J.C. Agarwal, eds., American Institute of Mining, Metallurgical and Petroleum Engineers, 1976.

34. P.B. Munoz, J.D. Miller, and M.E. Wadsworth, "Reaction Mechanism for the Acid Ferric Sulfate Leaching of Chalcopyrite," Met. Trans., 1979, vol. 10B, pp. 149-158.

35. D.L. Jones, Ph.D. Thesis, "The Leaching of Chalcopyrite," University of British Columbia, 1974.

36. M.B. Shirts, J.K. Winter, P.A. Bloom, and G.M. Potter, "Aqueous Reduction of Chalcopyrite Concentrate with Metals," U.S. Dept. of Interior, Bureau of Mines, RI 7953, 1974.

37. J.B. Hiskey and M.E. Wadsworth, "Galvanic conversion of Chalcopyrite," Met. Trans., 1975, vol. 6B, pp. 183-190.

38. M.J. Nicol, "Mechanism of Aqueous Reduction of Chalcopyrite by Copper, Iron and Lead," IIM, 1975, vol. 84, pp. C206-C209.

39. D.G. Chae, Ph.D. Thesis, "Part II, Cathodic Conversion of Chalcopyrite by Lead," Department of Metallurgy, University of Utah, 1980.

40. H.J. Sohn, Ph.D. Thesis, "Chemical and Electrochemical Conversion of Chalcopyrite into Copper Sulfide," Department of Metallurgy, University of Utah, 1980.

41. H.J. Sohn and M.E. Wadsworth, "Reduction of Chalcopyrite with SO_2 in the Presence of Cupric Ions," submitted for publication.

42. L.W. Beckstead and J.D. Miller, "Ammonia, Oxidation Leaching of Chalcopyrite - Reaction Kinetics," Met. Trans. B, 1977, Vol. 8B, pp. 19-38.

43. F.A. Forward and V.N. Mackiw, "Chemistry of the Ammonia Pressure Process for Leaching Ni, Cu and Co from Sherritt-Gordon Sulphide Concentrates," Trans, AIME, 1955, Vol. 203, pp. 457-463.

44. M.C. Kuhn, N. Arbiter, and H. Kling, "Anaconda's Arbiter Process for Copper," Can. Min. Met. Bull. 1974, Vol. 67, pp. 62-73.

Chapter 27

ELECTROCHEMISTRY IN SILVER CATALYSED FERRIC SULFATE LEACHING OF CHALCOPYRITE

J.D. Miller, P.J. McDonough and H.Q. Portillo

Department of Metallurgy and Metallurgical Engineering
University of Utah

ABSTRACT

Previous investigations have demonstrated the catalytic effect of silver additions in the ferric sulfate leaching of chalcopyrite.

$$CuFeS_2 + 4Fe^{+3} \rightarrow Cu^{++} + 5Fe^{++} + 2S^o$$

The enhanced rate of leaching was found to be due to the formation of an intermediate Ag_2S film which forms on the $CuFeS_2$ surface by an exchange reaction. Under these conditions, unlike the uncatalysed ferric sulfate leach, the elemental sulfur forms a nonprotective reaction product on the Ag_2S crystallites. As a result, the rate appears to become limited by an intermediate electrochemical reaction in the Ag_2S film rather than by transport through the elemental sulfur reaction product.

$$Ag_2S_{(Surface\ CuFeS_2)} + 2Fe^{+3} \rightarrow 2Ag^+ + 2Fe^{++} + S^o$$

The reaction rate was found to have a complex dependence on the Ag^+, Fe^{++} and Fe^{+3} concentrations, but independent of the Cu^{++} concentration. Electrochemical measurements with a Ag_2S-coated chalcopyrite electrode provide furtner information regarding the nature of the reaction mechanism. The kinetics of the rate limiting, intermediate electrochemical reaction are analyzed in terms of the Butler-Volmer equation.

ELECTROCHEMISTRY IN SILVER CATALYSED FERRIC SULFATE LEACHING OF CHALCOPYRITE

J.D. Miller, P.J. McDonough and H.Q. Portillo

Department of Metallurgy and Metallurgical Engineering
University of Utah

Previous investigations have demonstrated the catalytic effect of silver additions in the ferric sulfate leaching of chalcopyrite.

$$CuFeS_2 + 4Fe^{+3} \xrightarrow{Ag^+} Cu^{++} + 5Fe^{++} + 2S^o$$

The enhanced rate of leaching was found to be due to the formation of an intermediate Ag_2S film which forms on the $CuFeS_2$ surface by an exchange reaction. Under these conditions, unlike the uncatalysed ferric sulfate leach, the elemental sulfur forms a nonprotective reaction product on the Ag_2S crystallites. As a result, the rate appears to become limited by an intermediate electrochemical reaction in the Ag_2S film rather than by transport through the elemental sulfur reaction product.

$$Ag_2S_{(Surface\ CuFeS_2)} + 2Fe^{+3} \rightarrow 2Ag^+ + 2Fe^{++} + S^o$$

The reaction rate was found to have a complex dependence on the Ag^+, Fe^{++} and Fe^{+3} concentrations, but independent of the Cu^{++} concentration. Electrochemical measurements with a Ag_2S-coated chalcopyrite electrode provide further information regarding the nature of the reaction mechanism. The kinetics of the rate limiting, intermediate electrochemical reaction are analyzed in terms of the Butler-Volmer equation.

INTRODUCTION

In sulfuric acid solutions of ferric sulfate, chalcopyrite dissolves according to the following reaction,

$$CuFeS_2 + 4Fe^{+3} \rightarrow Cu^{+2} + 5Fe^{++} + 2S^o \qquad (1)$$

This reaction stoichiometry has been observed in a number of investigations. (1-5) Exceptions to this consensus are the early work of Sullivan (6) and the more recent results reported by Jones and Peters. (7) The topochemical nature of the leaching reaction is clearly illustrated in Figure 1 which shows the cross section of partially reacted chalcopyrite particle surrounded by a dense, tenacious sulfur layer.

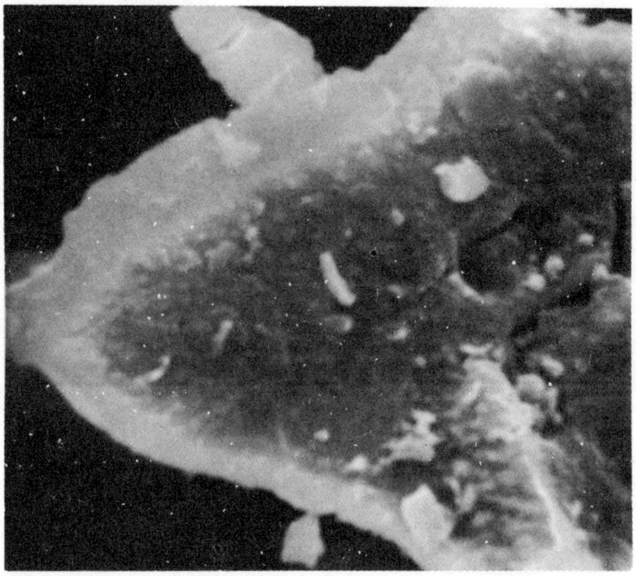

Figure 1. SEM photograph of the cross section of a partially leached chalcopyrite particle (10 µm) showing the sulfur layer and topochemical nature of the uncatalysed ferric sulfate leach. (5)

The formation of an elemental sulfur layer on the chalcopyrite surface may significantly influence the reaction kinetics by establishing a diffusion barrier. The details of rate control for this particular reaction have not been well established. Some investigators (1,7) attribute rate control to a surface reaction. Other investigators (2-5,8,9) report that the reaction rate is limited by transport in the chalcopyrite lattice or through the elemental sulfur reaction product layer. Differences in leaching behavior reported in the literature (5) may be due to variations in the intrinsic electrochemical properties of the chalcopyrites used in the various studies. These electrochemical differences are illustrated by the anodic polarization curves presented in Figure 2. (10)

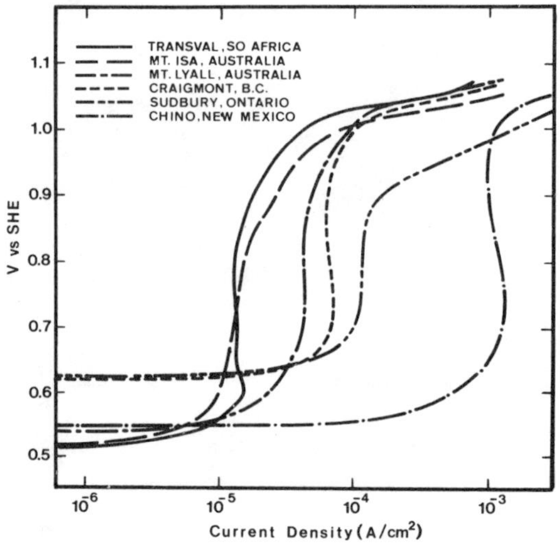

Figure 2. Anodic polarization curves for $CuFeS_2$ from six different locations in 1 M H_2SO_4, 30 mv/min, 25°C. (10)

Notice that the current density for a fixed potential can differ by as much as two orders of magnitude for different samples of chalcopyrite.

Most recently, leaching data for passive chalcopyrites which exhibit parabolic reaction kinetics have been reconsidered together with new experimental data and the results interpreted using Wagner's theory of oxidation. (5) Analysis of the ferric sulfate leaching system in this context suggests that the rate limiting process may be the transport of electrons through the elemental sulfur layer as depicted schematically in Figure 3.

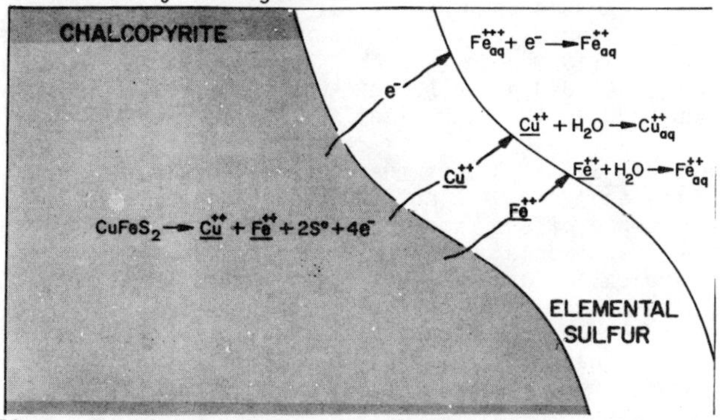

Figure 3. Schematic representation of the transport process in the ferric sulfate leaching of chalcopyrite. Note that the hydrated ferric ion does not advance through the sulfur product layer, but rather is discharged at the sulfur-solution interface. The analysis does not distinguish Cu^{++} and Fe^{++} from Cu^{+} and Fe^{+++} as the transported species. (5)

Predicted reaction rates calculated from first principles using the physicochemical properties of the system (conductivity of elemental sulfur, t σ, and the free energy change for the reaction, ΔG°)agree satisfactorily with experimentally determined rates. This analysis is summarised in Table I.

Table I. Ferric Sulfate Leaching of Chalcopyrite in the Absence of Silver (5)

The rate of acid ferric sulfate leaching of monosize chalcopyrite particles in a stirred reactor appears to be limited mainly by the transport of electrons through the elemental sulfur reaction product. This position has been analysed in terms of Wagner's theory of oxidation and is supported by:

1) Independence of rate on both reactant and product concentrations.
2) Parabolic reaction kinetics.
3) A rate constant equivalent in magnitude ($7.6 \times 10^{-13} ohm^{-1} cm^{-1}$) to the electrical conductivity of elemental sulfur ($10^{-13} ohm^{-1} cm^{-1}$).
4) An apparent activation energy, 20 kcal/mole (83.7 kJ/mole) equivalent to the temperature coefficient for electrical conduction in elemental sulfur, 23 kcal/mole (96.2 kJ/mole).

Silver Catalysis

It has been demonstrated that silver addition in acid ferric sulfate leaching of chalcopyrite and similar systems (11-13) accelerates greatly the rate

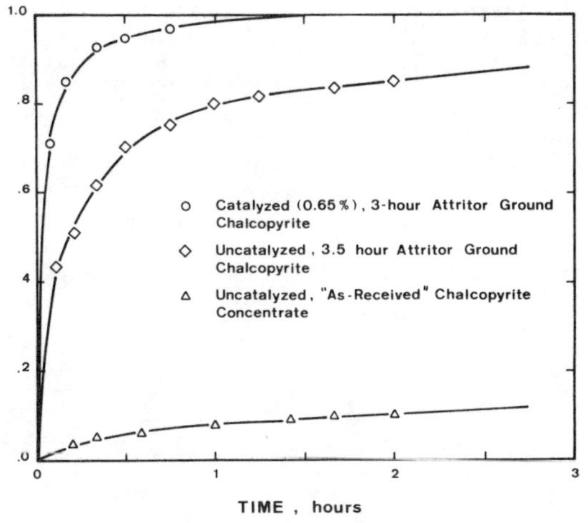

Figure 4. Fraction of copper extraction as a function of time for different products at 1 M Fe^{+3} and 90°C. (14)

of reaction. For example, the catalysed and uncatalysed leaching rates for an attritor-ground chalcopyrite are contrasted in Figure 4. (14)

Some mechanistic information on this catalytic effect has been reported. However, further research is required to elucidate details of the intrinsic reaction kinetics.

Silver Exchange Reaction

The reaction of silver ion with chalcopyrite in the absence of oxidant has been found to be described by the following exchange reaction (14)

$$CuFeS_2 + 4Ag^+ \rightarrow 2Ag_2S_{(Surface)} + Cu^{++} + Fe^{++} \quad (2)$$

Chalcopyrite when reacted with silver ion rapidly forms a blue-black product on the chalcopyrite surface. Analysis on this reaction product by x-ray diffraction showed that the reaction product consisted of Ag_2S (argentite) and unreacted $CuFeS_2$ with all reported "d" spacings identified. Solution analyses, however, showed some deviation from the indicated stoichiometry. The results obtained seem to be independent of acid strength (0.1 M to 1.0 M H_2SO_4) at 90°C.

The initial kinetics of the exchange were determined by following the depletion of silver ion with a specific ion electrode and were studied as a function of initial silver concentration, particle size, and temperature. (14) Generally the initial kinetics were first order with respect to silver concentration until the thickness of the Ag_2S film exceeded approximately 50A°. For the case of first order reaction kinetics, the reaction velocity constant can be found from a semi-log plot and should be independent of the initial silver concentration. Such was found to be the case for a wide range of concentrations (over 2 orders of magnitude) and for two different monosize samples (4.8 μm and 51.2 μm) at 25° as shown in Figure 5.

The effect of particle size was also considered in the analysis of the initial reaction kinetics. Generally the reaction velocity constant would be expected to be largely independent of the particle size even for rate control by diffusion in the mass transfer boundary layer. It was found that the reaction velocity constants for the 4.8 μm and 51.2 μm samples do not differ greatly, 2.67×10^{-2} cm/sec and 1.28×10^{-2} cm/sec respectively. The initial high rate of this first order reaction would suggest that the rate may be limited by diffusion in the mass transfer boundary layer. To test this hypothesis, the experimental reaction velocity constants were compared to predicted mass transfer coefficients for stirred reactors estimated using Harriot's approach (15,16) which is based on the Sherwood correlation. (17) Excellent agreement between experimental and predicted values was obtained.

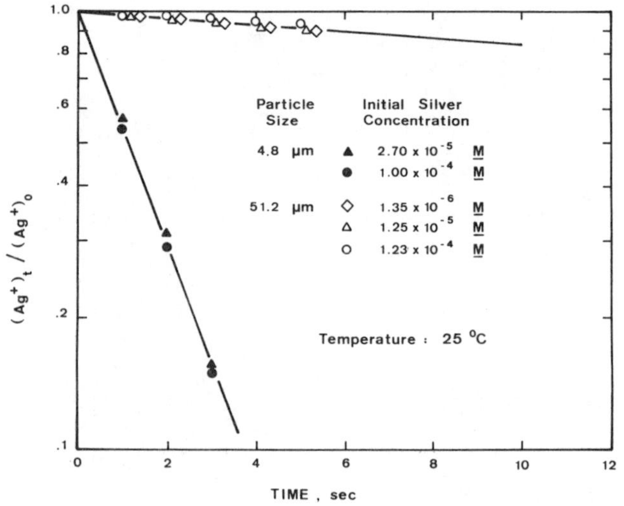

Figure 5. First order rate plot of fraction of silver reacted for two monosize chalcopyrite samples at 0.25$\underline{M}$ H_2SO_4, 0.2 percent solids, 600 rpm and various initial silver concentrations. (14)

Finally, if the initial reaction kinetics are controlled by a mass transfer process in the aqueous phase, then the temperature coefficient for the rate phenomenon should typically correspond to an activation energy of less than 5.0 kcal/mole (20.9 kJ/mole). An apparent activation energy of 4.2 kcal/mole (17.6 kJ/mole) was found for the 51.2 μm material which supports the hypothesis of rate control by mass transfer in the aqueous phase. Similar results were obtained for the 4.8 μm sample with an apparent activation energy of 3.1 kcal/mole (13.0 kJ/mole).

Silver Catalysed Ferric Sulfate Leach

The silver catalysed reaction appears to exhibit the same stoichiometry as the uncatalysed reaction presented in Equation 1. (14) In the absence of the chloride salts, elemental sulfur together with Ag_2S constituted the reaction product as determined by x-ray diffraction. Solution analyses for copper and ferrous revealed a Fe^{++}/Cu^{++} mole ratio of 5.08 (average of 18 values from 4 different experiments) at all stages of the reaction which closely corresponds to the 5/1 ratio specified by the stoichiometry of the suggested reaction in Equation 1.

The leaching response was found to be sensitive to the experimental procedure used for reaction initiation. As might be expected when the chalcopyrite was allowed to react first with the ferric sulfate the reaction kinetics were slow due to the formation of the impervious elemental sulfur layer during the initial time interval prior to the addition of silver. On the other hand, when the silver was added first a much more rapid leaching response was observed.

Further, the rate and extent of reaction were found to be dependent on the initial silver concentration. The rate increased with an increase in initial silver concentration up to $1 \times 10^{-3}$$\underline{M}$. Further additions beyond this level did not enhance the rate of reaction and even reduced the rate significantly.

Particle Size. One of the most diagnostic features in the analysis of kinetic data is the particle size dependence of the reaction rate. An inverse first order dependence on the initial particle size could be indicative of rate control by diffusion in the aqueous phase or by surface reaction. An inverse second order dependence on the initial particle size would suggest rate control by diffusion through a growning reaction product layer. Neither case was observed in the silver catalysed ferric sulfate leaching of chalcopyrite. Surprisingly, the rate, up to approximately 50% reacted, was essentially independent of initial particle size as shown in Figure 6.

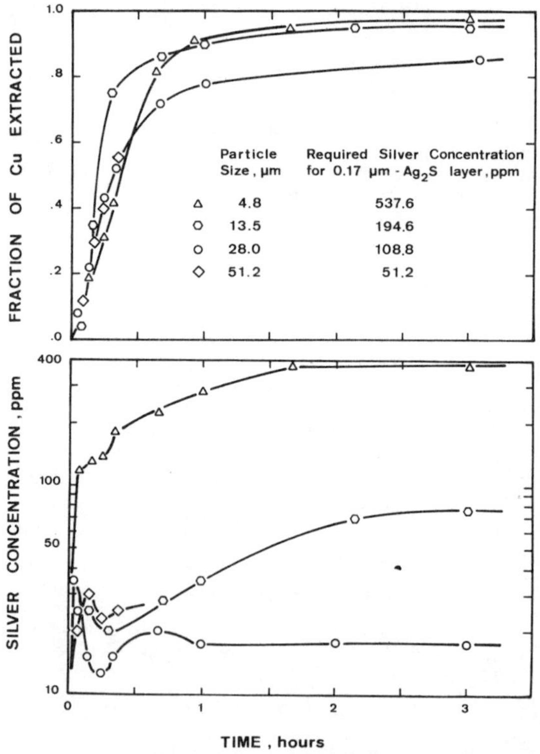

Figure 6. The effect of particle size on the rate and extent of the catalysed reaction at 0.25 $\underline{M}$ H_2SO_4 1.0 $\underline{M}$ Fe^{+3}, 0.3 percent solids, 600 rpm and 90°C. Also the corresponding plot of silver concentration in the aqueous phase as a function of reaction time is shown for the respective experiments.

These experiments were done by first completing the exchange reaction and establishing a silver sulfide layer of fixed thickness (0.174 μm) assuming spherical particles of a specified median particle size.

It is clear from the data that the Ag_2S film reacts rapidly with the ferric sulfate solution as revealed by the plots of silver concentration in the aqueous phase as a function of time.

$$Ag_2S_{(Surface)} + 2Fe^{+3} \rightarrow 2Ag^{+} + 2Fe^{2+} + S^{o} \qquad (3)$$

The silver sulfide layer is not completely removed and a rapid rate of chalcopyrite leaching, independent of initial particle size was observed. Under certain circumstances for the 51.2 μm sample the Ag_2S film dissolved completely during the first few minutes and the reaction became passivated due to sulfur formation at the chalcopyrite surface.

Only after extensive reaction ($\gtrsim$ 50% reacted), does the sulfur layer, which appears to form in the silver film, become protective. The very distinct difference between the sulfur which forms in the silver catalysed reaction and that which forms during the uncatalysed reaction is revealed by the SEM photographs presented in Figure 7. Note that the surface sulfur created in the case of the silver catalysed reaction is very porous with well defined sulfur crystallites seen. In the case of the uncatalysed reaction a smooth tenacious layer of elemental sulfur forms similar to that shown in cross section in Figure 1.

Concentration Effects. Unlike the uncatalysed reaction, a definite Fe^{+3} dependence of the rate was observed in the case of the silver catalysed reaction. In addition, the reaction kinetics were extremely sensitive to the Fe^{+2} concentration as shown in Figure 8. Note that the addition of 0.5 M $FeSO_4$ essentially passivates the system and very little reaction is achieved. On the other hand, the addition of 0.5 M $CuSO_4$ or $MgSO_4$ does not alter the reaction rate. These results are somewhat anomalous in that the rate is dependent on the concentration of one of the reaction products (Fe^{++}) but independent of the concentration of another reaction product (Cu^{++}).

On the basis of these observed results it was suggested (14) that the rate may be controlled by the intermediate electrochemical reaction of Ag_2S with Fe^{+3}.

$$Ag_2S_{(Surface)} + 2Fe^{+3} \rightarrow 2Ag^{+} + 2Fe^{++} + S^{o} \qquad (3)$$

It was hypothesized that this electrochemical reaction occurs at the surface of silver sulfide crystallites throughout the Ag_2S film. The crystallites act as short circuited microelectrodes allowing for the discharge of Fe^{+3} and the release of Ag^{+}. Under these circumstances, the kinetics could show a dependence on the Ag^{+}, Fe^{+3}, and Fe^{++} concentrations and would be independent of the initial particle size, i.e. the rate is dependent on the internal area of the crystallites rather than on the external area of the particle. It was suggested that elemental sulfur replaces the Ag_2S crystallites forming a porous, non-protective layer. The silver ion generated acts as a transfer agent while the silver sulfide film, necessary to prevent extensive sulfur formation at the chalcopyrite surface, is continually restored. A schematic representation of this mechanism is presented in Figure 9.

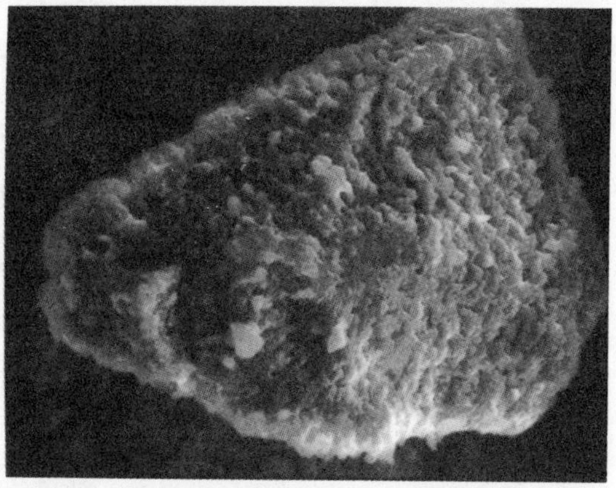

Partially Reacted 28 μm Particle -- Catalized

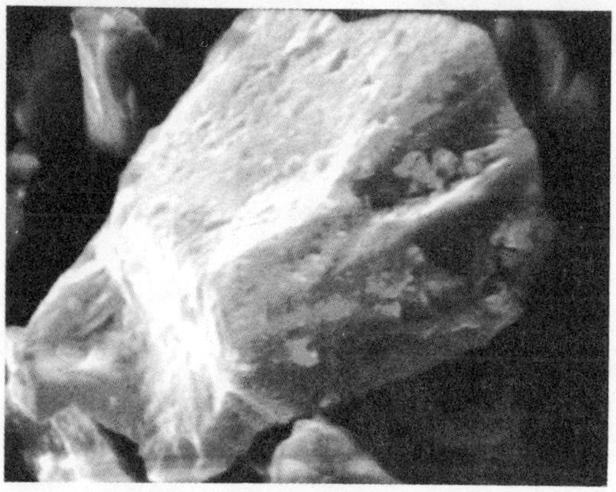

Partially Reacted 28 μm Particle -- Uncatalized

Figure 7. SEM photographs contrasting the nature of the elemental sulfur reaction product layer which forms during the catalysed reaction with that which forms during the uncatalysed reaction.

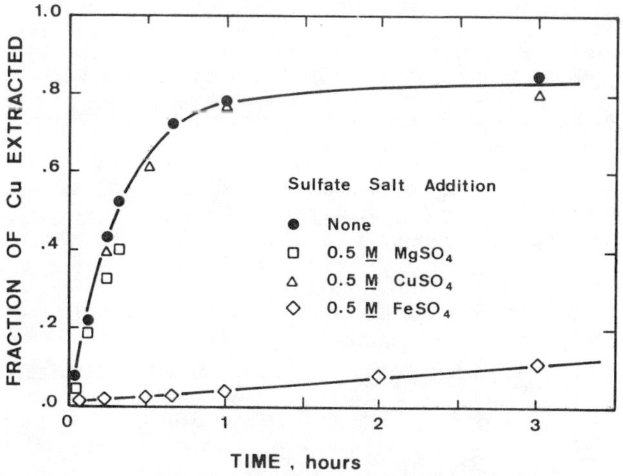

Figure 8. The effect of ferrous sulfate addition on the rate and extent of reaction for 28 μm particles at 0.25 M H_2SO_4, 1 M Fe^{+3}, 0.3 percent solids, 600 rpm and 90°C. (0.17 μm Ag_2S layer). (14)

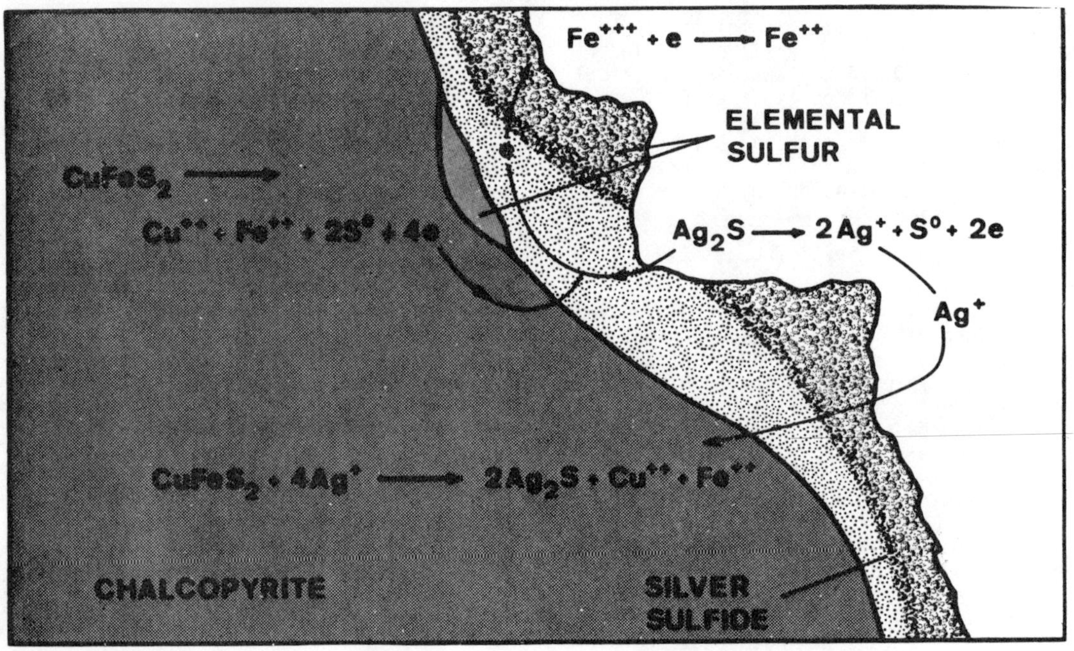

Figure 9. Schematic representation of the silver catalyzed ferric sulfate leaching of chalcopyrite with possible reactions of kinetic importance included.

EXPERIMENTAL TECHNIQUES

Two types of experiments were done in this study; leaching experiments in a stirred reactor and rest potentials measurements for a chalcopyrite electrode using a nulling circuit with a high impedance voltameter.

Materials

The chemicals used in this investigation were of reagent grade quality and the water used was distilled and deionized. For the silver catalysed ferric sulfate leaching experiments, monosize particles from a Pima flotation concentrate were prepared by wet screening and sizing with a Warman Cyclosizer. Elimination of particle aggregates was accomplished by dispersion with an ultrasonic probe. Gangue particles, primarily quartz, silicates and pyrite were removed by magnetic separation. The prepared samples were found to have the properties reported in Table II. (14,15)

Table II. Physical and Chemical Properties of Monosize Chalcopyrite Samples

Median Size (microns)	Chemical Analysis % Cu	% Fe	Specific Surface Area (m^2/gr)	Specific Gravity
4.8	26.5	29.5	0.480	3.75
13.5	30.5	28.5	0.195	3.92
28.0	31.5	30.5	0.092	3.92
51.2	33.5	30.5	0.031	3.92

The chalcopyrite sample used for rest potential measurements was from Transvaal, South Africa. The massive specimen is reported to be void of other sulfide phases such as cubanite ($CuFe_2S_3$) and pyrrhotite (FeS) both from mineralogical analysis and by x-ray analysis. (10) The sample of Transvaal Chalcopyrite was found to contain only minor amounts of quartz. A portion of this sample was cut into a rectangular parallelepiped the dimensions of which were 1 cm x 1 cm x 0.1 cm. One side of this parallelepiped was sputtered with a gold film to provide for low ohmic resistance and the sample was mounted in thermal plastic with provision for electrical contact via a mercury and platinum wire junction. The chemical reacting surface was polished with 1 μm diamond paste to clean the surface and establish a known area for subsequent current density determinations.

Leaching Experiments

Leaching experiments were carried-out under an atmospheric pressure of nitrogen using a four-necked unbaffled cylindrical reactor submerged in a constant temperature oil bath. The stirring speed was kept constant at 600 rpm, controlled with a Dyna-Mix variable speed stirrer. All experiments were carried-out in a total volume of 1 liter. Temperature of the bath was maintained within ±0.2°C. An initial two-hour acid wash (0.25 M H_2SO_4) of the monosize samples was necessary to remove and correct the kinetic data for acid soluble copper.

The following procedure was used to follow the silver catalysed acid ferric sulfate leaching of chalcopyrite.

1) A specified amount of ferric sulfate was dissolved with a minimum of the 0.25 M sulfuric acid solution. The ferric solution was heated separately to the desired temperature before pouring into the reactor.

2) The reactor was filled with the remaining amount of 0.25 M sulfuric acid solution and heated to the desired temperature at which time the monosize sample was introduced into the reactor.

3) After two hours of washing a 3 cc sample was withdrawn just before introducing the silver solution (~2 cc) necessary to achieve the desired silver concentration.

4) After the addition of the silver solution (30 sec), the ferric solution (prepared in step 1) was poured into the reactor. This was the starting point (time zero) for determination of the rate of reaction.

It should be pointed out that the sequence; solids, silver solution, and ferric solution was selected because it gave the most rapid kinetic response. All samples (3 cc) taken at timed intervals were cooled immediately in a container of ice. After cooling, the samples were centrifuged and the aqueous phase diluted as necessary and analysed for copper and/or ferrous concentrations. The extent of reaction was calculated in terms of the fraction of the copper extracted from the acid-washed sample.

Rest Potential Measurements

The rest potential measurements were made in a typical electrochemical cell (Figure 10) where the calomel reference electrode was separated by a Luggin capillary filled with Na_2SO_4 solution and the platinum button counter electrode was separated from the working Ag_2S-coated chalcopyrite electrode by a fritted glass membrane. The Ag_2S-coated electrode was prepared by reacting the chalcopyrite surface with a sulfuric acid silver nitrate solution. Based on depletion of the Ag^+ from solution a thickness of Ag_2S was calculated to be 6.8×10^{-6} cm. The electrochemical cell was purged with argon and submerged in the constant temperature oil bath mentioned previously.

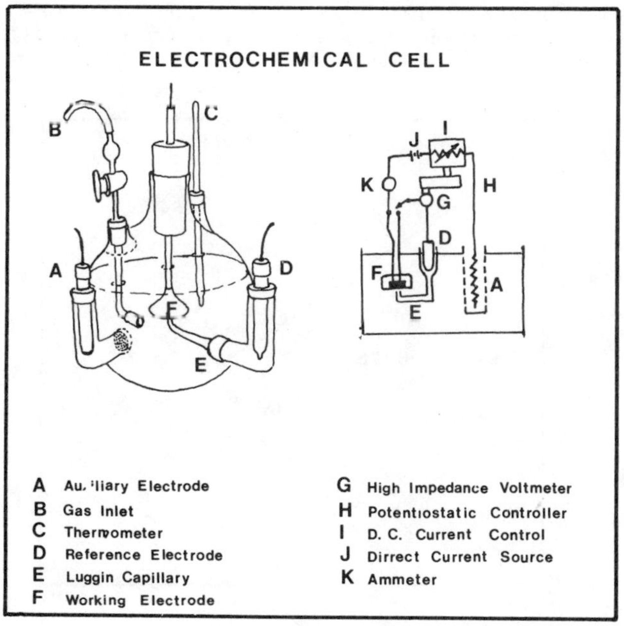

Figure 10. Schematic diagram of the electrochemical cell used for the rest potential measurements of silver coated chalcopyrite.

The electrode was rotated and controlled by a Dyna-Mix variable speed stirrer. The rest potentials were measured using a Princeton Applied Research Model 173 potentiostat and a Model 175 potential sweep generator.

EXPERIMENTAL RESULTS AND DISCUSSION

New rate data for the silver catalysed ferric sulfate leach of chalcopyrite is reported and analysed together with rest potential measurements in terms of an electrochemical reaction mechanism.

Leaching Experiments

In addition to previous experimental results (14) summarized in the introduction, other more recent experiments have further revealed the nature of the silver catalysed ferric sulfate leach. Rate data is presented as a function of concentration variations and as a function of temperature.

Concentration Effects

Unlike the uncatalysed ferric sulfate leach, (5) a significant dependence on ferric sulfate addition was observed for the silver catalysed system. This effect was not studied extensively. However, a reaction order of approximately 0.5 was estimated with respect to ferric ion. Anomalous results were observed at higher ferric concentrations (>1 M Fe^{+3}) which may involve transport problems or solution chemistry complications. Figure 11 shows the extraction of copper as function of time with ferric ion concentration as the parameter. Note that under these circumstances of higher silver addition, 4×10^{-3} M, the linear reaction kinetics observed previously (see Figure 6) do not prevail. These nonlinear results, besides demonstrating the ferric ion dependence, may

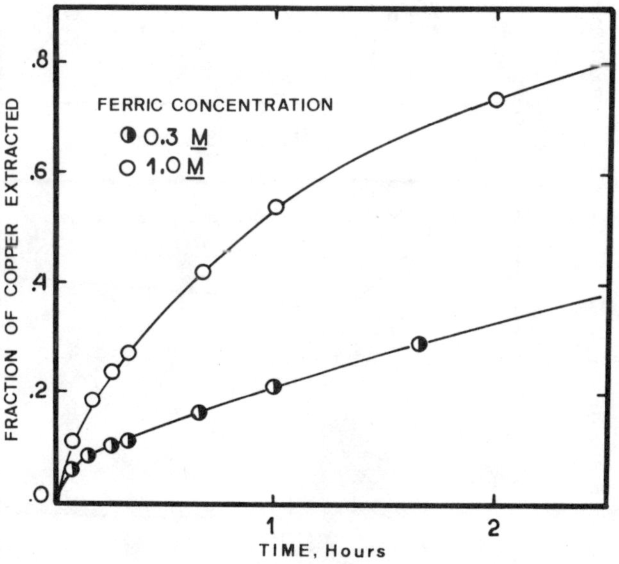

Figure 11. Influence of the initial ferric concentration on the extent of the catalysed reaction for 28 μm particles at 0.25 M H_2SO_4, 4×10^{-3} M Ag^+, 0.3 percent solids, 600 rpm and 90°C.

also reflect a reaction resistance due to transport through the reacting silver sulfide film which would be expected for a thicker layer of silver sulfide. However, the initial reaction kinetics for 1 M $Fe_2(SO_4)_3$ (open circles) are about the same as those observed for lower levels of silver addition, 10^{-3}M at 1 M $Fe_2(SO_4)_3$ (open circles) presented in Figure 6. The departure from linearity at the higher silver concentration and the slower rate of reaction also suggests that Ag^+ may inhibit the reaction via back reaction kinetics just as had been observed in the case of Fe^{++}. Recall that an addition of 0.5 M $FeSO_4$ essentially passivates the system and very little reaction is achieved as was shown in Figure 8. To test further the notion that the Ag^+ concentration can diminish the reaction rate, diagnostic experiments were attempted; the results of which are shown in Figure 12. Again, it can be noted that the initial rate is essentially independent of the amount of silver sulfide formed but for higher levels of silver addition (more silver sulfide on the surface initially), the rate decreases during intermediate reaction times (10 min. to 60 min.). Furthermore, the subsequent injection of silver (300 ppm) into the system after the catalysed ferric sulfate reaction had been initiated at 108 ppm silver (closed circles) reduces the rate considerably.

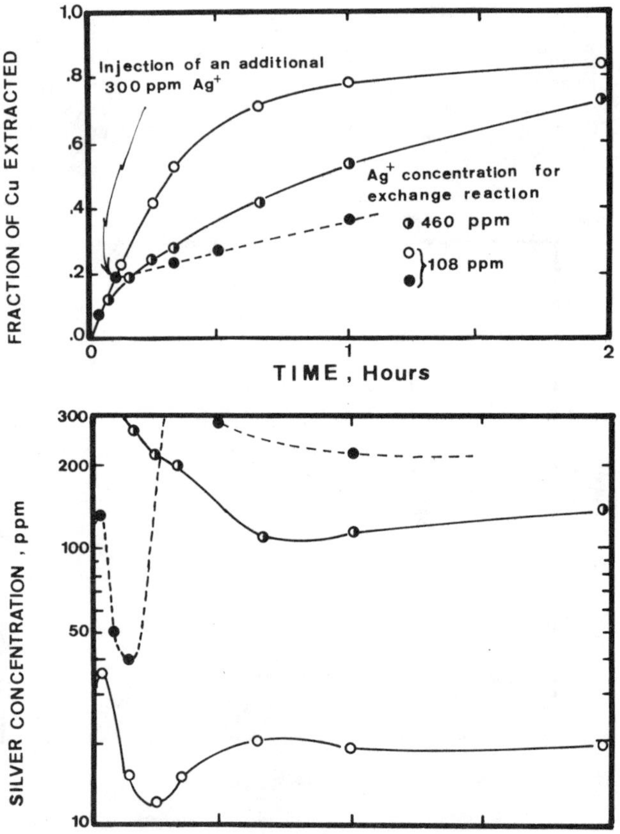

Figure 12. The effect of silver concentration on the rate and extent of reaction for 28 μm particles at 0.25 M H_2SO_4, 1 M Fe^{+3}, 0.3 percent solids, 600 rpm and 90°C.

Temperature

A significant temperature dependence on the reaction rate is observed for the catalysed ferric sulfate leach of the 4.8 μm sample as shown in Figure 13. In this particular set of data the ferric concentration is of sufficient excess so as to be considered constant which facilitates analysis of the experimental data. In addition, it is very important, as discussed earlier, to evaluate the temperature effect at constant, steady-state silver concentration (~25 ppm in aqueous phase). An Arrhenius plot of the data is presented in Figure 14 and an apparent activation energy of 15.7 kcal/mole (65.7 kJ/mole) is observed for this rate process.

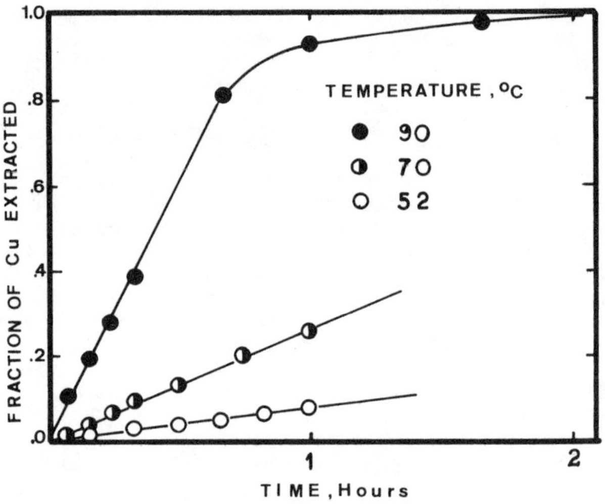

Figure 13. The influence of temperature on the rate and extent of the catalyzed reaction for 4.8 μm particles at 0.25 M H_2SO_4, 1 M Fe^{+3}, 0.3 percent solids and 600 rpm. (0.17 μm Ag_2S layer).

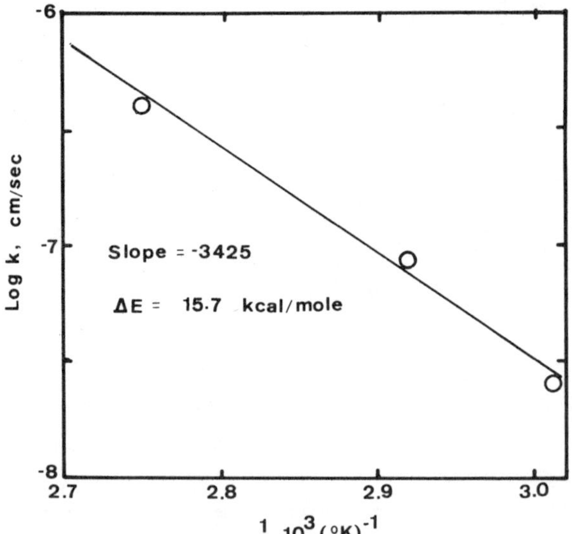

Figure 14. Arrhenius plot of the reaction velocity constant for the silver catalysed leaching of chalcopyrite with the 4.8 μm monosize chalcopyrite sample after having formed 0.17 μm layer of Ag_2S. Data taken from Figure 13.

The experimental reaction velocity constants used in construction of the Arrhenius plot were calculated assuming a first order rate process with respect to the reactant concentration, the Fe^{+3} concentration.

Values for the reaction velocity constant are on the order of 2.21×10^{-8} to 3.48×10^{-7} cm/sec, orders of magnitude less than the appropriate mass transfer coefficients ($\sim 10^{-2}$ cm/sec) which would be predicted if the rate were limited by diffusion in the mass transfer boundary layer.

These results together with those results published earlier (14) and summarized in the introduction (p.12) seem to sustain the original hypothesis that the silver catalysed reaction of chalcopyrite with ferric sulfate is limited by intermediate electrochemical reaction,

$$Ag_2S_{(Surface)} + 2Fe^{+3} \rightarrow 2Ag^{+}\ 2Fe^{++} + S^{o} \qquad (3)$$

Electrochemistry

The electrochemistry of the silver catalysed ferric sulfate leaching of chalcopyrite reaction is examined in terms of rest potential measurements used to explain the leaching kinetics in the context of the Butler-Volmer equation.

Rest Potential Measurements

The rest potential of the Ag_2S-coated chalcopyrite electrode was measured as a function of Ag^{+} concentration at 80°C. The linear relationship presented in Figure 15 shows that the electrode exhibits Nernstian behavior with an estimated standard potential of 787 mv and a slope of 69.8 mv per decade of concentration change. These results suggest that the electrode may be measuring the following half cell reaction,

$$Ag_2S \rightleftarrows 2Ag^{+} + S^{o} + 2e \qquad (4)$$

the standard potential for which is reported to vary from 799 mv to 1000 mv at 25°C. At higher temperatures lower rest potentials would be expected due to a negative temperature coefficient. (18) Similarly, the observed slope is close to that predicted at 80°C from the Nernst equation, 70.1 mv per decade change in concentration.

Other results shown in Figure 16 support this hypothesis inasmuch as the potential of the Ag_2S-coated chalcopyrite electrode is insensitive to Fe^{++} additions; whereas the potential of the uncoated chalcopyrite electrode exhibits a very definite dependence on the Fe^{++} addition. These results suggest that the Ag_2S film completely seals the chalcopyrite surface and a Nernstian half cell potential for Ag_2S is measured rather than a mixed potential involving a chalcopyrite half cell reaction.

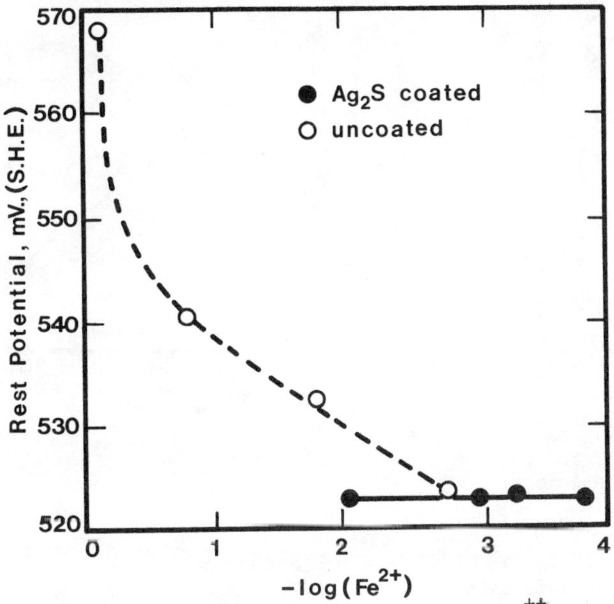

Figure 16. Rest potential as a function of Fe^{++} concentration at 0.25 M H_2SO_4 and 80°C.

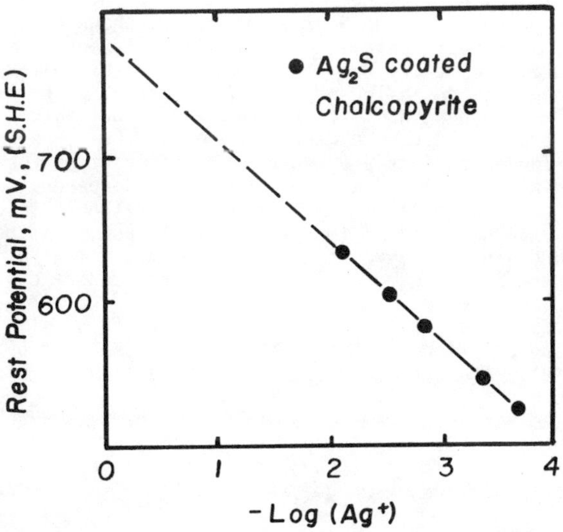

Figure 15. Rest potential of Ag_2S-coated chalcopyrite as a function of Ag^{+} concentration at 0.25 M H_2SO_4 and 80°C.

Reaction Mechanism

The key to understanding the rate enhancement which occurs during silver catalysis in ferric sulfate leaching of chalcopyrite is the very dramatic change in the nature of the sulfur deposit as shown by the photographs presented in Figure 7. The sulfur layer no longer inhibits the reaction progress as it does in the uncatalysed ferric sulfate leach. This observation together with the following data suggest that the leaching rate is controlled by the rate of an intermediate electrochemical reaction in the Ag_2S film;

$$Ag_2S + 2Fe^{+3} \rightarrow 2Ag^{+} + 2Fe^{++} + S^{o} \qquad (3)$$

1) magnitude of reaction velocity approximately 10^{-8} cm/sec, orders of magnitude less than predicted mass transfer coefficients (10^{-2} cm/sec)
2) independence of reaction rate on particle size
3) complex dependence of reaction rate on Ag^{+}, Fe^{+3}

and Fe^{++} concentrations and independence of reaction rate on Cu^{++} concentration.

4) Apparent activation energy 15.7 kcal/mole (65.7 kJ/mole).

Thermodynamics. The standard free energy change for this particular reaction (Equation 3) at 90°C is $\Delta G^\circ_{90°C} \simeq 5.0$ kcal/mole which was determined by interpolation since no thermodynamic data was available at this particular temperature. (19,20) A plot of the standard free energy of reaction vs. temperature is presented in Figure 17.

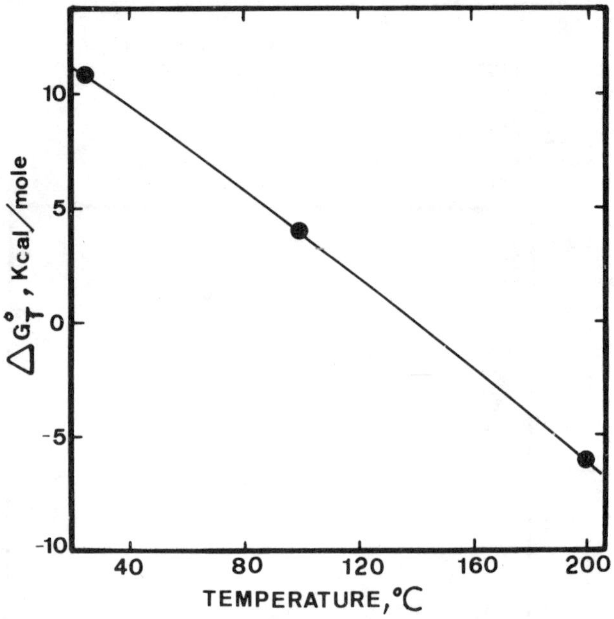

Figure 17. Standard free energy change as a function of temperature for the intermediate electrochemical reaction.

As can be seen under experimental conditions the reaction is close to equilibrium and as a result will be quite sensitive to slight changes in temperature and the composition of the aqueous phase. Complete evaluation of the possibility of rate control by this intermediate electrochemical reaction requires further thermodynamic analysis. In particular, the activities of the various species must be specified and the free energy of reaction determined at the temperatures of interest. By this approach conditions under which the intermediate reaction is thermodynamically feasible can be identified. The results of this thermodynamic analysis are presented in Figure 18. The set of surfaces for different temperatures specify the thermodynamic stability for various solution compositions.

Compositions which lie above a given surface correspond to conditions under which the intermediate electrochemical reaction should not occur. Whether or not the reaction should occur under specified temperature and aqueous phase composition can be determined if the activities of the reactants and products are known. These values can be determined using the University of Utah Solution Chemistry Program. For example, let's examine the experiment involving 0.5 M $FeSO_4$ which was presented in Figure 8. Computer analysis of the solution chemistry of this system results in the following activities:

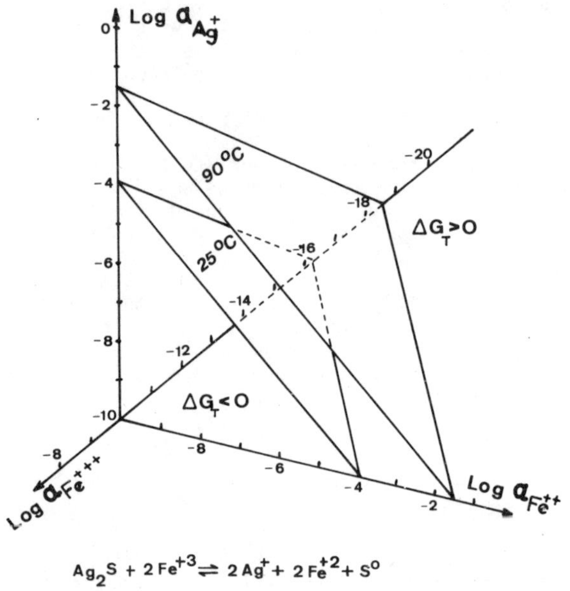

Figure 18. Equilibrium surfaces of the intermediate electrochemical reaction as a function of solution composition at 25°C and 90°C.

$$a_{Fe^{+3}} = 1.1 \times 10^{-3}$$
$$a_{Fe^{+2}} = 3.55 \times 10^{-2}$$
$$a_{Ag^{+}} \simeq 4.64 \times 10^{-4}$$

Evaluation of the free energy of reaction for these activities results in a value of -0.9 kcal/mole which indicates that the reaction should occur, but as can be seen in Figure 8 the reaction is exceedingly slow.

The independence of the rate on initial particle diameter suggests that the reaction kinetics are controlled by electrochemical discharge at the surface of the individual Ag_2S crystallites as shown by the schematic representation of the reaction mechanism in Figure 9. The crystallites act as short circuited microelectrodes allowing for the discharge of Fe^{+3} and the release of Ag^+. Under these circumstances the kinetics could show a dependence on the Fe^{+3}, Fe^{+2} and Ag^+ concentrations and would be independent of the initial particle size, i.e. the rate is dependent on the internal area of the crystallites rather than on the external area of the particle. In addition the strong sensitivity of the leaching rate to the Fe^{++} concentration suggests that either the back reaction kinetics are important or that the dependence arises from a mixed potential developed in the Ag_2S film during reaction. SEM photographs indicate that elemental sulfur replaces the Ag_2S crystallites forming a porous nonprotective layer.[2] The silver ions generated act as a transfer agent while the silver sulfide film, necessary to prevent extensive sulfur formation at the chalcopyrite surface, is continually restored in the exchange reaction.

The highly irreversible exchange reaction would not be rate limiting due to the rate dependence on ferric ion concentrations and the rate independence on the cupric ion concentration. Eventually, after ~50%

reaction, the rate becomes nonlinear (see Figure 6) and apparently enough sulfur forms to provide significant resistance to mass transport.

Butler-Volmer Equation. If the intermediate electrochemical reaction (Equation 3) is considered in terms of its respective half cell reactions:

anodic half cell,

$$Ag_2S \rightarrow 2Ag^+ + S^o + 2e^- \qquad (4)$$

cathodic half cell,

$$Fe^{++} \rightarrow Fe^{+3} + e \qquad (5)$$

which occur at distinct anodic and cathodic sites and are short-circuited through the Ag_2S film, then the reaction kinetics can be evaluated in terms of the Butler-Volmer equation. (18) If ohmic potentials are small then the system developes a mixed potential (E_m), a compromise potential, under which conditions the anodic current (i_a) equals the cathodic current (i_c). Assuming a first order dependence on reacting species as well as single electron transfer processes and transfer coefficients equal to 0.5, the partial currents for the respective half cells would be;

$$i_a = A_a z_a F \vec{k}_a \exp\left(\frac{FE_m}{2RT}\right) - A_a z_a F(Ag^+)\overleftarrow{k}_a \exp\left(-\frac{FE_m}{2RT}\right) \qquad (6)$$

and

$$i_c = A_c z_c F(Fe^{++})\vec{k}_c \exp\left(\frac{FE_m}{2RT}\right) - A_c z_c F(Fe^{+3})\overleftarrow{k}_c \exp\left(-\frac{FE_m}{2RT}\right) \qquad (7)$$

Since these half cell reactions have been assumed to be short-circuited,

$$i_a = -\, i_c \qquad (8)$$

it follows that

$$\exp\left(\frac{FE_m}{RT}\right) = \frac{A_c z_c (Fe^{+3})\overleftarrow{k}_c + A_a z_a (Ag^+)\overleftarrow{k}_a}{A_c z_c (Fe^{++})\vec{k}_c + A_a z_a \vec{k}_a} \qquad (9)$$

The reaction rate can then be formulated from the expression for either partial current in terms of the mixed potential as well as reactant and product concentrations,

$$-\frac{dn}{dt} = \frac{i_a}{Z_a F} = A_a \vec{k}_a \left[\frac{A_c z_c (Fe^{+3})\overleftarrow{k}_c + A_a z_a (Ag^+)\overleftarrow{k}_a}{A_c z_c (Fe^{++})\vec{k}_c + A_a z_a \vec{k}_a}\right]^{1/2} - A_a (Ag^+)\overleftarrow{k}_a \left[\frac{A_c z_c (Fe^{++})\vec{k}_c + A_a z_a \vec{k}_a}{A_c z_c (Fe^{+3})\overleftarrow{k}_c + A_a z_a (Ag^+)\overleftarrow{k}_a}\right]^{1/2} \qquad (10)$$

Apparent reaction orders are difficult to determine from such an expression which includes consideration of back reaction kinetics, which is reasonable in terms of the thermodynamic analysis presented previously. This complexity together with possible variation of anodic and cathodic areas during the course of the reaction makes it difficult to simplify the kinetic expression. However, it should be evident that, generally, increased Fe^{+++} concentration should enhance the rate of reaction, whereas increased Fe^{++} and Ag^+ concentrations should reduce the rate.

Evaluation of the Butler-Volmer equation for conditions under which back reaction kinetics are not significant, simplifies the rate expression to a certain extent;

$$-\frac{dn}{dt} = \frac{i_a}{z_a F} = A_a \vec{k}_a \frac{A_c z_c (Fe^{+3})\overleftarrow{k}_c}{.A_a z_a \vec{k}_a} \qquad (11)$$

in which case it is seen that the rate is independent of Fe^{++} and Ag^+ concentrations and has a one half order dependence on the Fe^{+3} concentration. In view of the experimental results, such a condition might only prevail for initial reaction kinetics and the limited data presented in Figure 11 would tend to support this premise.

SUMMARY AND CONCLUSIONS

Silver catalysis in acid ferric sulfate leaching of chalcopyrite has the following overall reaction stoichiometry:

$$CuFeS_2 + 4Fe^{+3} \overset{Ag^+}{\rightleftarrows} Cu^{+2} + 5Fe^{+2} + 2S^o$$

Two intermediate reactions in this catalytic reaction have been identified: first, an *exchange reaction* in which an argentite reaction product is deposited at the chalcopyrite surface;

$$CuFeS_2 + 4Ag^+ \rightleftarrows 2Ag_2S_{(Surface\ CuFeS_2)} + Cu^{+2} + Fe^{+2}$$

second, an *intermediate electrochemical reaction* of Fe^{+3} with the Ag_2S film producing elemental sulfur;

$$Ag_2S_{(Surface\ CuFeS_2)} + 2Fe^{+3} \rightleftarrows 2Ag^+ + 2Fe^{++} + S^o$$

The initial kinetics of the *exchange reaction* appear to be limited by transport through the mass transfer boundary layer as substantiated by the reaction order (1st order with respect to silver, the magnitude of the reaction velocity constant ($\sim 2x10^{-2}$ cm/sec), and the temperature dependence (activation energy less than 5 kal/mole).

However, the overall leaching rate appears to be controlled by the *intermediate electrochemical reaction* of Fe^{+3} with Ag_2S crystallites as evidenced by the fact that:

1. The rate of reaction has been shown to be dependent on Fe^{+3}, Fe^{+2} and Ag^+ concentrations, but independent of the Cu^{++} concentration.
2. The apparent reaction order with respect to the Fe^{+3} concentration)0.3 M to 1.0 M Fe^{+3} for initial kinetics was estimated to be one half.
3. The rate has been found to be independent of the initial chalcopyrite particle size apparently because the reaction occurs at the surface of the Ag_2S crystallites in the argentite surface film rather than at the surface of the particles.

4. An activation energy of 15.7 kcal/mole was calculated for the initial stage of the leaching reaction (up to 50% reacted), an activation energy consistent for rate control by an electrochemical reaction.

Unlike the uncatalysed ferric sulfate leach in which elemental sulfur forms a tight, tenacious, coherent film resistant to transport, the silver catalysed ferric sulfate leach results in a porous, non-protective sulfur film due to sulfur growth on Ag_2S crystallites. It seems that only after extensive reaction (≳50%) does sufficicent sulfur form to result in a significant resistance to the rate of reaction. The increased anodic rest potentials measured for the Ag_2S coated chalcopyrite electrode suggest a shift in the polarization curve such that the reaction would be driven by a larger mixed potential. Such a phenomenon may account for the porous sulfur reaction product as has been observed to form in the leaching of chalcopyrite with stronger oxidants.

ACKNOWLEDGEMENTS

The authors wish to acknowledge the LASPAU fellowship awarded to Mr. Portillo for his studies at the University of Utah as well as financial assistance from the University of Utah Research Committee for the purchase of supplies and equipment.

REFERENCES

1. Lowe, D.F., "The Kinetics of the Dissolution Reaction of Copper and Copper-Iron Sulfide Minerals Using Ferric Sulphate Solutions," Ph.D. thesis, University of Arizona, Tucson, Arizona, 1970.

2. Linge, H.G., "A Study of Chalcopyrite Dissolution in Acid Ferric Nitrate by Potentiometric Titration," Hydrometallurgy 2, 1976, p. 51-64. See also, Ling, H.G., "Reactivity Comparison of Australian Chalcopyrite Concentrates in Acidified Ferric Solution," Hydrometallurgy 2, 1976, p. 219-233.

3. Dutrizac, J.E. et. al., "The Kinetics of Dissolution of Synthetic Chalcopyrite in Aqueous Acid Ferric Sulfate Solutions," Metallurgical Transactions, 245, 1969, p. 955-959.

4. Beckstead, L.W., et. al., "Acid Ferric Sulfate Leaching of Attritor-Ground Chalcopyrite Concentrates," Extractive Metallurgy of Copper, Vol. 2, The Metallurgical Society of A.I.M.E., New York, 1976, p. 611-632.

5. Munoz, P.B., Miller, J.D. and Wadsworth, M.E., "Reaction Mechanism for the Acid Ferric Sulfate Leaching of Chalcopyrite," Metallurgical Transactions B, 10B, June 1979, p. 149-158.

6. Sullivan, J.D., "Chemical and Physical Features of Copper Leaching," Trans. AIME, 106, 1933, p.515-546.

7. Jones, D.L. and Peters, E., "The Leaching of Chalcopyrite with Ferric Sulfate and Ferric Chloride," Extractive Metallurgy of Copper, Vol. 2, The Metallurgical Society of A.I.M.E., New York, 1976, p.633-653.

8. Baur, J.P., Gibbs, H.L. and Wadsworth, M.E., "Initial-Stage Sulfuric Acid Leaching Kinetics of Chalcopyrite Using Radiochemical Techniques," U.S.B.M., R.I. 7823, 1974.

9. Peters, E., "The Physical Chemisty of Hydrometallurgy," Internat. Symp. on Hydromet. AIME New York, 1973, p. 205-228.

10. Warren, G. and Wadsworth, M.E., "Hydrometallurgy of Fine Mineral Particles," International Symposium of Fine Mineral Particles, A.I.M.E., Las Vegas, Nevada, Feb. 1979.

11. Pawlek, F.E., "The Influence of Grain Size and Mineralogical Composition on the Leachability of Copper Concentrates," Extractive Metallurgy of Copper, Editors Yannopoulos and Agarwal, A.I.M.E. 1976.

12. McElroy, R.O. and Duncan. D.W., British Columbia Research Council, Vancouver, Canada V6F 212, The Silver Institute Letter, 5, No. 8, Sept. 1975.

13. Snell, G.J. and Sze, M.C., "New Oxidative Leaching Process Uses Silver to Enhance Copper Recovery," EMJ, 178, No. 10, October 1977, p. 100-105.

14. Miller, J.D. and Portillo, H.W., "Silver Catalysis in Ferric Sulfate Leaching of Chalcopyrite," XIII International Mineral Processing Congress, Warsaw, Poland, 1979, p. 691-742.

15. Portillo, H.W., "Silver Catalysis in the Acid Ferric Sulfate Leaching of Chalcopyrite," M.S. Thesis, Dept. Metallurgy and Metallurgical Engineering, University of Utah, Salt Lake City, Utah 1978.

16. Wadsworth, M.E. and Miller, J.D., "Hydrometallurgical Processes." Rate Processes of Extractive Metallurgy, Plenum Press, 1978, p. 222.

17. Ranz, W.E.,and Marshall, W.R., Chem Eng. Progr., 48, 1952, p. 141.

18. Vesely, J., Jensen, O.J., and Nicolaisen, B., "Ion-Selective Electrodes Based on Silver Sulfide," Analytica Chimica Acta, 62, 1972, p. 1-13.

19. Helgeson, H.C., "Thermodynamics of Hydrothermal Systems at Elevated Temperatures and Pressures," American Journal of Science, 267, Summer 1969, p. 729-834.

20. Robie, R.A. and Waldbaum, D.R., "Thermodynamic Properties of Minerals and Related Substances at 25°C and One Atmosphere Pressure and at Higher Temperatures," Geological Survey Bulletin 1269.

21. Vetter, K.J., Electrochemical Kinetics Theoretical and Experimental Aspects, Academic Press, N.Y., 1967.

Chapter 28

A MODEL OF THE CHEMISTRY OF THE DUMP LEACHING OF CHALCOPYRITE

KNona C. Liddell* and Renato G. Bautista

Ames Laboratory USDOE and Department of Chemical Engineering
Iowa State University, Ames, Iowa 50011

ABSTRACT

A model of the reaction of chalcopyrite, $CuFeS_2$, with Fe^{3+} and with O_2 is considered. Homogeneous solution phase reactions are assumed to be always at equilibrium throughout the dissolution of the mineral. The dissociation of HSO_4^- and several complexation reactions involving Cu^{2+}, Fe^{2+} and Fe^{3+} are incorporated in the model equations. The relative importance of Fe^{3+}, O_2, and H^+ as oxidants and the role of solution equilibria are assessed and the implications for leaching practice discussed.

*Department of Chemical Engineering, Montana State University, Bozeman, Montana 59717

A MODEL OF THE CHEMISTRY OF THE DUMP LEACHING OF CHALCOPYRITE

KNona C. Liddell* and Renato G. Bautista

Ames Laboratory USDOE and Department of Chemical Engineering

Iowa State University, Ames, Iowa 50011

A model of the reaction of chalcopyrite, $CuFeS_2$, with Fe^{3+} and with O_2 is considered. Homogeneous solution phase reactions are assumed to be always at equilibrium throughout the dissolution of the mineral. The dissociation of HSO_4^- and several complexation reactions involving Cu^{2+}, Fe^{2+} and Fe^{3+} are incorporated in the model equations. The relative importance of Fe^{3+}, O_2, and H^+ as oxidants and the role of solution equilibria are assessed and the implications for leaching practice discussed.

INTRODUCTION

Chalcopyrite has recently received much attention because it is the most abundant copper sulfide mineral and is consequently extremely important to the copper industry. Moreover, processing chalcopyrite ore poses special problems because this mineral is generally less reactive than other copper sulfides. We have been particularly interested in the ambient temperature reactions by which chalcopyrite dissolves in leach dumps and have developed a mathematical model of the changes in solution concentrations during its dissolution by Fe^{3+}, O_2, and H^+. We focus our attention here primarily on the relative efficacy of chalcopyrite dissolution by these oxidants.

The derivation and application of the model equations are discussed elsewhere[1,2,3] in greater detail but we summarize a few highlights here. Leach liquor contains $Fe_2(SO_4)_3$, $FeSO_4$, H_2SO_4, and $CuSO_4$. For such solutions, two reactions have been identified by which chalcopyrite is dissolved at ambient temperatures[4,5]:

$$CuFeS_2 + 4Fe^{3+} \rightarrow Cu^{2+} + 5Fe^{2+} + 2S^0 \qquad (1)$$

$$CuFeS_2 + O_2 + 4H^+ \rightarrow Cu^{2+} + Fe^{2+} + 2S^0 + 2H_2O \qquad (2)$$

When the equilibrium constants for these reactions are compared with mass action quotients computed using data on dump effluent concentrations, it is evident that neither reaction is at equilibrium under dump leaching conditions. There are a large number of acid-base and complexation reactions going on in the leach liquor, however, and such aqueous phase reactions are very rapid with respect to the heterogeneous dissolution reactions. As the mineral dissolves, the concentrations in the solution can be considered to adjust themselves in such a way that chemical equilibrium within the solution is maintained. This is the physical reasoning behind the model. For each of the equilibrium reactions, the model contains an equation of the form

$$0 = \sum_j \frac{\nu_j n_j}{m_j} \qquad (3)$$

where ν_j is the stoichiometric coefficient of the jth solution phase species, n_j is the number of moles of the jth species formed per mole of $CuFeS_2$ reacting and m_j is the molality of the jth species. We have considered independent equilibria among H^+, HSO_4^-, SO_4^{2-}, Fe^{3+}, $FeOH^{2+}$, $Fe(OH)_2^+$, $Fe_2(OH)_2^{4+}$, $FeSO_4^+$, $Fe(SO_4)_2^-$, $FeHSO_4^{2+}$, Fe^{2+}, $FeSO_4^0$, $FeHSO_4^+$, Cu^{2+}, and $CuSO_4^0$. These equations are supplemented by material balances for Fe(III), Fe(II), Cu(II), and SO_4^{2-}, a charge balance and a kinetic equation based on the best available reaction rate data.[5] Following an initial computation of the equilibrium concentrations before any dissolution has taken place, the model equations are solved for successive small increments of chalcopyrite dissolution. The fraction of the total amount of copper dissolved due to reaction (2) was treated as an adjustable parameter and specified for each increment.

RESULTS AND DISCUSSION

At the start of each computation, analytical concentrations (molality) of $Fe_2(SO_4)_3$, $FeSO_4$, H_2SO_4 and $CuSO_4$ were specified. These concentrations correspond physically to the total concentrations in the leach liquor as it is applied to a leach dump just before dissolution begins. Trivalent iron was taken to be depleted when the sum of the concentrations of all the Fe(III) species had dropped to 10^{-4}m.

Although estimates of the dissolved oxygen content of the leach liquor in the interiors of dumps have been made[6,7], this quantity has apparently not actually been measured. It is likely, however, to be highly variable even within a single dump. Accordingly, we have made the simplifying assumption that oxygen never becomes the limiting reagent in chalcopyrite dissolution.

Throughout the dissolution process, the fraction of the $CuFeS_2$ dissolution due to reaction (2) was taken to be a constant. The development of the kinetic equation used is discussed in detail elsewhere.[2]

Precipitation of Fe^{3+} is an important problem in dump leaching since it causes reductions in dump

*Department of Chemical Engineering, Montana State University, Bozeman, Montana 59717

permeability and fouling of pumps and piping. Thermodynamic information on compounds that could form as Fe^{3+} precipitates from an acidic sulfate solution is sketchy but K_{sp} values are available for several such compounds, among them $KFe_3(SO_4)_2(OH)_6$,[8] amorphous $Fe(OH)_3$[9] and α-FeOOH.[9] Jarosite formation is critically dependent on the K^+ concentration, which must be very small if precipitation of this compound from leach liquor is to be avoided. We have treated the K^+ concentration as a second adjustable parameter. The details of the conditions under which jarosite precipitates in leaching operations are presented in other publications.[2,10] Goethite (α-FeOOH) is formed by the aging of amorphous $Fe(OH)_3$[11] and is not discussed further here. In the computations whose results are discussed below, the dissolution process is considered to be halted by precipitation of amorphous ferric hydroxide (K_{sp} = 2.51 x 10^{-39} at 25°C and ionic strength 3).[9]

The model equations were solved for successive dissolution increments until either Fe(III) was depleted or the K_{sp} for $Fe(OH)_3$ was exceeded. A total Fe(III) concentration of 10^{-4}m was chosen as the criterion for depletion of this reactant.

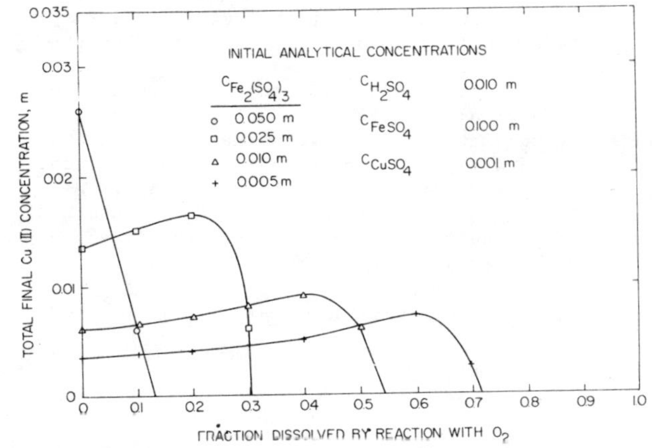

FIGURE 1. Final Cu(II) concentration as a function of F_{O_2} for various initial $Fe_2(SO_4)_3$ concentrations.

Fig. 1 shows the final total concentration of Cu(II) for several values of the initial ferric sulfate concentration. The final copper concentrations have been plotted against the fraction of $CuFeS_2$ dissolved according to reaction (2). At 0.025 m $Fe_2(SO_4)_3$, for example, the final copper concentration increases as this fraction increases from 0.0 to 0.2. Fe(III) is depleted under these conditions. This trend is to be expected because O_2 has been assumed to be available in unlimited amounts and the effect of this is to "extend the life" of the Fe(III), achieving a higher fractional conversion of the mineral before the trivalent iron is used up. Increasing the fraction to 0.3, however, leads to rapid precipitation of $Fe(OH)_3$. As reaction (2) proceeds, the H^+ concentration drops and the OH^- concentration rises in order to keep the solution in a state of chemical equilibrium. With the Fe^{3+} concentration still fairly high, the K_{sp} value for $Fe(OH)_3$ is exceeded. Thus at higher values of F_{O_2} (the fraction of the chalcopyrite reacted in reaction (2) or the ratio of the number of moles of O_2 reacted in reaction (2) to the total number of moles of $CuFeS_2$ reacted), precipitation of $Fe(OH)_3$ occurs before much of the Fe(III) reacts and before much of the chalcopyrite dissolves. The same trend is seen at lower values of the initial ferric sulfate concentration. As the initial $Fe_2(SO_4)_3$ level is decreased, however, the maxima in the curves occur at lower Cu(II) concentrations and precipitation occurs at greater F_{O_2} values. Increasing the initial $Fe_2(SO_4)_3$ concentration to 0.05 m gives a curve on which precipitation occurs at an F_{O_2} value of only 0.1. Thus having high Fe(III) concentrations, and simultaneously, large amounts of O_2 reacting is not advantageous and will lead to rapid clogging of a dump. If high levels of trivalent iron are to be used, low levels of dissolved oxygen are desirable. Increasing the starting ferric sulfate concentration necessitates closer control of the dissolved oxygen concentration, a requirement difficult to achieve.

The initial analytical concentration of sulfuric acid also dramatically affects the formation of $Fe(OH)_3$. If the H_2SO_4 concentration is reduced to 0.005 m, a solution that also contains 0.1 m $FeSO_4$, 0.001 m $CuSO_4$ and $Fe_2(SO_4)_3$ is not stable unless the ferric sulfate concentration is quite small; precipitation occurs if the $Fe_2(SO_4)_3$ concentration is 0.01 m or higher. Fig. 2 shows the total final copper

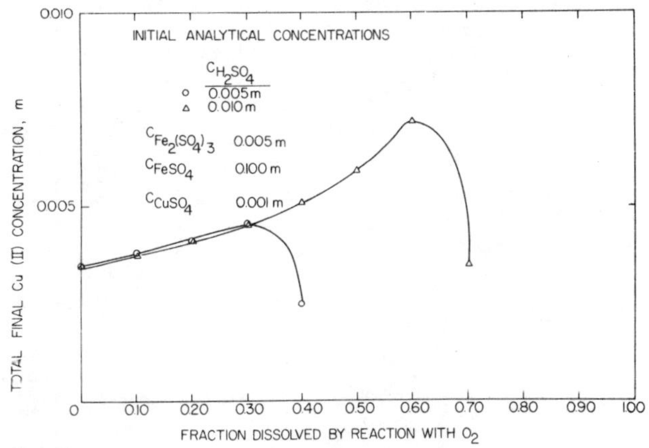

FIGURE 2. Final Cu(II) concentration. Shows effect of initial H_2SO_4 concentration.

concentration for two different values of the sulfuric acid concentration; the dependence of the Cu(II) recovery on F_{O_2} is shown for an initial $Fe_2(SO_4)_3$ concentration of 0.005 m. The initial portions of the curves are almost identical, but at the lower value of the sulfuric acid concentration, 0.005 m, $Fe(OH)_3$ precipitates when between 30 and 40% of the copper dissolution is due to the reaction of chalcopyrite with O_2 and H^+. With an initial sulfuric acid concentration of 0.01 m, the precipitation takes place when the F_{O_2} value is between 0.6 and 0.7. Doubling the initial H_2SO_4 concentration approximately doubles the maximum amount of copper that can be dissolved before precipitation of ferric hydroxide sets in. High levels of sulfuric acid are therefore valuable in preventing precipitation but the acid must not be consumed in dissolving chalcopyrite by reaction (2) or its protective effect is lost.

Changing the initial $FeSO_4$ concentration indirectly affects the dissolution process through the sulfate and charge balances and the various equilibria that involve the sulfate ion. For example, a solution containing 0.05 m $Fe_2(SO_4)_3$, 0.001 m $CuSO_4$, 0.01 m H_2SO_4 and 0.2 m $FeSO_4$ cannot be prepared because $Fe(OH)_3$ would precipitate from it. However, if the analytical concentration of $FeSO_4$ were reduced to 0.1 m, precipitation would not take place. The major effect of the initial ferrous sulfate level on the copper recovery is to shift the point at which $Fe(OH)_3$ precipitates to smaller F_{O_2} values. With 0.025 m $Fe_2(SO_4)_3$, 0.001 m $CuSO_4$, 0.01 m H_2SO_4 and 0.1 m $FeSO_4$, precipitation occurs between F_{O_2} = 0.2 and 0.3. This is shown in Fig. 1. Increasing the $FeSO_4$ concentration to 0.2 m causes this critical range of F_{O_2} values to decrease to 0.1 to 0.2. The total copper recovered when Fe(III) is depleted is virtually the same in the two cases, however. Our results show that, with the other analytical concentrations the same, an increase in the $FeSO_4$ concentration leads to a decrease in the Fe^{3+} equilibrium concentration by causing more $FeSO_4^+$ and $Fe(SO_4)_2^-$ to be formed. More $FeSO_4$ also produces a drop in the equilibrium concentration of H^+, principally by shifting the SO_4^{2-}/HSO_4^- equilibrium. The second effect is the larger and ferric hydroxide precipitates more readily.

The solution's pH at Fe(III) depletion gives a good indication of how the starting conditions affect the H^+ concentration. The effect of the initial H_2SO_4 concentration is shown in Fig. 3. The pH

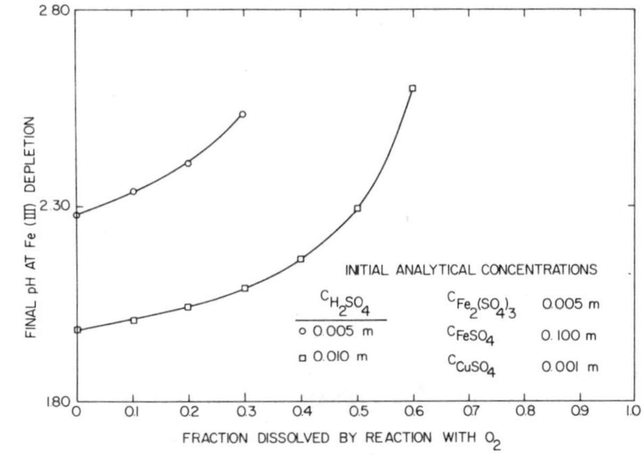

FIGURE 3. pH at depletion of Fe(III). Shows effect of initial H_2SO_4 concentration.

values at the beginning of the dissolution are 2.25 for an initial sulfuric acid concentration of 0.005 m and 1.97 for 0.010 m. When reaction (2) does not occur, the pH changes only slightly during dissolution due to the involvement of H^+ in the sulfate-bisulfate equilibrium and the equilibria between Fe^{3+} and its hydroxo complexes. As the dissolution of chalcopyrite by O_2 and H^+ becomes more important (increased fraction of total dissolution, F_{O_2}) the pH rises during leaching. The pH increase is steeper at the lower value of the initial acid concentration, corresponding to the fact that in this case leaching is terminated by $Fe(OH)_3$ precipitation at a lower value of F_{O_2}. Fig. 4 illustrates how the initial ferric sulfate concentration affects the final pH under conditions of Fe(III) depletion. With H^+ not consumed in chalcopyrite dissolution,

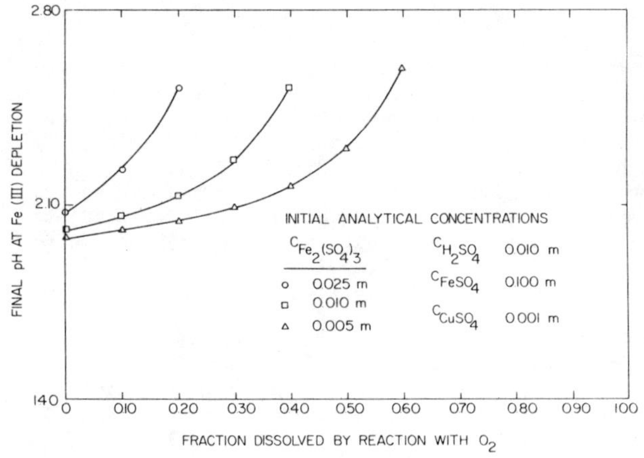

FIGURE 4. pH at depletion of Fe(III) when C_{FeSO_4} = 0.1 m.

the final pH does not depend strongly on the starting $Fe_2(SO_4)_3$ concentration. By a F_{O_2} value of 0.1, however, the curves have begun to spread apart. The separation is most pronounced at the lowest ferric sulfate concentration. The initial conditions for Fig. 5 differ from those of Fig. 4 only in that the $FeSO_4$ concentration in Fig. 5 is twice that in Fig. 4. At a F_{O_2} value of 0.0, the final pH is about 0.1 unit higher at the higher $FeSO_4$ value. $Fe(OH)_3$ precipitation occurs at a lower value of F_{O_2} in a solution with a higher starting $FeSO_4$ concentration. The curves spread apart in Fig. 5 qualitatively as they do in Fig. 4, but for particular F_{O_2} values and initial ferric sulfate concentrations, the final pH values are higher in Fig. 5 than in Fig. 4.

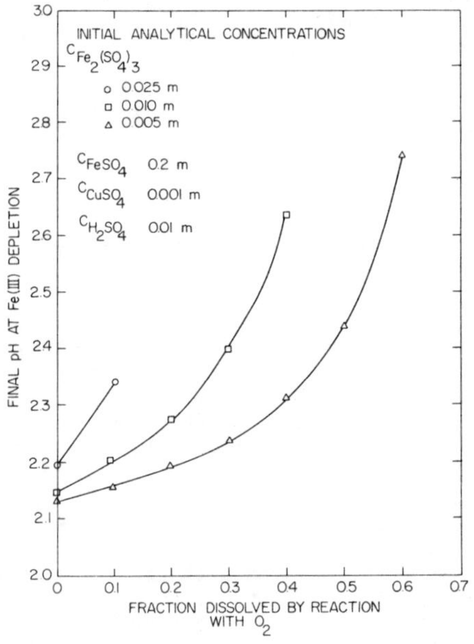

FIGURE 5. pH at depletion of Fe(III) when C_{FeSO_4} = 0.2 m.

The H^+ concentration of the leach solution at a certain point during $CuFeS_2$ dissolution results from the subtle interactions of a number of factors. The effects of the initial concentrations of $Fe_2(SO_4)_3$, $FeSO_4$ and H_2SO_4 and the fraction of the dissolution due to reaction (2) have already been discussed.

The partial equilibrium model has been developed to account for the response of a solution initially at equilibrium to perturbations imposed on it by irreversible reactions that involve some of the solution species. The solution concentrations change in such a way that the chemical equilibria in the solution are maintained while the slower irreversible process goes on. The major effect of this type is the shift of the bisulfate-sulfate equilibrium to replace protons as they are consumed by reaction (2). In Figs. 6 and 7 are shown the changes in H^+, SO_4^{2-} and HSO_4^- concentrations during dissolution for $F_{O_2} = 0.0$ and 0.2, respectively. In Fig. 6, the H^+ concentration declines slightly and the HSO_4^- concentration rises slightly. In Fig. 7, both these concentrations decline. The two curves cross; the H^+ concentration, initially at a higher level, declines more steeply than does the bisulfate concentration. This indicates that the bisulfate ion dissociation only partially replenishes the concentration of protons.

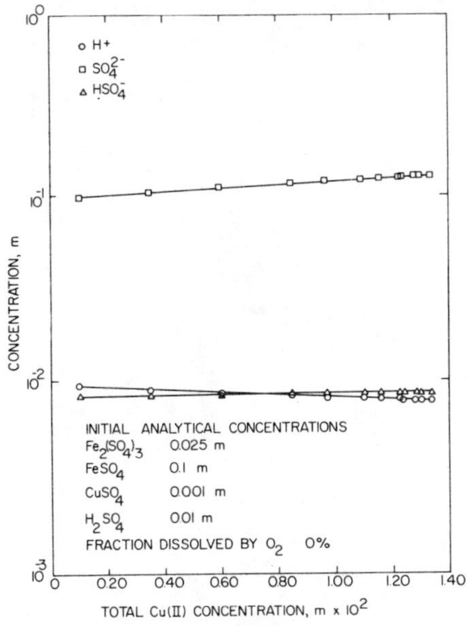

FIGURE 6. Concentration changes for H^+, SO_4^{2-} and HSO_4^- during $CuFeS_2$ leaching. $F_{O_2} = 0.0$.

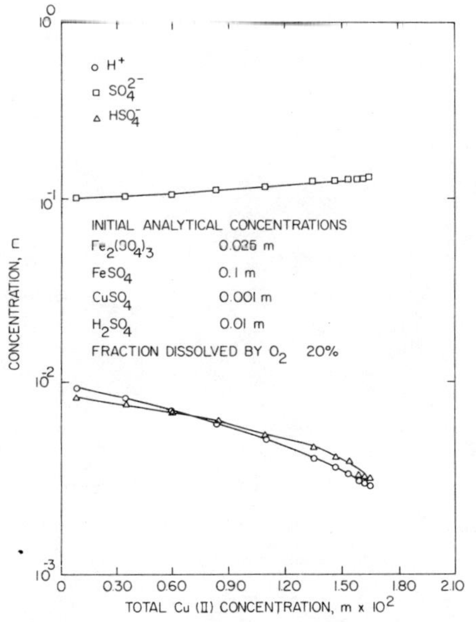

FIGURE 7. Concentration changes for H^+, SO_4^{2-} and HSO_4^- during $CuFeS_2$ leaching. $F_{O_2} = 0.2$.

Keeping the pH of the leach liquor low will delay $Fe(OH)_3$ precipitation; supplying additional H_2SO_4 during leaching will be one way to accomplish this. If the Fe(III) concentration has been substantially reduced by its consumption in reaction (1), it will also be beneficial to supply additional trivalent iron. Both objectives could be accomplished if $Fe(HSO_4)_3$ were provided in the influent liquor. This would often be preferable to introducing $Fe_2(SO_4)_3$ to raise the iron concentration because increases in the SO_4^{2-} level lead indirectly to precipitation.

CONCLUSION

A partial equilibrium model of the chemical changes occurring in the ambient temperature dissolution of $CuFeS_2$ by ferric ion and by oxygen and acid has been developed. The conclusions drawn from it should be valuable in understanding the chemistry of dump leaching. High copper recovery is favored by high ferric ion concentrations and by high acid concentrations. The acid, however, does not exert this effect by causing chalcopyrite dissolution according to reaction (2). Rather, it acts to prevent $Fe(OH)_3$ precipitation. Because 4 moles of H^+ are consumed in reaction (2) for every mole of O_2, high levels of dissolved oxygen have a deleterious influence on the acid concentrations and consequently on the likelihood of precipitation. High levels of ferrous iron shift the solution equilibria in such a way that precipitation occurs at higher sulfuric acid concentrations than would otherwise be the case. Maximum copper dissolution without precipitation is favored by high $Fe_2(SO_4)_3$ concentrations, high H^+ concentrations, low $FeSO_4$ concentrations and an absence of dissolved oxygen. If oxygen is present, H^+ is used up in reaction (2) but it is partially replenished by shifts in the equilibria in the solution, especially by dissociation of HSO_4^-.

ACKNOWLEDGMENT

This research was supported in part through the financial support of Monsanto, Union Carbide and Phillips Petroleum Corporations in the form of graduate fellowships, the National Science Foundation Energy Traineeship, and Graduate Assistantship from the Engineering Research Institute and the Ames Laboratory, USDOE to one of the authors (K.C.L.). This work was also supported in part by the U.S. Department of Energy, under contract No. W-7405-ENG-82, by the Assistant Secretary for Energy Research, Office of Basic Energy Sciences, WPAS-KC-03-02-02.

REFERENCES

1. Liddell, K. C., and Bautista, R. G., 1980, "Modeling the Chemical Equilibria in Solid Liquid Reactions - Application to Leaching of Ores", Thermodynamics of Aqueous Systems with Industrial Applications, Newman, S. A., Barner, H. E., Klein, M., and Sandler, S., eds., American Chemical Society, Washington, D.C.

2. Liddell, K. C., 1979, "A Mathematical Model of the Chemistry of the Dump Leaching of Chalcopyrite", Ph.D. Thesis, Iowa State University, 168 pp.

3. Helgeson, H. C., 1968, "Evaluation of Irreversible Reactions in Geochemical Processes Involving Minerals and Aqueous Solutions. I. Thermodynamic Relations", Geochimica et Cosmochimica Acta, Vol. 32, No. 8, Aug., pp. 853-877.

4. Dutrizac, J. E., 1978, "The Kinetics of Dissolution of Chalcopyrite in Ferric Ion Media", Metallurgical Transactions B, Vol. 9B, No. 3, Sept., pp. 431-439.

5. Baur, J. P., Gibbs, H. L., and Wadsworth, M. E., 1974, "Initial-Stage Sulfuric Acid Leaching Kinetics of Chalcopyrite Using Radiochemical Techniques", Report of Investigations 7823, U.S. Bureau of Mines.

6. Woodcock, J. T., 1967, "Copper Waste Dump Leaching", Proceedings, Australasian Institution of Mining and Metallurgy, No. 224, pp. 47-66.

7. Cathles, L. M., and Apps, J. A., "A Model of the Dump Leaching Process that Incorporates Oxygen Balance, Heat Balance, and Air Convection", Metallurgical Transactions B, Vol.6B, No. 4, Dec., pp. 617-624.

8. Brown, J. B., 1970, "A Chemical Study of Some Synthetic Potassium-Hydronium Jarosites", Canadian Mineralogist, Vol. 10, Pt. 4, pp. 696-703.

9. Smith, R. M., and Martell, A. E., 1976, Critical Stability Constants, Vol. 4, Plenum Press, New York.

10. Liddell, K. C., and Bautista, R. G., 1981, "A Chemically Based Model Characterizing the Precipitation of Ferric Ion in Leach Dumps", Metallurgical Transactions B, submitted.

11. Hsu, P. H., 1973, "Appearance and Stability of Hydrolyzed $Fe(ClO_4)_3$ Solutions", Clays and Clay Minerals, Vol. 21, No. 5, Oct., pp. 267-277.

Chapter 29

KINETICS OF Cl_2-O_2 LEACHING OF LEAD-ZINC FLOTATION CONCENTRATES

R. B. Dix and J. L. Hendrix

University of Nevada, Reno
Reno, Nevada 89557

ABSTRACT

A study was made of the United States Bureau of Mines chlorine-oxygen leaching process for complex sulfide concentrates. The investigation sought to determine the kinetic order and mechanisms involved in the reactions between galena and sphalerite with chlorine. The influence of copper on the extraction rates was also studied.

It has been shown that the rates of reaction of chlorine in aqueous solutions with sphalerite and galena are both zero order at 373 K with rate limiting step being the chemical reaction. Addition of cupric chloride increased the extractions of both lead and zinc.

KINETICS OF Cl_2-O_2 LEACHING OF LEAD-ZINC FLOTATION CONCENTRATES

R. B. Dix and J. L. Hendrix

University of Nevada, Reno
Reno, Nevada 89557

ABSTRACT

A study was made of the United States Bureau of Mines chlorine-oxygen leaching process for complex sulfide concentrates. The investigation sought to determine the kinetic order and mechanisms involved in the reactions between galena and sphalerite with chlorine. The influence of copper on the extraction rates was also studied.

It has been shown that the rates of reaction of chlorine in aqueous solutions with sphalerite and galena are both zero order at 373 K with rate limiting step being the chemical reaction. Addition of cupric chloride increased the extractions of both lead and zinc.

Introduction

The use of chlorine to recover metals from their ores is not new and was first used over 400 years ago in the treatment of silver ores (Varley, et al., 1923). Its use in the production of rare metals such such as titanium and zirconium today is well known (Hampel, 1967). Recently there has been a renewed interest in chlorine hydrometallurgy for the treatment of complex sulfide ores such as the New Brunswick deposits in Canada, the McArthur River deposits in Australia, and the "black ores" of Japan. These deposits are polymetallic pyritic ores in which the non-ferrous metals are so finely disseminated in the pyritic matrix that current mineral separation techniques, e.g. flotation, can not produce high grade concentrates. As high grade, easily concentrated ores are becoming more scarce, the metals industry will eventually have to use complex sulfide ores to meet world demands. Conventional smelting techniques however, have proved to be economically prohibitive with rising energy costs and the additional costs encountered in complying with environmental regulations. These and other factors have led to the renewed interest in hydrometallurgical processes.

Chlorine Processes

There are two main categories of chlorine metallurgy: 1) chloridization, and 2) chlorination. The first category refers to a process in which the chlorine is supplied indirectly through hydrogen chloride gas or a salt such as ferric or cupric chloride while the second category refers to a process in which chlorine is supplied directly as chlorine gas.

Example of chloridization processes are the Duval Clear process and the Cyprus Cymet process which are used to leach sulfide copper ores (Anon., 1976; Kruesi, et al., 1973). Recently, Minemet Recherche in France has developed a chloridization process for the treatment of complex lead-zinc sulfide ores (Demarthe and Georgeaux, 1978).

Chlorination processes in the past, such as the Swinburne-Ashcroft and Baker-Burwell processes, used a dry chlorination step followed by aqueous dissolution of the metal chlorides and subsequent separation steps (Parsons, 1978). Recently, the United States Bureau of Mines has developed a chlorine-oxygen leaching process for treating complex sulfides in which the chlorination step takes place in an aqueous phase (Scheiner, et al., 1977; Scheiner, et al., 1975).

In the USBM process the leaching is carried out batchwise in two steps. First a predetermined amount of chlorine gas is added to the pulp, the exact amount dependent on the concentrate being treated. After chlorine addition is complete, oxygen is then added to complete the oxidation of the remaining metal sulfides and convert dissolved iron to insoluble oxides. The purpose of this study was to collect kinetic data on the chlorine addition step in the leach process.

KINETIC CONSIDERATIONS

Hydrometallurgical processes are in general carried out under favorable thermodynamic conditions, i.e., there is a large decrease in free energy associated with the process. However, whether or not a thermodynamically possible reaction does in fact proceed is determined by kinetic factors.

In a three-phase leaching system the reaction proceeds by the following consecutive steps: 1) transfer of gas to the liquid phase, 2) transport of reactants to the solid-liquid phase boundary, 3) reaction at the solid-liquid phase boundary and, 4) transport of products away from the reaction zone. The rate of the first step is dependent on two factors: 1) gas solubility and 2) the efficiency of agitation. The rates of steps two and four are both dependent on the efficiency of agitation.

In a process where gaseous reagents are used in leaching reactions the gas that reacts must be replaced as rapidly as possible. The rate of gas transfer can be enhanced by increasing the partial pressure of the gas and by using vigorous agitation to increase the surface area of the liquid-gas interface. Vigorous agitation will also shorten the diffusion path of reactants and products to and from the solid surface so that in most cases the rate will be determined by the chemical reactions at the solid-liquid phase boundary.

EXPERIMENTAL

A zinc flotation concentrate was used for all leaching experiments. The concentrate was composed mainly of sphalerite with minor amounts of galena, pyrite and other impurities. A partial chemical analysis of the concentrate is given in Table I, and a typical screen analysis of the sample is shown in Table II.

TABLE I. Chemical analysis of Zn concentrate

Element	Percent
Zn	52.58
Pb	2.17
Fe	4.70
Cu	0.13
Cd	0.33
S	29.50

TABLE II. Screen analysis of Zn concentrate

Size (um)	Wt. Percent
+150	--
-150+75	6.10
-75+45	15.42
-45	78.48

Equipment

A 10 liter reactor with a teflon lining, shown in Figure 1, was used for all leaching experiments. The reactor top, stirring shaft, temperature well, gas sparger and cooling coil were made from titanium. All valves and gas lines were made of either teflon or polyethylene. The flowrates and amount of chlorine added was measured by a totalizing flowmeter. The temperature of the reactor was controlled by use of an external water bath and the cooling coil located inside the reactor. Agitation was produced by two 10 cm impellers driven by a 186 W (1/4 horsepower) motor.

Analytical Procedure

Periodic samples of approximately 10 ml were taken through the sampling tube and tested for pH and metal ions. Leach residues were filtered and washed thoroughly with water to remove any soluble zinc present. The residues were then leached with 20 percent brine solution at 363 K to extract the $PbCl_2$ present. The final residue was then dried, weighed and analyzed for Zn, Pb, Fe and Cu. Atomic absorption techniques were used for all metal analyses using a Perkin-Elmer model 360 absorption spectrophotometer.

Leach Conditions

In order to eliminate any transport effects on the leaching rate and to insure a sufficient gas transfer rate, preliminary experiments were conducted at various stirring speeds. These tests showed that at speeds above 900 rpm there was no increase in the extraction rates. Based on these results a stirring speed of 1,082 rpm was used in all tests.

Tests were also conducted using 5, 10 and 20 percent solids with no significant change in the extraction of lead and zinc. Based on these results all future tests were conducted using 10 percent solids mainly for ease of handling.

All tests were run at a temperature of 373 K. This temperature, which is called for in the USBM process, is used for two reasons: 1) to minimize the production of sulfate, and 2) to avoid elemental sulfur from melting and coating the solid particles.

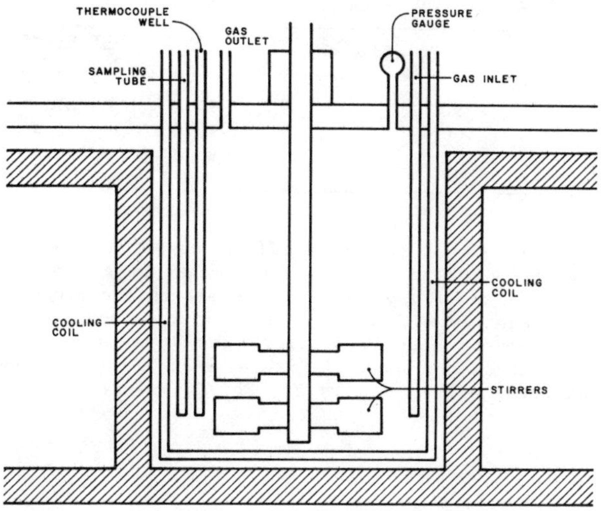

FIGURE 1 • DIAGRAM OF REACTOR

Leach Procedure

A standard leach procedure was followed. The concentrate was added to the reactor with the required amount of water to produce a pulp containing 10 percent solids. It was then heated to the reaction temperature, agitation started, and the chlorine added over a two and a half to three hour period. Tests were run for a period of four hours during which samples were taken at 15 minute intervals. After four hours the agitation was stopped and the reactor was rapidly cooled to room temperature using the external water bath and internal cooling coil. The cooled charge was then removed from the reactor and filtered. The volume of leach liquor was measured and analyzed. The residue was washed, leached with brine and then analyzed.

EXPERIMENTAL RESULTS

Batch leaching tests were performed on the concentrate to determine the reaction order and mechanisms involved. The variables studied to date include the amount and rate of chlorine addition and the addition of copper in the form of cupric chloride. Temperature, stirring speed and percent solids were held constant.

Experiments were conducted using 90, 100, and 150 percent of the stoichiometric chlorine required for 100 percent extraction of the Zn, Pb, Cu and Cd in the concentrate. The results of these experiments are shown in Figures 2-7. Table III shows the total extraction of Zn, Pb, Cu and Fe obtained. The differences in the total lead extractions shown in Table III and those which appear in Figures 2, 4, and 6 is due to the limited solubility of lead chloride in solution.

TABLE III. Metal Extractions with Cl_2

Percent Cl_2 Added	Percent Metal Extraction			
	Zn	Pb	Fe	Cu
90	79.3	97.1	38.0	62.1
100	70.3	97.6	36.8	50.4
150	88.8	98.4	54.8	63.4

TABLE IV. Metal extractions using Cl_2 with 10 g/L Cu

Percent Cl_2 Added	Percent Metal Extraction		
	Zn	Pb	Fe
90	90.3	98.8	43.5
100	95.5	98.7	46.8

In order to see what effect copper had on the extraction of Zn and Pb, experiments were conducted using 90 and 100 percent of the stoichiometric chlorine required with 10 grams per liter copper added in the form of cupric chloride. The results of these experiments are shown in Figures 8-11. Table IV shows the total extractions of Zn, Pb, Cu and Fe obtained.

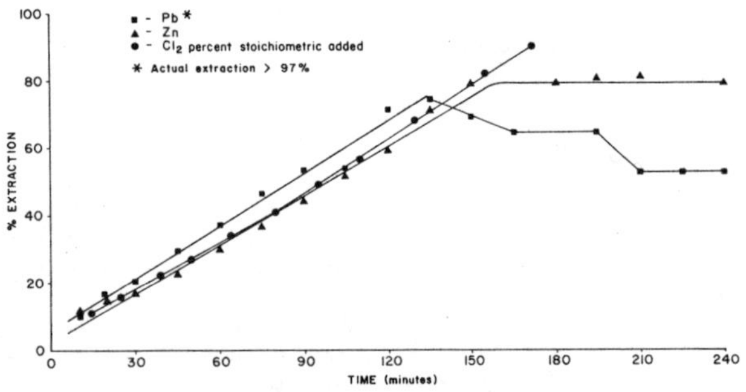

FIGURE 2 • LEACH OF Zn CONCENTRATE WITH Cl_2 AT 373 K

Conditions: 10 weight percent solids; 90 percent stoichiometric chlorine; 1082 RPM stirring speed.

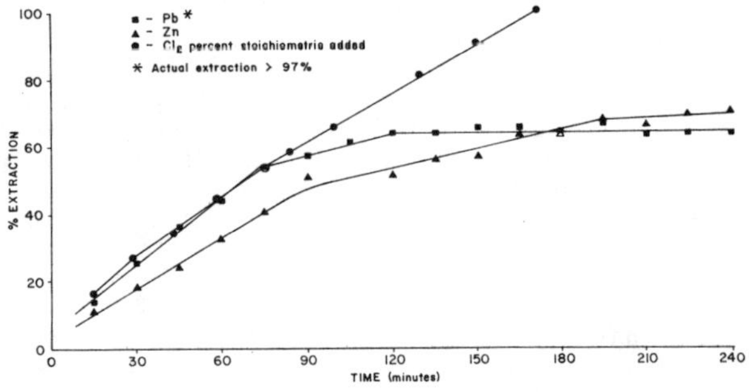

FIGURE 4 • LEACH OF Zn CONCENTRATE WITH Cl_2 AT 373 K

Conditions: 10 weight percent solids; 100 percent stoichiometric chlorine; 1082 RPM stirring speed.

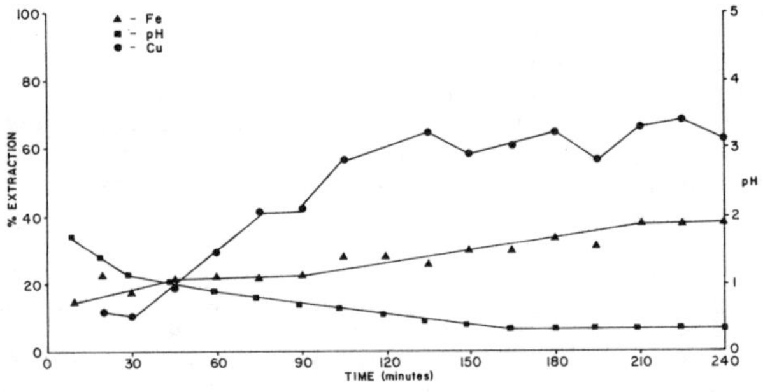

FIGURE 3 • LEACH OF Zn CONCENTRATE WITH Cl_2 AT 373 K

Conditions 10 weight percent solids; 90 percent stoichiometric Cl_2; 1082 RPM stirring speed.

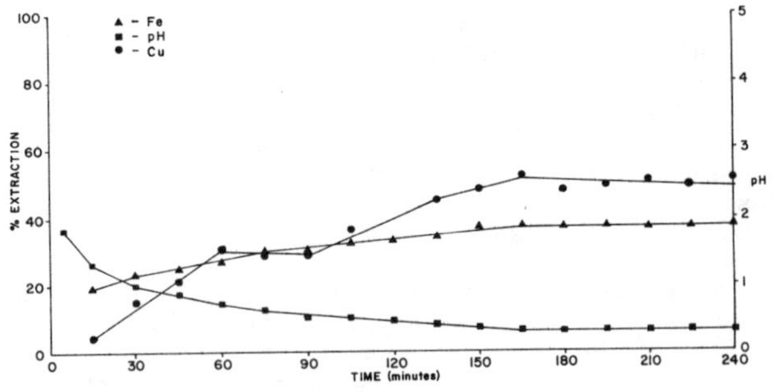

FIGURE 5 • LEACH OF Zn CONCENTRATE WITH Cl_2 AT 373 K

Conditions: 10 weight percent solids; 100 percent stoichiometric Cl_2; 1082 RPM stirring speed.

Discussion

Preliminary experiments were conducted in which a flowrate of 5,766 cm^3/min was used to add the chlorine in a period of thirty minutes, which resulted in a rapid buildup of pressure and release of heat in the reactor. Muir, et al. (1976) have reported that when a vigorous flowrate of chlorine is used, zinc is leached rapidly, but large amounts of iron and acid also report to the solution. It was therefore decided to reduce the flowrate of chlorine which resulted in chlorine addition times of two and a half to three hours.

Sherman and Strickland (1957), in studying the reaction rate of galena in aqueous chlorine solutions at temperatures ranging from 281 to 325 K, found the reaction to be first order. At room temperature they found the reaction to be mixed chemical and transport control while at temperatures above 318 K the reaction was transport controlled. Jan, et al. (1976) in studying the kinetics of pressure leaching sphalerite with sulfuric acid and oxygen found that the reaction was zero order. The extraction curves obtained for lead and zinc shown in Figures 2, 4, 6, 8, and 10 are all straight lines showing that the extraction rates for each exhibit zero order kinetics. Table V shows the rate constants obtained using linear regression analysis.

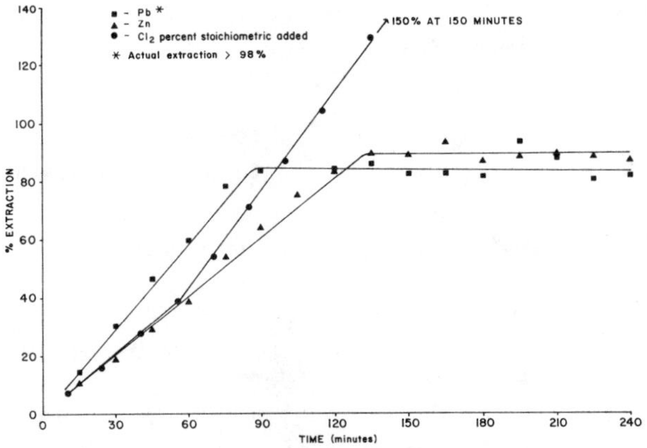

FIGURE 6 • LEACH OF Zn CONCENTRATE WITH Cl_2 AT 373 K

Conditions: 10 weight percent solids; 150 percent stoichiometric Cl_2; 1082 RPM stirring speed.

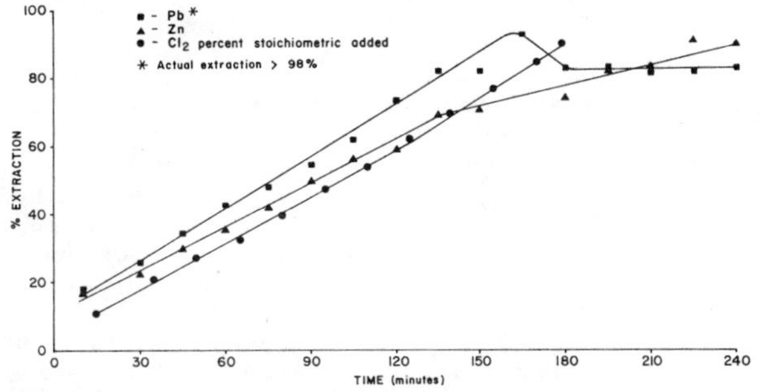

FIGURE 8 • LEACH OF Zn CONCENTRATE WITH Cl_2 AND $CuCl_2$ AT 373 K

Conditions: 10 weight percent solids; 10 gL^{-1} Cu as $CuCl_2$; 90 percent stoichiometric chlorine; 1082 RPM stirring speed.

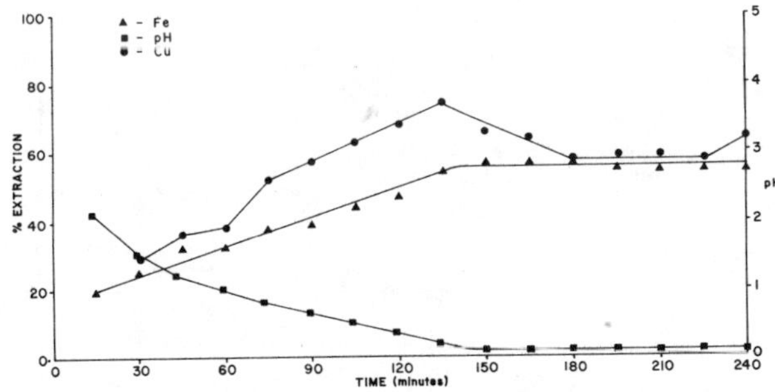

FIGURE 7 • LEACH OF Zn CONCENTRATE WITH Cl_2 AT 373 K

Conditions: 10 weight percent solids; 150 percent stoichiometric Cl_2; 1082 RPM stirring speed.

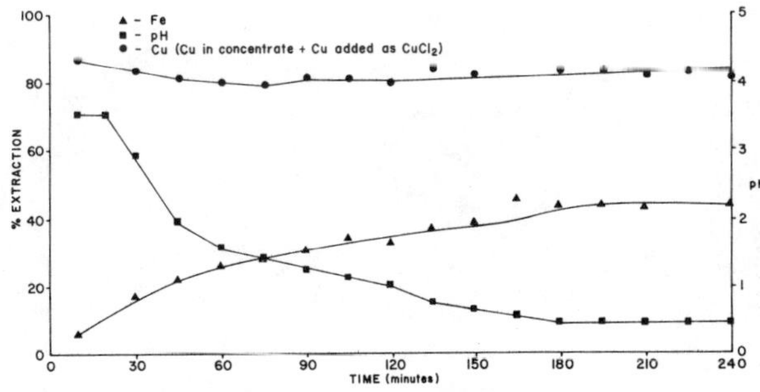

FIGURE 9 • LEACH OF Zn CONCENTRATE WITH Cl_2 AND $CuCl_2$ AT 373 K

Conditions: 10 weight percent solids; 10 gL^{-1} Cu as $CuCl_2$; 90 percent stoichiometric Cl_2; 1082 RPM stirring speed.

TABLE V. Rate constants obtained for Pb-Zn Extractions

Test	Element	$\frac{K}{C_{Ao}}$	Correlation Coefficient
90% Cl_2 added	Zn	.48	.99
	Pb	.51	.99
100% Cl_2 added	Zn	.52	.99
	Pb	.48	.97
150% Cl_2 added	Zn	.69	.99
	Pb	.95	.99
90% Cl_2 + 10 g/L Cu ($CuCl_2H_2O$)	Zn	.41	.99
	Pb	.50	.99
100% Cl_2 + 10 g/L Cu	Zn	.52	.98
	Pb	.61	.99

During the tests using 100 and 150 percent of the stoichiometric chlorine required, the flowrate of chlorine was adjusted to see what effect this had on the extraction rates. Figure 4 shows that decreasing the chlorine flowrate from 1,185 cm^3/min to 830 cm^3/min decreased the rate of extraction for both lead and zinc. Figure 6 shows that increasing the chlorine flowrate from 1,221 cm^3/min to 2,065 cm^3/min had virtually no effect on the extraction rates for lead and zinc. This indicates that the rate limiting step is the chemical reaction. Once the flowrate of chlorine has reached a certain value, which in these experiments was approximately 1,200 cm^3/min, increasing the flow has no effect on the rates of extraction of lead and zinc and the excess chlorine present reacts to from hypochlorous and hydrochloric acid which in turn leads to higher iron extractions. By controlling the flowrate of chlorine carefully it was possible to keep the temperature of the reactor at 373 K with only intermittent use of the cooling coil in addition to controlling the extraction rates of lead and zinc at chlorine flowrates below 1,221 cm^3/min.

Figures 3, 5, 7, 9, and 11 show the extraction curves obtained for iron and copper. Iron extractions in all tests were dependent on the pH of the solution. Extraction ceased after the pH reached a constant value. The higher iron extractions obtained with cupric chloride were probably due to the iron contained in the concentrate as (Zn,Fe)S being released as zinc extractions increased. The extraction curves obtained for copper do not exhibit any definite pattern and additional testing will have to be performed before any conclusions can be made.

Scott and Dyson (1968) have reported that 1 part Cu in 2,000 parts ZnS was sufficient to exert a pronounced catalytic effect on the oxidation of sphalerite. Muir, et al. (1976) found that addition of 20 g/L copper as cupric chloride approximately doubled the rate of extraction of zinc. Figures 8 and 10 show that addition of 10 g/L copper as cupric chloride using 90 percent of the stoichiometric chlorine required increased extraction of zinc from 75 to 90 and from 70 to 95 percent when 100 percent stoichiometric chlorine was used.

CONCLUSIONS

From the data obtained, the following conclusions can be made:

1) The main mechanism for the reaction of lead and zinc with chlorine is:

$$MS + Cl_2 \rightarrow MCl_2 + S^o \quad M = Pb, Zn$$

and is zero order with the rate limiting step being the chemical reaction.

2) The extraction rates of lead and zinc can be controlled by the chlorine flowrate up to a limiting value. Increases in the flowrate above this value results in chlorine being added faster then it is consumed in the reactions with lead and zinc. In addition, the flowrate of chlorine can also be used to control the temperature of the reactor, since the reactions involving lead and zinc are exothermic and by controlling the extraction rates the amount of heat liberated will also be controlled.

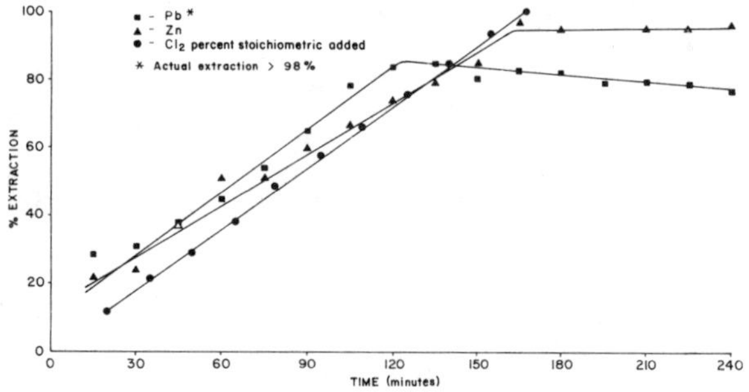

FIGURE 10 • LEACH OF Zn CONCENTRATE WITH Cl_2 AND $CuCl_2$ AT 373 K

Conditions: 10 weight percent solids; 10 g L^{-1} Cu as $CuCl_2$; 100 percent stoichiometric Cl_2; 1082 RPM stirring speed.

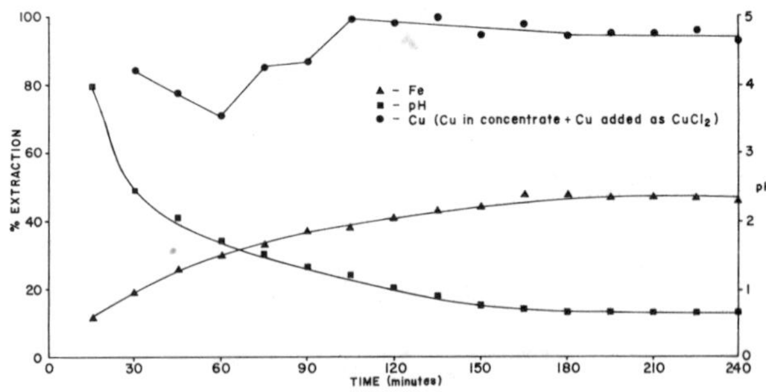

FIGURE 11 • LEACH OF Zn CONCENTRATE WITH Cl_2 AND $CuCl_2$ AT 373 K

Conditions: 10 weight percent solids; 10 gL^{-1} Cu as $CuCl_2$; 100 percent stoichiometric Cl_2; 1082 RPM stirring speed.

3) Addition of copper increases zinc and iron extraction with the amount of additional iron extracted dependent on the amount present as (Zn,Fe)S.

4) Iron extraction is a function of the pH with extraction increasing as pH decreases.

Acknowledgment

The authors wish to acknowledge the financial assistance of the United States Bureau of Mines through Grant No. G0284004, and to thank Renee' Webber for her assistance in running the reactor during these tests.

REFERENCES

Anon., 1976, "Hydrometallurgy: New Processes Move to Commercialization," Engineering and Mining Journal Vol. 177, No. 6, June, pp. 244-245.

Demarthe, J. M., and Georgeaux, A., 1978, "Hydrometallurgical Treatment of Complex Sulphides," Complex Metallurgy '78, Jones, M. J., ed., The Institution of Mining and Metallurgy, London, pp. 113-120.

Hampel, C. A., 1967, Rare Metals Handbook, Reinhold, New York.

Jan, R. J., Hepworth, M. T., and Fox, V. G., 1976, "A Kinetic Study on the Pressure Leaching of Sphalerite," Metallurgical Transactions B, Vol. 7B, No. 3, Sept., pp. 353-361.

Kruesi, P. R., Allen, E. S., and Lake, J. L., 1973, "Cymet Process: Hydrometallurgical Conversion of Base-Metal Sulphides to Pure Metals," The Canadian Mining and Metallurgical Bulletin, Vol. 66, No. 734, June, pp. 81-87.

Muir, D. M., et al., 1976, "Leaching of the McArthur River Zinc-Lead Sulphide Concentrate in Aqueous Chloride and Chlorine Systems," Proceedings Australasian Institute of Mining and Metallurgy, No. 259, Sept., pp. 23-35.

Parsons, H. W., 1978, "Low-Temperature Dry Chlorination of Sulphide Ores-A Review," The Canadian Mining and Metallurgical Bulletin, Vol. 71, No. 791, Mar., pp. 196-204.

Scheiner, B. J., et al., 1977, "Chlorine-Oxygen Leaching of Complex Sulfide Concentrates," TMS Paper Selection, No. A-77-86, TMS-AIME, New York, 10 pp.

Scheiner, B. J., Smyres, G. A. and Lindstrom, R. E., 1975, "Lead-Zinc Extraction from Flotation Concentrates by Chlorine-Oxygen Leaching," Paper presented at SME fall meeting, Salt Lake City, Utah, Sept. 10-11, Preprint No. 75-B-314.

Scott, T. R., and Dyson, N. F., 1968, "The Catalyzed Oxidation of Zinc Sulfide under Acid Pressure Leaching Conditions," Trans. SME-AIME, Vol. 242, No. 9, Sept., pp. 1815-1821.

Sherman, M. I., and Strickland, J. D. H., 1957, "Dissolution of Lead Sulfide Ores in Acid Chlorine Solutions," Trans. AIME, Vol. 209, June, pp. 795-800.

Varley, T., et al., 1923, "The Chloride Volatilization Process of Ore Treatment," Bulletin 211, U.S. Bureau of Mines, pp. 5-6.

Chapter 30

FLUIDIZED BED ELECTROWINNING OF COBALT

M. Dubrovsky and J. W. Evans

Department of Materials Science and Mineral Engineering
University of California, Berkeley
Berkeley, California

ABSTRACT

A laboratory scale cell (approximately 50 amps current) equipped with a fluidized bed cathode was used in a study of the electrodeposition of cobalt from acidified aqueous sulfate solutions. The effect of current density (in the range 0.1 to 1 amp/cm^2 of diaphragm), cobalt concentration (4.8 to 100 g/liter) and catholyte pH (1.35 to 6.2) on current efficiency and energy consumption were determined. Generally the current efficiency increased with increasing current density and cobalt concentration. Current efficiencies were comparable to those reported in the literature for conventional electrowinning.

FLUIDIZED BED ELECTROWINNING OF COBALT

M. Dubrovsky and J. W. Evans

Department of Materials Science and Mineral Engineering
University of California, Berkeley
Berkeley, California

INTRODUCTION AND PREVIOUS INVESTIGATIONS

High capital costs are associated with the electrowinning of metals from aqueous solutions.[1] The fluidized bed electrode, invented[2] in the late 60's, appears to be a means of significantly reducing such capital costs and there have been numerous investigations of the application of this electrode to electrodeposition of metals, particularly copper (see Table I). Other advantages that fluidized bed electrowinning appears to have are the possibility of automation of equipment and entrapment of oxygen generated at the anodes (with subsequent amelioration of the "acid mist" problem suffered by conventional cells). Apparent disadvantages of this approach to electrowinning include the lack of knowledge and experience in scaling up these electrodes, together with the higher power consumptions reported by several investigators.

TABLE 1
Investigation of Fluidized Bed Electrowinning of Metals

Investigators	Reference Number	Metal Deposited	Cell Configuration	Electrolyte	Superficial Current Density A/M^2
Goodridge et al.	3	Zn	rectangular side by side	20-80 g/liter zinc sulfate 0.2-5 g/liter H_2SO_4	2000-6500
Kreysa and Heitz	4	Ag	cylindrical plane parallel	3-5 g/liter silver as nitrate	250
Flett	5	Cu	cylindrical side by side	2 g/liter Cu 100 g/liter H_2SO_4	------
Fleischmann and Kelsal	6	Cu	cylindrical plane parallel	----------	------
Monhemius and Costa	7	Cu	cylindrical plane parallel	2 g/liter Cu 100 g/liter H_2SO_4	------
Carbin and Gabe	8	Cu	cylindrical plane parallel	0.2-4 g/liter 50 g/liter H_2SO_4	------
Germain and Goodridge	9	Cu	rectangular side by side	2 g/liter Cu 100 g/liter H_2SO_4	220-3000
Simpson	10	Cu	rectangular side by side	2 g/liter Cu 100 g/liter H_2SO_4	550-1100
Fleischmann et al.	11	Cu	cylindrical side by side	1 g/liter Cu 100 g/liter H_2SO_4	------
Wilkinson and Haines	12	Cu	rectangular side by side	2 g/liter Cu 100 g/liter H_2SO_4	1000-5000
Coeuret and Hutin	13	Cu	cylindrical plane parallel	5-40 g/liter Cu 10-100 g/liter H_2SO_4	450-3500
Evans and Sabacky	14	Cu	cylindrical side by side	20 g/liter Cu 100 g/liter H_2SO_4	2200-6600
Goodridge and Vance	15	Cu	rectangular side by side	2.5 g/liter Cu 100 g/liter H_2SO_4	1000-5000
Sherwood et al.	16	Ni	cylindrical side by side	2-4 g/liter Cu 60 g/liter Na_2SO_4 pH 3.0	220-2000

Two types of cell appear in Table I, the "plane parallel" cell wherein the direction of electrolyte flow is parallel to the direction of current flow (i.e. vertical) and the "side by side" cell wherein these flows are perpendicular. Due to the difficulty of scaling up a cell in the direction of current flow[17], the former cell design is impractical for an industrial operation although convenient for preliminary laboratory tests. The latter design suffers from the disadvantage that a cell diaphragm must be used to prevent collapse of the fluidized cathode on to the adjacent anode. The "superficial current density" appearing in Table I is the current per unit area normal to the flow, i.e. per unit area of cell diaphragm in the side by side design. Since the cell dimension in the direction of current flow is approximately the same in conventional and fluidized bed electrowinning, comparison of this superficial current density with the current density of conventional electrowinning(50-200A/m^2) gives an immediate impression of the compactness of fluidized bed electrowinning.

Hitherto there has been no report of an investigation into the use of fluidized bed electrodes to electrowin cobalt. Conventional electrowinning of cobalt has been reviewed by Löwe at al.[18], who have reported that comparatively low current densities (200-300 A/m^2) are used; current efficiencies fall in the range 65-88% and power consumptions in the range 3.2-6.4 kwh/kg cobalt. Low current densities are necessary to avoid excessive hydrogen evolution at the cathodes and to prevent formation of metal hydroxides at the cathode by depletion of hydrogen ions from solution[19,20].

EXPERIMENTAL APPARATUS AND PROCEDURE

The cell and associated system appear in Fig. 1. The cell is of the rectangular side by side type, constructed from Lucite. The anode and cathode compartments both had inside dimentions of 35 x 6½ x 2 cm. The fluidized cathode consisted of copper particles (645-850 μm) which had been coated with cobalt in preliminary electrodeposition experiments. The current feeder, positioned at the part of the bed most remote from the anode, consisted of a thin copper sheet. The anode and cathode compartments were separated by a diaphragm measuring 18 x 6 cms or 9 x 5 cms. The dimensionally stable anode was supplied by Diamond Shamrock Corporation, in the form of a coated titanium mesh, and measured 6 x 40 cm. It was located 0.6 cm from the cell diaphragm. Power to the cell was supplied by a Hewlett Packard constant current/voltage power supply and current; total cell voltage and voltage difference between reference electrodes and the current feeder were recorded on strip chart recorders. The saturated calomel reference electrodes were connected to the cell by Luggin capillaries whose tips were located on opposite sides of the cell diaphragm. The usual mode of operation was to maintain separate anolyte and catholytes. In this mode each electrolyte was

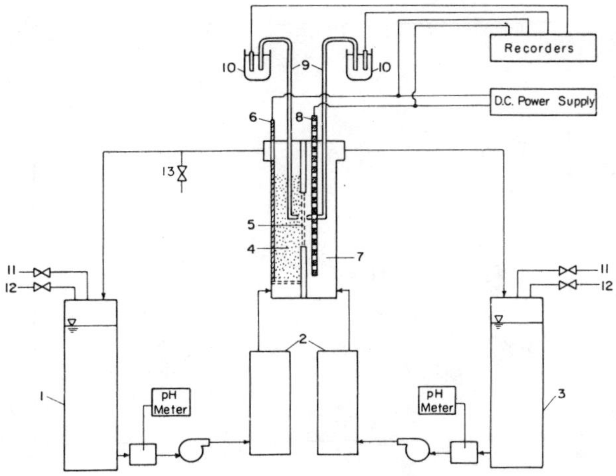

FIGURE 1 System layout for the investigation of the performance of fluidized bed electrode in the electrowinning of cobalt. (1) catholyte reservoir; (2) constant heat bath; (3) anolyte reservoir; (4) fluidized bed cathode; (5) diaphragm; (6) current feeder; (7) anode chamber; (8) DSA; (9) luggin capillaries; (10) reference electrode; (11) hydrogen or oxygen sampling port; (12) auxiliary exit/entry; (13) catholyte sampling port.

supplied to the cell from a separate system consisting of a reservoir, pump and heat exchanger coil immersed in a constant temperature bath.

Preliminary experiments suggested that a material supplied by Grace Corporation, "Daramic", was suitable as a diaphragm material. This was mounted on a cast acrylic frame by solvent welding. Electrolytes were prepared from Fisher Reagent Grade sulfuric acid, technical grade cobalt sulfate (21% Co, 0.1% Ni, 0.01% Fe, 0.001% Cu), and distilled water. Deposition experiments were carried out at constant temperature, current and bed expansion (25-30%). The pH of both electrolytes were controlled by intermittent addition of Fisher Reagent Grade sulfuric acid or sodium hydroxide solutions.

In preliminary experiments the current efficiency was measured in two ways. Samples of catholyte were analyzed for cobalt by atomic absorption spectrophotometry and the current efficiency calculated from the current passed through the cell, the drop in cobalt concentration and the catholyte volume. The second method consisted of measuring the volume of hydrogen generated at the cathode. This was done by connecting a gas burette to port 11 in Fig. 1. The current efficiency could then be calcualted by assuming that any inefficiency was solely due to hydrogen evolution. The two methods agreed to better than 3% and, since the second method is rapid and convenient, the results reported below were obtained by the second method.

EXPERIMENTAL RESULTS AND DISCUSSION

Fig. 2 shows the effect of superficial current density on current efficiency for various catholyte pH's during runs carried out at 24°C with 100 g/l of cobalt in the catholyte and an anolyte consisting of sulfuric acid at 40 g/l. The current efficiency is

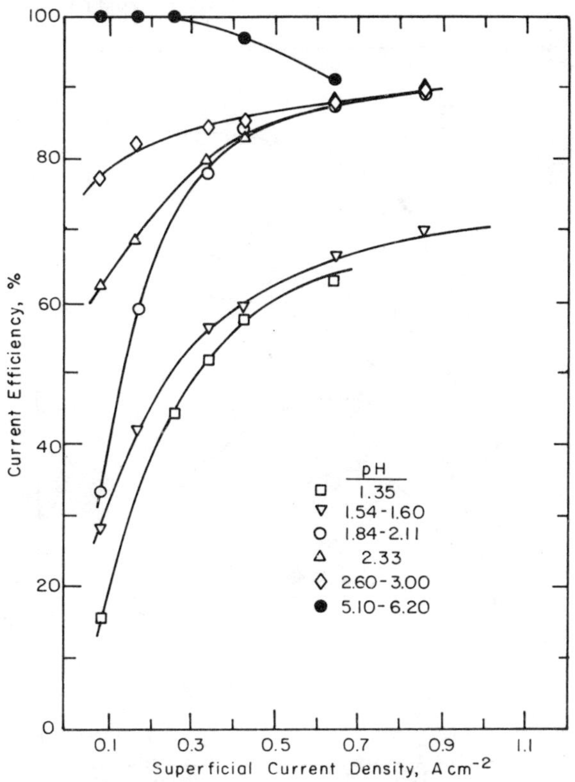

FIGURE 2 Current efficiency vs superficial current density at different catholyte solution pH. Temp 24°C, 100 g/l Co^{++}.

seen to increase with increasing current density (except for measurements at pH 5.1 - 6.2). This increase in current efficiency with durrent density has been observed in previous investigations (of copper electrodeposition), and is to be contrasted with the drop in current efficiency with increasing current density that is observed in conventional electrowinning. An explanation for the experimental results is as follows. At higher current densities, both the cobalt deposition and the hydrogen tend to approach their limiting currents, i.e. deposition of both elements is proceeding under mass transfer control. Since the diffusivities of the cobalt and hydrogen ion differ by less than an order of magnitude while the cobalt ion concentration is higher by a few orders of magnitude than that of the hydrogen ion, then the limiting current for cobalt deposition is greater than that for hydrogen evolution and the current efficiency is high. At low current densities the electrode is operating at close to electrochemical equilibrium, which in the case of the pH's under discussion, would correspond to hydrogen evolution rather than cobalt deposition. Therefore at low current densities the current efficiency is low. The precise shape of the curves between these two extremes is determined by the relative electrochemical kinetics for the two reactions and by mass transport. The anomalous behavior of the curve for pH 5.1 - 6.2 is thought to be due to migration of hydrogen ions from the anode chamber through the slightly permeable diaphragm to the cathode chamber, a phenomenon reported by Goodridge et al.[3] in their work on zinc electrodeposition.

Fig. 3 shows the effects of cobalt concentration

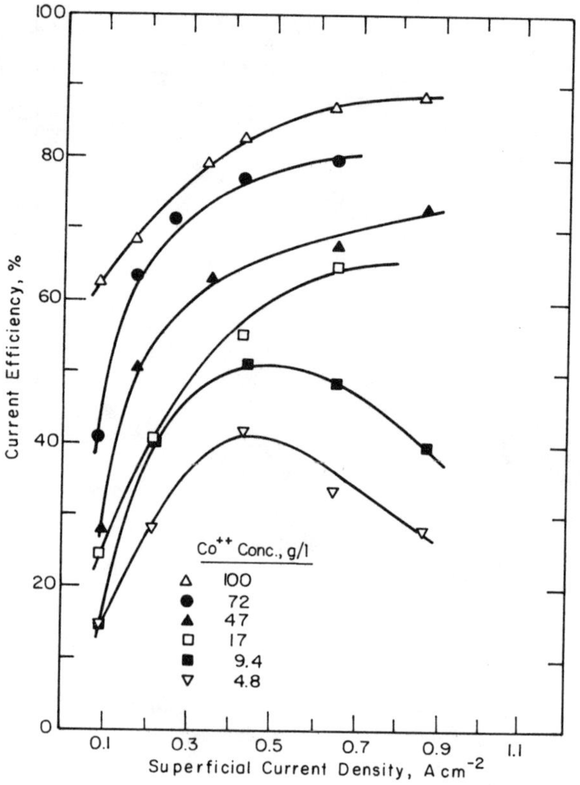

FIGURE 3 Effect of cobalt concentrations in electrolyte. Temp 24°C, pH 2.20 - 2.50.

on current efficiency for acidic catholytes (pH 2.2 - 2.5). The increase in current efficiency with increasing cobalt concentration is consistent with the explanation of electrode behavior reported in the above paragraph.

Voltage and current efficiency measurements permit the calculation of energy consumption rates for each current. Energy consumptions determined in this way appear in Fig. 4. The minimum in two of the curves is a consequence of low current efficiency at low current densities and excessive cell voltage at high current densities. The minimum power consumptions are well above 3.2 - 6.4 kwh/kg Co cited by Lowe et al. The cause of such high power consumptions can be readily seen from Fig. 5 wherein the results of voltage distribution measurements at various current densities are presented. The major portion of the total voltage drop across the cell can be ascribed to the anode chamber and diaphragm rather than to the fluidized cathode. It should be recognized that the anode chamber, which differs little from that of a conventional anode, is being driven at current densities many times that of conventional electrowinning and consequently shows a very high voltage drop. There is scope for improvements in anodes for use in conjunction with fluidized cathodes and some success has been achieved in this direction at Berkeley in connection with fluidized bed copper electrowinning.[21]

The cobalt deposits formed in the experiments

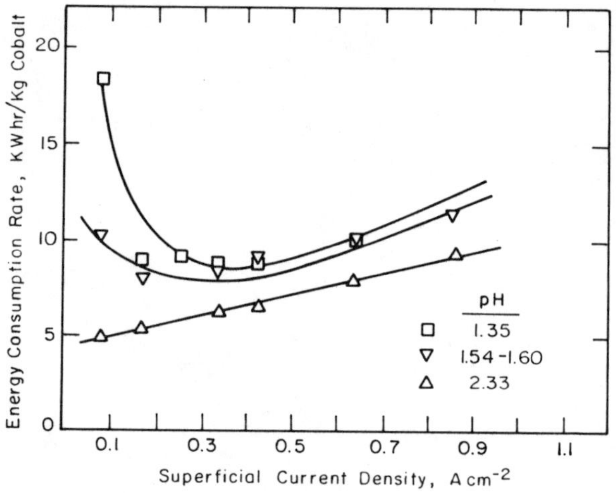

FIGURE 4 Energy consumption in fluidized bed electrowinning cobalt at different superficial current densities. Temp 24°C, 100 g/l Co^{++}.

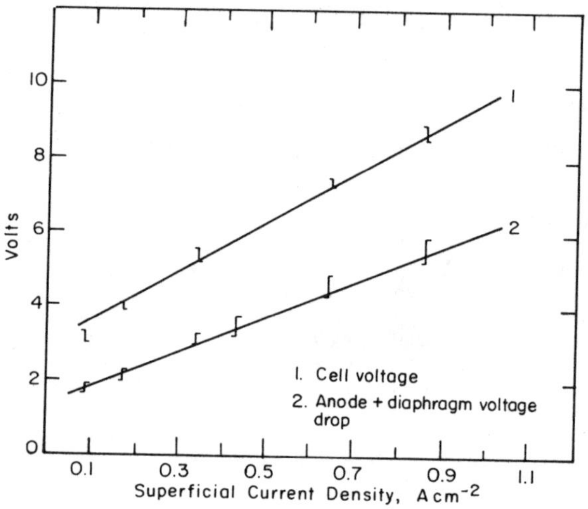

FIGURE 5 Voltage distribution in the cell used for cobalt electrowinning by a fluidized bed cathode. Temp 24°C, 100 g/l Co^{++}.

described above were bright and, presumably therefore, without significant contamination by hydroxide, even at the highest superficial current densities. Dull and discolored deposits were found in runs that are not reported here, wherein cobalt was deposited from near neutral solutions.

CONCLUDING REMARKS

An apparatus has been described in which current efficiencies for metal deposition in fluidized bed electrodes can be rapidly and conveniently measured.

Despite the proximity of the equilibrium potentials for hydrogen evolution and cobalt deposition, it has been shown that cobalt can be electrowon from acidified aqueous sulfate solutions at reasonable current efficiencies using a fluidized bed electrode. Current efficiency was found to increase with current density under acid conditions and bright adherent deposits were formed, even at high current densities. The energy consumption observed in this study was found to be high, compared to conventional electrowinning, but this high energy consumption might be significantly reduced by modifications to the anode chamber.

ACKNOWLEDGEMENT

Support of this investigation by NSF under Grant Number DAR77-24298 is gratefully acknowledged.

REFERENCES

1. Ammann, P.R., et al., 1976, "Direct Electrowinning of Copper from Dilute Leach Solutions," Extractive Metallurgy of Copper, Vol. 2, TMS, p. 994.
2. British Patent, 1,194,181.
3. Goodridge, F., and Vance, C.J., 1977, Electrochimica Acta, Vol. 22, p. 1073.
4. Kreyza, G., and Heitz, E.A., 1975, Chemical Industries, Vol. 19, p. 332.
5. Flett, D.S., 1972, Chemical Industries, Vol. 16, p. 983.
6. Fleischmann, M. and Kelsall, G.M., 1975, Chemical Industries, No. 8, p. 329.
7. Monhemius, A.J., and Costa, P.L.N., 1975, Hydrometallurgy, Vol. 1, p. 183.
8. Carbin, D.C., and Gabe, D.R., 1975, Journal of Applied Electrochemistry, Vol. 5, p. 137.
9. Germain, S., and Goodridge, R., 1976, Electrochimica Acta, Vol. 21, p. 545.
10. Simpson, C.C., 1977, Journal of Metals, Vol. 29, p. 6.
11. Fleischmann, M., Oldfield, J.W., and Tennakoo, L. 1971, Journal of Applied Electrochemistry, Vol. 1 p. 103.
12. Wilkinson, J.A.E., and Haines, K.P., 1972, Transactions of the Institute of Mining and Metallurgy Vol. 81, p. C157.
13. Hutton, D., and Coeuret, F., 1977, Journal of Applied Electrochemistry, Vol. 7, p. 463.
14. Sabacky, B.J., and Evans, J.W., 1979, Journal of the Electrochemical Society, Vol. 126, No. 7, p. 1180.
15. Goodrige, F., and Vance, C.J., 1979, Electrochimica Acta, Vol. 24, p. 1237.
16. Sherwood, W.G., et al., 1979, Metallurgical Transactions, Vol. 10B, p. 659.
17. Sabacky, B.J., and Evans, J.W., 1977, Metallurgical Transactions, Vol. 8B, p. 5.
18. Löwe, D., Muller, L., and Ufer, H., 1968, Neue Hutte, Vol. 13, p. 281.
19. Kublamovskii, V.S., and Belinskii, V.N., 1976, Elektzokhimiya, Vol. 12, No. 9, p. 448.
20. Nakahara, S., and Mahajan, S., 1980, Journal of the Electrochemical Society, Vol. 127, No. 2, 283.
21. Ziegler, D., 1980, M.Sc. Thesis, University of California, Berkeley.

Chapter 31

APPLICATION OF ACTIVITY-ACTIVITY DIAGRAMS TO AMMONIA HYDROMETALLURGY II. THE COPPER-, NICKEL-, COBALT-AMMONIA-WATER SYSTEMS AT ELEVATED TEMPERATURES

K. Osseo-Asare
Associate Professor of Metallurgy

The Pennsylvania State University
University Park, PA 16802

ABSTRACT

The thermodynamics of solid-aqueous solution equilibria up to 200°C in the systems Me-NH_3H_2O (Me=Cu, Ni, Co) are summarized with predominance area diagrams generated with the DIAGRAM computer program. The implications of the diagrams to ammonia hydrometallurgy are discussed.

APPLICATION OF ACTIVITY-ACTIVITY DIAGRAMS TO AMMONIA HYDROMETALLURGY II. THE COPPER-, NICKEL-, COBALT-AMMONIA-WATER SYSTEMS AT ELEVATED TEMPERATURES

K. Osseo-Asare
Associate Professor of Metallurgy

The Pennsylvania State University
University Park, PA 16802

INTRODUCTION

Increasingly, metals have to be extracted from ores of rapidly decreasing grade and under ever-stringent environmental constraints. In the development of processing technologies to meet these new challenges, a thorough knowledge of mineral decomposition chemistry is indispensable. The thermodynamic basis of complex leaching and solution treatment operations is best presented in terms of predominance area diagrams such as Eh-pH (or Pourbaix), log{Me}-pH, and log{Ligand}-pH diagrams (1-6).

In the first paper in this series (4) stability relations in the copper-, nickel-, and cobalt-ammonia water systems at 25°C were described with a variety of log(activity) versus log(activity) diagrams. The object of the present paper is to extend this earlier discussion to elevated temperatures. This effort is in recognition of the fact that hydrometallurgical processing in ammoniacal solutions often involves above-ambient temperatures. For example, in the Sherritt Gordon process, pressure leaching (using compressed air at 100 psig) is conducted at 70-90°C, while solution treatment operations such as hydrogen reduction of nickel and oxyhydrolysis are performed at temperatures in the range 200-250°C (7,8). The Arbiter process operates with leaching temperatures in the 60-90°C range (9), and temperatures up to 150°C have been investigated for the oxidative leaching of chalcopyrite in ammoniacal solutions (10,11). At Nicaro, Surigao, and Greenvale, steamstripping is used to regenerate ammonia and to precipitate a basic nickel carbonate product (8,12-14).

METHOD

In order to generate predominance area diagrams, the equilibrium constants for the pertinent chemical equations must be determined at the temperatures of interest. High temperature equilibrium constants can be extrapolated from low temperature values by means of the thermodynamic relations:

$$\ln K(T) = \frac{Tr}{T}\ln K(Tr) + \frac{1}{RT}\Big[(T-Tr)\Delta S^\circ(Tr) + \int_{Tr}^{T}\Delta C_p^\circ(T)\,d\ln T - \int_{Tr}^{T}\Delta C_p^\circ(T)\,dt\Big] \quad (1a)$$

or

$$\ln K(T) = \ln K(Tr) + \frac{\Delta H^\circ(Tr)}{R}\Big[\frac{1}{Tr} - \frac{1}{T}\Big] + \frac{1}{RT}\Big[T\int_{Tr}^{T}\Delta C_p^\circ(T)\,d\ln T - \int_{Tr}^{T}\Delta C_p\,(T)\,dt\Big] \quad (1b)$$

where Tr is some reference temperature, ΔH° and ΔS° are respectively the changes in standard enthalpy and entropy, and ΔC_p° is the standard molal heat capacity change for the reaction.

Tables 1, 2, and 3 present the pertinent equilibrium constants estimated with the aid of Equations 1a and 1b. The reference temperature used was generally 298.15°K. Values of $\Delta S^\circ(Tr)$, $\Delta H^\circ(Tr)$ and K(Tr) have been determined from values of standard entropy (S°), enthalpy of formation (ΔH_f°) and Gibbs free energy of formation (ΔG_f°) and stability constants taken from the literature (15-19).

For most of the solids and gases, heat capacity Kelley power functions are available (19),

$$C_p^\circ(t) = a + bT + cT^{-2} \quad (2)$$

Where the heat capacity parameters (a,b,c) or the standard entropies are not available from published sources, they have been estimated by the method of summation (20,21) taking H_2O as liquid and ice respectively for the aqueous and solid species. In the case of the simple cations, simple anions, and hydroxo species, C_p° values have been estimated as $C_p^\circ = bT$ with the Correspondence Principle (22,23) as modified by Lewis (24). The standard entropies of the aqueous ionic hydroxo species were estimated with the method of Couture and Laidler (25).

It has recently been shown by Osseo-Asare and Asihene (26) that the Van't Hoff Isobar, Helgeson's electrostatic-nonelectrostatic model (21) and the Correspondence Principle overestimate stability constants for metal ammines. The heat capacities of the metal ammine complexes have therefore been estimated with the "Nickel Model" (26) based on the

Table 1. High Temperature Equilibrium Constants for the NH_3-H_2O and Cu-NH_3-H_2O Systems.

Reaction	Log K(T) 25°C	60°C	100°C	200°C
$H_2 = 2H^+ + 2e$	0.0	0.0	0.0	0.0
$O_2 + 2H_2 = 2H_2O$	83.08	72.58	63.03	46.35
$OH^- + H^+ = H_2O$	14.00	13.02	12.27	11.24
$NH_3 + H^+ = NH_4^+$	9.25	8.26	7.41	5.76
$Cu + 2H^+ = Cu^{2+} + H_2$	-11.48	-10.28	- 9.18	- 7.22
$CuO + 2H^+ = Cu^{2+} + H_2O$	7.64	6.44	5.31	3.26
$Cu_2O + 4H^+ = 2Cu^{2+} + H_2 + H_2O$	- 7.36	- 7.09	- 6.87	- 6.53
$Cu(OH)_2(s) + 2H^+ = Cu^{2+} + 2H_2O$	8.70	7.51	6.38	4.23
$Cu^+ + H^+ = Cu^{2+} + \frac{1}{2} H_2$	- 2.64	- 2.79	- 2.96	- 3.37
$Cu(OH)^+ + H^+ = Cu^{2+} + H_2O$	7.48	6.92	6.39	5.37
$Cu(OH)_2(aq) + 2H^+ = Cu^{2+} + 2H_2O$	14.23	12.35	10.66	7.74
$HCuO_2^- + 3H^+ = Cu^{2+} + 2H_2O$	27.39	24.48	21.85	17.32
$CuO_2^{2-} + 4H^+ = Cu^{2+} + 2H_2O$	40.78	37.61	35.17	32.47
$Cu(NH_3)^+ = Cu^+ + NH_3$	- 5.79	- 5.08	- 4.30	- 2.42
$Cu(NH_3)_2^+ = Cu^+ + 2NH_3$	-10.57	- 9.22	- 7.78	- 4.55
$Cu(NH_3)^{2+} = Cu^{2+} + NH_3$	- 4.24	- 3.75	- 3.16	- 1.58
$Cu(NH_3)_2^{2+} = Cu^{2+} + 2NH_3$	- 7.83	- 6.89	- 5.83	- 3.20
$Cu(NH_3)_3^{2+} = Cu^{2+} + 3NH_3$	-10.80	- 9.41	- 7.88	- 4.23
$Cu(NH_3)_4^{2+} = Cu^{2+} + 4NH_3$	-13.00	-11.17	- 9.19	- 4.55
$Cu(NH_3)_5^{2+} = Cu^{2+} + 5NH_3$	-12.43	-10.27	- 7.95	- 2.57

experimental stability constant measurements of Letowski (27). According to this model (26), the partial molal heat capacity corresponding to the nth ammine complex ($Me(NH_3)_n^{z+}$) may be written as

$$C^\circ_{p,n} = b_n T \qquad (3)$$

where

$$b_n = e + gS^\circ_{Tr,n}(abs) \qquad (4)$$

and $S^\circ_{Tr,n}$ (abs) is the absolute entropy as defined by the Correspondence Principle (22,23). The parameters e and g are constants specific to metal ammines and with values (26):

$$e = 0.19 \ J/mol\cdot K^2 \ (0.045 \ cal/mol\cdot deg^2) \qquad (5a)$$

$$g = -1.46 \times 10^{-3} \qquad (5b)$$

The corresponding values for simple cations are (24):

$$e = 0.56 \ J/mol\cdot K^2 \ (0.133 \ cal/mol\cdot deg^2) \qquad (6a)$$

$$g = - 1.65 \times 10^{-3} \qquad (6b)$$

In order to eliminate the uncertainties concerning the thermodynamic properties of the electron, equilibrium constants for reactions involving

Table 2. High Temperature Equilibrium Constants for the $Ni-NH_3-H_2O$ System.

Reaction	Log K(T) 25°C	60°C	100°C	200°C
$Ni + 2H^+ = Ni^{2+} + H_2$	7.99	7.00	6.12	4.61
$NiO + 2H^+ = Ni^{2+} + H_2O$	12.45	10.63	8.99	6.21
$Ni_3O_4 + 6H^+ + H_2 = 3Ni^{2+} + 4H_2O$	65.43	56.75	48.96	35.66
$Ni_2O_3 + 4H^+ + H_2 = 2Ni^{2+} + 3H_2O$	58.32	50.91	44.24	32.81
$NiO_2 + 2H^+ + H_2 = Ni^{2+} + 2H_2O$	53.39	47.01	41.25	31.33
$Ni(OH)_2 + 2H^+ = Ni^{2+} + 2H_2O$	12.73	11.00	9.47	6.96
$Ni(OH)_3 + 2H^+ + 1/2\ H_2 = Ni^{2+} + 3H_2O$	37.71	33.43	29.60	23.12
$Ni(OH)^+ + H^+ = Ni^{2+} + H_2O$	9.14	8.25	7.42	5.92
$Ni(OH)_2(aq) + 2H^+ = Ni^{2+} + 2H_2O$	17.40	15.06	12.95	9.33
$HNiO_2^- + 3H^+ = Ni^{2+} + 2H_2O$	30.58	27.25	24.42	20.16
$Ni(NH_3)^{2+} = Ni^{2+} + NH_3$	- 2.81	- 2.43	- 1.93	- 0.53
$Ni(NH_3)_2^{2+} = Ni^{2+} + 2NH_3$	- 5.08	- 4.38	- 3.55	- 1.32
$Ni(NH_3)_3^{3+} = Ni^{2+} + 3NH_3$	- 6.85	- 5.81	- 4.60	- 1.52
$Ni(NH_3)_4^{2+} = Ni^{2+} + 4NH_3$	- 8.12	- 6.78	- 5.25	- 1.40
$Ni(NH_3)_5^{2+} = Ni^{2+} + 5NH_3$	- 8.93	- 7.32	- 5.49	- 0.92
$Ni(NH_3)_6^{2+} = Ni^{2+} + 6NH_3$	- 9.08	- 7.05	- 4.79	0.68

electrons (eg. Eq. 7) were rewritten in terms of the hydrogen half-cell reaction (Eq. 8). The equilibrium constants for the resulting equations (eg. Eq. 9) were then estimated for various temperatures using Equations 1a and 1b.

$$Cu = Cu^{2+} + 2e \tag{7}$$

$$2H^+ + 2e = H_2 \tag{8}$$

$$Cu + 2H^+ = Cu^{2+} + H_2 \tag{9}$$

A common convention (28-30) was used according to which the free energy change associated with the standard hydrogen electrode (SHE) is zero at all temperatures. According to this convention, Equations 7 and 9 are equivalent to each other, i.e., the free energy changes for the two chemical equations have the same value.

The DIAGRAM computer program (6) was used to calculate and plot the stability relationships. In the figures which follow, the symbols {Me} and {N} are used to denote respectively, the activities of total dissolved metal and total free ammonia $\{NH_3+NH_4^+\}$.

RESULTS AND DISCUSSION

Figures 1a-c, 2a-c, and 3a-c present respectively Eh-pH diagrams for the systems $Cu-NH_3-H_2O$, $Ni-NH_3-H_2O$, and $Co-NH_3-H_2O$ at 60, 100 and 200°C. In these diagrams, {Me} and {N} have been maintained constant at values of 0.1 and 5.0 respectively. The broken lines which run diagonally across these figures represent the boundaries of water stability. Above the upper broken line, water decomposes to oxygen:

$$H_2O = \frac{1}{2}O_2(g) + 2e + 2H^+(aq) \tag{10}$$

while below the lower line, hydrogen is the decomposition product (see Equation 8).

The diagrams show that as the temperature increases the locus of the metal ammine regions shift leftwards from pH 9.3 at 25°C (4) to pH 8.3 at 60°C (Fig. 1a, 2a, 3a), pH 7.4 at 100°C (Fig. 1b,2b,3b) and to pH 5.8 at 200°C (Fig. 1c, 2c). These pH values also represent (for the respective temperatures) the conditions at which NH_4^+ and NH_3 are present at equal activities as can be seen from the equilibrium constants for ammonia protonation (see Table 1). In addition, the sizes of the ammine stability regions decrease with increase in temperature. For example, at 60°C, the cupric tetrammine stability field extends across almost

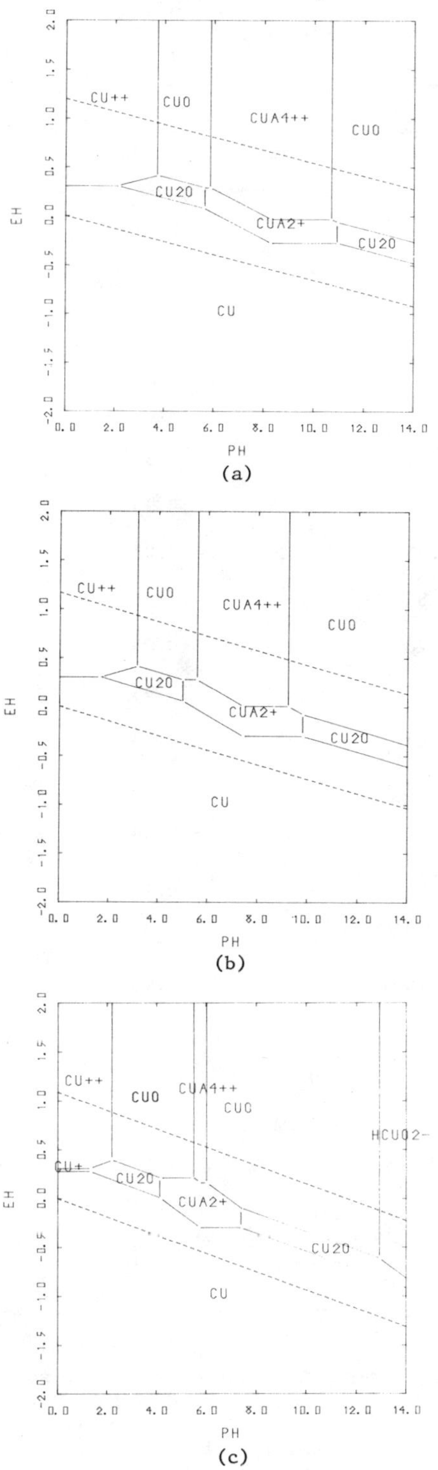

FIGURE 1. Eh-pH diagram for the Cu-NH_3-H_2O system at (a) 60°C, (b) 100°C, (c) 200°C; {Me}=0.1, {N}=5.0.

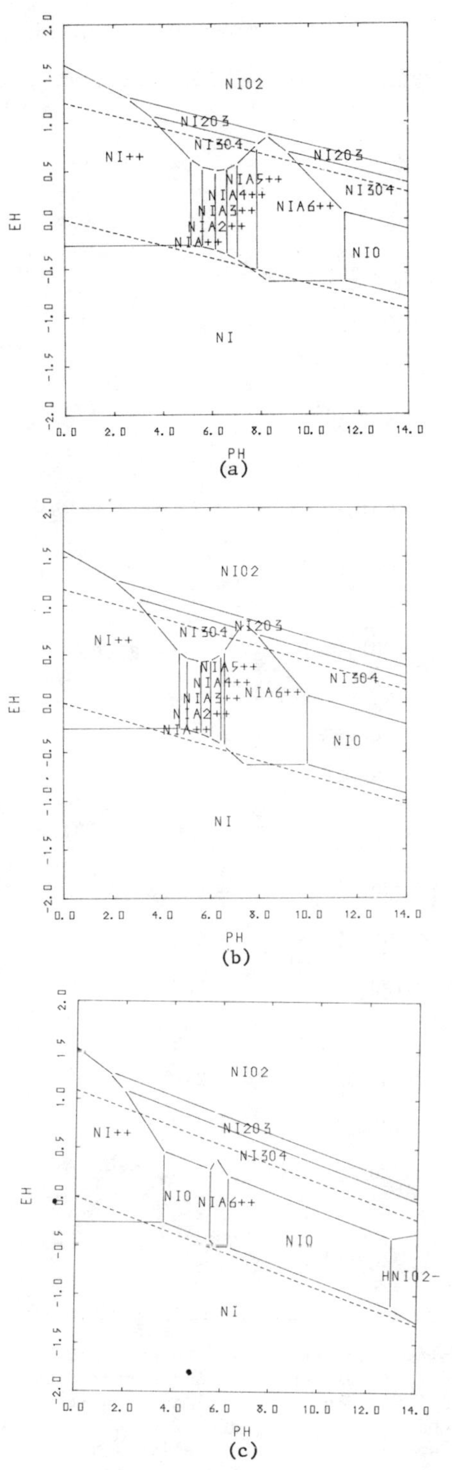

FIGURE 2. Eh-pH diagram for the Ni-NH_3-H_2O system at (a) 60°C, (b) 100°C, (c) 200°C; {Me}=0.1, {N}=5.0.

Table 3. High Temperature Equilibrium Constants for the Co-NH_3-H_2O System.

Reaction	Log K(T)			
	25°C	60°C	100°C	200°C
$Co + 2H^+ = Co^{2+} + H_2$	9.53	8.47	7.50	5.85
$CoO + 2H^+ = Co^{2+} + H_2O$	13.55	11.61	9.86	6.86
$Co_3O_4 + 6H^+ + H_2 = 3Co^{2+} + 4H_2O$	9.17	51.41	44.43	32.51
$Co_2O_3 + 4H^+ + H_2 = 2Co^{2+} + 3H_2O$	59.31	52.05	45.51	34.28
$CoO_2 + 2H^+ + H_2 = Co^{2+} + 2H_2O$	54.62	48.37	42.73	33.00
$Co(OH)_2 + 2H^+ = Co^{2+} + 2H_2O$	13.02	11.39	9.94	7.55
$Co(OH)_3 + 2H^+ + \frac{1}{2} H_2 = Co^{2+} + 3H_2O$	29.64	26.28	23.29	18.29
$Co^{3+} + \frac{1}{2} H_2 = Co^{2+} + H^+$	32.98	30.21	27.68	23.23
$Co(OH)^+ + H^+ = Co^{2+} + H_2O$	9.69	8.82	8.02	6.54
$Co(OH)_2(aq) + 2H^+ = Co^{2+} + 2H_2O$	18.68	16.20	13.96	10.09
$HCoO_2^- + 3H^+ = Co^{2+} + 2H_2O$	31.37	28.04	25.21	20.93
$Co(NH_3)^{2+} = Co^{2+} + NH_3$	- 2.14	- 1.84	- 1.43	- 0.17
$Co(NH_3)_2^{2+} = Co^{2+} + 2NH_3$	- 3.75	- 3.19	- 2.47	- 0.48
$Co(NH_3)_3^{2+} = Co^{2+} + 3NH_3$	- 4.89	- 4.07	- 3.06	- 0.34
$Co(NH_3)_4^{2+} = Co^{2+} + 4NH_3$	- 5.68	- 4.59	- 3.28	0.15
$Co(NH_4)_5^{2+} = Co^{2+} + 5NH_3$	- 5.94	- 4.58	- 2.99	1.14
$Co(NH_3)_6^{2+} = Co^{2+} + 6NH_3$	- 5.37	- 3.75	- 1.86	2.93
$Co(NH_3)^{3+} = Co^{3+} + NH_3$	- 7.43	- 6.53	- 5.56	- 3.31
$Co(NH_3)_2^{3+} = Co^{3+} + 2NH_3$	-14.27	-12.49	-10.67	- 6.72
$Co(NH_3)_3^{3+} = Co^{3+} + 3NH_3$	-20.50	-17.89	-15.22	- 9.61
$Co(NH_3)_4^{3+} = Co^{3+} + 4NH_3$	-26.23	-22.73	-19.24	-11.97
$Co(NH_3)_5^{3+} = Co^{3+} + 5NH_3$	-31.41	-27.06	-22.73	-13.84
$Co(NH_3)_6^{3+} = Co^{3+} + 6NH_3$	-35.94	-30.80	-25.71	-15.31

five pH units from pH6 to 11. However, at 200°C the tetrammine field is confined to a narrow region between pH 5.5 and 6.0. Nickel and cobalt show similar behavior. In the case of cobalt, both the cobaltous and cobaltic ammine regions, present at 60°C and 100°C, are completely absent in the 200°C diagram.

These figures also indicate that the higher the positive charge on an aqueous species, the more its aqueous stability decreases with increase in temperature. This effect is demonstrated by the Cu-NH_3-H_2O system which shows the appearance of Cu^+ and $HCuO_2^-$ in the 200°C diagram (Fig. 1c); these species are absent from the 60°C and 100°C diagrams (Figs. 1a,1b). Further, a comparison of Figures 1a,b and c shows that the proportion of the total ammine region occupied by the cuprous ammine species ($Cu(NH_3)_2^+$) increases with temperature. The cobalt system shows the same tendency. For example, at 150°C (figure not shown in this paper) a cobaltous ammine region exists even though a cobaltic ammine region is absent. This charge effect has also been noted by other investigators (29,30) and it is due to the fact that the greater the cationic charge, the more intense is the hydrolytic cation-hydroxyl ion interaction. As a result of the decrease of the dielectric constant of water with temperature, the tendency towards hydrolysis increases with temperature (21).

Figures 4a-c present log{Me}-pH diagrams at 100°C. In Figures 5a-c, the log{Me}-pH diagrams for 25, 60, 100, 200°C have been superimposed. These diagrams

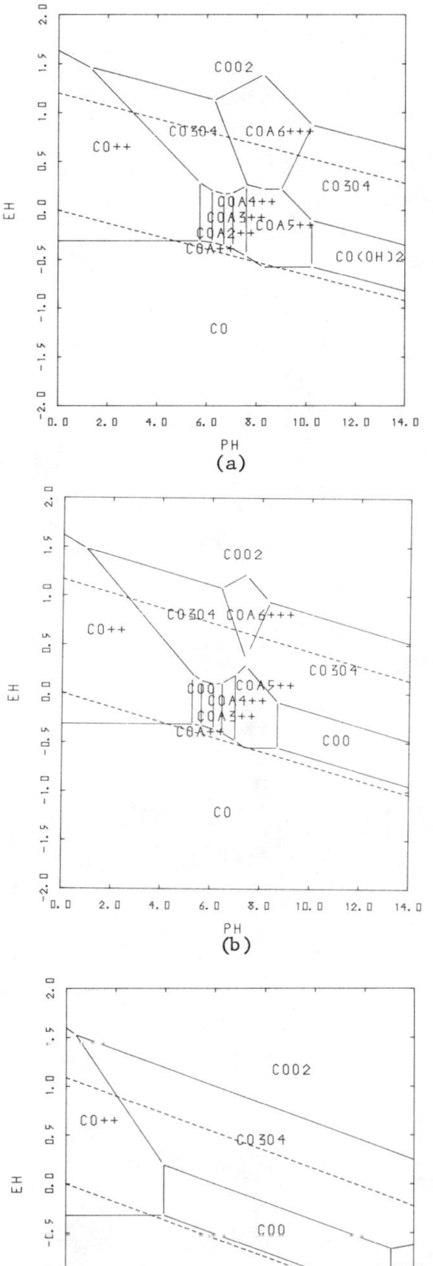

FIGURE 3. Eh-pH diagram for the $Co-NH_3-H_2O$ system at (a) 60°C, (b) 100°C, (c) 200°C; {Me}=0.1, {N}=5.0.

have been calculated for {N}=5.0, Eh=0.5V. The translation of the ammine stability fields towards low pH regions is again apparent. These diagrams further stress the limited solubility of metal ammines as temperature is raised.

Experiments by Morioka and Shimakage (31) on the dissolution of metallic copper and nickel powders at elevated temperatures showed that raising the temperature increases the initial rate of metal extraction but can have an adverse effect on the extent of dissolution. For example, using an oxygen partial pressure of 1 atmosphere, and a solution containing 1 $mol \cdot dm^{-3}$ each of NH_3 and NH_4^+, the copper powder gave 100% extraction ([Cu]=0.31 $mol \cdot dm^{-3}$) at 48 and 98°C, but at 192°C, the extraction was only 63% ([Cu]=0.20 $mol \cdot dm^{-3}$). Lack of activity coefficient data prevents detailed quantitative comparisons of the findings of Morioka and Shimakage with the stability diagrams presented here. However, examination of Figures 4a and 5a shows that the constraints imposed by the solution equilibria in the $Cu-NH_3-H_2O$ system can explain their experimental observations. At 100°C, the leaching window for $Cu(NH_3)_4^{2+}$ remains open at pH 7.4 ($\{NH_3\}=\{NH_4^+\}$ at 100°C) even up to {Cu}=10.0. On the other hand at 200°C, the cupric ammine stability region at pH 5.8 ($\{NH_3\}=\{NH_4^+\}$ at 200°C) terminates at log{Cu}≈-0.5.

The important role which thermodynamic factors play in high temperature ammonia leaching systems has also been noted by Reilly and Scott (32). These investigators found the final dissolved copper concentration to be the same irrespective of whether the starting material was covellite (CuS) or cupric sulfate. In either case, the solid reaction product was identified as CuO. Other workers (33,34) have drawn attention to the competing roles of hydrolysis and metal ammine formation at elevated temperatures.

The effect of total ammonia concentration on the stability relations is illustrated by the log{N}-pH diagrams presented in Figures 6a-c and 7a-c. In these diagrams, the Eh has been fixed at 0.5V and total dissolved metal activity has been kept constant at 0.1. It can be seen from these diagrams that as the temperature is raised, higher and higher ammonia concentrations are needed in order to keep the metal ammines in solution. It is this factor which allows ammonia stripping to be used as a means of regenerating ammonia from leach liquors and recovering a solid product (7-9, 11-14).

It has been noted above that Cu, Ni and Co follow similar trends. Nevertheless, it is of interest to compare the variations in the behavior of these three metals. In general, the size of the ammine stability regions seem to increase in the order Cu>Ni>Co. This is emphasized by the temperature versus pH diagrams shown in Figures 8a-c. For the specific conditions used in Figure 8 (Eh=0.5V, {N}=5.0, {Me}=0.1), the disappearance of the metal ammine regions occurs respectively at >200°C for copper, ~160°C for nickel, and ~110°C for cobalt. It can also be seen that at each temperature, the cupric ammine field spans the largest pH range followed by nickel. Cobalt has the least pH range for ammine stability, therefore this metal requires more careful pH and temperature control for effective dissolution to be achieved.

Cobalt extraction with ammonia is often found to be poor. According to Slobtsov et al. (35), the Nicaro laterite plant in Cuba gives only 10-20% cobalt

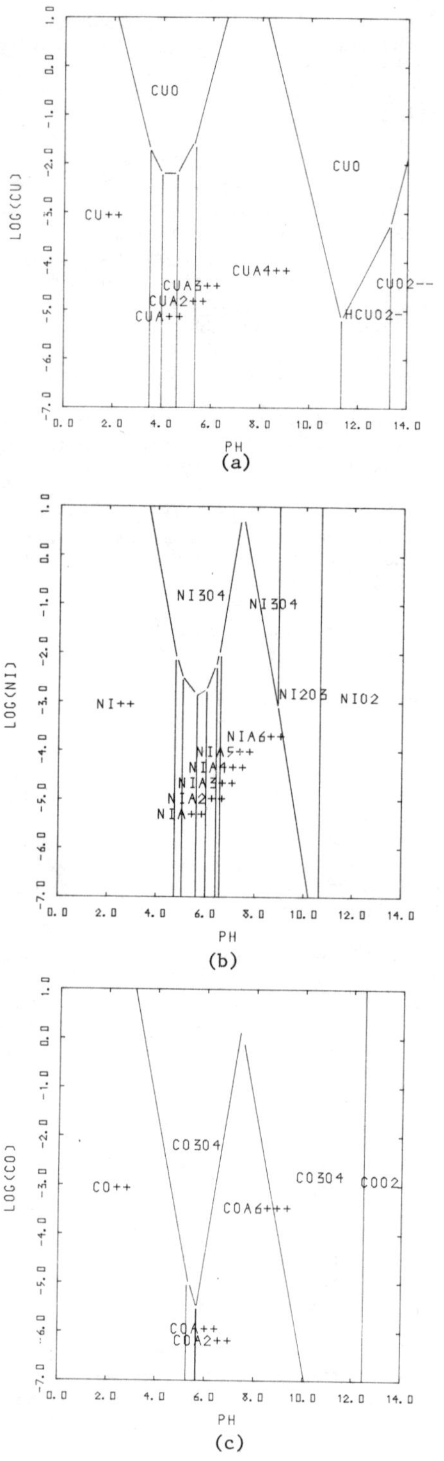

FIGURE 4. Log{Me} versus pH diagrams at 100°C for the systems (a) Cu-NH_3-H_2O, (b) Ni-NH_3-H_2O, (c) Co-NH_3-H_2O; Eh=0.5V, {N}=5.0.

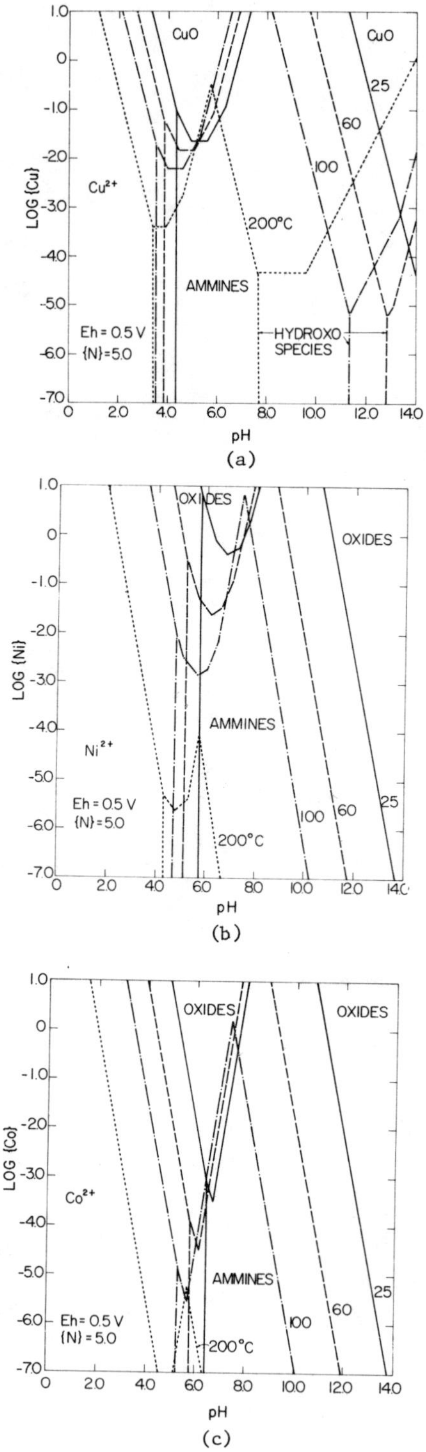

FIGURE 5. Superposition of log{Me} versus pH diagrams at 25, 60, 100, and 200°C for the systems (a) Cu-NH_3-H_2O, (b) Ni-NH_3-H_2O, (c) Co-NH_3-H_2O; Eh=0.5V, {N}=5.0.

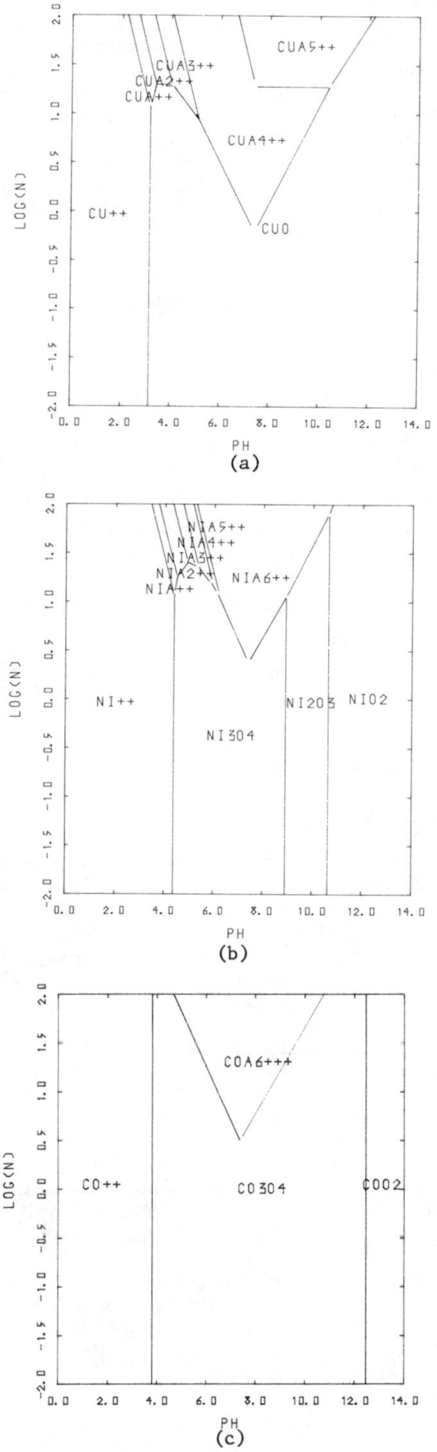

FIGURE 6. Log{N} versus pH diagrams at 100°C for the systems (a) Cu-NH_3-H_2O, (b) Ni-NH_3-H_2O, (c) Co-NH_3-H_2O; Eh=0.5V, {Me}=0.1.

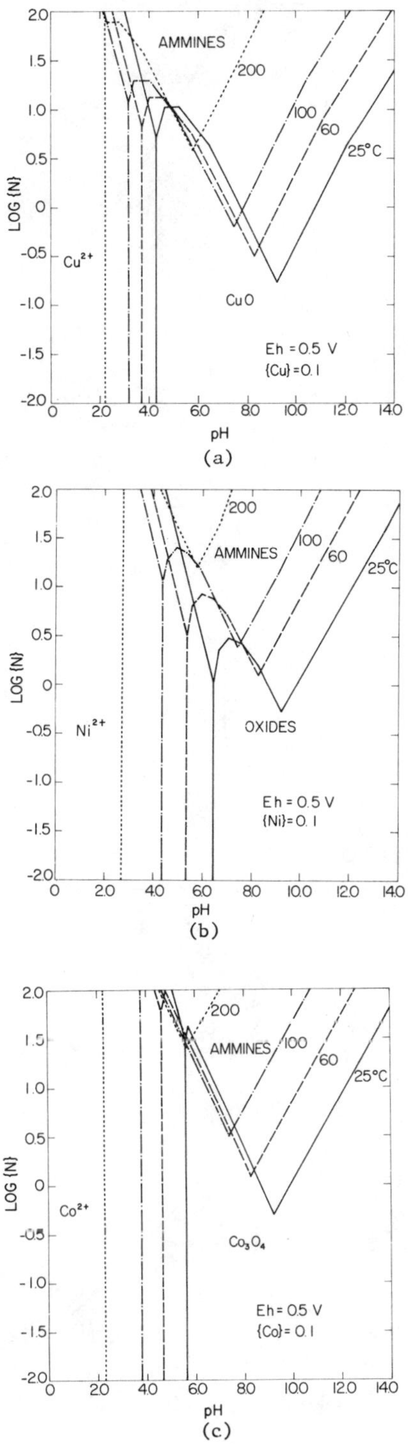

FIGURE 7. Superposition of log {N} versus pH diagrams at 25, 60, 100, and 200°C for the systems (a) Cu-NH_3-H_2O, (b) Ni-NH_3-H_2O, (c) Co-NH_3-H_2O; Eh=0.5V, {Me}=0.1.

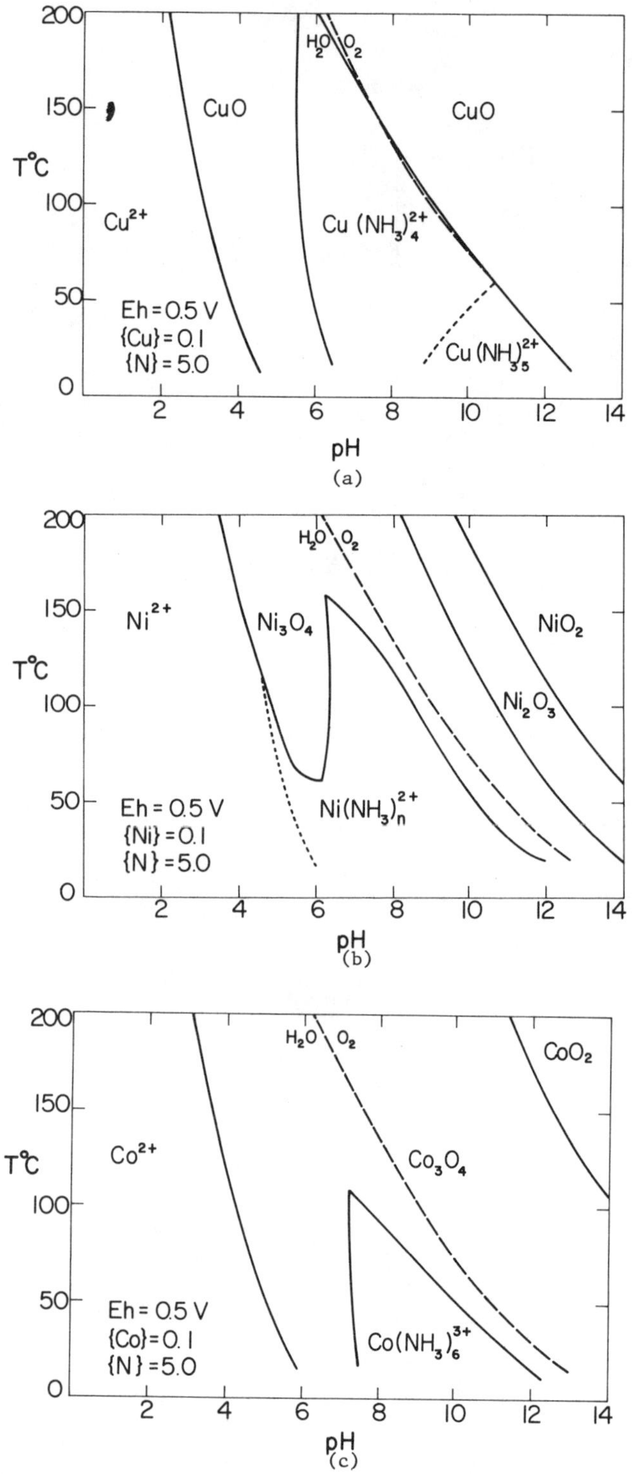

FIGURE 8. Temperature versus pH diagram for the systems (a) $Cu-NH_3-H_2O$, (b) $Ni-NH_3-H_2O$, (c) $Co-NH_3-H_2O$; Eh=0.5V, {Me}=0.1, {N}=5.0.

extraction compared with 75-80% for nickel. Han et al. (36) have also noted that ammonium-ammonia leaching of reduced manganese nodules does not give more than 50% cobalt extraction while copper and nickel extractions exceeding 80% individually can be obtained. The differences in the size of the ammine stability fields with respect to the corresponding oxides and hydroxides, may partly account for the observed differences in the extraction of copper, nickel, and cobalt from laterites and ocean nodules.

ACKNOWLEDGEMENTS

This work was supported in part by the National Science Foundation. Computer time was provided by the Experiment Station of the College of Earth and Mineral Sciences, The Pennsylvania State University. The assistance of S. W. Asihene with the data estimation is gratefully acknowledged.

REFERENCES

1. Pourbaix, M., 1966, Atlas of Electrochemical Equilibria in Aqueous Solutions, Pergamon, London, 644 pp.

2. Garrels, R. M. and Christ, C. L., 1965, Solutions, Minerals, and Equilibria, Harper and Row, New York, 450 pp.

3. Peters, E., 1976, "Direct Leaching of Sulfides: Chemistry and Application", Met. Trans. B. Vol. 7B, pp. 505-517.

4. Osseo-Asare, K., and Fuerstenau, D. W., 1978, "Application of Activity-Activity Diagrams to Ammonia Hydrometallurgy: The Systems $Cu-NH_3-H_2O$, $Ni-NH_3-H_2O$ and $Co-NH_3-H_2O$ at 25°C", Fundamental Aspects of Hydrometallurgical Processes, Chapman, T. W. et al., eds. AIChE Symposium Series, Vol. 74, No. 173, pp. 1-13.

5. Osseo-Asare, K., and Asihene, S. W., 1979, "Heterogeneous Equilibria in Ammonia/Laterite Leaching Systems", Proceedings International Laterite Symposium, AIME, pp. 585-609

6. Osseo-Asare, K., and Brown T. H., 1979, "A Numerical Method for Computing Hydrometallurgical Activity-Activity Diagrams", Hydrometallurgy, Vol. 4, pp. 217-232.

7. Forward, F. A., and Mackiw, V. N., 1955, "Chemistry of the Ammonia Pressure Process for Leaching Ni, Cu, and Co from Sherritt Gordon Sulfide Concentrates", Trans. AIME, Vol. 203, pp. 457-463.

8. Boldt, J. R. and Queneau, P., 1966, The Winning of Nickel, Van Nostrand Princeton, 487 pp.

9. Kuhn, M. C., Arbiter, N., and Kling, H., 1974, "Anaconda's Arbiter Process for Copper", Can. Min. Met. Bull., Vol. 67, pp. 62-72.

10. Beckstead, L. W., and Miller, J. D., 1977, "Ammonia, Oxidative Leaching of Chalcopyrite-Reaction Kinetics", Met. Trans. B, Vol. 8B, pp. 19-29.

11. Fonseca, A. G., 1974, "Ammonia-Oxidative Leach of Chalcopyrite", CIM Bull., July, pp. 105-110.

12. Power, L. F. and Geiger, G. H., 1977, "The Application of the Reduction Roast-Ammoniacal Carbonate Leach to Nickel Laterites", Miner. Sci. Eng., Vol. 9, (1), pp. 32-51.

13. Colvin, N. and Gulyas, J. W., 1979, "The Marinduque Surigao Nickel Refinery", Procedings International Laterite Symposium, Evans, D. J. I., Shoemaker, R. S., and Veltman, H., eds. AIME, New York, pp. 346-356.

14. Reid, J. G., 1979, "Operations at the Greenvale Nickel Project Mine and Refinery", Evans, D. J. I., Shoemaker, R. S., and Veltman, H., eds. Procedings International Laterite Symposium, AIME, New York, pp. 368-381.

15. Duby, P., 1977, The Thermodynamic Properties of Aqueous Inorganic Copper Systems. International Copper Research Association.

16. Sillen, L. G. and Martell, A. E., 1964, 1971, Stability Constants of Metal-Ion Complexes, Special Publication No. 17, No. 25, The Chemical Society, London.

17. Martell, A. E., and Smith, R. M., 1977, Critical Stability Constants, Vol. 4, Plenum, New York.

18. Wagman, D. D. et al., 1966-1971, Selected Values of Chemical Thermodynamic Properties, NBS Technical Note 270, parts 1-6, NBS, Washington.

19. Kelley, K. K., 1960, "Contributions to the Data on Theoretical Metallurgy. XIII 'High Temperature Heat Content, Heat Capacity and Entropy Data for the Elements and Inorganic Compounds'", U.S. Bur. Mines Bull, 584, 232 pp; Kelley, K. K. and King, E. G., 1961, "Contributions to the Data on Theoretical Metallurgy. XIV. Entropies of the Elements and Inorganic Compounds", U.S. Bur. Mines Bull., 592, 149 pp.

20. Latimer, W. M., 1952, The Oxidation States of the Elements and Their Potentials in Aqueous Solutions, Second Edition, Prentice-Hall, Englewood Cliffs, N.J., 392 pp.

21. Helgeson, H. C., 1969, "Thermodynamics of Hydrothermal Systems at Elevated Temperatures and Pressures", Am. J. Sci., 267. pp. 729-804.

22. Criss, C. N., and Cobble, J. W., 1964a, "The Thermodynamic Properties of High Temperature Aqueous Solutions. IV. Entropies of the Ions up to 200°C and the Correspondence Principle", J. Am. Chem. Soc., Vol. 86, pp. 5385-5390.

23. Criss, C. N., and Cobble, J. W., 1964b, "The Thermodynamics of High Temperature Aqueous Solutions. V. The Calculation of Ionic Heat Capacities up to 200°C. Entropies and Heat Capacities Above 200°C", J. Am. Chem. Soc., Vol. 86, pp. 5390-5393.

24. Lewis, D., 1971, "Studies of Redox Equilibria at Elevated Temperatures I. The Estimation of Equilibrium Constants and Standard Potentials for Aqueous Systems up to 374°C", Arkiv for Kemi, Vol. 32. pp. 385-404.

25. Couture, A. M., and Laidler, K. J., 1957, "The Entropies of Ions in Aqueous Solution II. An Empirical Equation for Oxy-Anions", Canad. J. Chem., Vol. 35, pp. 202-206.

26. Osseo-Asare, K., and Asihene, S. W., "Metal Ammine Formation at Elevated Temperatures", Submitted.

27. Letowski, F., 1970, "Etudes Thermodynamiques des Solutions Aqueuses des Ammoniacates de Metaux aux Temperatures Elevees. II. Constantes de Stabilite des Ammoniacates de Nickel en Temperatures jusqu'a 180°C", Roczniki Chem., Vol. 44, pp. 1665-1673.

28. Robins, R. G., 1968, The Application of Potential-pH Diagrams to the Prediction of Reactions in Pressure Hydrothermal Processes, Warren Spring Laboratory, L R 90 (MSF), Stevenage.

29. Ferreira, R. C. H., 1975, "High-temperature E-pH Diagrams for the Systems $S-H_2O$, $Cu-S-H_2O$ and $Fe-S-H_2O$", Leaching and Reduction in Hydrometallurgy, Burkin, A. R., ed., Instn. Min. Metal. London, pp. 67-83.

30. Townsend, H. E., 1970, "Potential-pH Diagrams at Elevated Temperature for the System $Fe-H_2O$", Corros. Sci., Vol. 10, 343-358.

31. Morioka, S., and Shimakage, K., 1971, "Studies on the Ammonia Pressure Leaching of Metallic Nickel and Copper Powders", Trans. Jap. Inst. Met., Vol. 12, pp. 198-205.

32. Reilly, I. G., and Scott, D. S., 1976, "The Hydrolysis of Cupric Ammines", Met. Trans B., Vol. 7B, pp. 718-720.

33. Ono, K., and Matsushima, T., 1969, "NH_3-CO_2 Leaching of Nickel from Iron-Rich Laterite", Studies in Metallurgy for Sachio Matoba, Tohoku University, Sendai, Japan, pp. 167-178.

34. Tozawa, K., Umetsu, Y., and Sato, K., 1976, "On Chemistry of Ammonia Leaching of Copper Concentrates", Extractive Metallurgy of Copper, Vol. II., Hydrometallurgy and Electrowinning, Yannopoulos, J. C., and Agarwal, J. C., eds., AIME, New York, pp. 706-721.

35. Slobtsov, L. E., et al., 1975, "Sorption Extraction of Nickel and Cobalt from Ammonia-Carbonate Pulp", Russ. J. Non-Ferrous Metals, Vol. 16, No. 5, pp. 16-18.

36. Han, K. N., Hoover, M., and Fuerstenau, D. W., 1974, "Ammonia-Ammonium Leaching of Deep-Sea Manganese Nodules", Int. J. Miner. Process., Vol. 1, pp. 215-230.

Chapter 32

EQUILIBRIUM RELATIONSHIPS IN THE EXTRACTION OF METALS FROM AMMONIACAL SOLUTION WITH LOW RANK COALS

G. V. Cullen

G. V. Cullen and Associates
36 Harrington Avenue
North Balwyn, Victoria 3104

and

N. G. Siviour

Division of Mineral Engineering
Commonwealth Scientific and Industrial Research Organization
Clayton, Victoria 3168

ABSTRACT

It has previously been reported that metal loadings of up to 6.8 m. eq/g. of dry coal can be obtained by contacting brown coal or lignite with ammoniacal solutions of ammine forming metals.

Application of equilibrium expressions to further experimental results indicates that the extraction is by ion exchange, that the exchange is divalent-univalent for Zn, Cd and Ni, but is mainly univalent-univalent for Cu, and that to obtain reasonably uniform values for the equilibrium constants correction formulae must be applied to compensate for multiple functionality of the coal and its pH dependence in ion exchange. The results are shown to be consistent with the hypothesis that metal ammine complex cations are exchanged for protons derived from carboxyl and/or phenolic hydroxyl groups in the coal.

EQUILIBRIUM RELATIONSHIPS IN THE EXTRACTION OF METALS FROM AMMONIACAL SOLUTION WITH LOW RANK COALS

G.V. Cullen and N.G. Siviour

ABSTRACT

It has previously been reported that metal loadings of up to 6.8 m. eq/g. of dry coal can be obtained by contacting brown coal or lignite with ammoniacal solutions of ammine forming metals.

Application of equilibrium expressions to further experimental results indicates that the extraction is by ion exchange, that the exchange is divalent-univalent for Zn, Cd and Ni, but is mainly univalent-univalent for Cu, and that to obtain reasonably uniform values for the equilibrium constants correction formulae must be applied to compensate for multiple functionality of the coal and its pH dependence in ion exchange. The results are shown to be consistent with the hypothesis that metal ammine complex cations are exchanged for protons derived from carboxyl and/or phenolic hydroxyl groups in the coal.

INTRODUCTION

Coals known as lignite or brown coal are classified by ASTM, 1979, as having a calorific value of less than 8300 BTU/lb (19.3 MJ/kg) on a mineral free but natural moisture containing basis, and as such these materials are the lowest ranking coals. For different deposits of these coals Herman, 1952, reports natural moisture contents ranging from about 30 to 67% and it is considered this indicates that the coals contain extensive porosity and consequently a large internal surface area. Fig. 1 shows a suggested structural formula, Anon, 1959, for brown coal containing peripheral carboxyl (COOH) and phenolic hydroxyl (OH) groups which have a capacity for reaction with other chemical species, probably facilitated greatly by their accessibility at the pore surfaces. The above formula indicates that brown coal contains 1.62 and 6.47 m. eq/g. of carboxyl and phenolic hydroxyl groups, respectively, whereas Brooks and Sternhill, 1957, obtained 0.7 - 1.8, 6.4 - 8.7 and 4.9 - 7.3 m. eq/g, respectively, for the carboxyl, total acidity and phenolic hydroxyl groups, the latter being determined by difference. However, Schafer, 1970, with several methods obtained a range of about 3.0 - 4.0 m. eq/g. for the carboxyl group alone in one sample of brown coal, these generally higher but variable results being obtained after refinement of the technique used in the earlier determinations. The work of Schafer, 1970, also demonstrated that the hydrogen in both the functional groups is capable of ion exchange with calcium and barium ions in solution, and that exchange for the carboxyl group predominates up to a pH of about 8.3 and for the phenolic hydroxyl group at higher pH, thus indicating that the coal contains acid groups with a range of strengths.

Cullen, Siviour and Pearson, 1978, have shown that loadings, by weight, of 22% copper, 15% nickel, 17% zinc, 31% cadmium, 37% silver and 13% cobalt could be obtained on brown coal by contacting it with ammoniacal solutions of the metals which are known to form ammine complex cations of the type $Me(NH_3)^{+}_{n}$ for univalent metal and $Me(NH_3)^{++}_{n}$ for divalent metals. Similar findings have been reported for nickel by Altekar, Shahani and Saha, 1974, and for cobalt by Saha, Shahani and Altekar, 1977.

It is important to note that the work reported by Cullen, Siviour and Pearson, 1978, indicated that approximately two ammonium ions are formed and released into solution for each atom of divalent metal loading on the coal, that substantial reduction in percentage loading and extraction occurs as the ammonium ion concentration increases, and for the metals discussed in this paper virtually maximum loading is obtained with contact times of 1 000 to 4 000 minutes.

G.V. CULLEN, formerly Manager, Institute of Materials Research, Department of Metallurgy (currently Department of Mining and Metallurgy), University of Melbourne, Parkville, Victoria 3052, Australia, is now Consultant, G.V. Cullen and Associates, 36 Harrington Avenue, North Balwyn, Victoria 3104.

N.G. SIVIOUR, formerly Senior Research Officer, Institute of Materials Research, Department of Metallurgy (currently Department of Mining and Metallurgy), University of Melbourne, is now Experimental Officer, Division of Mineral Engineering, Commonwealth Scientific and Industrial Research Organization, Clayton, Victoria 3168.

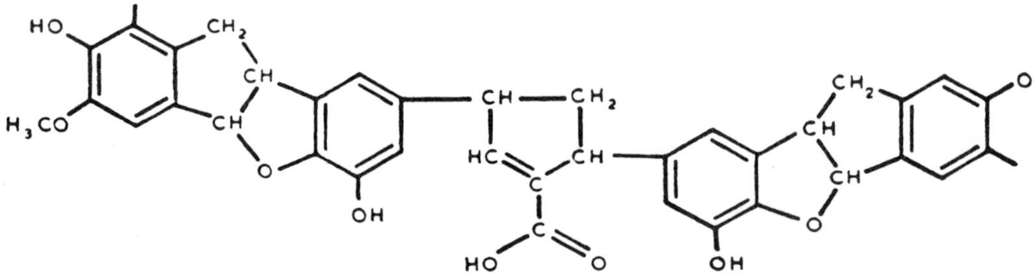

FIGURE 1. Tentative suggested formula for brown coal

In the light of the previous work by Schafer, 1970, it was considered likely that the extraction from ammoniacal solution involves ion exchange of a complex metal-ammine cation for two protons derived from either the carboxyl or phenolic hydroxyl groups, or from both of them.

The present paper gives experimental measurements which provide a basis for testing this hypothesis.

EXPERIMENTAL

Strategy

It was assumed that the extraction of metal from ammonical solution by brown coal may be expressed as follows :

$$\{Me(NH_3)_n\}^{++} + 2(coal - H) \rightleftarrows \{coal - Me(NH_3)_{n-2} - coal\} + 2(NH_4)^+ \quad (1)$$

where the reactants are the divalent metal ammine complex cation and the proton sites on the coal, respectively, and the products are the coal loaded with the stabilized metal ammine and the ammonium ions formed, respectively.

The above reaction can be written as follows :

$$K = \frac{\{activity(coal\text{-}Me(NH_3)_{n-2} - coal)\} \cdot \{activity(NH_4)^+\}^2}{\{activity(coal - H)\}^2 \cdot \{activity\ Me(NH_3)^{++}{}_n\}} \quad (2)$$

where K is the equilibrium constant.

If, as shown later, changes in the activity coefficients may be disregarded, Equation (2) can be rewritten, with square brackets indicating molar concentrations :

$$K = \frac{[coal_2 - Me(NH_3)_{n-2}] \cdot [NH^+{}_4]^2}{[coal - H]^2 \cdot [Me(NH_3)^{++}{}_n]} \quad (3)$$

If for a particular initial metal concentration a series of equilibrium contacts is made with increasing liquid to solid ratios, ie. (m L solution/g dry coal), a set of results is obtained in which the metal loadings increase and the percentage metal removed from solution decreases with each separate contact, and the extraction tends to zero as the loadings tend to the maximum possible by filling all the available functional sites on the coal. These results can be represented in Equation (3) as follows :

$$K = \frac{[q] \cdot [2(C_o - C)]^2}{[C] \cdot [2(Q - q)]^2} \quad (4)$$

where, for each contact at equilibrium, the representation is as follows :

q = metal equivalent of the complex loading $Me(NH_3)_{n-2}$ (see also below).
C_o = initial metal concentration in the solution (mol L^{-1}).
C = metal concentration in the solution at equilibrium (mol L^{-1}).
$2(C_o-C)$ = metal removed from solution as equivalent to the ammonium ions formed.
Q = metal equivalent of total available functional sites on the coal. The available sites may vary with the experimental conditions of each contact.
q = also equals the metal equivalent of the occupied available functional sites.
$2(Q-q)$ = the metal equivalent of the unoccupied available functional sites.

Equation (4) may be arranged to give :

$$K = \frac{[C_o]}{[Q]} \cdot \frac{[q/Q]}{[(1 - q/Q)^2]} \cdot \frac{[(1 - C/C_o)^2]}{[C/C_o]} \quad (5)$$

$$\text{or} \quad K\frac{Q}{C_o} = \frac{[q/Q]}{[(1 - q/Q)^2]} \cdot \frac{[(1 - C/C_o)^2]}{[C/C_o]} \quad (6)$$

Equation (6) is virtually the same as that which has been developed and adapted by others, for example Frisch and McGarvey, 1959, who presented the equation for ion exchange of divalent ions in solution for univalent ions in a resin matrix, and suggested that the solution activity ratios could usually be taken as near unity for engineering applications, and that the activity of the resin-phase could be combined into the equilibrium constant.

Similarly, the equation for univalent-univalent exchange can be written as :

$$K = \frac{[1 - C/C_o]}{[C/C_o]} \cdot \frac{[q/Q]}{[1 - q/Q]} \quad (7)$$

Frisch and McGarvey, 1959, also showed that by plotting q/Q versus C/C_o for divalent-univalent ion exchange, a series of characteristic curves, as shown in Fig. 2, can be obtained, each curve representing a particular value of $K\ Q/C_o$ for resins having single functionality. It can be shown, however, if such a plot were made for univalent-univalent exchange, only one curve would be obtained. Therefore K values and q/Q versus C/C_o plots are used to gather information on the particular characteristics of the system involved in the present work.

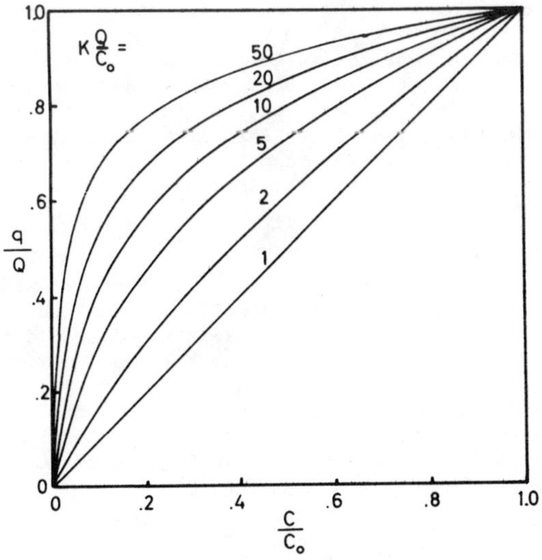

FIGURE 2. General form of equilibrium plots for divalent-univalent ion exchange

This preliminary study was confined to cadmium, zinc, nickel and copper because for the other ammine forming metals the previous work by Cullen et al., 1978, had shown that chemical reduction of silver by the coal caused problems, and there were difficulties in maintaining cobalt in solution.

Materials

The work was done with brown coal from Yallourn, Victoria, Australia, the coal being dried from its natural moisture of about 67% to 15% and crushed to minus 10 Tyler mesh for the zinc solutions and to minus 48 Tyler mesh for the other metal solutions.

The solutions were prepared by dissolving the metal sulfates in water and adding ammonium hydroxide in amounts slightly in excess of that required to stabilize the metal ammine complex cations, this resulting in the use of progressively higher ammonia to metal mole ratios with decreasing initial metal concentration. In all cases analytical grade chemicals were used.

Procedure

Brown coal was contacted with ammoniacal solutions of the metals by continuous agitation in closed vessels on a set of rolls. The coal was then separated by filtration and the metal loading on it was calculated from chemical analysis of the solutions before and after the contacting and the liquid to solid ratios. Contact time was 1 000 min for zinc and 4 000 min for the other metals. The contacts were made with a range of initial concentrations for each metal and with liquid to solid ratios selected from a range of 4 to 2 000 to give an appropriate range of equilibrium loadings for each of the solutions.

The shorter contact time and the larger particle size of the coal for the zinc solutions were used because under those conditions the results could also be used for another part of the overall program, the previous work by Cullen et al., 1978, having indicated that either set of conditions should be satisfactory for obtaining virtually maximum loadings consistent with the other specific conditions used.

In order to apply the equilibrium equations it was first necessary to find the value of the maximum metal loading for each of the solutions having different initial concentration of metal. This was done by plotting the loading for each equilibrium contact versus the fractions of the initial metal concentration $\frac{C_o - C}{C_o}$ left in solution, and then extrapolating to give the maximum equlibrium loading where $\frac{C_o - C}{C_o} = 0$ ie. at zero extraction.

The value for q at each equilibrium contact was determined by calculating the moles of metal on the amount of coal in contact with 1 litre of solution. The values for Q were obtained by calculating the moles of metal per gram of coal at the maximum loading for each solution, and multiplying this by the number of grams of coal in contact with 1 litre of solution for each equilibrium contact. The values for C and C_o were also calculated as mol L^{-1}.

RESULTS

Fig. 3, as well as showing the loadings and extraction values for all the equilibrium contacts made, also indicates the maximum possible loadings for each solution.

The groups of curves in Figs. 4 and 5 give the results obtained by plotting q/Q versus C/C_o for all the equilibrium contacts. It can be seen that for nickel, zinc and cadmium a set of distinctly separate curves progress away from each other as the values of C_o are increased, indicating compliance in part with the characteristic form for divalent-univalent exchange shown in Fig. 2. For copper, however, the curves for increasing values of C_o fall very nearly into one curve, indicating substantially univalent-univalent exchange. Furthermore, it will be noted that there does appear to be a small significant separation of the copper curves, but in the reverse order with respect to C_o, compared with the other metals. Clearly, therefore Equation (7) should be used for the results for copper.

It will also be noted that all the curves contain a more skewed form than is shown in Fig. 2, thus indicating that each curve represents a series of different values for $K\ Q/C_o$.

Similarly, it was found that the values of K, derived from Equation (5) for cadmium, zinc and nickel, and from Equation (7) for copper vary with C for each metal. It was recognized that these deviations could most probably be due to multiple functionality in the coal and changes in this at different values of C and C_o, and that these would be difficult to determine experimentally. Alternatively, however, it has been shown that correction formulae can be used to produce more constant K values, and it is assumed that these correction factors can be taken to represent the above changes in functionality. Thus for cadmium, zinc and nickel the K values from Equation (5) are first divided by $(1+x^3)/(1-0.81x^3)$ where $x = C/C_o$ and a limit on C/C_o is set such that $0.1 < C/C_o < 0.9$, to produce one set of corrected values for K, and then these were multiplied by C_o^n where n is 0.50 for cadmium and 0.33 for zinc and nickel, to produce a second set of values. For copper the K values derived from Equation (7) are multiplied by C_o to the power 0.67. In selecting the correction formulae the aim was simply to find the best compromise correction for the zinc, cadmium and nickel solutions, and another for the copper solutions. Table I shows that the first correction formula substantially narrows the range of values, and by application of the additional correction a set of similar equilibrium constants are obtained for each metal, except for zinc at 20.0 g L^{-1} for which it is speculated that the low value obtained may be due to non-equilibrium conditions which have appeared at that concentration with the shorter contact time and larger coal particle size used, despite the earlier predictions.

DISCUSSION OF RESULTS

The general decrease in loading associated with increase in initial metal concentration shown in Fig. 3 is to be expected because of the previously mentioned inhibiting effect of the concentration of ammonium ions formed, this being proportional to the amount of metal extracted and not to the loadings which are themselves dependent jointly on the amounts of metal extracted and the liquid to solid ratios. In addition, the loading for solutions based on any given initial metal concentration will reflect the fact that the different ammonia to metal mole ratios

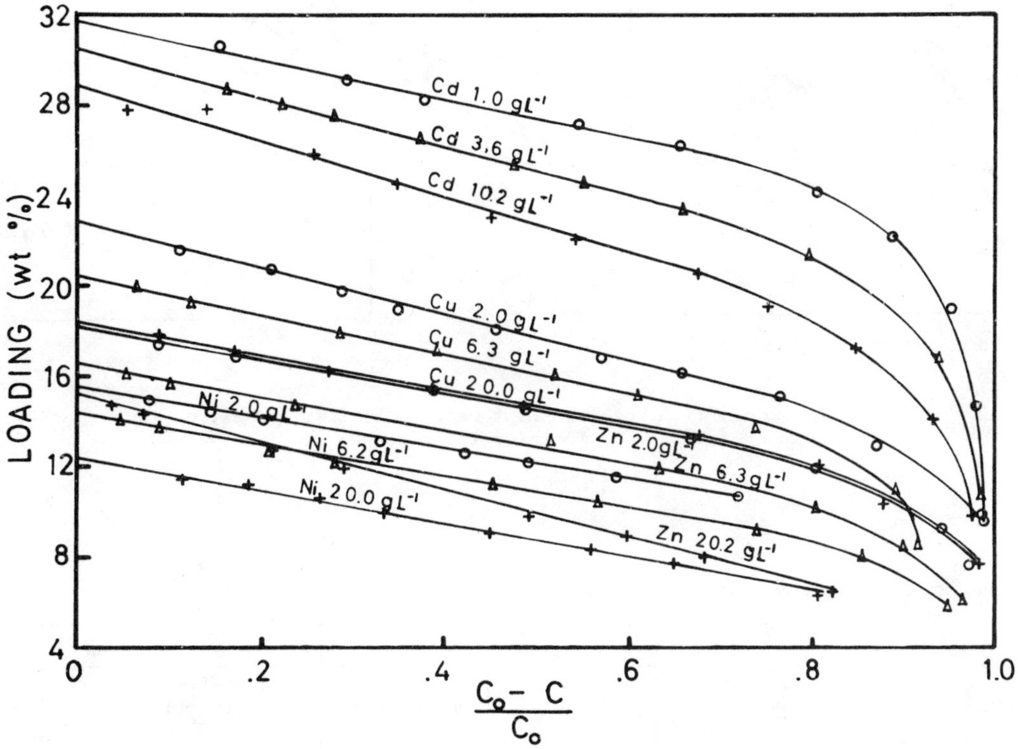

FIGURE 3. Data for determining maximum metal loadings

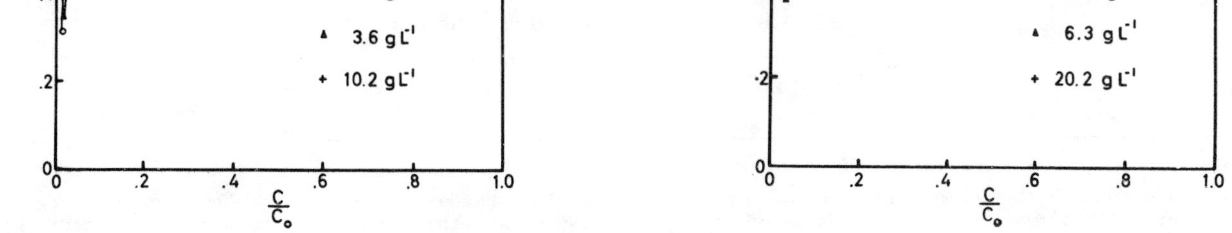

FIGURE 4. Equilibrium relationships between loading and extraction for cadmium and zinc

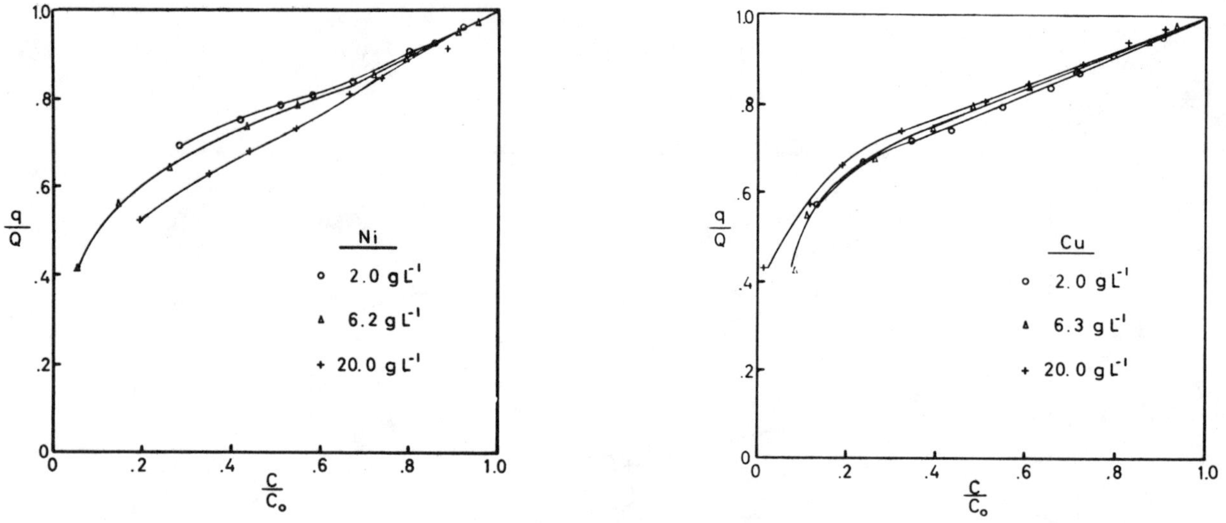

FIGURE 5. Equilibrium relationships between loading and extraction for nickel and copper

TABLE I. Equilibrium constants before and after corrections for multiple functionality of coal

Metal	Initial Concentr-ation $g\ L^{-1}$	K from Equation (5): Full Range	Mean of Full Range	Restricted Range*	Mean of Restricted Range	Adjusted K: After Correction** Range	After Correction** Mean	After Correction*** Range	After Correction*** Mean
Cd	1.0	38.8-129	53.7	41.2-129	56.7	26.4-43.1	36.8	2.5-4.1	3.5
	3.6	17.9-54.7	28.7	21.0-54.7	30.6	16.9-19.5	18.4	3.0-3.8	3.3
	10.2	8.8-429	56.5	11.9-63.1	21.5	9.9-19.0	12.5	3.0-5.7	3.7
Zn	2.0	13.5-56.5	26.2	15.5-48.5	24.1	13.2-16.7	15.2	4.1-5.2	4.8
	6.3	9.7-83.5	25.4	9.7-27.6	15.8	7.0-12.6	10.4	3.2-5.8	4.8
	20.2	2.7-19.2	6.1	2.7- 6.8	3.9	2.4- 2.8	2.6	1.6-1.9	1.8
Ni	2.0	13.3-59.5	22.8	13.3-28.8	17.6	8.4-13.2	10.9	2.7-4.3	3.5
	6.2	8.6-75.3	21.6	9.1-11.2	10.1	8.2- 9.6	8.8	3.9-4.6	4.2
	20.0	5.1-20.2	9.8	5.1-20.2	9.1	4.8- 5.3	5.0	3.4-3.7	3.5
Cu	2.0	2.1-49.2	8.7	2.7- 9.1	4.5	-	-	2.1-2.5	2.3
	6.3	2.1-10.0	4.6	2.1-10.0	4.3	-	-	1.9-2.7	2.3
	20.0	3.0-46.7	9.2	3.0-10.0	5.4	-	-	2.4-2.8	2.6

* Where $0.1 < C/C_o < 0.9$

** $(1 + x^3)/(1 - 0.81x^3)$ for Cd, Zn and Ni, where $x = C/C_o$

*** $(1 + x^3)/(1 - 0.81x^3)\ C_o^{\ n}$ where n = 0.5 for Cd and 0.33 for Zn and Ni. $C_o^{\ 0.67}$ for Cu.

that had to be used will produce differences in pH which in turn will affect the net activation of the functional groups on the coal which themselves have different pK values.

In considering the form of the curves in Figs. 4 and 5, and the progressive change in values of the terms containing the equilibrium constants, two aspects deserve attention. Firstly, the curves presented by

Frisch and McGarvey, 1959, were for virtually ideal systems, and they suggested that more complications result when resins possess more than a single functionality, or when weakly dissociated resins which exhibit a pH dependent exchange capacity are involved. Consequently, it may be expected that such complications should arise when the equilibrium expressions designed for single functioning ion exchange resins are applied to a low rank coal which depends for its exchange properties on at least two functional groups, each of which may possess varying pK values, and other chemical and structural inhomogeneity. The results of this method of evaluation, however, do support the initial hypothesis that ion exchange is the mechanism of extraction, and do indicate the modes of exchange involved.

Secondly, as evidence of additional complications for copper, it has been observed that copper is reduced from the cupric to cuprous state during contacting, and this seems to offer a basis for explaining not only the different form of the curves obtained, but also the fact that the m.eq/g. loadings for copper, which ranged from 5.8 to 7.1, were significantly higher than the range of 4.2 to 5.6 for the other metals.

SUMMARY AND CONCLUSIONS

1) Equilibrium expressions have been derived for the extraction of cadmium, zinc, nickel and copper from ammoniacal solution by brown coal or lignite. These expressions are essentially the same as those developed by previous workers for use with conventional cation exchange resins.

2) Application of the expressions to the results of equilibrium contacting between the coal and metal solutions has indicated the following :

   a) The metals are extracted by ion exchange of metal ammine complex cations for protons derived from the coal, with subsequent release of ammonium ions into solution.

   b) The complex cations of cadmium, zinc and nickel are exchanged as divalent ions, but the complex for copper behaves principally as a univalent ion, which appears to be due to concurrent reduction of copper from the cupric to cuprous state by the coal, particularly for long contact times.

   c) To obtain reasonably similar equilibrium constants over a range of metals and initial and equilibrium metal concentrations it is necessary to apply corrections for multiple functionality of the coal.

3) The general correspondence between the maximum metal loadings obtained and the available ion exchange capacity reported in the literature for the carboxyl and phenolic hydroxyl groups in the coal, support the hypothesis that the protons with which the metal complex cations exchange are derived from either the phenolic groups or from them and the carboxyl groups in the coal.

ACKNOWLEDGMENTS

This project was part of a program of research by the Institute of Materials Research in the Department of Metallurgy, University of Melbourne. The work was sponsored jointly by Conzinc Riotinto of Australia Limited and the Electrolytic Zinc Company of Australasia Limited.

The authors gratefully acknowledge the interest and co-operation from these organizations and contributions to the project by D. McDonald and G.M. Pearson, both former Experimental Officers in the Institute.

REFERENCES

Anon, 1959, "Chemistry of Brown Coals", CSIRO Coal Research, No.6, p.13.

Altekar, V.A., Shahani, M.J., and Saha, A.K., 1974, "Adsorption of Nickel by Lignite", Fuel, Vol.53, No.1, pp. 29-31.

ASTM, 1979, "Standard Specification for Classification of Coals by Rank, ANSI/ASTM D388-77", Annual Book of ASTM Standards, Part 26, American Society for Testing and Materials, Philadelphia, pp. 220-224.

Brooks, J.D., and Sternhill, S., 1957, "Chemistry of Brown Coals", Aust. J. App. Sc., Vol.8, No.3, pp. 206-221.

Cullen, G.V., Siviour, N.G., and Pearson, G.M., 1978, "Extracting Metals from Ammoniacal Solution with Low Rank Coals", Met. Trans. B, Vol.9B, pp. 409-413.

Frisch, N.W., and McGarvey, F.X., 1959, "Application of Ion Exchange Equilibrium Relationships to Process Design", Chem. Eng. Prog. Symposium Series No. 24, Vol.55, pp. 51-59.

Herman, H., 1952, Brown Coal, State Electricity Commission of Victoria, Melbourne, pp. 4-7.

Saha, A.K., Shahani, M.J., and Altekar, V.A., 1977, "Adsorption of Cobalt by Lignite", Hydrometallurgy, Vol.2, No.3, pp. 285-292.

Schafer, H.N.S., 1970, "Carboxyl Groups and Ion Exchange in Low Rank Coals", Fuel, Vol. 49, pp. 197-212.

Chapter 33

ANODIC DEPOLARIZATION BY COAL LOWERS ELECTRIC ENERGY CONSUMPTION DURING ELECTRODEPOSITION OF METAL

Robert W. Coughlin

Department of Chemical Engineering
University of Connecticut
Storrs, Connecticut 06268

ABSTRACT

Pulverized coal has been introduced at the anode where it is oxidized during the electrowinning of metals from their aqueous solutions. This substitutes an electrochemical reaction between coal and water for the usual oxygen-evolving half-cell reaction at the anode, with the result that overall cell potential is significantly lowered and the consumption of electrical energy is correspondingly reduced. Experimental results are reported for electrowinning copper as an example of this novel modification of the electrowinning process. In this example, the use of coal lowered the consumption of electricity by two-thirds of the requirement of the conventional process. When the potential electrical energy of the coal consumed is considered, the new, coal-assisted process consumes about 45% less energy, and the energy costs (combined costs of coal and electricity) are reduced by about 50%.

ANODIC DEPOLARIZATION BY COAL LOWERS ELECTRIC ENERGY CONSUMPTION DURING ELECTRODEPOSITION OF METAL

Robert W. Coughlin

Department of Chemical Engineering
University of Connecticut
Storrs, Connecticut 06268

INTRODUCTION

Recent publications (1-4) show a new method for electrochemically converting slurries of coal in aqueous electrolyte to oxides of carbon at the anode and hydrogen at the cathode. This process might be called electrochemical gasification of coal, or coal-assisted or coal-depolarized water electrolysis. This has now been extended to the electrochemical deposition of metal, instead of hydrogen evolution, at the cathode. Significant savings in electrical energy are thereby realized over conventional electrodeposition or electrowinning.

The metals which can be extracted in large amounts from aqueous solutions of their salts by electrowinning are Cr, Mn, Co, Ni, Cu, Zn, Ga, Cd, In and Tl. Gilroy (5) specifies an annual electrowinning production for 1971 of 2,400,000 tons of Zn and 540,000 tons of Cu, 53% and 10% respectively of total production at that time. We describe here a modification to the established electrowinning technique which lowers the energy consumption of the process.

The reactions for conventional electrowinning of a metal M are:

$$\text{Anodic:} \quad xH_2O \rightarrow \frac{1}{2}xO_2 + 2xH^+_{aq} + 2xe^- \qquad (1)$$

$$\text{Cathodic:} \quad M^{2x+}_{aq} + 2xe^- \rightarrow M \qquad (2)$$

The reversible thermodynamic potential, E°, of reaction (1) is 1.229 V. It and associated polarization overpotentials, E_{op}, constitute a large portion of the total potential E_{tot} of the practical electrowinning cell:

$$E_{tot} = E^\circ + E_{op} + E_m$$

where E_m is the half-cell potential for reducing the metal ion, reaction (2). Browall et al. (6) introduced the reducing agent, CO, at the anode to lower the cell potential for electrolytic production of hydrogen. Using coal as the reducing agent, we devised a new electrochemical process which converts coal and water into two separate gaseous products -- one at the anode, comprising essentially gaseous oxides of carbon, and the other at the cathode which is essentially pure hydrogen. We called this process electrochemical gasification of coal or, alternatively, coal-assisted water electrolysis. Coal contains a wide variety of organic molecules rich in carbon. Focusing only on the carbon of the coal the reactions of the electrochemical coal-gasification process, in simplified stoichiometry, would be:

$$\text{Anode:} \quad \frac{1}{2}C(s) + H_2O(l) \rightarrow \frac{1}{2}CO_2(g) + 2H^+ + 2e^- \qquad (3)$$

$$\text{Cathode:} \quad 2H^+ + 2e^- \rightarrow H_2(g) \qquad (4)$$

The reversible thermodynamic potential, E°, of half-cell reaction (3) is only about 0.21 V. We have now extended electrochemical coal gasification to the electrowinning of metals by substituting the half-cell reaction (3) (which consumes coal) for the half-cell reaction (1) which usually takes place during conventional electrowinning of metals from aqueous solutions of their salts. By this coal-based innovation the total overall cell potential of the resulting metal electrowinning process is lowered by about 1.0 V with a corresponding significant reduction in the consumption of electrical energy.

EXPERIMENTAL

Preliminary results (7) for coal-assisted electrowinning are illustrated by experiments with copper. Experiments were conducted in the simple cell shown in Figure 1. The external e.m.f. was controlled by a potentiostat (PAR 371) and the electrodes (both anode and cathode) were Pt mesh, gauge 52 (0.004-inch diameter wire, Mathey Bishop Inc.). Aqueous electrolytes were prepared with Baker Analyzed reagent grade sulphuric acid and Baker $CuSO_4 \cdot 5H_2O$ crystals. Coal slurries were prepared from pulverized North Dakota lignite coal; the ultimate analyses are described elsewhere (2). Metal analysis was carried out on a Perkin-Elmer 460 atomic absorption spectrophotometer.

For comparison Cu was first deposited by conventional electrowinning without using coal. Accordingly, using the cell of Fig. 1, Cu was deposited on

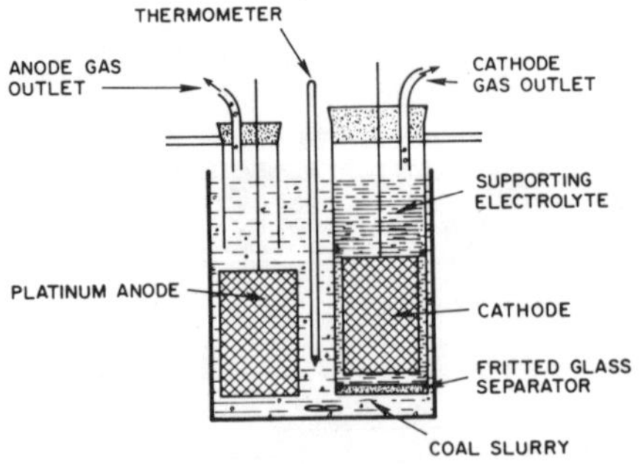

FIGURE 1. Coal gasification cell: electrolyte volume, 80 cm^3; anode, 5.3 cm^2 (geometrical) platinum mesh, gauge 52; cathode, 8.4 cm^2 (geometrical) platinum mesh, gauge 52; stirring, magnetic, $\frac{7}{8}$-inch magnet.

a platinum-mesh cathode from aqueous electrolyte (0.125 M $CuSO_4$ in 0.5 M H_2SO_4) at 60°C, and at four different galvanostatic currents: 5.9, 7.95, 10 and 12 mA. The corresponding change in cell potential was recorded and typical cell potentials corresponding to the galvanostatic currents of 5.9 and 12 mA are plotted in Figure 2. The conventional copper electrowinning reaction took place at corresponding cell potentials of about 1.65 and 1.73 V respectively. Figure 2 also shows the results obtained in parallel experiments, identical except that the coal was simultaneously oxidized at the anode according to equation (3) while Cu cations were reduced at the cathode according to equation (2). In these latter experiments, North Dakota lignite (NDL) coal slurry (0.15 g cm^{-3}) was introduced into the anode compartment of the cell and was anodically oxidized while copper deposited on the cathode in identical galvanostatic experimental conditions. As a result of the oxidation or gasification of the coal at the anode, the overall cell potential was lowered by ~1.1 V, compared to the conventional process. This is evident from Figure 2. It is seen that the simultaneous anodic oxidation of coal lowers the potential by ~1.10 V, a value which is very close to the difference in half-cell potential between reactions (1) and (3).

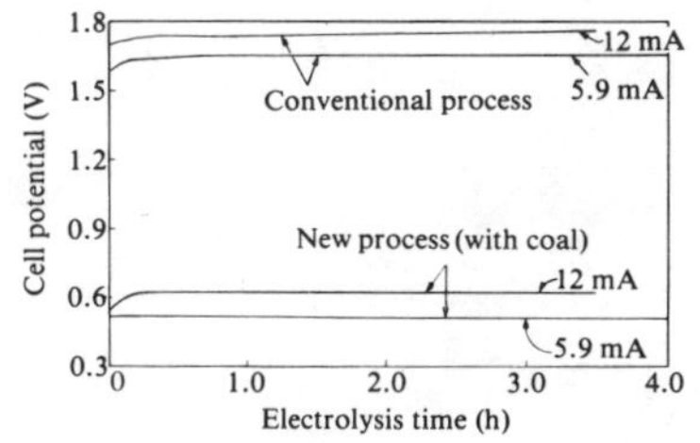

FIGURE 2. Comparison of cell potentials of the copper electrowinning processes in galvanostatic conditions: Electrolyte, 0.125 M $CuSO_4$ and 0.5 M H_2SO_4; coal sample, N. Dakota lignite; slurry concentration, 0.15 g cm^{-3}; particle size, 74-88 μm; temperature, 60°C.

Data on the current, electricity consumption and weight of Cu deposited are given in Table 1 for the conventional process, and in Table 2 for the new,

TABLE 1. Current efficiency and electrical energy consumption during copper deposition in the conventional process.

Reaction rate I (mA)	Total coulomb consumed It (C)	Theoretical Cu equivalent to the charge passed W_1 (g)	Measured amount of Cu deposited W_2 (g)	Current efficiency $(W_2/W_1) \times 100$ (%)	Electrical power consumed $E_1 = (I \int V\,dt)/W_2$ (kWh kg^{-1})
5.90	100.79	0.03319	0.03027	91.2	1.52
7.95	147.39	0.04853	0.04620	95.2	1.51
10.00	144.00	0.04741	0.04425	93.3	1.52
12.00	133.90	0.0442	1.04288	97.2	1.622

Electrolyte, 0.125 M $CuSO_4$ and 0.5 M H_2SO_4; Temperature, 60 °C.

TABLE 2. Current efficiency and electrical energy consumption during copper deposition: new coal-assisted process.

Reaction rate I (mA)	Total coulomb consumed It (C)	Theoretical Cu equivalent to the charge passed W_1 (g)	Measured amount of Cu deposited W_2 (g)	Current efficiency $(W_2/W_1) \times 100$ (%)	Electrical power consumed $E_2 = (I \int V\,dt)/W_2$ (kWh kg^{-1})
5.90	105.84	0.03485	0.031115	89.4	0.468
7.95	132.61	0.04366	0.04062	93.0	0.428
10.00	142.80	0.04702	0.04407	93.3	0.524
12.00	129.60	0.04267	0.04063	95.2	0.552

Electrolyte, 0.125 M $CuSO_4$ and 0.5 M H_2SO_4; Temperature, 60 °C; North Dakota lignite coal concentration 0.15 g cm^{-3}; particle size, 74–88 μm.

coal-assisted process. Current efficiencies are also given computed from the ratio of the amount of Cu actually deposited to the equivalents of the charge passed during the experiment. The introduction of coal into the anode compartment during electrolysis evidently reduces current efficiency only slightly, at the worst about 2%. Comparing the right-hand columns of Tables 1 and 2 reveals that the simultaneous anodic oxidation of coal during Cu electrowinning reduces the electrical energy consumption by a factor of about two-thirds or more. Whereas the conventional process uses only electrical energy, our new process uses both electrical energy (E_2) and the chemical energy (E_3') of coal directly. For a reasonable comparison of the energy consumption of the two processes, the chemical energy of the coal consumed should be converted into equivalent electrical energy. By the stoichiometry of reaction (3), the mass of carbon consumed to produce 1 kg of copper is approximately:

$$\frac{12 \text{ kg C}}{\text{kg atom C}} \times \frac{1 \text{ kg atom C}}{2 \text{ kg atom Cu}} \times \frac{1}{63.54} \frac{\text{kg atom Cu}}{\text{kg Cu}} = 0.0944 \frac{\text{kg C}}{\text{kg Cu}}$$

Based on the enthalpy of combustion of carbon, this corresponds to a chemicothermal energy consumption of

$$E_3' = 0.0944 \frac{\text{kg C}}{\text{kg Cu}} \times \frac{94{,}052 \text{ kcal}}{12 \text{ kg C}} \times \frac{4180 \text{ J}}{\text{kcal}} \times \frac{1 \text{ kWs}}{10^3 \text{J}} \times \frac{1 \text{ h}}{3{,}600 \text{ s}} = 0.859 \frac{\text{kWh}}{\text{kg Cu}}$$

Assuming a practical efficiency of 35% for conversion of the combustion enthalpy of carbon (coal) into electrical energy, the electrical energy equivalent to the coal consumed is:

$$E_3 = (0.859)(0.35) = 0.300 \frac{\text{kWh}}{\text{kg Cu}}$$

Using these values of E_3' and E_3 with the values of E_1 and E_2 from Tables 1 and 2, the total energy requirements and also the net energy savings have been calculated for the new process and are reported in Table 3. Compared to the conventional process, the new coal-consuming process requires about 45% less net equivalent electrical energy for producing the same amount of copper.

A more telling comparison of these processes might be on the basis of relative energy costs based on assumed costs of US $0.02 per kWh for electricity

TABLE 3. Savings in energy consumption.

Galvano-static reaction rate I (mA)	Total energy consumption in the new process E_2+E_3 (kWh per kg Cu)	Net equivalent electrical energy consumed in this new process E_2+E_3 (kWh per kg Cu)	Net savings in equivalent electrical energy $E_1-E_2-E_3$ (kWh per kg Cu)	Net equivalent energy savings $\frac{(E_1-E_2-E_3)}{E_1}\times 100$ (%)
5.9	1.328	0.77	0.75	49.3
7.95	1.29	0.73	0.78	51.6
10.0	1.384	0.826	0.694	45.7
12.0	1.412	0.852	0.768	47.3

and \$35.00 per ton (\$0.0514 per kg C) for coal assumed to be 75% carbon. Using mean values for E_1 and E_2 these costs are:

For the conventional process: Cost = (\$0.02) $\cdot\ \bar{E}_1$ = \$0.031 per kg Cu.
For the new coal-assisted process: Cost = (\$0.02) $\cdot\ \bar{E}_2$ + (0.0944)(\$0.0514) = \$0.0147 per kg Cu.
The savings in energy cost appears, therefore, to be ~50%.

In the experiments discussed so far, NDL coal which contains about 11.7% ash has been used as a reactant. The ash composition suggests the possibility of the presence of impurities such as Al, Fe, Ti, Ca, Mg, Na and K ions in the electrolytes. Consideration of the thermodynamic potential of the $M\text{-}M^{x+}$ couple of these metals indicates that only iron might interfere with the copper deposition for which half-cell potentials have been reported (8) as follows:

$$Cu^{2+} + 2e^- = Cu \qquad E^\circ = 0.337\ V$$

$$Cu^{+} + e^- = Cu \qquad E^\circ = 0.521\ V$$

$$Cu^{2+} + e^- = Cu^{+} \qquad E^\circ = 0.153\ V$$

Reported (8) potentials of the different iron couples are:

$$Fe^{2+} + 2e^- = Fe \qquad E^\circ = -0.440\ V$$

$$Fe^{3+} + e^- = Fe^{2+} \qquad E^\circ = 0.771\ V$$

$$Fe^{3+} + 3e^- = Fe \qquad E^\circ = -0.036\ V$$

The ash of NDL coal contains 12.5% Fe_2O_3 (other coals contain even less iron). Therefore iron is the only large-scale impurity which should be removed in a continuous process because reduction from the ferric to the ferrous state would reduce current efficiency for copper deposition and would increase the power consumption. Note that iron is generally also the only large-scale impurity present in copper ore itself and the technology for iron removal is well established, for example, by air oxidation of Fe^{2+} to Fe^{3+} which can easily be made to precipitate as the hydroxide. In some industrial processes ferric ion concentration is kept high to promote faster leaching of sulphide ores.

The purity of the electrodeposited copper was determined by dissolution in HNO_3 followed by atomic absorption spectrophotometry. In all experiments the purity of the deposited copper was >99%. The use of NDL or the presence of its ash in the electrolyte does not seem to have any significant effect on the copper purity. While our results, showing significant reduction in the energy requirement, pertain to one specific reaction system, similar observations are expected for other metal electrowinning reactions in aqueous electrolytes. Anode materials other than platinum have also been investigated. Graphite rods and graphite felt anode (Union Carbide Corp.) performed as well as platinum; copper plate cathodes have also been successfully tested.

For actual application questions remain as to what fraction of the coal can be consumed in practice. As reported elsewhere (2,3,4) surface oxygen compounds build up on the surface of the coal, requiring increased potential to achieve the same current density. This behavior is shown in Figure 3 taken from reference (2). It has been shown that the original activity of the coal can be restored in part by certain treatments (3).

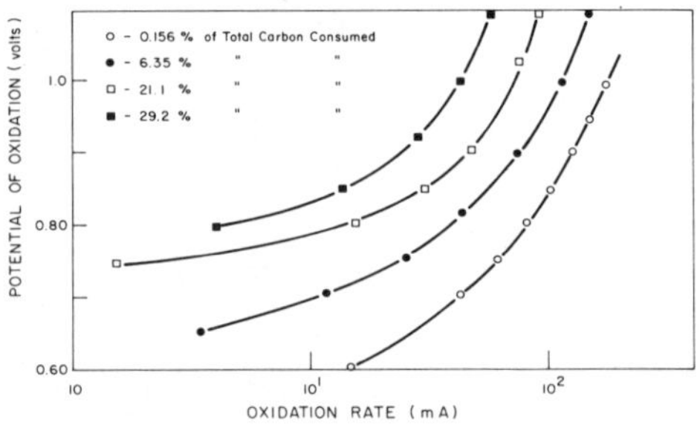

FIGURE 3. Effect of potential on the oxidation rate as the reaction proceeds; (North Dakota lignite; coal slurry concentration, 0.069 g/cm^3; supporting electrolyte, 5.60 M H_2SO_4; particle size: 125-149 μm; temperature 114°C).

It has also been shown (3) that the depolarization of the anode requires that contact be made between the coal particles and the electrode. This requirement, illustrated by Figure 4, will demand care and ingenuity in engineering design and scale-up of the process.

Although, in an energy conscious environment this method looks promising, the practical applicability of the process depends on: (1) Achieving current densities sufficiently large to keep the capital cost economical; in our simple laboratory experiments no attempt has been made to increase current densities which were far lower than those used in industrial processes. (2) Almost total consumption of the carbon-value of the coal particles. (3) The influence of the particular coal and its ash content on electrowinning kinetics and on the properties and morphology of the deposited metal. (4) Preventing build-up in a continuous process of impurity cations which might compete with copper ions in electrolysis; such impurities would have to be removed continuously, for example by air oxidation of Fe^{2+} to precipitate ferric hydroxide.

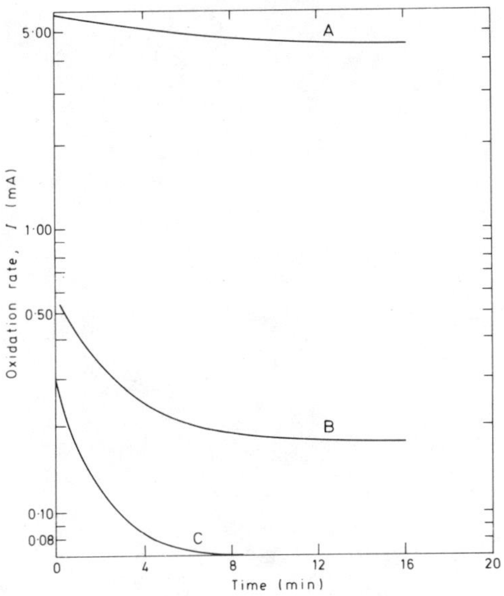

FIGURE 4. Effects of contact between the coal and the anode (coal sample, fresh NDL; slurry concentration, 0.0125 kg/dm^3; particle size, 44 μm and below; electrolyte, 4.13 mol/dm^3; oxidation potential, 1.0 V; anode area, 0.065 dm^2 (geometrical); temperature, 59°C). A, Anode directly in coal slurry; B, Coal slurry isolated from anode by fritted glass compartment; C, Blank, i.e. no coal.

This research was supported by the University of Connecticut Research Foundation and the U.S. Department of Energy.

REFERENCES

1. Coughlin, R. W. and Farooque, M., 1979, Nature 279, 301-303.

2. Coughlin, R. W. and Farooque, M., 1980, Ind. & Eng. Chem. Process Des. & Dev. 19, 211-219.

3. Farooque, M. and Coughlin, R. W., 1979, Fuel 58, 705-712.

4. Coughlin, R. W. and Farooque, M., J. Appl. Electrochem. (in press).

5. Gilroy, D., 1971, The Electrowinning of Metals, Industrial Electrochemical Processes, (ed. A. Kuhn), Chap. 6, Elsevier, New York.

6. Browall, K. and Hanneman, R. E., 1975, General Electrical Technical Information Ser. Rept. No. 75CRD012.

7. Farooque, M. and Coughlin, R. W., 1979, Nature 280, 666-668.

8. Latimer, W. M., 1952, Oxidation Potentials, Prentice-Hall, Englewood Cliffs.

INDEX

D

M

N

O

P

T

U

V